TRANSPORT PHENOMENA AND MATERIALS PROCESSING

TRANSPORT PHENOMENA AND MATERIALS PROCESSING

SINDO KOU
Department of Materials Science and Engineering
University of Wisconsin, Madison, Wisconsin

A Wiley-Interscience Publication
JOHN WILEY & SONS, INC.
New York / Chichester / Brisbane / Toronto / Singapore / Weinheim

This text is printed on acid-free paper.

Library of Congress Cataloging in Publication Data:

Kou, Sindo.
Transport phenomena and materials processing / by Sindo Kou.
p. cm.
Includes index.
ISBN 0-471-07667-8 (cloth : alk. paper)
1. Manufacturing processes—Mathematical models. 2. Mass transfer—Mathematical models. 3. Heat transfer—Mathematical models. 4. Transport theory. 5. Fluid mechanics. I. Title.
TS183.K68 1996
670—dc20 95-49988
CIP

Printed in the United States of America

10 9 8 7 6 5 4 3

To Tina, Nancy, and Katharine

CONTENTS

PART I INTRODUCTION TO TRANSPORT PHENOMENA

PREFACE

This book has been developed for the dual purpose of introducing transport phenomena to materials and chemical engineering students, and helping those mechanical, chemical, and materials engineering students already familiar with the subject apply it to materials processing. It is intended for undergraduate or graduate students. It can also be used as a reference book for practicing engineers and research workers interested in materials processing. It deals with fluid flow (momentum transport), heat transfer (energy transport), mass transfer (species transport), and their critical roles in materials processing.

The subject of transport phenomena in materials processing differs significantly from that in conventional mechanical or chemical engineering in one important aspect, that is, phase transformations or conversions are often involved. Typical examples are the liquid-to-solid transformation in crystal growth, casting and welding, solid-to-solid transformations in heat treating, and the vapor-to-solid conversion in chemical vapor deposition. It is well known in materials processing that transport phenomena have a significant effect on the shape and velocity of the interface between two phases, which, in turn, have a significant effect on the product quality and the production rate.

The book, which is divided into two parts, differs from existing transport phenomena books in several ways, which are described as follows.

Part I, an introduction to transport phenomena, is intended to be readable and compact. It is not intended to be a comprehensive treatment of transport phenomena, as excellent books of this kind are already available in both chemical and mechanical engineering. An overview of Part I is shown in Figure 1 (of this Preface).

Overview of Part I Introduction to Transport Phenomena[1]

	Momentum Transfer	Heat Transfer	Mass Transfer
Basic Concepts[2]	1.1	2.1	3.1
Overall Balance	1.2; 1.4; 1.9	2.2	3.2
Differential Balance	1.3; 1.5; 1.6	2.3; 2.4	3.3; 3.4
Turbulence	1.7	2.5	3.5
Correlations	1.8	2.6	3.6
Miscellaneous		2.7[3]	
Similarities & Coupling	4.1; 4.2		
Boundary Conditions[4]	5.1; 5.2; 5.3; 5.4		

1. Numbers in boxes refer to section numbers in the text,
2. Mechanisms, laws, boundary layers, transfer coefficients, etc.,
3. Radiation, and
4. For phase transformations.

Fig. 1.

On the basis of the conservation laws of mass, momentum, and energy, governing equations are derived in both the overall (input/output) form and the integral form in Chapters 1 through 3. As illustrated in Figure 2, the former is convenient for finding average (spatial) transport properties, such as the average velocity, temperature, and species concentration. The latter is readily converted into the differential form, which is useful for determining the spatial variations of the properties. Instead of the straight differential approach, I have decided to take this different approach, since it is less tedious and since overall and differential balance equations both become available.

Both the step-by-step procedure for problem solving and the commonly encountered boundary conditions are summarized in charts for ready reference. These, as I have observed in classrooms, are very useful for students in problem solving.

In Chapter 4 fluid flow, heat transfer and mass transfer are discussed together, first their similarities then their coupling. The boundary conditions at the

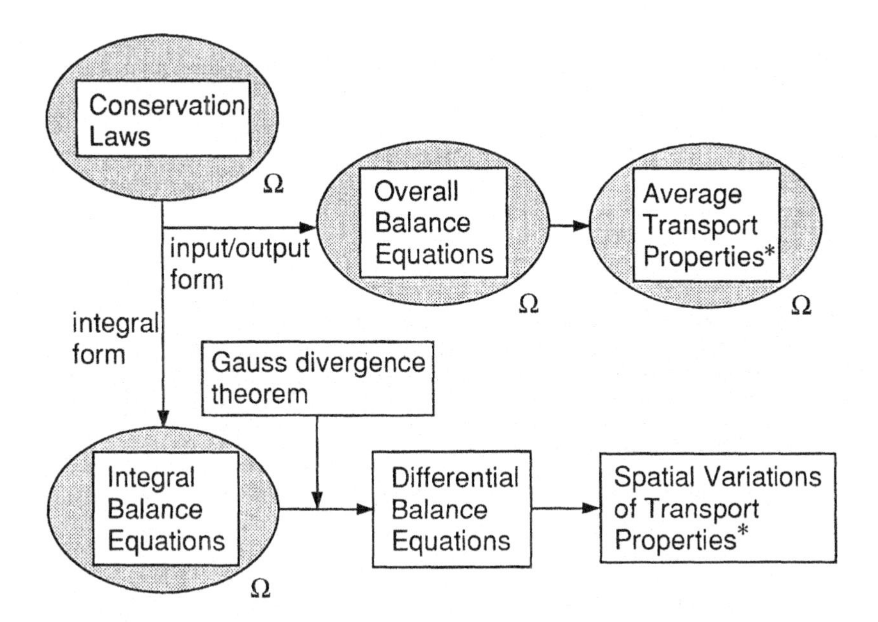

Fig. 2.

interfaces between different phases involved in phase transformations are described in Chapter 5. These interfaces are often encountered in materials processing, as mentioned previously.

Part II is devoted to the applications of transport phenomena in materials processing. An overview of Part II is shown in Figure 3. For mechanical or chemical engineering students and anyone not yet familiar with materials processing, a brief description of the selected processes is provided in Chapter 6. Crystal growth, casting, welding, heat treating, semiconductor device fabrication, powder processing, and fiber processing are selected as examples of the applications. They are organized in Chapters 7 through 9, rather than scattered all over the book. This, I believe, is more convenient for readers.

In the first section of Chapters 7 through 9, the overall balance equations are applied to materials processing. In the following sections the differential balance equations are applied. Analytical solutions are presented throughout the three chapters. Numerical solutions are also presented, especially in the last section of each chapter. Numerous excellent computational studies have been published and I apologize for not being able to mention most of them. Because of space limitation I have selected only a few that are most coherent with both the focus and level of this book.

Overview of Part II Transport Phenomena in Materials Processing[1]

	Momentum Transfer	Heat Transfer	Mass Transfer
Processing Technologies	6.1; 6.2; 6.3		
Overall Balance	7.1	8.1	9.1
Differential Balance (1-D)	7.2	8.2; 8.3	9.2; 9.3
Differential Balance (2,3-D)	7.3	8.4	
Differential Balance (coupled transfer)		8.5	9.4

1. Numbers in boxes refer to section numbers in the text.

Fig. 3.

Photographs of flow visualization, material processing in action and the materials produced are presented throughout the book. This, I hope, will stimulate the interest of readers.

Appendix A, which is a brief review of mathematics, includes 23 analytical solutions to differential equations encountered in this book. This is designed to separate equation solving from the main focus of understanding transport phenomena, specifically, to minimize detraction caused by lengthy mathematical manipulation. This, as I have experienced in classrooms, helps students enormously.

Some computer codes and publications relevant to transport phenomena in materials processing are suggested in Appendixes B and C, respectively.

The book can be used by following either the serial or parallel approach shown in Figure 4. It can be used in a three-credit course in either introduction to transport phenomena or transport phenomena in materials processing, as also shown in the figure. In the former, Appendix A should be reviewed before Part I is started, and many excellent examples of applications can be selected from Part II (see Sections 7.1, 7.2, 7.3, 8.1, 8.2, 8.3, 9.1, and 9.2). In the latter, Chapters 4 and 5 should be covered and the rest of Part I quickly reviewed before Part II is started.

In recent years, materials processing has been recognized worldwide as an important area in engineering. In fact, casting, welding, heat treating, powder processing, and fiber processing are being studied extensively in mechanical engineering, and so are crystal growth and chemical vapor deposition in

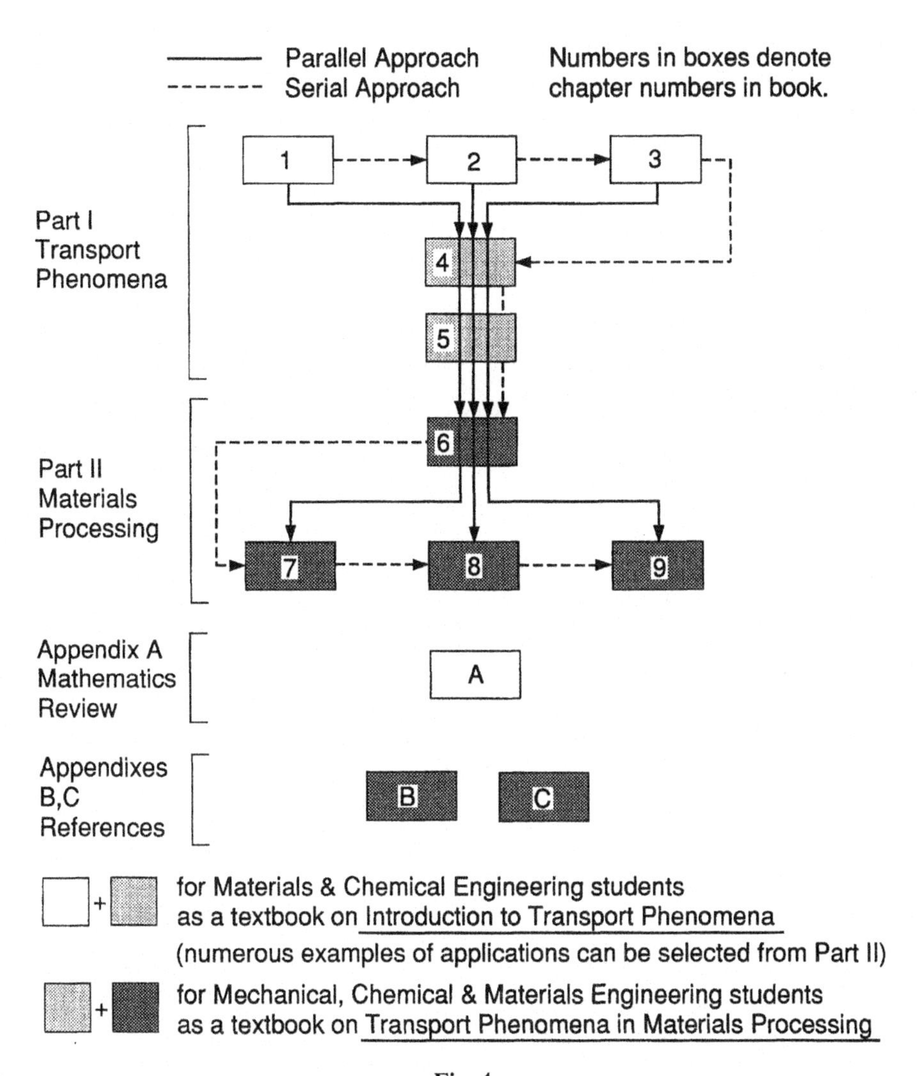

Fig. 4.

chemical engineering. The growing emphasis on materials processing, research and education alike, is expected to continue through the next two decades. To improve the effectiveness of materials processing and the quality of the resultant materials, it is essential to understand transport phenomena and their roles in materials processing. In view of this need, I have decided to prepare this book.

This book has grown in part from my research in crystal growth, welding, casting, and heat treating in the past 20 years, including those at the University of Wisconsin and Carnegie Mellon University. I would like to thank the National Science Foundation and National Aeronautics and Space Administration for support. I would also like to thank Lynn Neis and Peter Gunawa for manuscript typing, Todd Swanson, Bin Xiong, and Shuenn-Ching Tsaur for computer drawing, and Dr. C. R. Kao and Mr. Chau-Yun Yang for helpful discussion.

S. Kou

Madison, WI
March 1996

TRANSPORT PHENOMENA AND MATERIALS PROCESSING

PART I

INTRODUCTION TO TRANSPORT PHENOMENA

CHAPTER 1

INTRODUCTION TO FLUID FLOW

1.1. BASIC CONCEPTS

1.1.1. Laminar Flow and Turbulent Flow

The flow of a fluid can be in the regime of laminar flow or turbulent flow. Laminar flow is a well-ordered flow and is characterized by the smooth sliding of adjacent fluid layers (or lamina) over one another, with mixing between layers occurring only on a molecular level. Figure 1.1-1 shows a short section of two large parallel plates and the fluid in between them. The upper plate is stationary while the lower one moves at the velocity V. The thin layer of fluid immediately adjacent to the moving plate acquires a velocity in the z-direction and hence a certain z-momentum (momentum = mass × velocity). By the mechanism of **molecular diffusion**, this fluid imparts some of its momentum to the adjacent layer of fluid and causes it to also move in the z-direction. By the same mechanism this adjacent layer of fluid imparts some of its momentum to the next adjacent layer and causes it to also move in the z-direction and so on. The molecular diffusion across the boundary between two adjacent fluid layers in laminar flow is illustrated in the lower left of Fig. 1.1-1. In this way, the z-momentum is transmitted through the fluid in the y-direction. In fact, this is why fluid flow is considered as **momentum transfer**.

Turbulent flow, on the other hand, is a chaotic flow and is characterized by the transfer of small packets of fluid particles between layers, as illustrated in the lower right of Fig. 1.1-1. As expected, when fluid flow is intensified, it tends to switch from laminar to turbulent flow.

The transition from laminar to turbulent flow was studied by Reynolds in 1883, using an apparatus such as that shown in Fig. 1.1-2. As shown, a dye

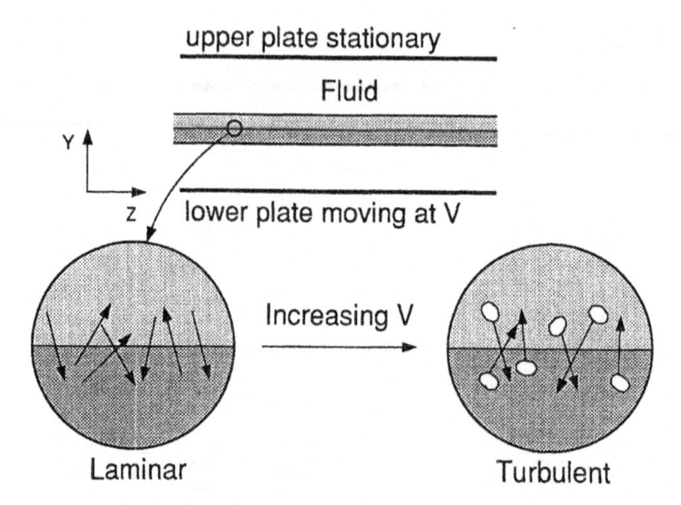

Fig. 1.1-1 Momentum transfer by motion of fluid molecules in laminar flow and by motion of fluid packets in turbulent flow.

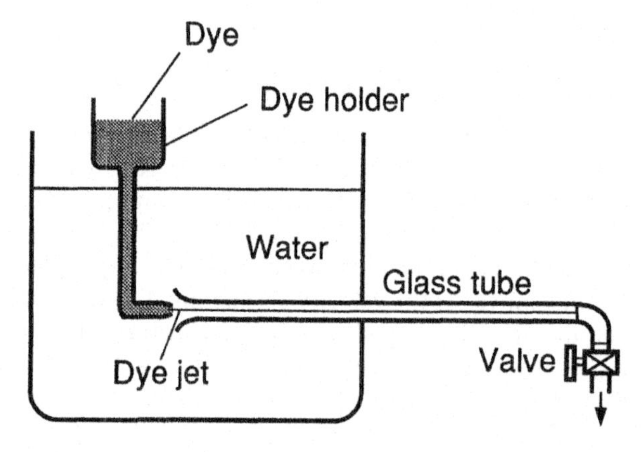

Fig. 1.1-2 Reynolds apparatus for studying transition from laminar to turbulent flow.

is injected along the axis of a glass tube through which water is flowing. Figure 1.1-3 shows representative patterns of the dye in water flowing through the glass tube.[1] At low water flow rates, the flow is laminar, as evidenced by the undisturbed single line of color in the water (top photo). The water layers or lamina are in the form of cylindrical shells of various diameters. At high water flow rates, the flow becomes turbulent, as evidenced by the dispersion of the dye in the water (bottom photo).

[1] H. Rouse, *Elementary Mechanics of Fluids*, Wiley, New York, 1946.

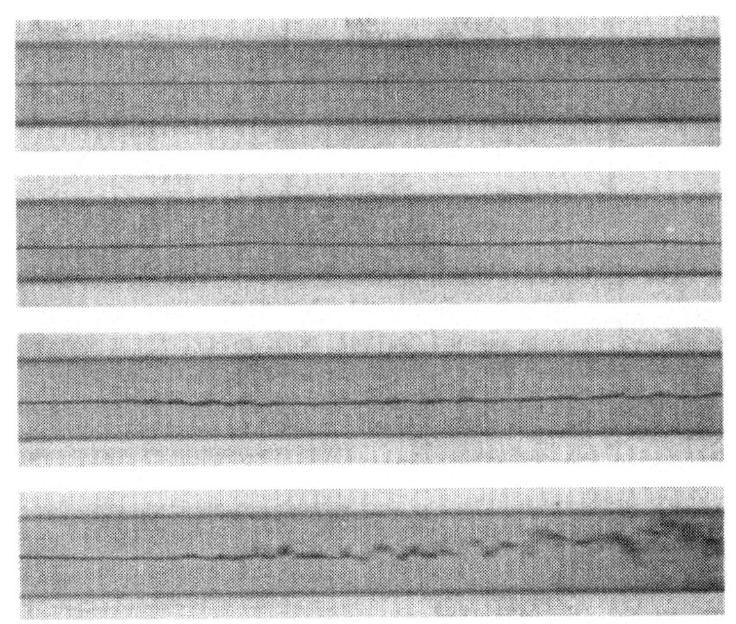

Fig. 1.1-3 Visualization of laminar-to-turbulent transition in water flowing through a glass tube. (From H. Rouse, *Elementary Mechanics of Fluids*, Wiley, New York, 1946.)

The transition from laminar to turbulent flow is visible at intermediate flow rates (center photos). Besides the fluid velocity (v), Reynolds found that the nature of flow is also affected by three other variables: the tube inner diameter (D), fluid density (ρ), and fluid viscosity (μ). The fluid velocity v here refers to that averaged over the cross-sectional area of the tube, specifically the volume of fluid collected at the exit per unit time divided by the inner cross-sectional area of the tube. These four variables can, in fact, be combined into a single dimensionless parameter:

$$\mathrm{Re}_D = \frac{D\rho v}{\mu} \quad \text{(Reynolds number)} \qquad [1.1\text{-}1]$$

This dimensionless parameter Re_D is called the **Reynolds number** in honor of Reynolds. The subscript D is to indicate that this particular Reynolds number is based on a diameter D. It has been observed that fluid flow in a circular tube is laminar if Re_D is less than approximately 2100. Beyond this, that is, the **critical Reynolds number**, fluid flow becomes turbulent. Although laminar flows have been observed in circular tubes with Re_D much larger than 2100, these flows are not stable and slight disturbances will cause a transition to turbulent flow.

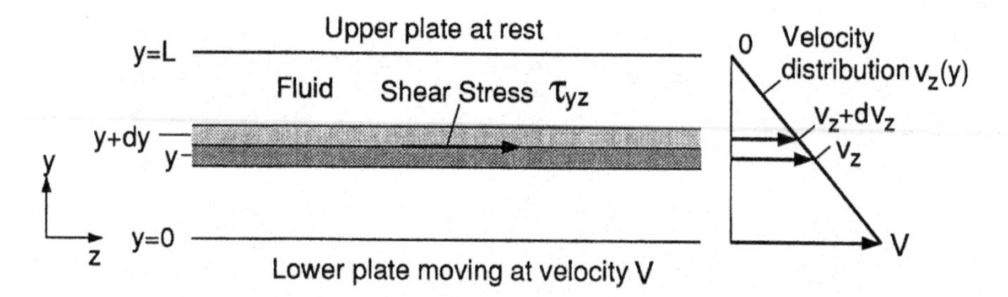

Fig. 1.1-4 One-dimensional flow of a fluid between two parallel plates caused by the motion of the lower plate.

1.1.2. Newton's Law of Viscosity

1.1.2.1. One-Dimensional

A fluid is contained between two large parallel plates separated by a small distance L, as shown in part in Fig. 1.1-4. The lower plate is pulled at a constant velocity V in the z direction, whereas the upper plate is kept stationary. A steady velocity profile $v_z(y)$ is established in the fluid, where v_z is the fluid velocity in the z direction and y the vertical distance form the lower plate. As shown, the fluid at the lower plate moves with it at the same velocity (i.e., $v_z = V$ at $y = 0$), whereas the fluid at the upper plate is motionless (i.e., $v_z = 0$ at $y = L$).

Let us consider the two thin layers of fluid shown in the figure. The lower layer at y has a velocity v_z, and the upper one at $y + dy$ has a slightly smaller velocity $v_z + dv_z$ ($dv_z < 0$). Since the lower layer is faster, it tends to exert a shear stress (stress = force/area) on the upper one in the y direction. This shear stress is denoted as τ_{yz} since it acts on a y plane (a plane of constant y) and in the z direction. The greater the velocity difference between the two layers – or, more appropriately, the steeper the velocity gradient dv_z/dy – is, the greater the shear stress τ_{yz} becomes. Furthermore, the more viscous the fluid is, the greater the shear stress τ_{yz} becomes. In fact, it is observed that

$$\boxed{\tau_{yz} = -\mu \frac{dv_z}{dy}} \qquad [1.1\text{-}2]$$

where μ is the **viscosity** of the fluid. This equation is **Newton's law of viscosity** for one-dimensional flow in the z direction. The minus sign in the equation is due to the fact that the velocity gradient dv_z/dy is negative. (*Note*: Throughout the text, boxing of equations is used for emphasis; equations that are boxed require closer attention.)

It can be shown from Eq. [1.1-2] that the unit of a shear stress is identical to that of a **momentum flux**, that is, the amount of momentum transferred per unit area per unit time (Problem 1.1-1). The minus sign before the velocity gradient suggests that a momentum flux occurs in the direction of decreasing velocity. The shear stress τ_{yz} is, in fact, the **flux of z-momentum in the y-direction**.

Newton's law of viscosity Eq. [1.1-2] applies in the regime of laminar flow. It does not apply in the regime of turbulent flow, which will be discussed later in Section 1.7.

Fluids that obey Eq. [1.1-2] are called **Newtonian fluids**. All gases and simple liquids (e.g., water, molten metals, molten semiconductors, and many molten salts) are Newtonian, while pastes, slurries and polymeric melts are non-Newtonian.

The cgs (centimeter-gram-second) unit of stress is dyn cm^{-2} (dyn/cm^2) or g cm^{-1} s^{-2}. Therefore, from Eq. [1.1-2] it can be shown that the cgs unit of viscosity is g cm^{-1} s^{-1}, which is called the **poise**. Most viscosity data are

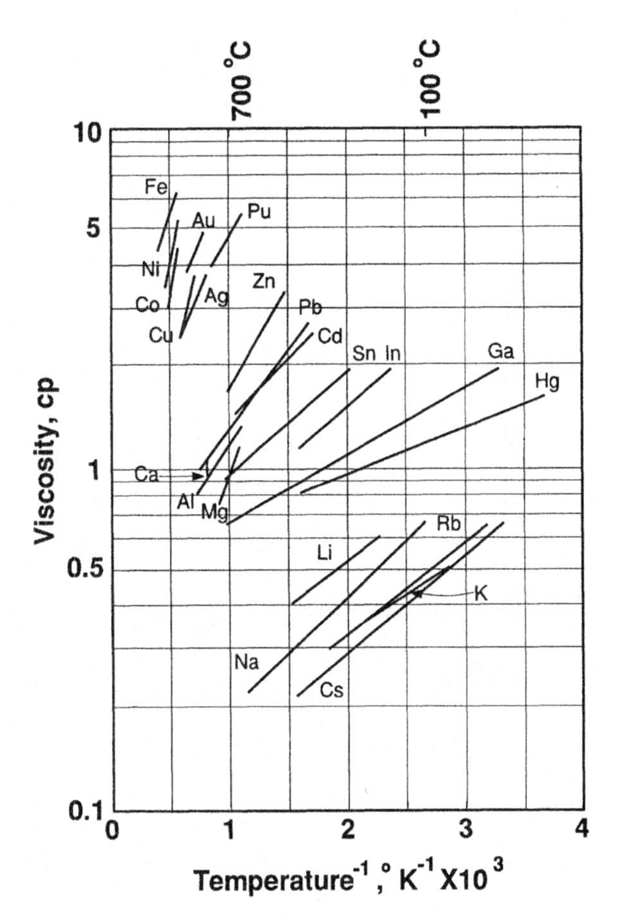

Fig. 1.1-5 Viscosities of some molten metals. (From T. Chapman, *AIChE Journal*, **12**, 1966, p. 395. Reproduced with permission of the American Institute of Chemical Engineers.)

reported either in this unit or in **centipoises** [1 cP = 0.01 P (poise)]. The viscosity of water at 20°C and 1 atm is 1.00 cP. The metric or mks (meter-kilogram-second) unit of stress is Pa (pascals) or kg $m^{-1} s^{-2}$. The corresponding unit of viscosity is Pa s (Pa s) or kg $m^{-1} s^{-1}$, and is equivalent to 10^3 cP.

The viscosities of some molten materials are shown in Figs. 1.1-5 through 1.1-7.

1.1.2.2. Three-Dimensional

In Eq. [1.1-2] Newton's law of viscosity is presented for one-directional (unidirectional) flow in the z direction. In general, fluid flow can be three-dimensional. Furthermore, each velocity component has three gradients, for example, $\partial v_x/\partial x$, $\partial v_x/\partial y$, and $\partial v_x/\partial z$ for the velocity component v_x. Therefore, one can expect to see nine stress terms: τ_{xx}, τ_{xy}, τ_{xz}, τ_{yx}, τ_{yy}, τ_{yz}, τ_{zx}, τ_{zy}, and τ_{zz}.

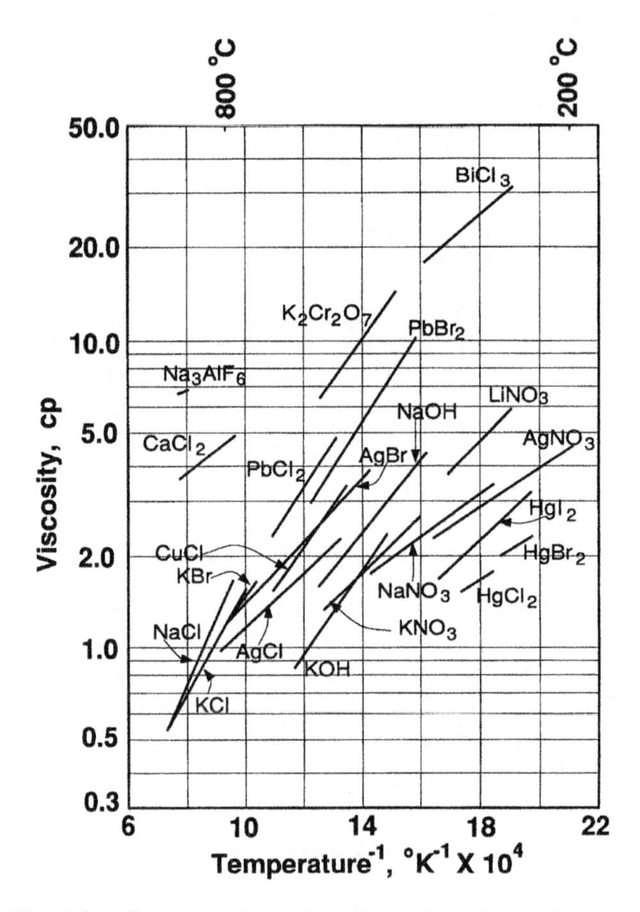

Fig. 1.1-6 Viscosities of some molten salts. (From C. J. Smithells, *Metals Reference Book*, 4th ed., vol. 1, Plenum Press, New York, 1967.)

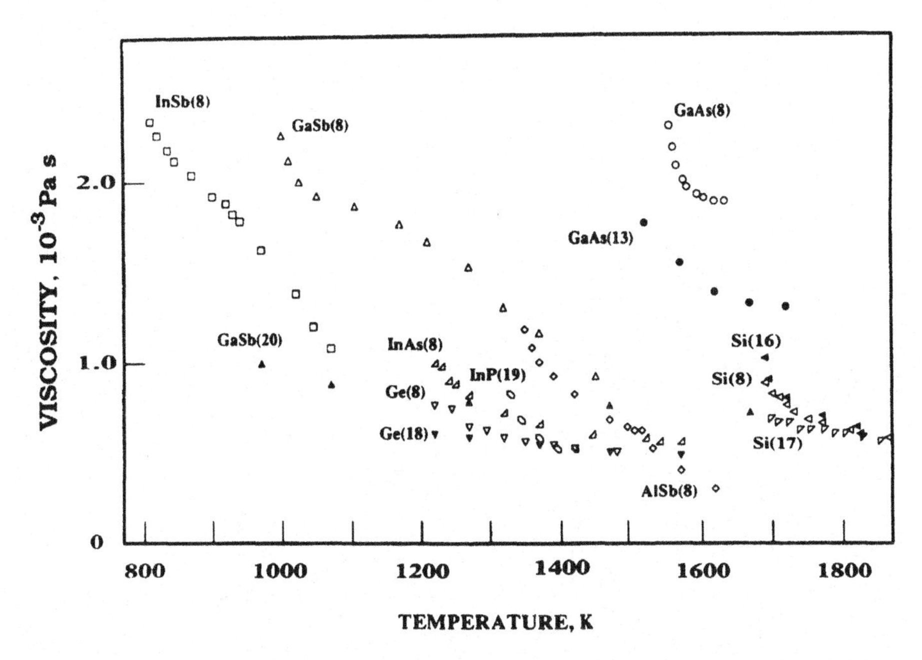

Fig. 1.1-7 Viscosities of some molten semiconductors. (S. Nakamura and T. Hibiya, *International Journal of Thermophysics*, **13**, 1992, p. 1061.)

As shown in Eq. [A.2-1], the stress tensor τ in rectangular coordinates is as follows:

$$\tau = \begin{bmatrix} \tau_{xx} & \tau_{xy} & \tau_{xz} \\ \tau_{yx} & \tau_{yy} & \tau_{yz} \\ \tau_{zx} & \tau_{zy} & \tau_{zz} \end{bmatrix} \qquad [1.1\text{-}3]$$

The planes and directions associated with these stress components are shown in Fig. 1.1-8. Notice that a stress component points in one direction on one surface of the cube and in the opposite direction on the opposite surface. For example, τ_{zy} points in the y direction on the top surface but in the negative y direction on the bottom surface. Suppose that v_y increases with increasing z and that the fluid in the cube moves more slowly in the y direction than the fluid above it and hence experiences a pulling shear at the top surface in the y direction. On the other hand, the fluid in the cube moves faster in the y direction than the fluid below it and hence experiences a dragging shear at the bottom surface in the negative y direction.

It has been shown[2] that for three-dimensional fluid flow, Newton's law of viscosity can be written as follows:

[2] H. Lamb, *Hydrodynamics*, Dover, New York, 1945.

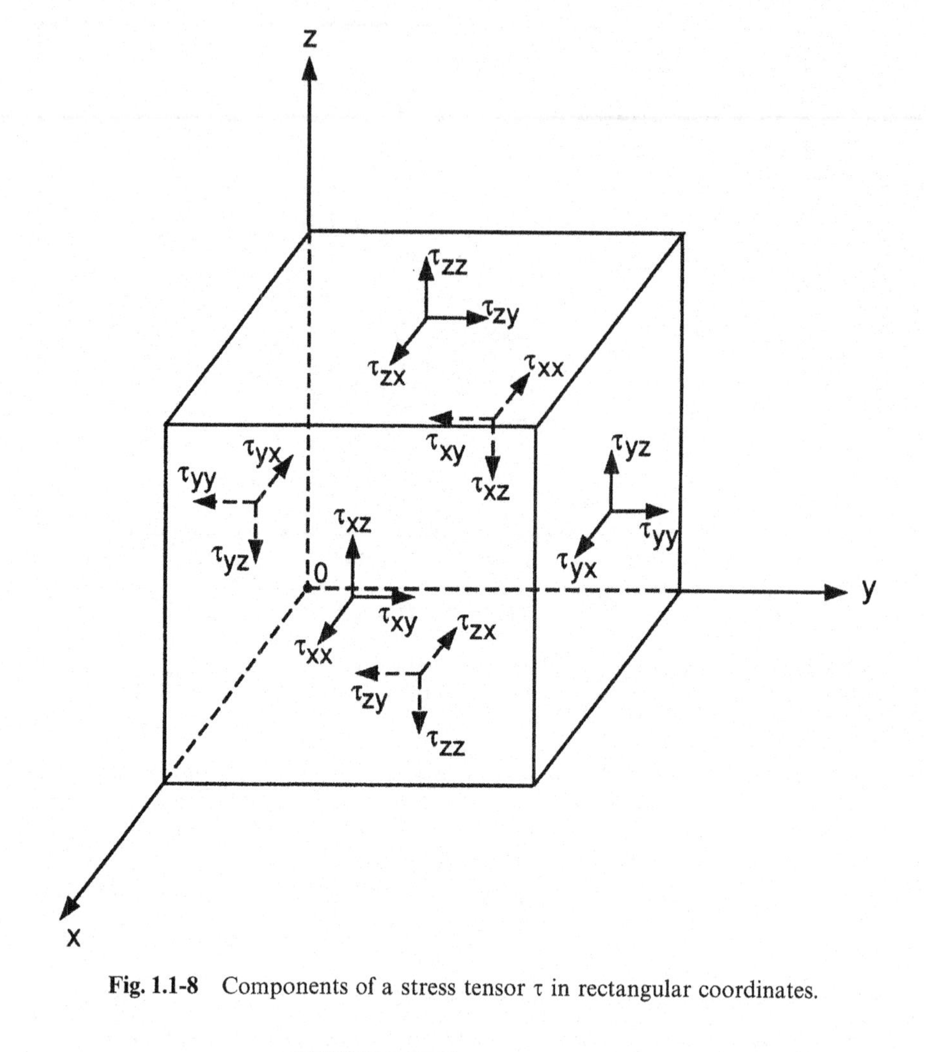

Fig. 1.1-8 Components of a stress tensor τ in rectangular coordinates.

$$\tau_{xx} = -2\mu \frac{\partial v_x}{\partial x} + \frac{2}{3}\mu (\nabla \cdot \mathbf{v}) \quad [1.1\text{-}4]$$

$$\tau_{yy} = -2\mu \frac{\partial v_y}{\partial y} + \frac{2}{3}\mu (\nabla \cdot \mathbf{v}) \quad [1.1\text{-}5]$$

$$\tau_{zz} = -2\mu \frac{\partial v_z}{\partial z} + \frac{2}{3}\mu (\nabla \cdot \mathbf{v}) \quad [1.1\text{-}6]$$

$$\tau_{xy} = \tau_{yx} = -\mu \left[\frac{\partial v_x}{\partial y} + \frac{\partial v_y}{\partial x} \right] \quad [1.1\text{-}7]$$

$$\tau_{yz} = \tau_{zy} = -\mu\left[\frac{\partial v_y}{\partial z} + \frac{\partial v_z}{\partial y}\right] \quad [1.1\text{-}8]$$

$$\tau_{zx} = \tau_{xz} = -\mu\left[\frac{\partial v_z}{\partial x} + \frac{\partial v_x}{\partial z}\right] \quad [1.1\text{-}9]$$

As shown, $\boldsymbol{\tau}$ is a symmetric tensor: $\tau_{ij} = \tau_{ji}$; $\mathbf{v}$ is the velocity vector, and the quantity $(\nabla \cdot \mathbf{v})$, as shown in Eq. [A.3-5], is given by

$$\nabla \cdot \mathbf{v} = \frac{\partial v_x}{\partial x} + \frac{\partial v_y}{\partial y} + \frac{\partial v_z}{\partial z} \quad [1.1\text{-}10]$$

For fluids with a constant density ρ, that is, **incompressible fluids**, $\nabla \cdot \mathbf{v} = 0$, as will be shown in Section 1.3. As such, from Eqs. [1.1-3] through [1.1-9], for incompressible Newtonian fluids

$$\boldsymbol{\tau} = -\mu\begin{bmatrix} \frac{\partial v_x}{\partial x} + \frac{\partial v_x}{\partial x} & \frac{\partial v_y}{\partial x} + \frac{\partial v_x}{\partial y} & \frac{\partial v_z}{\partial x} + \frac{\partial v_x}{\partial z} \\ \frac{\partial v_x}{\partial y} + \frac{\partial v_y}{\partial x} & \frac{\partial v_y}{\partial y} + \frac{\partial v_y}{\partial y} & \frac{\partial v_z}{\partial y} + \frac{\partial v_y}{\partial z} \\ \frac{\partial v_x}{\partial z} + \frac{\partial v_z}{\partial x} & \frac{\partial v_y}{\partial z} + \frac{\partial v_z}{\partial y} & \frac{\partial v_z}{\partial z} + \frac{\partial v_z}{\partial z} \end{bmatrix} \quad [1.1\text{-}11]$$

or

$$\boldsymbol{\tau} = -\mu\begin{bmatrix} \frac{\partial v_x}{\partial x} & \frac{\partial v_y}{\partial x} & \frac{\partial v_z}{\partial x} \\ \frac{\partial v_x}{\partial y} & \frac{\partial v_y}{\partial y} & \frac{\partial v_z}{\partial y} \\ \frac{\partial v_x}{\partial z} & \frac{\partial v_y}{\partial z} & \frac{\partial v_z}{\partial z} \end{bmatrix} - \mu\begin{bmatrix} \frac{\partial v_x}{\partial x} & \frac{\partial v_x}{\partial y} & \frac{\partial v_x}{\partial z} \\ \frac{\partial v_y}{\partial x} & \frac{\partial v_y}{\partial y} & \frac{\partial v_y}{\partial z} \\ \frac{\partial v_z}{\partial x} & \frac{\partial v_z}{\partial y} & \frac{\partial v_z}{\partial z} \end{bmatrix} \quad [1.1\text{-}12]$$

or, from Eq. [A.3-12].

$$\boldsymbol{\tau} = -\mu(\nabla\mathbf{v} + \nabla\mathbf{v}^T) \quad [1.1\text{-}13]$$

where T denotes the transpose of a tensor, specifically, that tensor formed by interchanging the subscripts on each of the elements.

Tables 1.1-1, 1.1-2, and 1.1-3 list the components of the stress tensor $\boldsymbol{\tau}$ in rectangular, cylindrical, and spherical coordinates, respectively.

TABLE 1.1-1 Components of the Stress Tensor τ for Newtonian Fluids in Rectangular Coordinates (x, y, z)

$\tau_{xx} = -\mu\left[2\dfrac{\partial v_x}{\partial x} - \dfrac{2}{3}(\nabla\cdot\mathbf{v})\right]$	[A]
$\tau_{yy} = -\mu\left[2\dfrac{\partial v_y}{\partial y} - \dfrac{2}{3}(\nabla\cdot\mathbf{v})\right]$	[B]
$\tau_{zz} = -\mu\left[2\dfrac{\partial v_z}{\partial z} - \dfrac{2}{3}(\nabla\cdot\mathbf{v})\right]$	[C]
$\tau_{xy} = \tau_{yx} = -\mu\left[\dfrac{\partial v_x}{\partial y} + \dfrac{\partial v_y}{\partial x}\right]$	[D]
$\tau_{yz} = \tau_{zy} = -\mu\left[\dfrac{\partial v_y}{\partial z} + \dfrac{\partial v_z}{\partial y}\right]$	[E]
$\tau_{zx} = \tau_{xz} = -\mu\left[\dfrac{\partial v_z}{\partial x} + \dfrac{\partial v_x}{\partial z}\right]$	[F]
$(\nabla\cdot\mathbf{v}) = \dfrac{\partial v_x}{\partial x} + \dfrac{\partial v_y}{\partial y} + \dfrac{\partial v_z}{\partial z}$	[G]

1.1.2.3. Kinematic Viscosity

The viscosity μ is sometimes called the **dynamic viscosity**. The so-called **kinematic viscosity** ν is defined as follows:

$$\boxed{\nu = \frac{\mu}{\rho}} \qquad [1.1\text{-}14]$$

where ρ is the density of the fluid. The cgs and mks unit of kinematic viscosity are cm^2/s and m^2/s, respectively.

1.1.2.4. Stress Calculation Procedure

According to Newton's law of viscosity (e.g., Eqs. [1.1-4]–[1.1-9]), the stresses acting by a moving fluid on a solid can be determined if the velocity distribution in the fluid is known. The procedure, which is illustrated in Examples 1.1-1, 1.1-2, and later in 1.3-2, is as follows:

1. Choose a proper coordinate system: rectangular coordinates (Table 1.1-1), cylindrical coordinates (Table 1.1-2), and spherical coordinates (Table 1.1-3).

TABLE 1.1-2 Components of the Stress Tensor τ for Newtonian Fluids in Cylindrical Coordinates (r, θ, z)

$$\tau_{rr} = -\mu\left[2\frac{\partial v_r}{\partial r} - \frac{2}{3}(\nabla\cdot\mathbf{v})\right] \qquad [A]$$

$$\tau_{\theta\theta} = -\mu\left[2\left(\frac{1}{r}\frac{\partial v_\theta}{\partial \theta} + \frac{v_r}{r}\right) - \frac{2}{3}(\nabla\cdot\mathbf{v})\right] \qquad [B]$$

$$\tau_{zz} = -\mu\left[2\frac{\partial v_z}{\partial z} - \frac{2}{3}(\nabla\cdot\mathbf{v})\right] \qquad [C]$$

$$\tau_{r\theta} = \tau_{\theta r} = -\mu\left[r\frac{\partial}{\partial r}\left(\frac{v_\theta}{r}\right) + \frac{1}{r}\frac{\partial v_r}{\partial \theta}\right] \qquad [D]$$

$$\tau_{\theta z} = \tau_{z\theta} = -\mu\left[\frac{\partial v_\theta}{\partial z} + \frac{1}{r}\frac{\partial v_z}{\partial \theta}\right] \qquad [E]$$

$$\tau_{zr} = \tau_{rz} = -\mu\left[\frac{\partial v_z}{\partial r} + \frac{\partial v_r}{\partial z}\right] \qquad [F]$$

$$(\nabla\cdot\mathbf{v}) = \frac{1}{r}\frac{\partial}{\partial r}(r v_r) + \frac{1}{r}\frac{\partial v_\theta}{\partial \theta} + \frac{\partial v_z}{\partial z} \qquad [G]$$

2. Determine which stress component is relevant to the problem and pick the corresponding equation from the table chosen in step 1, for instance, Eq. [F] of Table 1.1-2 for τ_{rz}.

3. Substitute the given velocity distribution(s) into the equation chosen in step 2 and determine the stress component.

The velocity distribution, which is given here tentatively for the purpose of illustration, can be found by solving the equation of motion, as will be described in Section 1.5.

Example 1.1-1 Shear Stress Due to Laminar Flow through a Tube

Consider the steady-state laminar flow of an incompressible Newtonian fluid through a tube of inner radius R and length L at a constant volume flow rate Q. Except near the entrance and the exit, the flow is one-dimensional along the axis. A short section of the tube is shown in Fig. 1.1-9. The velocity distribution is

TABLE 1.1-3 Components of the Stress Tensor τ for Newtonian Fluids in Spherical Coordinates (r, θ, ϕ)

$$\tau_{rr} = -\mu\left[2\frac{\partial v_r}{\partial r} - \frac{2}{3}(\nabla\cdot\mathbf{v})\right] \qquad \text{[A]}$$

$$\tau_{\theta\theta} = -\mu\left[2\left(\frac{1}{r}\frac{\partial v_\theta}{\partial\theta} + \frac{v_r}{r}\right) - \frac{2}{3}(\nabla\cdot\mathbf{v})\right] \qquad \text{[B]}$$

$$\tau_{\phi\phi} = -\mu\left[2\left(\frac{1}{r\sin\theta}\frac{\partial v_\phi}{\partial\phi} + \frac{v_r}{r} + \frac{v_\theta\cot\theta}{r}\right) - \frac{2}{3}(\nabla\cdot\mathbf{v})\right] \qquad \text{[C]}$$

$$\tau_{r\theta} = \tau_{\theta r} = -\mu\left[r\frac{\partial}{\partial r}\left(\frac{v_\theta}{r}\right) + \frac{1}{r}\frac{\partial v_r}{\partial\theta}\right] \qquad \text{[D]}$$

$$\tau_{\theta\phi} = \tau_{\phi\theta} = -\mu\left[\frac{\sin\theta}{r}\frac{\partial}{\partial\theta}\left(\frac{v_\phi}{\sin\theta}\right) + \frac{1}{r\sin\theta}\frac{\partial v_\theta}{\partial\phi}\right] \qquad \text{[E]}$$

$$\tau_{\phi r} = \tau_{r\phi} = -\mu\left[\frac{1}{r\sin\theta}\frac{\partial v_r}{\partial\phi} + r\frac{\partial}{\partial r}\left(\frac{v_\phi}{r}\right)\right] \qquad \text{[F]}$$

$$(\nabla\cdot\mathbf{v}) = \frac{1}{r^2}\frac{\partial}{\partial r}(r^2 v_r) + \frac{1}{r\sin\theta}\frac{\partial}{\partial\theta}(v_\theta\sin\theta) + \frac{1}{r\sin\theta}\frac{\partial v_\phi}{\partial\phi} \qquad \text{[G]}$$

as follows[3]

$$v_z = \frac{2Q}{\pi R^2}\left[1 - \left(\frac{r}{R}\right)^2\right] \qquad \text{[1.1-15]}$$

where r is the radial coordinate. Calculate the shear stress and friction force at the tube wall.

Solution Clearly cylindrical coordinates are most appropriate for the problem. At the tube wall, the shear stress acts on a plane of constant r ($r = R$) and points in the z direction, thus suggesting that τ_{rz} is the component of the stress tensor $\boldsymbol{\tau}$ to be considered. Newton's law of viscosity, according to Eq. [F] of Table 1.1-2,

$$\tau_{rz} = -\mu\left[\frac{\partial v_z}{\partial r} + \frac{\partial v_r}{\partial z}\right] \qquad \text{[1.1-16]}$$

[3] Derived in Example 1.5-2.

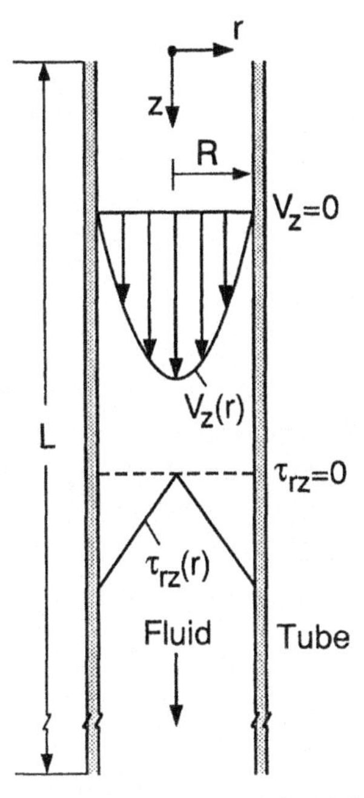

Fig. 1.1-9 Steady-state, one-dimensional fluid flow in a round tube.

Since fluid flow is in the z direction only, $v_r = 0$. Also, from Eq. [1.1-15] v_z is a function of r only. Therefore, the above equation reduces to

$$\tau_{rz} = -\mu \frac{\partial v_z}{\partial r} = -\mu \frac{dv_z}{dr} \qquad [1.1\text{-}17]$$

Substituting Eq. [1.1-15] into Eq. [1.1-17], we obtain

$$\tau_{rz} = -\frac{2\mu Q}{\pi R^2}\frac{d}{dr}\left[1 - \frac{r^2}{R^2}\right] = \frac{4\mu Q r}{\pi R^4} \qquad [1.1\text{-}18]$$

Therefore, τ_{rz} is a linear function of r. At the tube wall, $r = R$ and

$$\tau_{rz}|_{r=R} = \frac{4\mu Q}{\pi R^3} \qquad [1.1\text{-}19]$$

The friction force exerted by the fluid on the tube wall is

$$F_z = (2\pi R L)\,\tau_{rz}|_{r=R} \qquad [1.1\text{-}20]$$

Substituting Eq. [1.1-19] into Eq. [1.1-20]

$$F_z = \frac{8\mu L Q}{R^2} \qquad [1.1\text{-}21]$$

Example 1.1-2 Shear Stress Due to Creeping Flow around a Sphere

An incompressible, Newtonian fluid approaches a solid sphere of radius R at a constant and uniform velocity of v_∞, as illustrated in Fig. 1.1-10. The flow and the size of the sphere are such that the Reynolds number $2R\rho v_\infty/\mu$ is significantly less than one; this is the **creeping flow** condition. It has been found analytically that in creeping flow around a sphere[4]

$$v_r = v_\infty \left[1 - \frac{3}{2}\left(\frac{R}{r}\right) + \frac{1}{2}\left(\frac{R}{r}\right)^3 \right] \cos\theta \qquad [1.1\text{-}22]$$

$$v_\theta = -v_\infty \left[1 - \frac{3}{4}\left(\frac{R}{r}\right) - \frac{1}{4}\left(\frac{R}{r}\right)^3 \right] \sin\theta \qquad [1.1\text{-}23]$$

Calculate the shear stress acting on the sphere by the fluid.

Solution Clearly spherical coordinates are most appropriate for this problem. The relevant shear stress acting on the surface of the sphere (an r plane) are $\tau_{r\theta}$ and $\tau_{r\phi}$. The normal stress τ_{rr} will be considered in Example 1.2-1. Let us first examine $\tau_{r\theta}$. Newton's law of viscosity, according to Eq. [D] of Table 1.1-3, is as follows:

$$\tau_{r\theta} = -\mu \left[r\frac{\partial}{\partial r}\left(\frac{v_\theta}{r}\right) + \frac{1}{r}\frac{\partial v_r}{\partial \theta} \right] \qquad [1.1\text{-}24]$$

From Eqs. [1.1-22] and [1.1-23],

$$-\mu r\frac{\partial}{\partial r}\left(\frac{v_\theta}{r}\right) = \mu v_\infty r\frac{\partial}{\partial r}\left[\frac{1}{r} - \frac{3}{4}\frac{R}{r^2} - \frac{1}{4}\frac{R^3}{r^4}\right]\sin\theta$$

$$= \mu v_\infty \left[-\frac{1}{r} + \frac{3}{2}\frac{R}{r^2} + \frac{R^3}{r^4} \right] \sin\theta \qquad [1.1\text{-}25]$$

Similarly,

$$-\frac{\mu}{r}\frac{\partial v_r}{\partial \theta} = -\mu v_\infty \frac{1}{r}\left[1 - \frac{3}{2}\frac{R}{r} + \frac{1}{2}\frac{R^3}{r^3}\right]\frac{\partial}{\partial \theta}(\cos\theta)$$

$$= \mu v_\infty \left[\frac{1}{r} - \frac{3}{2}\frac{R}{r^2} + \frac{1}{2}\frac{R^3}{r^4} \right] \sin\theta \qquad [1.1\text{-}26]$$

[4] C. G. Stokes, *Trans. Cambridge Phil. Soc.*, **9**, 8, 1850.

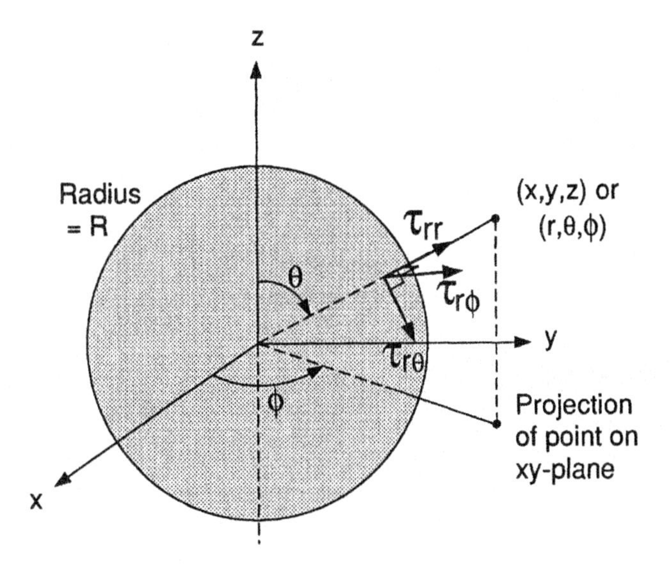

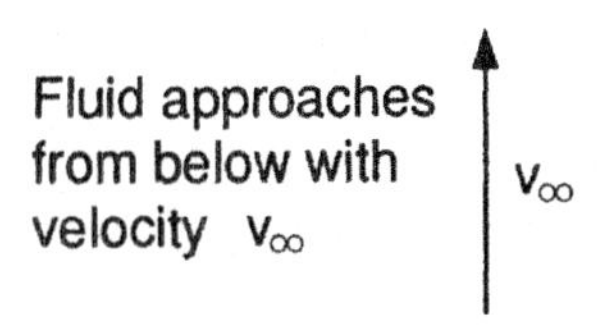

Fig. 1.1-10 Creeping flow around a sphere.

Substituting Eqs. [1.1-25] and [1.1-26] into Eq. [1.1-24]

$$\tau_{r\theta} = \frac{3}{2}\frac{\mu v_\infty}{R}\left[\frac{R}{r}\right]^4 \sin\theta \qquad [1.1\text{-}27]$$

and so

$$\tau_{r\theta}|_R = \frac{3}{2}\frac{\mu v_\infty}{R} \sin\theta \qquad [1.1\text{-}28]$$

Similarly, Newton's law of viscosity, according to Eq. [F] of Table 1.1-3, is as follows

$$\tau_{r\phi} = -\mu\left[\frac{1}{r\sin\theta}\frac{\partial v_r}{\partial\phi} + r\frac{\partial}{\partial r}\left(\frac{v_\phi}{r}\right)\right] \qquad [1.1\text{-}29]$$

Since there is no flow in the ϕ direction, $v_\phi = 0$. Furthermore, since the flow field is symmetric with respect to the z axis, v_r should not vary with ϕ (i.e., $\partial v_r/\partial\phi = 0$). In fact, it can also be shown from Eq. [1.1-22] that $\partial v_r/\partial\phi = 0$. As such, Eq. [1.1-29] reduces to

$$\tau_{r\phi} = 0 \qquad [1.1\text{-}30]$$

Equations [1.1-28] can be used to calculate the friction force exerted on the sphere by the fluid, as will be shown in Example 1.4-5.

1.1.3. Streamlines

Streamlines are a family of curves showing the pattern of fluid flow, which are drawn in such manner that the velocity vectors for all points along the curves would meet them tangentially. Figure 1.1-11*a* shows the streamlines in a fluid flowing past an egg-shaped body.

As illustrated in Fig. 1.1-11*b*

$$\frac{v_y}{v_x} = \frac{dy}{dx} = \text{slope of streamline} \qquad [1.1\text{-}31]$$

where v_x and v_y are the x component and y component of the velocity vector $\mathbf{v}$ respectively. As such, the velocity vector $\mathbf{v}$ is tangent to the streamline. In three dimensions the relation becomes

$$\frac{dx}{v_x} = \frac{dy}{v_y} = \frac{dz}{v_z} \qquad [1.1\text{-}32]$$

In steady flow, a streamline is also a **path line**, which is a line describing the path or trajectory of a given fluid element.

In fluid flow experiments called **flow visualization**, a tracer, such as a dye or smoke, is used to reveal streamlines in a transparent fluid. Figure 1.1-12*a* shows streamlines in water flowing through a two-dimensional channel with a boundary contraction, made visible by dye injection.[5]

Streamlines can also reveal variations in the fluid velocity, and this is particularly obvious in two-dimensional flows, such as that shown in Fig. 1.1-12. Convergence of streamlines represents accelerations, whereas divergence represents deceleration. Referring to the contraction in Fig. 1.1-12*a*, the amount of water passing by per unit time, that is, the water flow rate, is the same downstream as upstream. Since the cross-sectional area available for flow is smaller downstream than upstream, the average water velocity has to be higher

[5] H. Rouse, ibid.

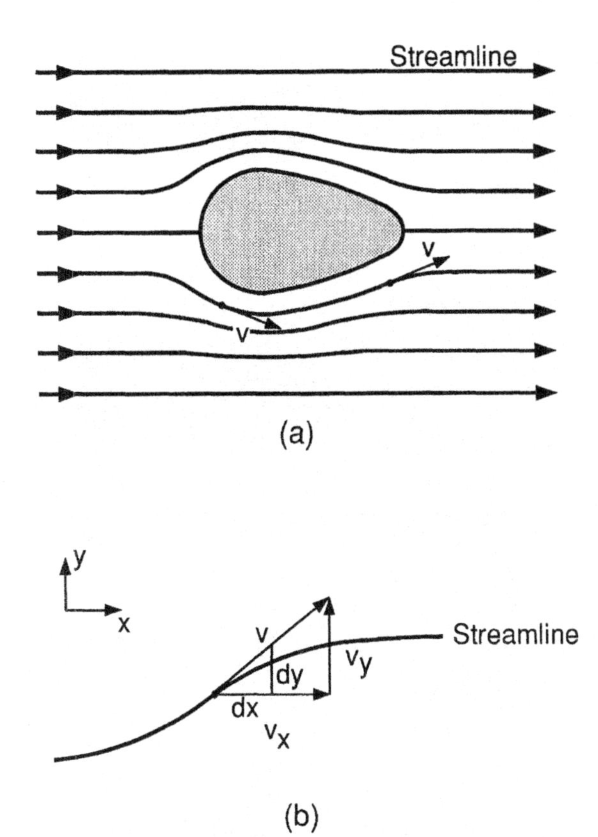

Fig. 1.1-11 Fluid flow past a submerged egg-shaped body: (*a*) streamlines; (*b*) a velocity vector **v** along a streamline.

downstream than upstream. This acceleration of water is reflected by the convergence of the streamlines at the contraction.

As the water channel in Fig. 1.1-12*a* is reversed in direction, water decelerates and the streamlines diverge at the enlargement.[6] However, as shown in Fig. 1.1-12*b*, the degree of divergence is significantly less than that of convergence shown previously. In fact, the stream is separated from the boundary of the channel by a region of backflow. Had the enlargement been made more gradual (i.e., streamlined), separation from the boundary could have been avoided.

It is worth noting that streamlines can still be used to represent the flow pattern when turbulence exists. These streamlines, however, represent only the average or time-smoothed pattern of fluid motion.

[6] H. Rouse, ibid.

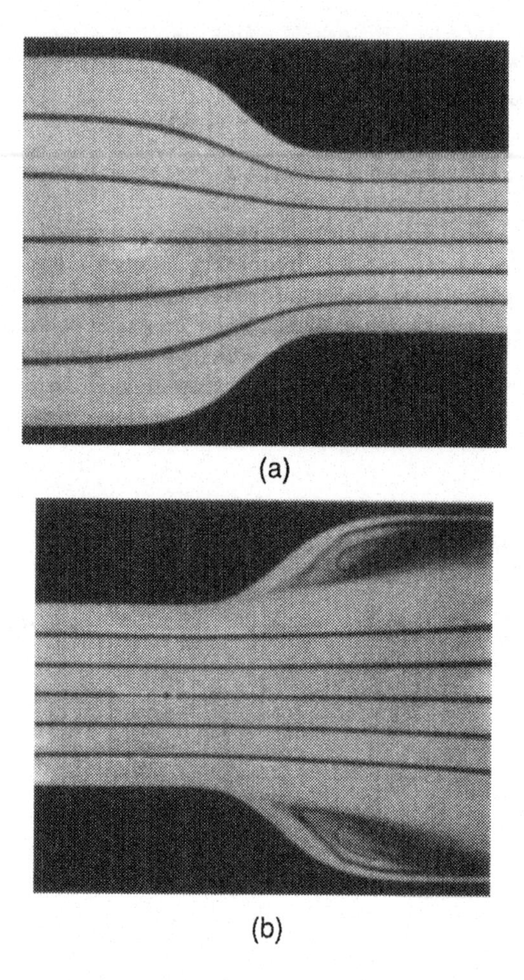

Fig. 1.1-12 Visualization of streamlines in two-dimensional flow of water at (*a*) a contraction; (*b*) an expansion. (From H. Rouse, *Elementary Mechanics of Fluids,* Wiley, New York, 1946).

1.1.4. Momentum Boundary Layer

1.1.4.1. External Flow

Consider first a fluid of uniform velocity v_∞ approaching a stationary flat plate in the direction parallel to the plate, as illustrated in Fig. 1.1-13. Because of the effect of viscosity, the velocity v_z of the fluid in the region near the plate is reduced by the plate, varying from zero at the plate surface to v_∞ in the stream. This region is called the **momentum boundary layer**. Its thickness δ is typically taken as the distance from the plate surface at which the dimensionless velocity v_z/v_∞ levels off to 0.99. In practice, it is usually specified that $v_z = v_\infty$ and

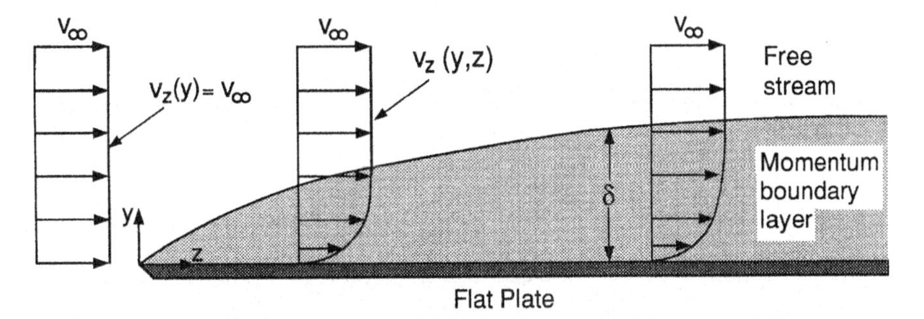

Fig. 1.1-13 Momentum boundary layer over a flat plate: laminar.

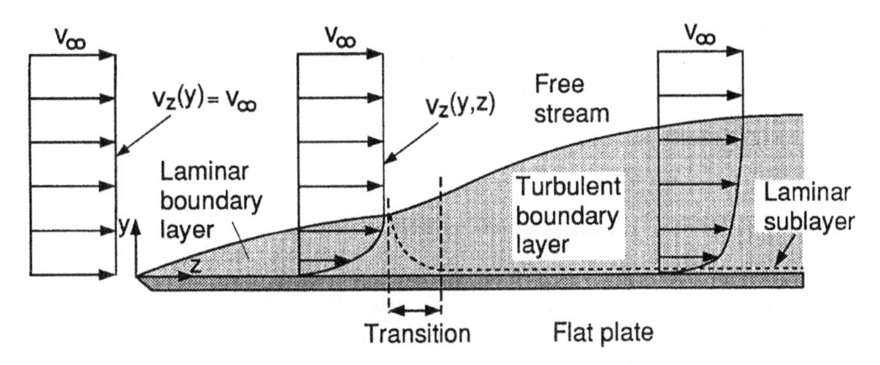

Fig. 1.1-14 Momentum boundary layer over a flat plate: laminar-to-turbulent transition.

$\partial v_z/\partial y = 0$ at $y = \delta$. With increasing distance from the leading edge of the plate, the effect of viscosity penetrates further into the stream and the boundary layer grows in thickness.

The concept of the momentum boundary layer was first proposed by Prandtl[7] in 1904. He suggested that fluid flows can be analyzed by dividing the flow into two regions: (1) a thin region adjacent to the solid boundary – the boundary layer, in which the flow is retarded because of the effect of viscosity; and (2) the region covering the rest of the fluid – the free stream, in which the velocity is uniform and the effect of viscosity is no longer significant.

If the flat plate is sufficiently long, the flow in the boundary layer can eventually change from laminar to turbulent. The higher the v_∞ is, the closer to the leading edge of the plate the laminar-to-turbulent transition occurs. This transition, as illustrated in Fig. 1.1-14, is not sharp but covers some distance along the plate. It is usually considered to occur at a length Reynolds number,

[7] L. Prandtl, in *Proceedings of the Third International Congress on Mathematics*, Heidelberg, 1904.

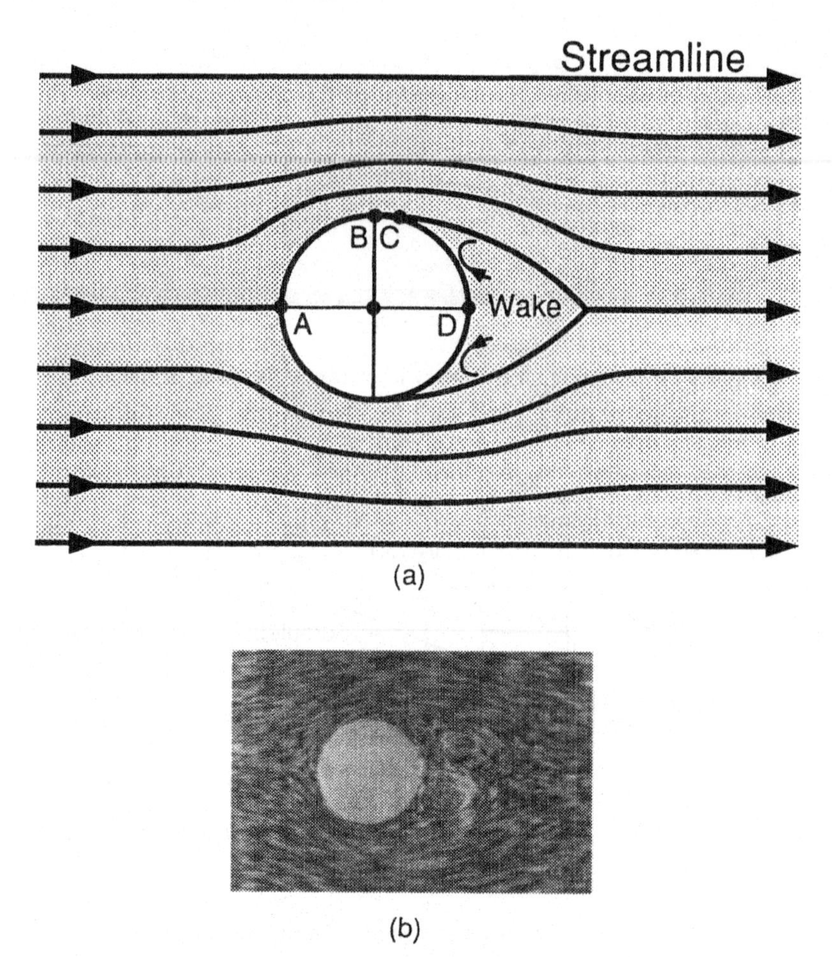

Fig. 1.1-15 Fluid flow past a cylinder: (*a*) schematic sketch; (*b*) flow visualization. (From H. Rouse, *Elementary Mechanics of Fluids*, Wiley, New York, 1946.)

specifically, $Re_z = \rho v_\infty z/\mu$, of about 2×10^5, where z is the distance or length from the leading edge.

Let us now consider fluid flow normal to a body, such as a cylinder or sphere. Figure 1.1-15 illustrates fluid flow past a cylinder. Point A is called the **stagnation point** where fluid comes to a stop at the cylinder. Since no fluid crosses the streamline passing through point A, we can consider only one half of the flow field, that is, the half on the side of $ABCD$.

The cylinder surface along AB acts as a contraction as in the case shown in Fig. 1.1-12*a*. As such, the adjacent streamlines converge and the fluid accelerates. In contrast, the cylinder surface along BD acts as an enlargement as in the case shown in Fig. 1.1-12*b*. Therefore, the adjacent streamlines diverge and the fluid decelerates. In fact, as illustrated in Fig. 1.1-15*a*, the flow is sufficiently fast that

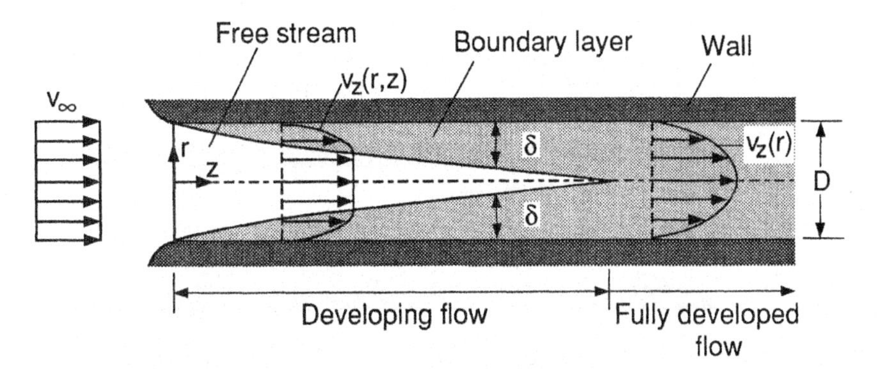

Fig. 1.1-16 Momentum boundary layers in a circular tube near its entrance: laminar.

the boundary layer separates itself from the cylinder at point *C* since surface *BD* is far from being streamlined. The region of slow backflow near surface *CD* is called a **wake**. Figure 1.1-15*b* shows boundary layer separation and backflow in water flowing past a cylinder.[8] Boundary layer separation was also shown in Fig. 1.1-12*b*.

1.1.4.2. Internal Flow

Consider a fluid that enters a circular tube of inner diameter *D* with a uniform velocity v_∞ parallel to its axis, as illustrated in Fig. 1.1-16. A laminar boundary layer begins to develop at the entrance, gradually expanding at the expense of the free stream until the layers from opposite sides approach the centerline at[9]

$$\frac{z}{D} \approx 0.05 \frac{\rho v_\infty D}{\mu} = 0.05 \text{ Re}_D \qquad [1.1\text{-}33]$$

The flow in the region where the boundary layer thickness δ and the velocity profile $v_z(r, z)$ vary with the axial distance z is called **developing flow**. On the other hand, the flow in the region where they no longer vary with z is called **fully developed flow**.

If the Reynolds number Re_D is increased significantly, however, fluid flow in the boundary layer can change to turbulent before the distance according to Eq. [1.1-33] is reached. This is illustrated in Fig. 1.1-17. As in the case of turbulent flow over the flat plate shown in Fig. 1.1-14, there exists a laminar sublayer along the wall.

[8] H. Rouse, ibid.

[9] H. L. Langhaar, *J. Appl. Mech.*, **64**, A-55, 1942.

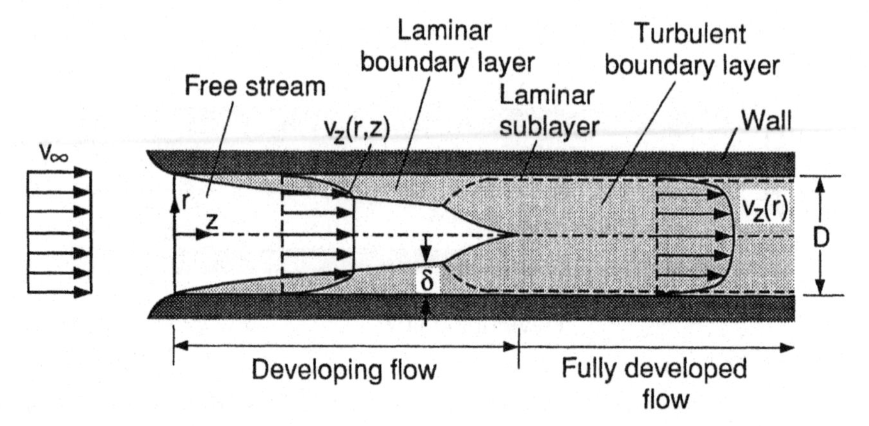

Fig. 1.1-17 Momentum boundary layer in a circular tube near its entrance: laminar-to-turbulent transition.

1.1.5. Momentum Transfer Coefficient and Other Coefficients

1.1.5.1. External Flow

Consider fluid flow over a flat plate such as that shown in Fig. 1.1-14. The shear stress at the plate is

$$\tau_{yz}\bigg|_{y=0} = -\mu \frac{\partial v_z}{\partial y}\bigg|_{y=0} \qquad [1.1\text{-}34]$$

This equation cannot be used to calculate the shear stress when the velocity gradient is an unknown. A convenient way to avoid this problem is to introduce a **momentum transfer coefficient**, defined as follows

$$C_f' = \left| \frac{\tau_{yz}|_{y=0}}{v_z|_{y=0} - v_\infty} \right| = \left| \frac{\tau_{yz}|_{y=0}}{0 - v_\infty} \right| \qquad [1.1\text{-}35]$$

The absolute values are used in order to keep C_f' always positive. The way this momentum transfer coefficient is defined is similar to the way the heat and mass transfer coefficients are defined, as will be shown in Sections 2.1.4 and 3.1.6. The **friction coefficient** is a dimensionless number defined by

$$\boxed{C_f = \frac{C_f'}{\frac{1}{2}\rho v_\infty} = \left\| \frac{\tau_{yz}|_{y=0}}{\frac{1}{2}\rho v_\infty^2} \right\|} \qquad [1.1\text{-}36]$$

Consider now fluid flow normal to a cylinder such as that shown in Fig. 1.1-15. Let F_D be the total drag force acting on the cylinder by the fluid. The cylinder not only presents a surface in contact with the fluid to cause the **friction**

drag, but also a physical body blocking the flow to cause the **form drag**. Apparently, the body of the cylinder is responsible for boundary layer separation and wake formation.

For flow normal to a cylinder, sphere, disk or plate, the **drag coefficient** is a dimensionless number defined by

$$C_D = \frac{F_D}{A_f(\rho v_\infty^2/2)} \qquad [1.1\text{-}37]$$

where A_f is the frontal area of the solid, that is, the area of the solid projected perpendicular to the free-stream velocity v_∞. For example, for a sphere of radius R, $A_f = \pi R^2$.

1.1.5.2. Internal Flow

For fluid flow through a tube, especially one that is long and small in diameter, the friction force between the fluid and the tube wall can be rather significant. For a horizontal tube the pressure at the entrance has to be greater than that at the exit in order to overcome the friction and force the fluid through the tube.

For fluid flow through a tube of length L and inner diameter D, the **friction factor**, is a dimensionless parameter defined by

$$f = \left[\frac{(p_0 + \rho g h_0) - (p_L + \rho g h_L)}{\rho v_{av}^2/2}\right]\frac{D}{L} \qquad [1.1\text{-}38]$$

where p is pressure, g is the gravitational acceleration, h is height, and v_{av} is the average fluid velocity. The subscripts 0 and L denote the entrance and exit of the tube, respectively. Height h is considered since fluid flow can be caused by the difference in height as well as in pressure. The average velocity is defined as follows

$$v_{av} = \frac{\iint_A \rho v \, dA}{\iint_A \rho \, dA} = \frac{m}{\rho A} \qquad [1.1\text{-}39]$$

Notice that the numerator is the mass flow rate. The density is considered as constant over the inner cross-sectional area A of the tube.

1.2. OVERALL MASS-BALANCE EQUATION

1.2.1. Derivation

Let us consider an arbitrary stationary volume element Ω in a fluid flow field, as illustrated in Fig. 1.2-1. This volume element, called the **control volume**,

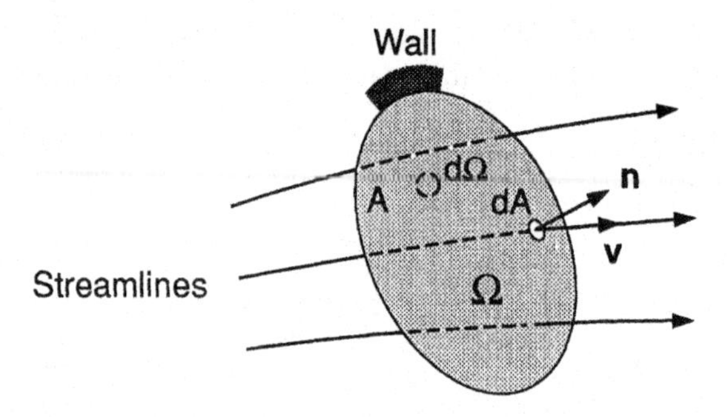

Fig. 1.2-1 Fluid flow through a volume element Ω.

represents the system being considered. The boundary of the control volume A is sometimes called the **control surface**. It presents no resistance to fluid flow except at the wall, where no fluid crosses. Let $d\Omega$ be a differential volume element in Ω and dA a differential surface element on A. The velocity vector $\mathbf{v}$ and the density ρ of the fluid are defined at every point in the control volume including its surface.

As shown in Fig. 1.2-1, $\mathbf{n}$ is the **outward unit normal vector** of a differential area dA on the control surface A, and $\mathbf{v}$ is the velocity of the fluid at dA. The control surface A can be considered to consist of three regions: A_{in} for the region where the fluid enters the control volume; A_{out}, where the fluid leaves; and A_{wall}, where the fluid is in contact with a wall. In other words,

$$A = A_{in} + A_{out} + A_{wall} \qquad [1.2\text{-}1]$$

Before deriving the integral mass-balance equation, let us first introduce the concept of the volume and mass flow rates of a fluid. Figure 1.2-2*a* shows the volume of a fluid flowing out of Ω per unit time through dA, namely, the outward volume flow rate through dA. As shown in the side view in Fig. 1.2-2*b*, it is represented by *abcd*, which equals *abef*, that is, $\mathbf{v}\cdot\mathbf{n}\,dA$. As such, the outward mass flow rate through dA is $\rho\mathbf{v}\cdot\mathbf{n}\,dA$, where ρ is the density of the fluid. Since $\mathbf{n}$ points outward

$$\text{inward mass flow rate through } dA = -\rho\mathbf{v}\cdot\mathbf{n}\,dA \qquad [1.2\text{-}2]$$

Let us now proceed to derive the integral mass-balance equation. For a homogeneous fluid of a pure material undergoing no chemical reactions, the following conservation law for mass can be applied to the control volume:

$$\underset{(1)}{\left\{\begin{matrix}\text{Rate of}\\ \text{mass}\\ \text{accumulation}\end{matrix}\right\}} = \underset{(2)}{\left\{\begin{matrix}\text{rate of}\\ \text{mass}\\ \text{in}\end{matrix}\right\}} - \underset{(3)}{\left\{\begin{matrix}\text{rate of}\\ \text{mass}\\ \text{out}\end{matrix}\right\}} \qquad [1.2\text{-}3]$$

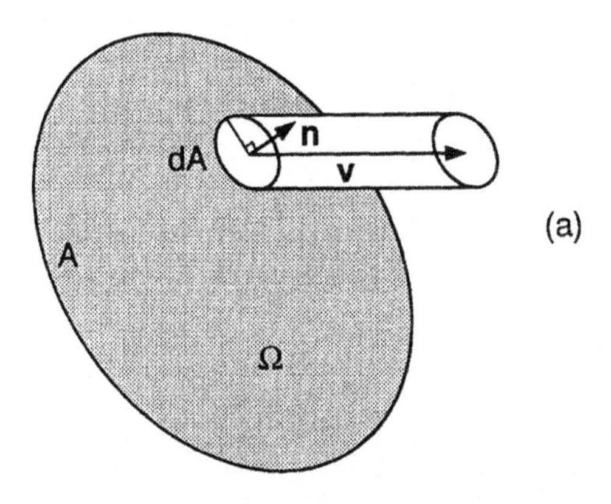

outward volume flow rate
= v·ndA

f
v·n
e
a
d
n
dA
v
b
c
(b)

Fig. 1.2-2 Volume flow rate through a differntial surface element dA: (a) three-dimensional view; (b) two-dimensional view.

Let us consider the terms in Eq. [1.2-3] one by one, starting from term (1). The mass of the fluid in the differential volume element $d\Omega$ is $dM = \rho\, d\Omega$. It can be integrated over Ω to obtain the mass of the fluid in the control volume:

$$\iiint_{\Omega} \rho\, d\Omega \quad ; \quad M$$

(integral) (overall)

and the rate of mass change in the control volume is thus

$$\frac{\partial}{\partial t}\iiint_{\Omega} \rho\, d\Omega \quad ; \quad \frac{dM}{dt} \qquad \text{term (1)}$$

(integral) (overall)

For convenience the first form is called the integral form and the second, the overall form. The same terminology will be applied to the remaining terms.

According to Eq. [1.2-2], the inward mass flow rate through a differential area dA is $-\rho\mathbf{v}\cdot\mathbf{n}\,dA$. It can be integrated over the control surface A to obtain the net (since A is a closed surface) inward mass flow rate into Ω, namely the rate of mass flow into Ω minus that out of Ω:

$$-\iint_A \rho\mathbf{v}\cdot\mathbf{n}\,dA$$

or, in view of Eq. [1.2-1],

$$-\iint_{A_{\text{in}}+A_{\text{out}}+A_{\text{wall}}} \rho\mathbf{v}\cdot\mathbf{n}\,d(A_{\text{in}}+A_{\text{out}}+A_{\text{wall}})$$

or, since $\mathbf{v}=0$ at the wall,

$$-\left(\iint_A \rho\mathbf{v}\cdot\mathbf{n}\,dA\right)_{\text{in}} - \left(\iint_A \rho\mathbf{v}\cdot\mathbf{n}\,dA\right)_{\text{out}} - (0)_{\text{wall}}$$

Let us define $v=|\mathbf{v}\cdot\mathbf{n}|$, namely, the velocity component along the normal to the control surface. At the inlet $\mathbf{v}$ is inward, $\mathbf{n}$ is outward and $v=|\mathbf{v}\cdot\mathbf{n}|=-\mathbf{v}\cdot\mathbf{n}$, while at the outlet $\mathbf{v}$ and $\mathbf{n}$ are both outward and $v=|\mathbf{v}\cdot\mathbf{n}|=\mathbf{v}\cdot\mathbf{n}$. Therefore, the rate of mass flow into Ω minus that out of Ω is

$$-\iint_A \rho\mathbf{v}\cdot\mathbf{n}\,dA \quad ; \quad \left[\left(\iint_A \rho v\,dA\right)_{\text{in}} - \left(\iint_A \rho v\,dA\right)_{\text{out}}\right]$$

(integral) (overall)

term (2)–term (3)

Substituting the integral form of terms (1) through (3) into Eq. [1.2-3]

$$\boxed{\frac{\partial}{\partial t}\iiint_\Omega \rho\,d\Omega = -\iint_A \rho\mathbf{v}\cdot\mathbf{n}\,dA} \quad \text{(integral mass-balance equation)} \qquad [1.2\text{-}4]$$

Now substituting the overall form of terms (1) through (3) into Eq. [1.2-3]

$$\boxed{\frac{dM}{dt} = \left(\iint_A \rho v\,dA\right)_{\text{in}} - \left(\iint_A \rho v\,dA\right)_{\text{out}} = m_{\text{in}} - m_{\text{out}}} \qquad [1.2\text{-}5]$$

where M is the mass in the control volume, that is, $\iiint_\Omega \rho\, d\Omega$. Substituting Eq. [1.1-39] into Eq. [1.2-5], we obtain

$$\boxed{\frac{dM}{dt} = (\rho v_{\rm av} A)_{\rm in} - (\rho v_{\rm av} A)_{\rm out} = m_{\rm in} - m_{\rm out}} \quad \text{(overall mass-balance equation)} \qquad [1.2\text{-}6]$$

where M = mass of fluid in control volume (= $\rho\Omega$ if ρ is uniform in Ω)
m = mass flow rate at inlet or outlet (= $\rho v_{\rm av} A$).

1.2.2. Examples

Example 1.2-1 Flow through a System of Constant Mass

Consider the steady-state flow of an incompressible fluid through the system shown in Fig. 1.2-3. How are the average velocities at the inlet and the outlet related to each other?

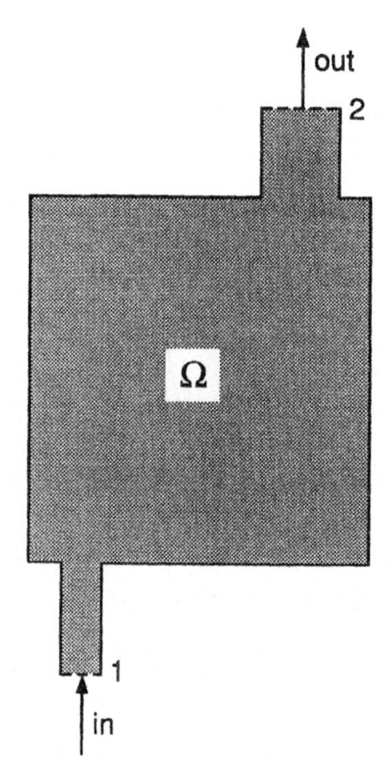

Fig. 1.2-3 Fluid flow through a container.

Solution From Eq. [1.2-6]

$$0 = (\rho v_{av} A)_{in} - (\rho v_{av} A)_{out} \tag{1.2-7}$$

Since ρ is constant, this equation reduces to

$$(v_{av} A)_{in} = (v_{av} A)_{out} \tag{1.2-8}$$

Example 1.2-2 Fluid Mass in a Mixing Tank

As illustrated in Fig. 1.2-4, a mixing tank receives a fluid from inlets 1 and 2 and discharges it from outlet 3 at the mass flow rates of m_1, m_2, and m_3, respectively. Determine the fluid mass M in the tank as a function of time. The initial fluid mass in the tank is M_0. Repeat for the case where inlet 1 is closed subsequently at time t_1.

Solution From Eq. [1.2-6]

$$\frac{dM}{dt} = m_1 + m_2 - m_3 \tag{1.2-9}$$

The initial condition is as follows

$$M = M_0 \quad \text{at} \quad t = 0 \tag{1.2-10}$$

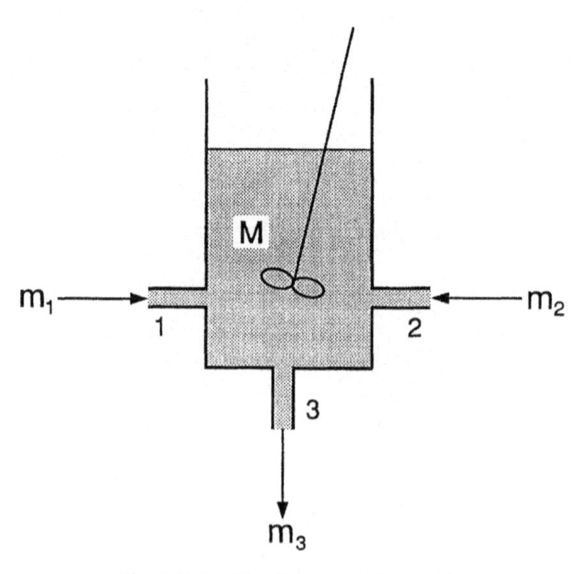

Fig. 1.2-4 Fluid in a mixing tank.

As such

$$\int_{M_0}^{M} dM = \int_0^t (m_1 + m_2 - m_3)\, dt \qquad [1.2\text{-}11]$$

and so

$$M = (m_1 + m_2 - m_3)t + M_0 \quad \text{for} \quad t < t_1 \qquad [1.2\text{-}12]$$

After time $t_1, m_1 = 0$ and Eq. [1.2-9] reduces to

$$\frac{dM}{dt} = (m_2 - m_3) \qquad [1.2\text{-}13]$$

On integration from M_1 at t_1

$$\int_{M_1}^{M} dM = \int_{t_1}^{t} (m_2 - m_3)\, dt \qquad [1.2\text{-}14]$$

and so

$$M = M_1 + (m_2 - m_3)(t - t_1) \qquad [1.2\text{-}15]$$

where M_1 is the fluid mass in the tank at time t_1. From Eq. [1.2-12]

$$M_1 = (m_1 + m_2 - m_3)t_1 + M_0 \qquad [1.2\text{-}16]$$

Substituting Eq. [1.2-16] into Eq. [1.2-15]

$$M = (m_2 - m_3)t + m_1 t_1 + M_0 \quad \text{for} \quad t \geq t_1 \qquad [1.2\text{-}17]$$

1.3. DIFFERENTIAL MASS-BALANCE EQUATION

1.3.1. Derivation

In Section 1.2 we derived the integral mass-balance equation, Eq. [1.2-4]. If the control volume Ω does not change with time, the equation can be rewritten as follows

$$\iiint_\Omega \frac{\partial \rho}{\partial t}\, d\Omega = -\iint_A \rho \mathbf{v} \cdot \mathbf{n}\, dA \qquad [1.3\text{-}1]$$

TABLE 1.3-1 Equation of Continuity

Rectangular coordinates (x, y, z):

$$\frac{\partial \rho}{\partial t} + \frac{\partial}{\partial x}(\rho v_x) + \frac{\partial}{\partial y}(\rho v_y) + \frac{\partial}{\partial z}(\rho v_z) = 0 \quad [A]$$

Cylindrical coordinates (r, θ, z):

$$\frac{\partial \rho}{\partial t} + \frac{1}{r}\frac{\partial}{\partial r}(\rho r v_r) + \frac{1}{r}\frac{\partial}{\partial \theta}(\rho v_\theta) + \frac{\partial}{\partial z}(\rho v_z) = 0 \quad [B]$$

Spherical coordinates (r, θ, ϕ):

$$\frac{\partial \rho}{\partial t} + \frac{1}{r^2}\frac{\partial}{\partial r}(\rho r^2 v_r) + \frac{1}{r \sin\theta}\frac{\partial}{\partial \theta}(\rho v_\theta \sin\theta) + \frac{1}{r \sin\theta}\frac{\partial}{\partial \phi}(\rho v_\phi) = 0 \quad [C]$$

The surface integral in Eq. [1.3-1] can be easily converted into a volume integral using the **Gauss divergence theorem** (i.e., Eq. [A.4-1]):

$$\iint_A \rho \mathbf{v} \cdot \mathbf{n}\, dA = \iiint_\Omega \nabla \cdot (\rho \mathbf{v})\, d\Omega \quad [1.3\text{-}2]$$

Substituting Eq. [1.3-2] into Eq. [1.3-1]

$$\iiint_\Omega \left\{ \frac{\partial \rho}{\partial t} + \nabla \cdot (\rho \mathbf{v}) \right\} d\Omega = 0 \quad [1.3\text{-}3]$$

The integrand, which is continuous, must be zero everywhere since the equation must hold for any arbitrary Ω.[10] Therefore, as shown by Theodore,[11]

$$\boxed{\frac{\partial \rho}{\partial t} + \nabla \cdot (\rho \mathbf{v}) = 0 \qquad \text{(general)}} \quad [1.3\text{-}4]$$

Equation [1.3-4] is the **differential mass-balance equation**, that is, the **equation of continuity**. In Table 1.3-1 this equation is given in expanded forms for rectangular, cylindrical, and spherical coordinates.

For an incompressible fluid, the density ρ is constant and Eq. [1.3-4] becomes

$$\boxed{\nabla \cdot \mathbf{v} = 0 \qquad (\rho = \text{constant})} \quad [1.3\text{-}5]$$

[10] W. Kaplan, *Advanced Calculus*, 2nd ed., Addison-Wesley, Reading, MA, 1973, p. 363.

[11] L. Theodore, *Transport Phenomena for Engineers*, International Textbook Company, Toronto, 1971, p. 78.

1.3.2. Examples

Examples 1.3-1 Fully Developed Flow in a Tube

Consider the steady-state, laminar flow of an incompressible fluid in a long tube, as illustrated in Fig. 1.1-16. In the region sufficiently far away from the entrance and exit of the tube, the flow is one-dimensional along the axis. Use the continuity equation to verify that in this region the flow is fully developed, that is, the velocity does not change with the axial distance z.

Solution Clearly, cylindrical coordinates are most appropriate for the problem. The continuity equation, according to Eq. [B] of Table 1.3-1, is as follows:

$$\frac{\partial \rho}{\partial t} + \frac{1}{r}\frac{\partial}{\partial r}(\rho r v_r) + \frac{1}{r}\frac{\partial}{\partial \theta}(\rho v_\theta) + \frac{\partial}{\partial z}(\rho v_z) = 0 \qquad [1.3\text{-}6]$$

At steady state

$$\frac{\partial \rho}{\partial t} = 0 \qquad [1.3\text{-}7]$$

Since the flow is one-dimensional along the z direction

$$v_r = v_\theta = 0 \qquad [1.3\text{-}8]$$

Therefore, Eq. [1.3-6] reduces to

$$\frac{\partial}{\partial z}(\rho v_z) = 0 \qquad [1.3\text{-}9]$$

Since ρ is constant, this equation can be divided by ρ to become

$$\frac{\partial v_z}{\partial z} = 0 \qquad [1.3\text{-}10]$$

Therefore, v_z is independent of z, and the flow is fully developed.

Example 1.3-2 Normal Stress Due to Creeping Flow around a Sphere

Repeat Example 1.1-2 but calculate the normal stress τ_{rr} acting on the sphere by the fluid.

Solution Newton's law of viscosity, according to Eq. [A] of Table 1.1-3, is as follows:

$$\tau_{rr} = -\mu\left[2\frac{\partial v_r}{\partial r} - \frac{2}{3}(\nabla\cdot\mathbf{v})\right] \quad [1.3\text{-}11]$$

Since the fluid is incompressible, $\nabla\cdot\mathbf{v} = 0$, according to Eq. [1.3-5], and Eq. [1.3-11] reduces to

$$\tau_{rr} = -2\mu\frac{\partial v_r}{\partial r} \quad [1.3\text{-}12]$$

From Eq. [1.1-22]

$$v_r = v_\infty\left[1 - \frac{3}{2}\left(\frac{R}{r}\right) + \frac{1}{2}\left(\frac{R}{r}\right)^3\right]\cos\theta \quad [1.3\text{-}13]$$

Therefore

$$\tau_{rr} = -2\mu v_\infty\left[\frac{3}{2}\frac{R}{r^2} - \frac{3}{2}\frac{R^3}{r^4}\right] \quad [1.3\text{-}14]$$

and so

$$\tau_{rr}|_{r=R} = -2\mu v_\infty\left[\frac{3}{2}\frac{R}{R^2} - \frac{3}{2}\frac{R^3}{R^4}\right] = 0 \quad [1.3\text{-}15]$$

1.4. OVERALL MOMENTUM-BALANCE EQUATION

1.4.1. Derivation

Let us consider an arbitrary stationary control volume Ω bounded by surface A through which a moving fluid is flowing, as shown in Fig. 1.4-1. The control surface A can be considered to consist of three parts: A_{in} for the region where the fluid enters the control volume; A_{out}, where the fluid leaves; and A_{wall}, where the fluid is in contact with a wall. As such

$$A = A_{\text{in}} + A_{\text{out}} + A_{\text{wall}} \quad [1.4\text{-}1]$$

For momentum transfer by convection, only A_{in} and A_{out} need to be considered as no fluid crosses the wall.

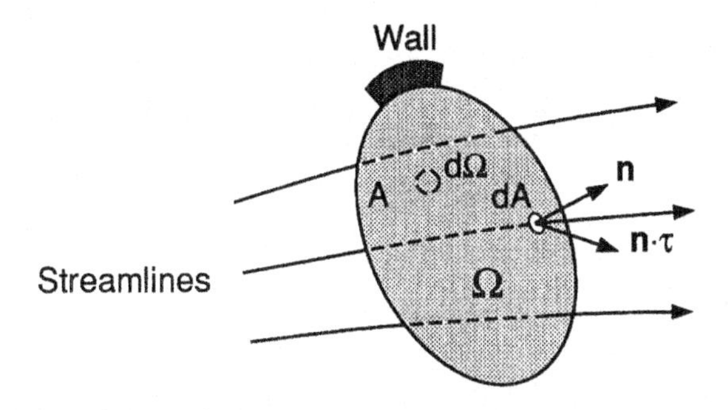

Fig. 1.4-1 Fluid flow through a volume element Ω.

The following conservation law for momentum can be applied to the control volume:

$$\left\{\begin{array}{l}\text{Rate of}\\ \text{momentum}\\ \text{accumulation}\end{array}\right\} = \left\{\begin{array}{l}\text{rate of}\\ \text{momentum}\\ \text{in}\end{array}\right\} - \left\{\begin{array}{l}\text{rate of}\\ \text{momentum}\\ \text{out}\end{array}\right\} + \left\{\begin{array}{l}\text{sum of forces}\\ \text{acting on}\\ \text{system}\end{array}\right\}$$

(1) (2) (3) (4)

[1.4-2]

This equation is consistent with **Newton's second law of motion**, that is, the rate of momentum change of a system equals the net force acting on the system. As can be shown in Problem 1.1-1, the rates of momentum change and transfer have the same unit as forces.

Let us now consider the terms in Eq. [1.4-2] one by one, starting from term (1). The mass contained in a differential volume element $d\Omega$ in the control volume is $\rho\, d\Omega$ and its momentum is $d\mathbf{P} = \rho\mathbf{v}\, d\Omega$. As such, the momentum of the fluid in Ω is

$$\iiint_{\Omega} \rho\mathbf{v}\, d\Omega \quad ; \quad \mathbf{P}$$

(integral) (overall)

and the rate of momentum change in Ω is

$$\frac{\partial}{\partial t}\iiint_{\Omega} \rho\mathbf{v}\, d\Omega \quad ; \quad \frac{d\mathbf{P}}{dt} \qquad \text{term (1)}$$

(integral) (overall)

For convenience the first form will be called the integral form and the second, the overall form. The same terminology will be applied to the remaining terms.

As shown in Eq. [1.2-2], the inward mass flow rate through dA is $-\rho(\mathbf{v}\cdot\mathbf{n})\,dA$. The inward momentum flow rate $-\rho\mathbf{v}(\mathbf{v}\cdot\mathbf{n})\,dA$ can be integrated over A to obtain the net (since A is a closed surface) inward momentum flow rate into Ω, that is, the rate of momentum flow into Ω minus that out of Ω:

$$-\iint_A \rho\mathbf{v}(\mathbf{v}\cdot\mathbf{n})\,dA$$

or, in view of Eq. [1.4-1],

$$-\iint_{A_{\text{in}}+A_{\text{out}}+A_{\text{wall}}} \rho\mathbf{v}(\mathbf{v}\cdot\mathbf{n})\,d(A_{\text{in}}+A_{\text{out}}+A_{\text{wall}})$$

or, since $\mathbf{v}=0$ at the wall,

$$-\left(\iint_A \rho\mathbf{v}(\mathbf{v}\cdot\mathbf{n})\,dA\right)_{\text{in}} - \left(\iint_A \rho\mathbf{v}(\mathbf{v}\cdot\mathbf{n})\,dA\right)_{\text{out}} - (0)_{\text{wall}}$$

As in Section 1.2, the velocity component along the normal to the control surface is defined as $v=|\mathbf{v}\cdot\mathbf{n}|$. At the inlet $\mathbf{v}$ is inward, $\mathbf{n}$ is outward and $v=|\mathbf{v}\cdot\mathbf{n}|=-(\mathbf{v}\cdot\mathbf{n})$, while at the outlet $\mathbf{v}$ and $\mathbf{n}$ are both outward and $v=|\mathbf{v}\cdot\mathbf{n}|$ $=(\mathbf{v}\cdot\mathbf{n})$. Therefore, the rate of momentum flow into Ω minus that out of Ω is

$$-\iint_A \rho\mathbf{v}\mathbf{v}\cdot\mathbf{n}\,dA \quad ; \quad \left[\left(\iint_A \rho v\mathbf{v}\,dA\right)_{\text{in}} - \left(\iint_A \rho v\mathbf{v}\,dA\right)_{\text{out}}\right]$$

(integral) (overall)

term (2)–term (3)

Note that $\mathbf{v}\mathbf{v}$ is a tensor and that $\mathbf{v}\mathbf{v}\cdot\mathbf{n}$ is a vector equal to $\mathbf{v}(\mathbf{v}\cdot\mathbf{n})$, as shown in Eq. [A.2-5].

Let us now consider the forces acting on the control volume. The pressure force acting on dA is $d\mathbf{F}_p=-p\mathbf{n}\,dA$; the minus sign results because the pressure force acts in the opposite direction of the outward normal vector $\mathbf{n}$. The pressure force acting on the entire control surface A is thus given by

$$-\iint_A p\mathbf{n}\,dA \quad ; \quad \mathbf{F}_p \qquad \text{term (4a)}$$

(integral) (overall)

Since the fluid in Ω exerts a stress $\boldsymbol{\tau}\cdot\mathbf{n}$ on its surroundings, the surroundings can be considered to exert a stress $-\boldsymbol{\tau}\cdot\mathbf{n}$ on the fluid in Ω. Note that $\boldsymbol{\tau}$ is a tensor and $\boldsymbol{\tau}\cdot\mathbf{n}$ is a vector, as can be seen by substituting $\mathbf{n}$ for the vector $\mathbf{v}$ in Eq.

[A.2-3]. Since the viscous force exerted on the fluid by the surrounding over dA is $d\mathbf{F}_v = -\boldsymbol{\tau}\cdot\mathbf{n}\,dA$, the viscous force exerted on the fluid by the surroundings over A is

$$-\iint_A \boldsymbol{\tau}\cdot\mathbf{n}\,dA \quad ; \quad \mathbf{F}_v \qquad \text{term (4b)}$$

(integral) (overall)

A force acting on the surface of the control volume, such as the pressure or viscous force, is a surface force. On the other hand, a force acting on the body of the control volume is a body force. Examples of body forces are the gravity force and the Lorentz (electromagnetic) force. The body force acting on the differential volume element $d\Omega$ is $d\mathbf{F}_b = \mathbf{f}_b\,d\Omega$, where $\mathbf{f}_b$ is the body force per unit volume. The body force acting on the entire control volume is thus given by

$$\iiint_\Omega \mathbf{f}_b\,d\Omega \quad ; \quad \mathbf{F}_b \qquad \text{term (4c)}$$

(integral) (overall)

Substituting the integral form of terms (1) through (4) into Eq. [1.4-2]

$$\boxed{\frac{\partial}{\partial t}\iiint_\Omega \rho\mathbf{v}\,d\Omega = -\iint_A \rho\mathbf{v}\mathbf{v}\cdot\mathbf{n}\,dA - \iint_A p\mathbf{n}\,dA - \iint_A \boldsymbol{\tau}\cdot\mathbf{n}\,dA + \iiint_\Omega \mathbf{f}_b\,d\Omega}$$

(integral momentum-balance equation) [1.4-3]

or

$$\frac{\partial}{\partial t}\iiint_\Omega \rho\mathbf{v}\,d\Omega = -\iint_A \rho\mathbf{v}\mathbf{v}\cdot\mathbf{n}\,dA + \sum\mathbf{F} \qquad [1.4\text{-}4]$$

where $\sum\mathbf{F}$ is the summation of the following forces acting on the control volume by the surroundings:

$$\text{Pressure force} \quad \mathbf{F}_p = -\iint_A p\mathbf{n}\,dA \qquad [1.4\text{-}5]$$

$$\text{Viscous force} \quad \mathbf{F}_v = -\iint_A \boldsymbol{\tau}\cdot\mathbf{n}\,dA \qquad [1.4\text{-}6]$$

$$\text{Body force} \quad \mathbf{F}_b = \iiint_\Omega \mathbf{f}_b\,d\Omega \qquad [1.4\text{-}7]$$

Substituting the overall form of terms (1) through (4) into Eq. [1.4-2]

$$\frac{d\mathbf{P}}{dt} = \left(\iint_A \rho v \mathbf{v}\, dA\right)_{\text{in}} - \left(\iint_A \rho v \mathbf{v}\, dA\right)_{\text{out}} + \mathbf{F}_p + \mathbf{F}_v + \mathbf{F}_b \qquad [1.4\text{-}8]$$

where $\mathbf{P}$ is the momentum in the control volume, that is, $\iiint_\Omega \rho \mathbf{v}\, d\Omega$.

If flow is perpendicular to the inlet and outlet and density is uniform

$$\begin{aligned}\frac{d\mathbf{P}}{dt} &= (a\rho v_{\text{av}} A \mathbf{v}_{\text{av}})_{\text{in}} - (a\rho v_{\text{av}} A \mathbf{v}_{\text{av}})_{\text{out}} + \mathbf{F}_p + \mathbf{F}_v + \mathbf{F}_b \\ &= (am\mathbf{v}_{\text{av}})_{\text{in}} - (am\mathbf{v}_{\text{av}})_{\text{out}} + \mathbf{F}_p + \mathbf{F}_v + \mathbf{F}_b\end{aligned}$$

(overall momentum-balance equation) [1.4-9]

where $\mathbf{P}$ = momentum of fluid in control volume ($= \rho\mathbf{v}\Omega = M\mathbf{v}$ if $\rho\mathbf{v}$ is uniform)
m = mass flow rate at inlet or outlet ($= \rho v_{\text{av}} A$)
$\mathbf{F}_p$ = pressure force acting **on** control volume by surroundings at inlet, outlet and wall
$\mathbf{F}_v$ = viscous force acting **on** control volume by surroundings (mainly at wall and negligible at inlet and outlet, where convective momentum transfer dominates), and
$\mathbf{F}_b$ = body force acting **on** control volume ($= \mathbf{f}_b\Omega$ if uniform $\mathbf{f}_b$)

The correction factor

$$a = \left(\iint_A v\mathbf{v}\, dA\right) \Big/ (A v_{\text{av}} \mathbf{v}_{\text{av}}) \qquad [1.4\text{-}10]$$

It can be shown (in Problem 1.9-5) that

$$a = \tfrac{4}{3} \text{ for laminar flow in round pipes} \qquad [1.4\text{-}11]$$

$$a \approx 1 \text{ for turbulent flow} \qquad [1.4\text{-}12]$$

1.4.2. Examples

Examples 1.4-1 Force Exerted by an Impinging Jet

A horizontal turbulent liquid jet is directed normal to a vertical plate and the subsequent flow is parallel to the plate[12], as shown in Fig. 1.4-2. The volume

[12] Eskinazi, *Principles of Fluid Mechanics*, Allyn and Bacon, Boston, 1962, p. 211.

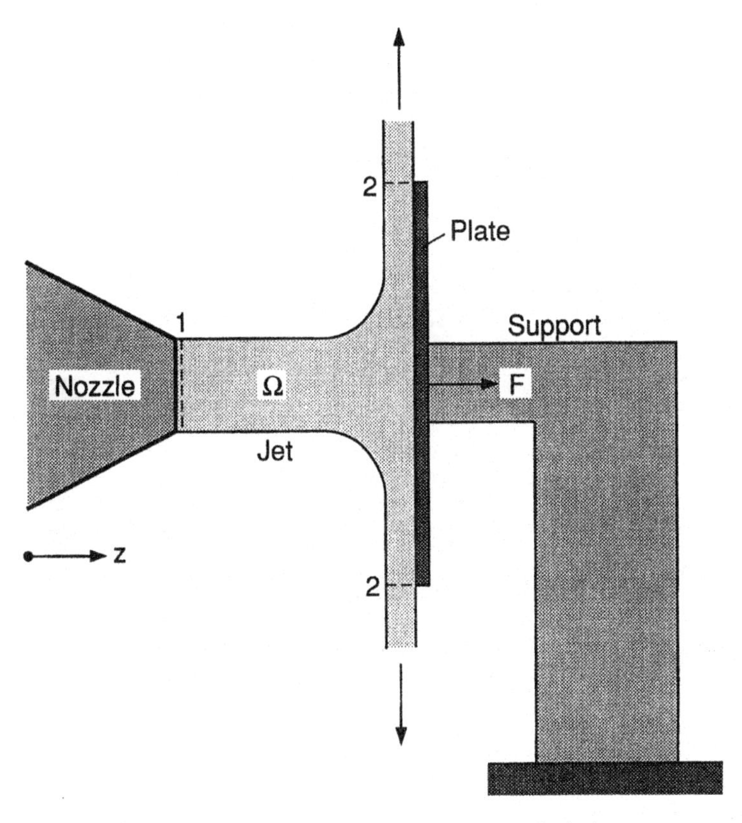

Fig. 1.4-2 A jet directed normal to a vertical plate.

flow rate of the liquid is Q and the cross-sectional area of the nozzle exit is A_n. Determine the pressure force F acting on the plate by the liquid at the steady state. As an approximation, neglect the viscous and gravity forces.

Solution Take the liquid between the nozzle and the plate as the control volume Ω. At the nozzle exit the average velocity of the liquid $v = Q/A_n$, $m = \rho Q$, and the z component of Eq. [1.4-9]

$$0 = (mv)_1 - (m0)_2 + F_{pz} + 0 + 0 \qquad [1.4\text{-}13]$$

The force acting on the plate by the liquid is:

$$\boxed{F = -F_{pz} = \frac{\rho Q^2}{A_n}} \qquad [1.4\text{-}14]$$

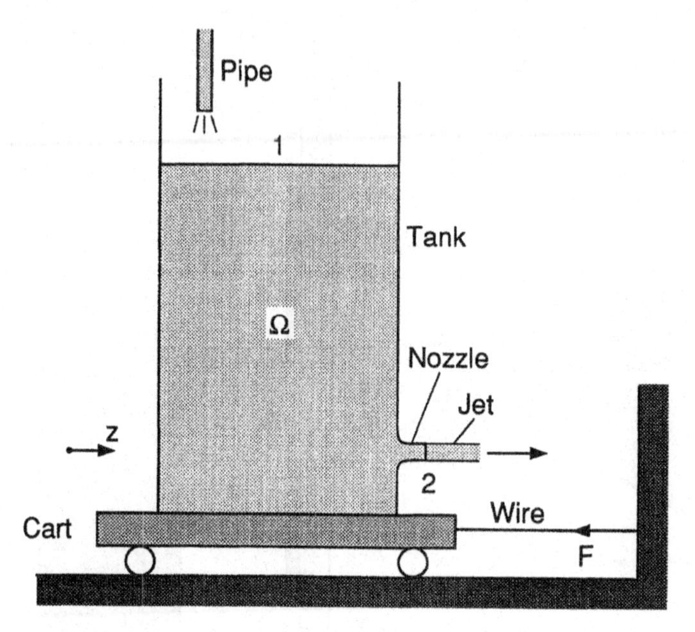

Fig. 1.4-3 A wire holding a tank from which a jet issues.

Example 1.4-2 Back Thrust of a Jet

A liquid tank is attached to a cart which is held stationary with a wire[13], as shown in Fig. 1.4-3. The liquid issues horizontally as a turbulent jet through a nozzle at the volume flow rate Q, and the cross-sectional area of the nozzle exit is A_n. Liquid is added through a vertical pipe to maintain a constant liquid level in the tank. Determine the force F acting on the wire by the jet. As an approximation, neglect the viscous force.

Solution Take the liquid in the tank as the control volume Ω. At the nozzle exit the average velocity of the liquid is $v = Q/A_n$ and the mass flow rate

$$m = \rho Q \qquad [1.4\text{-}15]$$

The z component of Eq. [1.4-9] is

$$0 = (m0)_1 - (mv)_2 + F_{pz} + 0 + 0 \qquad [1.4\text{-}16]$$

[13] Eskinazi, *Principles of Fluid Mechanics*, Allyn and Bacon, Boston, 1962, p. 209.

where F_{pz} is the pressure force acting on the liquid by the wire through the tank wall. The force acting on the wire by the liquid:

$$\boxed{F = -F_{pz} = -\frac{\rho Q^2}{A_n}} \qquad [1.4\text{-}17]$$

Example 1.4-3 Thrust on a Pipe Bend

Consider turbulent flow through a short horizontal pipe bend of uniform cross-section A_p at the constant volume flow rate of Q[14], as shown in Fig. 1.4-4. The angle of the bend is θ. Determine the pressure force F acting on the bend by the liquid at the steady state. As an approximation, neglect the viscous force.

Solution Take the liquid in the bend as the control volume. Let v be the average velocity:

$$v = \frac{Q}{A_p} \qquad [1.4\text{-}18]$$

and the mass flow rate

$$m = \rho Q \qquad [1.4\text{-}19]$$

The z component of Eq. [1.4-9] is

$$0 = (mv)_1 - (mv)_2 \cos\theta + F_{pz} + 0 + 0 \qquad [1.4\text{-}20]$$

and the y component is

$$0 = (m0)_1 - (mv)_2 \sin\theta + F_{py} + 0 + 0 \qquad [1.4\text{-}21]$$

where F_{pz} and F_{py} are the z and y components of the pressure force $\mathbf{F}_p$ acting on the liquid by the bend, respectively. Since $v_1 = v_2$ and $m_1 = m_2$, we have $(mv)_1 = (mv)_2 = mv$. From Eqs. [1.4-20] and [1.4-21], the force acting on the bend by the liquid is

$$\mathbf{F} = -\mathbf{F}_p = \mathbf{e}_z(-F_{pz}) + \mathbf{e}_y(-F_{py}) = mv[\mathbf{e}_z(1 - \cos\theta) + \mathbf{e}_y(-\sin\theta)] \qquad [1.4\text{-}22]$$

[14] S. Eskinazi, *Principles of Fluid Mechanics*, Allyn and Bacon, Boston, 1962, p. 205.

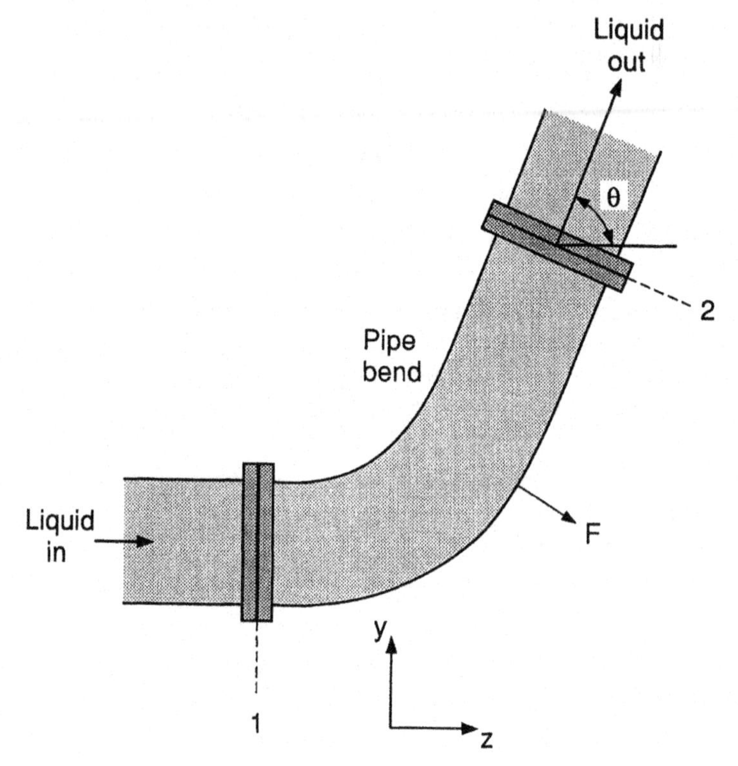

Fig. 1.4-4 Force induced by flow through a pipe bend.

Substituting Eqs. [1.4-18] and [1.4-19] into Eq. [1.4-22]

$$\mathbf{F} = \frac{\rho Q^2}{A_p}[\mathbf{e}_z(1 - \cos\theta) + \mathbf{e}_y(-\sin\theta)] \tag{1.4-23}$$

and so

$$\boxed{F = \frac{\rho Q^2}{A_p}\sqrt{2(1 - \cos\theta)}} \tag{1.4-24}$$

Example 1.4-4 Friction Force on a Pipe Wall

A liquid flows through a long vertical pipe of uniform inner radius R and length L, as shown in Fig. 1.4-5. Determine the viscous force F acting on the pipe wall by the liquid at the steady state.

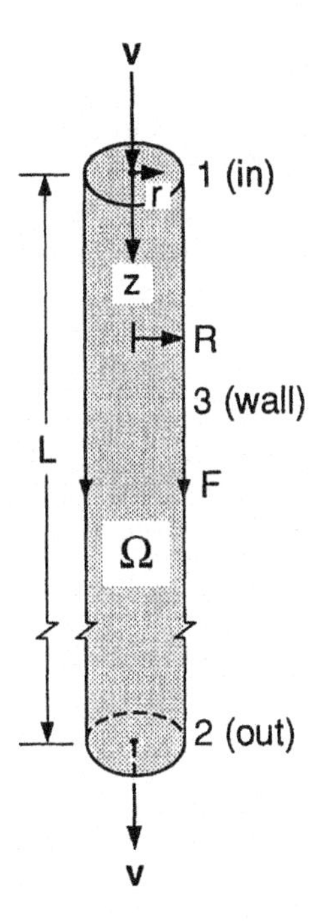

Fig. 1.4-5 Fluid flow through a vertical pipe.

Solution Let the liquid in the pipe be the control volume Ω and v be the average velocity of the liquid. The z component of Eq. [1.4-9] is

$$0 = (amv)_1 - (amv)_2 + p_1\pi R^2 - p_2\pi R^2 + F_{vz} + \pi R^2 L\rho g \qquad [1.4\text{-}25]$$

where F_{vz} is the viscous force acting on the liquid by the pipe wall. Since $m_1 = m_2$ and $v_1 = v_2$, we have $(amv)_1 = (amv)_2$. Substituting this into Eq. [1.4-25] and rearranging, we obtain the force acting on the pipe wall by the liquid:

$$\boxed{F = -F_{vz} = \pi R^2(p_1 - p_2 + \rho g L)} \qquad [1.4\text{-}26]$$

If the liquid is Newtonian

$$F = \tau_{rz}|_R(2\pi RL) = -\mu \left.\frac{dv_z}{dr}\right|_R (2\pi RL) = \pi R^2(p_1 - p_2 + \rho g L) \qquad [1.4\text{-}27]$$

which can be used to find the velocity gradient at the wall.

Example 1.4-5 Creeping Flow around a Sphere: Stokes' Law

An incompressible Newtonian fluid approaches a solid sphere of radius R at a constant and uniform velocity of v_∞, as illustrated in Fig. 1.4-6a. The Reynolds number $2R\rho v_\infty/\mu$ is significantly less than one. Under such a condition the pressure and velocity distributions have been found analytically[15] to be

$$p = p_0 - \rho g z - \frac{3}{2}\frac{\mu v_\infty}{R}\left(\frac{R}{r}\right)^2 \cos\theta \qquad [1.4\text{-}28]$$

$$v_r = v_\infty\left[1 - \frac{3}{2}\left(\frac{R}{r}\right) + \frac{1}{2}\left(\frac{R}{r}\right)^3\right]\cos\theta \qquad [1.4\text{-}29]$$

$$v_\theta = -v_\infty\left[1 - \frac{3}{4}\left(\frac{R}{r}\right) - \frac{1}{4}\left(\frac{R}{r}\right)^3\right]\sin\theta \qquad [1.4\text{-}30]$$

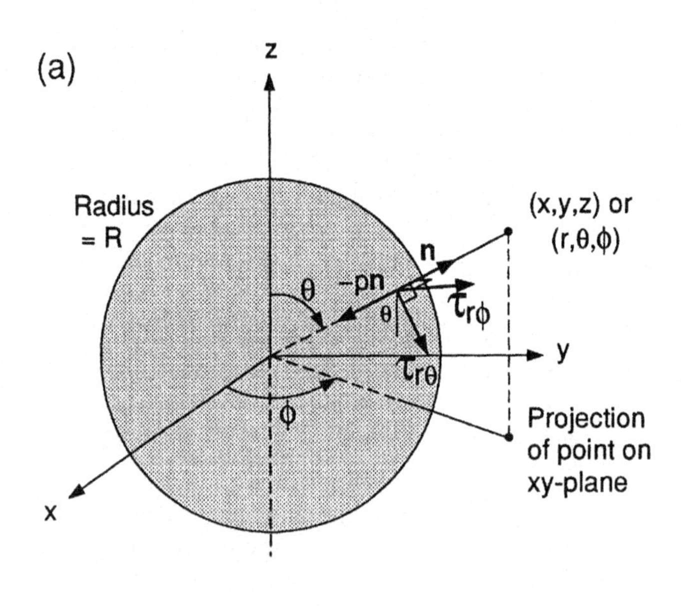

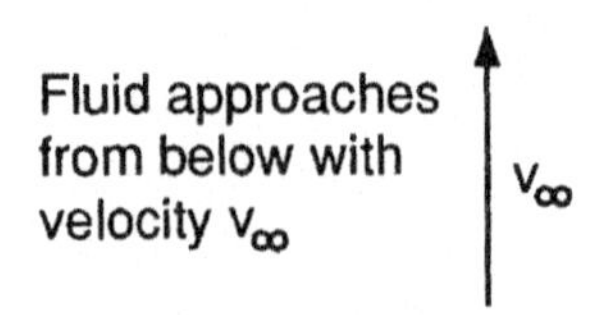

Fig. 1.4-6 Fluid flow past a sphere: (*a*) the sphere; (*b*) a surface element on the sphere.

[15] V. L. Streeter, *Fluid Dynamics*, McGraw-Hill, New York, 1948, p. 235.

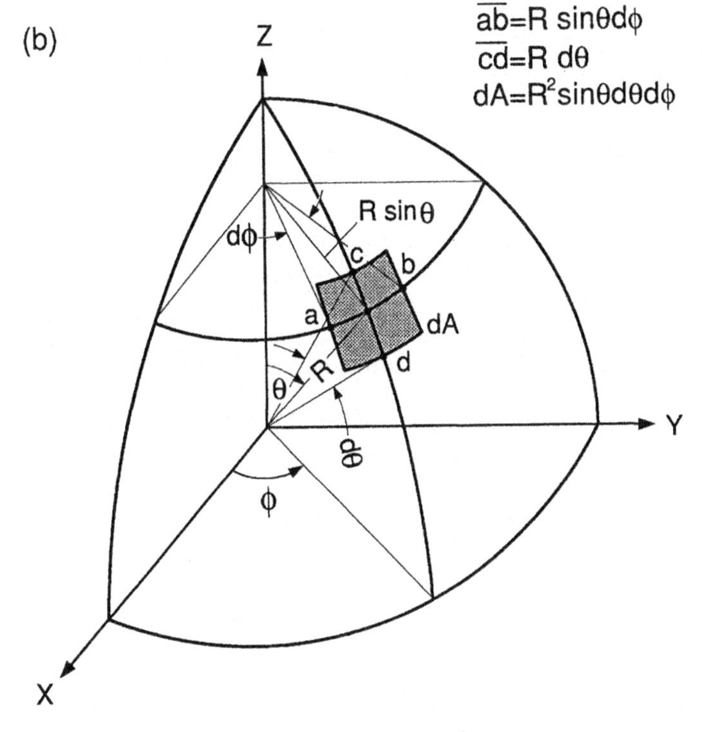

Fig. 1.4-6 (*Continued*).

Calculate the total force acting on the sphere by the fluid, that is, the pressure force and the viscous force.

Solution Let us first consider the pressure force acting on the sphere, $\mathbf{F}_p$. From Eq. [1.4-5] we have

$$\mathbf{F}_p = -\iint_A p\mathbf{n}\, dA \qquad [1.4\text{-}31]$$

where $\mathbf{n}$ is the outward-pointing unit normal vector, namely, $\mathbf{n} = \mathbf{e}_r$ (see Fig. A.5-1*c*). From the direction of $-p\mathbf{n}$ shown in Fig. 1.4-6*a*, the z component of the pressure force is as follows:

$$F_{pz} = -\iint_A p\cos\theta\, dA \qquad [1.4\text{-}32]$$

As illustrated in Fig. 1.4-6*b*, the differential area dA on the surface of the sphere is given by

$$dA = \overline{ab}\,\overline{cd} = R^2 \sin\theta\, d\theta\, d\phi \qquad [1.4\text{-}33]$$

Substituting Eqs. [1.4-28] and [1.4-33] into Eq. [1.4-32], and recognizing that $r = R$ and $z = R\cos\theta$ (see Eqn. [A.5-6]) at the surface of the sphere

$$F_{pz} = -\int_0^{2\pi}\int_0^{\pi}\left(p_0 - \rho g R\cos\theta - \frac{3}{2}\frac{\mu v_\infty}{R}\cos\theta\right)\cos\theta R^2 \sin\theta \, d\theta \, d\phi$$

$$= -2\pi\int_0^{\pi}\left(p_0 - \rho g R\cos\theta - \frac{3}{2}\frac{\mu v_\infty}{R}\cos\theta\right)R^2\sin\theta\cos\theta\, d\theta \qquad [1.4\text{-}34]$$

The integration can be carried out with the help of the following formulae:

$$\int_0^{\pi}\cos\theta\sin\theta\, d\theta = 0 \qquad [1.4\text{-}35]$$

$$\int_0^{\pi}\cos^2\theta\sin\theta\, d\theta = \tfrac{2}{3} \qquad [1.4\text{-}36]$$

Therefore,

$$F_{pz} = \tfrac{4}{3}\pi R^3 \rho g + 2\pi\mu R v_\infty \qquad [1.4\text{-}37]$$

Let us now consider the viscous force acting on the sphere, $\mathbf{F}_v$. From Eq. [1.4-6]

$$\mathbf{F}_v = -\iint_A \boldsymbol{\tau}\cdot\mathbf{n}\, dA \qquad [1.4\text{-}38]$$

From Eqn. [1.3-15] $\tau_{rr}|_{r=R} = 0$ and from Eq. [1.1-30] $\tau_{r\phi} = 0$. Since $\mathbf{n} = \mathbf{e}_r$, from Eq. [A.5-12] the θ component of $\boldsymbol{\tau}\cdot\mathbf{n}$ is $\tau_{r\theta}$ (the same as $\tau_{\theta r}$ since $\boldsymbol{\tau}$ is a symmetric tensor as shown in Table 1.1-3). From the direction of $\tau_{r\theta}$ shown in Fig. 1.4-6*a*, the z component of the viscous force is as:

$$F_{vz} = -\iint_A \tau_{r\theta}(-\sin\theta)\, dA \qquad [1.4\text{-}39]$$

From Eq. [1.1-28], at $r = R$,

$$\tau_{r\theta}|_R = \frac{3}{2}\frac{\mu v_\infty}{R}\sin\theta \qquad [1.4\text{-}40]$$

Substituting this equation and Eq. [1.4-33] into Eq. [1.4-39]

$$F_{vz} = \int_0^{2\pi} \int_0^{\pi} \frac{3}{2} \frac{\mu v_\infty}{R} \sin^2\theta \, R^2 \sin\theta \, d\theta \, d\phi$$

$$= 3\pi \int_0^{\pi} \mu v_\infty R \sin^3\theta \, d\theta \qquad [1.4\text{-}41]$$

The integration in this equation can be carried out with the help of the following formula:

$$\int_0^{\pi} \sin^3\theta \, d\theta = \tfrac{4}{3} \qquad [1.4\text{-}42]$$

Therefore

$$F_{vz} = 4\pi\mu R v_\infty \qquad [1.4\text{-}43]$$

From Eqs. [1.4-37] and [1.4-43] the total force F_z acting by the fluid on the sphere, that is, $F_{vz} + F_{pz}$, is given by

$$\boxed{F_z = \tfrac{4}{3}\pi R^3 \rho g + 6\pi\mu R v_\infty} \qquad [1.4\text{-}44]$$

The first term on the right-hand side (RHS) of Eq. [1.4-44] represents the buoyancy force and it exists even if the fluid is stationary. As such, it is designated as F_S. The second term, on the other hand, represents the force associated with the fluid movement – the kinetic contribution – and is designated as F_k. Therefore

$$\boxed{F_S = \tfrac{4}{3}\pi R^3 \rho g} \qquad [1.4\text{-}45]$$

and

$$\boxed{F_k = 6\pi\mu R v_\infty} \qquad [1.4\text{-}46]$$

Equation [1.4-46], known as **Stokes' law**, is valid for a Reynolds number less than one.

According to Eq. [1.1-37] the **drag coefficient**

$$C_D = \frac{F_D}{A_f(\rho v_\infty^2/2)} \qquad [1.4\text{-}47]$$

where F_D is the total drag force acting on the solid by the fluid, that is, F_k, A_f is the frontal area πR^2, that is, the area projected perpendicular to v_∞. Substituting Eq. [1.4-46] into Eq. [1.4-47]

$$C_D = \frac{6\pi\mu R v_\infty/\pi R^2}{(\frac{1}{2})\rho v_\infty^2} \qquad [1.4\text{-}48]$$

or

$$\boxed{c_D = \frac{24\mu}{2R\rho v_\infty} = \frac{24}{\text{Re}}} \qquad [1.4\text{-}49]$$

where Re is the Reynolds number $D\rho v_\infty/\mu$.

Example 1.4-6 Laminar Flow over a Flat Plate

Consider the steady-state laminar flow of an incompressible Newtonian fluid over a flat plate, as illustrated in Fig. 1.4-7*a*. The fluid approaches the plate with a uniform velocity v_∞ in the z direction. The width of the plate w (in the x direction) is large such that there are no significant velocity variations in the x direction. The pressure is uniform in the fluid. The concept of the momentum boundary layer has been described in Section 1.1.4. We shall determine the thickness of the momentum boundary layer, δ, as a function of the distance from the leading edge of the flat plate z using an integral approach.

Solution The control volume Ω is enlarged in Fig. 1.4-7*b*. Its top surface allows the fluid to enter at a mass flow rate of m_4 per unit width of the plate. At the steady state Eq. [1.2-5] reduces to

$$0 = \underset{(1)}{\left[\iint_A \rho v_z\, dA\right]\Bigg|_z} - \underset{(2)}{\left[\iint_A \rho v_z\, dA\right]\Bigg|_{z+\Delta z}} + \underset{(4)}{m_4 w} \qquad [1.4\text{-}50]$$

where the numbers below the terms indicate the surfaces of the control volume. The equation can be divided by the width of the plate w and rearranged to give

$$m_4 = \left[\int_0^\delta \rho v_z\, dy\right]\Bigg|_{z+\Delta z} - \left[\int_0^\delta \rho v_z\, dy\right]\Bigg|_z \qquad [1.4\text{-}51]$$

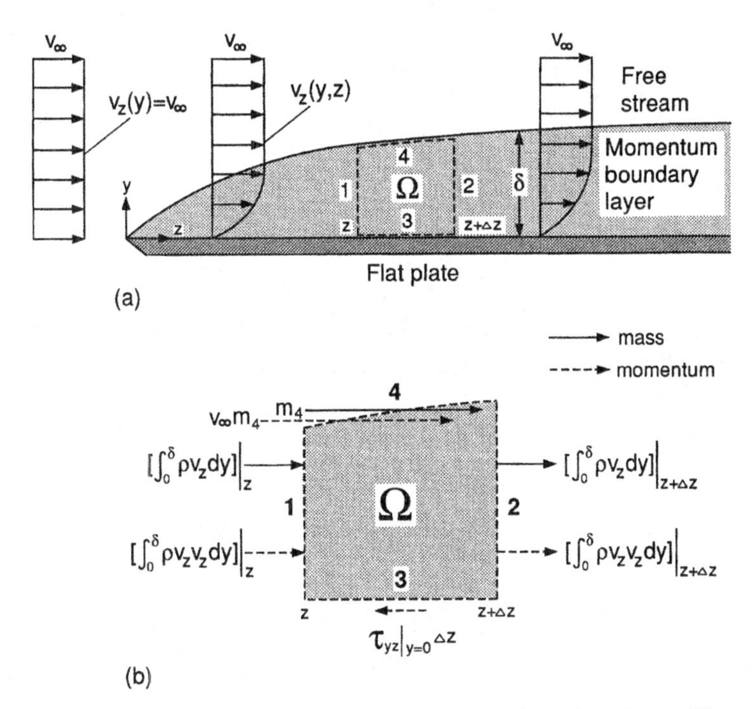

Fig. 1.4-7 Fluid flow over a flat plate: (*a*) momentum boundary layer; (*b*) a volume element in the layer.

noting that at surfaces 1 and 2, $dA = w\,dy$. From the definition of derivative and the fact that Δz is small

$$m_4 = \frac{d}{dz}\left[\int_0^\delta \rho v_z\,dy\right]\Delta z \qquad [1.4\text{-}52]$$

Let us consider the overall momentum-balance equation, Eq. [1.4-8]. The fluid entering the control volume through the top surface carries a z momentum of $m_4 v_\infty$ per unit width of the plate. The pressure force does not have to be considered since pressure is uniform. By the definition of the momentum boundary layer, $\partial v_z/\partial y$ and hence τ_{yz} are zero at surface 4. There is no gravity force in the z direction. At the steady state the z component of Eq. [1.4-8] reduces to

$$0 = \underset{(1)}{\left[\int_0^\delta \rho v_z v_z\,dy\right]\Bigg|_z} - \underset{(2)}{\left[\int_0^\delta \rho v_z v_z\,dy\right]\Bigg|_{z+\Delta z}} + \underset{(3)}{\tau_{yz}\Big|_{y=0}\Delta z} + \underset{(4)}{m_4 v_\infty}$$

[1.4-53]

Substituting Eq. [1.4-52] into this equation, dividing it by Δz and setting $\Delta z \to 0$

$$0 = -\frac{d}{dz}\left[\int_0^\delta \rho v_z v_z \, dy\right] + \frac{d}{dz}\left[\int_0^\delta \rho v_z v_\infty \, dy\right] + \tau_{yz}\bigg|_{y=0} \qquad [1.4\text{-}54]$$

From Newton's law of viscosity

$$\tau_{yz}|_{y=0} = -\mu \frac{\partial v_z}{\partial y}\bigg|_{y=0} \qquad [1.4\text{-}55]$$

Substituting this equation into Eq. [1.4-54]

$$\mu \frac{\partial v_z}{\partial y}\bigg|_{y=0} = \rho \frac{d}{dz}\left[\int_0^\delta v_z(v_\infty - v_z)\, dy\right] \qquad [1.4\text{-}56]$$

Since this equation is difficult to solve directly, the following approximate velocity profile is assumed:[16]

$$\boxed{\frac{v_z}{v_\infty} = \frac{3}{2}\left(\frac{y}{\delta}\right) - \frac{1}{2}\left(\frac{y}{\delta}\right)^3} \qquad [1.4\text{-}57]$$

This velocity profile is reasonable since it satisfies the following boundary conditions:

$$v_z = 0 \quad \text{at} \quad y = 0 \qquad [1.4\text{-}58]$$

$$v_z = v_\infty \quad \text{and} \quad \frac{\partial v_z}{\partial y} = 0 \quad \text{at} \quad y = \delta \qquad [1.4\text{-}59]$$

Substituting Eq. [1.4-57] into Eq. [1.4-56] and intergrating

$$\delta \, d\delta = \frac{140}{13} \frac{\mu \, dz}{\rho v_\infty} \qquad [1.4\text{-}60]$$

Integrating from $\delta = 0$ at $z = 0$

$$\boxed{\delta = 4.64 \sqrt{\frac{\mu z}{\rho v_\infty}}} \qquad [1.4\text{-}61]$$

or

$$\frac{\delta}{z} = \frac{4.64}{\sqrt{\mathrm{Re}_z}} \qquad [1.4\text{-}62]$$

[16] E. Pohlhausen, *Zamm*, **1**, 115, 1921.

where the **local Reynolds number** is defined as

$$\mathrm{Re}_z = \frac{z\rho v_\infty}{\mu} \qquad [1.4\text{-}63]$$

Equation [1.4-62] compares favorably with the exact solution,[17] which has a proportional constant of 5.0 instead of 4.64. As shown by Eq. [1.4-61], δ increases with $z^{1/2}$, that is, a parabolic relationship between the two.

The velocity profile $v_z(y, z)$ can be determined by substituting Eq. [1.4-61] into Eq. [1.4-57]:

$$\boxed{\frac{v_z}{v_\infty} = \frac{3}{2}\left(\frac{y}{4.64\sqrt{\mu z/\rho v_\infty}}\right) - \frac{1}{2}\left(\frac{y}{4.64\sqrt{\mu z/\rho v_\infty}}\right)^3} \qquad [1.4\text{-}64]$$

Substituting Eq. [1.4-64] into Eq. [1.4-55]

$$\tau_{yz}|_{y=0} = -0.323\, v_\infty \sqrt{\frac{\mu\rho v_\infty}{z}} \qquad [1.4\text{-}65]$$

The local friction coefficient C_{fz}, according to Eq. [1.1-36], is as follows:

$$C_{fz} = \left|\frac{\tau_{yz}|_{y=0}}{\frac{1}{2}\rho v_\infty^2}\right| \qquad [1.4\text{-}66]$$

Substituting Eq. [1.4-65] into Eq. [1.4-66]

$$\boxed{C_{fz} = \frac{0.646}{\sqrt{\mathrm{Re}_z}}} \qquad [1.4\text{-}67]$$

This equation compares favorably with the exact solution, which has a proportional constant of 0.664 instead of 0.646.

1.5. DIFFERENTIAL MOMENTUM-BALANCE EQUATION

1.5.1. Derivation

In Section 1.4 we derived the integral momentum-balance equation, Eq. [1.4-3]. If the control volume Ω does not change with time, the equation can be written

[17] H. Blasius, *Z. Math. U. Phys. Sci.* **1**, 1908.

as follows:

$$\iiint_{\Omega} \frac{\partial}{\partial t}(\rho \mathbf{v})\, d\Omega = -\iint_{A} \rho \mathbf{v}\mathbf{v}\cdot \mathbf{n}\, dA - \iint_{A} p\mathbf{n}\, dA - \iint_{A} \boldsymbol{\tau}\cdot \mathbf{n}\, dA + \iiint_{\Omega} \mathbf{f}_b\, d\Omega \qquad [1.5\text{-}1]$$

The surface integrals in Eq. [1.5-1] can be converted into their corresponding volume integrals. From Eq. [A.4-4]

$$\iint_{A} \rho \mathbf{v}\mathbf{v}\cdot \mathbf{n}\, dA = \iiint_{A} \nabla\cdot(\rho \mathbf{v}\mathbf{v})\, d\Omega \qquad [1.5\text{-}2]$$

Similarly, from Eqs. [A.4-2] and [A.4-3]

$$\iint_{A} p\mathbf{n}\, dA = \iiint_{\Omega} \nabla p\, d\Omega \qquad [1.5\text{-}3]$$

and

$$\iint_{A} \boldsymbol{\tau}\cdot \mathbf{n}\, dA = \iiint_{\Omega} \nabla\cdot \boldsymbol{\tau}\, d\Omega \qquad [1.5\text{-}4]$$

Equation [1.5-1] can now be rewritten as follows:

$$\iiint_{\Omega} \left\{ \frac{\partial}{\partial t}(\rho \mathbf{v}) + \nabla\cdot(\rho \mathbf{v}\mathbf{v}) + \nabla p + \nabla\cdot \boldsymbol{\tau} - \mathbf{f}_b \right\} d\Omega = 0 \qquad [1.5\text{-}5]$$

The integrand, which is continuous, must be zero everywhere since the equation must hold for any arbitrary region Ω.[18] Therefore

$$\frac{\partial}{\partial t}(\rho \mathbf{v}) + \nabla\cdot(\rho \mathbf{v}\mathbf{v}) + \nabla p + \nabla\cdot \boldsymbol{\tau} - \mathbf{f}_b = 0 \qquad [1.5\text{-}6]$$

The first two terms of this equation can be expanded; the second term, with the help of Eq. [A.3-15]. They can then be simplified with the help of the equation of continuity (Eq. [1.3-4]) as follows:

$$\rho \frac{\partial \mathbf{v}}{\partial t} + \mathbf{v}\frac{\partial \rho}{\partial t} + \mathbf{v}(\nabla\cdot \rho \mathbf{v}) + \rho \mathbf{v}\cdot \nabla \mathbf{v} = \rho \frac{\partial \mathbf{v}}{\partial t} + \rho \mathbf{v}\cdot \nabla \mathbf{v} \qquad [1.5\text{-}7]$$

[18] W. Kaplan, *Advanced Calculus*, 2nd ed., Addison-Wesley, Reading, MA, 1973, p. 363.

Furthermore, for a Newtonian fluid, the stress tensor τ is expressed in terms of velocity gradients, as shown in Tables 1.1-1 through 1.1-3. If the fluid is also incompressible, it can be shown that $-\nabla \cdot \tau$ becomes $\mu \nabla^2 \mathbf{v}$ (this is left as an exercise in Problem 1.5-14). Substituting this and Eq. [1.5-7] into Eq. [1.5-6]

$$\rho \frac{\partial \mathbf{v}}{\partial t} + \rho \mathbf{v} \cdot \nabla \mathbf{v} = -\nabla p + \mu \nabla^2 \mathbf{v} + \mathbf{f}_b \qquad \text{(constant } \rho \text{ and } \mu) \qquad [1.5\text{-}8]$$

Equation [1.5-8] is the **differential momentum-balance equation**, that is, the **equation motion.**

The body force per unit folume $\mathbf{f}_b = \rho \mathbf{g}$ if the gravity force is the only body force involved, that is

$$\rho \frac{\partial \mathbf{v}}{\partial t} + \rho \mathbf{v} \cdot \nabla \mathbf{v} = -\nabla p + \mu \nabla^2 \mathbf{v} + \rho \mathbf{g} \qquad \text{(constant } \rho \text{ and } \mu) \qquad [1.5\text{-}9]$$

Equation [1.5-9] is the special case of the differential momentum-balance equation for a fluid with a constant density and viscosity. This special form of the equation of motion, called the **Navier–Stokes equation**, was developed by Navier in 1827 and independently by Stokes in 1845.

In Tables 1.5-1 through 1.5-3, Eq. [1.5-9] is given in expanded forms for rectangular, cylindrical, and spherical coordinates, with the gravity force being the only body force involved. These tables are useful for ordinary **forced convection**, when flow is caused by external means, as, by a pump or fan.

TABLE 1.5-1 Equation of Motion in Rectangular Coordinates (x, y, z) (for a Newtonian Fluid with Constant ρ and μ)

x component:

$$\rho\left[\frac{\partial v_x}{\partial t} + v_x \frac{\partial v_x}{\partial x} + v_y \frac{\partial v_x}{\partial y} + v_z \frac{\partial v_x}{\partial z}\right] = -\frac{\partial p}{\partial x} + \mu\left[\frac{\partial^2 v_x}{\partial x^2} + \frac{\partial^2 v_x}{\partial y^2} + \frac{\partial^2 v_x}{\partial z^2}\right] + \rho g_x \qquad [A]$$

y component:

$$\rho\left[\frac{\partial v_y}{\partial t} + v_x \frac{\partial v_y}{\partial x} + v_y \frac{\partial v_y}{\partial y} + v_z \frac{\partial v_y}{\partial z}\right] = -\frac{\partial p}{\partial y} + \mu\left[\frac{\partial^2 v_y}{\partial x^2} + \frac{\partial^2 v_y}{\partial y^2} + \frac{\partial^2 v_y}{\partial z^2}\right] + \rho g_y \qquad [B]$$

z component:

$$\rho\left[\frac{\partial v_z}{\partial t} + v_x \frac{\partial v_z}{\partial x} + v_y \frac{\partial v_z}{\partial y} + v_z \frac{\partial v_z}{\partial z}\right] = -\frac{\partial p}{\partial z} + \mu\left[\frac{\partial^2 v_z}{\partial x^2} + \frac{\partial^2 v_z}{\partial y^2} + \frac{\partial^2 v_z}{\partial z^2}\right] + \rho g_z \qquad [C]$$

TABLE 1.5-2 Equation of Motion in Cylindrical Coordinates (r, θ, z) (for a Newtonian Fluid with Constant ρ and μ)

r component:

$$\rho\left[\frac{\partial v_r}{\partial t} + v_r\frac{\partial v_r}{\partial r} + \frac{v_\theta}{r}\frac{\partial v_r}{\partial \theta} - \frac{v_\theta^2}{r} + v_z\frac{\partial v_r}{\partial z}\right]$$

$$= -\frac{\partial p}{\partial r} + \mu\left[\frac{\partial}{\partial r}\left(\frac{1}{r}\frac{\partial}{\partial r}(rv_r)\right) + \frac{1}{r^2}\frac{\partial^2 v_r}{\partial \theta^2} - \frac{2}{r^2}\frac{\partial v_\theta}{\partial \theta} + \frac{\partial^2 v_r}{\partial z^2}\right] + \rho g_r \qquad \text{[A]}$$

θ component:

$$\rho\left[\frac{\partial v_\theta}{\partial t} + v_r\frac{\partial v_\theta}{\partial r} + \frac{v_\theta}{r}\frac{\partial v_\theta}{\partial \theta} + \frac{v_r v_\theta}{r} + v_z\frac{\partial v_\theta}{\partial z}\right]$$

$$= -\frac{1}{r}\frac{\partial p}{\partial \theta} + \mu\left[\frac{\partial}{\partial r}\left(\frac{1}{r}\frac{\partial}{\partial r}(rv_\theta)\right) + \frac{1}{r^2}\frac{\partial^2 v_\theta}{\partial \theta^2} + \frac{2}{r^2}\frac{\partial v_r}{\partial \theta} + \frac{\partial^2 v_\theta}{\partial z^2}\right] + \rho g_\theta \qquad \text{[B]}$$

z component:

$$\rho\left[\frac{\partial v_z}{\partial t} + v_r\frac{\partial v_z}{\partial r} + \frac{v_\theta}{r}\frac{\partial v_z}{\partial \theta} + v_z\frac{\partial v_z}{\partial z}\right]$$

$$= -\frac{\partial p}{\partial z} + \mu\left[\frac{1}{r}\frac{\partial}{\partial r}\left(r\frac{\partial v_z}{\partial r}\right) + \frac{1}{r^2}\frac{\partial^2 v_z}{\partial \theta^2} + \frac{\partial^2 v_z}{\partial z^2}\right] + \rho g_z \qquad \text{[C]}$$

The equation of motion is often expressed in terms of the **Stokes derivative** D/Dt, which is defined as follows:

$$\frac{D}{Dt} = \frac{\partial}{\partial t} + v_x\frac{\partial}{\partial x} + v_y\frac{\partial}{\partial y} + v_z\frac{\partial}{\partial z} = \frac{\partial}{\partial t} + \mathbf{v}\cdot\nabla \qquad \text{[1.5-10]}$$

The equation of motion (Eq. [1.5-9]) becomes

$$\underset{(1)}{\rho\frac{D\mathbf{v}}{Dt}} = \underset{(2)}{-\nabla p} + \underset{(3)}{\mu\nabla^2\mathbf{v}} + \underset{(4)}{\rho\mathbf{g}} \qquad \text{[1.5-11]}$$

In this equation term (1) is the inertia force per unit volume, namely, the mass per unit volume times acceleration. Terms (2), (3) and (4) are the pressure force, viscous force, and gravity force per unit volume, respectively. This equation is, in fact, a statement of **Newton's second law of motion**, which is mass times acceleration equal to the sum of forces. For the so-called creeping flow (Example 1.4-5), the acceleration term can be neglected.

TABLE 1.5-3 Equation of Motion in Spherical Coordinates (r, θ, ϕ) **(for a Newtonian Fluid with Constant** ρ **and** μ**)**[a]

r component:

$$\rho\left[\frac{\partial v_r}{\partial t} + v_r\frac{\partial v_r}{\partial r} + \frac{v_\theta}{r}\frac{\partial v_r}{\partial \theta} + \frac{v_\phi}{r\sin\theta}\frac{\partial v_r}{\partial \phi} - \frac{v_\theta^2 + v_\phi^2}{r}\right]$$

$$= -\frac{\partial p}{\partial r} + \mu\left[\nabla^2 v_r - \frac{2}{r^2}v_r - \frac{2}{r^2}\frac{\partial v_\theta}{\partial \theta} - \frac{2}{r^2}v_\theta\cot\theta - \frac{2}{r^2\sin\theta}\frac{\partial v_\phi}{\partial \phi}\right] + \rho g_r \qquad \text{[A]}$$

θ component:

$$\rho\left[\frac{\partial v_\theta}{\partial t} + v_r\frac{\partial v_\theta}{\partial r} + \frac{v_\theta}{r}\frac{\partial v_\theta}{\partial \theta} + \frac{v_\phi}{r\sin\theta}\frac{\partial v_\theta}{\partial \phi} + \frac{v_r v_\theta}{r} - \frac{v_\phi^2\cot\theta}{r}\right]$$

$$= -\frac{1}{r}\frac{\partial p}{\partial \theta} + \mu\left[\nabla^2 v_\theta + \frac{2}{r^2}\frac{\partial v_r}{\partial \theta} - \frac{v_\theta}{r^2\sin^2\theta} - \frac{2\cos\theta}{r^2\sin^2\theta}\frac{\partial v_\phi}{\partial \phi}\right] + \rho g_\theta \qquad \text{[B]}$$

ϕ component:

$$\rho\left[\frac{\partial v_\phi}{\partial t} + v_r\frac{\partial v_\phi}{\partial r} + \frac{v_\theta}{r}\frac{\partial v_\phi}{\partial \theta} + \frac{v_\phi}{r\sin\theta}\frac{\partial v_\phi}{\partial \phi} + \frac{v_\phi v_r}{r} + \frac{v_\theta v_\phi}{r}\cot\theta\right]$$

$$= -\frac{1}{r\sin\theta}\frac{\partial p}{\partial \phi} + \mu\left[\nabla^2 v_\phi - \frac{v_\phi}{r^2\sin^2\theta} + \frac{2}{r^2\sin\theta}\frac{\partial v_r}{\partial \phi} + \frac{2\cos\theta}{r^2\sin^2\theta}\frac{\partial v_\theta}{\partial \phi}\right] + \rho g_\phi \qquad \text{[C]}$$

[a] In these equations:

$$\nabla^2 = \frac{1}{r^2}\frac{\partial}{\partial r}\left(r^2\frac{\partial}{\partial r}\right) + \frac{1}{r^2\sin\theta}\frac{\partial}{\partial \theta}\left(\sin\theta\frac{\partial}{\partial \theta}\right) + \frac{1}{r^2\sin^2\theta}\left(\frac{\partial^2}{\partial \phi^2}\right)$$

1.5.2. Dimensionless Form

The equation of motion, and the continuity equation as well, can be presented in the dimensionless form; the main advantage is that the solutions can be more general.

In forced convection a **characteristic velocity** V is available as a reference, such as the velocity of a moving wall, the average velocity of flow in a pipe, or the velocity of a fluid approaching a solid surface or body. A **characteristic length** L and hence a **characteristic time** L/V are also available.

From Eq. [1.5-11], the characteristic velocity V, characteristic length L, and characteristic time L/V, the forces per unit volume in forced convection can be expressed qualitatively as follows

$$\text{inertia force:} \qquad \rho\frac{Dv}{Dt} = \rho\frac{V}{L/V} = \frac{\rho V^2}{L}$$

pressure force: $\nabla p = \dfrac{p}{L}$

viscous force: $\mu \nabla^2 v = \mu \dfrac{V}{L^2}$

gravity force: ρg

The expressions for these forces have been obtained from Eq. [1.5–11] by substituting V for v, $(L/V)^{-1}$ for D/Dt, L^{-1} for ∇ (Eq. [A.3–1]), and L^{-2} for ∇^2 (Eq. [A.3–10]). For forced covection the following dimensionless variables can be defined

$$\mathbf{v}^* = \frac{\mathbf{v}}{V} \qquad \text{(dimensionless velocity)} \qquad [1.5\text{-}12]$$

$$t^* = \frac{tV}{L} \qquad \text{(dimensionless time)} \qquad [1.5\text{-}13]$$

$$p^* = \frac{p - p_0}{\rho V^2} \qquad \text{(dimensionless pressure)} \qquad [1.5\text{-}14]$$

$$x^*, y^*, z^* = \frac{x}{L}, \frac{y}{L}, \frac{z}{L} \qquad \text{(dimensionless coordinates)} \qquad [1.5\text{-}15]$$

$$\nabla^* = L\nabla = \mathbf{e}_x \frac{\partial}{\partial x^*} + \mathbf{e}_y \frac{\partial}{\partial y^*} + \mathbf{e}_z \frac{\partial}{\partial z^*} \qquad \text{(dimensionless operator)} \qquad [1.5\text{-}16]$$

$$\nabla^{*2} = L^2\nabla^2 = \frac{\partial^2}{\partial x^{*2}} + \frac{\partial^2}{\partial y^{*2}} + \frac{\partial^2}{\partial z^{*2}} \qquad \text{(dimensionless operator)} \qquad [1.5\text{-}17]$$

For an incompressible Newtonian fluid and with gravity as the only body force involved, the equations of continuity and motion are

$$\text{Continuity:} \quad \nabla \cdot \mathbf{v} = 0 \qquad [1.5\text{-}18]$$

$$\text{Motion:} \quad \rho \frac{\partial \mathbf{v}}{\partial t} + \rho \mathbf{v} \cdot \nabla \mathbf{v} = -\nabla p + \mu \nabla^2 \mathbf{v} + \rho g \mathbf{e}_g \qquad [1.5\text{-}19]$$

where $\mathbf{e}_g$ is the unit vector in the direction of gravity:

$$\mathbf{g} = g\mathbf{e}_g \qquad [1.5\text{-}20]$$

Substituting Eqs. [1.5-12] through [1.5-17] into Eqs. [1.5-18] and [1.5-19]

$$\text{Continuity:} \quad \frac{1}{L}\nabla^* \cdot \mathbf{v}^* V = 0 \qquad [1.5\text{-}21]$$

$$\text{Motion:} \quad \rho \frac{V}{L}\frac{\partial}{\partial t^*}(\mathbf{v}^* V) + \rho V \mathbf{v}^* \cdot \nabla^* \mathbf{v}^* \frac{V}{L}$$

$$= -\frac{1}{L}\nabla^* p^* \rho V^2 + \mu \frac{1}{L^2}\nabla^{*2}\mathbf{v}^* V + \rho g \mathbf{e}_g \qquad [1.5\text{-}22]$$

Multiplying Eq. [1.5-21] by L/V and Eq. [1.5-22] by $L/(\rho V^2)$

$$\text{Continuity:} \quad \nabla^* \cdot \mathbf{v}^* = 0 \qquad [1.5\text{-}23]$$

$$\text{Motion:} \quad \frac{\partial \mathbf{v}^*}{\partial t^*} + \mathbf{v}^* \cdot \nabla^* \mathbf{v}^* = -\nabla^* p^* + \left(\frac{\nu}{LV}\right)\nabla^{*2}\mathbf{v}^* + \left(\frac{Lg}{V^2}\right)\mathbf{e}_g \qquad [1.5\text{-}24]$$

Therefore, for forced convection

$$\text{Continuity:} \quad \nabla^* \cdot \mathbf{v}^* = 0 \qquad [1.5\text{-}25]$$

$$\text{Motion:} \quad \frac{\partial \mathbf{v}^*}{\partial t^*} + \mathbf{v}^* \cdot \nabla^* \mathbf{v}^* = -\nabla^* p^* + \frac{1}{\text{Re}}\nabla^{*2}\mathbf{v}^* + \frac{1}{\text{Fr}}\mathbf{e}_g \qquad [1.5\text{-}26]$$

where

$$\text{Re} = \frac{LV}{\nu} \quad \left(\text{Reynolds number} = \frac{\text{inertia force } \rho V^2/L}{\text{viscous force } \mu V/L^2}\right) \qquad [1.5\text{-}27]$$

$$\text{Fr} = \frac{V^2}{gL} \quad \left(\text{Froude number} = \frac{\text{inertia force } \rho V^2/L}{\text{gravity force } \rho g}\right) \qquad [1.5\text{-}28]$$

1.5.3. Boundary Conditions

Boundary conditions commonly encountered in fluid flow are summarized in Figs. 1.5-1 and 1.5-2. The boundary conditions are explained as follows:

1. At the plane or axis of symmetry, the velocity gradient in the transverse direction is zero, as in case 1 in Figs. 1.5-1 and 1.5-2.

Commonly Encountered Fluid Flow Boundary Conditions in Rectangular Coordinates

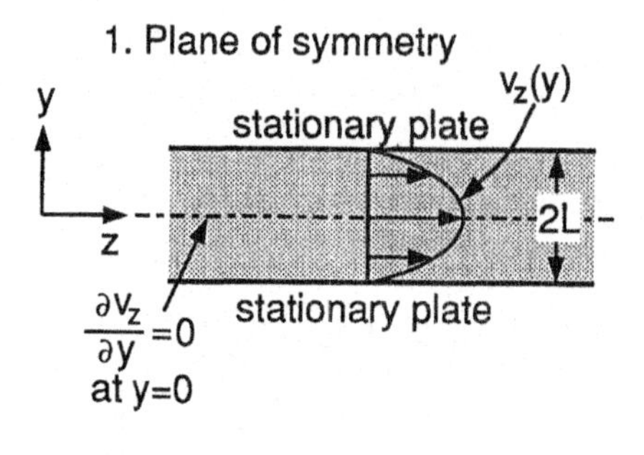

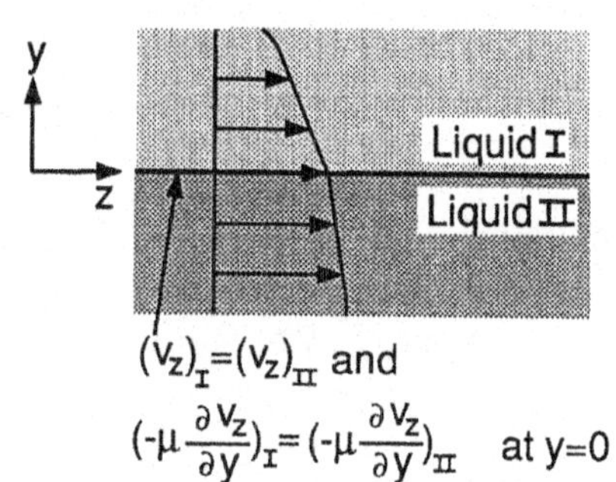

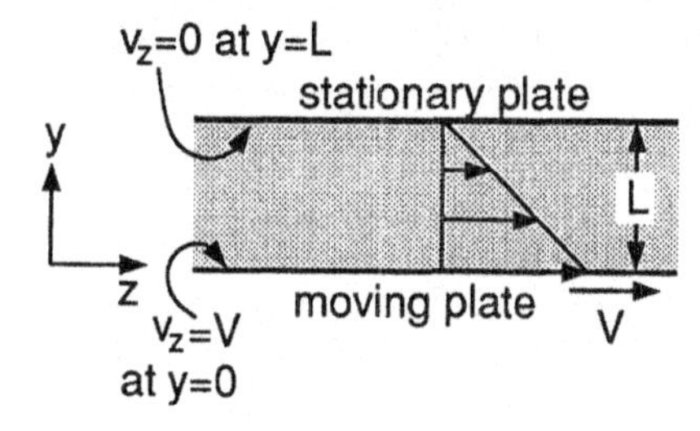

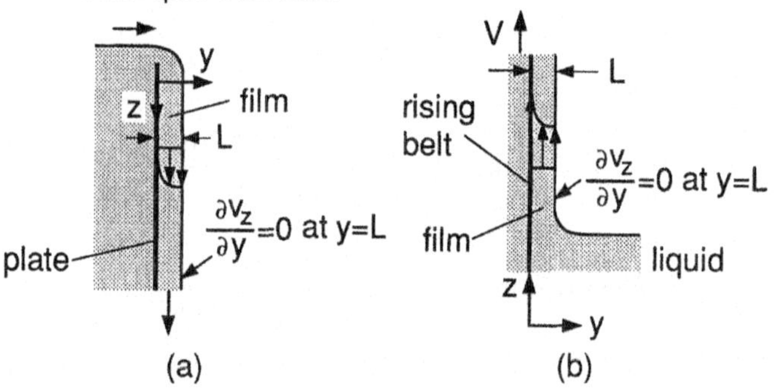

Fig. 1.5-1 Fluid flow boundary conditions: rectangular coordinates.

2. At a stationary wall of a tube or plate, the fluid velocity is zero. At a moving wall, on the other hand, the fluid velocity equals the velocity at which the wall is moving. These conditions are called the **no-slip condition**, as in case 2 in Figs. 1.5-1 and 1.5-2.

Commonly Encountered Fluid Flow Boundary Conditions in Cylindrical Coordinates

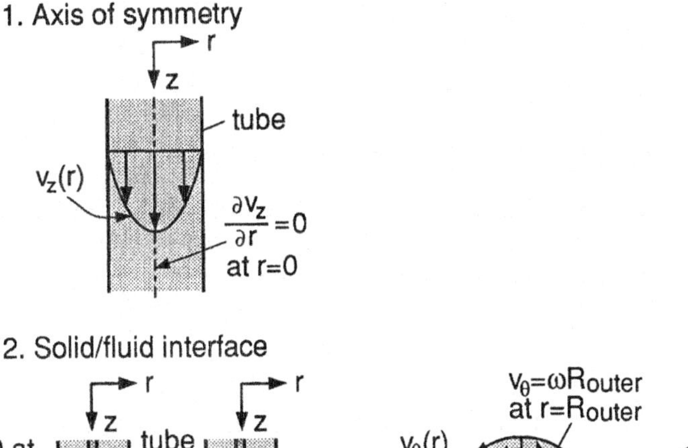

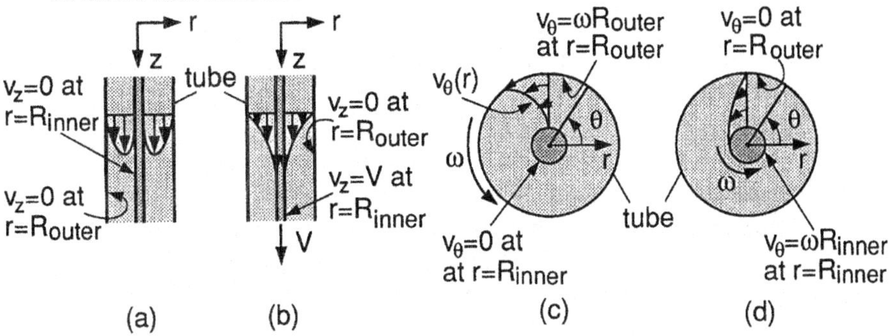

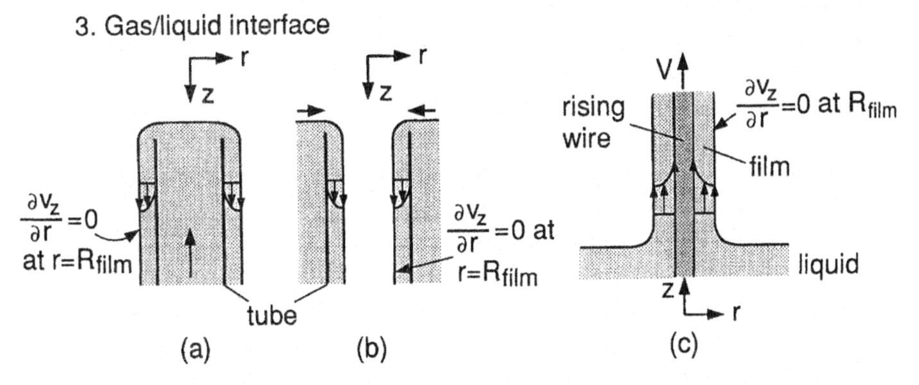

Fig. 1.5-2 Fluid flow boundary conditions: cylindrical coordinates.

3. At the free surface of a liquid, i.e., the liquid/gas interface, the momentum flux and hence the velocity gradient in the liquid are zero, as in case 3 in Figs. 1.5-1

and 1.5-2. Surface tension related boundary conditions will be discussed in Chapter 5.

4. For two immiscible liquids the velocity and momentum flux are both continuous across the interface, that is they are the same on both sides of the interface, as in case 4 in Fig. 1.5-1.
5. The boundary conditions at solid/liquid interfaces with phase transformations will be described in Chapter 5.

1.5.4. Solution Procedure

The purpose of the equation of motion, as will be illustrated in the following examples, is to described the motion of the fluid, that is, to determine its velocity distribution. The step-by-step procedure for solving the equation of motion, which is illustrated in Fig. 1.5-3 by the route represented by solid arrows, is as follows:

1. Choose a coordinate system that best describes the physical system geometrically and pick the corresponding form of the equation of continuity from Table 1.3-1.
2. Eliminate the zero term(s) from the equation of continuity.
3. Determine which velocity component is relevant to the problem and pick the corresponding component of the equation of motion. For ordinary forced convection the following tables are useful: Table 1.5-1 (rectangular coordinates), Table 1.5-2 (cylindrical coordinates), and Table 1.5-3 (spherical coordinates). Similar tables are available for thermosolutal convection, which will be discussed in Chapter 4.
4. Eliminate the zero term(s) from the equation of motion. The equation of continuity resulting from step 2 should be applied here.
5. Set up the boundary and/or initial conditions.
6. Solve the equation of motion, subject to the boundary and/or initial conditions in step 5, for the velocity distribution. For convenience, some solutions are provided in Appendix A.
7. Check to see if the velocity distribution satisfies the boundary and/or initial conditions in step 5.

After the velocity distribution is determined, it can be substituted into Newton's law of viscosity to determine the shear stress distribution. This is illustrated in Fig. 1.5-3 by the route represented by broken arrows.

It should be mentioned that once the velocity distribution is obtained, it can be used to determine the stress acting by the fluid at any solid/fluid interfaces. This can be done by following steps 2 and 3 of the procedure described in Section 1.1.2.4.

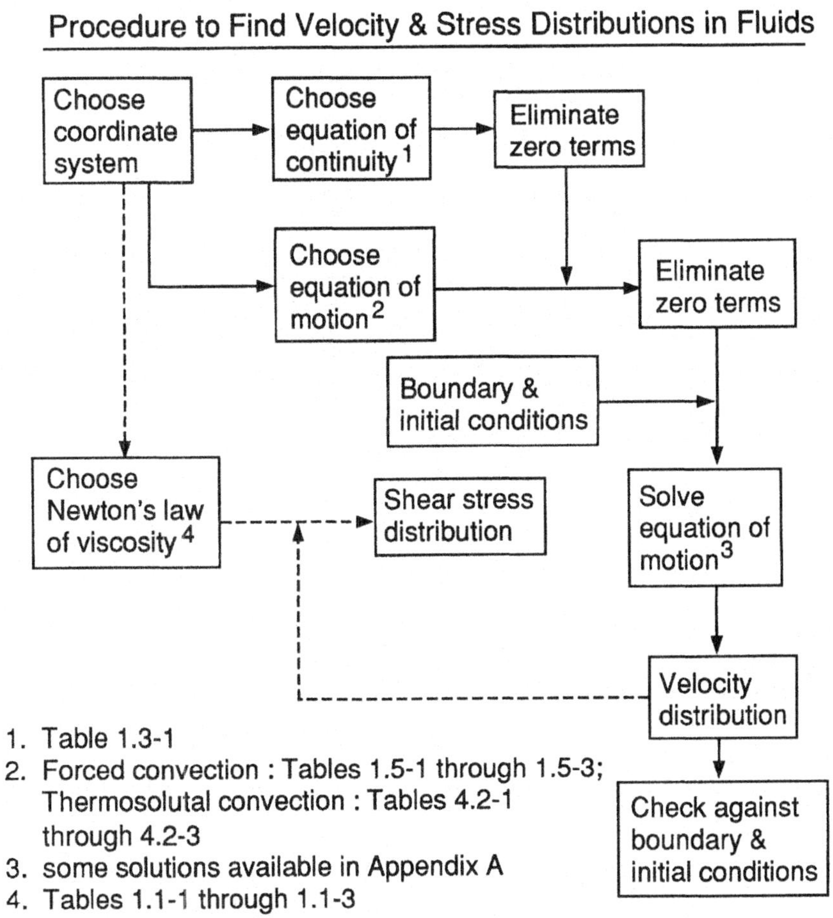

Procedure to Find Velocity & Stress Distributions in Fluids

Choose coordinate system

Choose equation of continuity[1]

Eliminate zero terms

Choose equation of motion[2]

Eliminate zero terms

Boundary & initial conditions

Choose Newton's law of viscosity[4]

Shear stress distribution

Solve equation of motion[3]

Velocity distribution

Check against boundary & initial conditions

1. Table 1.3-1
2. Forced convection : Tables 1.5-1 through 1.5-3; Thermosolutal convection : Tables 4.2-1 through 4.2-3
3. some solutions available in Appendix A
4. Tables 1.1-1 through 1.1-3

Fig. 1.5-3 Procedure for solving fluid flow problems by the equation of motion.

1.5.5. Examples

Example 1.5-1 Tangential Annular Flow

A fluid is held in the annular space between a stationary vertical cylinder of radius r_1 and a coaxial cylindrical container of radius r_2, which rotates at an angular velocity ω_2, as illustrated in Fig. 1.5-4. Determine the steady-state velocity and shear stress distributions in the fluid. The fluid is incompressible and Newtonian, and the flow is laminar. Assume that the fluid is deep and the end effects are thus negligible.

Solution As shown in Fig. 1.5-4, cylindrical coordinates are most appropriate for this problem. At the steady state all time-dependent terms should be zero.

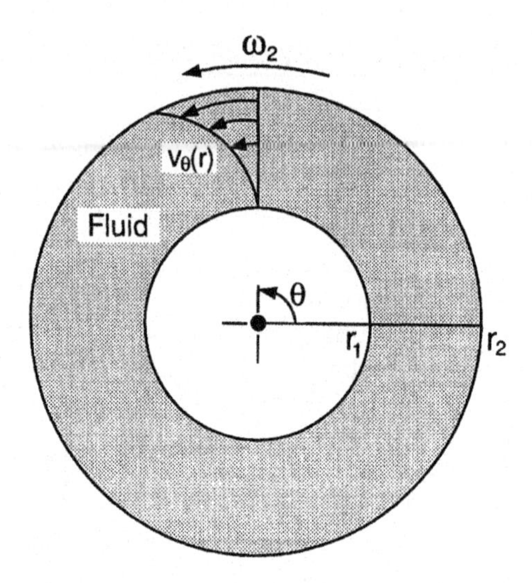

Fig. 1.5-4 Top view of a tangential flow between a stationary cylinder and a rotating container.

Since fluid flow is in the θ direction only, $v_r = v_z = 0$. Obviously, $g_\theta = 0$ since the θ component (horizontal) of g(vertical) is zero. Also, since fluid flow is axisymmetric, there should be no pressure variations in the θ direction. With negligible end effects, $\partial v_\theta/\partial z = 0$.

The equation of continuity in cylindrical coordinates, Eq. [B] of Table 1.3-1, is as follows:

$$\frac{\partial \rho}{\partial t} + \frac{1}{r}\frac{\partial}{\partial r}(\rho r v_r) + \frac{1}{r}\frac{\partial}{\partial \theta}(\rho v_\theta) + \frac{\partial}{\partial z}(\rho v_z) = 0 \qquad [1.5\text{-}29]$$

For the reasons states at the beginning, this equation is reduced to

$$\frac{\partial v_\theta}{\partial \theta} = 0 \qquad [1.5\text{-}30]$$

To determine the velocity distribution $v_\theta(r)$, the θ component of the equation of motion, Eq. [B] of Table 1.5-2, is shown below:

$$\rho\left[\frac{\partial v_\theta}{\partial t} + v_r\frac{\partial v_\theta}{\partial r} + \frac{v_\theta}{r}\frac{\partial v_\theta}{\partial \theta} + \frac{v_r v_\theta}{r} + v_z\frac{\partial v_\theta}{\partial z}\right]$$

$$= -\frac{1}{r}\frac{\partial p}{\partial \theta} + \mu\left\{\frac{\partial}{\partial r}\left[\frac{1}{r}\frac{\partial}{\partial r}(r v_\theta)\right] + \frac{1}{r^2}\frac{\partial^2 v_\theta}{\partial \theta^2} + \frac{2}{r^2}\frac{\partial v_r}{\partial \theta} + \frac{\partial^2 v_\theta}{\partial z^2}\right\} + \rho g_\theta \qquad [1.5\text{-}31]$$

For the reasons already stated and with the substitution of Eq. [1.5-30]

$$\frac{d}{dr}\left[\frac{1}{r}\frac{d}{dr}(rv_\theta)\right] = 0 \qquad [1.5\text{-}32]$$

The boundary conditions are

$$v_\theta = 0 \quad \text{at} \quad r = r_1 \qquad [1.5\text{-}33]$$

$$v_\theta = r_2\omega_2 \quad \text{at} \quad r = r_2 \qquad [1.5\text{-}34]$$

The solution, from Case L of Appendix A, is as follows:

$$v_\theta = \frac{1}{r_2^2 - r_1^2}\left(r\omega_2 r_2^2 - \frac{r_1^2 r_2^2}{r}\omega_2\right) \qquad [1.5\text{-}35]$$

As shown, v_θ is a nonlinear function of r.

From Newton's law of viscosity, Eq. [D] of Table 1.1-2, and the fact that $v_r = 0$

$$\tau_{r\theta} = -\mu r\frac{d}{dr}\left(\frac{v_\theta}{r}\right) \qquad [1.5\text{-}36]$$

Substituting Eq. [1.5-35] into Eq. [1.5-36]

$$\tau_{r\theta} = \frac{-\mu r\omega_2 r_2^2}{r_2^2 - r_1^2}\frac{d}{dr}\left(1 - \frac{r_1^2}{r^2}\right) \qquad [1.5\text{-}37]$$

or

$$\tau_{r\theta} = \frac{-2\mu\omega_2}{r_2^2 - r_1^2}\frac{r_1^2 r_2^2}{r^2} \qquad [1.5\text{-}38]$$

Let L be the length of the cylinder immersed in the fluid. The force F required to turn the cylindrical container is

$$F = 2\pi r_2 L(-\tau_{r\theta}|_{r=r_2}) = 4\pi\mu L\frac{\omega_2 r_1^2 r_2}{r_2^2 - r_1^2} \qquad [1.5\text{-}39]$$

and the corresponding torque is

$$\mathcal{T} = r_2 F = 4\pi\mu L\frac{\omega_2 r_1^2 r_2^2}{r_2^2 - r_1^2} \qquad [1.5\text{-}40]$$

Equation [1.5-40] can be used to measure the viscosity of the fluid.

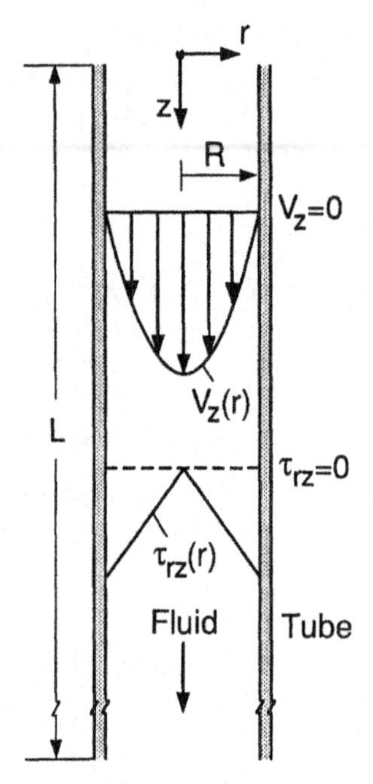

Fig. 1.5-5 Fluid flow through a circular tube.

Example 1.5-2 Laminar Flow through a Vertical Tube

Consider the steady-state, laminar flow of an incompressible, Newtonian fluid through a long, vertical tube of radius R and length L, as illustrated in Fig. 1.5-5. Determine the velocity distribution, the volume flow rate, and the shear stress acting on the tube wall in terms of the pressure drop across the tube.

Solution Like in the previous example cylindrical coordinates are most appropriate for this problem. At the steady state all time-dependent terms should be zero. Since the tube is long, we ignore the end effects, specifically, the fact that at the tube entrance and exit the flow is not necessarily parallel to the tube axis. Therefore, we set $v_r = v_\theta = 0$. Finally, since the flow is symmetric with respect to the z axis, v_z is independent of θ.

The equation of continuity in cylindrical coordinates, Eq. [B] of Table 1.3-1.

$$\frac{\partial \rho}{\partial t} + \frac{1}{r}\frac{\partial}{\partial r}(\rho r v_r) + \frac{1}{r}\frac{\partial}{\partial \theta}(\rho v_\theta) + \frac{\partial}{\partial z}(\rho v_z) = 0 \qquad [1.5\text{-}41]$$

This equation reduces to the following one for the reasons stated at the beginning:

$$\frac{\partial v_z}{\partial z} = 0 \qquad [1.5\text{-}42]$$

Let us now consider the z component of the equation of motion since we want the velocity distribution $v_z(r)$. According to Eq. [C] of Table 1.5-2, we have

$$\rho\left[\frac{\partial v_z}{\partial t} + v_r\frac{\partial v_z}{\partial r} + \frac{v_\theta}{r}\frac{\partial v_z}{\partial \theta} + v_z\frac{\partial v_z}{\partial z}\right]$$

$$= -\frac{\partial p}{\partial z} + \mu\left[\frac{1}{r}\frac{\partial}{\partial r}\left(r\frac{\partial v_z}{\partial r}\right) + \frac{1}{r^2}\frac{\partial^2 v_z}{\partial \theta^2} + \frac{\partial^2 v_z}{\partial z^2}\right] + \rho g_z \qquad [1.5\text{-}43]$$

Since the z direction is vertically downward, $g_z = g$, where g is gravitational acceleration 980 cm/sec^2. Following the same reasons and substituting Eq. [1.5-42] and $g_z = g$, Eq. [1.5-43] reduces to

$$\mu\frac{1}{r}\frac{d}{dr}\left(r\frac{dv_z}{dr}\right) = \frac{dp}{dz} - \rho g \qquad [1.5\text{-}44]$$

The left-hand side (LHS) of Eq. [1.5-44] is either a function of r or a constant, but the RHS is either a function of z or a constant. Therefore, we conclude that both must be a constant and the LHS can be written in the following finite form:

$$\frac{dp}{dz} - \rho g = \frac{p_L - p_0}{L - 0} - \rho g = \frac{(p_L - p_0) - \rho g L}{L} \qquad [1.5\text{-}45]$$

Equation [1.5-44] now becomes

$$\frac{1}{r}\frac{d}{dr}\left(r\frac{dv_z}{dr}\right) = \frac{(p_L - p_0) - \rho g L}{\mu L} \qquad [1.5\text{-}46]$$

The boundary conditions are

$$\frac{dv_z}{dr} = 0 \quad \text{at} \quad r = 0 \qquad [1.5\text{-}47]$$

$$v_z = 0 \quad \text{at} \quad r = R \qquad [1.5\text{-}48]$$

The solution, from Case I of Appendix A, is as follows:

$$\boxed{v_z = \left[\frac{(p_0 - p_L) + \rho g L}{4\mu L}\right](R^2 - r^2)} \qquad [1.5\text{-}49]$$

where $(p_0 - p_L)$ is the pressure drop across the tube. This equation shows that v_z is a parabolic function of r.

The average velocity v_{av}, as in Eq. [1.1-39], is defined by

$$v_{av} = \frac{\iint_A \rho v_z \, dA}{\iint_A \rho \, dA} = \frac{\int_0^R v_z 2\pi r \, dr}{\pi R^2} \qquad [1.5\text{-}50]$$

Substituting Eq. [1.5-49] into Eq. [1.5-50], we get

$$v_{av} = \frac{[(p_0 - p_L) + \rho g L] R^2}{8\mu L} \qquad [1.5\text{-}51]$$

Since the average velocity is simply the volume flow rate Q divided by the inner cross-sectional area πR^2 of the tube

$$\boxed{Q = \frac{\pi[(p_0 - p_L) + \rho g L] R^4}{8\mu L}} \qquad [1.5\text{-}52]$$

This equation is called the **Hagen–Poiseuille law.**

Since $v_z = v_{max}$ at $r = 0$, from Eq. [1.5-49]

$$v_{max} = \frac{[(p_0 - p_L + \rho g L] R^2}{4\mu L} = 2v_{av} = \frac{2Q}{\pi R^2} \qquad [1.5\text{-}53]$$

Thus the maximum velocity is twice the average velocity for laminar flow in a round tube. Substituting Eq. [1.5-53] into Eq. [1.5-49]

$$v_z = v_{max}\left[1 - \left(\frac{r}{R}\right)^2\right] = \frac{2Q}{\pi R^2}\left[1 - \left(\frac{r}{R}\right)^2\right] \qquad [1.5\text{-}54]$$

From Newton's law of viscosity, Eq. [F] of Table 1.1-2, and the fact that $v_r = 0$,

$$\tau_{rz} = -\mu \frac{dv_z}{dr} \qquad [1.5\text{-}55]$$

Substituting Eq. [1.5-49] into the equation

$$\boxed{\tau_{rz} = \left[\frac{(p_0 - p_L) + \rho g L}{2L}\right] r} \qquad [1.5\text{-}56]$$

Therefore, τ_{rz} is a linear function of r. At the tube wall $r = R$ and

$$\tau_{rz}\big|_{r=R} = \left[\frac{(p_0 - p_L) + \rho g L}{2L}\right] R \qquad [1.5\text{-}57]$$

This equation is consistent with Eq. [1.4-27], which has been derived previously from integral momentum balance.

Before leaving the subject, let us consider the friction factor f. From Eq. [1.5-51]

$$(p_0 - p_L) + \rho g L = \frac{32\,\mu L v_{av}}{D^2} \qquad [1.5\text{-}58]$$

where D is the tube inner diameter. From Eq. [1.1-38] the friction factor

$$f = \left[\frac{(p_0 + \rho g h_0) - (p_L + \rho g h_L)}{\rho v_{av}^2/2}\right]\frac{D}{L} = \left[\frac{(p_0 - p_L) + \rho g L}{\rho v_{av}^2/2}\right]\frac{D}{L}. \qquad [1.5\text{-}59]$$

Substituting Eq. [1.5-58] into Eq. [1.5-59]

$$\boxed{f = \frac{64\mu}{D\rho v_{av}} = \frac{64}{\mathrm{Re}_D}} \qquad [1.5\text{-}60]$$

where the Reynolds number

$$\mathrm{Re}_D = \frac{D\rho v_{av}}{\mu} \qquad [1.5\text{-}61]$$

Example 1.5-3 Flow of a Rising Film

A continuous belt moves vertically upward at a constant velocity V and passes through an incompressible Newtonian fluid. As shown in Fig. 1.5-6, the belt picks up a film with it; the film thickness L is uniform in the x direction. The movement of the belt and the viscosity of the fluid keep it from running off the belt completely. Find the steady-state velocity distribution in the film. Neglect end effects.

Solution As shown in Fig. 1.5-6, rectangular coordinates are most appropriate for the problem. At the steady state all time-dependent terms should be zero. Since the wall is wide and long as compared to the film thickness, the end effects can be neglected: $v_x = v_y = 0$, and v_z does not vary with x.

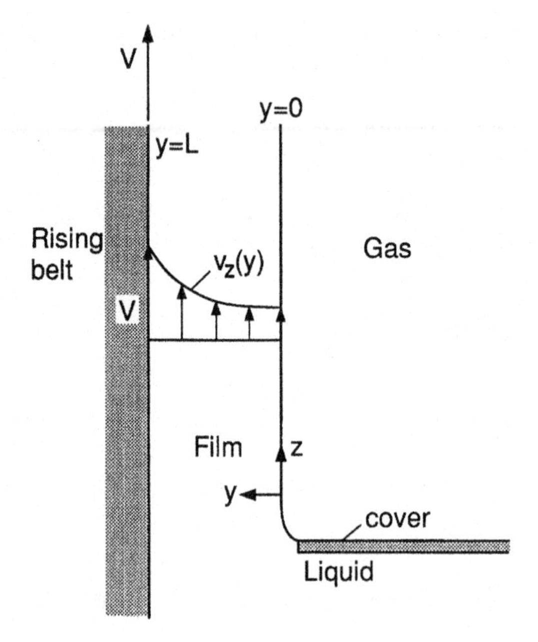

Fig. 1.5-6 Fluid flow in a liquid film on a rising belt.

The equation of continuity in rectangular coordinates, Eq. [A] of Table 1.3-1,

$$\frac{\partial \rho}{\partial t} + \frac{\partial}{\partial x}(\rho v_x) + \frac{\partial}{\partial y}(\rho v_y) + \frac{\partial}{\partial z}(\rho v_z) = 0 \qquad [1.5\text{-}62]$$

This equation reduces to the following one for the reasons stated at the beginning

$$\frac{\partial v_z}{\partial z} = 0 \qquad [1.5\text{-}63]$$

Let us now consider the z component of the equation of motion, Eq. [C] of Table 1.5-1:

$$\rho\left[\frac{\partial v_z}{\partial t} + v_x \frac{\partial v_z}{\partial x} + v_y \frac{\partial v_z}{\partial y} + v_z \frac{\partial v_z}{\partial z}\right] = -\frac{\partial p}{\partial z} + \mu\left[\frac{\partial^2 v_z}{\partial x^2} + \frac{\partial^2 v_z}{\partial y^2} + \frac{\partial^2 v_z}{\partial z^2}\right] + \rho g_z \qquad [1.5\text{-}64]$$

Following the same reasons and substituting Eq. [1.5-63] and $g_z = -g$, Eq. [1.5-64] reduces to

$$\frac{d^2 v_z}{dy^2} = \frac{\rho}{\mu} g \qquad [1.5\text{-}65]$$

Note that no pressure gradient can be applied in the z direction. Since the fluid is not enclosed but has a free surface. Also, $g_z = -g = -980\ \text{cm/s}^2$ since g is vertically downward but g_z is vertically upward. The boundary conditions are

$$\frac{dv_z}{dy} = 0 \quad \text{at} \quad y = 0 \qquad [1.5\text{-}66]$$

$$v_z = V \quad \text{at} \quad y = L \qquad [1.5\text{-}67]$$

The solution, from Case D of Appendix A, is as follows:

$$\boxed{v_z = V + \frac{\rho g}{2\mu}(y^2 - L^2)} \qquad [1.5\text{-}68]$$

which shows that v_z is a parabolic function of y.

The minimum velocity is at $y = 0$ and is

$$v_{\min} = V - \frac{\rho g L^2}{2\mu} \qquad [1.5\text{-}69]$$

From Newton's law of viscosity

$$\tau_{yz} = -\mu\frac{dv_z}{dy} \qquad [1.5\text{-}70]$$

Substituting Eq. [1.5-68]

$$\boxed{\tau_{yz} = -\rho g y} \qquad [1.5\text{-}71]$$

Example 1.5-4 Laminar Flow over a Flat Plate

Consider the steady-state, laminar flow of an incompressible, Newtonian fluid over a wide flat plate. Determine the velocity distribution in and the thickness of the momentum boundary layer δ. This problem was solved in Example 1.4-6 using an integral momentum-balance approach. Solve the problem again using the differential approach.

Solution From Eq. [A] of Table 1.3-1 and Eq. [C] of Table 1.5-1

$$\frac{\partial \rho}{\partial t} + \frac{\partial}{\partial x}(\rho v_x) + \frac{\partial}{\partial y}(\rho v_y) + \frac{\partial}{\partial z}(\rho v_z) = 0 \qquad \text{(continuity)} \qquad [1.5\text{-}72]$$

$$\rho\left[\frac{\partial v_z}{\partial t} + v_x\frac{\partial v_z}{\partial x} + v_y\frac{\partial v_z}{\partial y} + v_z\frac{\partial v_z}{\partial z}\right] = -\frac{\partial p}{\partial z} + \mu\left[\frac{\partial^2 v_z}{\partial x^2} + \frac{\partial^2 v_z}{\partial y^2} + \frac{\partial^2 v_z}{\partial z^2}\right] + \rho g_z$$

$$(z \text{ momentum}) \qquad [1.5\text{-}73]$$

Since the flow is not great in the y direction, we do not consider the y component of the equation of motion. Under the steady state $\partial\rho/\partial t$ and $\partial v_z/\partial t$ are both zero. Since there is no flow nor velocity gradient in the x direction, v_x and $\partial^2 v_z/\partial x^2$ are both zero. The terms $\partial p/\partial z$ and ρg_z are also both zero since there is no pressure gradient or gravity along the plate. Since momentum transfer in the z direction is mainly by convection rather than by molecular diffusion, $\mu\,\partial^2 v_z/\partial z^2$ is far smaller than $\rho v_z\,\partial v_z/\partial z$ and can thus be neglected as an approximation. As such, the two equations reduce to

$$\frac{\partial v_y}{\partial y} + \frac{\partial v_z}{\partial z} = 0 \qquad \text{(continuity)} \qquad [1.5\text{-}74]$$

and

$$v_y \frac{\partial v_z}{\partial y} + v_z \frac{\partial v_z}{\partial z} = \nu \frac{\partial^2 v_z}{\partial y^2} \qquad (z \text{ momentum}) \qquad [1.5\text{-}75]$$

where the kinematic viscosity $\nu = \mu/\rho$.

From Eq. [1.5-74]

$$v_y = -\int_0^y \frac{\partial v_z}{\partial z}\,dy \qquad [1.5\text{-}76]$$

which is substituted into Eq. [1.5-75] to give

$$-\left(\int_0^y \frac{\partial v_z}{\partial z}\,dy\right)\frac{\partial v_z}{\partial y} + v_z \frac{\partial v_z}{\partial z} = \nu \frac{\partial^2 v_z}{\partial y^2} \qquad [1.5\text{-}77]$$

Since this equation is difficult to solve directly, the following simple velocity profile can be assumed as an approximation:

$$\boxed{\frac{v_z}{v_\infty} = \frac{3}{2}\left(\frac{y}{\delta}\right) - \frac{1}{2}\left(\frac{y}{\delta}\right)^3} \qquad [1.5\text{-}78]$$

where δ is the thickness of the momentum boundary layer. As shown in Example 1.4-6, this velocity profile is reasonable since it satisfies the following boundary conditions

$$v_z = 0 \quad \text{at} \quad y = 0 \qquad [1.5\text{-}79]$$

and

$$v_z = v_\infty \quad \text{and} \quad \frac{\partial v_z}{\partial y} = 0 \quad \text{at} \quad y = \delta \qquad [1.5\text{-}80]$$

From Eq. [1.5-78]

$$\frac{\partial v_z}{\partial z} = \frac{v_\infty}{2}\left(\frac{-3y}{\delta^2} + \frac{3y^3}{\delta^4}\right)\frac{d\delta}{dz} \qquad [1.5\text{-}81]$$

$$\frac{\partial v_z}{\partial y} = \frac{v_\infty}{2}\left(\frac{3}{\delta} - \frac{3y^2}{\delta^3}\right) \qquad [1.5\text{-}82]$$

$$\frac{\partial^2 v_z}{\partial y^2} = \frac{-3v_\infty y}{\delta^3} \qquad [1.5\text{-}83]$$

Substituting Eqs. [1.5-78] and [1.5-81] through [1.5-83] into Eq. [1.5-77], rearranging, and then integrating with respect to y from 0 to δ, we get

$$\int_0^\delta \left[\frac{-3v_\infty}{4}\frac{d\delta}{dz}\left(\frac{y^2}{2} - \frac{7y^4}{12\delta^2} + \frac{y^6}{12\delta^4}\right) + \nu y\right] dy = 0 \qquad [1.5\text{-}84]$$

which yields

$$\delta\frac{d\delta}{dz} = \frac{140}{13}\frac{\nu}{v_\infty} \qquad [1.5\text{-}85]$$

On integration

$$\int_0^\delta 2\delta\, d\delta = \frac{280}{13}\frac{\nu}{v_\infty}\int_0^z dz \qquad [1.5\text{-}86]$$

or

$$\boxed{\delta = 4.64\sqrt{\frac{\nu z}{v_\infty}}} \qquad [1.5\text{-}87]$$

which is identical to Eq. [1.4-61] derived previously on the basis of integral momentum balance.

Example 1.5-5 Adjacent Flow of Two Immiscible Fluids

Two immiscible fluids of equal thickness are held between two large horizontal parallel plates separated from each other by a distance of $2b$, as shown in Fig. 1.5-7. The upper plate is stationary, while the lower one moves at a constant velocity V in the z direction. Find the velocity distributions in each fluid. The fluids are incompressible and Newtonian, and the flow is steady and laminar.

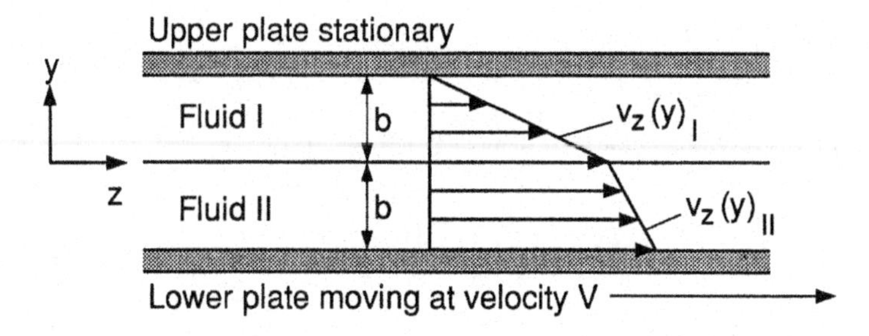

Fig. 1.5-7 Fluid flow in two adjacent immiscible fluids.

Solution The equation of continuity, Eq. [A] of Table [1.3-1], is as follows:

$$\frac{\partial \rho}{\partial t} + \frac{\partial}{\partial x}(\rho v_x) + \frac{\partial}{\partial y}(\rho v_y) + \frac{\partial}{\partial z}(\rho v_z) = 0 \qquad [1.5\text{-}88]$$

At the steady state $\partial\rho/\partial t = 0$. Since the flow is in the z direction only, $v_x = v_y = 0$. As such, Eq. [1.5-88] reduces to

$$\frac{\partial v_z}{\partial z} = 0 \qquad [1.5\text{-}89]$$

The equation of motion, Eq. [C] of Table [1.5-1], is as follows:

$$\rho\left[\frac{\partial v_z}{\partial t} + v_x\frac{\partial v_z}{\partial x} + v_y\frac{\partial v_z}{\partial y} + v_z\frac{\partial v_z}{\partial z}\right] = -\frac{\partial p}{\partial z} + \mu\left[\frac{\partial^2 v_z}{\partial x^2} + \frac{\partial^2 v_z}{\partial y^2} + \frac{\partial^2 v_z}{\partial z^2}\right] + \rho g_z \qquad [1.5\text{-}90]$$

At the steady state $\partial v_z/\partial t = 0$. As already stated, $v_x = v_y = 0$. From continuity $\partial v_z/\partial z$ and hence $\partial^2 v_z/\partial z^2$ are zero. In view of the fact that the flow is induced by moving the lower plate and not by pressure gradients, $\partial p/\partial z = 0$. Since there are no velocity gradients in the x direction, $\partial^2 v_z/\partial x^2 = 0$; and since the z direction is horizontal, $g_z = 0$. As such, Eq. [1.5-90] reduces to

$$\frac{d^2 v_z}{dy^2} = 0 \qquad [1.5\text{-}91]$$

The boundary conditions for fluid I are

$$-\mu_I \left.\frac{dv_z}{dy}\right|_{0^+} = -\mu_{II}\left.\frac{dv_z}{dy}\right|_{0^-} \quad \text{at} \quad y = 0 \qquad [1.5\text{-}92]$$

$$v_z = 0 \quad \text{at} \quad y = b \qquad [1.5\text{-}93]$$

The solution, from Case D of Appendix A, is as follows

$$(v_z)_{\mathrm{I}} = \frac{\mu_{\mathrm{II}}}{\mu_{\mathrm{I}}} \left.\frac{dv_z}{dy}\right|_{0^-} (y - b) \qquad [1.5\text{-}94]$$

The boundary conditions for fluid II are

$$-\mu_{\mathrm{II}} \left.\frac{dv_z}{dy}\right|_{0^-} = -\mu_{\mathrm{I}} \left.\frac{dv_z}{dy}\right|_{0^+} \quad \text{at} \quad y = 0 \qquad [1.5\text{-}95]$$

$$v_z = V \quad \text{at} \quad y = -b \qquad [1.5\text{-}96]$$

The solution, from Case D of Appendix A, is as follows

$$(v_z)_{\mathrm{II}} = V + \frac{\mu_{\mathrm{I}}}{\mu_{\mathrm{II}}} \left.\frac{dv_z}{dy}\right|_{0^+} (y + b) \qquad [1.5\text{-}97]$$

At the interface the continuity of velocity requires that

$$(v_z)_{\mathrm{I}} = (v_z)_{\mathrm{II}} \quad \text{at} \quad y = 0 \qquad [1.5\text{-}98]$$

Substituting Eqs. [1.5-94] and [1.5-97] into Eq. [1.5-98]

$$-b \frac{\mu_{\mathrm{II}}}{\mu_{\mathrm{I}}} \left.\frac{dv_z}{dy}\right|_{0^-} = V + b \frac{\mu_{\mathrm{I}}}{\mu_{\mathrm{II}}} \left.\frac{dv_z}{dy}\right|_{0^+} \qquad [1.5\text{-}99]$$

Substituting Eq. [1.5-95] into Eq. [1.5-99]

$$\left.\frac{dv_z}{dy}\right|_{0^+} = -\frac{V}{b}\left(\frac{\mu_{\mathrm{II}}}{\mu_{\mathrm{I}} + \mu_{\mathrm{II}}}\right) \qquad [1.5\text{-}100]$$

Substituting this equation into Eq. [1.5-95]

$$\left.\frac{dv_z}{dy}\right|_{0^-} = -\frac{V}{b}\left(\frac{\mu_{\mathrm{I}}}{\mu_{\mathrm{I}} + \mu_{\mathrm{II}}}\right) \qquad [1.5\text{-}101]$$

So, substituting Eq. [1.5-101] into Eq. [1.5-94]

$$\boxed{(v_z)_{\mathrm{I}} = V\left(\frac{\mu_{\mathrm{II}}}{\mu_{\mathrm{I}} + \mu_{\mathrm{II}}}\right)\left(1 - \frac{y}{b}\right)} \qquad [1.5\text{-}102]$$

and substituting Eq. [1.5-100] into Eq. [1.5-97]

$$\boxed{(v_z)_{\mathrm{II}} = V\left[1 - \left(\frac{\mu_{\mathrm{I}}}{\mu_{\mathrm{I}} + \mu_{\mathrm{II}}}\right)\left(1 + \frac{y}{b}\right)\right]} \qquad [1.5\text{-}103]$$

These linear velocity distributions are shown in Fig. 1.5-7.

1.6. DIFFERENTIAL MOMENTUM-BALANCE EQUATION IN STREAM FUNCTION AND VORTICITY

As discussed in the previous section, the equations of continuity and motion are expressed in terms of the velocity components involved. In the present section, the equation of motion will be expressed in terms of the so-called stream function and vorticity, and the advantage will be described. This approach is often used in two-dimensional and axisymmetric problems, as will be seen in Chapters 8 and 9.

1.6.1. Two-Dimensional Problems

The **stream function** ψ can be defined as follows:

$$v_x = \frac{\partial \psi}{\partial y} \qquad [1.6\text{-}1]$$

$$v_y = -\frac{\partial \psi}{\partial x} \qquad [1.6\text{-}2]$$

Other similar definitions have also been used, in which the RHSs of the two equations can be divided by ρ or -1 or both. Therefore

$$d\psi = \frac{\partial \psi}{\partial x} dx + \frac{\partial \psi}{\partial y} dy = -v_y\, dx + v_x\, dy \qquad [1.6\text{-}3]$$

Consider a path of constant ψ in the xy plane. Since $d\psi = 0$ along the path, from Eq. [1.6-3]

$$\left.\frac{dy}{dx}\right|_{\psi = \text{constant}} = \frac{v_y}{v_x} \qquad [1.6\text{-}4]$$

Substituting Eqn. [1.1-31] into Eq. [1.6-4]

$$\left.\frac{dy}{dx}\right|_{\psi = \text{constant}} = \text{slope of streamline} \qquad [1.6\text{-}5]$$

As such, a path of constant ψ is a streamline, and the tangent at every point on this path gives this direction of the velocity at that point. A flow field is usually presented by series of streamlines, with a constant $\Delta\psi$ between each two neighboring streamlines. In two-dimensional flow, as evident from Eq. [1.6-1] and [1.6-2], fluid flow is faster where the streamlines are more closely spaced and slower where they are more separated from each other.

The **vorticity** ω is defined as

$$\omega = \frac{\partial v_y}{\partial x} - \frac{\partial v_x}{\partial y} \qquad [1.6\text{-}6]$$

which is the z component of $\nabla \times \mathbf{v}$, as shown in Eq. [C] of Table A.5-1. It is, in fact, twice the angular velocity of the rotational motion of the fluid in the xy plane.

For an incompressible fluid the equation of continuity (i.e., Eq. [A] of Table 1.3-1) reduces to the following equation for two-dimensional flow:

$$\frac{\partial v_x}{\partial x} + \frac{\partial v_y}{\partial y} = 0 \qquad [1.6\text{-}7]$$

Substituting Eqs. [1.6-1] and [1.6-2] into the LHS of this equation

$$\frac{\partial}{\partial x}\left(\frac{\partial \psi}{\partial y}\right) - \frac{\partial}{\partial y}\left(\frac{\partial \psi}{\partial x}\right) = \frac{\partial^2 \psi}{\partial x \, \partial y} - \frac{\partial^2 \psi}{\partial y \, \partial x} = 0 \qquad [1.6\text{-}8]$$

Therefore, for incompressible fluid the equation of continuity is **automatically satisfied** and thus no longer required.

From the equation of motion (i.e., Eqs. [A] and [B] in Table 1.5-1)

$$\rho\left[\frac{\partial v_x}{\partial t} + v_x \frac{\partial v_x}{\partial x} + v_y \frac{\partial v_x}{\partial y}\right] = -\frac{\partial p}{\partial x} + \mu\left[\frac{\partial^2 v_x}{\partial x^2} + \frac{\partial^2 v_x}{\partial y^2}\right] + \rho g_x \qquad [1.6\text{-}9]$$

and

$$\rho\left[\frac{\partial v_y}{\partial t} + v_x \frac{\partial v_y}{\partial x} + v_y \frac{\partial v_y}{\partial y}\right] = -\frac{\partial p}{\partial y} + \mu\left[\frac{\partial^2 v_y}{\partial x^2} + \frac{\partial^2 v_y}{\partial y^2}\right] + \rho g_y \qquad [1.6\text{-}10]$$

By differentiating Eq. [1.6-9] with respect to y and Eq. [1.6-10] with respect to x and subtracting one from the other, we can eliminate the pressure term.

Substituting Eqs. [1.6-1], [1.6-2], and [1.6-6] into the resulting equation, we can obtain the following **vorticity equation**:

$$\rho\left(\frac{\partial \omega}{\partial t}+\frac{\partial \psi}{\partial y}\frac{\partial \omega}{\partial x}-\frac{\partial \psi}{\partial x}\frac{\partial \omega}{\partial y}\right)=\mu\left(\frac{\partial^2 \omega}{\partial x^2}+\frac{\partial^2 \omega}{\partial y^2}\right) \quad [1.6\text{-}11]$$

Substituting Eqs. [1.6-1] and [1.6-2] into Eq. [1.6-6], we obtain the following **stream-function equation**:

$$\left(\frac{\partial^2 \psi}{\partial x^2}+\frac{\partial^2 \psi}{\partial y^2}\right)=-\,\omega \quad [1.6\text{-}12]$$

In summary, instead of having to solve three governing equations, – Eqs. [1.6-7], [1.6-9], and [1.6-10], – we now only have to solve two governing equations: Eqs. [1.6-11] and [1.6-12]. This, in fact, is a significant advantage of the streamfunction/vorticity approach. Since pressure is eliminated, we have one less variable to deal with, too.

The boundary conditions should be converted from those expressed in terms of v_x and v_y to those in terms of ψ and ω.

1.6.2. Axisymmetric Problems

The stream function and vorticity can be defined in a similar way as follows:

$$v_r=\frac{1}{r}\frac{\partial \psi}{\partial z} \quad [1.6\text{-}13]$$

$$v_z=\frac{-1}{r}\frac{\partial \psi}{\partial r} \quad [1.6\text{-}14]$$

and

$$\omega=\frac{\partial v_r}{\partial z}-\frac{\partial v_z}{\partial r} \quad [1.6\text{-}15]$$

This ω is the θ component of $\nabla\times\mathbf{v}$ as shown in Eq. [C] of Table A.5-2.

For an incompressible fluid the equation of continuity (i.e., Eq. [B] of Table 1.3-1) reduces to the following equation for axisymmetric flow:

$$\frac{1}{r}\frac{\partial}{\partial r}(rv_r)+\frac{\partial v_z}{\partial z}=0 \quad [1.6\text{-}16]$$

Substituting Eqs. [1.6-13] and [1.6-14] into the LHS of this equation

$$\frac{1}{r}\frac{\partial}{\partial r}\left(\frac{\partial \psi}{\partial z}\right) - \frac{\partial}{\partial z}\left(\frac{1}{r}\frac{\partial \psi}{\partial r}\right) = \frac{1}{r}\left(\frac{\partial^2 \psi}{\partial r\,\partial z} - \frac{\partial^2 \psi}{\partial r\,\partial z}\right) = 0 \qquad [1.6\text{-}17]$$

Again, the equation of continuity is automatically satisfied.

From the equation of motion (i.e., Eqs. [A]-[C] in Tables 1.5-2), for axisymmetric flow

$$\rho\left[\frac{\partial v_r}{\partial t} + v_r\frac{\partial v_r}{\partial r} - \frac{v_\theta^2}{r} + v_z\frac{\partial v_r}{\partial z}\right] = -\frac{\partial p}{\partial r} + \mu\left(\frac{\partial}{\partial r}\left[\frac{1}{r}\frac{\partial}{\partial r}(rv_r)\right] + \frac{\partial^2 v_r}{\partial z^2}\right)$$

(r momentum) [1.6-18]

$$\rho\left[\frac{\partial v_z}{\partial t} + v_r\frac{\partial v_z}{\partial r} + v_z\frac{\partial v_z}{\partial z}\right] = -\frac{\partial p}{\partial z} + \mu\left(\frac{1}{r}\frac{\partial}{\partial r}\left[r\frac{\partial v_z}{\partial r}\right] + \frac{\partial^2 v_z}{\partial z^2}\right) + \rho g_z$$

(z momentum) [1.6-19]

$$\rho\left[\frac{\partial v_\theta}{\partial t} + \frac{v_r}{r}\frac{\partial}{\partial r}(rv_\theta) + v_z\frac{\partial v_\theta}{\partial z}\right] = \mu\left(\frac{\partial}{\partial r}\left[\frac{1}{r}\frac{\partial}{\partial r}(rv_\theta)\right] + \frac{\partial^2 v_\theta}{\partial z^2}\right)$$

(θ momentum) [1.6-20]

In Eq. [1.6-20] $\partial p/\partial \theta = 0$, due to axisymmetry, and since the z direction is taken as the vertical direction, $g_\theta = g_r = 0$.

By differentiating Eq. [1.6-18] with respect to z and Eq. [1.6-19] with respect to r and substracting one from the other, we can eliminate the pressure term. Substituting Eqs. [1.6-13] through [1.6-15] into the resulting equation, we obtain the following **vorticity equation**:

$$\rho\left(\frac{\partial \omega}{\partial t} - \frac{\partial \psi}{\partial z}\frac{\partial}{\partial r}\left[\frac{\omega}{r}\right] + \frac{\partial \psi}{\partial r}\frac{\partial}{\partial z}\left[\frac{\omega}{r}\right]\right) = \mu\left(\frac{\partial}{\partial r}\left[\frac{1}{r}\frac{\partial (r\omega)}{\partial r}\right] + \frac{\partial}{\partial z}\left[\frac{1}{r}\frac{\partial (r\omega)}{\partial z}\right]\right) + \rho\frac{\partial}{\partial z}\left(\frac{v_\theta^2}{r}\right) \qquad [1.6\text{-}21]$$

Substituting Eqs. [1.6-13] and [1.6-14] into Eqs. [1.6-15] and [1.6-20], we obtain the following **stream-function equation**:

$$\frac{\partial}{\partial r}\left(\frac{1}{r}\frac{\partial \psi}{\partial r}\right) + \frac{\partial}{\partial z}\left(\frac{1}{r}\frac{\partial \psi}{\partial z}\right) = \omega \qquad [1.6\text{-}22]$$

and **circulation equation**:

$$\rho\left[\frac{\partial v_\theta}{\partial t}+\frac{1}{r^2}\frac{\partial \psi}{\partial z}\frac{\partial}{\partial r}(rv_\theta)-\frac{1}{r}\frac{\partial \psi}{\partial r}\frac{\partial v_\theta}{\partial z}\right]=\mu\left(\frac{\partial}{\partial r}\left[\frac{1}{r}\frac{\partial}{\partial r}(rv_\theta)\right]+\frac{\partial^2 v_\theta}{\partial z^2}\right) \qquad [1.6\text{-}23]$$

1.7. TURBULENT FLOW

The subject of turbulent flow has been introduced in Section 1.1.1. It was mentioned that as the critical Reynolds number is exceeded, for example, by increasing the fluid velocity, fluid flow becomes turbulent. Turbulent flow is characterized by the random and chaotic motion of fluid molecules and packets (Fig. 1.1-1).

1.7.1 Time-Smoothed Variables

Figure 1.7-1 illustrates steady and unsteady laminar and turbulent flows. The velocity refers to that at a fixed position in the fluid. As shown, turbulent flow differs from laminar flow in that the velocity fluctuates all the time. A turbulent flow can still be considered as steady if the time-smoothed velocity remains constant, such as water flowing through a pipe at a high but constant flow rate.

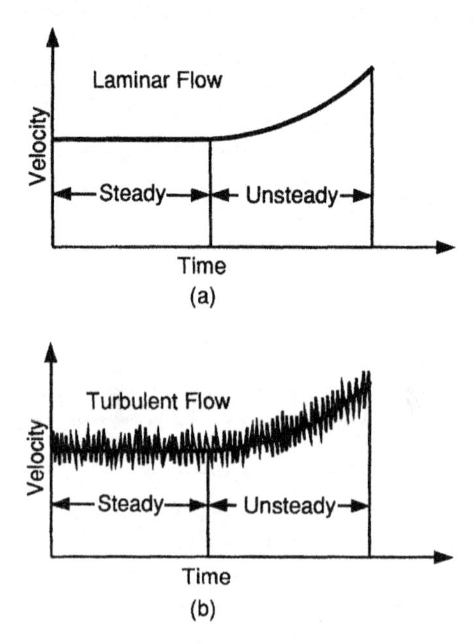

Fig. 1.7-1 Velocity variations with time: (*a*) laminar flow; (*b*) turbulent flow.

As illustrated in Fig. 1.7-2, we assume that the fluid velocity $\mathbf{v}$ at a fixed point in space over a given finite time interval t_0 can be resolved into the time-smoothed velocity $\bar{\mathbf{v}}$ and a fluctuation velocity term $\mathbf{v}'$ that accounts for the turbulent motion:

$$\mathbf{v} = \bar{\mathbf{v}} + \mathbf{v}' \qquad [1.7\text{-}1]$$

where

$$\bar{\mathbf{v}} = \frac{1}{t_0}\int_t^{t+t_0} \mathbf{v}\, dt \qquad [1.7\text{-}2]$$

The time interval t_0 is small with respect to the time over which $\bar{\mathbf{v}}$ varies but large with respect to the time of turbulent fluctuations. From Eqs. [1.7-1] and [1.7-2], it is obvious that

$$\bar{\mathbf{v}}' = \frac{1}{t_0}\int_t^{t+t_0} \mathbf{v}'\, dt = 0 \qquad [1.7\text{-}3]$$

Since the local pressure is affected by the velocity, we can write a similar expression for the pressure at a fixed point in space over the same time interval:

$$p = \bar{p} + p' \qquad [1.7\text{-}4]$$

The significance of turbulence can be illustrated by fluid flow through a circular pipe. From Eqs. [1.5-53] and [1.5-54], for laminar flow in a

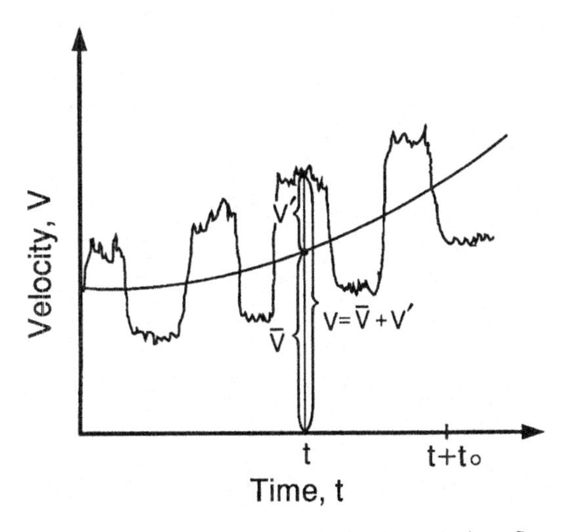

Fig. 1.7-2 Time-smoothed velocity in turbulent flow.

circular pipe

$$\boxed{\frac{v_z}{v_{max}} = 1 - \left(\frac{r}{R}\right)^2} \qquad [1.7\text{-}5]$$

and

$$\frac{v_{av}}{v_{max}} = \frac{1}{2} \qquad [1.7\text{-}6]$$

In contrast, for turbulent flow in a circular pipe, it has been proposed and verified experimentally that

$$\boxed{\frac{\bar{v}_z}{\bar{v}_{max}} = \left(1 - \frac{r}{R}\right)^{1/7}} \qquad [1.7\text{-}7]$$

and

$$\frac{\bar{v}_{av}}{\bar{v}_{max}} \approx \frac{4}{5} \qquad [1.7\text{-}8]$$

Equation [1.7-7] is known as the **Blasius one-seventh power law**. The velocity distributions according to Eqs. [1.7-5] and [1.7-7] are shown in Fig. 1.7-3. As shown, the velocity distribution is significantly more uniform in the case of turbulent flow, as a result of better mixing in the bulk fluid. Because of the no-slip condition at the tube wall, however, the velocity drops sharply near the wall, making the velocity gradient $d\bar{v}_z/dr$ at the tube wall much higher than that in laminar flow.

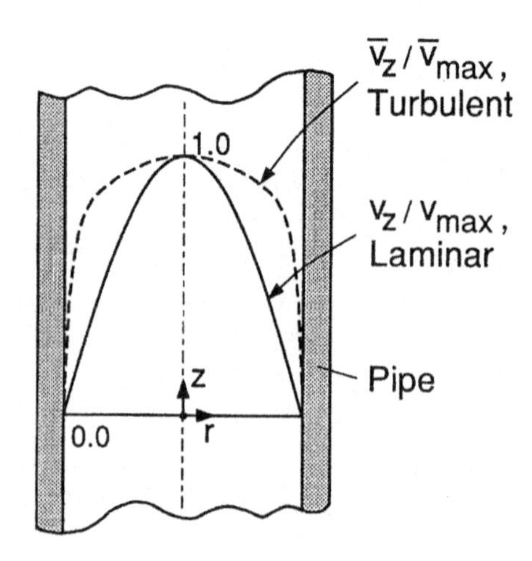

Fig. 1.7-3 Velocity distributions for flow in a circular tube.

1.7.2. Time-Smoothed Governing Equations

Substituting Eqs. [1.7-1] into Eq. [1.3-5] and then taking the time average according to Eqs. [1.7-2] and [1.7-3], we see that

$$\nabla \cdot \bar{\mathbf{v}} = 0 \qquad [1.7\text{-}9]$$

which is the **time-smoothed equation of continuity** for incompressible fluids. This equation is the same as the equation of continuity for laminar flow (Eq. [1.3-5]), except that the time-smoothed velocity replaces the instantaneous velocity.

Similar, substituting Eqs. [1.7-1] and [1.7-4] into Eq. [1.5-6] and then taking the time average according to Eqs. [1.7-2] and [1.7-3], we get the following **time-smoothed equation of motion**:

$$\boxed{\rho \frac{\partial \bar{\mathbf{v}}}{\partial t} + \rho \bar{\mathbf{v}} \cdot \nabla \bar{\mathbf{v}} = -\nabla \bar{p} - \nabla \cdot \bar{\tau} - \nabla \cdot \overline{\tau'} + \mathbf{f}_b} \qquad [1.7\text{-}10]$$

where $\tau = -\nabla(\mu \bar{\mathbf{v}})$ and

$$\overline{\tau'} = \rho \begin{pmatrix} \overline{v'_x v'_x} & \overline{v'_x v'_y} & \overline{v'_x v'_z} \\ \overline{v'_y v'_x} & \overline{v'_y v'_y} & \overline{v'_y v'_z} \\ \overline{v'_z v'_x} & \overline{v'_z v'_y} & \overline{v'_z v'_z} \end{pmatrix} \qquad [1.7\text{-}11]$$

$\overline{\tau'}$, which is due to turbulent velocity fluctuations, is sometimes collectively called the **Reynolds stresses**. Note that the nine components of $\overline{\tau'}$ are not zero even though $\overline{v'_x} = \overline{v'_y} = \overline{v'_z} = 0$ according to Eq. [1.7-3]. Equation [1.7-10] is the same as the equation of motion for laminar flow, Eq. [1.5-6], except that the time-smoothed velocity and pressure replace the instantaneous velocity and pressure, and that a new term $\nabla \cdot \overline{\tau'}$ arises.

1.7.3. Turbulent Momentum Flux

Several semiempirical relations have been proposed for the Reynolds stresses τ', in order to solve Eq. [1.7-10] for velocity distributions in turbulent flow as shown below.

1.7.3.1. Eddy Viscosity

Boussinesq proposed the following simplest form:

$$\overline{\tau'_{yz}} = -\mu' \frac{d\bar{v}_z}{dy} \qquad [1.7\text{-}12]$$

by analogy with Eq. [1.1-2], Newton's law of viscosity. The coefficient μ' is a turbulent or eddy viscosity and is usually position-dependent.

1.7.3.2. Prandtl's Mixing Length

Based on the assumption that eddies move about in a fluid like molecules do in a gas, Prandtl proposed the concept of the mixing length. The mixing length plays a role somewhat similar to that of the mean free path in the gas kinetic theory. This led him to the following relation

$$\overline{\tau'_{yz}} = -\rho\ell^2 \left|\frac{d\bar{v}_z}{dy}\right| \frac{d\bar{v}_z}{dy} \qquad [1.7\text{-}13]$$

and from Eq. [1.7-12]

$$\mu' = \rho\ell^2 \left|\frac{d\bar{v}_z}{dy}\right| \qquad [1.7\text{-}14]$$

when ℓ, called the **mixing length**, is a function of position, For example, $\ell = ay$, where y is the distance from the solid surface.

There are still other forms proposed, including the popular **two-equation model** of Launder and Spalding.[19] The two equations are one for the turbulence kinetic energy K and the other for its dissipation rate ε. For this reason this model is also called the K-ε model.

1.8. MOMENTUM TRANSFER CORRELATIONS

Momentum transfer, as described in Section 1.5, can be determined by solving the governing equations for fluid flow. Correlations that are derived theoretically but verified experimentally or are based on experimental data alone are useful for studying fluid flow. Some of these correlations will be presented in this section in dimensionless numbers.

1.8.1. External Flow

1.8.1.1. Flow over a Flat Plate

Flow over a flat plate is laminar for local Reynolds number $\text{Re}_z < 2 \times 10^5$. As described in Example 1.4-6, the following theoretical correlation can be used for laminar flow over a flat plate:

$$\boxed{C_{fz} = \frac{0.664}{\sqrt{\text{Re}_z}}} \qquad (\text{Re}_z < 2 \times 10^5) \qquad [1.8\text{-}1]$$

[19] B. E. Launder and D. B. Spalding, *Comp. Meth. Appl. Mech. Eng.*, **3**, 269, 1974.

where, as shown in Eq. [1.1-6],

$$C_{fz} = \left| \frac{\tau_{yz}|_{y=0}}{\frac{1}{2}\rho v_\infty^2} \right| \qquad \text{(local friction coefficient)} \qquad [1.8\text{-}2]$$

and

$$\mathrm{Re}_z = \frac{z v_\infty}{\nu} = \frac{z \rho v_\infty}{\mu} \qquad \text{(local Reynolds number)} \qquad [1.8\text{-}3]$$

From these equations the friction coefficient averaged over a distance L from the leading edge of the plate is

$$C_{fL} = \frac{1}{L} \int_0^L C_{fz}\, dz \qquad [1.8\text{-}4]$$

Substituting Eq. [1.8-1] into Eq. [1.8-4]

$$C_{fL} = 0.664 \left(\frac{1}{L}\right) \left(\frac{\nu}{v_\infty}\right)^{1/2} \int_0^L z^{-1/2}\, dz = 1.328 \left(\frac{\nu}{L v_\infty}\right)^{1/2} \qquad [1.8\text{-}5]$$

Hence

$$\boxed{C_{fL} = \frac{1.328}{\sqrt{\mathrm{Re}_L}}} \qquad (\mathrm{Re}_L < 2 \times 10^5) \qquad [1.8\text{-}6]$$

where

$$\mathrm{Re}_L = \frac{L v_\infty}{\nu} = \frac{L \rho v_\infty}{\mu} \qquad \text{(average Reynolds number)} \qquad [1.8\text{-}7]$$

For turbulent flow over a flat plate the following empirical correlation has been suggest:[20]

$$\boxed{C_{fz} = 0.0592\, \mathrm{Re}_z^{-(1/5)}} \qquad (5 \times 10^5 < \mathrm{Re}_z < 10^7) \qquad [1.8\text{-}8]$$

and from Eq. [1.8-4]

$$\boxed{C_{fL} = 0.074\, \mathrm{Re}_L^{-(1/5)}} \qquad (5 \times 10^5 < \mathrm{Re}_L < 10^7) \qquad [1.8\text{-}9]$$

[20] H. Schlichting, *Boundary Layer Theory*, 6th ed., McGraw-Hill, New York, 1968.

1.8.1.2. Flow Normal to a Cylinder

For incompressible flow normal to a cylinder of diameter D as that shown previously in Fig. 1.1-15, the drag coefficient is a function of Reynolds number as follows:[21]

$$\boxed{C_D = C_D\,(\mathrm{Re}_D)} \qquad \text{(Fig. 1.8-1a)}^{22} \qquad [1.8\text{-}10]$$

where, according to Eq. [1.1-37],

$$C_D = \frac{F_D}{A_f\,(\rho v_\infty^2/2)} \qquad \text{(drag coefficient)} \qquad [1.8\text{-}11]$$

and

$$\mathrm{Re}_D = \frac{D v_\infty}{\nu} = \frac{D \rho v_\infty}{\mu} \qquad \text{(Reynolds number)} \qquad [1.8\text{-}12]$$

1.8.1.3. Flow Past a Sphere

As described in Example 1.4-5, the following theoretical correlation can be used for creeping incompressible flow around a sphere of diameter D:

$$\boxed{C_D = \frac{24}{\mathrm{Re}_D}} \qquad \mathrm{Re}_D < 1 \qquad [1.8\text{-}13]$$

where C_D and Re_D are as defined in Eqs. [1.8-11] and [1.8-12].

The following empirical correlations[23] can be used for higher Re_D:

$$\boxed{C_D = \frac{18.5}{\mathrm{Re}_D^{3/5}}} \qquad (2 < \mathrm{Re}_D < 5 \times 10^2) \qquad [1.8\text{-}14]$$

and

$$\boxed{C_D \doteq 0.44} \qquad (5 \times 10^2 < \mathrm{Re}_D < 2 \times 10^5) \qquad [1.8\text{-}15]$$

The experimental data[24] of C_D as a function of Re_D are shown in Fig. 1.8-1*b*.[22]

[21] H. Schlichting, ibid.

[22] Curve taken from R. W. Fox and A. T. McDonald, *Introduction to Fluid Mechanics*, 4th ed., Wiley, New York, 1992.

[23] R. B. Bird, W. E. Stewart and E. N. Lightfoot, *Transport Phenomena*, Wiley, New York, 1960.

[24] H. Schlichting, *Boundary Layer Theory*, 6th ed., McGraw-Hill, 1968.

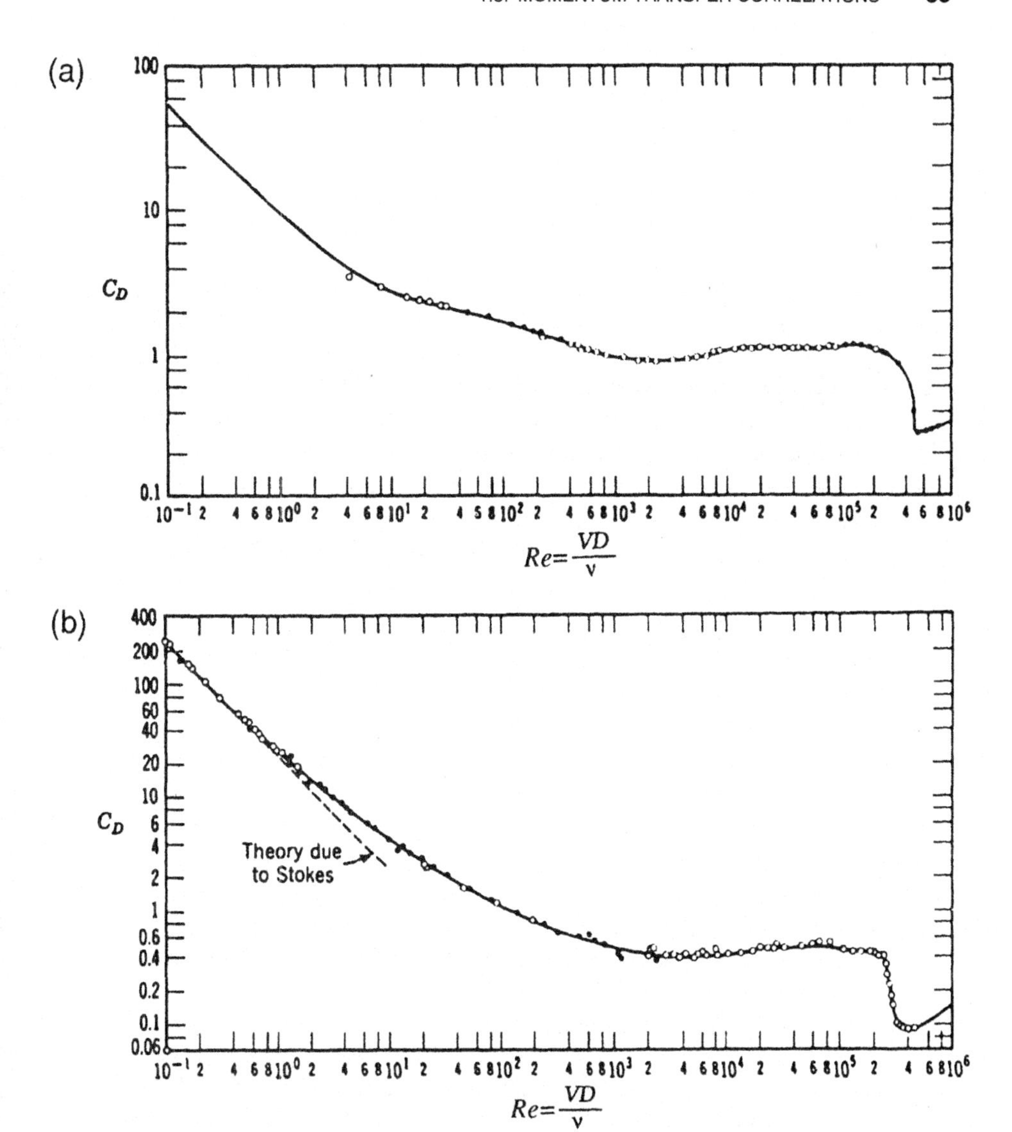

Fig. 1.8-1 Drag coefficients as a function of Reynolds number: (*a*) smooth cylinder; (*b*) smooth sphere. (Curves taken from R. W. Fox and A. T. McDonald, *Introduction to Fluid Mechanics*, 4th ed., Wiley, New York, copyright 1992. Reprinted by permission of John Wiley and Sons, Inc.)

1.8.2. Internal Flow

1.8.2.1. Flow through a Circular Tube

As described in Example 1.5-2, the following theoretical correlation can be used for fully developed laminar flow in a circular tube of inner diameter D:

$$\boxed{f = \frac{64}{\mathrm{Re}_D}} \qquad (\mathrm{Re}_D < 2100) \tag{1.8-16}$$

where, according to Eq. [1.1-38], the friction factor f is defined by

$$f = \left[\frac{(p_0 + \rho g h_0) - (p_L + \rho g h_L)}{\rho v_{av}^2/2}\right]\frac{D}{L} \tag{1.8-17}$$

and the Reynolds number is as defined in Eq. [1.8-12].

In fully developed turbulent flow in pipes, the experimental results of the friction factor f called the **Moody diagram** are as shown by Moody[25] in Fig. 1.8-2. As illustrated, f is a function of both Re_D and e/D, where e is the pipe roughness and e/D the relative pipe roughness.

For fully developed incompressible turbulent flow in a smooth tube the following empirical correlations[26] can be used

$$\boxed{f = 0.316\,\mathrm{Re}_D^{-(1/4)}} \qquad \mathrm{Re}_D < 2 \times 10^4 \tag{1.8-18}$$

and

$$\boxed{f = 0.184\,\mathrm{Re}_D^{-(1/5)}} \qquad \mathrm{Re}_D > 2 \times 10^4 \tag{1.8-19}$$

Equation [1.8-18] is known as the **Blasius formula**.

When the fluid velocity is known, it can be used to find the Reynolds number and, from Fig. 1.8-2, the friction factor. From the friction factor the pressure drop $p_0 - p_L$ along a pipe of length L can be found. To find the velocity from a known pressure drop, however, is less straightforward. One approach to this problem is described below.

From Eq. [1.8-17], for a horizontal pipe,

$$\mathrm{Re}_D\sqrt{f} = \frac{D\rho v_{av}}{\mu}\sqrt{\frac{p_0 - p_L}{L}\frac{2D}{\rho v_{av}^2}} = \frac{D\rho}{\mu}\sqrt{\frac{2(p_0 - p_L)D}{\rho L}} \tag{1.8-20}$$

noting that v_{av} drops out from the equation. From Fig. 1.8-2, $\log \mathrm{Re}_D$ can be plotted against $\log(\mathrm{Re}_D\sqrt{f})$ for various e/D values. From such a plot, Re_D and hence the velocity can be found from $\mathrm{Re}_D\sqrt{f}$. For example, for turbulent flow in smooth pipes a $\log \mathrm{Re}_D/\log(\mathrm{Re}_D\sqrt{f})$ plot is as shown in Fig. 1.8-3. The

[25] L. F. Moody, *Transact. ASME*, **66**, 671, 1944.

[26] F. P. Incropera and D. P. DeWitt, *Introduction to Heat Transfer*, 3rd ed., Wiley, New York, 1990.

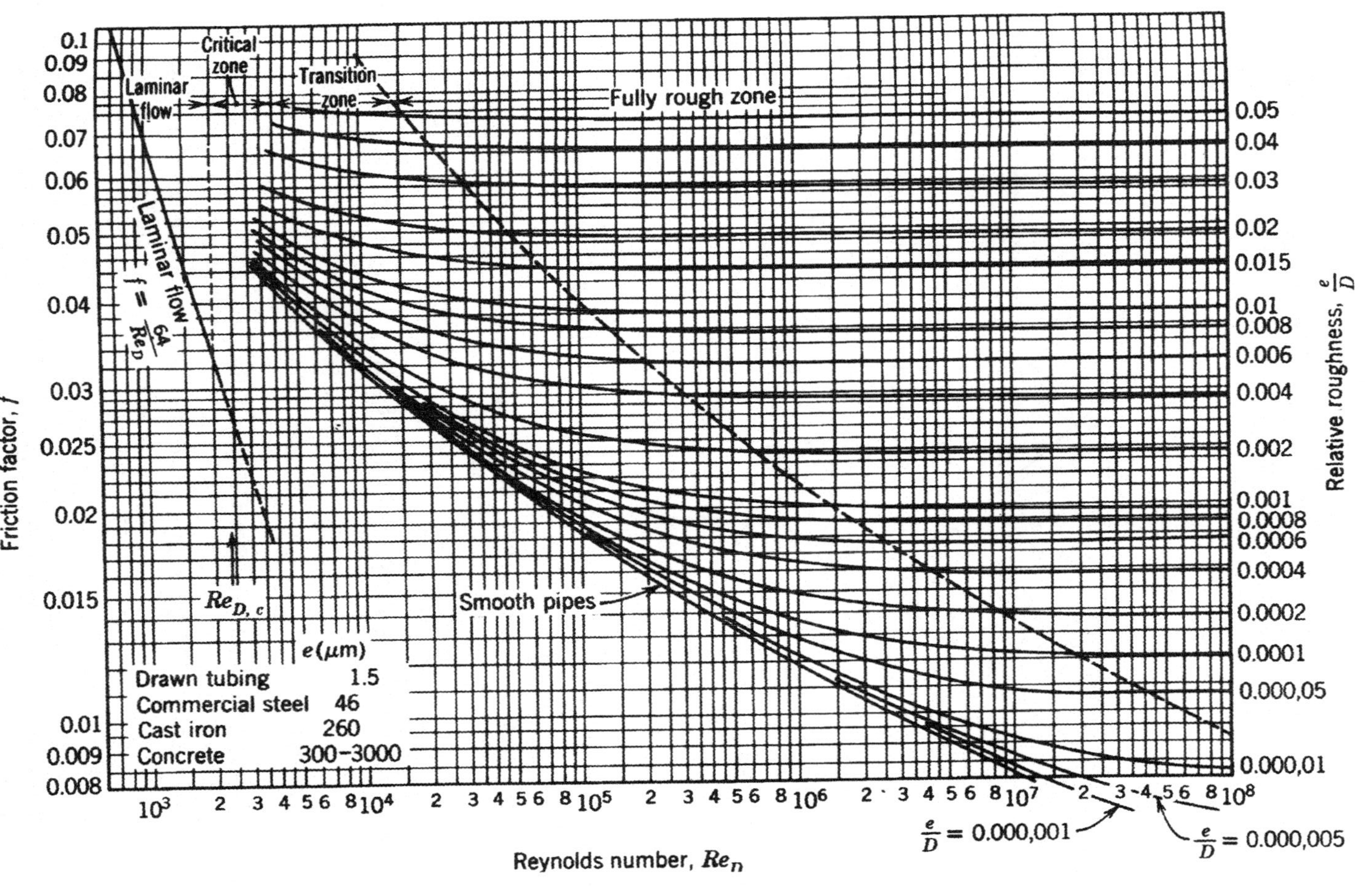

Fig. 1.8-2 Friction factor f for fully developed flow in circular pipes. (From L. F. Moody, *Trans. ASME*, **66**, 671, 1944. Courtesy of ASME.)

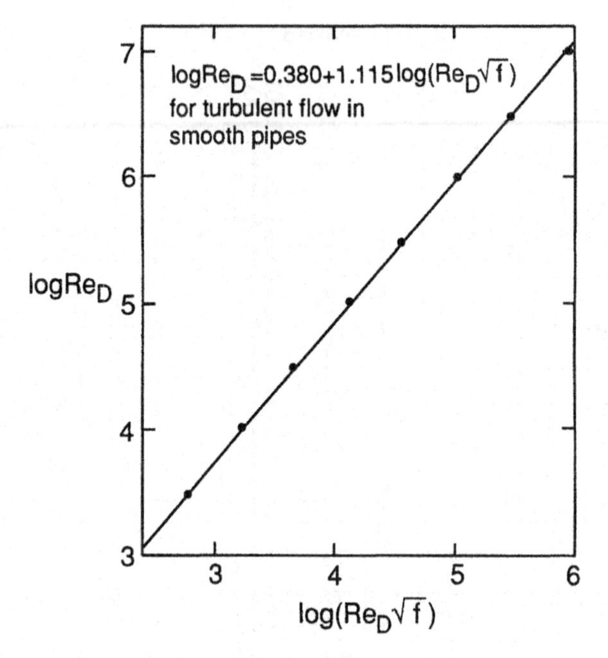

Fig. 1.8-3 $\log \mathrm{Re}_D$ as a function of $\log(\mathrm{Re}_D\sqrt{f})$ for turbulent flow in smooth circular pipes.

relationship happens to be close to a linear one in this particular case:

$$\boxed{\log \mathrm{Re}_D = 0.380 + 1.115 \log(\mathrm{Re}_D\sqrt{f})} \quad \text{(turbulent flow in smooth pipes)} \tag{1.8-21}$$

where $\mathrm{Re}_D\sqrt{f}$ can be obtained from Eq. [1.8-20].

1.8.3. Examples

Example 1.8-1 Flow through a Cooling Tube

A vacuum chamber of 0.5 m high is cooled by running cooling water through a copper tubing soldered to the chamber wall, as shown in Fig. 1.8-4. The cooling water is supplied at $2.758 \times 10^6 \, \mathrm{g\,s^{-2}\,cm^{-1}}$ above the atmospheric pressure, The copper tubing is 0.4 cm ID (inner diameter) and 20 m long. The water viscosity and density are $1 \times 10^{-2} \, \mathrm{g\,cm^{-1}\,s^{-1}}$ and $1 \, \mathrm{g/cm^3}$, respectively. Find the friction factor and the water flow rate through the copper tubing, assuming that its inner wall is smooth as an approximation.

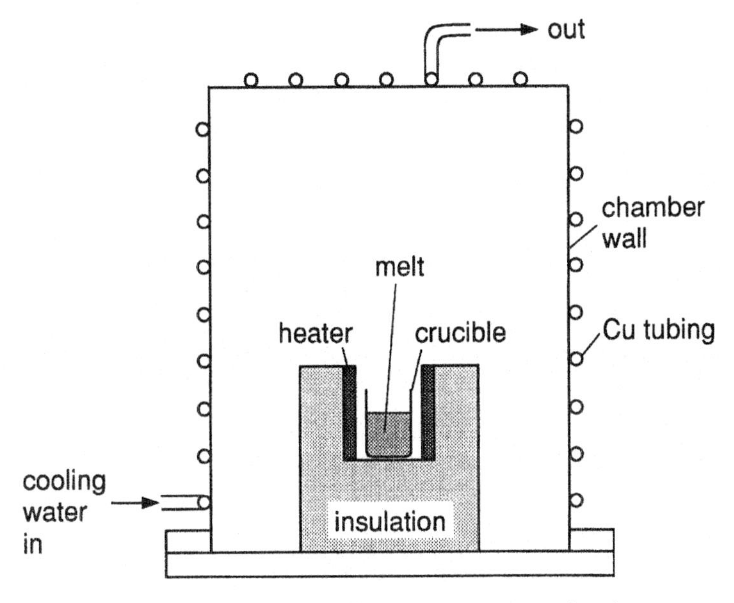

Fig. 1.8-4 Copper cooling tube on a vacuum chamber.

Solution The length (20 m) of the copper tubing is much greater than the height (0.5 m) of the chamber water. As such, the copper tubing is essentially horizontal. From Eq. [1.8-20]

$$\begin{aligned} \mathrm{Re}_D\sqrt{f} &= \frac{0.4\,\mathrm{cm}\times 1\,\mathrm{g\,cm^{-3}}}{1\times 10^{-2}\,\mathrm{g\,cm^{-1}\,s^{-1}}}\times\sqrt{\frac{2\times 2.758\times 10^6\,\mathrm{g\,s^{-2}\,cm^{-1}}\times 0.4\,\mathrm{cm}}{1\,\mathrm{g\,cm^{-3}}\times 2000\,\mathrm{cm}}} \\ &= 1328.6 \end{aligned} \qquad [1.8\text{-}22]$$

From Eq. [1.8-21]

$$\log \mathrm{Re}_D = 0.380 + 1.115\log(1328.6) = 3.836 \qquad [1.8\text{-}23]$$

or

$$\mathrm{Re}_D = 7288 \qquad \text{(turbulent flow)} \qquad [1.8\text{-}24]$$

Substituting into Eq. [1.8-22]

$$f = \left(\frac{1328.6}{7288}\right)^2 = 0.033 \qquad [1.8\text{-}25]$$

The water flow rate

$$m = \frac{\pi}{4} D^2 \rho v_{av} = \frac{\pi}{4} D\mu \left(\frac{D\rho v_{av}}{\mu} \right) \qquad [1.8\text{-}26]$$

as such

$$m = \frac{\pi}{4} D\mu \,\mathrm{Re}_D = \frac{\pi}{4} \times 0.4 \text{ cm} \times 1 \times 10^{-2} \text{ g cm}^{-1} \text{s}^{-1} \times 7288$$

$$= 22.90 \text{ g/s} \qquad [1.8\text{-}27]$$

1.9. OVERALL MECHANICAL ENERGY BALANCE

1.9.1. Derivation

Let us consider the flow of a fluid through a pipe and a turbine, as shown in Fig. 1.9-1*a*. Let the fluid in the pipe and the turbine be the control volume Ω, that is, the system. The control surfaces at the inlet A_1 and the outlet A_2 are chosen to be perpendicular to the pipe walls. The fluids, pipe, and turbine are assumed to be at the same temperature, that is, **isothermal** condition. As such, only the mechanical energy and not the thermal energy will be considered.

The following conservation law can be applied to the control volume:

$$\left\{ \begin{array}{l} \text{Rate of mechanical} \\ \text{energy} \\ \text{accumulation} \end{array} \right\} = \left\{ \begin{array}{l} \text{rate of mechanical} \\ \text{energy in} \\ \text{by convection} \end{array} \right\} - \left\{ \begin{array}{l} \text{rate of mechanical} \\ \text{energy out} \\ \text{by convection} \end{array} \right\}$$

(1) (2) (3)

$$- \left\{ \begin{array}{l} \text{rate of work} \\ \text{done by system} \\ \text{on surroundings} \end{array} \right\} \qquad [1.9\text{-}1]$$

(4)

Let us consider the case of steady-state flow of an incompressible fluid. At the steady state, the rate of mechanical energy change in Ω is

zero (term 1)

The mass flow rate of the fluid entering Ω through a differential area dA_1 at the entrance is $\rho v\, dA_1$, where v is the velocity component along the normal to the control surface A_1, namely, $v = |\mathbf{v} \cdot \mathbf{n}|$. The kinetic and potential energy per unit mass of the fluid are $\frac{1}{2}v^2$ and gz, respectively. As such, the kinetic and

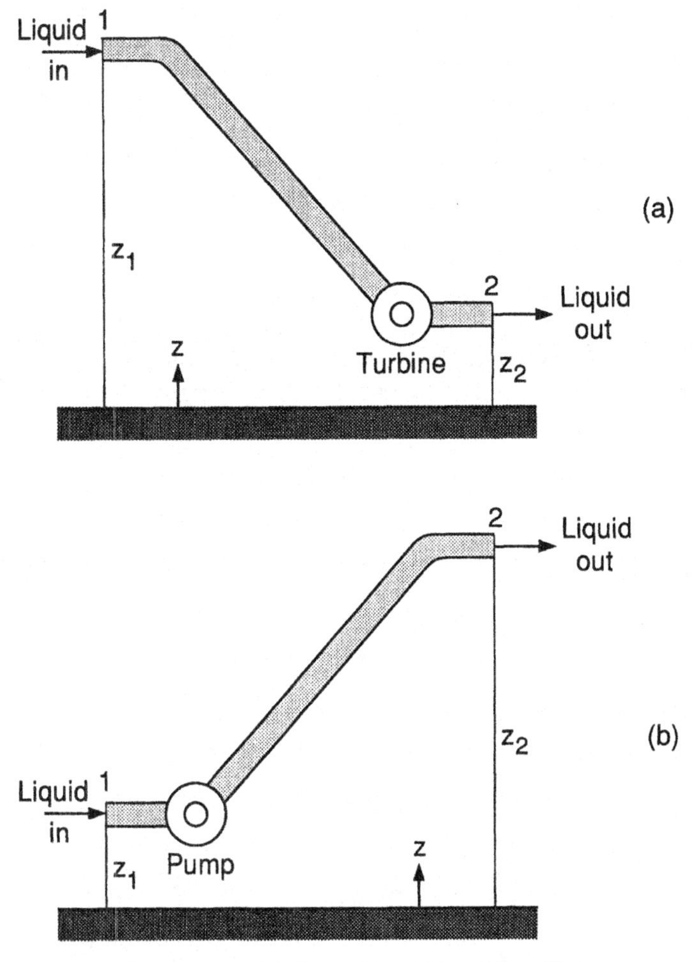

Fig. 1.9-1 Shaft work in pipelines: (*a*) turbine; (*b*) pump.

potential energy of the fluid enter Ω through the inlet A_1 at the rate of:

$$\iint_{A_1} (\tfrac{1}{2}v^2 + gz)\rho v\, dA_1 \qquad \text{(term 2)}$$

Similarly, the kinetic and potential energy leave Ω through the outlet A_2 at the rate of:

$$\iint_{A_2} (\tfrac{1}{2}v^2 + gz)\rho v\, dA_2 \qquad \text{(term 3)}$$

The volume flow rate of the fluid entering Ω through dA_1 is $v\,dA_1$. At the inlet the surrounding fluid has to work against the pressure of the system in order to enter Ω. Since the pressure work equals pressure times the volume of the fluid, the rate of the pressure work equals pressure times the volume flow rate of the fluid. This means that at the inlet A_1 the rate of the pressure work done by the surrounding fluid on the system is $\iint_{A_1} pv\,dA_1$. As such, the system can be considered to do work on the surroundings at the rate of:

$$-\iint_{A_1} pv\,dA_1 \qquad \text{(term 4a)}$$

At the outlet A_2 the system has to work against the pressure of the surrounding fluid in order to leave Ω. The rate of this pressure work is:

$$\iint_{A_2} pv\,dA_2 \qquad \text{(term 4b)}$$

The rate of the shaft work done by the fluid in the system on the surroundings through the turbine is

$$W_S \qquad \text{(term 4c)}$$

It should be noted that if a pump is used, as shown in Fig. 1.9-1*b*, the shaft work W_S is considered negative in value. This is because the work is done by the surroundings on the system.

Finally, the rate of the viscous work done by the fluid in the system on the pipe wall, that is, the friction loss, is

$$W_V \qquad \text{(term 4d)}$$

Now substituting all the above terms into Eq. [1.9-1]

$$0 = \iint_{A_1} \tfrac{1}{2}\rho v^3\,dA_1 + \iint_{A_1} \rho gzv\,dA_1 - \iint_{A_2} \tfrac{1}{2}\rho v^3\,dA_2 - \iint_{A_2} \rho gzv\,dA_2 - \left(-\iint_{A_1} pv\,dA_1 + \iint_{A_2} pv\,dA_2 + W_S + W_V\right) \qquad [1.9\text{-}2]$$

On rearranging

$$\iint_{A_1} \tfrac{1}{2}\rho v^3\,dA_1 + \iint_{A_1} \rho gzv\,dA_1 + \iint_{A_1} pv\,dA_1 = \iint_{A_2} \tfrac{1}{2}\rho v^3\,dA_2 + \iint_{A_2} \rho gzv\,dA_2 + \iint_{A_2} pv\,dA_2 + W_S + W_V \qquad [1.9\text{-}3]$$

At the inlet and outlet the variations in the height z and the pressure p over the cross-sectional area can be assumed negligible as an approximation. However, the velocity v can still vary significantly, especially in the case of laminar flow. From Eq. [1.2-8], for steady-state flow of an incompressible fluid

$$(v_{av}A)_1 = (v_{av}A)_2 = v_{av}A \qquad [1.9\text{-}4]$$

where from Eq. [1.1-39] the average velocity v_{av} is as follows

$$v_{av} = \frac{\iint_A v\,dA}{A} \qquad [1.9\text{-}5]$$

As such, Eq. [1.9-3] can be rewritten for an incompressible fluid as follows

$$\begin{aligned}(\tfrac{1}{2}\rho v_{av}^2)_1(v_{av}A)_1\left(\frac{1}{A}\iint_A\left(\frac{v}{v_{av}}\right)^3 dA\right)_1 + \rho g z_1(v_{av}A)_1 + p_1(v_{av}A)_1 \\ = (\tfrac{1}{2}\rho v_{av}^2)_2(v_{av}A)_2\left(\frac{1}{A}\iint_A\left(\frac{v}{v_{av}}\right)^3 dA\right)_2 + \rho g z_2(v_{av}A)_2 + p_2(v_{av}A)_2 + W_S + W_V\end{aligned} \qquad [1.9\text{-}6]$$

Dividing by the volume flow rate Q $(= v_{av}A)$, the equation becomes

$$\boxed{\left(\frac{b}{2}\rho v_{av}^2\right)_1 + \rho g z_1 + p_1 = \left(\frac{b}{2}\rho v_{av}^2\right)_2 + \rho g z_2 + p_2 + w_S + w_V}$$

(overall mechanical-energy balance) [1.9-7]

where v_{av}^2 is meant to be $(v_{av})^2$ and the **correction factor**

$$b = \frac{1}{A}\iint_A\left(\frac{v}{v_{av}}\right)^3 dA \qquad [1.9\text{-}8]$$

and the shaft work per unit volume of the fluid

$$w_S = \frac{W_S}{Q} \qquad [1.9\text{-}9]$$

and the viscous work or loss per unit volume of the fluid

$$w_V = \frac{W_V}{Q} \qquad [1.9\text{-}10]$$

It can be shown, by substituting Eqs. [1.7-5] and [1.7-6] into Eq. [1.9-8], that for laminar flow through a circular pipe

$$\boxed{\alpha = 2 \qquad \text{(laminar flow)}} \tag{1.9-11}$$

Similarly, it can be shown, by substituting Eqs. [1.7-7] and [1.7-8] into Eq. [1.9-8], that for turbulent flow through a circular pipe

$$\boxed{\alpha \approx 1 \qquad \text{(turbulent flow)}} \tag{1.9-12}$$

For most practical applications involving shaft work, the flow is fast and turbulent.

A special case of Eq. [1.9-7] is the flow of an inviscid fluid ($w_V = 0$) with an essentially uniform velocity ($\alpha = 1$) and in the absence of any shaft work ($w_S = 0$). In other words

$$\boxed{\tfrac{1}{2}\rho v_1^2 + \rho g z_1 + p_1 = \tfrac{1}{2}\rho v_2^2 + \rho g z_2 + p_2} \tag{1.9-13}$$

This equation is called the **Bernoulli equation.** It can also be written as follows

$$\tfrac{1}{2}\rho v^2 + \rho g z + p = \text{constant} \tag{1.9-14}$$

The friction loss w_V in Eq. [1.9-7] can be considered to consist of two parts, namely, that associated with pipes w_{vp} and that associated with fittings and valves w_{vf}. The former can be determined from the friction factor f for flow in pipes. For a pipe of length L and diameter D, from Eq. [1.1-38],

$$f = \left[\frac{(p_o + \rho g h_o) - (p_L + \rho g h_L)}{\rho v_{av}^2/2}\right]\frac{D}{L} \tag{1.9-15}$$

where o and L denote the inlet and outlet of the pipe, respectively, and h denotes height. As such, the friction loss associated with a series of i pipes, which is reflected by the changes in the pressure and height, is as follows

$$\begin{aligned} w_{vp} &= \sum_i [(p_o + \rho g h_o) - (p_L + \rho g h_L)]_i \\ &= \sum_i \left(\tfrac{1}{2}\rho v_{av}^2 \frac{L}{D} f\right)_i \end{aligned} \tag{1.9-16}$$

The so-called **loss coefficient** K for a pipe fitting or valve is defined as

$$K = \frac{p_{in} - p_{out}}{\rho v_{av}^2/2} \quad [1.9\text{-}17]$$

where p_{in} is the pressure at the inlet of the fitting or valve and p_{out} that at the outlet. Note that unlike the definition of the friction factor f in Eq. [1.9-15], the height difference is not considered in the definition of the loss coefficient K in Eq. [1.9-17]. Obviously, a fitting or valve is much shorter than a pipe, and the height difference between its inlet and outlet is negligible. As such, the friction loss associated with a series of i fittings and/or valves, which is reflected by the pressure drops through them, is as follows

$$w_{vf} = \sum_i (p_{in} - p_{out})_i = \sum_i \left(\tfrac{1}{2}\rho v_{av}^2 K\right)_i \quad [1.9\text{-}18]$$

Substituting Eq. [1.9-16] and [1.9-18] into Eq. [1.9-7], we obtain the following overall mechanical-energy balance for turbulent flow ($b \approx 1$) in pipe lines:

$$\left(\tfrac{1}{2}\rho v_{av}^2\right)_1 + \rho g z_1 + p_1 = \left(\tfrac{1}{2}\rho v_{av}^2\right)_2 + \rho g z_2 + p_2 + w_S + \sum_i \left(\tfrac{1}{2}\rho v_{av}^2 \frac{L}{D} f\right)_i + \sum_i \left(\tfrac{1}{2}\rho v_{av}^2 K\right)_i \quad [1.9\text{-}19]$$

The friction factor f can be obtained from Fig. 1.8-2, or from Eqs. [1.8-18] and [1.8-19]. The loss coefficient K has been provided for various fittings and valves[27]. Here, two equations are given for the sudden expansion and contraction shown in Fig. 1.9-2:

$$K = \left(1 - \frac{A_{in}}{A_{out}}\right)^2 = \frac{p_{in} - p_{out}}{(\rho v_{av}^2/2)_{in}} \quad \text{(sudden expansion)} \quad [1.9\text{-}20]$$

where A_{in} and A_{out} are the cross-sectional areas of the inlet (smaller) and outlet (larger), respectively, and

$$K \approx 0.45\left(1 - \frac{A_{out}}{A_{in}}\right) = \frac{p_{in} - p_{out}}{(\rho v_{av}^2/2)_{out}} \quad \text{(sudden contraction)} \quad [1.9\text{-}21]$$

[27] F. M. White, *Fluid Mechanics*, 3rd ed., McGraw-Hill, New York, 1994, pp. 335–342.

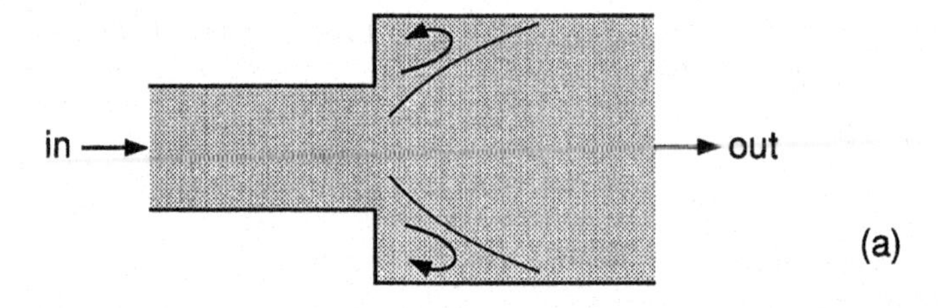

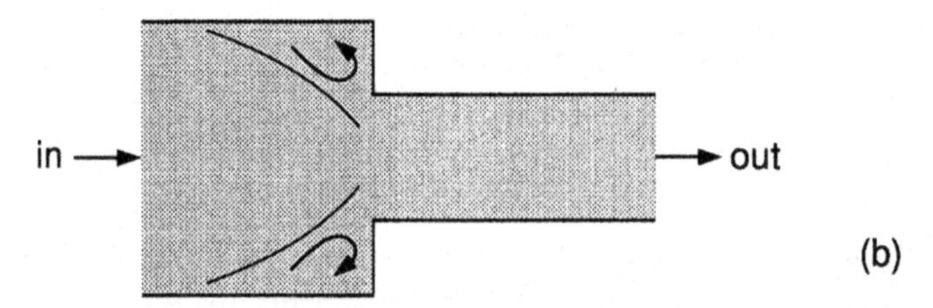

Fig. 1.9-2 Sudden changes in pipe cross-section: (*a*) sudden expansion; (*b*) sudden contraction.

where A_{out} and A_{in} are the cross-sectional areas of the outlet (smaller) and inlet (larger), respectively.

1.9.2. Examples

Example 1.9-1 Flow through a Siphon

A liquid is discharged through a siphon of inner radius R, as illustrated in Fig. 1.9-3. How is the volume flow rate Q related to the height difference between the liquid level in the reservoir and the exit of the siphon? The siphon is much smaller in cross-section than the reservoir.

Solution From Bernoulli's equation (Eq. [1.9-13])

$$\tfrac{1}{2}\rho v_1^2 + p_1 + \rho g z_1 = \tfrac{1}{2}\rho v_2^2 + p_2 + \rho g z_2 \qquad [1.9\text{-}22]$$

Since the reservoir is much larger in cross-section, $v_1 \approx 0$. Also, $p_1 = p_2 = 1$ atm. As such

$$\boxed{v_2 = \sqrt{2g(z_1 - z_2)}} \qquad [1.9\text{-}23]$$

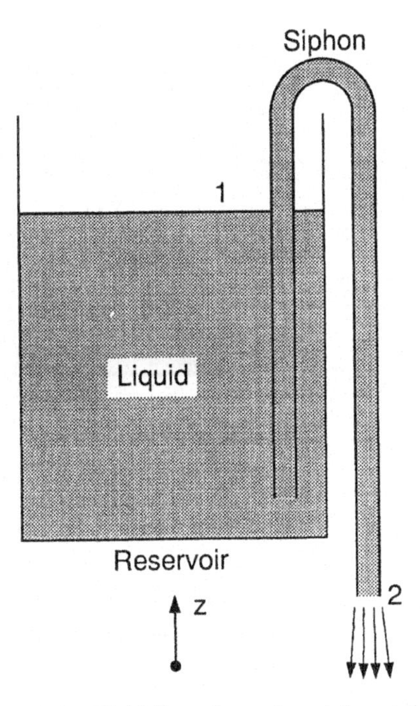

Fig. 1.9-3 Fluid flow through a siphon tube.

and so

$$Q = \pi R^2 \sqrt{2g(z_1 - z_2)} \qquad [1.9\text{-}24]$$

Example 1.9-2 Friction Loss in a Sudden Expansion

Consider the steady-state flow of an incompressible Newtonian fluid through a sudden expansion from tube 1 of cross-sectional area A_1 to tube 2 of cross-sectional area A_2, as shown in Fig. 1.9-4. Use the suggested control volume Ω to find the loss coefficient K. Neglect the friction at the wall.

Solution

1. *Mass balance*: From Eq. [1.2-8]

$$\frac{v_2}{v_1} = \frac{A_1}{A_2} = \beta \qquad [1.9\text{-}25]$$

where v_1 and v_2 are the average velocities in tube 1 at surface 1 and in tube 2 at surface 2, respectively.

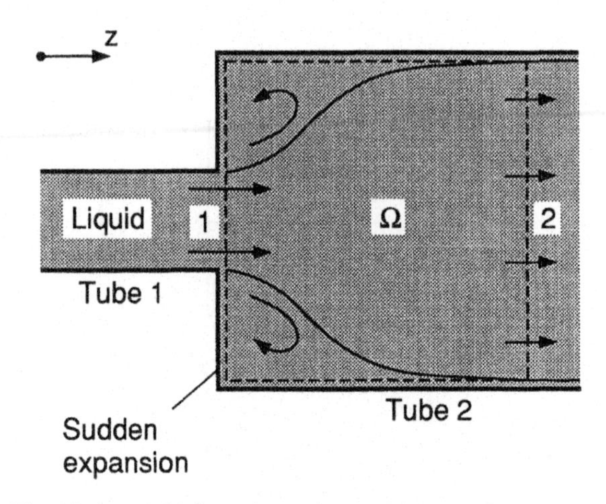

Fig. 1.9-4 Fluid flow through a sudden expansion.

2. *Momentum balance*: The pressure varies primarily in the z direction, from p_1 at surface 1 to p_2 at surface 2, but only slightly over each surface. As such, the z component of Eq. [1.4-9] is

$$0 = \rho v_1^2 A_1 - \rho v_2^2 A_2 + p_1 A_2 - p_2 A_2 + 0 + 0 \tag{1.9-26}$$

Substituting Eq. [1.9-25] to eliminate the areas

$$p_2 - p_1 = \rho v_1^2 \beta - \rho v_2^2 \tag{1.9-27}$$

Substituting Eq. [1.9-25] again to eliminate v_2

$$p_2 - p_1 = \rho v_1^2 \beta (1 - \beta) \tag{1.9-28}$$

3. *Mechanical-energy balance*: From Eq. [1.9-19], and with $z_1 = z_2$, $w_S = 0$ and $f = 0$,

$$\tfrac{1}{2} \rho v_1^2 K = \tfrac{1}{2} \rho (v_1^2 - v_2^2) + (p_1 - p_2) \tag{1.9-29}$$

Substituting Eqs. [1.9-25] and [1.9-28] into Eq. [1.9-29]

$$\boxed{K = (1 - \beta)^2 = \left(1 - \frac{A_1}{A_2}\right)^2} \tag{1.9-30}$$

which is identical to Eq. [1.9-20].

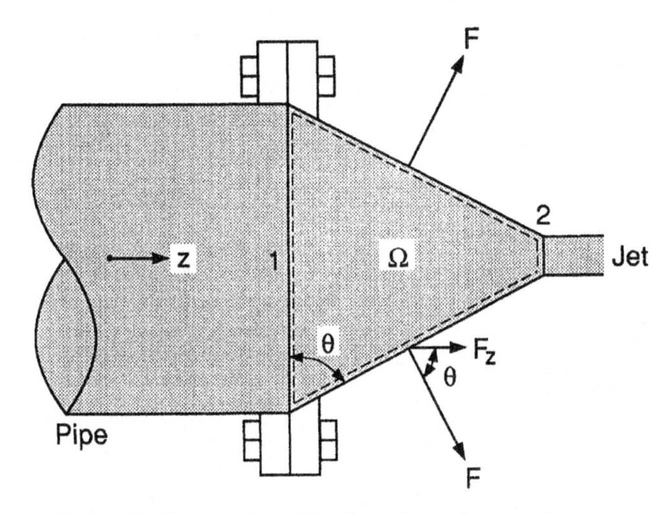

Fig. 1.9-5 Force induced by flow through a nozzle.

Example 1.9-3 Pressure Force Acting on a Nozzle

As shown in Fig. 1.9-5, consider the steady-state flow of a liquid through a horizontal nozzle[28] whose inner cross-sectional area converges from A_1 to A_2 at an angle θ. The volume flow rate is Q. Determine the force F acting on the nozzle by the liquid. Neglect the viscous force.

Solution Let the fluid in the nozzle be the control volume Ω.

1. *Mass balance*: From Eq. [1.2-6]

$$0 = \rho v_1 A_1 - \rho v_2 A_2 \qquad [1.9\text{-}31]$$

where the velocities refer to the average velocities of the fluid. As such

$$v_1 A_1 = v_2 A_2 = Q \qquad [1.9\text{-}32]$$

2. *Momentum balance*: The z-component of Eq. [1.4-9] is as follows

$$0 = \rho v_1^2 A_1 - \rho v_2^2 A_2 + (p_1 A_1 - p_a A_2 + F_{pz}) + 0 + 0 \qquad [1.9\text{-}33]$$

where p_a is the atmospheric pressure and F_{pz} is the z component of the pressure force $\mathbf{F}_p$ acting on the liquid by the nozzle. Substituting Eq. [1.9-32]

[28] S. Eskinazi, *Principles of Fluid Mechanics*, Allyn and Bacon, Boston, 1962, p. 208.

into Eq. [1.9-33] and rearranging, we obtain the z component of the force F acting on the nozzle by the liquid:

$$F_z = -F_{pz} = \rho v_1^2 A_1 - \rho v_2^2 A_2 + p_1 A_1 - p_a A_2 \qquad [1.9\text{-}34]$$

3. *Mechanical-energy balance*: From Eq. [1.9-13] and with $z_1 = z_2$

$$\tfrac{1}{2}\rho v_1^2 + p_1 = \tfrac{1}{2}\rho v_2^2 + p_a \qquad [1.9\text{-}35]$$

and so

$$p_1 = p_a + \frac{\rho}{2}(v_2^2 - v_1^2) \qquad [1.9\text{-}36]$$

Substituting Eq. [1.9-36] into Eq. [1.9-34] and rearranging, we obtain

$$F_z = \rho v_1^2 A_1 - \rho v_2^2 A_2 + \frac{\rho}{2}(v_2^2 - v_1^2)A_1 + p_a(A_1 - A_2) \qquad [1.9\text{-}37]$$

Substituting Eq. [1.9-32] into Eq. [1.9-37] and simplifying, we obtain

$$F_z = \left(\frac{\rho Q^2(A_1 - A_2)}{2A_1 A_2^2} + p_a\right)(A_1 - A_2) \qquad [1.9\text{-}38]$$

Since $F_z = F\cos\theta$, the force acting on the nozzle by the liquid is

$$\boxed{F = \left(\frac{\rho Q^2(A_1 - A_2)}{2A_1 A_2^2} + p_a\right)\left(\frac{A_1 - A_2}{\cos\theta}\right)} \qquad [1.9\text{-}39]$$

Example 1.9-4 Power Requirement for Pipe-Line Flow

Water is pumped from a lower reservoir to a higher one, through the pipe line shown in Fig. 1.9-6. The inner wall of the pipe is 10 cm in diameter and is smooth. Each of the three 90° elbows has a loss coefficient $K \approx 0.6$. What is the power required to pump the water at a flow rate $Q = 6$ liters/s? The efficiency of the pump is 70%. The water viscosity and density are 1×10^{-2} g cm^{-1} s^{-1} and 1 g/cm^3, respectively.

Solution The average velocity in the pipe is

$$v_{av} = \frac{Q}{\pi R^2} = \frac{6000 \text{ cm}^3/\text{s}}{\pi(5 \text{ cm})^2} = 76 \text{ cm/s} \qquad [1.9\text{-}40]$$

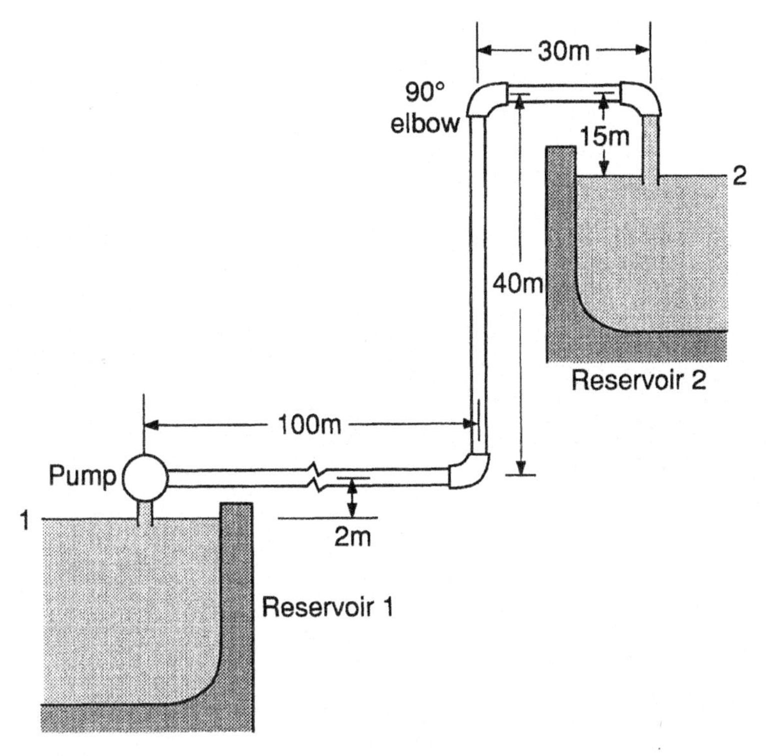

Fig. 1.9-6 Pipe line for pumping water from a lower reservoir to a higher one.

and the Reynolds number is

$$\mathrm{Re} = \frac{D\rho v_{\mathrm{av}}}{\mu} = \frac{10\ \mathrm{cm} \times 1\ \mathrm{g/cm^3} \times 76\ \mathrm{cm/s}}{1 \times 10^{-2}\ \mathrm{g\,cm^{-1}\,s^{-1}}} = 7.6 \times 10^4 \qquad [1.9\text{-}41]$$

The flow is, therefore, turbulent. From Eq. [1.8-19] the friction factor

$$f = 0.184\ \mathrm{Re}^{-(1/5)} = 0.184\ (7.6 \times 10^4)^{-(1/5)} = 0.0194 \qquad [1.9\text{-}42]$$

The friction loss associated with the various lengths of the pipe is

$$\sum_i \left(\frac{1}{2}\rho v_{\mathrm{av}}^2 \frac{L}{D} f\right) = \frac{1}{2}\left[1\ \mathrm{g/cm^3} \times (76\ \mathrm{cm/s})^2 \frac{0.0194}{10\ \mathrm{cm}}\right]$$
$$\times (200\ \mathrm{cm} + 10000\ \mathrm{cm} + 4000\ \mathrm{cm} + 3000\ \mathrm{cm} + 1500\ \mathrm{cm})$$
$$= 1.05 \times 10^5\ \mathrm{g\,cm^{-1}\,s^{-2}} \qquad [1.9\text{-}43]$$

For the sudden contraction at the point 1, from Eq. [1.9-21] and with the pipe cross section A_{out} much less than the water cross section A_{in},

$$K \approx 0.45\left(1 - \frac{A_{out}}{A_{in}}\right) \approx 0.45(1 - 0) = 0.45 \qquad [1.9\text{-}44]$$

Similarly, for the sudden expansion at point 2, from Eq. [1.9-20] and with the pipe cross section A_{in} much less than the water cross-section A_{out},

$$K = \left(1 - \frac{A_{in}}{A_{out}}\right)^2 \approx (1 - 0)^2 = 1 \qquad [1.9\text{-}45]$$

The friction loss associated with the sudden contraction, three 90° elbows and sudden expansion is then

$$\sum_i \left(\tfrac{1}{2}\rho v_{av}^2 K\right)_i = \tfrac{1}{2}[1\ \text{g/cm}^3 \times (76\ \text{cm/s})^2] \times (0.45 + 3 \times 0.6 + 1)$$

$$= 9.39 \times 10^3\ \text{g cm}^{-1}\,\text{s}^{-2} \qquad [1.9\text{-}46]$$

From Eq. [1.9-19], and with $(v_{av})_1 = (v_{av})_2$ and $p_1 = p_2 = 1$ atm,

$$\rho g z_1 = \rho g z_2 + w_S + \sum_i \left(\frac{1}{2}\rho v_{av}^2 \frac{L}{D} f\right)_i + \sum_i \left(\tfrac{1}{2}\rho v_{av}^2 K\right)_i \qquad [1.9\text{-}47]$$

Substituting Eqs. [1.9-40], [1.9-43], and [1.9-46] into Eq. [1.9-47]

$$0 = (1\ \text{g/cm}^3)(980\ \text{cm/s}^2)(200\ \text{cm} + 4000\ \text{cm} - 1500\ \text{cm})$$

$$+ w_S + 1.05 \times 10^5\ \text{g cm}^{-1}\,\text{s}^{-2} + 9.39 \times 10^3\ \text{g cm}^{-1}\,\text{s}^{-2} \qquad [1.9\text{-}48]$$

or

$$w_S = 2.76 \times 10^6\ \text{g cm}^{-1}\,\text{s}^{-2} \qquad [1.9\text{-}49]$$

As such,

$$W_S = Q w_S = 6000 \frac{\text{cm}^3}{\text{s}} \times 2.76 \times 10^6 \frac{\text{g}}{\text{cm s}^2} = 1.66 \times 10^{10}\ \text{erg/s} = 1.66 \times 10^3\ \text{Watt} \qquad [1.9\text{-}50]$$

The actual power requirement is 1.66×10^3 Watt/0.70 or 2.37×10^3 Watt.

PROBLEMS[29]

1.1-1 Show that a stress and a momentum flux have the same unit by showing that a force and the rate of momentum change or transfer have the same unit.

1.1-2 Show that Eq. [1.1-8] reduces to Eq. [1.1-2] for the one-dimensional flow in Fig. 1.1-4.

1.2-1 An incompressible fluid is flowing through a circular conduit, as shown in Fig. P1.2-1. The velocity profile of the fluid in the 8-cm-diameter pipe is $v_z = 0.1\,[1 - (r^2/16)]$ cm/s. What is the average velocity in the 2-cm-diameter pipe?

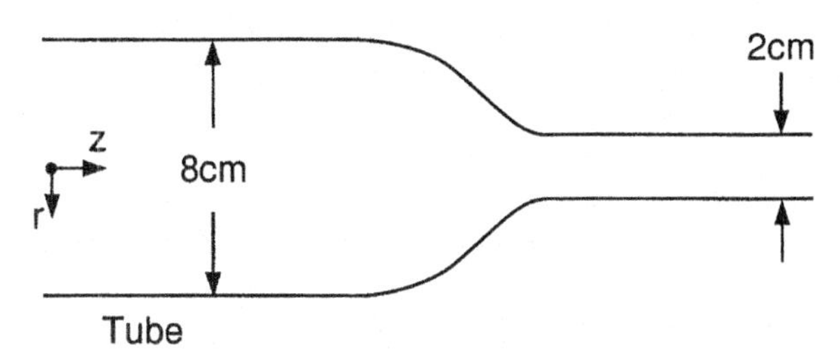

Fig. P1.2-1

1.3-1 A fluid is flowing through a round tube and its velocity profile is $v_z = 0.1\,[1 - (r^2/16)]$ cm s^{-1} at the steady state. Is the fluid incompressible?

1.3-2 Referring to Fig. 1.1-4, show that v_z is independent of z if the plates are long enough for the end effects to be neglected.

1.4-1 A horizontal turbulent liquid jet of diameter D_1 and average velocity v_j strikes concentrically a vertical plate having an orifice at the center[30], thus exerting a force F on the plate, as shown in Fig. P1.4-1 (see p. 104). The jet leaving the orifice has a diameter D_2 and the same average velocity v_j. Determine the force needed at the steady state to hold the plate in place. Neglect the viscous and gravity forces.

1.4-2 A horizontal turbulent liquid jet of diameter D_1 and average velocity v_j strikes concentrically a circular dish having an orifice at the center[30], thus exerting a force F on the disk, as shown in Fig. P1.4-2 (see p. 104). The jet leaving the orifice has a diameter D_2 and the same average velocity v_j. The remaining liquid is deflected at the same average velocity v_j and at an angle θ. Determine the force needed at the steady state to hold the dish in place. Neglect the viscous and gravity forces.

[29] The first two digits in problem numbers refer to section numbers in text.

[30] R. W. Fox and A. T. McDonald, *Introduction to Fluid Mechanics*, 4th ed., Wiley, New York, 1992, pp. 164–178.

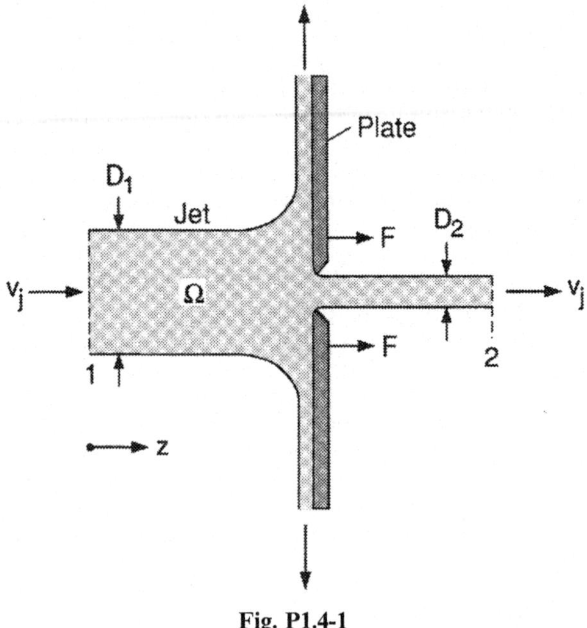

Fig. P1.4-1

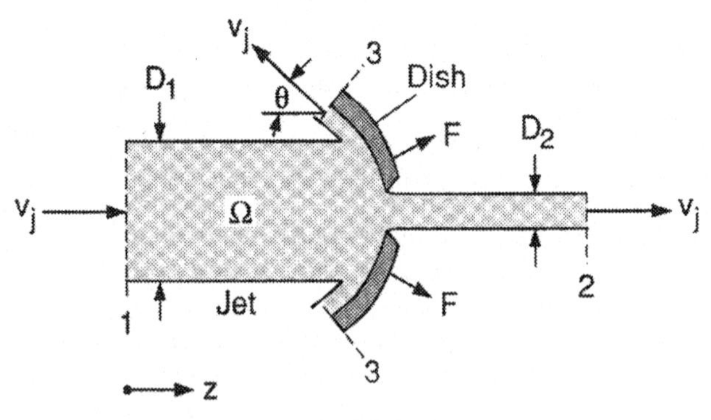

Fig. P1.4-2

1.4-3 A horizontal turbulent liquid jet of diameter D and average velocity v_j impinges on a vertical plate mounted on a cart[31], thus exerting a force F on the plate, as shown in Fig. P1.4-3 (see p. 105). Determine the opposing force needed to keep the plate moving at a constant velocity v_p ($<v_j$). Neglect the viscous and gravity forces. (*Hint*: Let the control volume Ω move with the plate.)

[31] F. M. White, *Fluid Mechanics*, 3rd ed., McGraw-Hill, New York, 1994, p. 132.

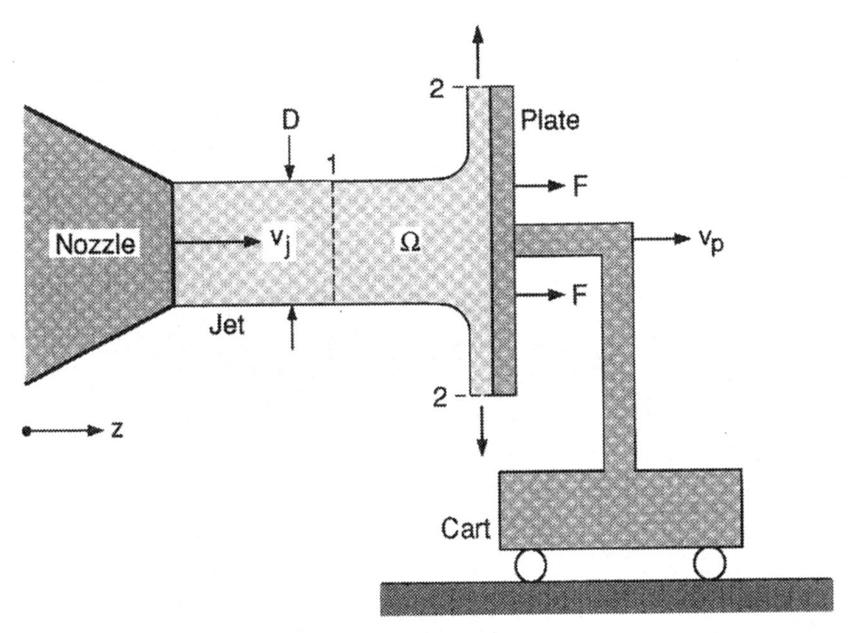

Fig. P1.4-3

1.4-4 A horizontal turbulent liquid jet of diameter D and average velocity v_j is deflected at an angle θ by a vane mounted on a cart[32], thus exerting a force F on the vane, as shown in Fig. P1.4-4 (see p. 106). Determine the opposing force needed to keep the vane moving at a constant velocity v_v $(< v_j)$. Neglect the viscous and gravity forces. (*Hint*: Let the control volume Ω move with the vane.)

1.4-5 A horizontal turbulent jet from the nozzle of a liquid tank is deflected by a vane at an angle θ. The tank and the vane are attached to a cart which is held stationary by a wire[32], as shown in Fig. P1-4-5 (see p. 106). Liquid is added through a vertical pipe to keep the water level in the tank constant. The cross-sectional area of the nozzle exit is A_n and the volume flow rate of the liquid is Q. Determine the force F acting on the wire by the liquid. Neglect the viscous and gravity forces.

1.4-6 A horizontal turbulent liquid jet of diameter D_1 and average velocity v_j strikes concentrically a circular cone having an orifice at the center, thus exerting a force F on the cone, as shown in Fig. P1.4-6 (see p. 107). The jet leaving the orifice has a diameter D_2 and the same average velocity v_j. The remaining liquid is deflected at the same average velocity v_j and at an angle θ. Determine the force needed at the steady state to hold the cone in place. Neglect the viscous and gravity forces.

[32] R. W. Fox and A. T. McDonald, *Introduction to Fluid Mechanics*, 4th ed., Wiley, New York, 1992, pp. 164–178.

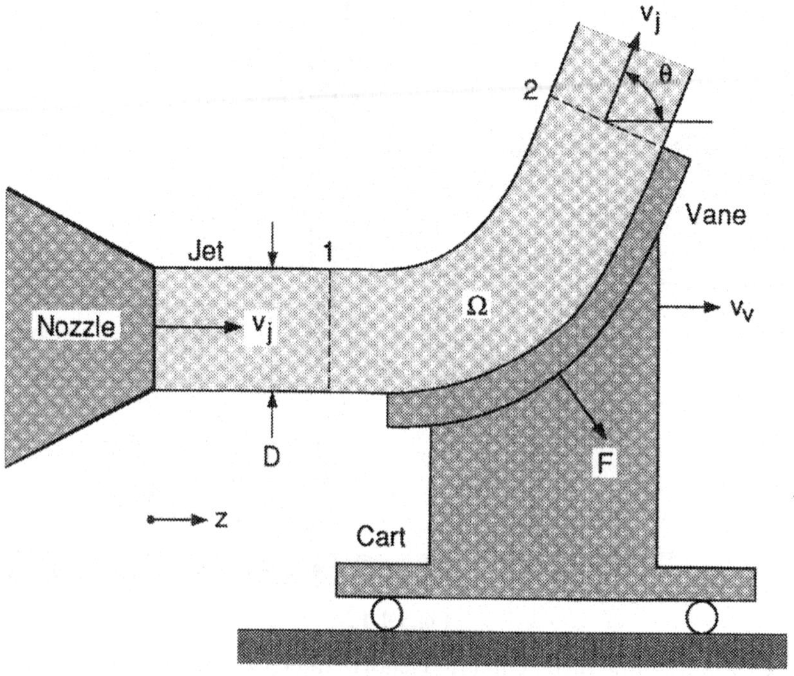

Fig. P1.4-4

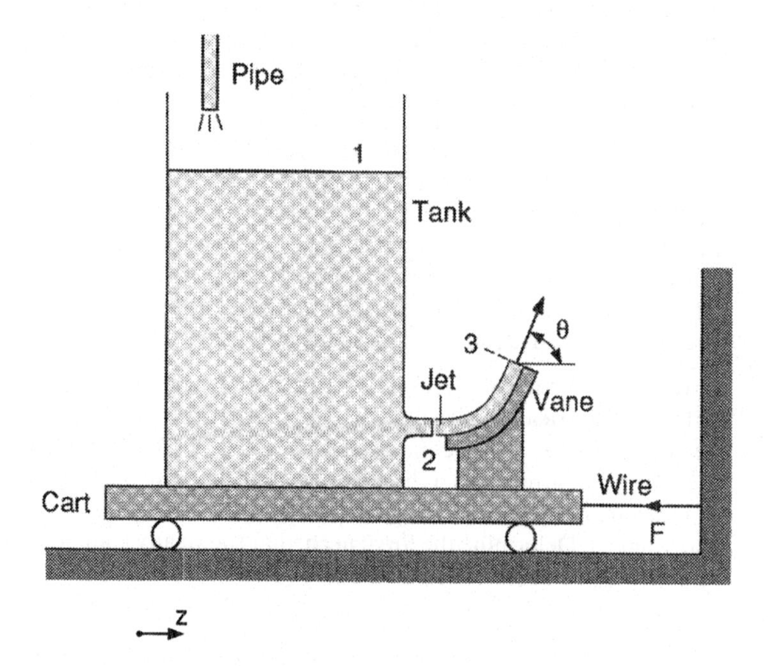

Fig. P1.4-5

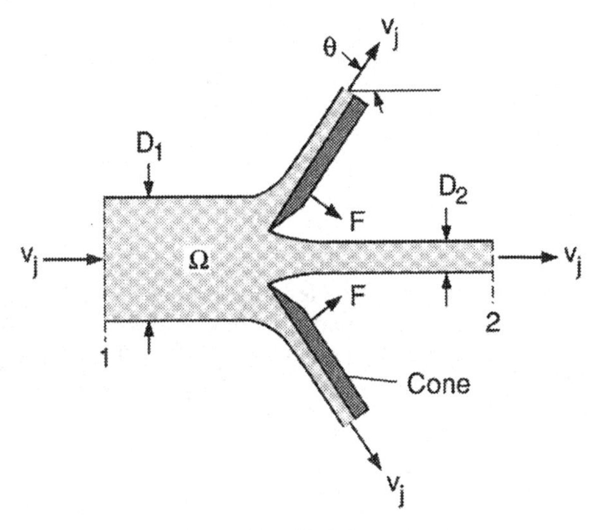

Fig. P1.4-6

1.4-7 Referring to Fig. P1.4-1, what is the force needed to keep the plate moving at a constant velocity v_p ($<v_j$) in the z direction? (*Hint*: Let the control volume Ω move with the plate.)

1.4-8 Referring to Fig. P1-4-1, what is the force needed to keep the plate moving at a constant velocity v_p in the negative z direction? (*Hint*: Let the control volume Ω move with the plate.)

1.4-9 Referring to Fig. P1-4-2, what is the force needed to keep the dish moving at a constant velocity v_d in the z direction? (*Hint*: Let the control volume Ω move with the dish.)

1.4-10 Suppose water is flowing over a flat plate with a free-stream velocity of 1 cm/s. What is the velocity profile 2 cm away from the leading edge of the plate?

1.5-1 Referring to Fig. 1.1-4, use the equation of motion to determine the steady-state velocity distribution $v_z(y)$ and the force required to keep the lower plate moving. The flow is due to the motion of the lower plate alone: $\partial p/\partial z = 0$. Assume that the fluid is Newtonian and incompressible, and the flow is laminar. Neglect end effects.

1.5-2 Consider the incompressible Newtonian fluid flowing, under the pressure gradient dp/dz, between two stationary horizontal plates separated by a gap of $2L$, as shown in Fig. P1.5-2 (see p. 108). The volume flow rate Q is known. Find the steady-state velocity distribution and the shear force acting on the two plates by the fluid.

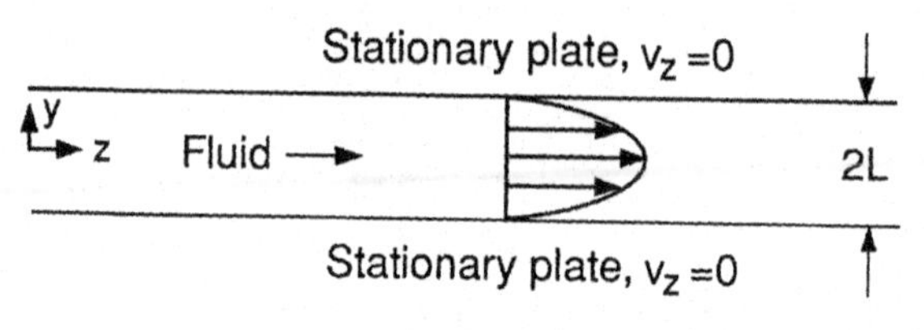

Fig. P1.5-2

1.5-3 A film of an incompressible Newtonian liquid and thickness L is held between two large vertical plates, as shown in Fig. P1.5-3. The left plate is stationary, while the right plate moves vertically upward at a constant speed V. Find the steady-state velocity distribution in the film.

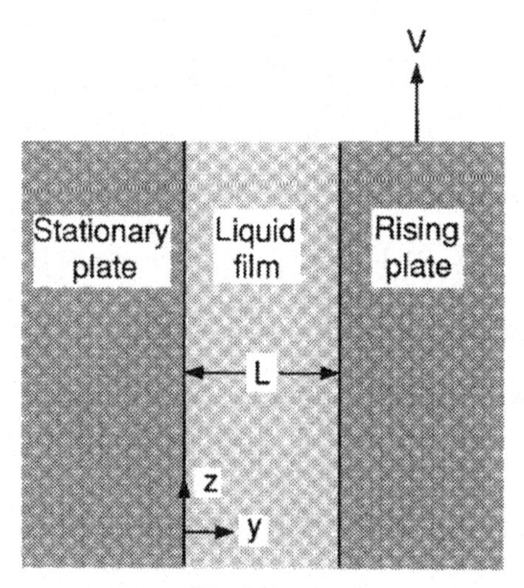

Fig. P1.5-3

1.5-4 Consider the steady-state overflow of an incompressible Newtonian fluid down a vertical plate, as shown in Fig. P1.5-4 (see p. 109). The thickness of the film is L. Find the velocity distribution in the film.

1.5-5 Consider the steady-state overflow of an incompressible Newtonian liquid along an inclined plate of angle θ, as shown in Fig. P1.5-5 (see p. 109). The thickness of the liquid film is L. Find the velocity distribution in the film.

1.5-6 As shown in Fig. P1.5-6 (see p. 110), a fluid overflows down the outside of a vertical tube and forms a film; the outer radii of the tube and the film are r_1 and r_2, respectively. Determine the steady-state velocity distribution in the film. The fluid is incompressible and Newtonian, and the flow is laminar. Neglect end effects.

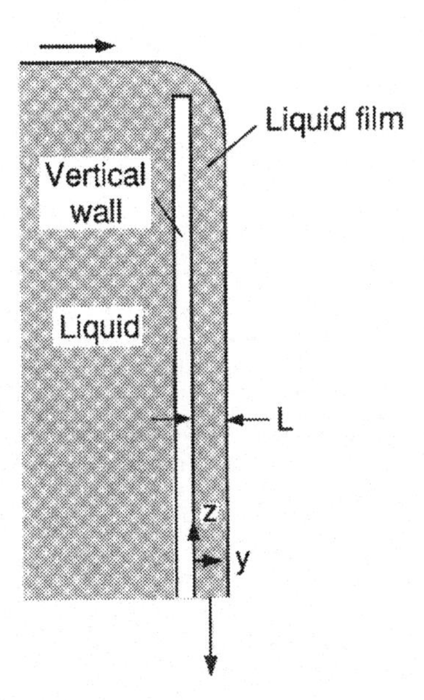

Fig. P1.5-4

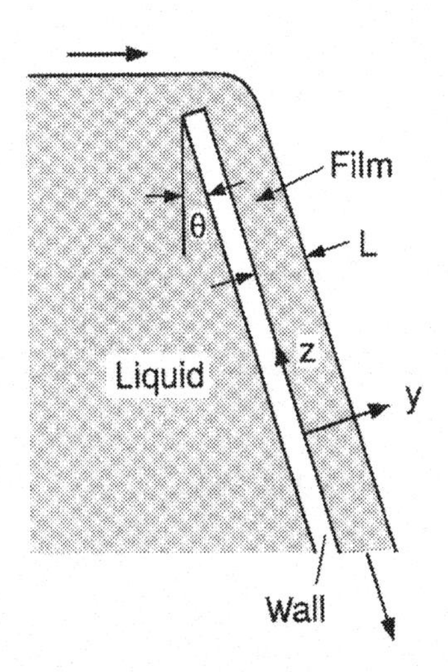

Fig. P1.5-5

1.5-7 Consider the steady-state flow of an incompressible Newtonian fluid through a horizontal tube of inner radius R and length L, as shown in Fig. P1.5-7 (see p. 110). Find the velocity distribution in the liquid. Neglect end effects.

1.5-8 A wire of radius r_1 is being pulled coaxially at a constant velocity V through an incompressible Newtonian fluid in a horizontal tube of

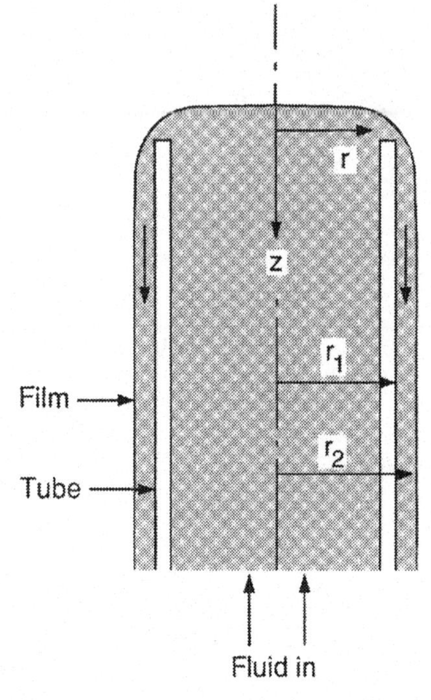

Fig. P1.5-6

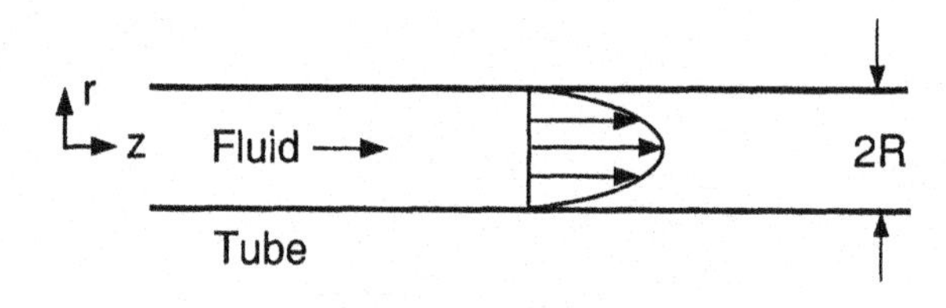

Fig. P1.5-7

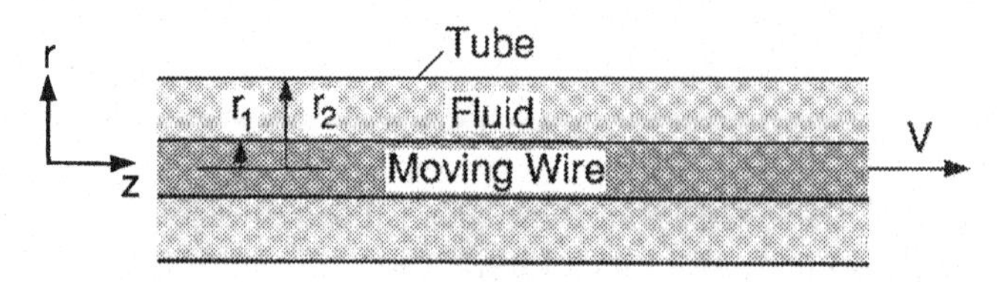

Fig. P1.5-8

inner radius r_2, as shown in Fig. P1.5-8. Find the steady-state velocity distribution in the fluid. Neglect end effects.

1.5-9 A wire of radius r_1 is being pulled coaxially at a constant velocity V through an incompressible Newtonian fluid in a vertical tube of inner radius r_2, as shown in Fig. P1.5-9. Find the steady-state velocity distribution in the fluid. Neglect end effects.

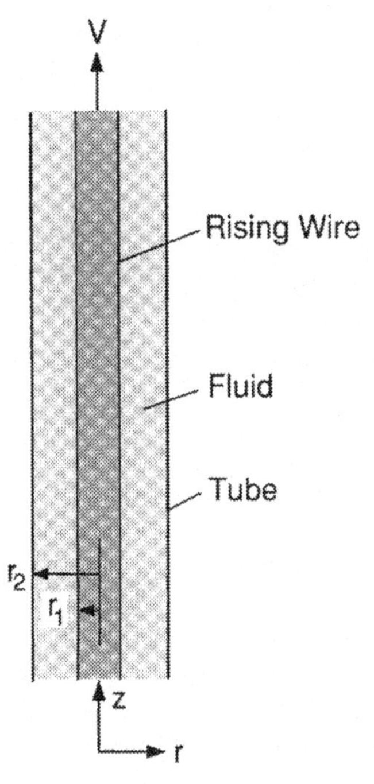

Fig. P1.5-9

1.5-10 Consider the steady-state flow of an incompressible Newtonian fluid through the annular space between two stationary horizontal coaxial tubes of length L, as shown in Fig. P1.5-10. The outer radius of the smaller tube is r_1, and the inner radius of the larger tube is r_2. Find the velocity distribution in the fluid. Neglect end effects.

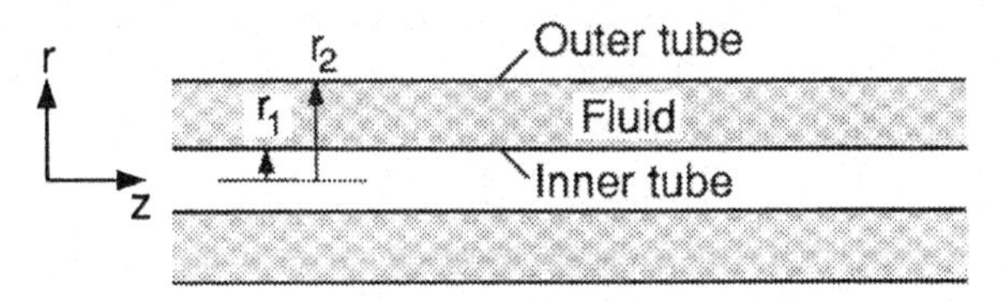

Fig. P1.5-10

1.5-11 Consider the steady-state overflow of an incompressible Newtonian liquid into a vertical tube to form a film on its inner wall, as shown in Fig. P1.5-11 (see p. 112). The radii of the film and the tube are r_1 and r_2, respectively. Find the velocity distribution in the film.

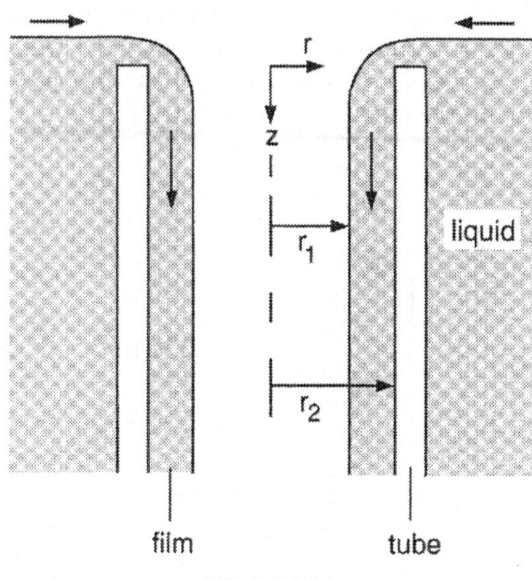

Fig. P1.5-11

1.5-12 A continuous wire of radius r_1 moves vertically upward at a constant speed V and passes through an incompressible Newtonian fluid. As shown in Fig. P1.5-12 (see p. 113), it picks up a liquid film with it, and the outer radius of the film is r_2. The movement of the wire and the viscosity of the fluid keep it from running off the wire completely. Find the steady-state velocity distribution in the film.

1.5-13 A semiinfinite body of an incompressible Newtonian liquid is bounded on one side by a large flat plate, as shown in Fig. P1.5-13 (see p. 113). Initially, both are at rest; but at time $t = 0$ the plate is suddenly set into motion with velocity V. Find the velocity as a function of time and position.

1.5-14 Prove that $-\nabla \cdot \boldsymbol{\tau} = \mu \nabla^2 \mathbf{v}$ for an incompressible Newtonian fluid.

1.5-15 Consider the flow over the flat plate shown in Fig. 1.4-7*a*. Prove that $\partial^2 v_z/\partial y^2 = 0$ at $y = 0$. Does the velocity distribution given by Eq. [1.4-57] satisfy this condition?

1.8-1 Consider the vacuum furnace shown in Fig. 1.8-4. If a water flow rate of 30 g/s is needed for proper cooling of the chamber wall, at what pressure should the cooling water be supplied? The copper tubing is 0.4 cm ID and 20 m long. Assume smooth inner surface of the copper tubing.

1.8-2 A pair of water skis 15 cm wide and 150 cm long is towed on still water of 20°C at 10 m/s. Find the drag force on the bottom of each ski. The viscosity $\mu = 1 \times 10^{-2}$ g cm^{-1} s^{-1}.

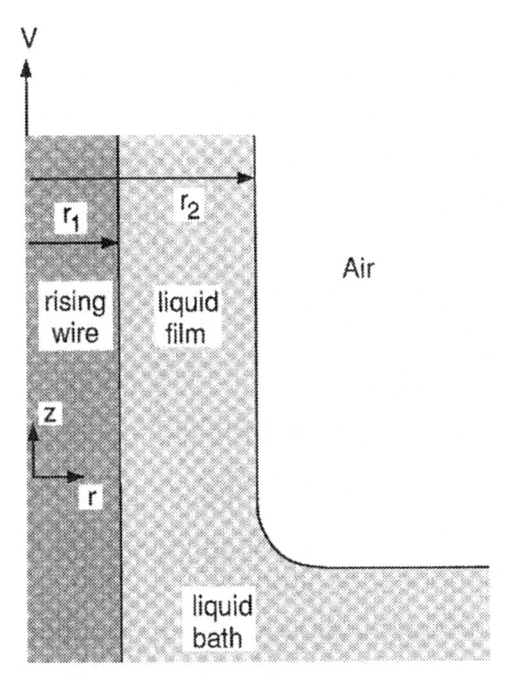

Fig. P1.5-12

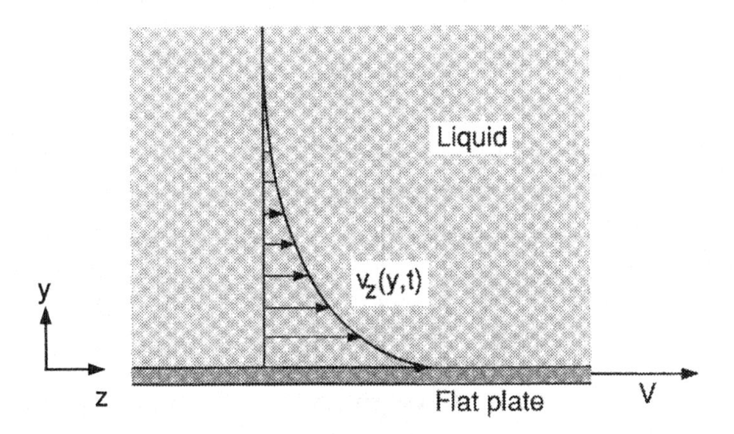

Fig. P1.5-13

1.8-3 A transmission cable 3 cm in diameter is struck at right angles by a wind blown at 30 m/s. Find the drag force per meter of the cable. Assume $\rho = 1.2 \times 10^{-5}$ g cm^3 and $\mu = 1.81 \times 10^{-4}$ g cm^{-1} s^{-1}.

1.9-1 Consider the velocity measurement using a Pitot tube, as shown in Fig. P1.9-1 (see p. 114). How is the fluid velocity at point 1 related to the

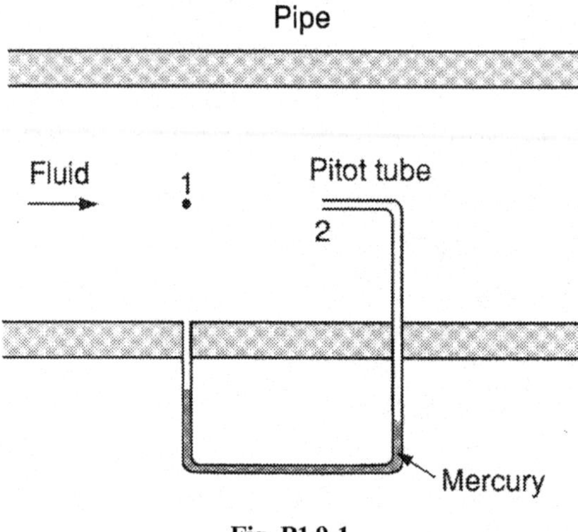

Fig. P1.9-1

pressure difference measured by the manometer. Assume inviscid incompressible flow as an approximation.

1.9-2 Consider the velocity measurement using a Venturi tube, as shown in Fig. P1.9-2. How is the fluid velocity at point 1 related to the pressure difference measured by the manometer. Assume inviscid incompressible flow as an approximation.

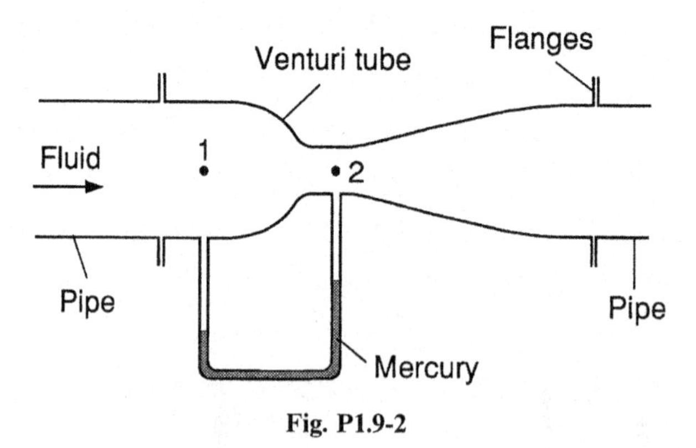

Fig. P1.9-2

1.9-3 Consider the hydroelectric power plant[33] shown in Fig. P1.9-3 (see p. 115). The water is discharged at 24 m^3/s through a channel of 8 m^2 cross section and the flow is turbulent. Determine the power output from the turbine assuming 70% efficiency. Neglect the friction loss.

[33] F. M. White, *Fluid Mechanics*, 3rd ed., McGraw-Hill, New York, 1994, p. 154.

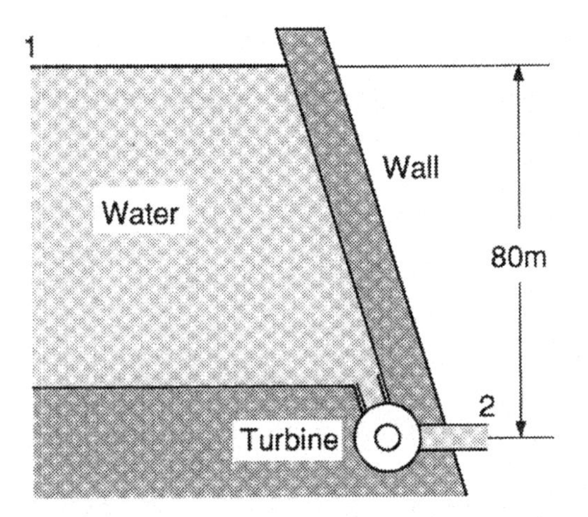

Fig. P1.9-3

1.9-4 As shown in Fig. P1.9-4, water is pumped at 1 liter/s through a 2 cm diameter smooth pipe line of total length 50 m. The loss efficiency K is about 0.5 for each of the three elbows, and 0.8 for the valve. How much power of the pump is required assuming an efficiency of 70%?

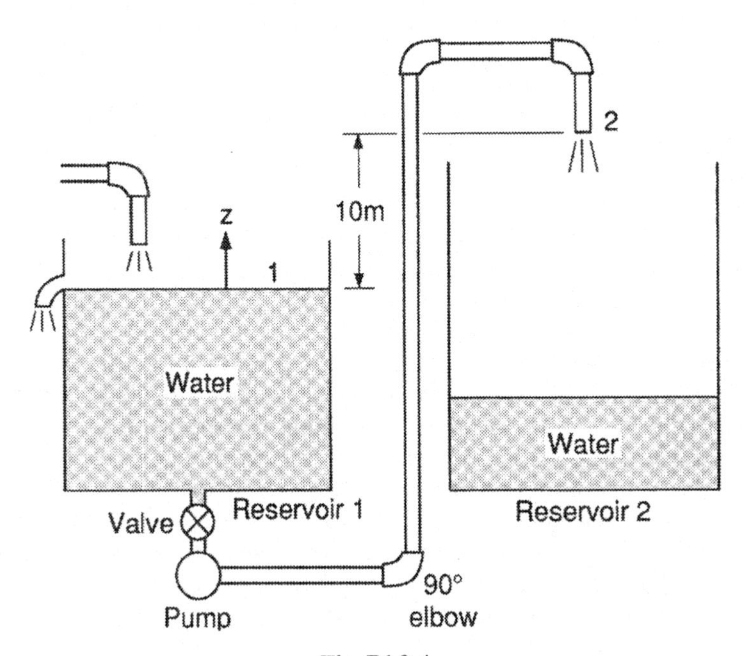

Fig. P1.9-4

1.9-5 Prove Eq. [1.4-11] using Eqs. [1.7-5] and [1.7-6].

1.9-6 Prove Eq. [1.9-11] using Eqs. [1.7-5] and [1.7-6].

CHAPTER 2

INTRODUCTION TO HEAT TRANSFER

2.1. BASIC CONCEPTS

2.1.1. Conduction, Convection, and Radiation

Heat transfer is the transfer of heat due to a temperature difference. Heat transfer can occur by three different mechanisms: conduction, convection, and radiation.

Conduction refers to heat transfer that occurs across a stationary solid or fluid in which a temperature gradient exists. In contrast, **convection** refers to the heat transfer that occurs across a moving fluid in which a temperature gradient exists. **Radiation**, specifically, thermal radiation, refers to the heat transfer between two surfaces at different temperatures separated by a medium transparent to the electromagnetic waves emitted by the surfaces.

An example in which all three mechanisms of heat transfer are significant is shown in Fig. 2.1-1. The charge material in the crucible, which has been melted, can be cast to form an ingot by metal casting or pulled to form a crystal in crystal growth. In any case, temperature decreases monotonically from the heater to the center of the crucible. Heat transfer from the heater to the crucible is by radiation, whereas that through the crucible is by conduction. Heat transfer through the melt is dominated by convection. Conduction, which occurs regardless of whether the melt is recirculating or stationary, can still play a role in heat transfer through the melt, especially in the case of metals, which conduct heat easily.

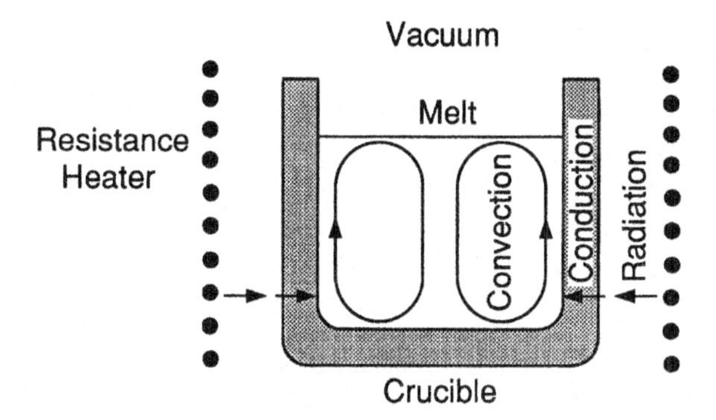

Fig. 2.1-1 Heat transfer mechanisms: conduction, convection, and radiation.

2.1.2. Fourier's Law of Conduction

2.1.2.1. One-Dimensional

Let us consider the conduction of heat through a slab of thickness L, as shown in part in Fig. 2.1-2. The lower surface is kept at a constant temperature T_1 and the upper surface at a constant but lower temperature T_2. A steady-state temperature profile $T(y)$ is established in the slab, where y is the distance from the lower surface.

Let us consider the two thin layers of the slab shown in the figure. The lower layer at y has a temperature T, whereas the upper one at $y + dy$ has a slightly lower temperature, $T + dT$ ($dT < 0$). Since the lower layer is hotter, a heat flux tends to flow from it to the cooler upper layer. A **heat flux** is defined as the amount of heat transferred per unit area per unit time. This heat flux is denoted as q_y since it is in the y direction. The greater the temperature difference between the two layers is, or more appropriately, the steeper the temperature gradient dT/dy is, the greater the heat flux q_y becomes. Furthermore, the more thermally conductive the slab is, the greater the heat flux q_y becomes. In fact, it is observed that

$$\boxed{q_y = -k\frac{dT}{dy}} \qquad [2.1\text{-}1]$$

where k is the **thermal conductivity** of the fluid. This equation is **Fourier's law of conduction** for one-dimensional heat conduction in the y direction. The minus sign in the equation indicates that the temperature gradient is negative. It also reminds us that heat conduction occurs in the direction of decreasing temperature.

The cgs and mks units of the heat flux are W/cm^2 and W/m^2, respectively. From Eq. [2.1-1] the cgs and mks units of the thermal conductivity are $W\,cm^{-1}\,K^{-1}$ and $W\,m^{-1}\,K^{-1}$, respectively.

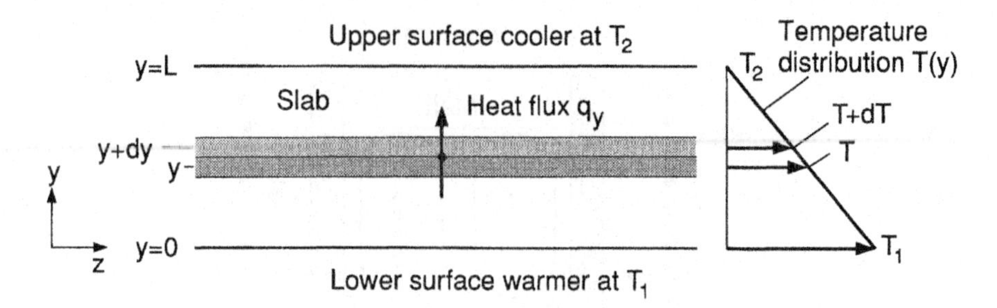

Fig. 2.1-2 One-dimensional heat conduction through a slab.

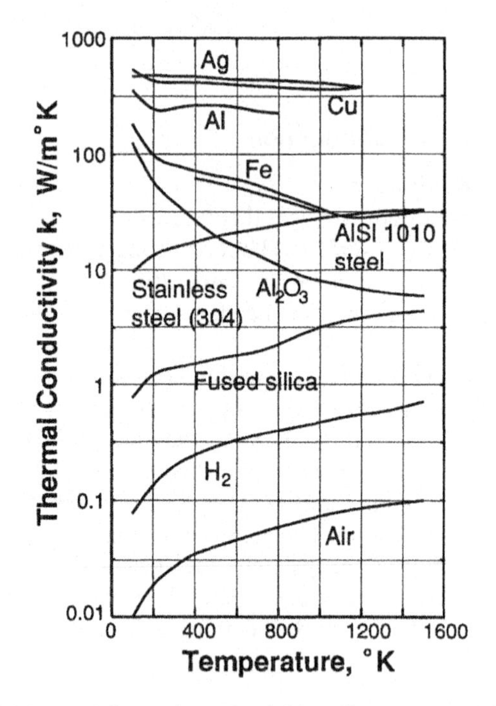

Fig. 2.1-3 Thermal conductivities of some materials.

The thermal conductivities of some common materials are given in Fig. 2.1-3. The thermal conductivities of some semiconductors are listed in Table 2.1-1.

2.1.2.2. Three-Dimensional

In the case of three-dimensional heat flow, an equation like Eq. [2.1-1] can be written for each of the three coordinate directions as follows:

$$q_x = -k\frac{\partial T}{\partial x} \qquad [2.1\text{-}2]$$

TABLE 2.1-1 Thermal Conductivities of Some Molten and Solid Semiconductor Materials at Their Melting Points

Semiconductor	Melting Point (K)	K_L (W cm^{-1} K^{-1})	K_S (W cm^{-1} K^{-1})
Si	1683	0.669	0.314
		0.515	0.216
Ge	1210	0.711	0.243
InSb	798	0.123	0.0474
GaSb	985	0.171	0.0781
GaAs	1511	0.178	0.0712
InP	1335	0.228	0.0911

Source: A. S. Jordan, *Journal of Crystal Growth*, **71**, 551, 1985.

$$q_y = -k\frac{\partial T}{\partial y} \qquad [2.1\text{-}3]$$

$$q_z = -k\frac{\partial T}{\partial z} \qquad [2.1\text{-}4]$$

Note that it has been assumed here that the material being considered is isotropic, that is, its physical properties are the same in all three directions. This assumption is valid for fluids and for most homogeneous solids. Some solid materials are nonisotropic, however, especially those with a fibrous or laminated structure. Equations [2.1-2] through [2.1-4] are components of the following equation:

$$\boxed{\mathbf{q} = -k\nabla T} \qquad [2.1\text{-}5]$$

Equation [2.1-5] is the three-dimensional form of Fourier's law of conduction. Its expanded forms are given in Table 2.1-2 for rectangular, cylindrical, and spherical coordinates.

2.1.2.3. Thermal Diffusivity

As already mentioned, the proportional constant k in Fourier's law of conduction is called the thermal conductivity. The **thermal diffusivity** α, on the other hand, is defined as follows:

$$\boxed{\alpha = \frac{k}{\rho C_v}} \qquad [2.1\text{-}6]$$

where ρ and C_v are the density and specific heat of the material, respectively. Like the kinematic viscosity, the cgs and mks units of the thermal diffusivity are cm^2/s and m^2/s, respectively.

TABLE 2.1-2 Components of the Heat Conduction Flux q

Rectangular Coordinates		Cylindrical Coordinates		Spherical Coordinates	
$q_x = -k\dfrac{\partial T}{\partial x}$	[A]	$q_r = -k\dfrac{\partial T}{\partial r}$	[D]	$q_r = -k\dfrac{\partial T}{\partial r}$	[G]
$q_y = -k\dfrac{\partial T}{\partial y}$	[B]	$q_\theta = -k\dfrac{1}{r}\dfrac{\partial T}{\partial \theta}$	[E]	$q_\theta = -k\dfrac{1}{r}\dfrac{\partial T}{\partial \theta}$	[H]
$q_z = -k\dfrac{\partial T}{\partial z}$	[C]	$q_z = -k\dfrac{\partial T}{\partial z}$	[F]	$q_\phi = -k\dfrac{1}{r\sin\theta}\dfrac{\partial T}{\partial \phi}$	[I]

2.1.3. Thermal Boundary Layer

Consider first a fluid of uniform temperature T_∞ approaching a flat plate of constant temperature T_s in the direction parallel to the plate, as illustrated in Fig. 2.1-4. At the solid/liquid interface the fluid temperature is T_s since the local fluid particles achieve thermal equilibrium at the interface temperature of the plate. As a result of conduction, the fluid temperature T in the region near the plate is affected by the plate, varying from T_s at the plate surface to T_∞ in the stream. This region is called the **thermal boundary layer**. Its thickness δ_T is typically taken as the distance from the plate surface at which the dimensionless temperature $(T - T_S)/(T_\infty - T_S)$ or $(T_S - T)/(T_S - T_\infty)$ reaches 0.99. In practice it is usually specified that $T = T_\infty$ and $\partial T/\partial y = 0$ at $y = \delta_T$. With increasing distance from the leading edge of the plate, the effect of conduction penetrates further into the stream and the boundary layer grows in thickness.

The effect of conduction is significant only in the boundary layer. Beyond it the temperature is uniform and the effect of conduction is no longer significant.

Let us now consider a fluid of uniform temperature T_∞ entering a circular tube of inner diameter D and uniform wall temperature T_S, as illustrated in Fig. 2.1-5. A thermal boundary layer begins to develop at the entrance, gradually expanding until the layers from oppositive sides approach the centerline. This occurs at[1]

$$\frac{z}{D} \approx 0.05\left(\frac{\rho v_\infty D}{\mu}\right)\left(\frac{C_v \mu}{k}\right) = 0.05\ \mathrm{Re}_D\,\mathrm{Pr} \qquad [2.1\text{-}7]$$

where

$$\mathrm{Re}_D = \frac{v_\infty D}{\nu} = \frac{\rho v_\infty D}{\mu} \qquad \text{(Reynolds number)} \qquad [2.1\text{-}8]$$

[1] W. M. Kay and M. E. Crawford, *Convective Heat and Mass Transfer*, McGraw-Hill, New York, 1980.

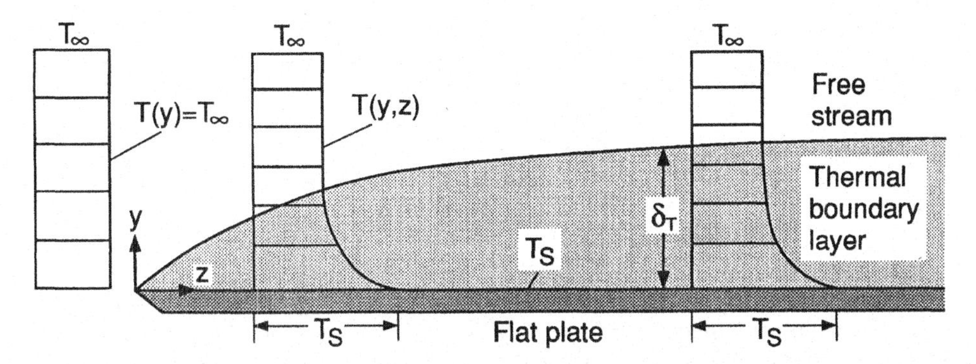

Fig. 2.1-4 Thermal boundary layer over a flat plate.

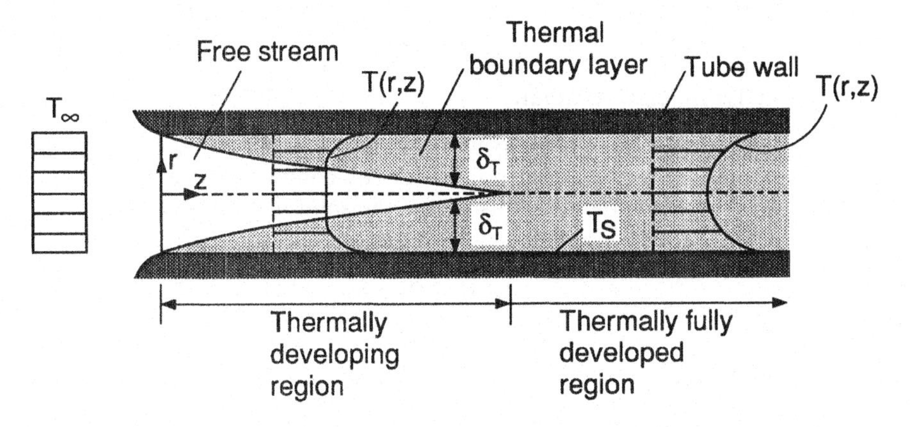

Fig. 2.1-5 Thermal boundary layer in a circular tube near its entrance.

and

$$\mathrm{Pr} = \frac{\nu}{\alpha} = \frac{C_v \mu}{k} \qquad \text{(Prandtl number)} \qquad [2.1\text{-}9]$$

Equation [2.1-7] differs from Eq. [1.1-33] by a factor of Pr.

Before we discuss the thermally fully developed temperature profile, let us first define the average temperature. For flow through a cross-sectional area A, such as that of a tube, the average temperature is defined as

$$T_{\mathrm{av}} = \frac{\iint_A \rho C_v T v \, dA}{\iint_A \rho C_v v \, dA} \qquad [2.1\text{-}10]$$

Notice that the numerator is the thermal-energy flow rate. If the specific heat is assumed constant over A, this equation becomes

$$T_{av} = \frac{\iint_A \rho C_v T v \, dA}{m C_v} = \frac{\iint_A \rho T v \, dA}{m} \quad [2.1\text{-}11]$$

where m is the mass flow rate of the fluid.

The **thermally fully developed** temperature profile in a tube is one with a dimensionless temperature $(T - T_S)/(T_{av} - T_S)$ or $(T_S - T)/(T_S - T_{av})$ independent of the axial position, that is

$$\frac{\partial}{\partial z}\left[\frac{T_S - T}{T_S - T_{av}}\right] = 0 \quad [2.1\text{–}12]$$

2.1.4. Heat Transfer Coefficient

Consider the thermal boundary layer shown previously in Fig. 2.1-4. At the solid/liquid interface heat transfer occurs only by conduction since there is no fluid motion. Therefore, the heat flux across the solid/liquid interface is

$$q_y|_{y=0} = -k \left.\frac{\partial T}{\partial y}\right|_{y=0} \quad [2.1\text{-}13]$$

This equation cannot be used to calculate the heat flux when the temperature gradient is an unknown. A convenient way to avoid this problem is to introduce a **heat transfer coefficient**, defined as follows:

$$h = \left|\frac{q_y|_{y=0}}{(T_S - T_\infty)}\right| = \left|\frac{-k(\partial T/\partial y)|_{y=0}}{(T_S - T_\infty)}\right| \quad [2.1\text{-}14]$$

The absolute values are used to keep h always positive. From Eq. [2.1-14]

$$|q_y|_{y=0}| = h|(T_S - T_\infty)| \quad [2.1\text{-}15]$$

This equation has been called **Newton's law of cooling**. For fluid flow through a tube of an inner radius R and wall temperature T_S, a similar equation can be used:

$$h = \left|\frac{q_r|_{r=R}}{(T_S - T_{av})}\right| = \left|\frac{-k(\partial T/\partial r)|_{r=R}}{(T_S - T_{av})}\right| \quad [2.1\text{-}16]$$

where T_{av} is the average fluid temperature over the cross-sectional area πR^2.

Let us consider the thermally fully developed region shown in Fig. 2.1-5 again. In the case of a **constant heat flux** $q_r|_{r=R}$, the heat transfer coefficient h is constant in the thermally fully developed region. From Eq. [2.1-16] we see that $(T_S - T_{av})$ is also constant. From this and Eq. [2.1-12], we see that

$$\frac{\partial T_S}{\partial z} = \frac{\partial T}{\partial z} = \frac{\partial T_{av}}{\partial z} \qquad (\text{constant } q_r|_{r=R}) \qquad [2.1\text{-}17]$$

Since T_S and T_{av} are independent of r, $\partial T/\partial z$ is also **independent of *r*.**

Let us now consider the case of a **constant wall temperature** T_S. Equation [2.1-12] can be expanded and solved for $\partial T/\partial z$ to give

$$\frac{\partial T}{\partial z} = \left(\frac{T_S - T}{T_S - T_{av}}\right)\frac{\partial T_{av}}{\partial z} \qquad (\text{constant } T_S) \qquad [2.1\text{-}18]$$

Since T is dependent on r, $\partial T/\partial z$ is also **dependent on *r*.**

2.2. OVERALL ENERGY-BALANCE EQUATION

2.2.1. Derivation

Let us consider an arbitrary stationary control volume Ω bounded by surface A through which a moving fluid is flowing, as illustrated in Fig. 2.2-1. The control surface A can be considered to consist of three different regions: A_{in} for the region where the fluid enters the control volume; A_{out}, where the fluid leaves; and A_{wall}, where the fluid is in contact with a wall. In other words

$$A = A_{in} + A_{out} + A_{wall} \qquad [2.2\text{-}1]$$

Before deriving the integral energy-balance equation, let us consider the outward heat transfer, for example, conduction, rate through dA shown in

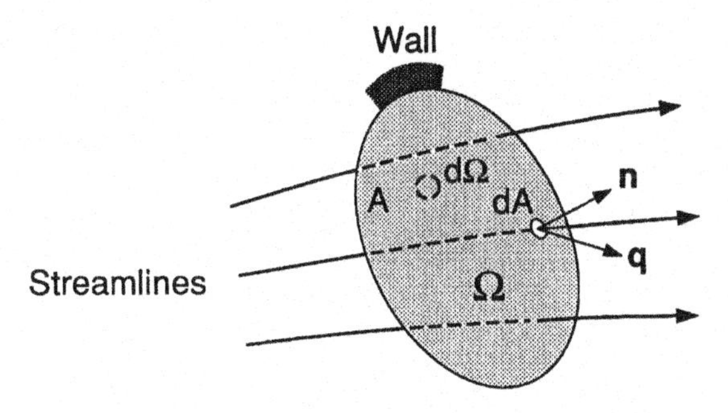

Fig. 2.2-1 Fluid flow through a volume element Ω with heat transfer.

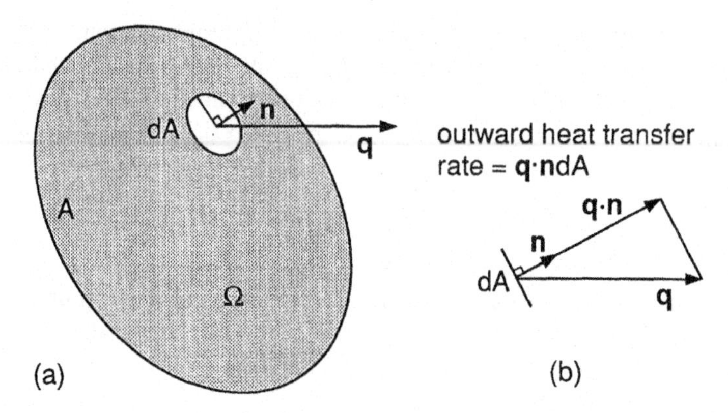

Fig. 2.2-2 Heat conduction rate through a differential surface element dA: (*a*) three-dimensional view; (*b*) two-dimensional view.

Fig. 2.2-2*a*. As shown in Fig. 2.2-2*b*, it is $\mathbf{q} \cdot \mathbf{n}\, dA$. Since $\mathbf{n}$ points outward

$$\text{inward heat transfer rate through area } dA = -\mathbf{q} \cdot \mathbf{n}\, dA \qquad [2.2\text{-}2]$$

The following energy conversation law can be applied to the control volume:

$$\underset{(1)}{\begin{Bmatrix} \text{rate of} \\ \text{energy} \\ \text{accumulation} \end{Bmatrix}} = \underset{(2)}{\begin{Bmatrix} \text{rate of} \\ \text{energy in by} \\ \text{mass inflow} \end{Bmatrix}} - \underset{(3)}{\begin{Bmatrix} \text{rate of} \\ \text{energy out by} \\ \text{mass outflow} \end{Bmatrix}}$$

$$+ \underset{(4)}{\begin{Bmatrix} \text{rate of other heat} \\ \text{transfer to system} \\ \text{from surroundings} \end{Bmatrix}} - \underset{(5)}{\begin{Bmatrix} \text{rate of work} \\ \text{done by system} \\ \text{on surroundings} \end{Bmatrix}} + \underset{(6)}{\begin{Bmatrix} \text{rate of heat} \\ \text{generation} \\ \text{in system} \end{Bmatrix}} \qquad [2.2\text{-}3]$$

This equation is, in fact, the **first law of thermodynamics** written for an open system under the unsteady-state condition. In terms (1) through (3) the energy includes the thermal, kinetic, and potential energies.

Let us now consider the terms in Eqn. [2.2-3] one by one, starting from term (1). The thermal, kinetic, and potential energy per unit mass of the fluid are $C_v T$, $v^2/2$, and ϕ, respectively, where C_v is the specific heat and T is the temperature. The total energy **per unit mass** of the fluid

$$e_t = C_v T + \tfrac{1}{2} v^2 + \phi \qquad [2.2\text{-}4]$$

Therefore, the total energy in the differential volume element $d\Omega$ is $dE_t = \rho e_t \, d\Omega$. This can be integrated over Ω to obtain the total energy in the control volume:

$$\iiint_\Omega \rho e_t \, d\Omega \quad ; \quad E_t$$

(integral) (overall)

and the rate of energy change in Ω is

$$\frac{\partial}{\partial t}\iiint_\Omega \rho e_t \, d\Omega \quad ; \quad \frac{dE_t}{dt} \qquad \text{term (1)}$$

(integral) (overall)

For convenience the first form will be called the integral form and the second the overall form. The same terminology will be applied to the remaining terms.

From Eq. [1.2-2] the inward mass flow rate through dA is $-\rho\mathbf{v}\cdot\mathbf{n}\,dA$. The inward energy flow rate $-\rho e_t \mathbf{v}\cdot\mathbf{n}\,dA$ can be integrated over A to obtain the net (since A is a closed surface), inward energy flow rate into Ω, that is, the rate of energy flow into Ω minus that out of Ω:

$$-\iint_A \rho e_t \mathbf{v}\cdot\mathbf{n}\,dA$$

or, in view of Eq. [2.2-1],

$$-\iint_{A_{\text{in}}+A_{\text{out}}+A_{\text{wall}}} \rho e_t \mathbf{v}\cdot\mathbf{n}\,d(A_{\text{in}}+A_{\text{out}}+A_{\text{wall}})$$

or, since $\mathbf{v} = 0$ at the wall,

$$-\left(\iint_A \rho e_t \mathbf{v}\cdot\mathbf{n}\,dA\right)_{\text{in}} - \left(\iint_A \rho e_t \mathbf{v}\cdot\mathbf{n}\,dA\right)_{\text{out}} - (0)_{\text{wall}}$$

As in Section 1.2, the velocity component along the normal to the control surface is defined as $v = |\mathbf{v}\cdot\mathbf{n}|$. At the inlet, $\mathbf{v}$ is inward, $\mathbf{n}$ is outward, and $v = |\mathbf{v}\cdot\mathbf{n}| = -\mathbf{v}\cdot\mathbf{n}$, while at the outlet $\mathbf{v}$ and $\mathbf{n}$ are both outward and $v = |\mathbf{v}\cdot\mathbf{n}| = \mathbf{v}\cdot\mathbf{n}$. Therefore, the rate of energy flow into Ω minus that out of Ω is

$$-\iint_A \rho e_t \mathbf{v}\cdot\mathbf{n}\,dA \quad ; \quad \left[\left(\iint_A \rho e_t v\,dA\right)_{\text{in}} - \left(\iint_A \rho e_t v\,dA\right)_{\text{out}}\right]$$

(integral) (overall)

term(2)–term(3)

According to Eq. [2.2-2] the rate of heat transfer, that is, conduction, into the control volume through dA is $dQ = -\mathbf{q}\cdot\mathbf{n}\,dA$. This can be integrated over A to obtain the rate of heat transfer into the control volume from the surroundings other than terms (2) and (3):

$$-\iint_A \mathbf{q}\cdot\mathbf{n}\,dA \quad ; \quad Q \qquad \text{term (4)}$$

(integral) (overall)

The rate of the shaft work done by the fluid in the control volume on the surroundings, that is, through a turbine or compressor, is

$$W_S \qquad \text{term (5a)}$$

In order to leave the control volume through dA, the fluid has to work against the pressure of the surrounding fluid. Since the pressure force is $p\mathbf{n}\,dA$, the rate of pressure work required is $dW_p = p\mathbf{n}\cdot\mathbf{v}\,dA = p\mathbf{v}\cdot\mathbf{n}\,dA$. Note that work equals force times displacement and the rate of work equals force times velocity. This can be integrated over A to obtain the rate of pressure work the fluid has to do to go through the control volume:

$$\iint_A p\mathbf{v}\cdot\mathbf{n}\,dA \quad ; \quad W_p \qquad \text{term (5b)}$$

(integral) (overall)

The fluid also has to do work to overcome the viscous force $\boldsymbol{\tau}\cdot\mathbf{n}\,dA$, and the rate of viscous work required is $dW_v = (\boldsymbol{\tau}\cdot\mathbf{n})\cdot\mathbf{v}\,dA = (\boldsymbol{\tau}\cdot\mathbf{v})\cdot\mathbf{n}\,dA$, noting that $(\boldsymbol{\tau}\cdot\mathbf{n})\cdot\mathbf{v} = (\boldsymbol{\tau}\cdot\mathbf{v})\cdot\mathbf{n}$ since $\boldsymbol{\tau}$ is a symmetrical tensor (Section 1.1.2.2). This can be integrated over A to obtain the rate of viscous work the fluid has to do:

$$\iint_A (\boldsymbol{\tau}\cdot\mathbf{v})\cdot\mathbf{n}\,dA \quad ; \quad W_v \qquad \text{term (5c)}$$

(integral) (overall)

Let s be the heat generation rate **per unit volume**, such as that due to Joule heating, phase transformations, or chemical reactions. The rate of heat generation in the differential volume element $d\Omega$ is $dS = s\,d\Omega$. This can be integrated over Ω to obtain the rate of heat generation in the control volume:

$$\iiint_\Omega s\,d\Omega \quad ; \quad S \qquad \text{term (6)}$$

(integral) (overall)

Substituting the integral form of terms (1) through (6) into Eq. [2.2-3]

$$\frac{\partial}{\partial t}\iiint_{\Omega} \rho e_t \, d\Omega = -\iint_A \rho e_t \mathbf{v}\cdot\mathbf{n}\, dA - \iint_A \mathbf{q}\cdot\mathbf{n}\, dA - \iint_A p\mathbf{v}\cdot\mathbf{n}\, dA$$

$$-\iint_A (\boldsymbol{\tau}\cdot\mathbf{v})\cdot\mathbf{n}\, dA + \iiint_{\Omega} s\, d\Omega - W_s \qquad [2.2\text{-}5]$$

In most problems, including those in materials processing, the kinetic and potential energies are negligible as compared to the thermal energy. Furthermore, the pressure, viscous and shaft work are usually negligible or even absent. As such, Eq. [2.2-5] reduces to

$$\boxed{\frac{\partial}{\partial t}\iiint_{\Omega} \rho C_v T \, d\Omega = -\iint_A \rho C_v T \mathbf{v}\cdot\mathbf{n}\, dA - \iint_A \mathbf{q}\cdot\mathbf{n}\, dA + \iiint_{\Omega} s\, d\Omega}$$

(integral energy-balance equation) [2.2-6]

Now substituting the overall form of terms (1) through (6) into Eq. [2.2-3] and neglecting the kinetic and potential energy and the pressure, viscous, and shaft work

$$\boxed{\frac{dE_T}{dt} = \left(\iint_A \rho C_v T v\, dA\right)_{\text{in}} - \left(\iint_A \rho C_v T v\, dA\right)_{\text{out}} + Q + S} \qquad [2.2\text{-}7]$$

where E_T is the thermal energy in the control volume, $\iiint_{\Omega} \rho C_v T\, d\Omega$. Substituting Eq. [2.1-11] into this equation and assuming constant C_v, we obtain:

$$\boxed{\begin{aligned}\frac{dE_T}{dt} &= (\rho C_v T_{\text{av}} v_{\text{av}} A)_{\text{in}} - (\rho C_v T_{\text{av}} v_{\text{av}} A)_{\text{out}} + Q + S \\ &= (m C_v T_{\text{av}})_{\text{in}} - (m C_v T_{\text{av}})_{\text{out}} + Q + S\end{aligned}}$$

(overall energy-balance equation) [2.2-8]

where E_T = thermal energy in the control volume ($= \rho C_v T\, \Omega = M C_v T$ if $\rho C_v T$ is uniform in Ω)

m = mass flow rate at inlet or outlet ($= \rho v_{\text{av}} A$)

Q = heat transfer rate **into** control volume from surroundings (other than the two $m C_v T_{\text{av}}$ terms), that is, by conduction, and

S = heat generation rate in the control volume ($= s\Omega$ if uniform s).

2.2.2. Bernoulli's Equation

Before leaving the overall energy-balance equation let us consider the Bernoulli equation. This equation has been derived previously in Section 1.9.1 from overall mechanical-energy balance. Here it will be derived from overall energy balance.

Let us consider the steady-state isothermal flow of an inviscid incompressible fluid without heat generation, heat conduction, shaft work, and viscous work. Substituting Eq. [2.2-4] into Eq. [2.2-5] and assuming uniform properties over the cross-sectional area A, we obtain

$$0 = -\rho \iint_A \left(e_t + \frac{p}{\rho}\right) \mathbf{v} \cdot \mathbf{n} \, dA$$

$$= \rho \left[\left(C_v T + \frac{1}{2}v^2 + \phi + \frac{p}{\rho}\right) vA\right]_1 - \rho \left[\left(C_v T + \frac{1}{2}v^2 + \phi + \frac{p}{\rho}\right) vA\right]_2 \tag{2.2-9}$$

Since $T_1 = T_2$ and $(\rho v A)_1 = (\rho v A)_2$, Eq. [2.2-9] reduces to

$$\frac{1}{2}v_1^2 + \phi_1 + \frac{p_1}{\rho} = \frac{1}{2}v_2^2 + \phi_2 + \frac{p_2}{\rho} \tag{2.2-10}$$

If the z direction is taken vertically upward, $\phi = gz$, where g is the gravitational acceleration. As such, Eq. [2.2-10], on multiplying by ρ, becomes

$$\tfrac{1}{2}\rho v_1^2 + \rho g z_1 + p_1 = \tfrac{1}{2}\rho v_2^2 + \rho g z_2 + p_2 \tag{2.2-11}$$

or simply

$$\tfrac{1}{2}\rho v^2 + \rho g z + p = \text{constant} \tag{2.2-12}$$

which is the Bernoulli equation given previously in Eq. [1.9-14].

2.2.3. Examples

Example 2.2-1 Fluid Temperature in a Mixing Tank

As illustrated in Fig. 2.2-3, a mixing tank receives fluid from two inlets and discharges it through one outlet. The mass of the fluid in the tank is M and its

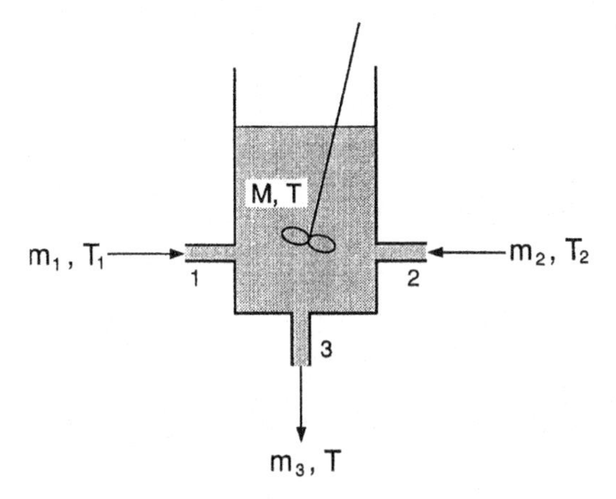

Fig. 2.2-3 Fluid temperature in a mixing tank.

temperature T is uniform as the result of stirring. Both the tank wall and the stirrer are very light in mass and a very good thermal insulator. As such, they can be ignored in the heat flow analysis. The initial fluid mass and temperature in the tank are M_0 and T_0, respectively. The mass flow rates and temperatures are respectively m_1 and T_1 for inlet 1, m_2 and T_2 for inlet 2, and m_3 and T for the outlet. Determine the fluid temperature in the tank as a function of time. The kinetic energy, potential energy, the pressure work of the fluid, and the work done on the fluid by the stirrer are negligible as compared to the thermal energy of the fluid. At the inlets and outlet, heat transfer due to conduction is negligible as compared to that due to convection.

Solution Since temperature is uniform, the thermal energy in the control volume

$$E_T = MC_vT \qquad [2.2\text{-}13]$$

From Eq. [2.2-8] with $Q = S = 0$

$$\frac{d}{dt}[MC_vT] = m_1C_vT_1 + m_2C_vT_2 - m_3C_vT \qquad [2.2\text{-}14]$$

From Eq. [1.2-12]

$$M = (m_1 + m_2 - m_3)t + M_0 = mt + M_0 \qquad [2.2\text{-}15]$$

where $m = m_1 + m_2 - m_3$.

Substituting Eq. [2.2-15] into Eq. [2.2-14] and dividing it by C_v, we obtain

$$\frac{d}{dt}[(mt + M_0)T] = m_1T_1 + m_2T_2 - m_3T \qquad [2.2\text{-}16]$$

or, since $d(xy) = x\,dy + y\,dx$,

$$(mt + M_0)\frac{dT}{dt} + mT = m_1T_1 + m_2T_2 - m_3T \qquad [2.2\text{-}17]$$

So

$$\frac{dT}{dt} = \frac{-(m + m_3)\{T - [(m_1T_1 + m_2T_2)/(m + m_3)]\}}{m[t + (M_0/m)]} \qquad [2.2\text{-}18]$$

The initial condition is

$$T = T_0 \quad \text{at} \quad t = 0 \qquad [2.2\text{-}19]$$

From Case B of Appendix A

$$\left(\frac{T - [(m_1T_1 + m_2T_2)/(m + m_3)]}{T_0 - [(m_1T_1 + m_2T_2)/(m + m_3)]}\right)^m = \left(\frac{t + (M_0/m)}{(M_0/m)}\right)^{-(m+m_3)} \qquad [2.2\text{-}20]$$

Example 2.2-2 Conduction through Cylindrical Composite Wall

The cylindrical composite wall shown in Fig. 2.2-4 is made of three different materials, A, B, and C, each having its own thermal conductivity – i.e., k_A, k_B, and k_C – respectively. The temperatures of the bulk fluids inside and outside the composite walls are T_a and T_b, respectively, and the heat transfer coefficients are h_1 and h_4, respectively. The **overall heat transfer coefficients** U_1 based on the inner surface and U_4 based on the outer surface, are defined as

$$Q_r = (2\pi r_1 L)U_1(T_a - T_b) = (2\pi r_4 L)U_4(T_a - T_b) \qquad [2.2\text{-}21]$$

where Q_r is the rate of heat flow through the composite wall, L is the length of the wall, and $(T_a - T_b)$ is the overall temperature difference. Determine at the steady state U_1 as a function of the thermal conductivities and heat transfer coefficients; to calculate Q_r from U_1 and $(T_a - T_b)$.

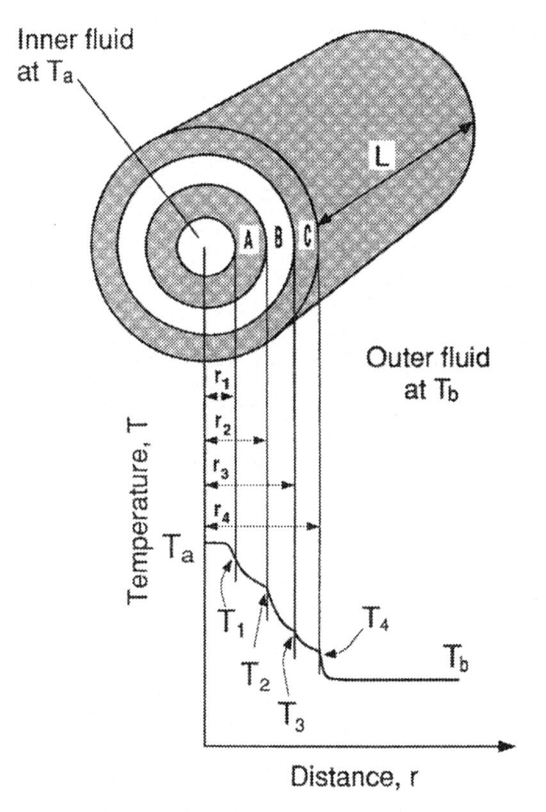

Fig. 2.2-4 Heat conduction through a composite cylindrical wall.

Solution Let us consider material A as the control volume Ω. Since there is no mass flow through nor heat generation within material A, at the steady state Eq. [2.2-8] reduces to

$$0 = Q = (2\pi r L q_r)|_{r_1^+} - (2\pi r L q_r)|_{r_2^-} \qquad [2.2\text{-}22]$$

and so

$$(2\pi r L q_r)|_{r_1^+} = (2\pi r L q_r)|_{r_2^-} = (2\pi r L q_r)|_A = Q_r \text{ (constant)} \qquad [2.2\text{-}23]$$

or

$$(r q_r)|_A = \frac{Q_r}{2\pi L} \text{ (constant)} \qquad [2.2\text{-}24]$$

Repeating for the other two materials, it can be shown that

$$[rq_r]|_A = [rq_r]|_B = [rq_r]|_C = rq_r = \frac{Q_r}{2\pi L} \qquad [2.2\text{-}25]$$

Therefore, from Fourier's law of conductivity (i.e., Eq. [D] of Table 2.1-2)

$$\left[-kr\frac{dT}{dr}\right]\bigg|_A = \left[-kr\frac{dT}{dr}\right]\bigg|_B = \left[-kr\frac{dT}{dr}\right]\bigg|_C = \frac{Q_r}{2\pi L} \qquad [2.2\text{-}26]$$

Integrating this equation over each individual material, we have

$$T_1 - T_2 = \frac{Q_r}{2\pi L k_A}\ln\left(\frac{r_2}{r_1}\right) \qquad [2.2\text{-}27]$$

$$T_2 - T_3 = \frac{Q_r}{2\pi L k_B}\ln\left(\frac{r_3}{r_2}\right) \qquad [2.2\text{-}28]$$

$$T_3 - T_4 = \frac{Q_r}{2\pi L k_C}\ln\left(\frac{r_4}{r_3}\right) \qquad [2.2\text{-}29]$$

At the two fluid/solid interfaces, from Newton's law of cooling Eq. [2.1-16]

$$T_a - T_1 = \frac{q_r|_{r_1^-}}{h_1} = \frac{Q_r}{2\pi L r_1 h_1} \qquad [2.2\text{-}30]$$

and

$$T_4 - T_b = \frac{q_r|_{r_4^+}}{h_4} = \frac{Q_r}{2\pi L r_4 h_4} \qquad [2.2\text{-}31]$$

Adding Eqs. [2.2-27] through [2.2-31]

$$T_a - T_b = \left[\frac{1}{r_1 h_1} + \frac{\ln(r_2/r_1)}{k_A} + \frac{\ln(r_3/r_2)}{k_B} + \frac{\ln(r_4/r_3)}{k_C} + \frac{1}{r_4 h_4}\right]\frac{Q_r}{2\pi L} \qquad [2.2\text{-}32]$$

Substituting Eq. [2.2-21] into Eq. [2.2-32]

$$\boxed{U_1 = r_1^{-1}\left[\frac{1}{r_1 h_1} + \frac{\ln(r_2/r_1)}{k_A} + \frac{\ln(r_3/r_2)}{k_B} + \frac{\ln(r_4/r_3)}{k_C} + \frac{1}{r_4 h_4}\right]^{-1}} \qquad [2.2\text{-}33]$$

From Eq. [2.2-21] U_4 can also be found since $U_4 = U_1 r_1/r_4$.

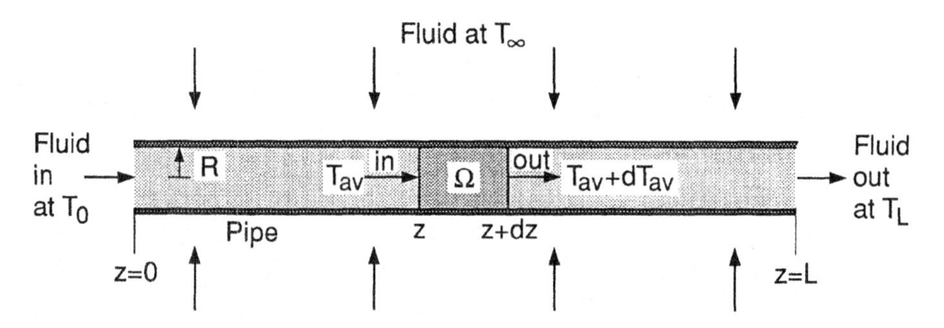

Fig. 2.2-5 Fluid flow through a pipe which is exposed to a fluid of temperature T_∞.

Example 2.2-3 Heat Transfer in Fluid Flow Through a Pipe

Consider a fluid of low thermal conductivity flowing through a pipe of length L and inner radius R. The outer surface of the pipe is exposed to a fluid of a higher temperature T_∞, as shown in Fig. 2.2-5, and the overall heat transfer coefficient based on the inner radius is U. The mass flow rate is m, and the average inlet temperature is T_0. Find the average outlet temperature T_L and the rate of heat exchange between the two fluids at the steady state.

Solution Let the volume element of length dz be the control volume Ω. From Eq. [2.2-8]

$$0 = \underset{\text{(in)}}{mC_vT_{\text{av}}} - \underset{\text{(out)}}{mC_v(T_{\text{av}} + dT_{\text{av}})} + \underset{\text{(wall)}}{Q} \qquad [2.2\text{-}34]$$

From U and the overall temperature difference $(T_\infty - T_{\text{av}})$

$$Q = (2\pi R\, dz)U(T_\infty - T_{\text{av}}) \qquad [2.2\text{-}35]$$

Substituting Eq. [2.2-35] into Eq. [2.2-34], we obtain

$$mC_v dT_{\text{av}} = 2\pi RU(T_\infty - T_{\text{av}})dz \qquad [2.2\text{-}36]$$

Since T_∞ is constant, $d(T_\infty - T_{\text{av}}) = -dT_{\text{av}}$ and Eq. [2.2-36] becomes

$$\frac{d(T_\infty - T_{\text{av}})}{(T_\infty - T_{\text{av}})} = d[\ln(T_\infty - T_{\text{av}})] = -\left(\frac{2\pi RU}{mC_v}\right)dz \qquad [2.2\text{-}37]$$

Integrating from $T_{av} = T_0$ at $z = 0$

$$\boxed{\ln\left(\frac{T_\infty - T_{av}}{T_\infty - T_0}\right) = -\frac{2\pi RUz}{mC_v}} \quad [2.2\text{-}38]$$

At the outlet $z = L$ and

$$\ln\left(\frac{T_\infty - T_L}{T_\infty - T_0}\right) = -\frac{2\pi RUL}{mC_v} \quad [2.2\text{-}39]$$

which is

$$T_\infty - T_L = (T_\infty - T_0)\exp\left(-\frac{2\pi RUL}{mC_v}\right) \quad [2.2\text{-}40]$$

or

$$T_L - T_0 = (T_\infty - T_0)\left[1 - \exp\left(-\frac{2\pi RUL}{mC_v}\right)\right] \quad [2.2\text{-}41]$$

Now applying Eq. [2.2-8] to the entire length of the pipe as the control volume Ω:

$$\underset{\text{(in)}}{0 = mC_vT_0} - \underset{\text{(out)}}{mC_vT_L} + \underset{\text{(wall)}}{Q} \quad [2.2\text{-}42]$$

As such, the rate of heat transfer into the pipe through the pipe wall is

$$Q = mC_v(T_L - T_0) \quad [2.2\text{-}43]$$

Substituting Eq. [2.2-41] into Eq. [2.2-43], we obtain

$$\boxed{Q = mC_v(T_\infty - T_0)\left[1 - \exp\left(-\frac{2\pi RUL}{mC_v}\right)\right]} \quad [2.2\text{-}44]$$

Example 2.2-4 Counterflow Heat Exchanger

Consider the simple double-pipe heat exchanger of length L shown in Fig. 2.2-6. The hot stream going through the inner pipe is cooled by the cold stream going through the outer pipe. The hot stream has an inlet temperature T_{h1}, an outlet temperature T_{h2}, and a mass flow rate of m_h. The cold stream has an inlet temperature T_{c2}, an outlet temperature T_{c1}, and a mass flow rate m_c. The inner pipe has a thin metal wall of radius R, and its thermal resistance can be ignored.

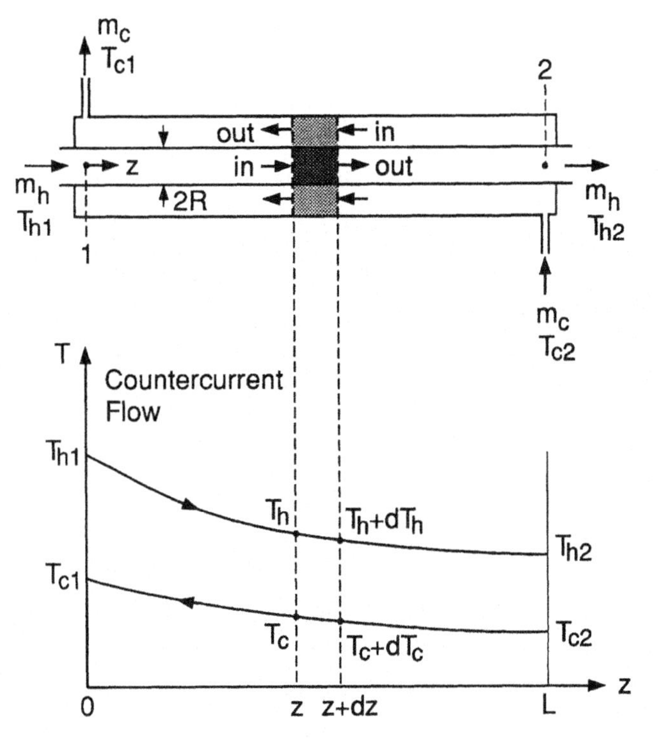

Fig. 2.2-6 Double-pipe counterflow heat exchanger.

Fluid flow is turbulent in both pipes and the overall heat transfer coefficient U can be assumed uniform in the z direction. The heat losses from the outer pipe to the surroundings can be ignored. The properties of the streams can be assumed uniform in the radial direction as an approximation. Find the steady-state heat exchange rate Q_e as a function of the inlet and outlet temperatures and U.

Solution Heat conduction is negligible except that through the inner pipe wall. From the overall energy-balance equation, Eq. [2.2-8], for the hot stream in the entire length of the inner pipe

$$0 = m_h C_v T_{h1} - m_h C_v T_{h2} + (-Q_e) \qquad [2.2\text{-}45]$$

and for the cold stream in the entire length of the outer pipe

$$0 = m_c C_v T_{c2} - m_c C_v T_{c1} + Q_e \qquad [2.2\text{-}46]$$

Therefore,

$$\frac{1}{m_h C_v} = -\frac{T_{h2} - T_{h1}}{Q_e} \qquad [2.2\text{-}47]$$

and

$$\frac{1}{m_c C_v} = -\frac{T_{c2} - T_{c1}}{Q_e} \qquad [2.2\text{-}48]$$

Let us now consider the volume element in the inner pipe. From Eq. [2.2-8]

$$\underset{\text{(in)}}{0 = m_h C_v T_h} - \underset{\text{(out)}}{m_h C_v (T_h + dT_h)} + \underset{\text{(wall)}}{[-(U 2\pi R\, dz)(T_h - T_c)]} \qquad [2.2\text{-}49]$$

and so

$$\frac{dT_h}{T_h - T_c} = -\frac{2\pi R U}{m_h C_v} dz \qquad [2.2\text{-}50]$$

Similarly, for the volume element in the outer pipe

$$0 = \underset{\text{(in)}}{m_c C_v (T_c + dT_c)} - \underset{\text{(out)}}{m_c C_v T_c} + \underset{\text{(wall)}}{(U 2\pi R\, dz)(T_h - T_c)} \qquad [2.2\text{-}51]$$

and so

$$\frac{dT_c}{T_h - T_c} = -\frac{2\pi R U}{m_c C_v} dz \qquad [2.2\text{-}52]$$

Subtracting Eq. [2.2-52] from Eq. [2.2-50]

$$\frac{d(T_h - T_c)}{T_h - T_c} = -2\pi R U \left(\frac{1}{m_h C_v} - \frac{1}{m_c C_v}\right) dz \qquad [2.2\text{-}53]$$

Substituting Eqs. [2.2-47] and [2.2-48] and integrating from $z = 0$ to L

$$\int_{T_{h1} - T_{c1}}^{T_{h2} - T_{c2}} d\ln(T_h - T_c) = \frac{2\pi R U}{Q_e}(T_{h2} - T_{h1} - T_{c2} + T_{c1}) \int_0^L dz \qquad [2.2\text{-}54]$$

On rearranging

$$\boxed{Q_e = (2\pi R L) U \left[\frac{(T_{h2} - T_{c2}) - (T_{h1} - T_{c1})}{\ln[(T_{h2} - T_{c2})/(T_{h1} - T_{c1})]}\right]} \qquad [2.2\text{-}55]$$

The terms in parentheses can be considered as the logarithmic mean temperature difference between the two streams.

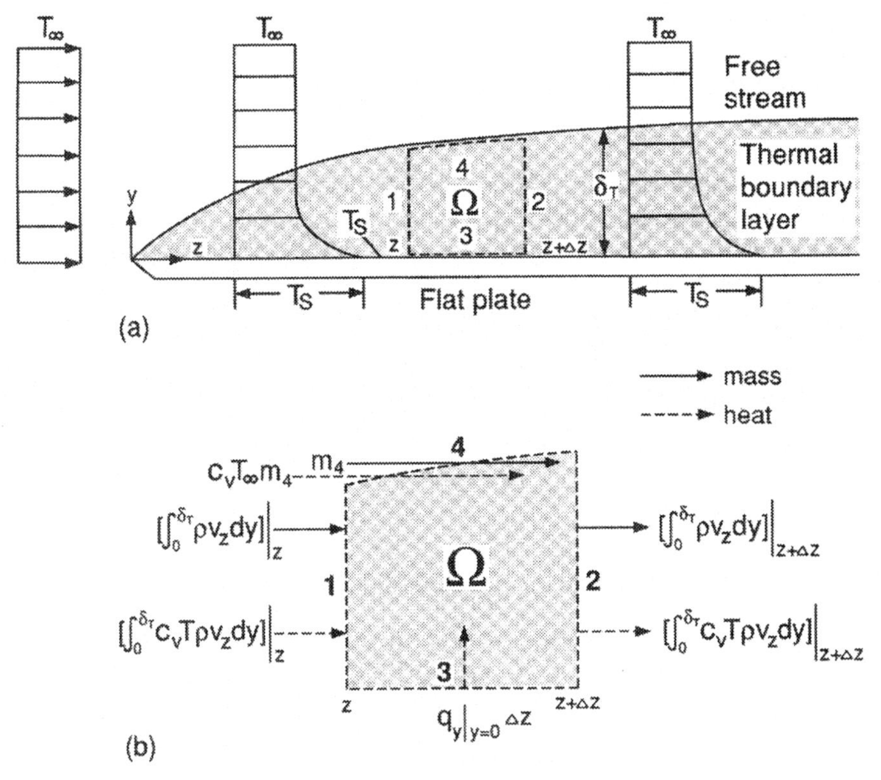

Fig. 2.2-7 Heat transfer in fluid flow over a heated flat plate: (*a*) thermal boundary layer; (*b*) volume element in the layer.

Example 2.2-5 Heat Transfer in Laminar Flow over a Flat Plate

In Example 1.4-6 we studied the momentum boundary layer over a flat plate using an integral approach. Here we shall use a similar approach to consider the thermal boundary layer over a flat plate. As illustrated in Fig. 2.2-7*a*, the temperatures of the fluid in the bulk stream and at the plate surface are T_∞ and T_s, respectively. The plate is wide in the x direction such that there are no significant velocity or temperature variations in this direction. The concept of the thermal boundary layer has been described in Section 2.1.3. We shall determine the thickness of the thermal boundary layer δ_T and the heat transfer coefficient as a function of the distance from the leading edge of the plate z. Assume steady state, constant physical properties, and no heat generation.

Solution The control volume Ω is enlarged in Fig. 2.2-7*b*. Similar to Eq. [1.4-52], the mass flow rate across the top surface per unit width of the plate is

$$m_4 = \frac{d}{dz}\left[\int_0^{\delta_T} \rho v_z \, dy\right]\Delta z \qquad [2.2\text{-}56]$$

The fluid entering the control volume through the top surface carries an energy $m_4 C_v T_\infty$ per unit width of the plate.

Since heat transfer in the z direction is mainly by convection, heat conduction in this direction can be neglected. At the top surface, $\partial T/\partial y = 0$ and hence $q_y = 0$. At the steady state the energy balance according to Eq. [2.2-7] reduces to

$$0 = \underset{(1)}{\left[\int_0^{\delta_T} C_v T \rho v_z \, dy\right]\Bigg|_z} - \underset{(2)}{\left[\int_0^{\delta_T} C_v T \rho v_z \, dy\right]\Bigg|_{z+\Delta z}} + \underset{(3)}{q_y|_{y=0}\,\Delta z} + \underset{(4)}{m_4 C_v T_\infty} \qquad [2.2\text{-}57]$$

Substituting Eq. [2.2-56] into this equation, dividing it by Δz and setting $\Delta z \to 0$

$$-q_y|_{y=0} = -\frac{d}{dz}\left[\int_0^{\delta_T} C_v T \rho v_z \, dy\right] + \frac{d}{dz}\left[\int_0^{\delta_T} C_v T_\infty \rho v_z \, dy\right] \qquad [2.2\text{-}58]$$

From Fourier's law of conduction

$$q_y|_{y=0} = -k\left.\frac{\partial T}{\partial y}\right|_{y=0} \qquad [2.2\text{-}59]$$

Substituting this equation into Eq. [2.2-58] and dividing it by $C_v \rho$

$$\alpha\left.\frac{\partial T}{\partial y}\right|_{y=0} = \frac{d}{dz}\int_0^{\delta_T} v_z (T_\infty - T)\, dy \qquad [2.2\text{-}60]$$

where the thermal diffusivity $\alpha = k/\rho C_v$. Since this equation is difficult to solve directly, the following approximate temperature profile is assumed:

$$\boxed{\frac{T - T_s}{T_\infty - T_s} = \frac{3}{2}\left(\frac{y}{\delta_T}\right) - \frac{1}{2}\left(\frac{y}{\delta_T}\right)^3} \qquad [2.2\text{-}61]$$

This temperature profile is reasonable since it satisfies the following boundary conditions:

$$T = T_s \quad \text{at} \quad y = 0 \qquad [2.2\text{-}62]$$

and

$$T = T_\infty \quad \text{and} \quad \frac{\partial T}{\partial y} = 0 \quad \text{at} \quad y = \delta_T \qquad [2.2\text{-}63]$$

Equation [2.2-61] is analogous to Eq. [1.4-57] for the momentum boundary layer:

$$\boxed{\frac{v_z}{v_\infty} = \frac{3}{2}\left(\frac{y}{\delta}\right) - \frac{1}{2}\left(\frac{y}{\delta}\right)^3} \qquad [2.2\text{-}64]$$

where v_∞ is the free-stream velocity of the fluid.

Equations [2.2-61] and [2.2-64] can be substituted into Eq. [2.2-60] to find the relationship between δ_T and δ. The integration in Eq. [2.2-60] yields a $(\delta_T/\delta)^2$ term and a $(\delta_T/\delta)^4$ term about 14 times smaller in coefficient. As an approximation, the latter can be neglected, if we assume that $\delta_T \leq \delta$. With further help from Eq. [1.4-61], the following equation is obtained:

$$\boxed{\frac{\delta_T}{\delta} = \mathrm{Pr}^{-(1/3)}} \qquad [2.2\text{-}65]$$

where the Prandtl number

$$\mathrm{Pr} = \frac{\nu}{\alpha} = \frac{C_v \mu}{k} \qquad [2.2\text{-}66]$$

From Eqn. [2.1-14], the heat transfer coefficient

$$h = \frac{-k(\partial T/\partial y)|_{y=0}}{(T_s - T_\infty)} \qquad [2.2\text{-}67]$$

sine $\partial T/\partial y < 0$ and $(T_s - T_\infty) > 0$. Substituting Eq. [2.2-61] into Eq. [2.2-67]

$$h = \frac{3}{2}\frac{k}{\delta_T} \qquad [2.2\text{-}68]$$

From Eq. [1.4-62]

$$\frac{\delta}{z} = \frac{4.64}{\sqrt{\mathrm{Re}_z}} \qquad [2.2\text{-}69]$$

where the local Reynolds number

$$\mathrm{Re}_z = \frac{z v_\infty}{\nu} = \frac{z \rho v_\infty}{\mu} \qquad [2.2\text{-}70]$$

From Eqs. [2.2-65], [2.2-68], and [2.2-69]

$$\boxed{\frac{hz}{k} = \mathrm{Nu}_z = 0.323\,\mathrm{Pr}^{1/3}\,\mathrm{Re}_z^{1/2}} \qquad [2.2\text{-}71]$$

where Nu_z is the local Nusselt number. This equation compares favorably with the exact solution,[2] which has a proportional constant of 0.332 instead of 0.323. Like the exact solution, Eqs. [2.2-65] and [2.2-71] apply in the range of $0.6 \leq \mathrm{Pr} \leq 50$. For liquid metals and semiconductors Pr is significantly less than 0.6 and the equations cannot be applied. With these materials of high thermal conductivity the thermal boundary layer is thick and $\delta_T > \delta$.

2.3. DIFFERENTIAL ENERGY-BALANCE EQUATION

2.3.1. Derivation

In materials processing the kinetic and potential energy are negligible as compared to the thermal energy, that is, the total energy per unit mass $e_t = C_v T$. Furthermore, the pressure, viscosity, and shaft work are usually negligible. The surface integrals in Eq. [2.2-6] can be converted into volume integrals. From Gauss' divergence theorem (i.e., Eq. [A.4-1])

$$\iint_A \rho C_v T \mathbf{v} \cdot \mathbf{n}\, dA = \iiint_\Omega \nabla \cdot (\rho C_v T \mathbf{v}) d\Omega \qquad [2.3\text{-}1]$$

$$\iint_A \mathbf{q} \cdot \mathbf{n}\, dA = \iiint_\Omega \nabla \cdot \mathbf{q}\, d\Omega \qquad [2.3\text{-}2]$$

Substituting Eqs. [2.3-1] and [2.3-2] into Eq. [2.2-6]

$$\frac{\partial}{\partial t} \iiint_\Omega \rho C_v T\, d\Omega + \iiint_\Omega \nabla \cdot (\rho C_v T \mathbf{v}) d\Omega + \iiint_\Omega \nabla \cdot \mathbf{q}\, d\Omega - \iiint_\Omega s\, d\Omega = 0$$

$$[2.3\text{-}3]$$

If the control volume Ω does not change with time, $\partial/\partial t$ in Eq. [2.3-3] can be moved inside the integration sign:

$$\iiint_\Omega \left\{ \frac{\partial}{\partial t} (\rho C_v T) + \nabla \cdot (C_v T \rho \mathbf{v}) + \nabla \cdot \mathbf{q} - s \right\} d\Omega = 0 \qquad [2.3\text{-}4]$$

[2] J. R. Welty, C. E. Wicks, and R. E. Wilson, *Fundamentals of Momentum, Heat and Mass Transfer*, 3rd ed., Wiley, New York, 1984, p. 595.

The integrand, which is continuous, must be zero everywhere since the equation must hold for any arbitrary region Ω:[3]

$$\frac{\partial}{\partial t}(\rho C_v T) + \nabla \cdot (C_v T \rho \mathbf{v}) + \nabla \cdot \mathbf{q} - s = 0 \qquad [2.3\text{-}5]$$

The first two terms in the equation can be expanded; the second term with the help of Eq. [A.3-7] as follows:

$$\rho \frac{\partial}{\partial t}(C_v T) + (C_v T)\frac{\partial \rho}{\partial t} + (C_v T)\nabla \cdot (\rho \mathbf{v}) + \rho \mathbf{v} \cdot \nabla (C_v T)$$

$$= \rho \frac{\partial}{\partial t}(C_v T) + \rho \mathbf{v} \cdot \nabla (C_v T) \qquad [2.3\text{-}6]$$

Note that the equation of continuity, Eq. [1.3-4], has been used.

Substituting Eqs. [2.1-5] and [2.3-6] into Eq. [2.3-5]

$$\boxed{\rho \frac{\partial}{\partial t}(C_v T) + \rho \mathbf{v} \cdot \nabla (C_v T) = \nabla \cdot (k \nabla T) + s} \qquad [2.3\text{-}7]$$

Assuming constant C_v and k

$$\boxed{\rho C_v \left[\frac{\partial T}{\partial t} + \mathbf{v} \cdot \nabla T \right] = k \nabla^2 T + s \qquad \text{(constant } C_v\text{, and } k\text{)}} \qquad [2.3\text{-}8]$$

Equation [2.3-8] is the **differential energy-balance equation**, or the **equation of energy**. The viscous dissipation is neglected. It is significant when the velocity gradients and viscosity are very high as in lubrication with engine oil. It is, however, negligible in most other applications.

In Table 2.3-1, Eq. [2.3-8] is given in expanded forms for rectangular, cylindrical, and spherical coordinates.

2.3.2. Dimensionless Form

Like the equation of motion, the equation of energy can be presented in the dimensionless form to make the solutions more general.

For forced convection the following dimensionless variables can be defined:

$$T^* = \frac{T - T_0}{T_1 - T_0} \qquad \text{(dimensionless temperature)} \qquad [2.3\text{-}9]$$

[3] W. Kaplan, *Advanced Calculus*, 2nd ed., Addison-Wesley, Reading, MA, 1973, p. 363

TABLE 2.3-1 Equation of Energy for Constant C_v and k

Rectangular coordinates

$$\rho C_v\left[\frac{\partial T}{\partial t}+v_x\frac{\partial T}{\partial x}+v_y\frac{\partial T}{\partial y}+v_z\frac{\partial T}{\partial z}\right]=k\left[\frac{\partial^2 T}{\partial x^2}+\frac{\partial^2 T}{\partial y^2}+\frac{\partial^2 T}{\partial z^2}\right]+s \quad \text{[A]}$$

Cylindrical coordinates

$$\rho C_v\left[\frac{\partial T}{\partial t}+v_r\frac{\partial T}{\partial r}+\frac{v_\theta}{r}\frac{\partial T}{\partial \theta}+v_z\frac{\partial T}{\partial z}\right]=k\left[\frac{1}{r}\frac{\partial}{\partial r}\left(r\frac{\partial T}{\partial r}\right)+\frac{1}{r^2}\frac{\partial^2 T}{\partial \theta^2}+\frac{\partial^2 T}{\partial z^2}\right]+s \quad \text{[B]}$$

Spherical coordinates

$$\rho C_v\left[\frac{\partial T}{\partial t}+v_r\frac{\partial T}{\partial r}+\frac{v_\theta}{r}\frac{\partial T}{\partial \theta}+\frac{v_\phi}{r\sin\theta}\frac{\partial T}{\partial \phi}\right]$$

$$=k\left[\frac{1}{r^2}\frac{\partial}{\partial r}\left(r^2\frac{\partial T}{\partial r}\right)+\frac{1}{r^2\sin\theta}\frac{\partial}{\partial \theta}\left(\sin\theta\frac{\partial T}{\partial \theta}\right)+\frac{1}{r^2\sin^2\theta}\frac{\partial^2 T}{\partial \phi^2}\right]+s \quad \text{[C]}$$

$$t^*=\frac{tV}{L} \quad \text{(dimensionless time)} \quad [2.3\text{-}10]$$

$$\mathbf{v}^*=\frac{\mathbf{v}}{V} \quad \text{(dimensionless velocity)} \quad [2.3\text{-}11]$$

$$p^*=\frac{p-p_0}{\rho V^2} \quad \text{(dimensionless pressure)} \quad [2.3\text{-}12]$$

$$x^*, y^*, z^*=\frac{x}{L},\frac{y}{L},\frac{z}{L} \quad \text{(dimensionless coordinates)} \quad [2.3\text{-}13]$$

$$\nabla^*, \nabla^{*2}=L\nabla, L^2\nabla^2 \quad \text{(dimensionless operators)} \quad [2.3\text{-}14]$$

where (T_1-T_0), V, and L are any characteristic temperature difference, velocity, and length, respectively. Equations [2.3-10] through [2.3-14] are identical to Eqs. [1.5-12] through [1.5-17] for forced convection in an isothermal system.

In the absence of a heat source, the energy equation Eq. [2.3-8] reduces to

$$\frac{\partial T}{\partial t}+\mathbf{v}\cdot\nabla T=\alpha\nabla^2 T \quad [2.3\text{-}15]$$

Substituting Eqs. [2.3-9] through [2.3-14] into Eq. [2.3-15]

$$\frac{V}{L}\frac{\partial}{\partial t^*}[T^*(T_1 - T_0)] + V\mathbf{v}^* \cdot \nabla^* T^* \frac{1}{L}(T_1 - T_0) = \alpha \frac{1}{L^2} \nabla^{*2} T^*(T_1 - T_0)$$

[2.3-16]

Multiplying Eq. [2.3-16] by $L/[V(T_1 - T_0)]$

$$\frac{\partial T^*}{\partial t^*} + \mathbf{v}^* \cdot \nabla^* T^* = \frac{\alpha}{LV} \nabla^{*2} T^* \qquad [2.3\text{-}17]$$

By combining Eq. [2.3-17] with Eq. [1.5-25] through [1.5-28], the following equations can be obtained for heat transfer in forced convection:

Continuity: $\nabla^* \cdot \mathbf{v}^* = 0$ [2.3-18]

Motion: $$\frac{\partial \mathbf{v}^*}{\partial t^*} + \mathbf{v}^* \cdot \nabla^* \mathbf{v}^* = -\nabla^* p^* + \frac{1}{\text{Re}} \nabla^{*2} \mathbf{v}^* + \frac{1}{\text{Fr}} \mathbf{e}_g \qquad [2.3\text{-}19]$$

Energy: $$\frac{\partial T^*}{\partial t^*} + \mathbf{v}^* \cdot \nabla^* T^* = \frac{1}{\text{Re}\,\text{Pr}} \nabla^{*2} T^* = \frac{1}{\text{Pe}_T} \nabla^{*2} T^* \qquad [2.3\text{-}20]$$

where

$$\text{Re} = \frac{LV}{\nu} \quad \left(\text{Reynold number} = \frac{\text{inertia force } \rho V^2/L}{\text{viscous force } \mu V/L^2}\right) \qquad [2.3\text{-}21]$$

$$\text{Fr} = \frac{V^2}{gL} \quad \left(\text{Froude number} = \frac{\text{inertia force } \rho V^2/L}{\text{gravity force } \rho g}\right) \qquad [2.3\text{-}22]$$

$$\text{Pr} = \frac{\nu}{\alpha} \quad \left(\text{Prandtl number} = \frac{\text{viscous diffusivity } \nu}{\text{thermal diffusivity } \alpha}\right) \qquad [2.3\text{-}23]$$

$$\text{Pe}_T = \text{Re}\,\text{Pr} = \frac{LV}{\alpha}$$

$$= \left(\text{thermal Peclet number} = \frac{\text{convection heat transport } \rho C_v V(T_1 - T_0)}{\text{conduction heat transport } k(T_1 - T_0)/L}\right)$$

[2.3-24]

The physical meanings of Re and Fr have been shown in Eqs. [1.5-27] and [1.5-28], respectively.

2.3.3. Boundary Conditions

Boundary conditions commonly encountered in heat transfer are summarized in Figs. 2.3-1 and 2.3-2. These boundary conditions are explained as follows:

Commonly Encountered Heat Flow Boundary Conditions in Rectangular Coordinates

1. Plane of symmetry

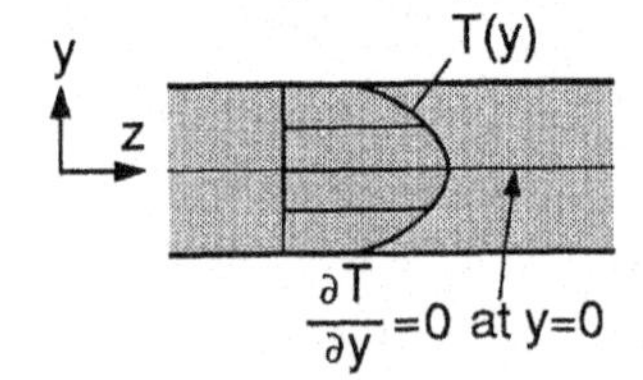

$\frac{\partial T}{\partial y} = 0$ at y=0

2. Constant surface temperature

$T = T_S$ at y=0

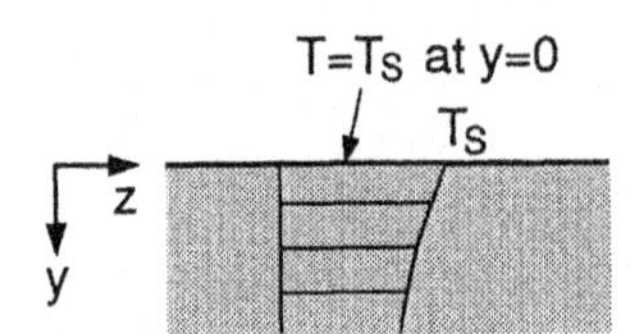

3. Adiabatic or insulated surface

$\frac{\partial T}{\partial y} = 0$ at y=0

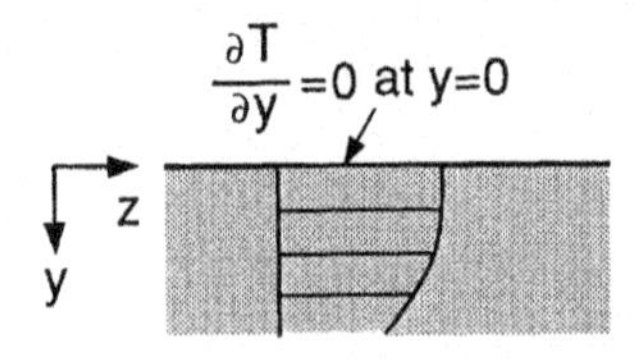

4. Constant surface heat flux

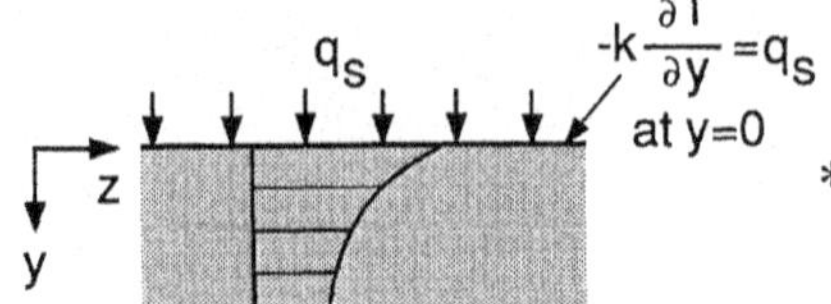

5. Convection exchange

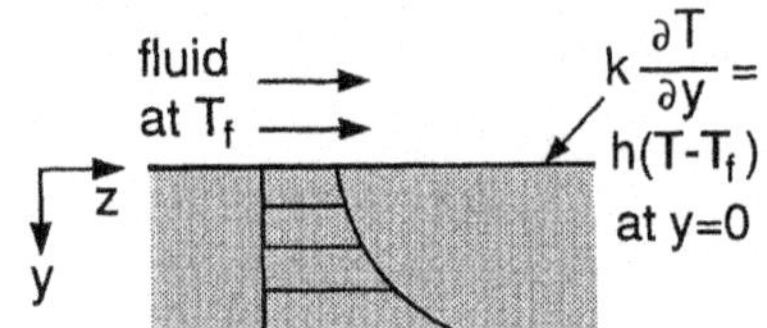

6.* Interface I/II: S/S,L/S,G/S,L/L,G/L

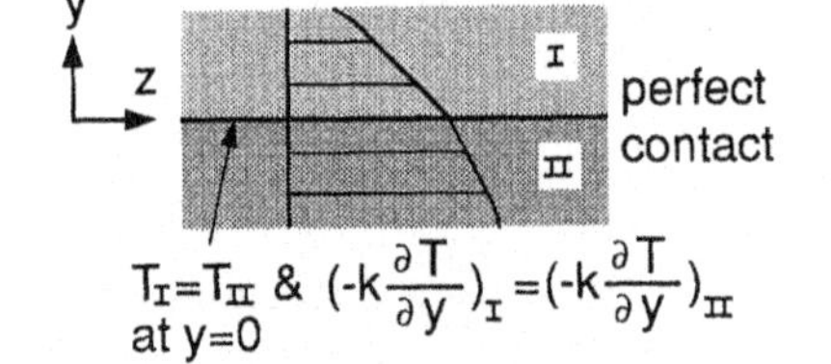

$T_I = T_{II}$ & $(-k\frac{\partial T}{\partial y})_I = (-k\frac{\partial T}{\partial y})_{II}$ at y=0

7.* Solid/solid contact

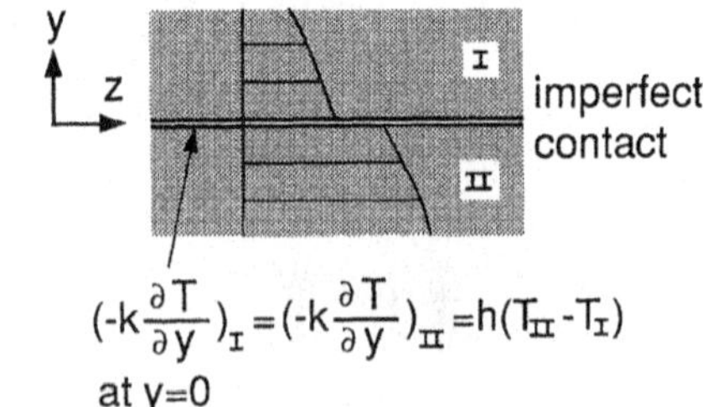

$(-k\frac{\partial T}{\partial y})_I = (-k\frac{\partial T}{\partial y})_{II} = h(T_{II} - T_I)$ at y=0

* When temperature distribution is to be found in both phases.

Fig. 2.3-1 Heat flow boundary conditions: rectangular coordinates.

Commonly Encountered Heat Flow Boundary Conditions in Cylindrical Coordinates

1. Axis of symmetry

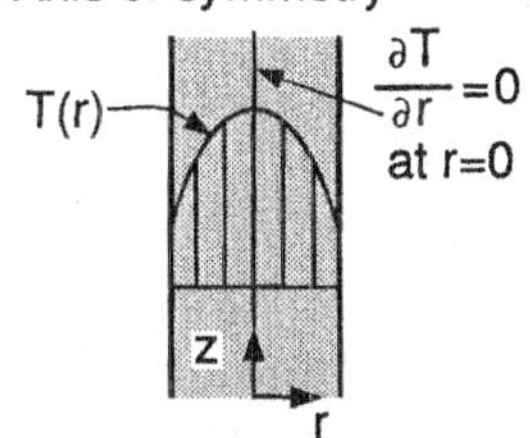

2. Constant surface temperature

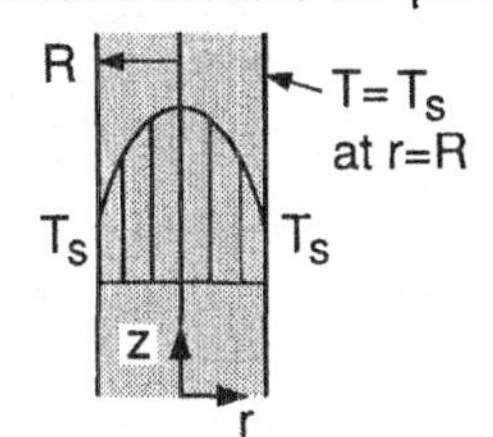

3. Adiabatic or insulated surface

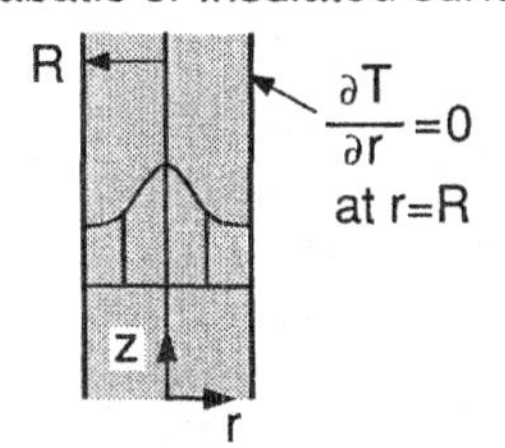

4. Constant surface heat flux

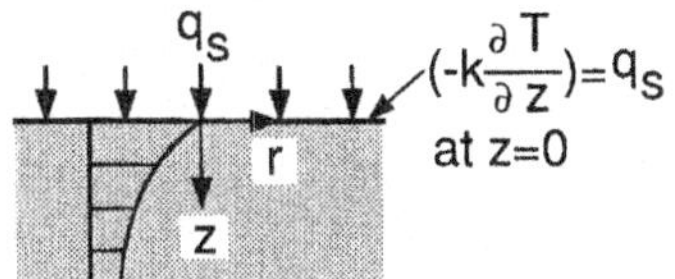

5. Convection exchange

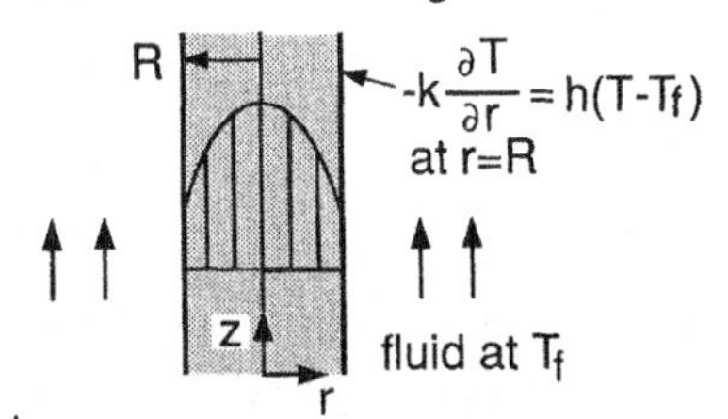

6.* Interface I/II : S/S, L/S, G/S

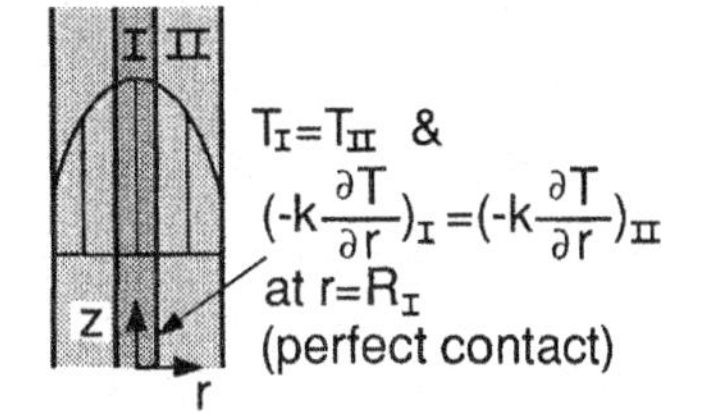

7.* Solid/solid contact

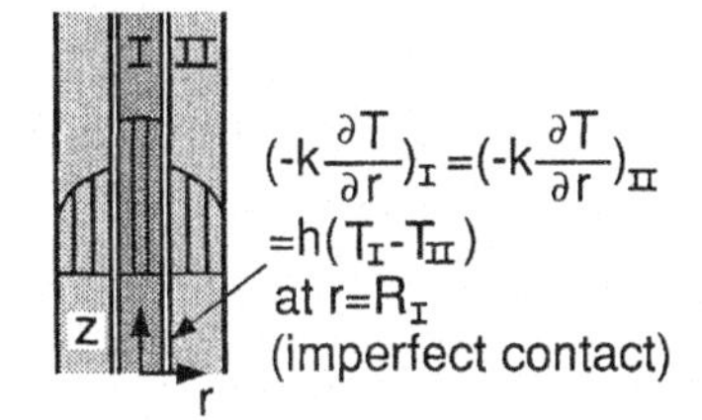

* When temperature distribution is to be found in both phases.

Fig. 2.3-2 Heat flow boundary conditions: cylindrical coordinates.

1. At the plane or axis of symmetry the temperature gradient in the transverse or radial direction is zero, as in case 1 of Figs. 2.3-1 and 2.3-2.
2. A wall in contact with a fluid or the surface of a material may be kept at a constant temperature, as in case 2 of Figs. 2.3-1 and 2.3-2.
3. The surface of a solid or liquid may be adiabatic or insulated, as in case 3 of Figs. 2.3-1 and 2.3-2. The heat flux and hence the temperature gradient are zero.

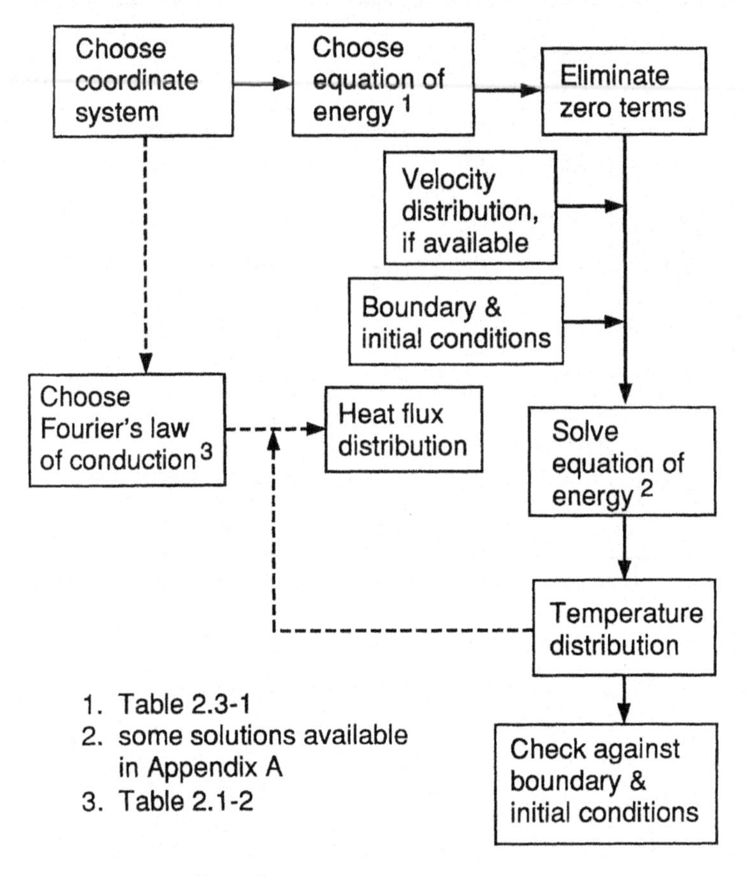

Fig. 2.3-3 Procedure for solving heat transfer problems by the equation of energy.

4. The surface of a solid or liquid may be exposed to a heat flux, as in case 4 of Figs. 2.3-1 and 2.3-2.
5. The free surface of a fluid may be exposed to a gas of bulk temperature T_f or the surface of a solid may be exposed to a gas or liquid of bulk temperature T_f. The boundary condition, which is consistent with Eq. [2.1-14], is

$$|q| = \left| -k\frac{\partial T}{\partial y} \right| = |h(T - T_f)| \qquad [2.3\text{-}25]$$

for rectangular coordinates, as in case 5 of Fig. 2.3-1. For cylindrical coordinates, as in case 5 of Fig. 2.3-2, y in Eq. [2.3-25] is replaced by r.

6. For two phases in perfect contact with each other, the temperature and the heat flux are both continuous across the interface; in other words they are the same on both sides of the interface, as in case 6 of Figs. 2.3-1 and 2.3-2.
7. If there is a small gap between two solids in contact with each other, the heat flux across the gap can be expressed as $|q| = |h(T_{\mathrm{I}} - T_{\mathrm{II}})|$, where h is a heat transfer coefficient, T_{I} the temperature of solid I at the gap, and T_{II} the temperature of solid II at the gap. Examples are case 7 of Figs. 2.3-1 and 2.3-2. Note that items 6 and 7 are needed when the temperature fields in **both** I and II are being determined.
8. The boundary conditions at solid/liquid interfaces with phase transformations will be described in Chapter 5.

The boundary conditions for heat transfer in spheres are identical to those for heat transfer in cylinders (case 4 excluded).

2.3.4. Solution Procedure

The purpose of the equation of energy, as will be illustrated in the following examples, is to determine the temperature distribution. The step-by-step procedure for solving the equation of energy, which is illustrated in Fig. 2.3-3 by the route represented by solid arrows, is as follows:

1. Choose a coordinate system that best describes the physical system geometrically.
2. Choose the equation of energy from Table 2.3-1 for the coordinate system.
3. Eliminate the zero terms from the equation of energy.
4. Substitute the velocity distribution, if it is available and temperature-independent, into the equation of energy.
5. Set up the boundary and/or initial conditions.
6. Solve the equation of energy, subject to the boundary and/or initial conditions in step 5, for the temperature distribution. For convenience, some solutions are provided in Appendix A.
7. Check to see if the temperature distribution satisfies the boundary and/or initial conditions in step 5.

If the velocity and temperature distributions depend on each other, that is, if they are coupled, the equations of continuity and motion should also be considered. In other words, the equation of continuity, motion, and energy need to be solved simultaneously.

After the temperature distribution is determined, it can be substituted into Fourier's law of conduction to determine the heat flux distribution due to conduction. This is illustrated in Fig. 2.3-1 by the route represented by broken arrows.

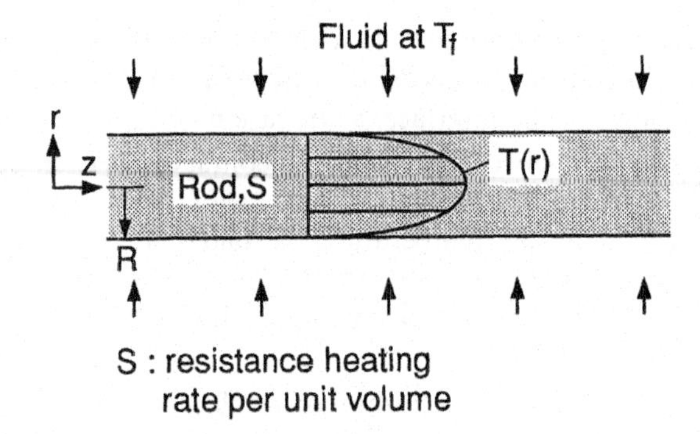

Fig. 2.3-4 Heat conduction in a resistance heated rod.

2.3.5. Examples

Example 2.3.1 Heat Conduction in a Resistance Heated Rod

A long rod of radius R and electrical conductivity k_e is resistance heated by passing an electric current I through it. The rate of heat generation per unit volume, s, is given by the expression

$$s = \frac{1}{k_e}\left[\frac{I}{\pi R^2}\right]^2 \tag{2.3-26}$$

The rod is cooled by directing a fluid of temperature T_f at its surface, as shown in Fig. 2.3-4. The heat transfer coefficient is h. Determine the steady-state temperature distribution in the rod and the heat flux across the wire surface. Assume constant thermal properties. Neglect radiation and end effects.

Solution For cylindrical coordinates the equation of energy, Eq. [B] of Table 2.3-1, becomes

$$\rho C_v\left[\frac{\partial T}{\partial t} + v_r\frac{\partial T}{\partial r} + \frac{v_\theta}{r}\frac{\partial T}{\partial \theta} + v_z\frac{\partial T}{\partial z}\right] = k\left[\frac{1}{r}\frac{\partial}{\partial r}\left(r\frac{\partial T}{\partial r}\right) + \frac{1}{r^2}\frac{\partial^2 T}{\partial \theta^2} + \frac{\partial^2 T}{\partial z^2}\right] + s \tag{2.3-27}$$

In this equation $\partial T/\partial t = 0$ in view of the steady-state condition. In the absence of convection in the rod $v_r = v_z = v_\theta = 0$. $\partial T/\partial\theta = 0$, due to axisymmetry of the temperature field. Also $\partial T/\partial z = 0$ in view of negligible end effects. With these conditions and the fact that T is a function of r only, Eq. [2.3-27] is reduced to

$$\frac{1}{r}\frac{d}{dr}\left[r\frac{dT}{dr}\right] = -\frac{s}{k} \tag{2.3-28}$$

The boundary conditions are

$$\frac{dT}{dr} = 0 \qquad \text{at} \quad r = 0 \tag{2.3-29}$$

$$-k\frac{dT}{dr} = h(T - T_f) \quad \text{at} \quad r = R \tag{2.3-30}$$

Equation [2.3-29] is due to axisymmetry of the temperature field.

The solution from Case K of Appendix A, is as follows:

$$\boxed{T = \frac{s}{4k}(R^2 - r^2) + \frac{sR}{2h} + T_f} \tag{2.3-31}$$

From Fourier's law of conduction

$$q_r = -k\frac{dT}{dr} \tag{2.3-32}$$

Substituting Eq. [2.3-31] into Eq. [2.3-32]

$$\boxed{q_r = \frac{sr}{2}} \tag{2.3-33}$$

and so

$$q_r|_{r=R} = \frac{sR}{2} \tag{2.3-34}$$

Example 2.3-2 Heat Conduction in a Cooling Fin

Consider the cooling fin of the hot wall shown in Fig. 2.3-5. Since the fin is thin, temperature variations in the thickness direction are negligible. For the same reason, convective heat losses to the air from the edges of the fin are negligible. The wall temperature is T_w, the ambient temperature T_a, and the heat transfer coefficient h. Find the steady-state temperature distribution along the fin and the heat loss to the air from the fin.

Solution The equation of energy, Eq. [A] of Table 2.3-1, is as follows:

$$\rho C_v\left[\frac{\partial T}{\partial t} + v_x\frac{\partial T}{\partial x} + v_y\frac{\partial T}{\partial y} + v_z\frac{\partial T}{\partial z}\right] = k\left[\frac{\partial^2 T}{\partial x^2} + \frac{\partial^2 T}{\partial y^2} + \frac{\partial^2 T}{\partial z^2}\right] + s \tag{2.3-35}$$

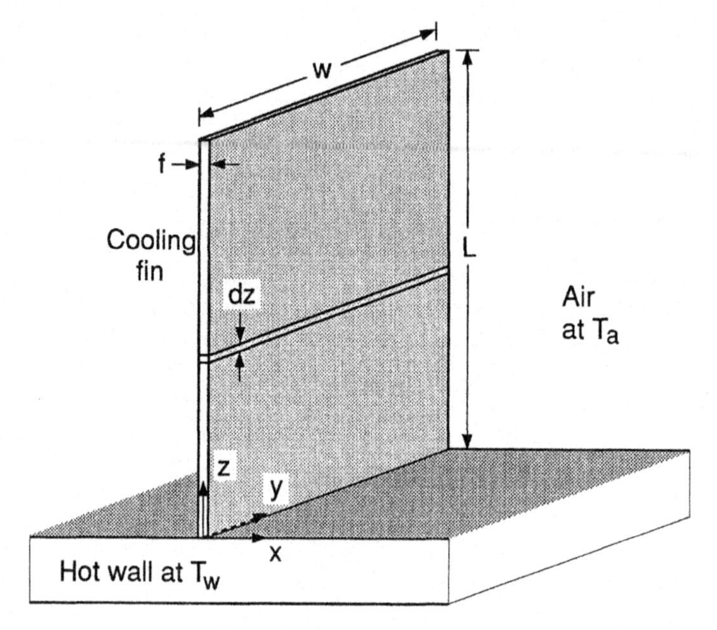

Fig. 2.3-5 Heat conduction in a cooling fin.

At the steady state $\partial T/\partial t = 0$. In the absence of convection, $v_x = v_y = v_z = 0$. Since T varies with z alone, $\partial^2 T/\partial x^2 = \partial^2 T/\partial y^2 = 0$.

The heat loss from the fin to the air cannot be expressed as a boundary condition in the form of $-k\,\partial T/\partial x = h(T - T_a)$ since $\partial T/\partial x = 0$ in this one-dimensional heat flow problem. Let us consider a small volume element $fw\,dz$ in the fin. The rate of heat loss from (both sides of) the element is $(2\,w\,dz)\,h\,(T - T_a)$. In other words, the element is a heat sink losing heat at the rate of $(2\,w\,dz)\,h(T - T_a)$. Since s is defined as the heat generation rate per unit volume,

$$s = -\frac{(2\,w\,dz)\,h(T - T_a)}{fw\,dz} = -\frac{2h}{f}(T - T_a) \qquad [2.3\text{-}36]$$

As such, Eq. [2.3-35] reduces to

$$\frac{d^2T}{dz^2} = a(T - T_a) \qquad [2.3\text{-}37]$$

where

$$a = \frac{2h}{fk} \qquad [2.3\text{-}38]$$

The boundary conditions are

$$\frac{dT}{dz} = 0 \quad \text{at} \quad z = L \qquad [2.3\text{-}39]$$

and

$$T = T_w \quad \text{at} \quad z = 0 \qquad [2.3\text{-}40]$$

The first boundary condition is the result of negligible heat loss from the tip of the fin.

The solution, from Case G of Appendix A, is as follows:

$$T = T_a + \frac{T_w - T_a}{e^{-2\sqrt{a}L} + 1}[e^{\sqrt{a}(z-2L)} + e^{-\sqrt{a}z}] \qquad [2.3\text{-}41]$$

and from Fourier's law of conduction

$$q_z = -k\frac{dT}{dz} \qquad [2.3\text{-}42]$$

Substituting Eq. [2.3-41] into Eq. [2.3-42]

$$q_z = \frac{-k(T_w - T_a)}{e^{-2\sqrt{a}L} + 1}[\sqrt{a}e^{\sqrt{a}(z-2L)} - \sqrt{a}e^{-\sqrt{a}z}] \qquad [2.3\text{-}43]$$

The heat loss from the fin, that is, the heat flow through the base of the fin, is

$$\boxed{Q_z = (fw)q_z|_{z=0} = \frac{-kfw(T_w - T_a)}{e^{-2\sqrt{a}L} + 1}[\sqrt{a}e^{-2\sqrt{a}L} - 1]} \qquad [2.3\text{-}44]$$

where the parameter a is defined in Eq. [2.3-38].

Example 2.3-3 Heat Conduction Into a Semiinfinite Solid

A solid body occupying the space from $x = 0$ to $x = \infty$ is initially at a uniform temperature T_i. At time $t = 0$, the surface at $x = 0$ is suddenly raised to a temperature T_s and maintained at that temperature for $t > 0$. This is illustrated in Fig. 2.3-6. Find the time-dependent temperature profile $T(x, t)$ and the heat flux q_x. Assume that the thermal properties are temperature-independent.

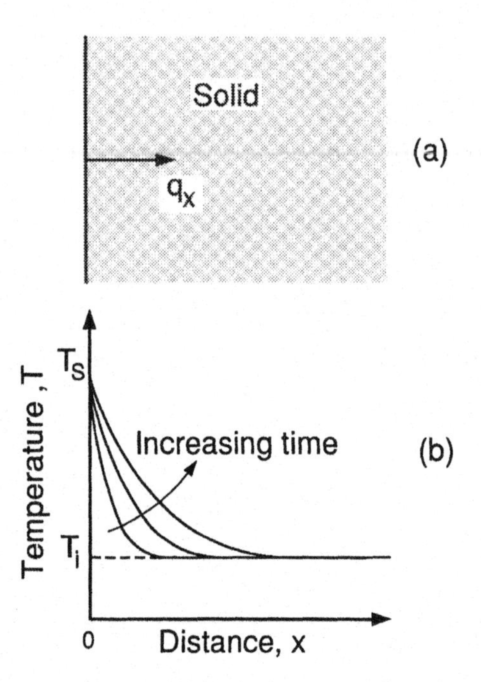

Fig. 2.3-6 Surface heating of a semiinfinite solid: (*a*) heat flux; (*b*) temperature distribution.

Solution Clearly rectangular coordinates are most appropriate for the problem. Without a heat source Eq. [A] of Table 2.3-1 becomes

$$\rho C_v \left[\frac{\partial T}{\partial t} + v_x \frac{\partial T}{\partial x} + v_y \frac{\partial T}{\partial y} + v_z \frac{\partial T}{\partial z} \right] = k \left[\frac{\partial^2 T}{\partial x^2} + \frac{\partial^2 T}{\partial y^2} + \frac{\partial^2 T}{\partial z^2} \right] \quad [2.3\text{-}45]$$

Since the solid is stationary, $v_x = v_y = v_z = 0$. Then $\partial^2 T/\partial y^2 = \partial^2 T/\partial z^2 = 0$ because temperature gradients exist in the x direction only. As such, Eq. [2.3-45] becomes

$$\frac{\partial T}{\partial t} = \alpha \frac{\partial^2 T}{\partial x^2} \quad [2.3\text{-}46]$$

where $\alpha = k/(\rho C_v)$.

The initial and boundary conditions are

$$T(x, 0) = T_i \quad [2.3\text{-}47]$$

$$T(0, t) = T_s \quad [2.3\text{-}48]$$

$$T(\infty, t) = T_i \quad [2.3\text{-}49]$$

The solution, from Case O of Appendix A, is as follows

$$\boxed{\frac{T - T_s}{T_i - T_s} = \operatorname{erf}\left[\frac{x}{\sqrt{4\alpha t}}\right]} \qquad [2.3\text{-}50]$$

where the error function, as shown (Eq. [A.7-23]), is defined as follows:

$$\operatorname{erf}(\eta) = \frac{2}{\sqrt{\pi}} \int_0^\eta e^{-\eta^2} \, d\eta \qquad [2.3\text{-}51]$$

The values of the error function can be found from Fig. 2.3-7. As shown

$$\operatorname{erf}(0) = 0 \qquad [2.3\text{-}52]$$

and

$$\operatorname{erf}(\infty) = 1 \qquad [2.3\text{-}53]$$

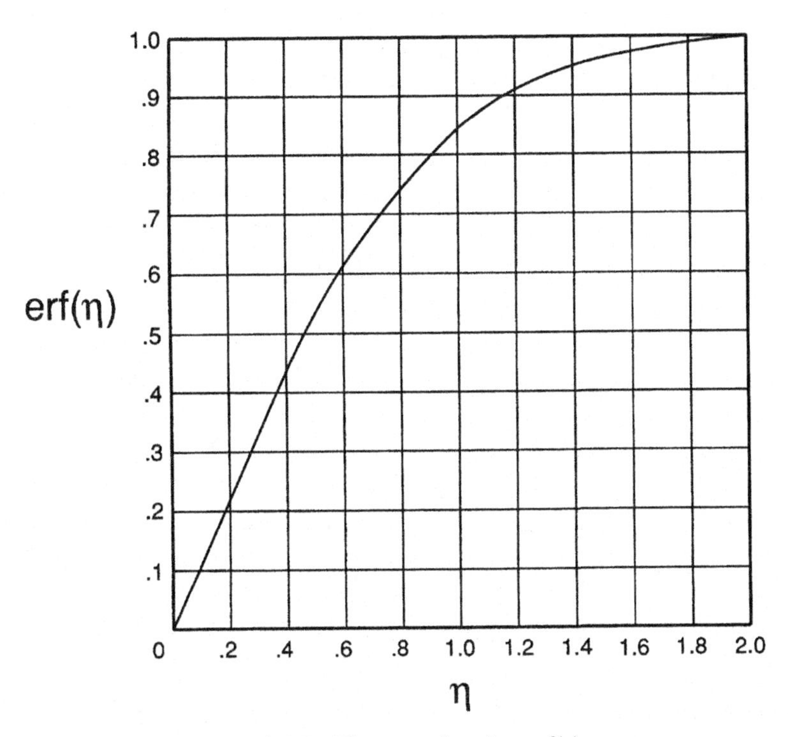

Fig. 2.3-7 The error function erf(η).

Since

$$\frac{\partial}{\partial x}[\text{erf}(\eta)] = \frac{2}{\sqrt{\pi}} e^{-\eta^2} \frac{\partial \eta}{\partial x} \qquad [2.3\text{-}54]$$

we have

$$\frac{\partial T}{\partial x} = \frac{1}{\sqrt{\pi \alpha t}} (T_i - T_s)\, e^{-x^2/4\alpha t} \qquad [2.3\text{-}55]$$

From Fourier's law of conduction

$$q_x = -k \frac{\partial T}{\partial x} \qquad [2.3\text{-}56]$$

Substituting Eq. [2.3-55] into Eq. [2.3-56],

$$q_x = -\rho C_v \sqrt{\frac{\alpha}{\pi t}} (T_i - T_s)\, e^{-x^2/4\alpha t} \qquad [2.3\text{-}57]$$

Example 2.3-4 Heat Loss from a Rising Film

As described in Example 1.5-3, a liquid film of thickness L forms on a continuous belt that moves upward at a constant velocity V passing through a liquid bath at temperature T_i. As the liquid film rises, it is exposed briefly to a gas at temperature T_f and the liquid near the free surface is cooled, as shown in Fig. 2.3-8. Find the steady-state temperature distribution in the film. Assume constant thermal properties and neglect end effects.

Solution The equation of energy in rectangular coordinates (Eq. [A] of Table 2.3-1)

$$\rho C_v \left[\frac{\partial T}{\partial t} + v_x \frac{\partial T}{\partial x} + v_y \frac{\partial T}{\partial y} + v_z \frac{\partial T}{\partial z} \right] = k \left[\frac{\partial^2 T}{\partial x^2} + \frac{\partial^2 T}{\partial y^2} + \frac{\partial^2 T}{\partial z^2} \right] + s \qquad [2.3\text{-}58]$$

At the steady state $\partial T/\partial t = 0$. Since fluid flow is in the z direction, $v_x = v_y = 0$. Since the wall is much wider than the thickness of the film L, the liquid temperature is uniform in the x direction and $\partial^2 T/\partial x^2 = 0$. Heat transfer in the z direction is mainly by convection, and the conduction term $k\,\partial^2 T/\partial z^2$ can be neglected. This is particularly true for high-Pr liquids such as nonmetals. Finally, $s = 0$ since there is no internal heat generation and since viscous

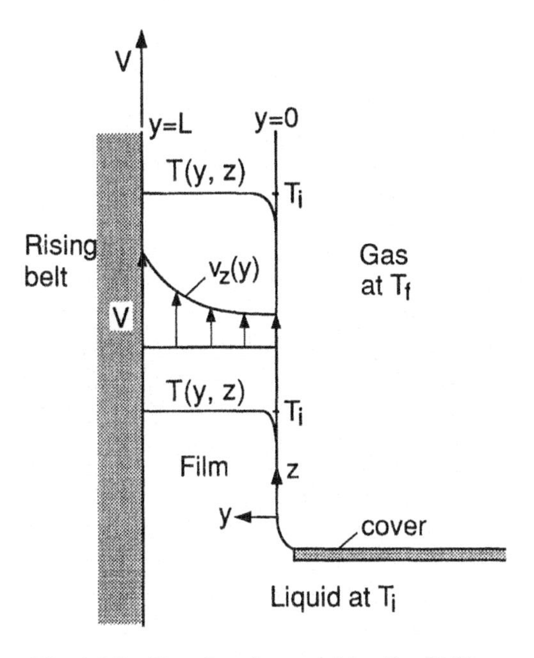

Fig. 2.3-8 Heat loss from a rising liquid film.

dissipation is negligible. As such, Eq. [2.3-58] reduces to

$$v_z \frac{\partial T}{\partial z} = \alpha \frac{\partial^2 T}{\partial y^2} \qquad [2.3\text{-}59]$$

where the thermal diffusivity $\alpha = k/(\rho C_v)$.

The velocity distribution $v_z(y)$ was solved in Example 1.5-3. From Eq. [1.5-68]

$$v_z = V + \frac{\rho g}{2\mu}(y^2 - L^2) \qquad [2.3\text{-}60]$$

Since the liquid film is exposed to the gas briefly, its temperature drops significantly only near the free surface. This is especially true for high-Pr liquids in which heat conduction is slow. Since $y \ll L$ near the free surface,

$$v_z = v_{\min} = V - \frac{\rho g}{2\mu} L^2 \qquad [2.3\text{-}61]$$

Substituting into Eq. [2.3-59]

$$\frac{\partial T}{\partial z} = \frac{\alpha}{v_{\min}} \frac{\partial^2 T}{\partial y^2} \qquad [2.3\text{-}62]$$

and the boundary conditions are

$$T(y,0) = T_i \qquad [2.3\text{-}63]$$

$$k\frac{\partial T(0,z)}{\partial y} = h[T(0,z) - T_f] \qquad [2.3\text{-}64]$$

$$T(y,z) = T_i \quad \text{as} \quad y \to \infty \qquad [2.3\text{-}65]$$

The solution, from Case S of Appendix A, is as follows:

$$\frac{T - T_f}{T_i - T_f} = \operatorname{erf}\left(\frac{y}{\sqrt{4\alpha z/v_{\min}}}\right) + \exp\left(\frac{hy}{k} + \frac{h^2\alpha z/v_{\min}}{k^2}\right) \times\left[1 - \operatorname{erf}\left(\frac{y}{\sqrt{4\alpha z/v_{\min}}} + \frac{h}{k}\sqrt{\alpha z/v_{\min}}\right)\right] \qquad [2.3\text{-}66]$$

This equation can be differential with respect to y, with the help of Eq. [A.7-25], to find the heat flux to the gas. It can then be integrated across the free surface to find the heat loss to the gas.

Example 2.3-5 Heat Transfer in Laminar Flow over a Flat Plate

Consider the steady-state, laminar flow of an incompressible Newtonian fluid over a wide flat plate of uniform temperature T_s, as illustrated in Fig. 2.3-9. Determine the temperature distribution in and the thickness of the thermal boundary layer. Assume that the physical properties of the fluid are temperature-independent so that fluid flow is not affected by heat transfer.

Solution Without a heat source and viscous dissipation, the equation of energy (i.e., Eq. [A] of Table 2.3-1), is as follows:

$$\rho C_v\left[\frac{\partial T}{\partial t} + v_x\frac{\partial T}{\partial x} + v_y\frac{\partial T}{\partial y} + v_z\frac{\partial T}{\partial z}\right] = k\left[\frac{\partial^2 T}{\partial x^2} + \frac{\partial^2 T}{\partial y^2} + \frac{\partial^2 T}{\partial z^2}\right] \qquad [2.3\text{-}67]$$

At the steady state $\partial T/\partial t$ is zero. Since there is no fluid flow nor temperature gradient in the x direction, v_x, $\partial T/\partial x$, and $\partial^2 T/\partial x^2$ are zero. Since heat transfer in the z direction is mainly by convection rather than by conduction, $k\,\partial^2 T/\partial z^2$ is far smaller than $\rho C_v v_z\,\partial T/\partial z$ and can thus be neglected. This approximation

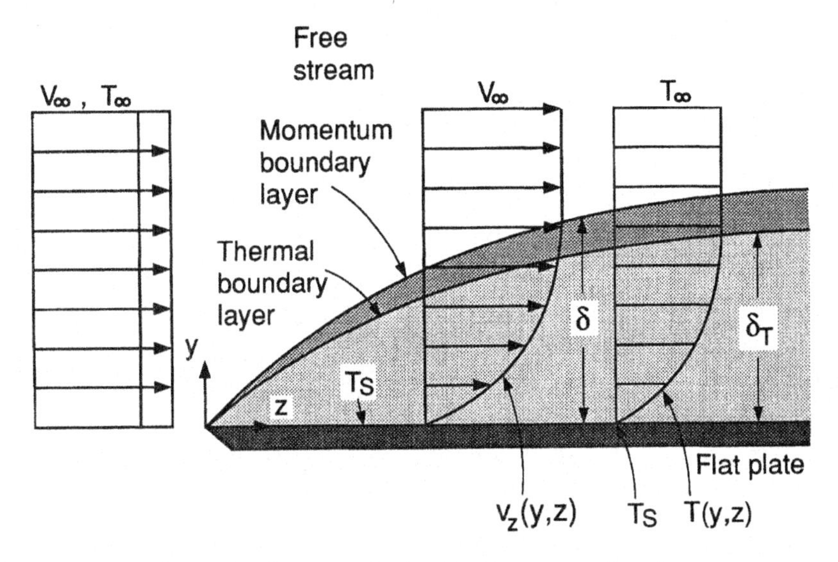

Fig. 2.3-9 Heat transfer in fluid flow over a flat plate.

is good except for molten metals and semiconductors in view of their high thermal conductivities.

The equations of continuity, motion, and energy are as follows:

$$\text{Continuity:} \quad \frac{\partial v_y}{\partial y} + \frac{\partial v_z}{\partial z} = 0 \qquad [2.3\text{-}68]$$

$$\text{Motion:} \quad v_y \frac{\partial v_z}{\partial y} + v_z \frac{\partial v_z}{\partial z} = \nu \frac{\partial^2 v_z}{\partial y^2} \qquad [2.3\text{-}69]$$

$$\text{Energy:} \quad v_y \frac{\partial T}{\partial y} + v_z \frac{\partial T}{\partial z} = \alpha \frac{\partial^2 T}{\partial y^2} \qquad [2.3\text{-}70]$$

where the thermal diffusivity $\alpha = k\rho/C_v$. From Eq. [2.3-68]

$$v_y = -\int_0^y \frac{\partial v_z}{\partial z}\, dy \qquad [2.3\text{-}71]$$

which is substituted into Eq. [2.3-70] to give

$$-\left(\int_0^y \frac{\partial v_z}{\partial z}\, dy\right)\frac{\partial T}{\partial y} + v_z \frac{\partial T}{\partial z} = \alpha \frac{\partial^2 T}{\partial y^2} \qquad [2.3\text{-}72]$$

Since this equation is difficult to solve directly, the following approximate temperature profile is assumed:

$$\boxed{\frac{T - T_s}{T_\infty - T_s} = \frac{3}{2}\left(\frac{y}{\delta_T}\right) - \frac{1}{2}\left(\frac{y}{\delta_T}\right)^3} \quad [2.3\text{-}73]$$

where δ_T is the thickness of the thermal boundary layer. This temperature profile is reasonable since it satisfies the following boundary conditions

$$T = T_s \quad \text{at} \quad y = 0 \quad [2.3\text{-}74]$$

and

$$T = T_\infty \quad \text{and} \quad \frac{\partial T}{\partial y} = 0 \quad \text{at} \quad y = \delta_T \quad [2.3\text{-}75]$$

The temperature profile shown in Fig. 2.3-9 differs from that shown previously in Fig. 2.2-7*a* in that $T_\infty > T_s$ in the former while $T_\infty < T_s$ in the latter. This difference, however, will not affect the conclusion of this example as either profile can be described by Eq. [2.3-73].

From Eq. [2.3-73]

$$\frac{\partial T}{\partial z} = \frac{(T_\infty - T_s)}{2}\left(\frac{-3y}{\delta_T^2} + \frac{3y^3}{\delta_T^4}\right)\frac{d\delta_T}{dz} \quad [2.3\text{-}76]$$

$$\frac{\partial T}{\partial y} = \frac{(T_\infty - T_s)}{2}\left(\frac{3}{\delta_T} - \frac{3y^2}{\delta_T^3}\right) \quad [2.3\text{-}77]$$

$$\frac{\partial^2 T}{\partial y^2} = \frac{-3(T_\infty - T_s)y}{\delta_T^3} \quad [2.3\text{-}78]$$

Equations [2.3-76] through [2.3-78] can be substituted into Eq. [2.3-72]. Since fluid flow is not affected by heat transfer, we can substitute Eq. [1.5-78] for v_z and Eq. [1.5-81] for $\partial v_z/\partial z$, into Eq. [2.3-72]. After rearranging and integrating with respect to y from 0 to δ_T, we get

$$\frac{8}{5}\delta\delta_T^3\left[-1 + \frac{1}{7}\left(\frac{\delta_T}{\delta}\right)^2\right]\frac{d\delta_T}{dz} + \frac{6}{5}\delta_T^4\left[\frac{2}{3} - \frac{1}{7}\left(\frac{\delta_T}{\delta}\right)^2\right]\frac{d\delta}{dz} = -\frac{8\alpha}{v_\infty}\delta^2\delta_T \quad [2.3\text{-}79]$$

This equation can be solved for δ_T as a function of δ, if z can be eliminated. From Eq. [1.5-87]

$$\delta = 4.64\sqrt{\frac{\nu z}{v_\infty}} \quad [2.3\text{-}80]$$

which yields

$$2\delta\, d\delta = \frac{280}{13}\frac{\nu}{v_\infty}\, dz \qquad [2.3\text{-}81]$$

Let us define

$$\frac{\delta_T}{\delta} = a \qquad [2.3\text{-}82]$$

where a is a constant depending only on the physical properties of the fluid. Substituting Eqs. [2.3-80] and [2.3-82] into Eq. [2.3-79]

$$a^5 - 14a^3 + \frac{13}{\mathrm{Pr}} = 0 \qquad [2.3\text{-}83]$$

where the Prandtl number

$$\boxed{\mathrm{Pr} = \frac{\nu}{\alpha} = \frac{C_v\mu}{k}} \qquad [2.3\text{-}84]$$

Let us consider the case where $a < 1$, that is, where the thermal boundary layer is thinner than the momentum boundary layer. Since $a^5 \ll 14a^3$, Eq. [2.3-83] reduces approximately to

$$a = \mathrm{Pr}^{-(1/3)} \qquad [2.3\text{-}85]$$

Substituting into Eq. [2.3-82]

$$\frac{\delta_T}{\delta} = \mathrm{Pr}^{-(1/3)} \qquad [2.3\text{-}86]$$

which is identical to Eq. [2.2-65] derived previously on the basis of integral energy balance. Substituting Eq. [2.3-80] into Eq. [2.3-86]

$$\boxed{\delta_T = 4.64\,\mathrm{Pr}^{-(1/3)}\sqrt{\frac{\nu z}{v_\infty}}} \qquad [2.3\text{-}87]$$

Therefore, the thickness of the thermal boundary layer increases with the square root of the distance from the leading edge of the plate. From Eqs. [2.3-73] and [2.3-87] the temperature profile can be determined.

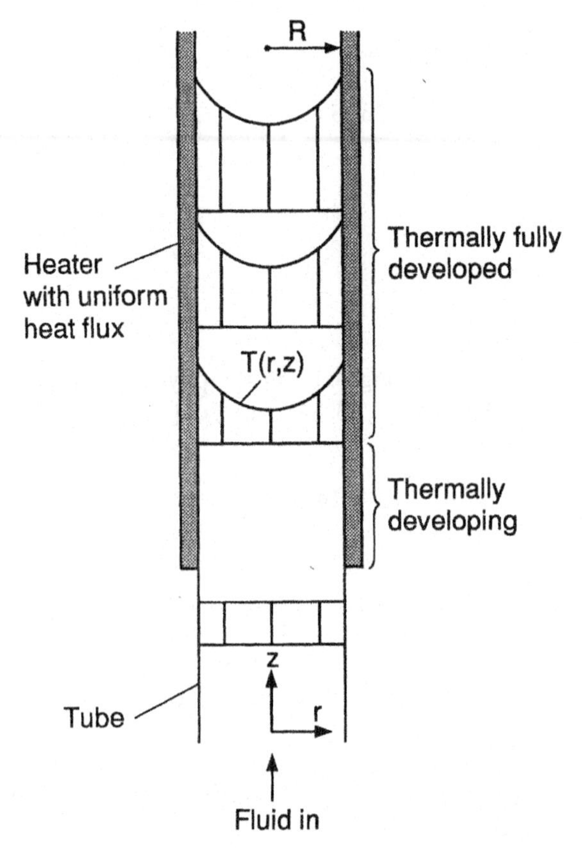

Fig. 2.3-10 Heat transfer in fluid flow through a circular tube.

Example 2.3-6 Heat Transfer with Laminar Flow in a Tube

Let us consider an incompressible, Newtonian fluid in steady state, laminar flow in a circular tube of radius R, as illustrated in Fig. 2.3-10. The heater produces a uniform heat flux q_R (taken as positive) at the wall. From Eq. [2.1-16] and the fact that q_R and $\partial T/\partial r$ are both positive

$$q_R = h(T_R - T_{av}) = k \left. \frac{\partial T}{\partial r} \right|_{r=R} \qquad [2.3\text{-}88]$$

where h is the heat transfer coefficient and T_R the fluid temperature at the wall. The average fluid temperature T_{av} has been defined in Eq. [2.1-11] as follows

$$T_{av} = \frac{2\pi \int_0^R \rho r v_z T \, dr}{m} = \frac{2\pi \int_0^R r v_z T \, dr}{Q} \qquad [2.3\text{-}89]$$

where m and Q are the mass and volume flow rates, respectively. As shown in Eq. [2.1-17], $\partial T/\partial z$ is independent of r when q_R is constant.

The physical properties of the fluid are assumed temperature-independent so that fluid flow is not affected by heat transfer. Also, fluid flow is significant enough that heat transfer in the z direction is dominated by convection. Determine the heat transfer coefficient h in the thermally fully developed region.

Solution From Eq. [2.3-88]

$$h = \frac{k}{(T_R - T_{\text{av}})} \frac{\partial T}{\partial r}\bigg|_{r=R} \quad \text{[2.3-90]}$$

In order to find h we must first find $\partial T/\partial r|_{r=R}$ and $(T_R - T_{\text{av}})$. To do this we must first find the temperature distribution. With no heat source and negligible heat dissipation, the equation of energy in cylindrical coordinates, Eq. [B] of Table 2.3-1, becomes

$$\rho C_v \left[\frac{\partial T}{\partial t} + v_r \frac{\partial T}{\partial r} + \frac{v_\theta}{r}\frac{\partial T}{\partial \theta} + v_z \frac{\partial T}{\partial z}\right] = k\left[\frac{1}{r}\frac{\partial}{\partial r}\left(r\frac{\partial T}{\partial r}\right) + \frac{1}{r^2}\frac{\partial^2 T}{\partial \theta^2} + \frac{\partial^2 T}{\partial z^2}\right] \quad \text{[2.3-91]}$$

In this equation $\partial T/\partial t = 0$ at the steady state. Since there is no fluid flow in the r or θ direction, $v_r = v_\theta = 0$. Because of axisymmetry of the temperature field, $\partial T/\partial \theta = 0$ and hence $\partial^2 T/\partial \theta^2 = 0$. Since heat transfer in the z direction is dominated by convection, the conduction term $k\,\partial^2 T/\partial z^2$ is negligible as compared to the convection term $\rho C_v v_z \partial T/\partial z$. The equation of energy now reduces to

$$v_z \frac{\partial T}{\partial z} = \alpha\left[\frac{1}{r}\frac{\partial}{\partial r}\left(r\frac{\partial T}{\partial r}\right)\right] \quad \text{[2.3-92]}$$

where

$$\alpha = \frac{k}{\rho C_v}. \quad \text{[2.3-93]}$$

Since fluid flow is not affected by heat transfer in the present problem, from Eq. [1.5-54]

$$v_z = \frac{2Q}{\pi R^2}\left[1 - \left(\frac{r}{R}\right)^2\right] \quad \text{[2.3-94]}$$

Substituting this equation into Eq. [2.3-92] and keeping in mind that $\partial T/\partial z$ is independent of r

$$\frac{1}{r}\frac{\partial}{\partial r}\left(r\frac{\partial T}{\partial r}\right) = \left(\frac{2Q}{\alpha\pi R^2}\frac{\partial T}{\partial z}\right)\left(1 - \frac{r^2}{R^2}\right) \quad [2.3\text{-}95]$$

The boundary conditions are

$$\frac{\partial T}{\partial r} = 0 \quad \text{at} \quad r = 0 \quad [2.3\text{-}96]$$

and

$$T = T_R \quad \text{at} \quad r = R \quad [2.3\text{-}97]$$

The solution, from Case I of Appendix A, is as follows:

$$T = T_R + \frac{Q}{2\alpha\pi R^2}\frac{\partial T}{\partial z}(r^2 - R^2) - \frac{Q}{8\alpha\pi R^4}\frac{\partial T}{\partial z}(r^4 - R^4) \quad [2.3\text{-}98]$$

or

$$\boxed{T = T_R - \frac{Q}{8\alpha\pi R^4}\frac{\partial T}{\partial z}(3R^4 - 4R^2r^2 + r^4)} \quad [2.3\text{-}99]$$

Differentiating with respect to r

$$\frac{\partial T}{\partial r} = -\frac{Q}{8\alpha\pi R^4}\frac{\partial T}{\partial z}(-8R^2r + 4r^3) \quad [2.3\text{-}100]$$

and so

$$\left.\frac{\partial T}{\partial r}\right|_{r=R} = \frac{Q}{2\alpha\pi R}\frac{\partial T}{\partial z} \quad [2.3\text{-}101]$$

Now let us find $(T_R - T_{av})$. Substituting Eqs. [2.3-94] and Eq. [2.3-99] into Eq. [2.3-89]

$$T_{av} - T_R = \frac{11Q}{48\alpha\pi}\frac{\partial T}{\partial z} \quad [2.3\text{-}102]$$

Substituting Eqs. [2.3-101] and [2.3-102] into Eq. [2.3-90]

$$\boxed{\mathrm{Nu}_D = \frac{hD}{k} = 4.36} \tag{2.3-103}$$

where Nu_D is the **Nusselt number** and D the inner diameter of the tube, specifically, $2R$.

2.4. DIFFERENTIAL ENERGY-BALANCE EQUATION IN STREAM FUNCTION

As discussed in the previous section, the differential governing equation for convective heat transfer is the equation of energy and is expressed in terms of temperature and the velocity components involved. In the present section a different type of differential governing equation, expressed in temperature and the stream function, will be introduced. This approach is often used in two-dimensional and axisymmetric problems, as will be seen in Chapters 8 and 9.

2.4.1. Two-Dimensional Problems

The equation of energy, Eq. [A] of Table 2.3-1, becomes

$$\rho C_v \left[\frac{\partial T}{\partial t} + v_x \frac{\partial T}{\partial x} + v_y \frac{\partial T}{\partial y} \right] = k \left[\frac{\partial^2 T}{\partial x^2} + \frac{\partial^2 T}{\partial y^2} \right] + s \tag{2.4-1}$$

From Eqs. [1.6-1] and [1.6-2]

$$v_x = \frac{\partial \psi}{\partial y} \tag{2.4-2}$$

and

$$v_y = -\frac{\partial \psi}{\partial x} \tag{2.4-3}$$

Substituting Eqs. [2.4-2] and [2.4-3] into Eq. [2.4-1]

$$\boxed{\rho C_v \left[\frac{\partial T}{\partial t} + \frac{\partial \psi}{\partial y}\frac{\partial T}{\partial x} - \frac{\partial \psi}{\partial x}\frac{\partial T}{\partial y} \right] = k \left[\frac{\partial^2 T}{\partial x^2} + \frac{\partial^2 T}{\partial y^2} \right] + s} \tag{2.4-4}$$

For two-dimensional convective heat transfer, this equation is solved simultaneously with Eqs. [1.6-11] and [1.6-12] for the temperature and velocity fields.

2.4.2. Axisymmetric Problems

The equation of energy, Eq. [B] of Table 2.3-1, is as follows:

$$\rho C_v\left[\frac{\partial T}{\partial t}+v_r\frac{\partial T}{\partial r}+v_z\frac{\partial T}{\partial z}\right]=k\left[\frac{1}{r}\frac{\partial}{\partial r}\left(r\frac{\partial T}{\partial r}\right)+\frac{\partial^2 T}{\partial z^2}\right]+s \qquad [2.4\text{-}5]$$

From Eqs. [1.6-13] and [1.6-14]

$$v_r=\frac{1}{r}\frac{\partial\psi}{\partial z} \qquad [2.4\text{-}6]$$

and

$$v_z=\frac{-1}{r}\frac{\partial\psi}{\partial r} \qquad [2.4\text{-}7]$$

Substituting Eqs. [2.4-6] and [2.4-7] into Eq. [2.4-5]

$$\boxed{\rho C_v\left[\frac{\partial T}{\partial t}+\frac{1}{r}\frac{\partial\psi}{\partial z}\frac{\partial T}{\partial r}-\frac{1}{r}\frac{\partial\psi}{\partial r}\frac{\partial T}{\partial z}\right]=k\left[\frac{1}{r}\frac{\partial}{\partial r}\left(r\frac{\partial T}{\partial r}\right)+\frac{\partial^2 T}{\partial z^2}\right]+s} \qquad [2.4\text{-}8]$$

For axisymmetric convective heat transfer, Eq. [2.4-8] is solved simultaneously with Eqs. [1.6-21] through [1.6-23] for the temperature and velocity fields.

2.5. TURBULENCE

2.5.1. Time-Smoothed Variables

As described in Section 1.7, the velocity fluctuations arising in turbulent flow affect the local pressure. The velocity fluctuations, in fact, also affect the local temperature. Therefore, in addition to time-smoothed velocity we need to use time-smoothed temperature.

As illustrated in Fig. 2.5-1, one may assume that the temperature T at a fixed point in space over a given finite time interval t_0 can be resolved into the time-smoothed temperature $\bar{T}$ and a fluctuation temperature term T' that accounts for the turbulent motion:

$$T=\bar{T}+T' \qquad [2.5\text{-}1]$$

where

$$\bar{T}=\frac{1}{t_0}\int_t^{t+t_0}T\,dt \qquad [2.5\text{-}2]$$

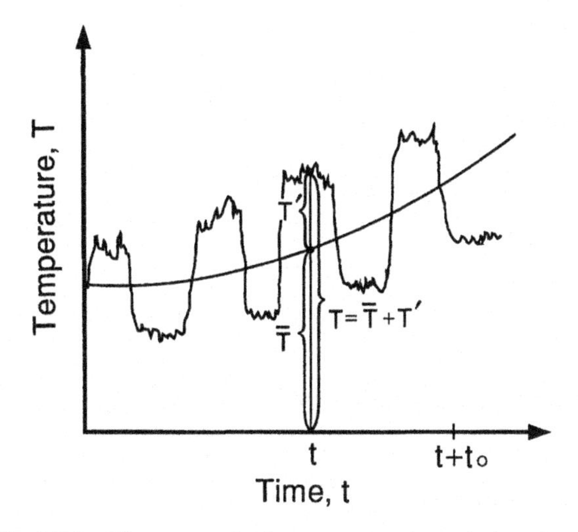

Fig. 2.5-1 Time-smoothed temperature in turbulent flow.

The time interval t_0 is small with respect to the time over which $\bar{T}$ may vary in an unsteady-state problem, but large with respect to the period of turbulent fluctuations. From these two equations, it is obvious that

$$\overline{T'} = \frac{1}{t_0} \int_t^{t+t_0} T' \, dt = 0 \tag{2.5-3}$$

2.5.2. Time-Smoothed Governing Equations

Substituting Eqs. [1.7-1] and [2.5-1] into Eq. [2.3-7] and then taking the time average according to Eqs. [1.7-2], [1.7-3], [2.5-2], and [2.5-3], we get the following **time-smoothed equation of energy**:

$$\rho C_v \left[\frac{\partial \bar{T}}{\partial t} + \bar{\mathbf{v}} \cdot \nabla \bar{T} \right] = -\nabla \cdot \bar{\mathbf{q}} - \nabla \cdot \overline{\mathbf{q}'} + s \tag{2.5-4}$$

where $\bar{\mathbf{q}} = -k \nabla \bar{T}$ and the turbulent heat flux

$$\overline{\mathbf{q}'} = \rho C_v (\overline{v_x' T'} + \overline{v_y' T'} + \overline{v_z' T'}) \tag{2.5-5}$$

Equation [2.5-4] is the same as the equation of energy for laminar flow Eq. [2.3-7], except that the time-smoothed velocity and temperature replace the instantaneous velocity and temperature, and that one new terms $\nabla \cdot \overline{\mathbf{q}'}$ arises.

2.5.3. Turbulent Heat Flux

Several semiempirical relations have been proposed for the turbulent heat flux $\overline{q'}$, in order to solve Eq. [2.5-4] for temperature distributions in turbulent flow.

2.5.3.1. Eddy Thermal Conductivity

By analogy with Eq. [2.1-1] Fourier's law of conduction, one may write

$$\boxed{\overline{q'_y} = -k' \frac{d\overline{T}}{dy}} \qquad [2.5\text{-}6]$$

The coefficient k' is a turbulent or eddy thermal conductivity and is usually position-dependent.

2.5.3.2. Prandtl's Mixing Length

By analogy with Eq. [1.7-13]

$$\overline{q'_y} = -\rho C_v \ell^2 \left|\frac{d\bar{v}_z}{dy}\right| \frac{d\overline{T}}{dy} \qquad [2.5\text{-}7]$$

and from Eq. [2.5-7]

$$k' = \rho C_v \ell^2 \left|\frac{d\bar{v}_z}{dy}\right| \qquad [2.5\text{-}8]$$

2.6. HEAT TRANSFER CORRELATIONS

Convective heat transfer, as described in Section 2.3, can be determined by solving the governing equations for fluid flow and heat transfer. Correlations that are derived theoretically but verified experimentally or are based on experimental data alone, are useful for studying convective heat transfer. Some of these correlations will be presented in this section in dimensionless numbers.

2.6.1. External Flow

2.6.1.1. Forced Convection over a Flat Plate

Flow over a flat plate is laminar for local Reynolds number $\mathrm{Re}_z < 2 \times 10^5$. As described in Example 2.2-5, the following theoretical correlation can be used for laminar flow of a fluid with a bulk temperature T_∞ over a flat plate with a surface temperature T_s:

$$\boxed{\mathrm{Nu}_z = 0.332\,\mathrm{Re}_z^{1/2}\,\mathrm{Pr}^{1/3}} \qquad 0.6 < \mathrm{Pr} < 50 \qquad [2.6\text{-}1]$$

where

$$\mathrm{Nu}_z \text{ (local Nusselt number)} = \frac{h_z z}{k} \qquad [2.6\text{-}2]$$

$$\mathrm{Re}_z \text{ (local Reynolds number)} = \frac{z v_\infty}{\nu} = \frac{z \rho v_\infty}{\mu} \qquad [2.6\text{-}3]$$

$$\mathrm{Pr} \text{ (Prandtl number)} = \frac{\nu}{\alpha} = \frac{C_v \mu}{k} \qquad [2.6\text{-}4]$$

From these equations, the heat transfer coefficient averaged over a distance L from the leading edge of the plate is

$$h_L = \frac{1}{L}\int_0^L h_z \, dz \qquad [2.6\text{-}5]$$

Substituting Eq. [2.6-1] into Eq. [2.6-5]

$$h_L = 0.332 \left(\frac{k}{L}\right) \mathrm{Pr}^{1/3} \left(\frac{v_\infty}{\nu}\right)^{1/2} \int_0^L z^{-1/2} \, dz \qquad [2.6\text{-}6]$$

or

$$\boxed{\mathrm{Nu}_L = 0.664\, \mathrm{Re}_L^{1/2}\, \mathrm{Pr}^{1/3}} \qquad (0.6 < \mathrm{Pr} < 50) \qquad [2.6\text{-}7]$$

where

$$\mathrm{Nu}_L \text{ (average Nusselt number)} = \frac{h_L L}{k} \qquad [2.6\text{-}8]$$

$$\mathrm{Re}_L \text{ (average Reynolds number)} = \frac{L v_\infty}{\nu} = \frac{L \rho v_\infty}{\mu} \qquad [2.6\text{-}9]$$

The fluid properties in Eqs. [2.6-1] and [2.6-7] are evaluated at the **film temperature**:

$$T_f = \frac{T_\infty + T_s}{2} \qquad [2.6\text{-}10]$$

For liquid metals and semiconductors $\mathrm{Pr} \ll 1$ and Eqs. [2.6-1] and [2.6-7] cannot be applied. For these materials the following theoretical correction has been suggested:[4]

$$\boxed{\mathrm{Nu}_z = 0.565(\mathrm{Re}_z\,\mathrm{Pr})^{1/2} = 0.565\mathrm{Pe}_z^{1/2}} \qquad (\mathrm{Pr} < 0.05,\ \mathrm{Pe} > 100) \qquad [2.6\text{-}11]$$

where

$$\mathrm{Pe}_z \text{ (local Peclet number)} = \mathrm{Re}_z\,\mathrm{Pr} = \frac{z v_\infty}{\alpha} \qquad [2.6\text{-}12]$$

For turbulent flow over a flat plate, the following theoretical correlation has been suggested:[5]

$$\boxed{\mathrm{Nu}_z = 0.0288\,\mathrm{Re}_z^{4/5}\,\mathrm{Pr}^{1/3}} \qquad (0.6 < \mathrm{Pr} < 60) \qquad [2.6\text{-}13]$$

and from Eq. [2.6-5]

$$\boxed{\mathrm{Nu}_z = 0.036\,\mathrm{Re}_L^{4/5}\,\mathrm{Pr}^{1/3}} \qquad (0.6 < \mathrm{Pr} < 60) \qquad [2.6\text{-}14]$$

2.6.1.2. *Forced Convection Normal to a Cylinder*

For the flow of air normal to a cylinder of diameter D, McAdams[6] has plotted the data from 13 studies and found excellent agreement when plotted as Nu_D against Re_D. A widely used empirical correlation for these data is[7]

$$\boxed{\mathrm{Nu}_D = a\mathrm{Re}_D^b\,\mathrm{Pr}^{1/3}} \qquad (0.1 < \mathrm{Re}_D < 3 \times 10^5, \mathrm{Pr} \geq 0.7) \qquad [2.6\text{-}15]$$

where

$$\mathrm{Nu}_D = \frac{hD}{k} \qquad [2.6\text{-}16]$$

$$\mathrm{Re}_D = \frac{D v_\infty}{\nu} = \frac{D \rho v_\infty}{\mu} \qquad [2.6\text{-}17]$$

and the constants a and b are as listed in Table 2.6-1.

[4] W. M. Kays and M. E. Crawford, *Convective Heat and Mass Transfer*, McGraw-Hill, New York, 1980.

[5] J. R. Welty, C. E. Wicks, and R. E. Wilson, *Fundamentals of Momentum, Heat and Mass Transfer*, 3rd ed., Wiley, New York, 1984, p. 370.

[6] W. H. McAdams, *Heat Transmission*, 3rd ed., McGraw-Hill, New York, 1949.

[7] R. Hilpert, *Forsch. Geb. Ingenieurwes.*, **4**, 215, 1933.

TABLE 2.6-1 Constants of Eq. [2.6-15] for Flow Normal to a Circular Cylinder*

Re_D	a	b
0.4–4	0.989	0.330
4–40	0.911	0.385
40–4000	0.683	0.466
4000–40,000	0.193	0.618
40,000–400,000	0.027	0.805

Source: R. Hilpert., *Forsch. Geb. Ingenieurwes.*, **4**, 215, 1933.

The following empirical correlation has also been suggested:[8]

$$\mathrm{Nu}_D = 0.3 + \frac{0.62\mathrm{Re}_D^{1/2}\,\mathrm{Pr}^{1/3}}{[1+(0.4/\mathrm{Pr})^{2/3}]^{1/4}}\left[1+\left(\frac{\mathrm{Re}_D}{282{,}000}\right)^{5/8}\right]^{4/5} \qquad [2.6\text{-}18]$$

where all properties are evaluated at the film temperature.

2.6.1.3. Forced Convection Past a Sphere

For the flow of air past a sphere of diameter D, McAdams[9] has also plotted, with good agreement, the experimental data from several studies relating Nu_D to Re_D. The following empirical correlation has been proposed:[10]

$$\mathrm{Nu}_D = 2 + (0.4\mathrm{Re}_D^{1/2} + 0.06\mathrm{Re}_D^{2/3})\,\mathrm{Pr}^{0.4}\left(\frac{\mu_\infty}{\mu_s}\right)^{1/4} \qquad [2.6\text{-}19]$$

$$(0.71 < \mathrm{Pr} < 380,\ 3.5 < \mathrm{Re}_D < 7.6\times 10^4,\ 1 < \mu/\mu_s < 3.2)$$

All fluid properties are evaluated at bulk T_∞ except for μ_s, which is evaluated at the surface temperature of the sphere. This correlation is accurate to within $\pm 30\%$ for the range of parameter values specified.

A special case of convective heat transfer from sphere is that of a freely falling drop. The following theoretical correlation has been proposed:[11]

$$\mathrm{Nu}_D = 2 + 0.6\mathrm{Re}_D^{1/2}\,\mathrm{Pr}^{1/3} \qquad [2.6\text{-}20]$$

[8] S. W. Churchill, and M. Bernstein, *J. Heat Transfer*, **99**, 300, 1977.

[9] W. H. McAdams, ibid.

[10] S. Whitaker, *AIChe J.*, **18**, 361, 1972.

[11] W. Ranz and W. Marshall, *Chem. Eng. Progr.*, **48**, 141, 1952.

2.6.2. Internal Flow

2.6.2.1. Forced Convection Inside a Circular Tube

For laminar fully developed flow inside a circular tube of diameter D and uniform surface heat flux, the following theoretical correlation (i.e., Eq. [2.3-103]) has been proposed:

$$\boxed{\mathrm{Nu}_D = 4.36} \qquad (\mathrm{Pr} \geq 0.6) \qquad [2.6\text{-}21]$$

For a uniform surface temperature rather than heat flux, the following theoretical correlation has been proposed:[12]

$$\boxed{\mathrm{Nu}_D = 3.66} \qquad (\mathrm{Pr} \geq 0.6) \qquad [2.6\text{-}22]$$

For turbulent fully developed flow inside a circular tube of diameter D and length L, the following theoretical correlation has been suggested:[13]

$$\mathrm{Nu}_D = 0.023\mathrm{Re}_D^{4/5}\mathrm{Pr}^{1/3} \qquad [2.6\text{-}23]$$

A slightly different and preferred correlation is as follows:[14]

$$\boxed{\mathrm{Nu}_D = 0.023\mathrm{Re}_D^{4/5}\mathrm{Pr}^n} \qquad \left(0.6 < \mathrm{Pr} < 160,\ \mathrm{Re}_D > 10{,}000,\ \frac{L}{D} > 10\right) \qquad [2.6\text{-}24]$$

where n is 0.4 if the fluid is being heated and 0.3 if it is being cooled.

Equation [2.6-24] is good for small to moderate temperature difference between the wall and the bulk fluid. The following equation is preferred for flows characterized by large property variations:[15]

$$\boxed{\mathrm{Nu}_D = 0.027\mathrm{Re}_D^{4/5}\mathrm{Pr}^{1/3}\left(\frac{\mu_b}{\mu_w}\right)^{0.14}}$$

$$\left(0.7 < \mathrm{Pr} < 16{,}700,\ \mathrm{Re}_D > 10{,}000,\ \frac{L}{D} > 10\right) \qquad [2.6\text{-}25]$$

All fluid properties are evaluated at the bulk temperature except μ_w, which is evaluated at the wall temperature.

[12] W. M. Kays and M. E. Crawford, *Convection Heat and Mass Transfer*, McGraw-Hill, New York, 1980.

[13] A. P. Colburn, *Trans. A.I.Ch.E.*, **29**, 174, 1933.

[14] F. W. Dittus and L. M. K. Boelter, University of California, *Publ. Eng.*, **2**, 443, 1930.

[15] E. N. Seider and G. E. Tate, *Ind. Eng. Chem.*, **28**, 1429, 1936.

Equations [2.6-23] through [2.6-25] are for $Re_D > 10^4$. The following correlation[16] can be used even if Re_D is below 10^4:

$$\boxed{Nu_D = \frac{(f/8)(Re_D - 1000)Pr}{1 + 12.7(f/8)^{1/2}(Pr^{2/3} - 1)}}$$

$$(0.5 < Pr < 2000,\ 2300 < Re_D < 5 \times 10^6) \qquad [2.6\text{-}26]$$

where the friction factor f can be obtained from the Moody diagram (Fig. 1.8-2) or, for smooth pipes, from

$$f = (0.79 \ln Re_D - 1.64)^{-2} \qquad [2.6\text{-}27]$$

2.6.3. Examples

Example 2.6-1 Flow through a Cooling Tube

Cooling water runs through a copper tubing 0.4 cm in diameter and 20 m long at a mass flow rate of 20 g/s. Find the heat transfer coefficient, assuming the inner surface is smooth. $\mu = 1 \times 10^{-2}$ g cm^{-1} s^{-1}, $\rho = 1$ g/cm^3, $C_p = 4.2$ J g^{-1} °C^{-1}, and $k = 6 \times 10^{-3}$ W cm^{-1} °C^{-1}.

Solution The Prandtl number

$$Re_D = \frac{D\rho v_{av}}{\mu} = \frac{\pi}{4} D^2 \rho v_{av} \times \frac{4}{\pi D \mu} = \frac{4m}{\pi D \mu} \qquad [2.6\text{-}28]$$

where m is the mass flow rate. As such

$$Re_D = \frac{4 \times 20 \text{ g/s}}{\pi \times 0.4 \text{ cm} \times 1 \times 10^{-2} \text{ g cm}^{-1} \text{s}^{-1}} = 6366 \qquad (\text{turbulent but } < 10^4) \qquad [2.6\text{-}29]$$

Substituting into Eq. [2.6-27]

$$f = (0.79 \ln 6366 - 1.64)^{-2} = 0.036 \qquad [2.6\text{-}30]$$

The Prandtl number

$$Pr = \frac{C_p \mu}{k} = \frac{4.2 \text{ J g}^{-1}\,°\text{C}^{-1} \times 1 \times 10^{-2} \text{ g cm}^{-1} \text{s}^{-1}}{6 \times 10^{-3} \text{ W cm}^{-1}\,°\text{C}^{-1}} = 7 \qquad [2.6\text{-}31]$$

[16] V. Gnielinski, *Int. Chem. Eng.*, **16**, 359, 1976.

From Eq. [2.6-26]

$$\mathrm{Nu}_D = \frac{(0.036/8)(6366 - 1000) \times 7}{1 + 12.7(0.036/8)^{1/2}(7^{2/3} - 1)} = 51.8 \qquad [2.6\text{-}32]$$

From Eq. [2.6-16] the heat transfer coefficient

$$h = \frac{\mathrm{Nu}_D k}{D} = \frac{51.8 \times 6 \times 10^{-3}\ \mathrm{W\,cm^{-1}\,^{\circ}C^{-1}}}{0.4\ \mathrm{cm}} = 0.78\ \mathrm{W\,cm^{-2}\,^{\circ}C^{-1}} \qquad [2.6\text{-}33]$$

2.7. RADIATION

2.7.1. Definitions and Laws

Heat transfer by conduction and convection, as described in the previous sections, requires the existence of a material medium, either a solid or a fluid. This, however, is not required in heat transfer by radiation. Radiation can travel through an empty space at the speed of light in the form of an electromagnetic wave. As shown in the electromagnetic spectrum in Fig. 2.7-1, **thermal radiation** covers approximately the range of wavelength from 10^{-7} to 10^{-4} m (i.e., 0.1–100 μm).

2.7.1.1. Absorptivity

Thermal radiation impinging on the surface of an opaque solid is either absorbed or reflected. The **absorptivity** is defined as the fraction of the incident

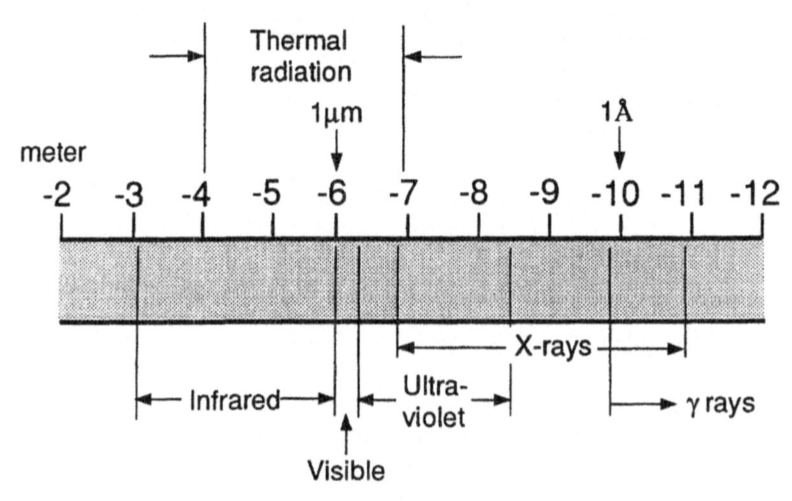

Fig. 2.7-1 Spectrum of electromagnetic radiation.

radiation that is absorbed:

$$\alpha = \frac{q^{(a)}}{q^{(i)}} \qquad [2.7\text{-}1]$$

where $q^{(a)}$ is the energy absorbed per unit area per unit time and $q^{(i)}$ is the energy impinging per unit area per unit time.

Let us define $q_\lambda^{(a)}$ and $q_\lambda^{(i)}$ such that $q_\lambda^{(a)}\,d\lambda$ and $q_\lambda^{(i)}\,d\lambda$ represent respectively the absorbed and incident energies per unit area per unit time in the wavelength range λ to $\lambda + d\lambda$. The **monochromatic absorptivity** α_λ is defined as follows:

$$\alpha_\lambda = \frac{q_\lambda^{(a)}\,d\lambda}{q_\lambda^{(i)}\,d\lambda} = \frac{q_\lambda^{(a)}}{q_\lambda^{(i)}} \qquad [2.7\text{-}2]$$

For any real body $\alpha_\lambda < 1$ and depends on λ. A **graybody** is a hypothetical one for which $\alpha_\lambda < 1$ but independent of λ and temperature. The limiting case of $\alpha_\lambda = 1$ for all λ and temperatures is known as a **blackbody**. In other words, a blackbody absorbs all the incident radiation.

2.7.1.2. Emissivity

The **emissivity** of a surface is defined as

$$\varepsilon = \frac{q^{(e)}}{q_b^{(e)}} \qquad [2.7\text{-}3]$$

where $q^{(e)}$ and $q_b^{(e)}$ are the energies emitted per unit area per unit time by a real body and a blackbody, respectively. Table 2.7-1 lists the emissivities of some materials. These averaged value have been used widely even though the emissivity actually depends on the wavelength and the angle of emission.

Let us define $q_\lambda^{(e)}$ and $q_{b\lambda}^{(e)}$ such that $q_\lambda^{(e)}\,d\lambda$ and $q_{b\lambda}^{(e)}\,d\lambda$ represent respectively the energies emitted per unit area per unit time in the wavelength range λ to $\lambda + d\lambda$ by a real body and a blackbody. The **monochromatic emissivity** ε_λ is defined as follows

$$\varepsilon_\lambda = \frac{q_\lambda^{(e)}\,d\lambda}{q_{b\lambda}^{(e)}\,d\lambda} = \frac{q_\lambda^{(e)}}{q_{b\lambda}^{(e)}} \qquad [2.7\text{-}4]$$

Since at any given temperature the radiant energy emitted by a blackbody represents an upper limit to the radiant energy emitted by a real body, the emissivity is less than unity for a real body and equal to unity of a blackbody.

2.7.1.3. Kirchhoff's Law

Let us consider the body enclosed in the cavity shown in Fig. 2.7-2. Suppose the body is at the same temperature as the wall of the cavity, that is, the two are at

TABLE 2.7-1 Emissivities of Some Surfaces

Surface	T (K)	Emissivity
Aluminum		
Highly polished, 98% pure	500–850	0.039–0.057
Oxidized at 865 K	470–865	0.11–0.19
Copper		
Polished	373	0.052
Oxidized at 865 K	470–865	0.57
Stainless steels		
Polished	370	0.074
Oxidized heavily	490–800	0.90–0.97
Asbestos		
Board	295	0.96
Paper	70–645	0.93
Lampblack	465–500	0.96

Source: W. H. McAdams ed., *Heat Transmission*, 3rd ed., McGraw-Hill, New York, 1954.

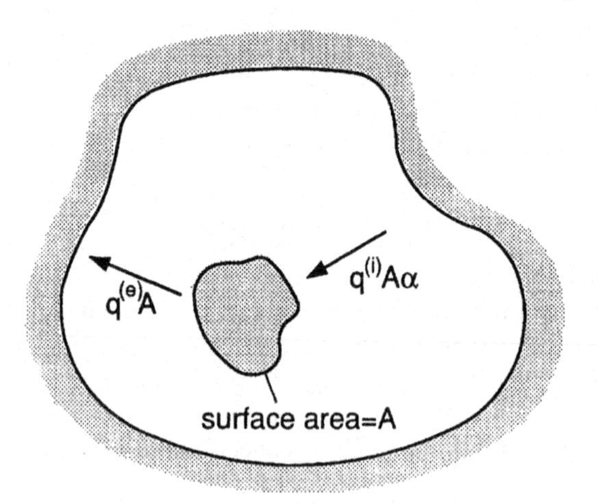

Fig. 2.7-2 A body enclosed in a cavity.

thermodynamic equilibrium with each other. Since there is no net heat transfer from the cavity wall to the body, the energy emitted by the body must be equal to the energy absorbed.

$$q^{(e)}A = q^{(i)}A\alpha \qquad [2.7\text{-}5]$$

where $q^{(e)}$ is the energy emitted per unit area of the body per unit time, A the surface area of the body, $q^{(i)}$ the energy impinging per unit area of the body per unit time, and α the absorptivity of the body. If the body is a blackbody, Eq. [2.7-5] becomes

$$q_b^{(e)} A = q^{(i)} A \qquad [2.7\text{-}6]$$

Dividing Eq. [2.7-5] by Eq. [2.7-6], we get

$$\frac{q^{(e)}}{q_b^{(e)}} = \alpha \qquad [2.7\text{-}7]$$

Substituting Eq. [2.7-7] into Eq. [2.7-3]

$$\boxed{\varepsilon = \alpha} \qquad [2.7\text{-}8]$$

This is **Kirchhoff's law**, which states that for a system in thermodynamic equilibrium the emissivity and absorptivity are the same. Following a similar procedure, we can show that

$$\boxed{\varepsilon_\lambda = \alpha_\lambda} \qquad [2.7\text{-}9]$$

2.7.1.4. Planck's Distribution Law

Planck[17] derived the following famous equation for the energy emitted by a blackbody as a function of the wavelength and temperature:

$$\boxed{q_{b\lambda}^{(e)} = \frac{2\pi c^2 h \lambda^{-5}}{\exp(ch/k\lambda T) - 1}} \qquad [2.7\text{-}10]$$

where T = temperature in K
c = speed of light = 3.0×10^{10} cm/s
h = Planck's constant = 6.624×10^{-27} erg s
k = Boltzmann's constant = 1.380×10^{-16} erg/K

The Planck distribution is shown in Fig. 2.7-3.

2.7.1.5. Stefan–Boltzmann Law

The Planck distribution law can be integrated over wavelengths from zero to infinity to determine the total emissive energy of a blackbody:

$$q_b^{(e)} = \int_0^\infty q_{b\lambda}^{(e)}\, d\lambda = \frac{2\pi^5 k^4 T^4}{15 c^2 h^3} \qquad [2.7\text{-}11]$$

[17]M. Planck, *Verh. d. dent. physik. Gesell.*, **2**, 237, 1900.

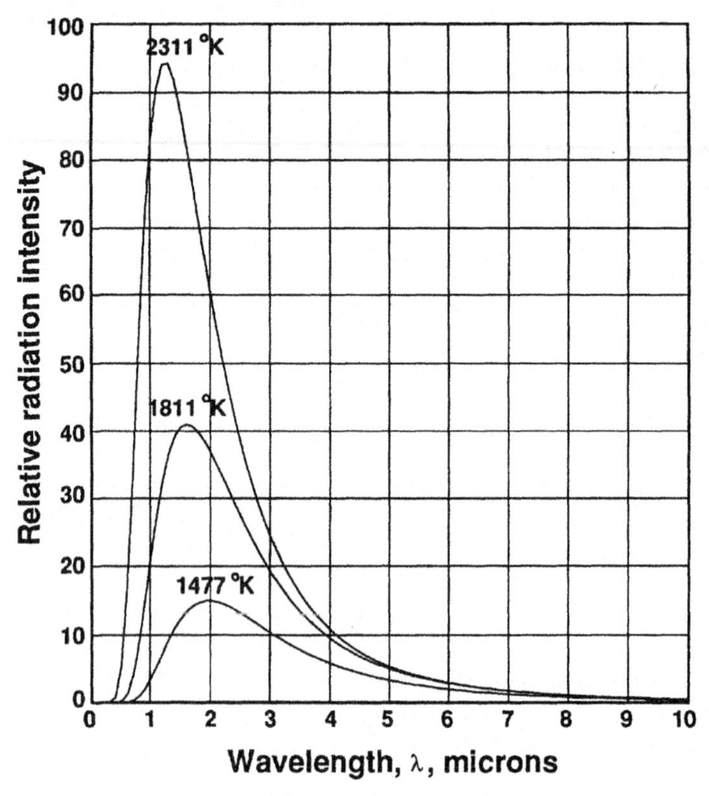

Fig. 2.7-3 Blackbody emissive power as a function of wavelength and temperature according to Planck's law.

or

$$\boxed{q_b^{(e)} = \sigma T^4} \qquad [2.7\text{-}12]$$

where σ (Stefan–Boltzmann constant) $= 5.676 \times 10^{-8}\ \mathrm{W\,m^{-2}\,K^{-4}}$.

2.7.1.6. The Solid Angle

Let us consider a hemisphere of radius r surrounding a differential area dA_1 at the center. For convenience, Fig. 2.7-4*a* shows only one-quarter of the hemisphere. As shown, on the hemisphere

$$dA_2 = (\overline{ab})(\overline{cd}) = (r \sin\theta\, d\phi)(r\, d\theta) \qquad [2.7\text{-}13]$$

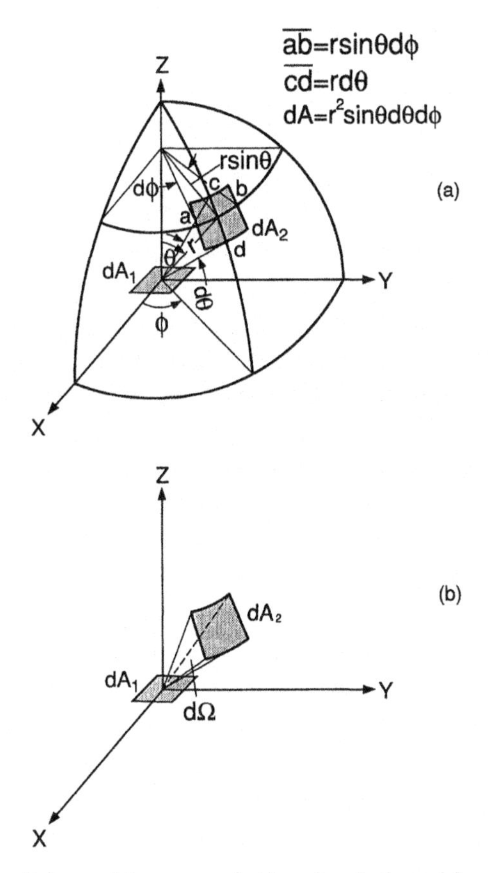

Fig. 2.7-4 A differential area dA_1 surrounded by a hemisphere: (*a*) a differential area dA_2 on the hemisphere; (*b*) the solid angle $d\Omega$.

The solid angle that intersects dA_2 on the hemisphere, shown in Fig. 2.7-4*b*, is defined as follows

$$\boxed{d\Omega = \frac{dA_2}{r^2} = \sin\theta\, d\theta\, d\phi} \qquad [2.7\text{-}14]$$

Figure 2.7-5 shows the emission from a differential surface area dA_1. Consider the direction that is at an angle θ from the normal to the surface. The projection of dA_1 along the direction is $\cos\theta\, dA_1$. Let I be the energy emitted per unit projected area per unit time per unit solid angle:

$$\boxed{I = \frac{d^2Q}{(dA_1 \cos\theta) d\Omega}} \qquad [2.7\text{-}15]$$

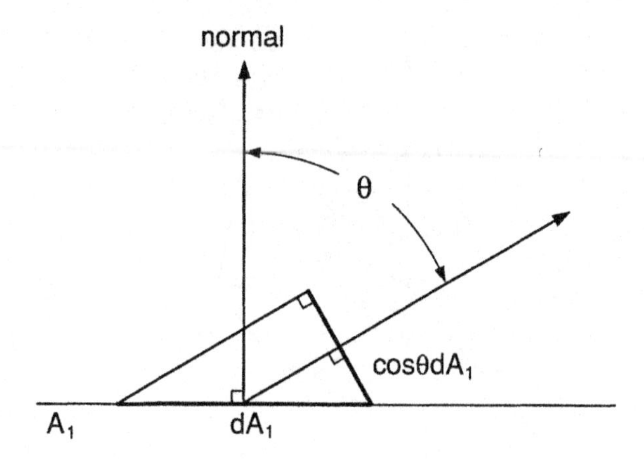

Fig. 2.7-5 Emission from a differential area dA_1.

so that

$$q^{(e)} = \frac{dQ}{dA_1} = \int I \cos\theta \, d\Omega \qquad [2.7\text{-}16]$$

where Q is the radiation heat transfer rate.

Substituting Eq. [2.7-14] into Eq. [2.7-16] and integrating over the entire hemisphere

$$q^{(e)} = I \int_{\phi=0}^{\phi=2\pi} \int_{\theta=0}^{\theta=\pi/2} \cos\theta \sin\theta \, d\theta \, d\phi \qquad [2.7\text{-}17]$$

This yields the following equation:

$$q^{(e)} = \pi I \qquad [2.7\text{-}18]$$

Substituting Eq. [2.7-18] into Eq. [2.7-15]

$$\frac{d^2Q}{\cos\theta \, dA_1 \, d\Omega} = \frac{q^{(e)}}{\pi} \qquad [2.7\text{-}19]$$

2.7.2. Radiation between Blackbodies

Let us consider the two black surfaces A_1 and A_2 shown in Fig. 2.7-6. In order to consider the heat exchange between the two surfaces, we need first to determine the amount of energy that leaves one surface and reaches the other. To do this

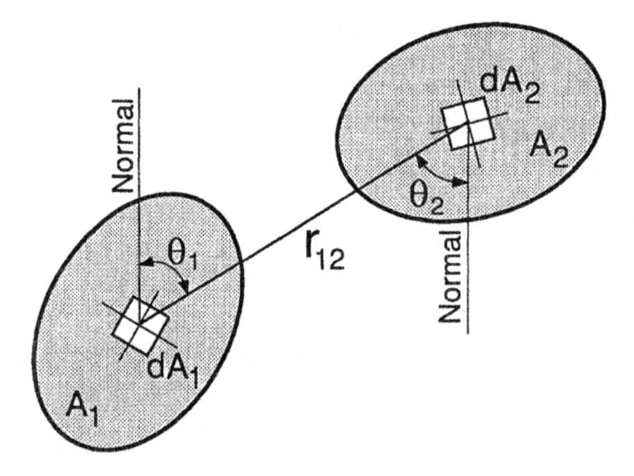

Fig. 2.7-6 Radiant heat transfer between two surfaces.

let us define the **view factors** as follows:

F_{12} = fraction of energy leaving surface 1 that reaches surface 2

F_{21} = fraction of energy leaving surface 2 that reaches surface 1

As such, the energy leaving surface 1 that reaches surface 2 per unit time is

$$Q_{1\to 2} = q_{b1}^{(e)} A_1 F_{12} \qquad [2.7\text{-}20]$$

Similarly, the energy leaving surface 2 that reaches surface 1 per unit time is

$$Q_{2\to 1} = q_{b2}^{(e)} A_2 F_{21} \qquad [2.7\text{-}21]$$

Therefore, the net energy exchange rate is as follows:

$$Q_{12} = Q_{1\to 2} - Q_{2\to 1} = q_{b1}^{(e)} A_1 F_{12} - q_{b2}^{(e)} A_2 F_{21} \qquad [2.7\text{-}22]$$

If both surfaces are at the same temperature T, there is no net energy exchange and $Q_{12} = 0$. Then, substituting Eq. [2.7-12] into Eq. [2.7-22]

$$Q_{12} = \sigma T^4 (A_1 F_{12} - A_2 F_{21}) = 0 \qquad [2.7\text{-}23]$$

so that

$$\boxed{A_1 F_{12} = A_2 F_{21}} \qquad [2.7\text{-}24]$$

which is called the **reciprocity relationship**. Substituting Eqs. [2.7-12] and [2.7-24] into Eq. [2.7-22]

$$\boxed{Q_{12} = A_1F_{12}\sigma(T_1^4 - T_2^4) = A_2F_{21}\sigma(T_1^4 - T_2^4)} \qquad [2.7\text{-}25]$$

Let us now proceed to determine the view factors F_{12} and F_{21} between the two black surfaces. From Eq. [2.7-19]

$$d^2Q_{1\to 2} = \frac{q_{b1}^{(e)}}{\pi}\cos\theta_1\, dA_1\, d\Omega_{12} \qquad [2.7\text{-}26]$$

where $d\Omega_{12}$ is the solid angle subtended by dA_2 as seen from dA_1. Let r_{12} be the distance between the centers of dA_1 and dA_2. Consider the hemisphere of radius r_{12} and centered at dA_1. The projection of dA_2 on the hemisphere is $\cos\theta_2\, dA_2$. Therefore, from Eq. [2.7-14]

$$d\Omega_{12} = \frac{\cos\theta_2\, dA_2}{r_{12}^2} \qquad [2.7\text{-}27]$$

Substituting Eq. [2.7-27] into Eq. [2.7-26]

$$d^2Q_{1\to 2} = q_{b1}^{(e)}\frac{\cos\theta_1\cos\theta_2\, dA_1\, dA_2}{\pi r_{12}^2} \qquad [2.7\text{-}28]$$

and so

$$Q_{1\to 2} = q_{b1}^{(e)}\int\int\frac{\cos\theta_1\cos\theta_2}{\pi r_{12}^2}\, dA_1\, dA_2 \qquad [2.7\text{-}29]$$

Similarly it can be shown that

$$Q_{2\to 1} = q_{b2}^{(e)}\int\int\frac{\cos\theta_1\cos\theta_2}{\pi r_{12}^2}\, dA_1\, dA_2 \qquad [2.7\text{-}30]$$

Substituting Eq. [2.7-29] and Eq. [2.7-30] respectively into Eqs. [2.7-20] and [2.7-21]

$$\boxed{F_{12} = \frac{1}{A_1}\int\int\frac{\cos\theta_1\cos\theta_2}{\pi r_{12}^2}\, dA_1\, dA_2} \qquad [2.7\text{-}31]$$

(a) large(infinite) parallel plates :
$F_{12} = 1$

(b) long(infinite) concentric cylinders :
$F_{12} = 1$

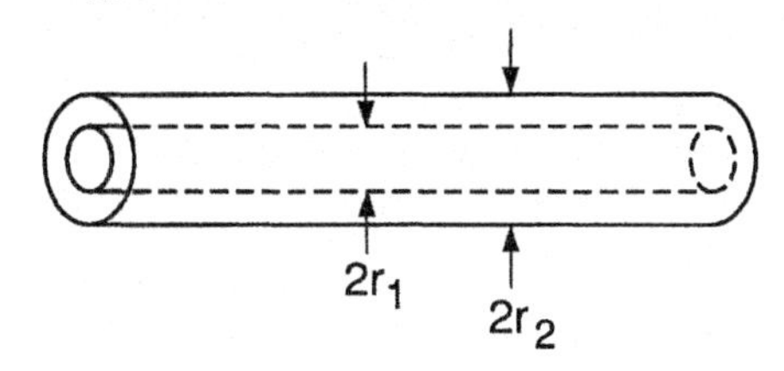

(c) concentric spheres :
$F_{12} = 1$

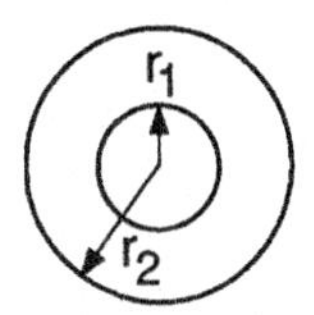

Fig. 2.7-7 Examples of view factors equal to one: (*a*) large parallel plates; (*b*) long concentric cylinders; (*c*) concentric spheres.

and

$$\boxed{F_{21} = \frac{1}{A_2} \int\!\!\int \frac{\cos\theta_1 \cos\theta_2}{\pi r_{12}^2}\, dA_1\, dA_2} \qquad [2.7\text{-}32]$$

The integration in Eq. [2.7-31] and [2.7-32] is often difficult and needs to be done numerically. Analytical equations are available for a number of special cases.[18] Some examples are given below.

For large (infinite) parallel plates, long (infinite) concentric cylinders and concentric spheres $F_{12} = 1$, as shown in Fig. 2.7-7.

[18] R. Siegel and J. R. Howell, *Termal Radiation Heat Transfer*, 2nd ed., Hemisphere Publishing, New York, 1981.

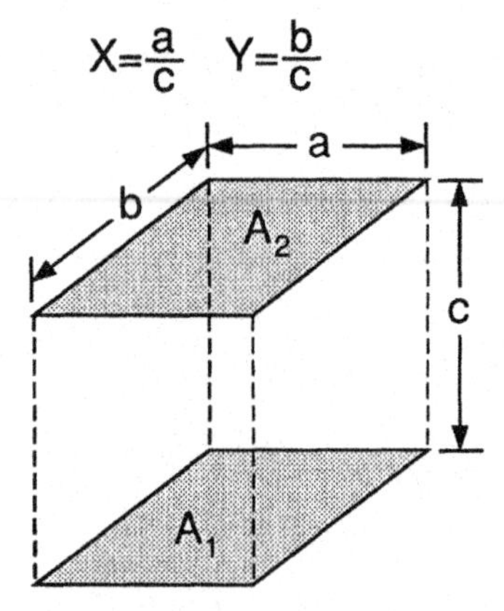

Fig. 2.7-8 Two identical rectangles parallel to each other.

For the two identical, parallel, directly opposed rectangles shown in Fig. 2.7-8

$$F_{12} = \frac{2}{\pi XY} \left\{ \ln \left[\frac{(1 + X^2)(1 + Y^2)}{1 + X^2 + Y^2} \right]^{1/2} + X\sqrt{1 + Y^2} \tan^{-1} \frac{X}{\sqrt{1 + Y^2}} + Y\sqrt{1 + X^2} \tan^{-1} \frac{Y}{\sqrt{1 + X^2}} - X \tan^{-1} X - Y \tan^{-1} Y \right\} \quad [2.7\text{-}33]$$

where X and Y are as defined in the figure.

For the two parallel concentric circular disks shown in Fig. 2.7-9,

$$F_{12} = \frac{1}{2} \left[X - \sqrt{X^2 - 4\left(\frac{R_2}{R_1}\right)^2} \right] \quad [2.7\text{-}34]$$

where X, R_1, and R_2 are as defined in the figure.

For the sphere and the disk shown in Fig. 2.7-10,

$$F_{12} = \frac{1}{2}\left(1 - \frac{1}{\sqrt{1 + R_2^2}}\right) \quad [2.7\text{-}35]$$

2.7.3. Radiation between Graybodies

In radiation heat transfer between blackbodies, all the radiant energy that strikes a surface is absorbed. In radiation heat transfer between nonblackbodies, however, all the energy striking a surface will not be absorbed; part will be reflected. Since the energy emitted and the energy reflected by a nonblack

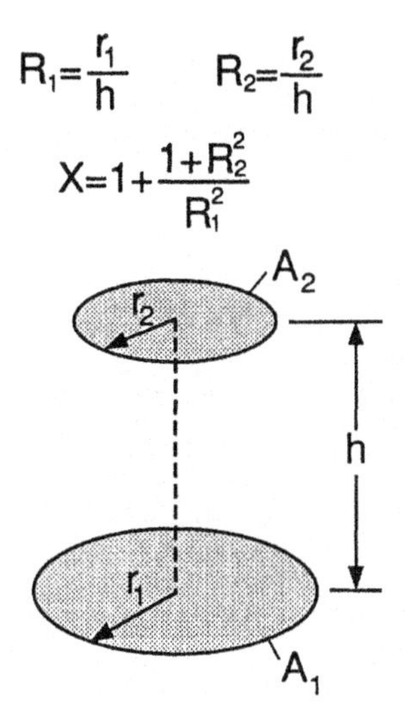

Fig. 2.7-9 Two parallel disks.

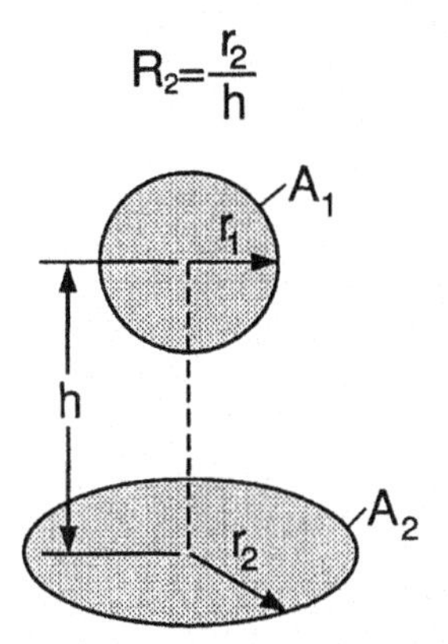

Fig. 2.7-10 A sphere over a disk.

surface both contribute to the total energy leaving the surface, we can write

$$J = q^{(e)} + \rho q^{(i)} = \varepsilon q_b^{(e)} + \rho q^{(i)} \qquad [2.7\text{-}36]$$

where J, called the **radiosity**, is the total energy leaving a surface per unit area per unit time and ρ, called the **reflectivity**, is the fraction of the incident energy reflected.

Note that the emissivity ε has been defined in Eq. [2.7-3].

From this definition for J, we see that

$$J - q^{(i)} = \text{net energy leaving a surface per unit area per unit time}$$

and

$$q^{(i)} - J = \text{net energy received at a surface per unit area per unit time}$$

Let us find an expression for $J - q^{(i)}$. For an opaque material the incident energy is either reflected or absorbed:

$$\rho + \alpha = 1 \qquad [2.7\text{-}37]$$

If Kirchhoff's law can be applied, that is, $\varepsilon = \alpha$ according to Eq. [2.7-8], Eq. [2.7-37] becomes

$$\rho + \varepsilon = 1 \qquad [2.7\text{-}38]$$

From Eqs. [2.7-36] and [2.7-38]

$$J - q^{(i)} = J - \frac{J - \varepsilon q_b^{(e)}}{\rho} = \frac{(1-\varepsilon)J - J + \varepsilon q_b^{(e)}}{1-\varepsilon} = \frac{q_b^{(e)} - J}{(1/\varepsilon) - 1} \qquad [2.7\text{-}39]$$

Let us now consider radiation heat transfer between two gray surfaces A_1 and A_2. Since the net energy transfer from A_1 to A_2 (i.e., Q_{12}) equals either the net energy leaving A_1 or the net energy received at A_2, we can write, with the help of Eq. [2.7-39]:

$$Q_{12} = A_1[J_1 - q_1^{(i)}] = A_1\left(\frac{q_{b1}^{(e)} - J_1}{(1/\varepsilon_1) - 1}\right) \qquad [2.7\text{-}40]$$

and

$$Q_{12} = A_2[q_2^{(i)} - J_2] = A_2\left(\frac{J_2 - q_{b2}^{(e)}}{(1/\varepsilon_2) - 1}\right) \qquad [2.7\text{-}41]$$

Similar to Eqs. [2.7-20] and [2.7-21] for two black surfaces, we can write the following expressions for two gray surfaces in terms of the view factors:

$$Q_{1\to 2} = J_1 A_1 F_{12} \qquad [2.7\text{-}42]$$

and

$$Q_{2\to 1} = J_2 A_2 F_{21} \qquad [2.7\text{-}43]$$

Since $A_1F_{12} = A_2F_{21}$ according to Eq. [2.7-24]

$$Q_{12} = Q_{1\to 2} - Q_{2\to 1} = A_1F_{12}(J_1 - J_2) \qquad [2.7\text{-}44]$$

From Eqs. [2.7-40], [2.7-44], and [2.7-41]

$$q_{b1}^{(e)} - J_1 = Q_{12}\left(\frac{(1/\varepsilon_1) - 1}{A_1}\right) \qquad [2.7\text{-}45]$$

$$J_1 - J_2 = Q_{12}\left(\frac{1}{A_1F_{12}}\right) \qquad [2.7\text{-}46]$$

$$J_2 - q_{b2}^{(e)} = Q_{12}\left(\frac{(1/\varepsilon_2) - 1}{A_2}\right) \qquad [2.7\text{-}47]$$

Adding these three equations, we have

$$q_{b1}^{(e)} - q_{b2}^{(e)} = Q_{12}\left[\frac{(1/\varepsilon_1) - 1}{A_1} + \frac{1}{A_1F_{12}} + \frac{(1/\varepsilon_2) - 1}{A_2}\right] \qquad [2.7\text{-}48]$$

Substituting Eq. [2.7-12] and rearranging, we obtain

$$\boxed{Q_{12} = A_1\mathscr{F}_{12}\sigma(T_1^4 - T_2^4) = A_2\mathscr{F}_{21}\sigma(T_1^4 - T_2^4)} \qquad [2.7\text{-}49]$$

where

$$\boxed{\frac{1}{A_1\mathscr{F}_{12}} = \frac{1}{A_2\mathscr{F}_{21}} = \frac{(1/\varepsilon_1) - 1}{A_1} + \frac{1}{A_1F_{12}} + \frac{(1/\varepsilon_2) - 1}{A_2}} \qquad [2.7\text{-}50]$$

Equation [2.7-49] is similar in form to Eq. [2.7-25] for blackbodies.

2.7.4. Examples

Example 2.7-1 Radiation between Two Black Disks

Two black disks of diameter 50 cm are placed directly opposite each other at a distance of 100 cm. The temperatures of disk 1 and disk 2 are 1000 and 500 K, respectively. Calculate the radiation heat transfer rate between the two disks.

Solution From Figure 2.7-9

$$R_1 = R_2 = \frac{50 \text{ cm}/2}{100 \text{ cm}} = 0.25 \qquad [2.7\text{-}51]$$

$$X = 1 + \frac{1 + (0.25)^2}{(0.25)^2} = 18 \qquad [2.7\text{-}52]$$

From Eq. [2.7-34]

$$F_{12} = \tfrac{1}{2}[18 - \sqrt{18^2 - 4}] = 0.0557 \qquad [2.7\text{-}53]$$

From Eq. [2.7-25]

$$Q_{12} = A_1 F_{12} \sigma (T_1^4 - T_2^4) = A_2 F_{21} \sigma (T_1^4 - T_2^4) \qquad [2.7\text{-}54]$$

and so

$$Q_{12} = \frac{\pi}{4}(0.5 \text{ m})^2 \times 0.0557 \times 5.676 \times 10^{-8} \frac{W}{\text{m}^2\,\text{K}^4}$$

$$[(1000 \text{ K})^4 - (500 \text{ K})^4] = 582 \text{ W} \qquad [2.7\text{-}55]$$

Example 2.7-2 Radiation between Two Gray Disks

The two black disks in Example 2.7-1 are now coated with a layer of copper and oxidized; the dimension and temperatures of the disks are unchanged. Calculate the radiation heat transfer rate between the two disks.

Solution From Table 2.7-1, the emissivity of the two copper disks, when oxidized in air, is estimated to be around 0.57%. From Eqs. [2.7-50] and [2.7-53]

$$\frac{1}{A_1 \mathscr{F}_{12}} = \frac{(1/0.57) - 1}{(\pi/4)(0.5 \text{ m})^2} + \frac{1}{(\pi/4)(0.5 \text{ m})^2 \times 0.0557} + \frac{(1/0.57) - 1}{(\pi/4)(0.5 \text{ m})^2} \qquad [2.7\text{-}56]$$

so that

$$A_1 \mathscr{F}_{12} = 0.0100 \text{ m}^2 \qquad [2.7\text{-}57]$$

From Eq. [2.7-49]

$$Q_{12} = A_1\mathscr{F}_{12}\sigma(T_1^4 - T_2^4) = A_2\mathscr{F}_{21}\sigma(T_1^4 - T_2^4) \qquad [2.7\text{-}58]$$

and so

$$\begin{aligned} Q_{12} &= 0.0100 \text{ m}^2 \times 5.676 \times 10^{-8} \text{ W m}^{-2}\text{K}^{-4} [(1000 \text{ K})^4 - (500 \text{ K})^4] \\ &= 532 \text{ W} \qquad [2.7\text{-}59] \end{aligned}$$

This is about 10% lower than that for the two black disks.

Example 2.7-3 Radiation Shields

Radiation shields of low-emissivity (high-reflectivity) materials are often used to reduce radiation heat transfer between two surfaces. Consider the radiation shield between two large parallel gray planes having the same area A, as shown in Fig. 2.7-11. Find the radiation heat transfer rate between the two parallel gray surfaces.

Solution Since $A_1 = A_2 = A_3$ and $F_{13} = F_{32} = 1$, from Eq. [2.7-50]

$$\frac{1}{A\mathscr{F}_{13}} = \frac{1}{A}\left(\frac{1}{\varepsilon_1} + \frac{1}{\varepsilon_3} - 1\right) \qquad [2.7\text{-}60]$$

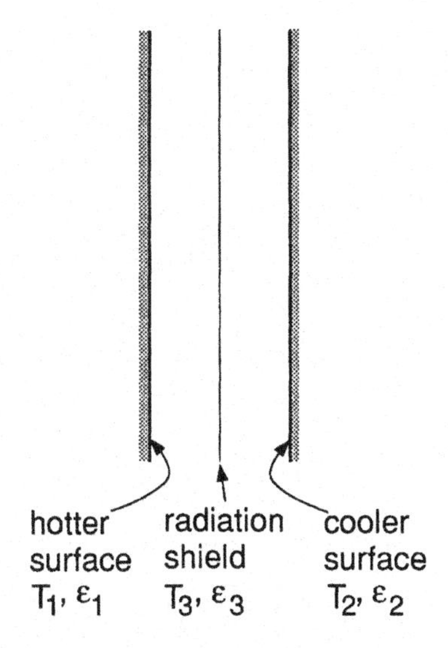

Fig. 2.7-11 Radiation shield between two large parallel surfaces.

and

$$\frac{1}{A\mathscr{F}_{32}} = \frac{1}{A}\left(\frac{1}{\varepsilon_3} + \frac{1}{\varepsilon_2} - 1\right) \qquad [2.7\text{-}61]$$

Substituting these two equations into Eq. [2.7-49]

$$Q_{13} = \frac{A\sigma(T_1^4 - T_3^4)}{(1/\varepsilon_1) + (1/\varepsilon_3) - 1} \qquad [2.7\text{-}62]$$

and

$$Q_{32} = \frac{A\sigma(T_3^4 - T_2^4)}{(1/\varepsilon_3) + (1/\varepsilon_2) - 1} \qquad [2.7\text{-}63]$$

Since $Q_{13} = Q_{32} = Q_{132}$, from Eqs. [2.7-62] and [2.7-63]

$$T_1^4 - T_3^4 = \frac{Q_{132}}{A\sigma}\left(\frac{1}{\varepsilon_1} + \frac{1}{\varepsilon_3} - 1\right) \qquad [2.7\text{-}64]$$

and

$$T_3^4 - T_2^4 = \frac{Q_{132}}{A\sigma}\left(\frac{1}{\varepsilon_3} + \frac{1}{\varepsilon_2} - 1\right) \qquad [2.7\text{-}65]$$

Adding the two equations

$$T_1^4 - T_2^4 = \frac{Q_{132}}{A\sigma}\left[\left(\frac{1}{\varepsilon_1} + \frac{1}{\varepsilon_3} - 1\right) + \left(\frac{1}{\varepsilon_3} + \frac{1}{\varepsilon_2} - 1\right)\right] \qquad [2.7\text{-}66]$$

and so

$$\boxed{Q_{132} = \frac{A\sigma(T_1^4 - T_2^4)}{[(1/\varepsilon_1) + (1/\varepsilon_3) - 1] + [(1/\varepsilon_3) + (1/\varepsilon_2) - 1]}} \qquad [2.7\text{-}67]$$

Without a radiation shield, from Eqs. [2.7-49] and [2.7-50]

$$Q_{12} = \frac{A\sigma(T_1^4 - T_2^4)}{(1/\varepsilon_1) + (1/\varepsilon_2) - 1} \qquad [2.7\text{-}68]$$

Therefore, the ratio of radiation heat transfer with a shield to that without one is

$$\frac{Q_{132}}{Q_{12}} = \frac{(1/\varepsilon_1) + (1/\varepsilon_2) - 1}{[(1/\varepsilon_1) + (1/\varepsilon_3) - 1] + [(1/\varepsilon_3) + (1/\varepsilon_2) - 1]} \qquad [2.7\text{-}69]$$

If all three surfaces have the same emissivity, that is, $\varepsilon_1 = \varepsilon_2 = \varepsilon_3 = \varepsilon$, this fraction is one-half. Multiple shields can be used to further reduce radiation heat transfer, as often practiced in high-temperature insulation.

PROBLEMS[19]

2-2.1 A wire of radius R, length L, and initial temperature T_i is cooled by a fluid of temperature T_f at time $t = 0$. The heat transfer coefficient is h. Find the temperature and the cooling rate of the wire as a function of time. Assume constant thermal properties and neglect the radial temperature gradients and end effects.

2.2-2 A thin-walled tube of inner radius R_1, outer radius R_2, and initial temperature T_i is cooled by a fluid of temperature T_f at time $t = 0$. The heat transfer coefficient is h. Find the temperature and cooling rate of the tube as a function of time. Assume constant thermal properties and neglect the radial temperature gradients and end effects.

2.2-3 A thin disk of radius R and thickness L and initial temperature T_i is cooled by a fluid of temperature T_f at time $t = 0$. The heat transfer coefficient is h. Find the temperature and cooling rate of the disk as a function of time. Assume constant thermal properties and neglect the temperature gradients in the disk.

2.2-4 Consider the steady-state heat conduction through a composite slab made of three different materials A, B, and C, each having its own thermal conductivity k_A, k_B, and k_C, respectively. A portion of the slab is shown in Fig. P2.2-4 (see p. 190). The slab is exposed to two fluids of temperature T_a and T_b; the heat transfer coefficients are h_1 and h_4, respectively. The overall heat transfer coefficient is defined by

$$Q_x = AU(T_a - T_b)$$

where Q_x is the heat flow rate through the slab. Find U as a function of the thermal conductivities and heat transfer coefficients.

2.2-5 Consider a fluid flowing through a tube of length L and inner radius R. The tube is heated from the outside and the heat flux at the inner surface q_s is constant, as shown in Fig. P2.2-5 (see p. 190). The mass flow rate is m, the average inlet temperature T_0, and the average outlet temperature T_L. Find the average fluid temperature T_{av} as a function of the distance from the inlet.

[19] The first two digits of problem numbers refer to section numbers in text.

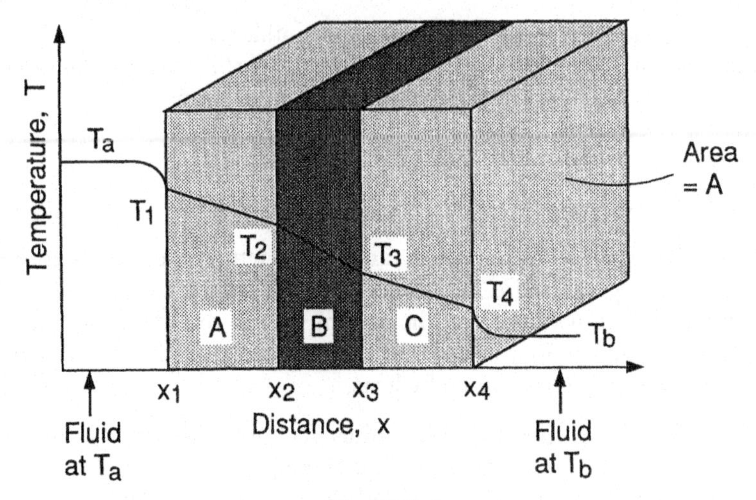

Fig. P2.2-4

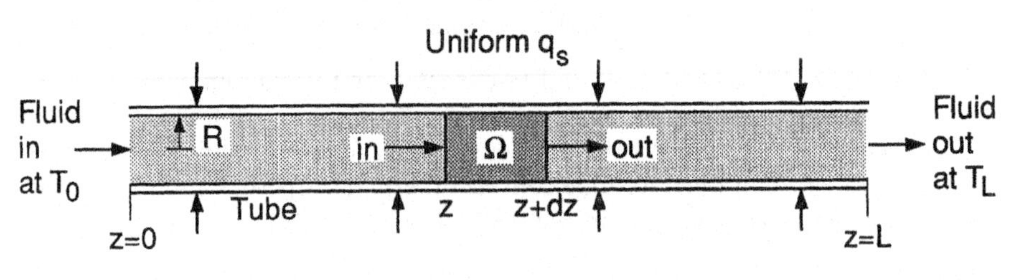

Fig. P2.2-5

2.2-6 A long wire of radius R and initial temperature T_i is resistance-heated at the rate of s per unit volume in an inert gas of temperature T_f, as shown in Fig. P2.2-6. The heat transfer coefficient is h. Find the temperature of the wire as a function of time. Assume constant thermal properties and neglect temperature gradients in the wire.

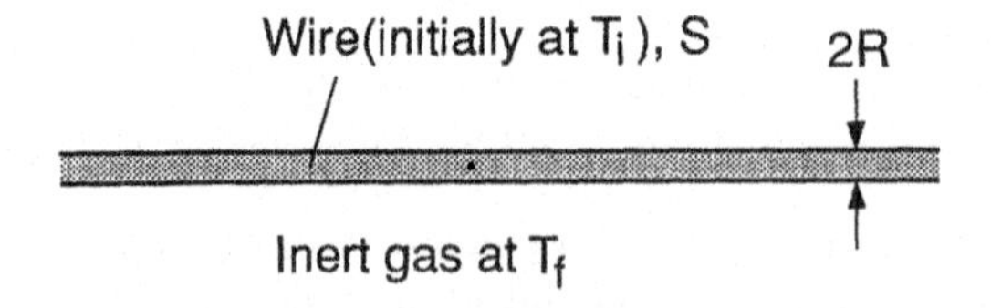

Fig. P2.2-6

2.3-1 Use the equation of energy to determine the steady-state temperature distribution and heat flux in the slab shown in Fig. 2.1-2. Assume constant thermal properties and neglect end effects.

2.3-2 A fluid is held between a lower plate of temperature T_1 and an upper plate of temperature T_2; the distance between the two plates is L. The lower plate moves at a constant velocity V while the upper one is at rest, as shown in Fig. P2.3-2. Find the steady-state temperature distribution in the fluid. Assume constant properties and neglect end effects.

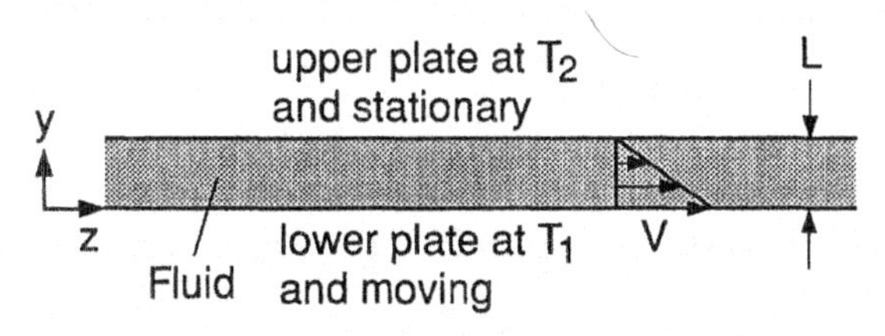

Fig. P2.3-2

2.3-3 A plate of thickness $2L$ is resistance-heated at the rate of s per unit volume; its surfaces are kept at a constant temperature T_1, as shown in Fig. P2.3-3. Find the steady-state temperature distribution in the plate. Assume constant thermal properties and neglect end effects.

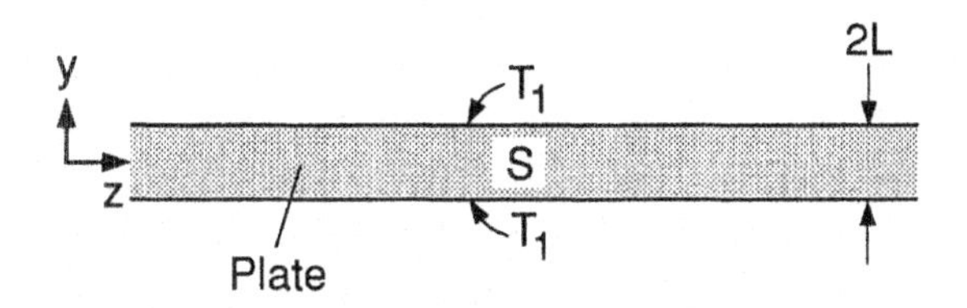

Fig. P2.3-3

2.3-4 A plate of thickness L is resistance-heated at the rate of s per unit volume; its bottom surface is kept at a constant temperature T_1 and its top surface at T_2, as shown in Fig. P2.3-4. Find the steady-state temperature distribution in the plate. Assume constant properties and neglect end effects.

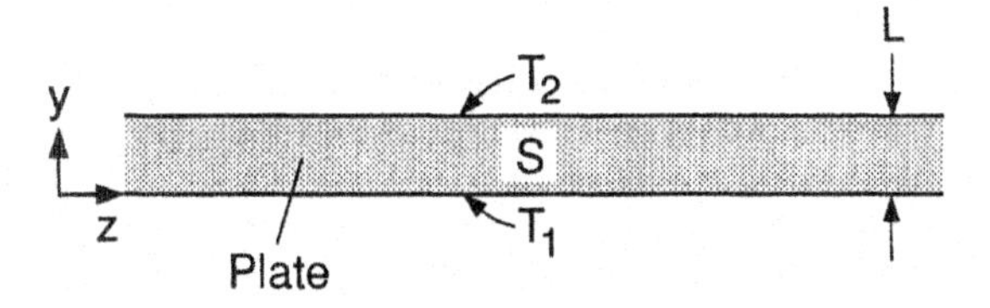

Fig. P2.3-4

2.3-5 A plate of thickness L is resistance-heated at the rate of s per unit volume; its bottom surface is kept at a constant temperature T_1 and its top surface is insulated, as shown in Fig. P2.3-5 (see p. 192). Find the steady-state temperature distribution in the plate. Assume constant properties and neglect end effects.

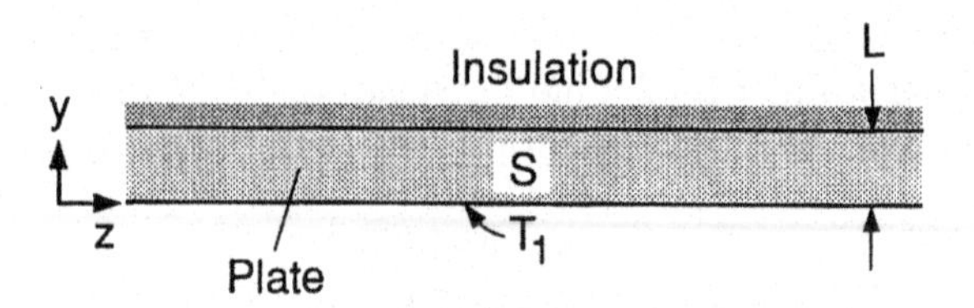

Fig. P2.3-5

2.3-6 A foil of thickness f and length $2L$ is resistance-heated at the rate of s per unit volume; its two ends are kept at a constant temperature T_1 and its surfaces are thermally insulated, as shown in Fig. P2.3-6. Find the temperature distribution along the foil. Assume constant thermal properties and neglect temperature gradients in the thickness direction.

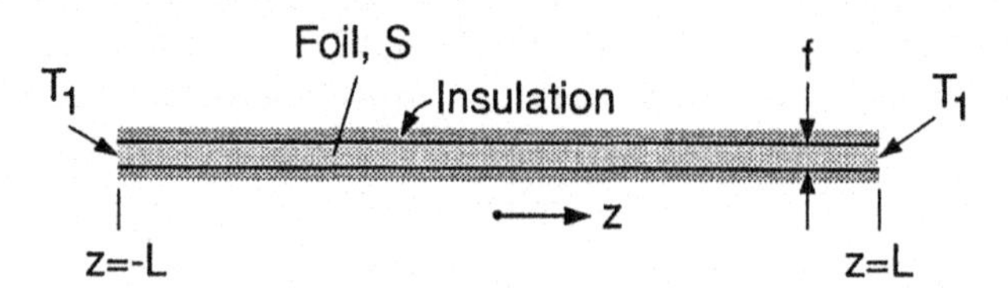

Fig. P2.3-6

2.3-7 A foil of thickness f and length L is resistance-heated at the rate of s per unit volume; its two ends are kept at constant temperature T_1 and T_2 and its surfaces are thermally insulated, as shown in Fig. P2.3-7. Find the temperature distribution along the foil. Assume constant thermal properties and neglect temperature gradients in the thickness direction.

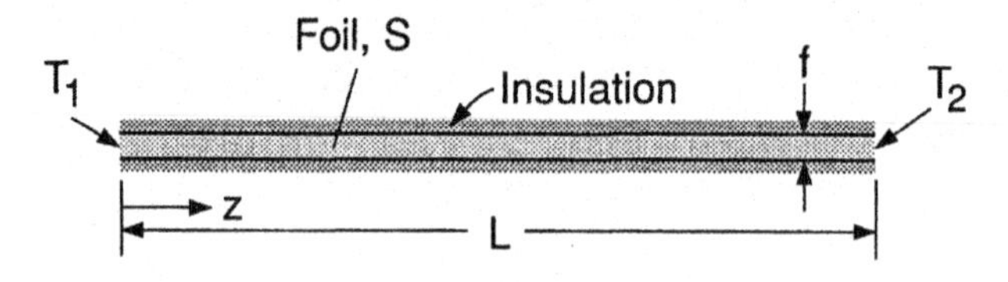

Fig. P2.3-7

2.3-8 A foil of thickness f and length $2L$ is resistance-heated at the rate of s per unit volume; its two ends are kept at a constant temperature T_1 and its surfaces are exposed to a fluid of temperature T_f, as shown in Fig. P2.3-8 (see p. 193). The heat transfer coefficient is h. Find the temperature distribution along the foil. Assume constant thermal properties and neglect temperature gradients in the thickness direction.

2.3-9 A rod of radius R is resistance-heated at the rate of s per unit volume; its surface is kept at a constant temperature T_R, as shown in Fig. P2.3-9 (see p. 193). Find the steady-state temperature distribution in the rod. Assume constant thermal properties and neglect end effects.

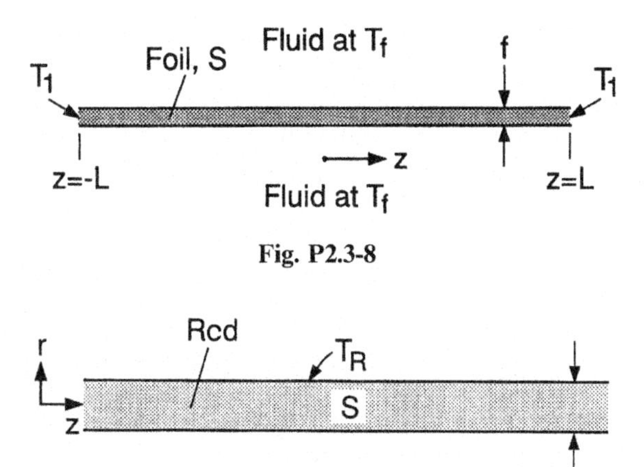

Fig. P2.3-9

2.3-10 A tube of inner radius r_1 and outer radius r_2 is kept at a constant temperature T_1 at its inner surface and T_2 at its outer surface, as shown in Fig. P2.3-10. Find the steady-state temperature distribution in the tube. Assume constant thermal properties and neglect end effects.

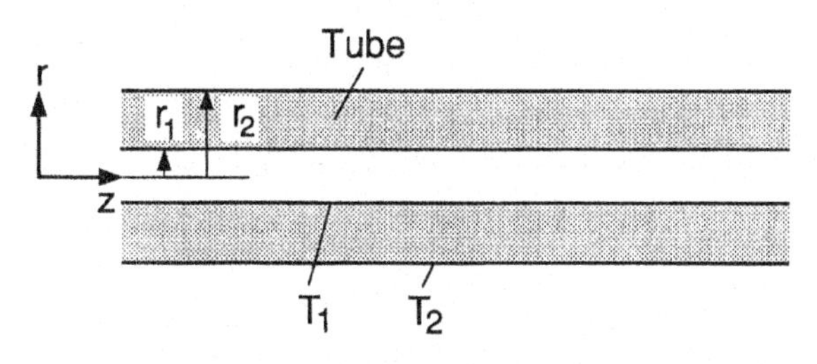

Fig. P2.3-10

2.3-11 A tube of inner radius r_1 and outer radius r_2 is resistance-heated at the rate of s per unit volume; its inner surface is kept at a constant temperature T_1 and its outer surface at T_2, as shown in Fig. P2.3-11. Find the steady-state temperature distribution in the tube. Assume constant thermal properties and neglect end effects.

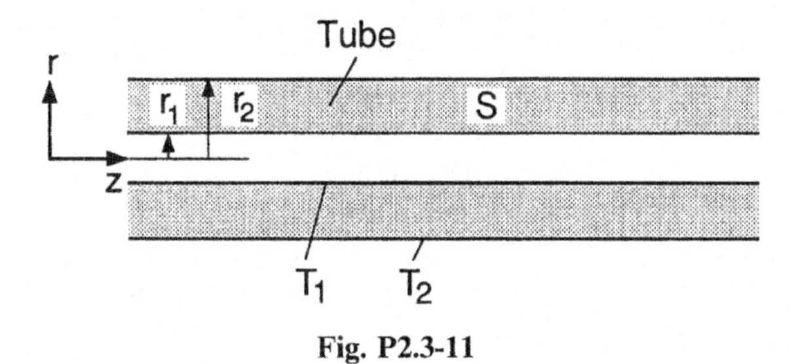

Fig. P2.3-11

2.3-12 A tube of inner radius r_1 and outer radius r_2 is resistance-heated at the rate of s per unit volume; its inner surface is kept at a constant temperature T_1 and its outer surface insulated, as shown in Fig. P2.3-12. Find the steady-state temperature distribution in the tube. Assume constant thermal properties and neglect end effects.

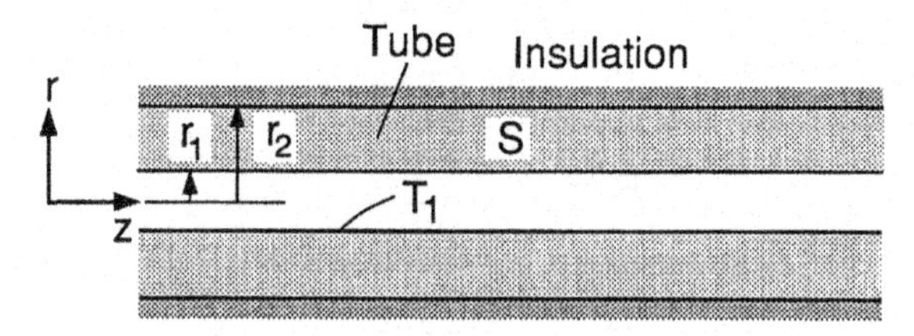

Fig. P2.3-12

2.3-13 A tube of inner radius r_1 and outer radius r_2 is resistance-heated at the rate of s per unit volume; its inner surface is insulated and its outer surface kept at a constant temperature T_2, as shown in Fig. P2.3-13. Find the temperature distribution in the tube. Assume constant thermal properties and neglect end effects.

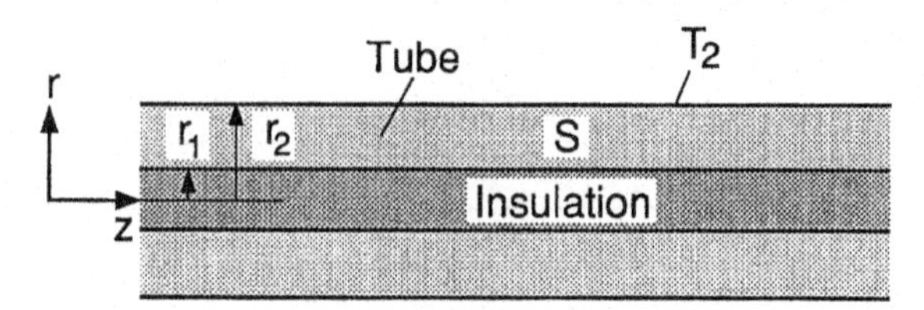

Fig. P2.3-13

2.3-14 A wire of radius R and length $2L$ is resistance-heated at the rate of s per unit volume; its two ends are kept at a constant temperature T_1 and its surface insulated, as shown in Fig. P2.3-14. Find the temperature distribution in the wire. Assume constant thermal properties and neglect temperature gradients in the radial direction.

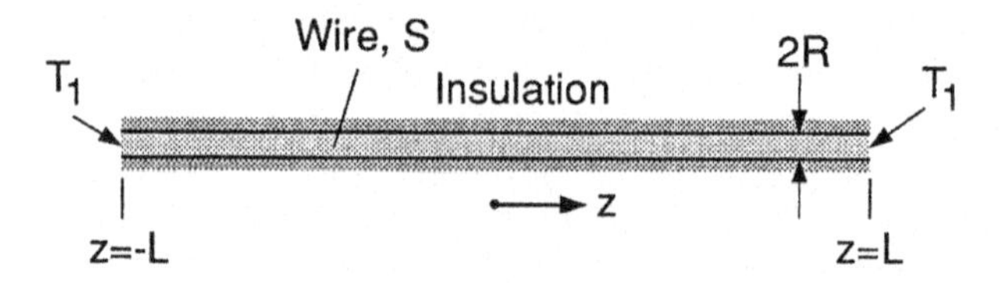

Fig. P2.3-14

2.3-15 A wire of radius R and length L is resistance-heated at the rate of s per unit volume; its two ends are kept at constant temperatures T_1 and T_2 and its surface is insulated, as shown in Fig. P2.3-15 (see p. 195). Find the temperature distribution in the wire. Assume constant thermal properties and neglect temperature gradients in the radial direction.

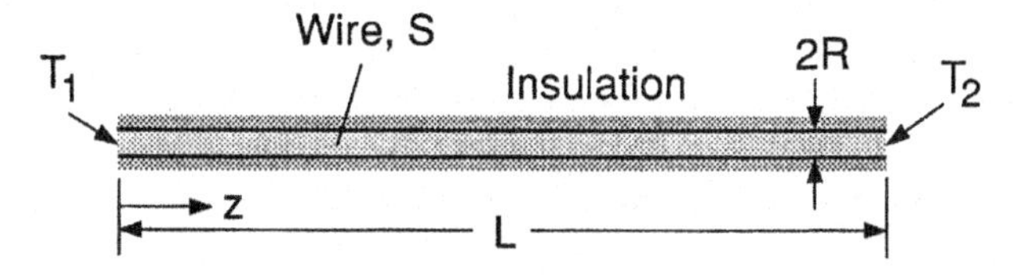

Fig. P2.3-15

2.3-16 A wire of radius R and length $2L$ is resistance-heated at the rate of s per unit volume; its two ends are kept at a constant temperature T_1 and its surface exposed to a fluid of temperature T_f, as shown in Fig. P2.3-16. The heat transfer coefficient is h. Find the steady-state temperature distribution along the wire. Assume constant thermal properties and neglect temperature gradients in the radial direction.

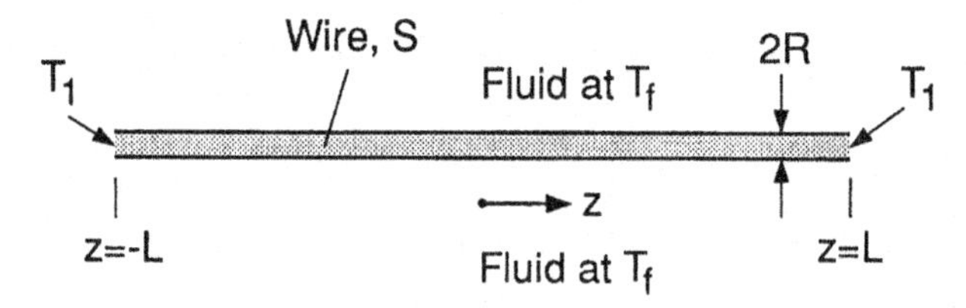

Fig. P2.3-16

2.3-17 Consider the pin fin of radius R and length L shown in Fig. P2.3-17. The temperature of the wall is T_w and the heat loss from the tip of the pin is negligible. Find the steady-state temperature distribution along the fin

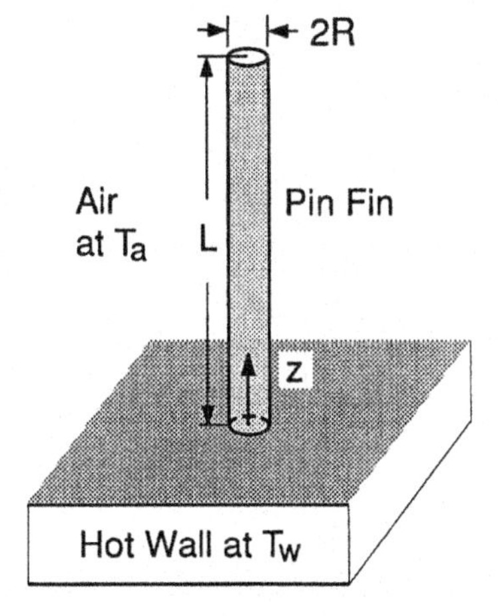

Fig. P2.3-17

and the heat loss from the pin to the ambient air of temperature T_a. The heat transfer coefficient is h.

2.3-18 A carbon steel plate ($\alpha = 0.1\ \text{cm}^2/\text{s}$) 3 cm thick is initially at 25°C. At $t = 0$, its temperature at the bottom surface is raised to 100°C (T_s) and kept at this level. At $t = 5$ s, what is the temperature at the top surface? Judging from your answer, can the plate be considered semiinfinite for this combination of T_s and t? What is the temperature at $x = 0.38$ cm and $t = 1$ s? What is the rate of temperature rise at this location and time?

2.3-19 Consider the fluid flow over the flat plate shown in Fig. 2.2-7*a*. Prove that $\partial^2(T - T_s)/\partial y^2 = 0$ at $y = 0$. Does the temperature distribution given by Eq. [2.2-61] satisfy this condition?

2.6-1 A thin-walled tube of diameter 3 cm is heated with a uniform heat flux of $0.1\ \text{W/cm}^2$ from the outside. A fluid flows through the tube at the flow rate of 2.5 g/s and an inlet temperature of 25°C. What is the length of the tube required to obtain a steady outlet temperature of 30°C? What is the inner wall temperature at the tube outlet? The physical properties of the fluid are $C_v = 4.2\ \text{J g}^{-1}\text{K}^{-1}$, $\mu = 1.1 \times 10^{-2}\ \text{g cm}^{-1}\text{s}^{-1}$, and $k = 6.7 \times 10^{-3}\ \text{W cm}^{-1}\text{K}^{-1}$.

2.7-1 Two identical black squares of 100×100 cm, parallel and directly opposed, are separated at 100 cm from each other. The temperatures of square 1 and square 2 are 2000 and 1000 K, respectively. Calculate the radiation heat transfer between the two squares.

2.7-2 Repeat Problem 2.7-1 for two gray squares with an emissivity of 0.1.

2.7-3 Find the radiation heat transfer rate between the two long, concentric cylinders shown in Fig. P2.7-3.

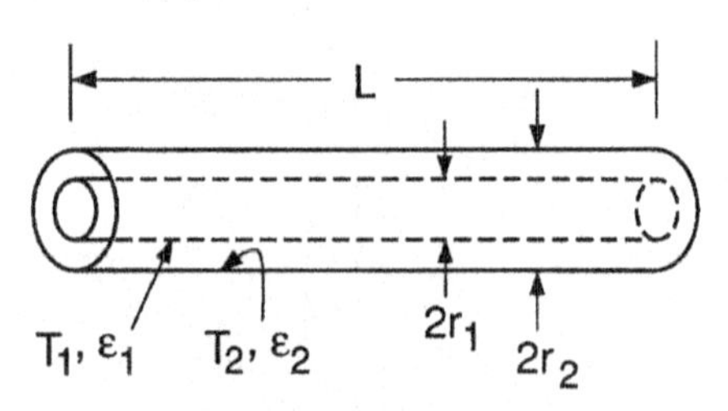

Fig. P2.7-3

2.7-4 Find the radiation heat transfer rate between two concentric spheres shown in Fig. P2.7-4 (see p. 197).

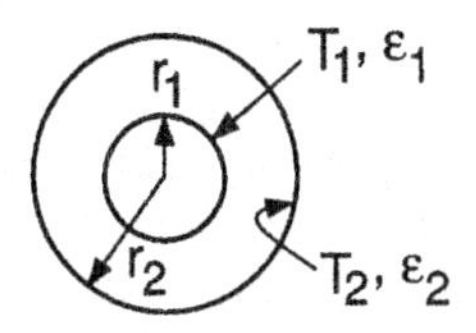

Fig. P2.7-4

2.7-5 A radiation shield is placed between two large parallel planes having the same area A, as shown in Fig. P2.7-5. The left surface of the shield has an emissivity of ε_{3L}; the right side, ε_{3R}. Find the radiation heat transfer rate between surfaces 1 and 2.

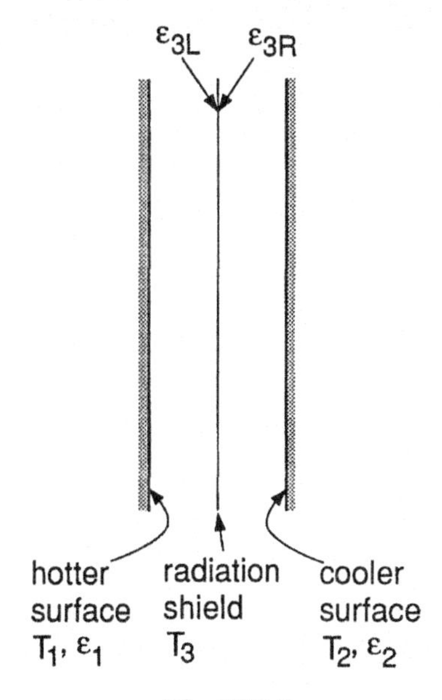

Fig. P2.7-5

2.7-6 A radiation shield is inserted between two long cylinders of radius r_1 and r_2, as shown in Fig. P2.7-6. Find the radiation heat transfer rate between the two cylinders.

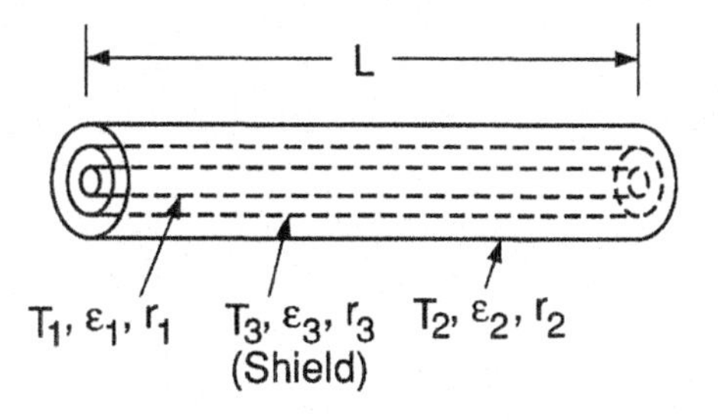

Fig. P2.7-6

2.7-7 A long wire of radius R held in a water-cooled tubular vacuum chamber of temperature T_a is resistance heated to a temperature T_i, as shown in Fig. P2.7-7 (see p. 198). At time $t = 0$ the electric current going through the wire is turned off. Find the wire temperature and cooling rate as a function of time. The wire temperatures of interest are at a level much higher than T_a. Assume constant thermal properties and neglect temperature gradients in the wire.

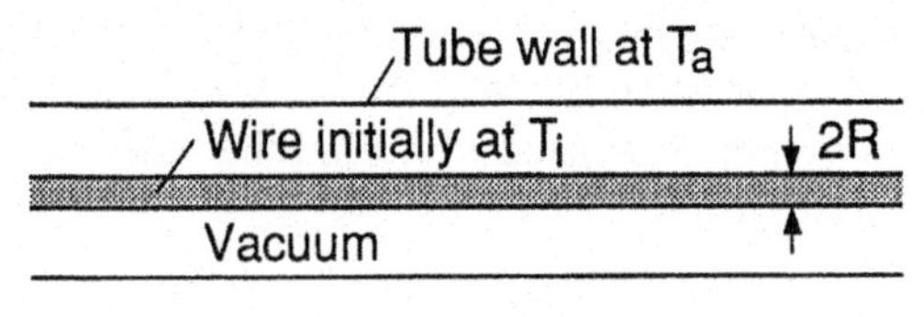

Fig. P2.7-7

CHAPTER 3

INTRODUCTION TO MASS TRANSFER

3.1. BASIC CONCEPTS

3.1.1. Diffusion and Convection

Mass transfer in this chapter refers to the transfer of a species in the presence of a concentration gradient of the species. Under a concentration gradient mass transfer can occur by either diffusion or convection.

Diffusion refers to the mass transfer that occurs across a stationary solid or fluid in which a concentration gradient exists. In contrast, **convection** refers to mass transfer that occurs across a moving fluid in which a concentration gradient exists.

The distinction between the two mechanisms of mass transfer can be appreciated by considering the dissolution of a sugar cube in water. As shown in Fig. 3.1-1*a*, the concentration of sugar molecules is significant only in the vicinity of the cube, since diffusion is a slow process. By stirring the water with a spoon to create forced convection, however, sugar molecules are transferred to the bulk water much faster, as illustrated in Fig. 3.1-1*b*. Diffusion, which occurs regardless of whether the water is stirred or not, still contributes to mass transfer but is overshadowed by convection.

3.1.2. Fick's Law of Diffusion

3.1.2.1. One-Dimensional

Let us consider the diffusion of species A through a thin sheet of thickness L, as shown in part in Fig. 3.1-2. Let w_A be the mass (or weight) fraction of species A in the sheet. The lower surface is kept at a constant concentration w_{A1} by being

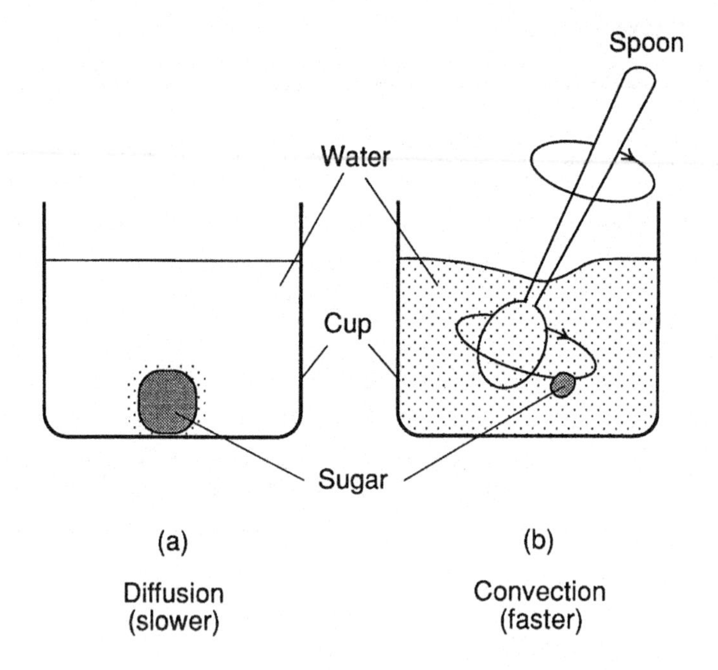

Fig. 3.1-1 Mass transfer mechanisms: (*a*) diffusion; (*b*) convection.

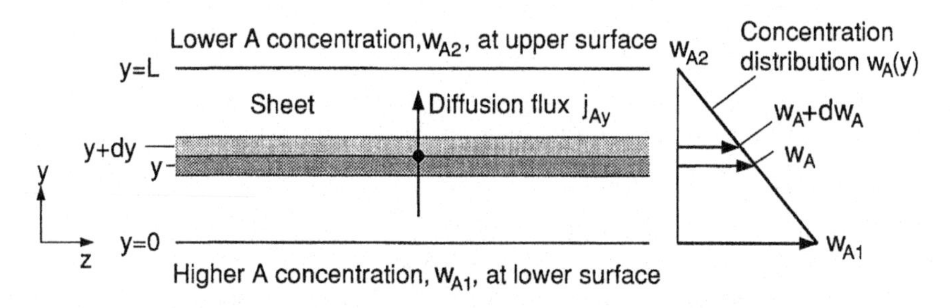

Fig. 3.1-2 One-dimensional diffusion through a thin sheet.

exposed to a fluid carrying A. The upper surface is kept at a constant but lower concentration w_{A2} by being exposed to a second fluid carrying less A per unit volume. A steady-state concentration profile $w_A(y)$ is established in the sheet, where y is the distance from the lower surface.

One such example is the diffusion of hydrogen atoms H through a sheet-steel partition of a chamber, which separates a higher-pressure H_2 at the bottom from a lower-pressure one at the top. At elevated temperatures both the solubility and diffusion coefficient of H in steel are appreciable. Another example is the diffusion of helium He through a sheet-glass partition of a chamber, which separates a natural gas containing He at the bottom and one containing no He at the top. It is well known that He diffuses through Pyrex glass far faster than any other gases.

Let us now consider two thin layers in the sheet. The lower layer at y has a concentration of w_A and the upper one at $y + dy$ has a slightly lower concentration of $w_A + dw_A$ ($dw_A < 0$). Since the lower layer is richer in A, a diffusion flux of A tends to occur from the lower layer to the upper one. A **diffusion flux** is defined as the amount of material diffused per unit area per unit time. This diffusion flux is denoted as j_{Ay} since species A diffuses in the y direction. The greater the concentration difference between the two layers is, or more appropriately, the steeper the concentration gradient dw_A/dy is, the greater the diffusion flux j_{Ay} becomes. Furthermore, the easier it is for species A to diffuse in the sheet, the greater the diffusion flux j_{Ay} becomes. In fact, it is observed that

$$\boxed{j_{Ay} = -\rho D_A \frac{dw_A}{dy}} \qquad [3.1\text{-}1]$$

where ρ is the mass density of the solution (i.e., the sheet), D_A the **diffusion coefficient** of species A in the solution, and w_A the mass (or weight) fraction of species A. This equation is **Fick's law of diffusion** for one-dimensional diffusion of species A in the y direction. The minus sign in the equation indicates that the concentration gradient is negative. It also reminds us that diffusion occurs in the direction of decreasing concentration.

The cgs and mks units of the mass diffusion flux are $\text{g cm}^{-2}\text{s}^{-1}$ and $\text{kg m}^{-2}\text{s}^{-1}$, respectively, and those of the diffusion coefficient are cm^2/s and m^2/s, respectively. The density ρ is required on the right-hand side (RHS) of the equation since w_A is the mass fraction of species A rather than the mass of species A per unit volume.

3.1.2.2. Three-Dimensional

In the case of three-dimensional diffusion, an equation such as Eq. [3.1-1] can be written for each of the three coordinates as follows:

$$j_{Ax} = -\rho D_A \frac{\partial w_A}{\partial x} \qquad [3.1\text{-}2]$$

$$j_{Ay} = -\rho D_A \frac{\partial w_A}{\partial y} \qquad [3.1\text{-}3]$$

$$j_{Az} = -\rho D_A \frac{\partial w_A}{\partial z} \qquad [3.1\text{-}4]$$

Note that it has been assumed that the material is isotropic, that is, ρ and D_A are direction-independent.

Equations [3.1-2] through [3.1-4] are the components of the following equation:

$$\mathbf{j}_A = -\rho D_A \nabla w_A \qquad [3.1\text{-}5]$$

where $\mathbf{j}_A$ is the **mass diffusion flux vector**. If the mass density ρ is constant, Eq. [3.1-5] can be written as follows:

$$\mathbf{j}_A = -D_A \nabla \rho_A \qquad [3.1\text{-}6]$$

where $\rho_A = \rho w_A$ is the mass of species A per unit volume of the solution, or the mass concentration of species A. In a dilute solution the mass density of the solution ρ is essentially constant.

Fick's law of diffusion can also be written as

$$\underline{\mathbf{j}}_A = -c D_A \nabla x_A \qquad [3.1\text{-}7]$$

where $\underline{\mathbf{j}}_A$ is the **molar diffusion flux vector,** c the molar density (i.e., moles per unit volume) of the solution, and x_A the mole fraction of species A. The underlining bar denotes a molar quantity. If the molar density of the solution is constant, Eq. [3.1-7] can be written as

$$\underline{\mathbf{j}}_A = -D_A \nabla c_A \qquad [3.1\text{-}8]$$

where $c_A = c x_A$ is the moles of species A per unit volume of the solution, that is, the molar concentration of species A. In a dilute solution the molar density of the solution c is essentially constant.

In Tables 3.1-1 and 3.1-2, Eqs. [3.1-5] and [3.1-7] are given in expanded forms for rectangular, cylindrical, and spherical coordinates.

3.1.3. Thermal Diffusion

In a nonisothermal system spatial temperature variations can induce the so-called thermal diffusion, and Fick's law of diffusion can be modified as follows:[1]

$$\mathbf{j}_A = -D_A\left(\nabla \rho_A + \frac{1}{\rho}\alpha_T \rho_A \rho_B \nabla \ln T\right) \qquad [3.1\text{-}9]$$

and dividing by M_A, the molecular weight of species A,

$$\underline{\mathbf{j}}_A = -D_A\left(\nabla c_A + \frac{1}{\rho}\underline{\alpha}_T c_A c_B \nabla \ln T\right) \qquad [3.1\text{-}10]$$

[1] R. B. Bird, W. E. Stewart, and E. N. Lightfoot, *Transport Phenomena*, Wiley, New York, 1960.

TABLE 3.1-1 Components of the Mass Diffusion Flux $\mathbf{j}_A$

Rectangular Coordinates	Cylindrical Coordinates	Spherical Coordinates
$j_{Ax} = -\rho D_A \frac{\partial w_A}{\partial x}$ [A]	$j_{Ar} = -\rho D_A \frac{\partial w_A}{\partial r}$ [D]	$j_{Ar} = -\rho D_A \frac{\partial w_A}{\partial r}$ [G]
$j_{Ay} = -\rho D_A \frac{\partial w_A}{\partial y}$ [B]	$j_{A\theta} = -\rho D_A \frac{1}{r}\frac{\partial w_A}{\partial \theta}$ [E]	$j_{A\theta} = -\rho D_A \frac{1}{r}\frac{\partial w_A}{\partial \theta}$ [H]
$j_{Az} = -\rho D_A \frac{\partial w_A}{\partial z}$ [C]	$j_{Az} = -\rho D_A \frac{\partial w_A}{\partial z}$ [F]	$j_{A\phi} = -\rho D_A \frac{1}{r\sin\theta}\frac{\partial w_A}{\partial \phi}$ [I]

TABLE 3.1-2 Components of the Molar Diffusion Flux $\underline{\mathbf{j}}_A$

Rectangular Coordinates	Cylindrical Coordinates	Spherical Coordinates
$\underline{j}_{Ax} = -cD_A \frac{\partial x_A}{\partial x}$ [A]	$\underline{j}_{Ar} = -cD_A \frac{\partial x_A}{\partial r}$ [D]	$\underline{j}_{Ar} = -cD_A \frac{\partial x_A}{\partial r}$ [G]
$\underline{j}_{Ay} = -cD_A \frac{\partial x_A}{\partial y}$ [B]	$\underline{j}_{A\theta} = -cD_A \frac{1}{r}\frac{\partial x_A}{\partial \theta}$ [E]	$\underline{j}_{A\theta} = -cD_A \frac{1}{r}\frac{\partial x_A}{\partial \theta}$ [H]
$\underline{j}_{Az} = -cD_A \frac{\partial x_A}{\partial z}$ [C]	$\underline{j}_{Az} = -cD_A \frac{\partial x_A}{\partial z}$ [F]	$\underline{j}_{A\phi} = -cD_A \frac{1}{r\sin\theta}\frac{\partial x_A}{\partial \phi}$ [I]

where α_T and $\underline{\alpha}_T$ are **thermal diffusion factors** based on mass and molar concentrations, respectively. The two factors are related to each other through $\underline{\alpha}_T = M_B \alpha_T$, there M_B is the molecular weight of species B.

3.1.4. Diffusion Boundary Layer

Consider a fluid approaching a flat plate in the direction parallel to the plate, as illustrated in Fig. 3.1-3. The plate is coated with a material containing species A, which has a limited solubility in the fluid. The composition of the approaching fluid is $w_{A\infty}$ and that of the fluid at the plate surface is w_{AS}, both of which are constant. Usually, w_{AS} is taken as the concentration of the fluid in equilibrium with the concentration of species A in the coating. Because of the effect of diffusion, the concentration of the fluid in the region near the plate is affected by the coating, varying from w_{AS} at the plate surface to $w_{A\infty}$ in the stream. This region is called the **diffusion** or **concentration boundary layer**. Its thickness δ_c is typically taken as the distance from the plate surface at which the dimensionless concentration $(w_A - w_{AS})/(w_{A\infty} - w_{AS})$ or $(w_{AS} - w_A)/(w_{AS} - w_{A\infty})$ levels off to 0.99. In practice it is usually specified that $w_A = w_{A\infty}$ and $\partial w_A/\partial y = 0$ at $y = \delta_c$. With increasing distance from the leading edge of the plate, the effect of diffusion penetrates farther into the stream and the boundary layer grows in thickness.

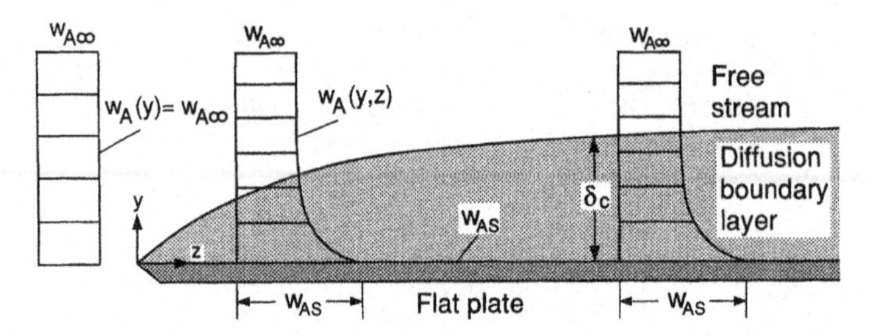

Fig. 3.1-3 Concentration boundary layer over a flat plate coated with species A.

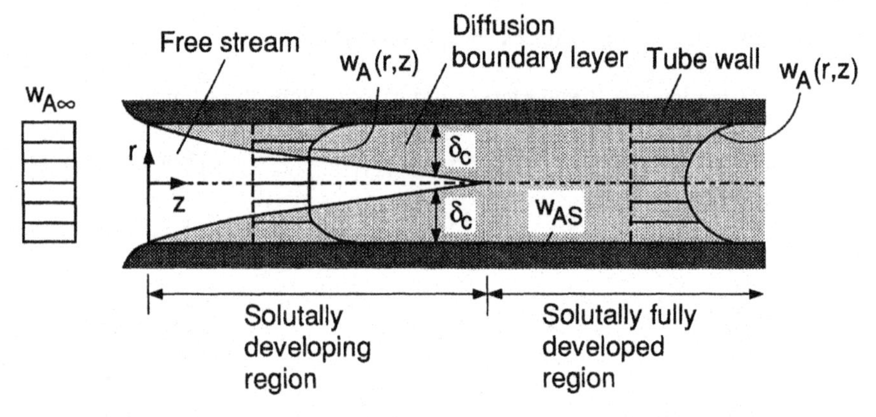

Fig. 3.1-4 Concentration boundary layer in a circular tube coated with species A.

The effect of diffusion is significant only in the boundary layer. Beyond it the concentration is uniform and the effect or diffusion is no longer significant.

Let us now consider a fluid of uniform concentration $w_{A\infty}$ entering a coated circular tube, as shown in Fig. 3.1-4. The fluid at the wall has a uniform concentration w_{AS}. A diffusion boundary layer begins to develop at the entrance, gradually expanding until the layer from opposite sides approach the centerline. The concept of the solutally fully developed region is similar to that of the thermally fully developed region described in Section 2.1-3. A tube with a coating of uniform w_{AS} is analogous to a pipe with uniform wall temperature T_S.

3.1.5. Mass Transfer Flux

Let $\mathbf{v}_A$ and $\mathbf{v}_B$ be the velocities of species A and B with respect to stationary coordinates, respectively. These species velocities result from both the bulk motion of the fluid at velocity $\mathbf{v}$ and the diffusion of the species superimposed on the bulk motion. Consequently, the **mass flux** of species A with respect to stationary coordinates, specifically, $\mathbf{n}_A = \rho_A \mathbf{v}_A$, can be considered to result from a mass flux due to the bulk motion of the fluid, $\rho_A \mathbf{v}$, and a mass flux due to the diffusion superimposed on the bulk motion, $\mathbf{j}_A$. In other words,

$$\boxed{\mathbf{n}_A \equiv \rho_A \mathbf{v}_A = \rho_A \mathbf{v} + \mathbf{j}_A} \tag{3.1-11}$$

For a binary system consisting of species A and B, which is the focus of the present chapter, the velocity $\mathbf{v}$ is a local **mass average velocity** defined by

$$\mathbf{v} = \frac{\rho_A \mathbf{v}_A + \rho_B \mathbf{v}_B}{\rho_A + \rho_B} = \frac{\rho_A \mathbf{v}_A + \rho_B \mathbf{v}_B}{\rho} \left(= \frac{\mathbf{n}_A + \mathbf{n}_B}{\rho} \right) = w_A \mathbf{v}_A + w_B \mathbf{v}_B \tag{3.1-12}$$

Substituting Eq. [3.1-12] into Eq. [3.1-11]

$$\boxed{\mathbf{n}_A = w_A \rho \mathbf{v} + \mathbf{j}_A = w_A(\mathbf{n}_A + \mathbf{n}_B) + \mathbf{j}_A} \tag{3.1-13}$$

Similarly, the **molar flux** of species A with respect to stationary coordinates, specifically, $\underline{\mathbf{n}}_A = c_A \mathbf{v}_A$, can be considered to result from a molar flux, $c_A \underline{\mathbf{v}}$, due to the bulk motion of the fluid at velocity $\underline{\mathbf{v}}$ and a molar flux due to the diffusion superimposed on the bulk motion, $\underline{\mathbf{j}}_A$. In other words,

$$\boxed{\underline{\mathbf{n}}_A \equiv c_A \mathbf{v}_A = c_A \underline{\mathbf{v}} + \underline{\mathbf{j}}_A} \tag{3.1-14}$$

where the local **molar average velocity**

$$\underline{\mathbf{v}} = \frac{c_A \mathbf{v}_A + c_B \mathbf{v}_B}{c_A + c_B} = \frac{c_A \mathbf{v}_A + c_B \mathbf{v}_B}{c} \left(= \frac{\underline{\mathbf{n}}_A + \underline{\mathbf{n}}_B}{c} \right) = x_A \mathbf{v}_A + x_B \mathbf{v}_B \tag{3.1-15}$$

Substituting Eq. [3.1-15] into Eq. [3.1-14]

$$\boxed{\underline{\mathbf{n}}_A = x_A c \underline{\mathbf{v}} + \underline{\mathbf{j}}_A = x_A(\underline{\mathbf{n}}_A + \underline{\mathbf{n}}_B) + \underline{\mathbf{j}}_A} \tag{3.1-16}$$

3.1.6. Mass Transfer Coefficient

Consider fluid flow over a flat plate such as that shown in Fig. 3.1-3. The mass diffusion flux across the solid/liquid interface is

$$j_{Ay}|_{y=0} = -\rho D_A \left.\frac{\partial w_A}{\partial y}\right|_{y=0} \tag{3.1-17}$$

This equation cannot be used to calculate the diffusion flux when the concentration gradient is an unknown. A convenient way to avoid this problem is to introduce a mass transfer coefficient.

At the solid/liquid (or liquid/gas) interface the mass transfer coefficient k_m is defined by

$$n_{Ay}|_{y=0} = k_m(\rho_{A0} - \rho_{A\infty}) \qquad [3.1\text{-}18]$$

where ρ_{A0} and $\rho_{A\infty}$ are the mass concentrations of species A in the fluid at the interface and in the bulk (or free-stream) fluid, respectively. If ρ is constant

$$n_{Ay}|_{y=0} = k_m \rho(w_{A0} - w_{A\infty}) \qquad [3.1\text{-}19]$$

Furthermore, if the solubility of species A in the fluid is limited so that v_y is essentially zero at the interface, the following approximation[2,3] can be made in view of Eq. [3.1-11]:

$$n_{Ay}|_{y=0} = j_{Ay}|_{y=0} \qquad [3.1\text{-}20]$$

Substituting Eqs. [3.1-17] and [3.1-19] into Eq. [3.1-20]

$$\boxed{k_m = \left|\frac{j_{Ay}|_{y=0}}{\rho(w_{AS} - w_{A\infty})}\right| = \left|\frac{-D_A(\partial w_A/\partial y)|_{y=0}}{(w_{AS} - w_{A\infty})}\right|} \qquad [3.1\text{-}21]$$

The absolute values are used in order to keep k_m always positive.

For fluid flow through a pipe such as that shown in Fig. 3.1-4, we can write

$$\boxed{k_m = \left|\frac{j_{Ar}|_{r=R}}{\rho(w_{AS} - w_{A,\text{av}})}\right| = \left|\frac{-D_A(\partial w_A/\partial r)|_{r=R}}{w_{AS} - w_{A,\text{av}}}\right|} \qquad [3.1\text{-}22]$$

where the average concentration is defined as follows

$$\boxed{w_{A,\text{av}} = \frac{\iint_A \rho w_A v\, dA}{\iint_A \rho v\, dA} = \frac{\iint_A \rho w_A v\, dA}{m} = \frac{\iint_A \rho w_A v\, dA}{\rho v_{\text{av}} A}} \qquad [3.1\text{-}23]$$

Notice that the numerator is the species mass flow rate.

Similar equations on the molar basis can be written for flow over a flat plate:

$$k_m = \left|\frac{j_{Ay}|_{y=0}}{c(x_{AS} - x_{A\infty})}\right| = \left|\frac{-D_A(\partial x_A/\partial y)|_{y=0}}{x_{AS} - x_{A\infty}}\right| \qquad [3.1\text{-}24]$$

[2] R. W. Fahien, *Fundamentals of Transport Phenomena*, McGraw-Hill, New York, 1983, p. 60.

[3] J. R. Welty, C. E. Wicks, and R. E. Wilson, *Fundamentals of Momentum, Heat and Mass Transfer*, 3rd ed., Wiley, New York, 1984, p. 598.

and for flow through a tube

$$k_m = \left| \frac{j_{Ar}|_{r=R}}{c(x_{AS} - x_{A,\mathrm{av}})} \right| = \left| \frac{-D_A (\partial x_A/\partial r)|_{r=R}}{x_{AS} - x_{A,\mathrm{av}}} \right| \qquad [3.1\text{-}25]$$

where the average concentration is defined as follows

$$\boxed{x_{A,\mathrm{av}} = \frac{\iint_A c x_A \underline{v}\, dA}{\iint_A c \underline{v}\, dA} = \frac{\iint_A c x_A \underline{v}\, dA}{\underline{m}} = \frac{\iint_A c x_A \underline{v}\, dA}{c \underline{v}_{\mathrm{av}} A}} \qquad [3.1\text{-}26]$$

3.1.7. Diffusion in Solids

3.1.7.1. Diffusion Mechanisms

Vacancy diffusion and interstitial diffusion are the two most frequently encountered diffusion mechanisms in solids, although other mechanisms have also been proposed.

In **vacancy diffusion** an atom in a solid jumps from a lattice position of the solid into a neighboring unoccupied lattice site or vacancy, as illustrated in Fig. 3.1-5*a*. At temperatures above absolute zero all solids contain some vacancies; the higher the temperature, the more the vacancies. The atom can continue to

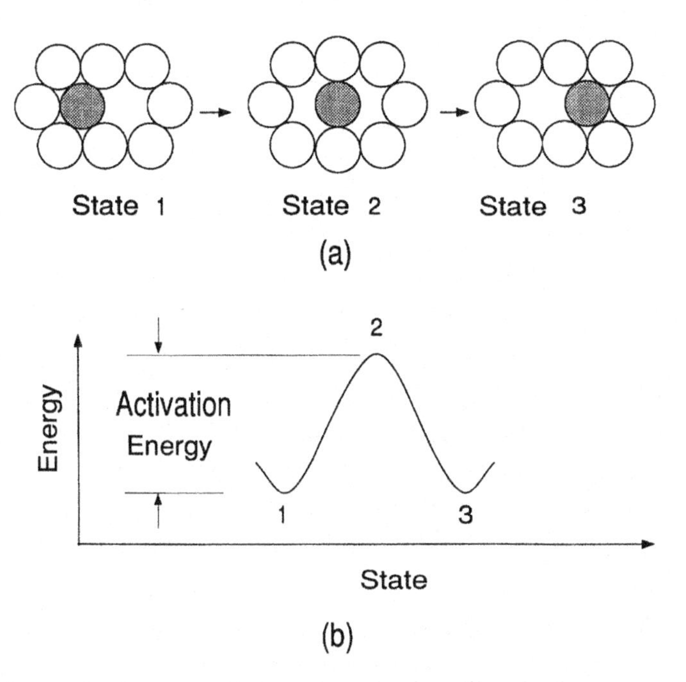

Fig. 3.1-5 Vacancy diffusion: (*a*) mechanism; (*b*) activation energy.

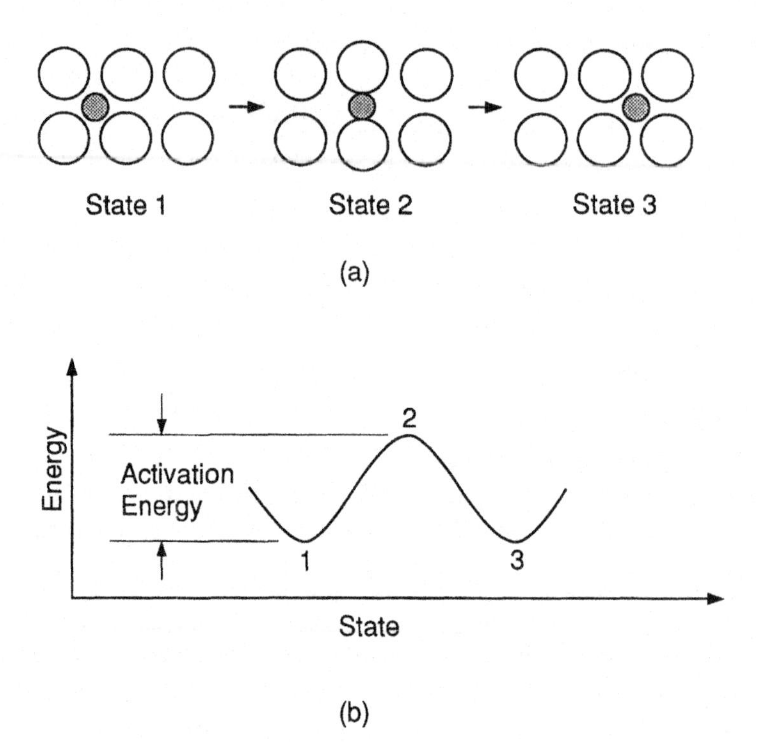

Fig. 3.1-6 Interstitial diffusion: (*a*) mechanism; (*b*) activation energy.

diffuse through the solid by a series of exchanges with vacancies that appear to be adjacent to it from time to time. Vacancy diffusion is usually the diffusion mechanism for substitutional solid solutions. In such materials the solute atoms, which are comparable to the solvent atoms in size, substitute the solvent atoms at their lattice sites. Examples of substitutional solid solutions are Cu–Zn alloy (brass) and Au–Ni alloy.

In **interstitial diffusion** an atom in a solid jumps from an interstitial site of the lattice to a neighboring one, as illustrated in Fig. 3.1-6*a*. The atom can continue to diffuse through the solid by a series of jumps to neighboring interstitial sites that are unoccupied. Interstitial diffusion is the diffusion mechanism for interstitial solid solutions. In such materials the solute atoms, which are significantly smaller than the solvent atoms, occupy the interstitial sites of the lattice. The most well known example of interstitial solid solution is the iron–carbon alloy, specifically, carbon steel, in which the small carbon atoms occupy the interstitial sites of the iron lattice.

3.1.7.2. Diffusion Coefficients

Several different types of diffusion coefficients have been used in studying diffusion in solids.

In the so-called self-diffusion experiment, a solute A in the form of a radioactive isotope, such as ^{63}Ni, is allowed to diffuse through the lattice of a nonradioactive solid of the same material, Ni. The diffusion coefficient D_{A^*} is known as the **self-diffusion coefficient**, in view of the absence of a chemical composition gradient as the driving force for diffusion.

In practical situations, however, diffusion usually occurs under the influence of a chemical composition gradient, such as the diffusion of carbon in steel from a higher-carbon-concentration region to the lower one. This diffusion coefficient D_A is known as the **intrinsic diffusion coefficient**. The so-called **interdiffusion coefficient** $\tilde{D}$ is often used to describe situations involving the interdiffusion of two different chemical species, such as, Au into Ni and Ni into Au as in an Au–Ni diffusion couple.

Quantitative discussion on these diffusion coefficients, which requires the species differential mass-balance equation, is given in Examples 3.3-1 through 3.3-5.

3.1.7.3. Effect of Temperature

The diffusion coefficient has been observed to increase with increasing temperature according to the following Arrhenius equation:

$$\boxed{D = D_0 e^{-Q/RT}} \qquad [3.1\text{-}27]$$

where D is any type of diffusion coefficient, D_0 a proportional constant, Q the activation energy, R the gas constant, and T the absolute temperature. As illustrated in Fig. 3.1-5*b*, a significant energy barrier has to be overcome before an atom can jump from one lattice site to a neighboring one by vacancy diffusion. Similarly, as illustrated in Fig. 3.1-6*b*, a smaller but still significant energy barrier has to be overcome before an interstitial atom can jump from one interstitial site to a neighboring one by interstitial diffusion. Equation [3.1-27] can, in fact, be derived theoretically.[4]

Tables 3.1-3 and 3.1-4 list the experimental data of D_0 and Q for substitutional diffusion and interstitial diffusion in some materials. As shown, Q is significantly lower for interstitial diffusion than for substitutional diffusion. As shown in Table 3.1-3, Q is smaller for substitutional self diffusion in body-center-cubic (bcc) iron than in face-center-cubic (fcc) iron. Since atoms are more loosely packed in a bcc structure than an fcc structure, Q is smaller in bcc iron than in fcc iron. For the same reason, Q is also smaller for interstitial diffusion of C, N, and H in bcc iron than in fcc iron. Figure 3.1-7 shows some diffusion coefficients as a function of temperature. As shown, the diffusion coefficient is much higher for interstitial diffusion, for example, C in fcc Fe, than for substitutional diffusion, such as Ni in fcc Fe.

[4] C. Zener, *J. Appl. Phys.* **22**, p. 372, 1951.

TABLE 3.1-3 Diffusion Data for Self-Diffusion in Pure Metals

Structure	Metal	D_0 (mm^2/s)	Q (kJ/mol)
Fcc	Au	10.7	176.9
Fcc	Cu	31	200.3
Fcc	Ni	190	279.7
Fcc	Fe(γ)	49	284.1
Bcc	Fe(α)	200	239.7
Bcc	Fe(δ)	190	238.5

TABLE 3.1-4 Diffusion Data for Interstitials in Iron

Structure	Solute	D_0 (mm^2/s)	Q (kJ/mol)
Bcc	C	2.0	84.1
Bcc	N	0.3	76.1
Bcc	H	0.1	13.4
Fcc	C	2.5	144.2

3.2. SPECIES OVERALL MASS-BALANCE EQUATION

3.2.1. Derivation

The overall mass-balance equation for single-component (i.e., pure-material) systems has been derived in Section 1.2. In this section we shall derive a similar equation for binary systems.

Let us consider an arbitrary stationary control volume Ω bounded by surface A through which a moving fluid is flowing, as illustraed in Fig. 3.2-1. The control surface A can be considered to consist of three different regions: A_{in} for the region where the fluid enters the control volume; A_{out}, where the fluid leaves; and A_{wall}, where the fluid is in contact with a wall. In other words

$$A = A_{in} + A_{out} + A_{wall} \qquad [3.2\text{-}1]$$

Before deriving the species integral mass-balance equation, let us consider the outward transfer (e.g., diffusion), rate of species A through dA shown in Fig. 3.2-2*a*. As shown in Fig. 3.2-2*b*, it is $\mathbf{j}_A \cdot \mathbf{n}\, dA$. Since $\mathbf{n}$ points outward

$$\text{inward mass transfer rate through area } dA = -\mathbf{j}_A \cdot \mathbf{n}\, dA. \qquad [3.2\text{-}2]$$

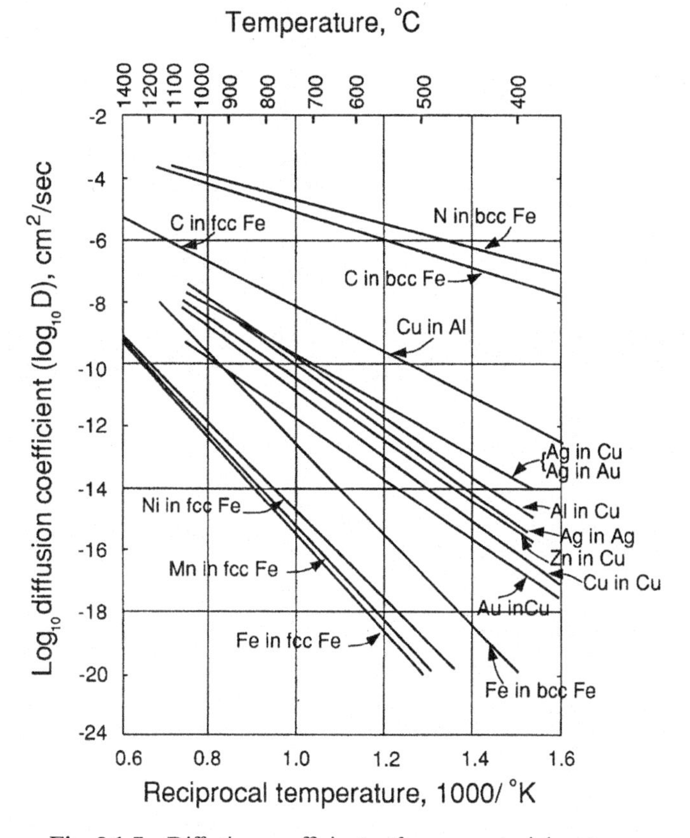

Fig. 3.1-7 Diffusion coefficients of some material systems.

The following mass conservation law for species A can be applied to the control volume in Fig. 3.2-1:

$$\begin{Bmatrix} \text{Rate of} \\ \text{species } A \\ \text{accumulation} \end{Bmatrix} = \begin{Bmatrix} \text{Rate of} \\ \text{species } A \text{ in} \\ \text{by mass inflow} \end{Bmatrix} - \begin{Bmatrix} \text{Rate of} \\ \text{species } A \text{ out} \\ \text{by mass outflow} \end{Bmatrix}$$
$$\qquad (1) \qquad\qquad\qquad (2) \qquad\qquad\qquad (3)$$

$$+ \begin{Bmatrix} \text{Rate of other species} \\ A \text{ transfer to system} \\ \text{from surroundings} \end{Bmatrix} + \begin{Bmatrix} \text{Rate of species} \\ A \text{ generation} \\ \text{in system} \end{Bmatrix}$$
$$\qquad (4) \qquad\qquad\qquad (5)$$

[3.2-3]

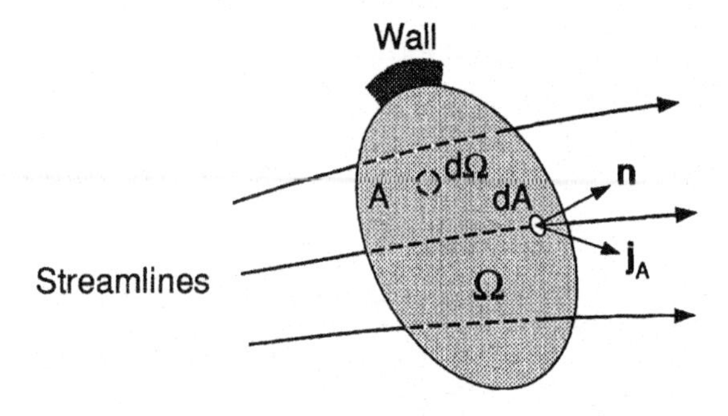

Fig. 3.2-1 Fluid flow through a volume element Ω with mass transfer.

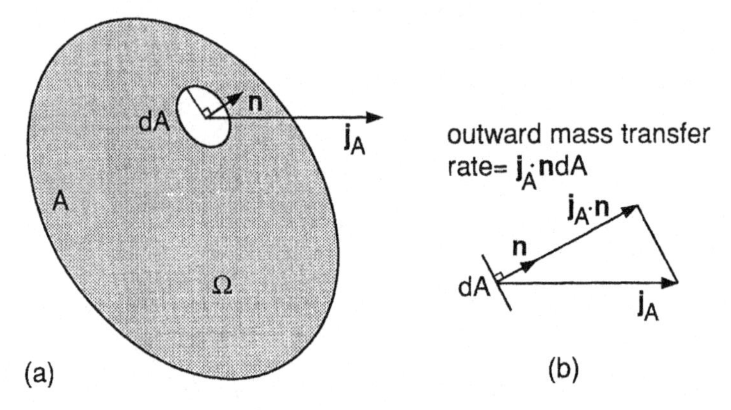

Fig. 3.2-2 Diffusion rate of species A through a differential surface element dA: (*a*) three-dimensional view; (*b*) two-dimensional view.

Let us now consider the terms in Eq. [3.2-3] one by one, starting from term (1). The mass of species A in a differential volume element $d\Omega$ is $dM_{A\Omega} = \rho_A d\Omega$, where ρ_A is the mass concentration of species A, that is, mass of A per unit volume of the fluid. It can be integrated over Ω to obtain the mass of species A in the control volume:

$$\iiint_{\Omega} \rho_A \, d\Omega \quad ; \quad M_{A\Omega}$$

(integral) (overall)

and so the rate of change in the mass of species A in Ω is

$$\frac{\partial}{\partial t} \iiint_{\Omega} \rho_A \, d\Omega \quad ; \quad \frac{dM_{A\Omega}}{dt} \qquad \text{term (1)}$$

(integral) (overall)

For convenience the first form will be called the integral form and the second the overall form. The same terminology will be applied to the remaining terms.

From Eq. [1.2-2], the inward mass flow rate through a differential area dA is $-\rho\mathbf{v}\cdot\mathbf{n}\,dA$. The inward mass flow of species A $-\rho_A\mathbf{v}\cdot\mathbf{n}\,dA$, can be integrated over the control surface A to obtain the net (since A is a closed surface) inward flow rate of species A into Ω, that is, the rate species A flows into Ω minus the rate it flows out of Ω:

$$-\iint_A \rho_A \mathbf{v}\cdot\mathbf{n}\,dA$$

or, in view of Eq. [3.2-1],

$$-\iint_{A_{\text{in}}+A_{\text{out}}+A_{\text{wall}}} \rho_A \mathbf{v}\cdot\mathbf{n}\,d(A_{\text{in}}+A_{\text{out}}+A_{\text{wall}})$$

or

$$-\left(\iint_A \rho_A \mathbf{v}\cdot\mathbf{n}\,dA\right)_{\text{in}} - \left(\iint_A \rho_A \mathbf{v}\cdot\mathbf{n}\,dA\right)_{\text{out}} - \left(\iint_A \rho_A \mathbf{v}\cdot\mathbf{n}\,dA\right)_{\text{wall}}$$

Note that if the wall is coated with species A, it can dissolve in the fluid and result in a slow inward flow from the wall. As in Section 1.2, the velocity component along the normal to the control surface is defined as $v = |\mathbf{v}\cdot\mathbf{n}|$. At the inlet and the wall $\mathbf{v}$ is inward, $\mathbf{n}$ is outward, and $v = |\mathbf{v}\cdot\mathbf{n}| = -\mathbf{v}\cdot\mathbf{n}$, while at the outlet $\mathbf{v}$ and $\mathbf{n}$ are both outward and $v = |\mathbf{v}\cdot\mathbf{n}| = \mathbf{v}\cdot\mathbf{n}$. Therefore, the rate species A flows into Ω minus the rate it flows out of Ω is

$$\underset{\text{(integral)}}{-\iint_A \rho_A \mathbf{v}\cdot\mathbf{n}\,dA;} \quad \underset{\text{(overall)}}{\left[\left(\iint_A \rho_A v\,dA\right)_{\text{in}} - \left(\iint_A \rho_A v\,dA\right)_{\text{out}} + \left(\iint_A \rho_A v\,dA\right)_{\text{wall}}\right]}$$

term (2)–term (3)

As already mentioned, if the wall is coated with a material containing species A, it can dissolve in the fluid. However, if the concentration of A in the coating is low or if the solubility of A in the fluid is low, convective mass transfer at the wall is negligible.

According to Eq. [3.2-2], the inward mass transfer (i.e., diffusion), rate through a differential area dA is $-\mathbf{j}_A\cdot\mathbf{n}\,dA$. This can be integrated over control surface A to obtain the rate species A goes into the control volume from the surroundings **other than** terms (2) and (3):

$$-\iint_A \mathbf{j}_A \cdot \mathbf{n}\, dA; \quad J_A \qquad \text{term (4)}$$

(integral) (overall)

Let r_A be the mass production rate of species A per unit volume by chemical reactions. The mass production rate of species A in the differential volume element $d\Omega$ is $dR_A = r_A\, d\Omega$. This can be integrated over Ω to obtain the mass generation rate of species A in Ω;

$$\iiint_A r_A\, d\Omega; \quad R_A \qquad \text{term (5)}$$

(integral) (overall)

Substituting the integral form of terms (1) through (5) into Eq. [3.2-3]

$$\frac{\partial}{\partial t}\iiint_\Omega \rho_A\, d\Omega = -\iint_A \rho_A \mathbf{v} \cdot \mathbf{n}\, dA - \iint_A \mathbf{j}_A \cdot \mathbf{n}\, dA + \iiint_\Omega r_A\, d\Omega$$

(species integral mass-balance equation) [3.2-4]

Note that the mass convection and diffusion terms can be combined into one through $\mathbf{n}_A = \rho_A \mathbf{v} + \mathbf{j}_A$ (i.e., Eq. [3.1-11]).

By following a similar approach, a similar equation can be derived on the basis of the molar density c_A, the molar flux $\underline{\mathbf{j}}_A$, the local molar average velocity $\underline{\mathbf{v}}$, and the molar production rate of A per unit volume $\underline{r}_A$. This equation is

$$\frac{\partial}{\partial t}\iiint_\Omega c_A\, d\Omega = -\iint_A c_A \underline{\mathbf{v}} \cdot \mathbf{n}\, dA - \iint_A \underline{\mathbf{j}}_A \cdot \mathbf{n}\, dA + \iiint_A \underline{r}_A\, d\Omega$$

(species integral mass-balance equation) [3.2-5]

Note that the molar convection and diffusion terms can be combined into one through $\underline{\mathbf{n}}_A = c_A \underline{\mathbf{v}} + \underline{\mathbf{j}}_A$ (i.e., Eq. [3.1-14]).

Now substituting the overall form of terms (1) through (5) into Eq. [3.2-3] and with $\rho_A = \rho w_A$

$$\frac{dM_{A\Omega}}{dt} = \left(\iint_A \rho w_A v\, dA\right)_{\text{in}} - \left(\iint_A \rho w_A v\, dA\right)_{\text{out}} + J_A + R_A$$
$$= (m_A)_{\text{in}} - (m_A)_{\text{out}} + J_A + R_A \qquad [3.2\text{-}6]$$

where $M_{A\Omega}$ is the mass of species A in the control volume; that is, $\iiint_\Omega \rho w_A\, d\Omega$. Note that convective mass transfer at the wall has been neglected as an approximation for the reasons already stated. Substituting Eq. [3.1-23] into Eq. [3.2-6], we obtain

$$\frac{dM_{A\Omega}}{dt} = (\rho\, w_{A,\text{av}}\, v_{\text{av}}\, A)_{\text{in}} - (\rho\, w_{A,\text{av}}\, v_{\text{av}}\, A)_{\text{out}} + J_A + R_A$$
$$= (m w_{A,\text{av}})_{\text{in}} - (m w_{A,\text{av}})_{\text{out}} + J_A + R_A$$

(species overall mass-balance equation) [3.2-7]

where

$M_{A\Omega}$ = mass of species A in control volume ($= \rho w_A \Omega = M w_A$ if uniform ρw_A)
m = mass flow rate at inlet or outlet ($= \rho v_{\text{av}} A$)
J_A = mass transfer rate of species A **into** control volume from surroundings (other than the two $m w_{A,\text{av}}$ terms), e.g., by **diffusion**,
R_A = mass generation rate of species A in control volume ($= \underline{r}_A \Omega$ if uniform $\underline{r}_A$).

Equations similar to [3.2-6] and [3.2-7] can be derived on the molar basis, i.e.,

$$\frac{d\underline{M}_{A\Omega}}{dt} = \left(\iint_A c x_A \underline{v}\, dA\right)_{\text{in}} - \left(\iint_A c x_A \underline{v}\, dA\right)_{\text{out}} + \underline{J}_A + \underline{R}_A$$
$$= (\underline{m}_A)_{\text{in}} - (\underline{m}_A)_{\text{out}} + \underline{J}_A + \underline{R}_A \qquad [3.2\text{-}8]$$

and

$$\frac{d\underline{M}_{A\Omega}}{dt} = (cx_{A,\text{av}}\underline{v}_{\text{av}}A)_{\text{in}} - (cx_{A,\text{av}}\underline{v}_{\text{av}}A)_{\text{out}} + \underline{J}_A + \underline{R}_A$$

$$= (\underline{m}x_{A,\text{av}})_{\text{in}} - (\underline{m}x_{A,\text{av}})_{\text{out}} + \underline{J}_A + \underline{R}_A \qquad [3.2\text{-}9]$$

where

$\underline{M}_{A\Omega}$ = moles of species A in the control volume ($= cx_A\Omega = \underline{M}x_A$ if uniform cx_A)
$\underline{m}$ = molar flow rate at inlet or outlet ($= c\underline{v}_{\text{av}}A$)
$\underline{J}_A$ = molar transfer rate of species A **into** control volume from surroundings (other than the two $\underline{m}x_{A,\text{av}}$ terms), that is, by **diffusion**,
$\underline{R}_A$ = molar generation rate of species A in control volume ($= \underline{r}_A\Omega$ if uniform $\underline{r}_A$).

3.2.2. Examples

Example 3.2-1 Fluid Composition in a Mixing Tank

As illustrated in Fig. 3.2-3, a mixing tank receives fluid from two inlets and discharges it from one outlet without any chemical reactions. The mass of the fluid in the tank is M and its solute concentration w_A is uniform as the result of stirring. The mass flow rates and solute concentrations are respectively m_1 and

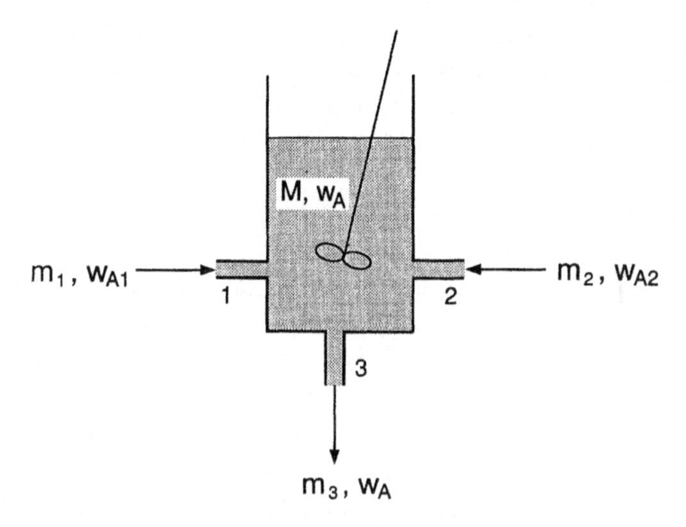

Fig. 3.2-3 Solute concentration in a mixing tank.

w_{A_1} for inlet 1, m_2 and w_{A_2} for inlet 2, and m_3 and w_A for the outlet. The initial fluid mass and solute concentration in the tank are M_0 and w_{A_0}, respectively. Determine the solute concentration in the tank as a function of time. At the inlets and outlet, mass transfer due to diffusion is negligible as compared to that due to convection.

Solution Since the solute concentration is uniform, the mass of A in the tank

$$M_{A\Omega} = Mw_A \tag{3.2-10}$$

also

$$J_A = 0 \tag{3.2-11}$$

and

$$R_A = 0 \tag{3.2-12}$$

Substituting Eqs. [3.2-10] through [3.2-12] into Eq. [3.2-7]

$$\frac{d}{dt}[Mw_A] = m_1 w_{A_1} + m_2 w_{A_2} - m_3 w_A \tag{3.2-13}$$

From Eq. [1.2-12]

$$M = mt + M_0 \tag{3.2-14}$$

where

$$m = m_1 + m_2 - m_3 \tag{3.2-15}$$

As such, Eq. [3.2-13] becomes

$$(mt + M_0)\frac{dw_A}{dt} + mw_A = m_1 w_{A_1} + m_2 w_{A_2} - m_3 w_A \tag{3.2-16}$$

or

$$\frac{dw_A}{dt} = \frac{-(m + m_3)\{w_A - [(m_1 w_{A_1} + m_2 w_{A_2})/(m + m_3)]\}}{m[t + (M_0/m)]} \tag{3.2-17}$$

The initial condition is

$$w_A = w_{A_0} \quad \text{at} \quad t = 0 \tag{3.2-18}$$

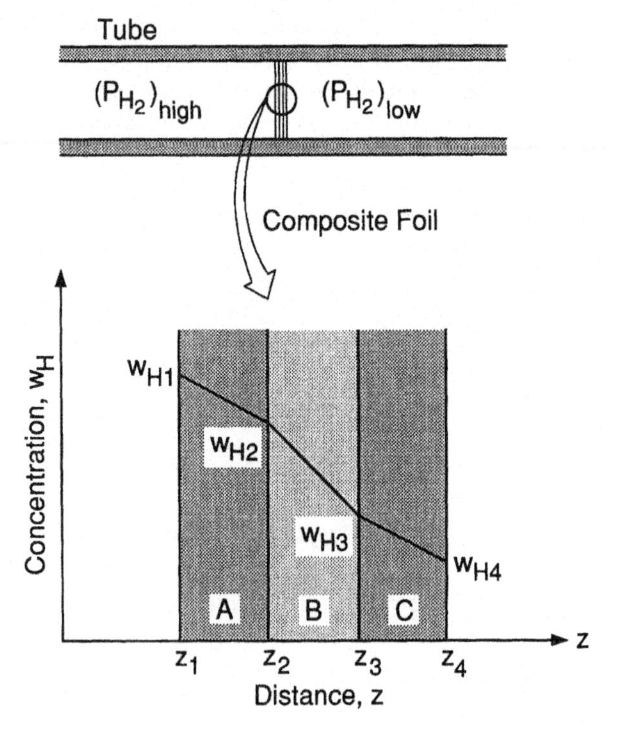

Fig. 3.2-4 Diffusion of hydrogen through a composite foil.

From Case B of Appendix A

$$\left(\frac{w_A - [(m_1 w_{A_1} + m_2 w_{A_2})/(m + m_3)]}{w_{A_0} - [(m_1 w_{A_1} + m_2 w_{A_2})/(m + m_3)]}\right)^m = \left(\frac{t + (M_0/m)}{(M_0/m)}\right)^{-(m+m_3)} \tag{3.2-19}$$

Example 3.2-2 Diffusion through Composite Foil

A tube of inner cross-sectional area A and impermeable to hydrogen is separated by a composite foil into a left chamber of a higher hydrogen pressure $(p_{H_2})_{high}$ and a right chamber of a lower hydrogen pressure $(p_{H_2})_{low}$, as shown in Fig. 3.2-4. The composite foil consists of three different materials A, B, and C, each with its own property of $(\rho D_H)_A$, $(\rho D_H)_B$, and $(\rho D_H)_C$, respectively. Determine the steady-state diffusion flux of atomic hydrogen j_H through the composite foil. The equilibrium concentration of atomic hydrogen at a gas/solid interface is determined by Sievert's law[5]:

$$w_H = K_p p_{H_2}^{1/2} \tag{3.2-20}$$

[5] L. S. Darken and R. W. Gurry, *Physical Chemistry of Metals*, McGraw-Hill, New York, 1953, p. 258.

where K_p is the equilibrium constant for the following reaction

$$\tfrac{1}{2}H_2(g) \rightarrow H \qquad \text{(in solid solution)} \qquad [3.2\text{-}21]$$

Solution Let us consider material A as the control volume Ω. Since there is no convection nor chemical reactions within it, at the steady state Eq. [3.2-7] reduces to

$$0 = J_H = (Aj_H)|_{z_1^+} - (Aj_H)|_{z_2^-} \qquad [3.2\text{-}22]$$

and so

$$j_H|_{z_1^+} = j_H|_{z_2^-} = j_H|_A = \text{constant} \qquad [3.2\text{-}23]$$

Repeating for the other two materials, it can be shown that

$$j_H|_A = j_H|_B = j_H|_C = j_H = \text{constant} \qquad [3.2\text{-}24]$$

Therefore, from Fick's law of diffusion (i.e., Eq. [F] of Table 3.1-1):

$$\left[-\rho D_H \frac{dw_H}{dz}\right]\bigg|_A = \left[-\rho D_H \frac{dw_H}{dz}\right]\bigg|_B = \left[-\rho D_H \frac{dw_H}{dz}\right]\bigg|_C = j_H \qquad [3.2\text{-}25]$$

Integrating this equation over each individual material, we have

$$w_{H1} - w_{H2} = \frac{j_H}{(\rho D_H)_A}(z_2 - z_1) \qquad [3.2\text{-}26]$$

$$w_{H2} - w_{H3} = \frac{j_H}{(\rho D_H)_B}(z_3 - z_2) \qquad [3.2\text{-}27]$$

$$w_{H3} - w_{H4} = \frac{j_H}{(\rho D_H)_C}(z_4 - z_3) \qquad [3.2\text{-}28]$$

Adding Eqs. [3.2-26] through [3.2-28]

$$w_{H1} - w_{H4} = \left[\frac{z_2 - z_1}{(\rho D_H)_A} + \frac{z_3 - z_2}{(\rho D_H)_B} + \frac{z_4 - z_3}{(\rho D_H)_C}\right] j_H \qquad [3.2\text{-}29]$$

or

$$j_H = \left[\frac{z_2 - z_1}{(\rho D_H)_A} + \frac{z_3 - z_2}{(\rho D_H)_B} + \frac{z_4 - z_3}{(\rho D_H)_C}\right]^{-1} (w_{H1} - w_{H4}) \qquad [3.2\text{-}30]$$

Substituting Eq. [3.2-20] into Eq. [3.2-30], we have

$$j_{\text{H}} = \left[\frac{z_2 - z_1}{(\rho D_{\text{H}})_A} + \frac{z_3 - z_2}{(\rho D_{\text{H}})_B} + \frac{z_4 - z_3}{(\rho D_{\text{H}})_C}\right]^{-1} K_p \left[(p_{\text{H}_2})^{1/2}_{\text{high}} - (p_{\text{H}_2})^{1/2}_{\text{low}}\right] \qquad [3.2\text{-}31]$$

Example 3.2-3 Diffusion of Gas Through Tube Wall

A methane–helium gas mixture with a low helium concentration x_{A_0} flows through a long quartz tube, as shown in Fig. 3.2-5. The tube has a length L, inner diameter D, and wall thickness l ($\ll D$). Helium can diffuse through quartz but not methane. Because of the low helium concentration, the molar flow rate $\underline{m}$ of the gas mixture remains essentially constant throughout the tube. The flow rate is fast enough that mass transfer in the axial direction is dominated by convection. Furthermore, since the diffusion coefficient is much smaller in the wall than

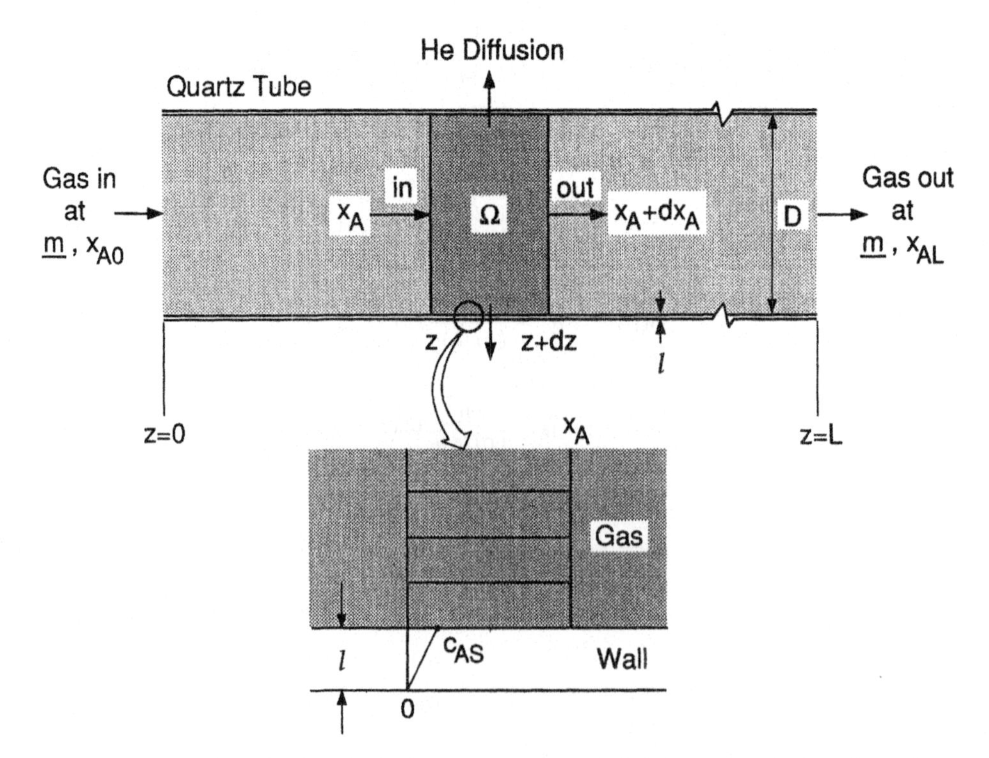

Fig. 3.2-5 Helium flowing through a quartz tube and diffusing through the tube wall.

in the gas, the radial composition gradient in the gas is negligible as compared to that in the wall. Determine the helium concentration x_{AL} at the tube exit and the rate of helium loss by diffusion at the steady state. The concentration of a monatomic gas A like helium in a solid at the solid/gas interface is

$$c_{AS} = K_x x_A \qquad [3.2\text{-}32]$$

were K_x is the equilibrium constant for the following reaction

$$A(g) = \underline{A} \text{ (in solid solution)} \qquad [3.2\text{-}33]$$

Solution Let the volume element of length dz be the control volume Ω. From Eq. [3.2-9] and with $\underline{R}_A = 0$

$$\underset{\text{(in)}}{0 = \underline{m} x_A} - \underset{\text{(out)}}{\underline{m}(x_A + dx_A)} + \underset{\text{(wall)}}{\underline{J}_A} \qquad [3.2\text{-}34]$$

where $\underline{J}_A$ (< 0) is the molar diffusion rate of helium into Ω from outside the tube. Since $l \ll D$, the tube wall can be considered as a flat sheet as an approximation (i.e., $\underline{j}_A = D_A \Delta c/l$). Therefore, from Eq. [3.2-32]

$$\underline{J}_A = (\pi D\, dz) D_A \left(\frac{0 - c_{AS}}{l}\right) = -(\pi D\, dz) D_A K_x x_A / l \qquad [3.2\text{-}35]$$

where c_{AS} is the helium concentration in the quartz tube at its inner wall. At the outer wall it is zero since the concentration of helium in air is essentially zero.

Substituting Eq. [3.2-35] into Eq. [3.2-34] and rearranging, we have

$$\frac{dx_A}{x_A} = d(\ln x_A) = -\left(\frac{\pi D D_A K_x}{\underline{m} l}\right) dz \qquad [3.2\text{-}36]$$

Integrating over the length of the tube

$$\ln\left(\frac{x_{AL}}{x_{A0}}\right) = -\frac{\pi D D_A K_x L}{\underline{m} l} \qquad [3.2\text{-}37]$$

or

$$x_{AL} = x_{A0} \exp\left(-\frac{\pi D D_A K_x L}{\underline{m} l}\right) \qquad [3.2\text{-}38]$$

Applying Eq. [3.2-9] to the entire length of the tube now as the control volume,

$$0 = \underline{m}x_{A_0} - \underline{m}x_{AL} + \underline{J}_A \tag{3.2-39}$$

Therefore,

$$\underline{J}_A = \underline{m}(x_{AL} - x_{A_0}) = -\underline{m}x_{A_0}\left[1 - \exp\left(-\frac{\pi D D_A K_x L}{\underline{m}l}\right)\right] \tag{3.2-40}$$

Example 3.2-4 Mass Transfer in Laminar Flow over a Flat Plate

In Examples 1.4-6 and 2.2-5 we studied the momentum and thermal boundary layers over a flat plate using an integral approach, respectively. Here we shall use a similar approach to consider the diffusion boundary layer over a flat plate. The plate is coated with a layer of material containing species A, which has a low solubility in the fluid. As illustrated in Fig. 3.2-6a, the compositions of the fluid in the bulk stream and at the plate surface are $w_{A\infty}$ and w_{AS}, respectively. The plate is wide in the x direction such that there are no significant velocity or the composition variations in this direction. The concept of the diffusion boundary layer has been described in Section 3.1.4. We shall determine the thickness of the diffusion boundary layer δ_c and the mass transfer coefficient k_m as a function of the distance from the leading edge of the plate z. Assume steady state, constant physical properties and no chemical reactions in the fluid.

Solution The control volume Ω is enlarged in Fig. 3.2-6b. Similar to Eq. [1.4-52], the mass flow rate across surface 4 per unit width of the plate is

$$m_4 = \frac{d}{dz}\left[\int_0^{\delta_c} \rho v_z\, dy\right]\Delta z \tag{3.2-41}$$

The fluid entering the control volume through the top surface carries $m_4 w_{A\infty}$ of A per unit width of the plate. Furthermore, at the top $\partial w_A/\partial y$ is essentially zero and, therefore, $j_{Ay} = 0$. Since mass transfer in the z direction is mainly by convection, diffusion in this direction can be neglected. At the steady state Eq. [3.2-6] reduces to

$$0 = \underbrace{\left[\int_0^{\delta_c} \rho w_A v_z\, dy\right]\Bigg|_z}_{(1)} - \underbrace{\left[\int_0^{\delta_c} \rho w_A v_z\, dy\right]\Bigg|_{z+\Delta z}}_{(2)} + \underbrace{j_{Ay}\Big|_{y=0}\Delta z}_{(3)} + \underbrace{m_4 w_{A\infty}}_{(4)} \tag{3.2-42}$$

The velocity v_y at the wall is essentially zero in view of the limited dissolution of species A from the wall. Substituting Eq. [3.2-41] into this equation, dividing it

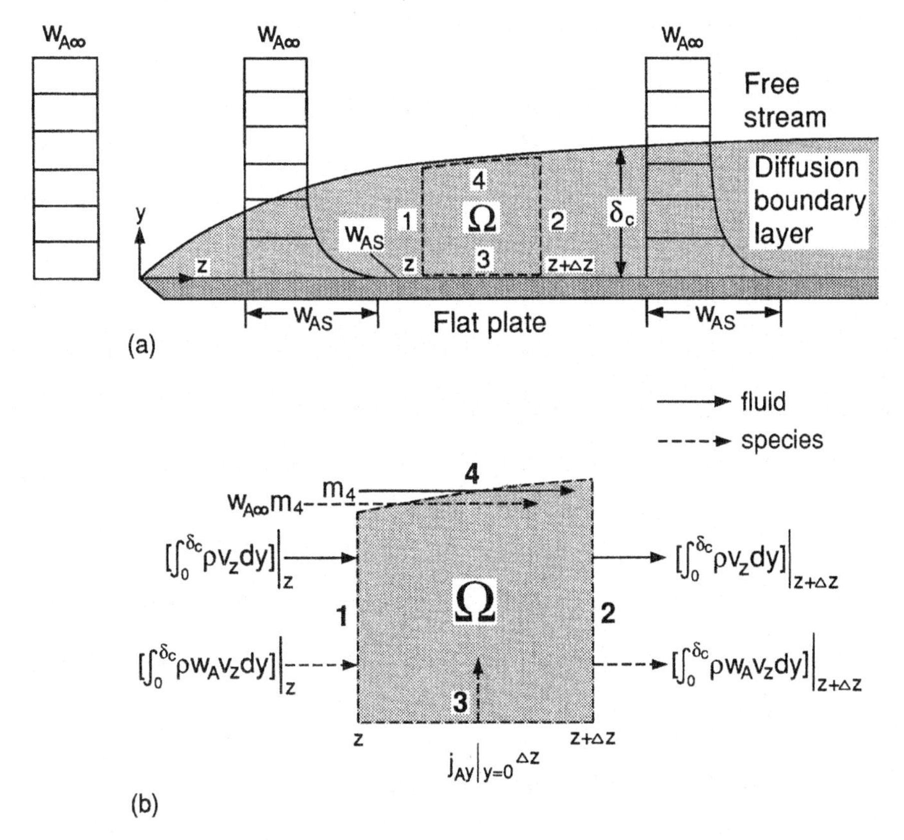

Fig. 3.2-6 Mass transfer in fluid flow over a flat plate coated with species A: (*a*) diffusion boundary layer; (*b*) volume element in the layer.

by Δz and letting $\Delta z \to 0$

$$- j_{Ay}|_{y=0} = -\frac{d}{dz}\left[\int_0^{\delta_c} \rho w_A v_z \, dy\right] + \frac{d}{dz}\left[\int_0^{\delta_c} \rho w_{A\infty} v_z \, dy\right] \qquad [3.2\text{-}43]$$

From Fick's law of diffusion

$$j_{Ay}|_{y=0} = -\rho D_A \left.\frac{\partial w_A}{\partial y}\right|_{y=0} \qquad [3.2\text{-}44]$$

Substituting this equation into Eq. [3.2-43] and dividing it by ρ

$$D_A \left.\frac{\partial w_A}{\partial y}\right|_{y=0} = \frac{d}{dz}\int_0^{\delta_c} v_z(w_{A\infty} - w_A)\, dy \qquad [3.2\text{-}45]$$

Since this equation is difficult to solve directly, the following approximate concentration profile is assumed:

$$\frac{w_A - w_{AS}}{w_{A\infty} - w_{AS}} = \frac{3}{2}\left(\frac{y}{\delta_c}\right) - \frac{1}{2}\left(\frac{y}{\delta_c}\right)^3 \quad [3.2\text{-}46]$$

This concentration profile is reasonable since it satisfies the following boundary conditions:

$$w_A = w_{AS} \quad \text{at } y = 0 \quad [3.2\text{-}47]$$

and

$$w_A = w_{A\infty} \quad \text{and} \quad \frac{\partial w_A}{\partial y} = 0 \quad \text{at} \quad y = \delta_c \quad [3.2\text{-}48]$$

Equation [3.2-46] is similar to Eq. [1.4-57] for the momentum boundary layer:

$$\frac{v_z}{v_\infty} = \frac{3}{2}\left(\frac{y}{\delta}\right) - \frac{1}{2}\left(\frac{y}{\delta}\right)^3 \quad [3.2\text{-}49]$$

where v_∞ is the free stream velocity of the fluid.

Equations [3.2-46] and [3.2-49] can be substituted into Eq. [3.2-45] to find the relationship between δ_c and δ. The integration in Eq. [3.2-45] yields a $(\delta_c/\delta)^2$ term and a $(\delta_c/\delta)^4$ term with a coefficient about 14 times smaller. As an approximation the latter can be neglected, if we assume that $\delta_c \leq \delta$. With further help from Eq. [1.4-61], the following equation is obtained:

$$\frac{\delta_c}{\delta} = \mathrm{Sc}^{-(1/3)} \quad [3.2\text{-}50]$$

where the **Schmidt number** is a dimensionless number defined by

$$\mathrm{Sc} = \frac{\nu}{D_A} = \frac{\mu}{\rho D_A} \quad [3.2\text{-}51]$$

From Eq. [3.1-21] the mass transfer coefficient

$$k_m = \frac{-[D_A(\partial w_A/\partial y)]_{y=0}}{(w_{AS} - w_{A\infty})} \quad [3.2\text{-}52]$$

since $\partial w_A/\partial y < 0$ and $(w_{AS} - w_{A\infty}) > 0$. Substituting Eq. [3.2-46] into Eq. [3.2-52]

$$k_m = \frac{3}{2}\frac{D_A}{\delta_c} \qquad [3.2\text{-}53]$$

From Eq. [1.4-62]

$$\frac{\delta}{z} = \frac{4.64}{\sqrt{\text{Re}_z}} \qquad [3.2\text{-}54]$$

where the local Reynolds number is defined as

$$Re_z = \frac{zv_\infty}{\nu} = \frac{z\rho v_\infty}{\mu} \qquad [3.2\text{-}55]$$

From Eq. [3.2-50], [3.2-53], and [3.2-54]

$$\boxed{\frac{k_m z}{D_A} = \text{Sh}_z = 0.323\ \text{Sc}^{1/3}\text{Re}_z^{1/2}} \qquad [3.2\text{-}56]$$

where Sh_z is the **local Sherwood number**. This equation compares favorably with the exact solution,[6] which has a proportional constant of 0.332 instead of 0.323. Like the exact solution, Eqs. [3.2-50] and [3.2-56] apply in the range of $0.6 \leq \text{Sc} \leq 50$.

3.3. SPECIES DIFFERENTIAL MASS-BALANCE EQUATION

3.3.1. Derivation

In Section 3.2 the species integral mass-balance equation Eq. [3.2-4] was derived. If the control volume Ω does not change with time, it can also be written as follows

$$\iiint_\Omega \frac{\partial \rho_A}{\partial t}\, d\Omega = -\iint_A \rho_A \mathbf{v}\cdot\mathbf{n}\, dA - \iint_A \mathbf{j}_A\cdot\mathbf{n}\, dA + \iiint_\Omega r_A\, d\Omega \qquad [3.3\text{-}1]$$

[6] H. Blasius, *Z. Math. U. Phys. Sci.*, **1**, 1908.

The surface integrals in Eq. [3.3-1] can be converted into volume integrals using the Gauss divergence theorem (i.e., Eq. [A.4-1]):

$$\iint_A \rho_A \mathbf{v} \cdot \mathbf{n} \, dA = \iiint_\Omega \nabla \cdot \rho_A \mathbf{v} \, d\Omega \tag{3.3-2}$$

$$\iint_A \mathbf{j}_A \mathbf{v} \cdot \mathbf{n} \, dA = \iiint_\Omega \nabla \cdot \mathbf{j}_A \, d\Omega \tag{3.3-3}$$

Substituting Eq. [3.3-2] and [3.3-3] into Eq. [3.3-1]

$$\iiint_\Omega \left[\frac{\partial \rho_A}{\partial t} + \nabla \cdot \rho_A \mathbf{v} + \nabla \cdot \mathbf{j}_A - r_A \right] d\Omega = 0 \tag{3.3-4}$$

The integrand, which is continuous, must be zero since the equation must hold for any arbitrary region Ω.[7] Therefore

$$\boxed{\frac{\partial \rho_A}{\partial t} = -\nabla \cdot \rho_A \mathbf{v} - \nabla \cdot \mathbf{j}_A + r_A = -\nabla \cdot \mathbf{n}_A + r_A \quad \text{(variable properties)}} \tag{3.3-5}$$

noting that $\mathbf{n}_A = \rho_A \mathbf{v} + \mathbf{j}_A$ from Eq. [3.1-11].

By following a similar approach, an equation can be derived based on the basis of molar density c_A (moles A per unit volume), the molar diffusion flux $\underline{\mathbf{j}}_A$, the local molar average velocity $\underline{\mathbf{v}}$, and the molar production rate of A per unit volume $\underline{r}_A$. This equation is

$$\boxed{\frac{\partial c_A}{\partial t} = -\nabla \cdot c_A \underline{\mathbf{v}} - \nabla \cdot \underline{\mathbf{j}}_A + \underline{r}_A = -\nabla \cdot \underline{\mathbf{n}}_A + \underline{r}_A \quad \text{(variable properties)}} \tag{3.3-6}$$

noting that $\underline{\mathbf{n}}_A = c_A \underline{\mathbf{v}} + \underline{\mathbf{j}}_A$ from Eq. [3.1-14].

Fick's law of diffusion, according to Eqs. [3.1-5] and [3.1-7], is

$$\mathbf{j}_A = -\rho D_A \nabla w_A \tag{3.3-7}$$

$$\underline{\mathbf{j}}_A = -c D_A \nabla x_A \tag{3.3-8}$$

Substituting Eqs. [3.3-7] and [3.3-8] into Eqs. [3.3-5] and [3.3-6], respectively, the following equations can be obtained:

$$\frac{\partial \rho_A}{\partial t} + \nabla \cdot (\rho_A \mathbf{v}) = \nabla \cdot (\rho D_A \nabla w_A) + r_A \tag{3.3-9}$$

[7] W. Kaplan, *Advanced Calculus*, 2nd edition, Addison-Wesley, Reading, MA, 1973, p. 363.

$$\frac{\partial c_A}{\partial t} + \nabla\cdot(c_A\underline{\mathbf{v}}) = \nabla\cdot(cD_A\nabla x_A) + \underline{r}_A \qquad [3.3\text{-}10]$$

Let us now consider the case of incompressible fluids. From the continuity equation $\nabla\cdot\mathbf{v} = 0$ and so $\nabla\cdot(\rho_A\mathbf{v}) = \mathbf{v}\cdot\nabla\rho_A + \rho_A\nabla\cdot\mathbf{v} = \mathbf{v}\cdot\nabla\rho_A$. Since $\rho_A = \rho w_A$, for constant ρ and D_A, Eq. [3.3-9] reduces to

$$\boxed{\frac{\partial \rho_A}{\partial t} + \mathbf{v}\cdot\nabla\rho_A = D_A\nabla^2\rho_A + r_A \quad (\text{constant } \rho \text{ and } D_A)} \qquad [3.3\text{-}11]$$

Let M_A be the molecular weight of species A, then $c_A = \rho_A/M_A$ and $\underline{r}_A = r_A/M_A$. As such, Eq. [3.3-11] can be divided by M_A to become

$$\boxed{\frac{\partial c_A}{\partial t} + \mathbf{v}\cdot\nabla c_A = D_A\nabla^2 c_A + \underline{r}_A \quad (\text{constant } \rho \text{ and } D_A)} \qquad [3.3\text{-}12]$$

Equation [3.3-11] or [3.3-12] is the species differential mass-balance equation or the **species continuity equation**. In Tables 3.3-1 and 3.3-2 Eqs. [3.3-11] and [3.3-12] are respectively given in expanded forms for rectangular, cylindrical, and spherical coordinates.

TABLE 3.3-1 The Species Continuity Equation in Terms of Mass Concentration (for a Fluid with Constant ρ and D_A)

Rectangular coordinates

$$\frac{\partial \rho_A}{\partial t} + \left(v_x\frac{\partial \rho_A}{\partial x} + v_y\frac{\partial \rho_A}{\partial y} + v_z\frac{\partial \rho_A}{\partial z}\right) = D_A\left(\frac{\partial^2 \rho_A}{\partial x^2} + \frac{\partial^2 \rho_A}{\partial y^2} + \frac{\partial^2 \rho_A}{\partial z^2}\right) + r_A \qquad [A]$$

Cylindrical coordinates

$$\frac{\partial \rho_A}{\partial t} + \left(v_r\frac{\partial \rho_A}{\partial r} + v_\theta\frac{1}{r}\frac{\partial \rho_A}{\partial \theta} + v_z\frac{\partial \rho_A}{\partial z}\right) = D_A\left[\frac{1}{r}\frac{\partial}{\partial r}\left(r\frac{\partial \rho_A}{\partial r}\right) + \frac{1}{r^2}\frac{\partial^2 \rho_A}{\partial \theta^2} + \frac{\partial^2 \rho_A}{\partial z^2}\right] + r_A \qquad [B]$$

Spherical coordinates

$$\frac{\partial \rho_A}{\partial t} + \left(v_r\frac{\partial \rho_A}{\partial r} + v_\theta\frac{1}{r}\frac{\partial \rho_A}{\partial \theta} + v_\phi\frac{1}{r\sin\theta}\frac{\partial \rho_A}{\partial \phi}\right)$$

$$= D_A\left[\frac{1}{r^2}\frac{\partial}{\partial r}\left(r^2\frac{\partial \rho_A}{\partial r}\right) + \frac{1}{r^2\sin\theta}\frac{\partial}{\partial \theta}\left(\sin\theta\frac{\partial \rho_A}{\partial \theta}\right) + \frac{1}{r^2\sin^2\theta}\frac{\partial^2 \rho_A}{\partial \phi^2}\right] + r_A \qquad [C]$$

TABLE 3.3-2 The Species Continuity Equation in Terms of Molar Concentration (for a Fluid with Constant ρ and D_A)

Rectangular coordinates

$$\frac{\partial c_A}{\partial t} + \left(v_x \frac{\partial c_A}{\partial x} + v_y \frac{\partial c_A}{\partial y} + v_z \frac{\partial c_A}{\partial z} \right) = D_A \left(\frac{\partial^2 c_A}{\partial x^2} + \frac{\partial^2 c_A}{\partial y^2} + \frac{\partial^2 c_A}{\partial z^2} \right) + \underline{r}_A \quad \text{[A]}$$

Cylindrical coordinates

$$\frac{\partial c_A}{\partial t} + \left(v_r \frac{\partial c_A}{\partial r} + v_\theta \frac{1}{r} \frac{\partial c_A}{\partial \theta} + v_z \frac{\partial c_A}{\partial z} \right) = D_A \left[\frac{1}{r} \frac{\partial}{\partial r} \left(r \frac{\partial c_A}{\partial r} \right) + \frac{1}{r^2} \frac{\partial^2 c_A}{\partial \theta^2} + \frac{\partial^2 c_A}{\partial z^2} \right] + \underline{r}_A \quad \text{[B]}$$

Spherical coordinates

$$\frac{\partial c_A}{\partial t} + \left(v_r \frac{\partial c_A}{\partial r} + v_\theta \frac{1}{r} \frac{\partial c_A}{\partial \theta} + v_\phi \frac{1}{r \sin\theta} \frac{\partial c_A}{\partial \phi} \right)$$

$$= D_A \left[\frac{1}{r^2} \frac{\partial}{\partial r} \left(r^2 \frac{\partial c_A}{\partial r} \right) + \frac{1}{r^2 \sin\theta} \frac{\partial}{\partial \theta} \left(\sin\theta \frac{\partial c_A}{\partial \theta} \right) + \frac{1}{r^2 \sin^2\theta} \frac{\partial^2 c_A}{\partial \phi^2} \right] + \underline{r}_A \quad \text{[C]}$$

3.3.2. Dimensionless Form

Like the equations of motion and energy, the species continuity equation can be presented in the dimensionless form to make the solutions more general.

For forced convection the following dimensionless variables can be defined

$$T^* = \frac{T - T_0}{T_1 - T_0} \quad \text{(dimensionless temperature)} \quad [3.3\text{-}13]$$

$$c_A^* = \frac{c_A - c_{A0}}{c_{A1} - c_{A0}} \quad \text{(dimensionless concentration)} \quad [3.3\text{-}14]$$

$$t^* = \frac{tV}{L} \quad \text{(dimensionless time)} \quad [3.3\text{-}15]$$

$$\mathbf{v}^* = \frac{\mathbf{v}}{V} \quad \text{(dimensionless velocity)} \quad [3.3\text{-}16]$$

$$p^* = \frac{p - p_0}{\rho V^2} \quad \text{(dimensionless pressure)} \quad [3.3\text{-}17]$$

$$x^*, y^*, z^* = \frac{x}{L}, \frac{y}{L}, \frac{z}{L} \quad \text{(dimensionless coordinates)} \quad [3.3\text{-}18]$$

$$\nabla^*, \nabla^{*2} = L\nabla, L^2\nabla^2 \quad \text{(dimensionless operators)} \quad [3.3.19]$$

where $(c_{A1} - c_{A0})$, V and L are any characteristic concentration difference, velocity, and length, respectively. Equations [3.3-15] through [3.3-19] are identical to Eqs. [1.5-12] through [1.5-17] for forced convection in an isothermal single-component system.

Let us assume that there are no chemical reactions involved. Equation [3.3-12] reduces to

$$\frac{\partial c_A}{\partial t} + \mathbf{v} \cdot \nabla c_A = D_A \nabla^2 c_A \qquad [3.3\text{-}20]$$

Substituting Eqs. [3.3-13] through [3.3-19] into Eq. [3.3-20]

$$\frac{V}{L}\frac{\partial}{\partial t^*}[c_A^*(c_{A1} - c_{A0})] + V\mathbf{v}^* \cdot \nabla^* c_A^* \frac{1}{L}(c_{A1} - c_{A0}) = D_A \frac{1}{L^2}\nabla^{*2} c_A^*(c_{A1} - c_{A0}) \qquad [3.3\text{-}21]$$

Multiplying Eq. [3.3-21] by $L/[V(c_{A1} - c_{A0})]$

$$\frac{\partial c_A^*}{\partial t^*} + \mathbf{v}^* \cdot \nabla^* c_A^* = \frac{D_A}{LV}\nabla^{*2} c_A^* \qquad [3.3\text{-}22]$$

By combining Eq. [3.3-22] with Eqs. [2.3-18] through [2.3-24], the following equations can be obtained for mass transfer in forced convection:

Continuity: $$\nabla^* \cdot \mathbf{v}^* = 0 \qquad [3.3\text{-}23]$$

Motion: $$\frac{\partial \mathbf{v}^*}{\partial t^*} + \mathbf{v}^* \cdot \nabla^* \mathbf{v}^* = -\nabla^* p^* + \frac{1}{\mathrm{Re}}\nabla^{*2}\mathbf{v}^* + \frac{1}{\mathrm{Fr}}\mathbf{e}_g \qquad [3.3\text{-}24]$$

Energy: $$\frac{\partial T^*}{\partial t^*} + \mathbf{v}^* \cdot \nabla^* T^* = \frac{1}{\mathrm{Re}\,\mathrm{Pr}}\nabla^{*2} T^* = \frac{1}{\mathrm{Pe}_T}\nabla^{*2} T^* \qquad [3.325]$$

Species: $$\frac{\partial c_A^*}{\partial t^*} + \mathbf{v}^* \cdot \nabla^* c_A^* = \frac{1}{\mathrm{Re}\,\mathrm{Sc}}\nabla^{*2} c_A^* = \frac{1}{\mathrm{Pe}_S}\nabla^{*2} c_A^* \qquad [3.3\text{-}26]$$

where

$$\mathrm{Re} = \frac{LV}{\nu} \qquad \left(\text{Reynolds number} = \frac{\text{inertia force } \rho V^2/L}{\text{viscous force } \mu V/L^2}\right) \qquad [3.3\text{-}27]$$

$$\mathrm{Fr} = \frac{V^2}{gL} \qquad \left(\text{Froude number} = \frac{\text{inertia force } \rho V^2/L}{\text{gravity force } \rho g}\right) \qquad [3.3\text{-}28]$$

$$\mathrm{Pr} = \frac{\nu}{\alpha} \qquad \left(\text{Prandlt number} = \frac{\text{viscous diffusivity } \nu}{\text{thermal diffusivity } \alpha}\right) \qquad [3.3\text{-}29]$$

$$\mathrm{Sc} = \frac{\nu}{D_A} \qquad \left(\text{Schmidt number} = \frac{\text{viscous diffusivity } \nu}{\text{species diffusivity } D_A}\right) \qquad [3.3\text{-}30]$$

$$\mathrm{Pe}_T = \mathrm{Re}\,\mathrm{Pr} = \frac{LV}{\alpha}$$

$$\left(\text{thermal Peclet number} = \frac{\text{convection heat transport } \rho C_v V(T_1 - T_0)}{\text{conduction heat transport } k(T_1 - T_0)/L}\right)$$

[3.3-31]

$$\mathrm{Pe}_S = \mathrm{Re}\,\mathrm{Sc} = \frac{LV}{D_A}$$

$$\left(\text{solutal Peclet number} = \frac{\text{convection species transport } \; V(c_{A1} - c_{A0})}{\text{diffusion species transport } D_A(c_{A1} - c_{A0})/L}\right)$$

[3.3-32]

The physical meanings of Re, Fr, Pr, and Pe_T have been shown in Eqs. [1.5-27], [1.5-28], [2.3-23], and [2.3-24], respectively.

3.3.3. Boundary Conditions

Boundary conditions commonly encountered in mass transfer are summarized in Figs. 3.3-1 and 3.3-2. These boundary conditions are explained as follows:

1. At the plane or axis of symmetry, the concentration gradient in the transverse direction is zero, as in case 1 of Figs. 3.3-1 and 3.3-2.
2. A wall in contact with a fluid or the surface of a solid or fluid may be kept at a given solute concentration, as in case 2 of Figs. 3.3-1 and 3.3-2.
3. A wall in contact with a fluid or the surface of a solid or liquid may allow no penetration, evaporation, or reactions. As such, the diffusion flux and hence the concentration gradient are zero, as in case 3 of Figs. 3.3-1 and 3.3-2.
4. The free surface of a fluid may be exposed to a gas of solute concentration w_{Af}, or the surface of a solid may be exposed to a gas or liquid of solute concentration w_{Af}. A boundary conditions consistent with Eqn. [3.1-21] is as follows:

$$\left|\frac{j_{Ay}}{\rho}\right| = \left|-D_A \frac{\partial w_A}{\partial y}\right| = |k_m(w_A - w_{Af})| \qquad [3.3\text{-}33]$$

Commonly Encountered Mass Transfer Boundary Conditions in Rectangular Coordinates

1. Plane of symmetry

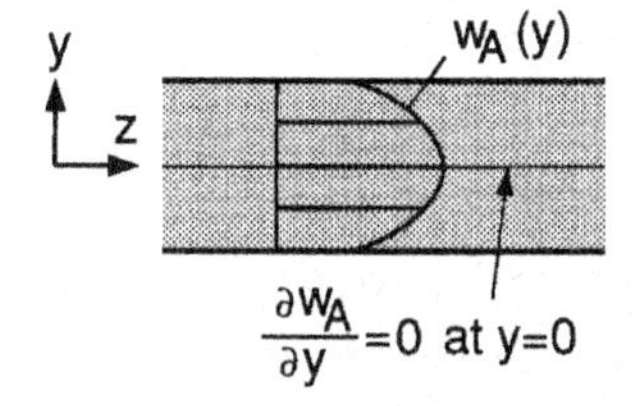

2. Constant surface concentration

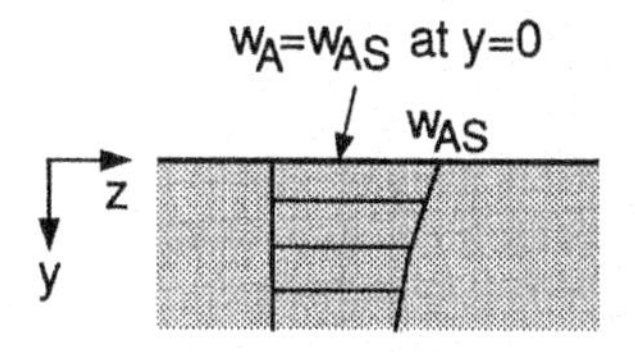

3. Sealed surface : no penetration, evaporation or reactions

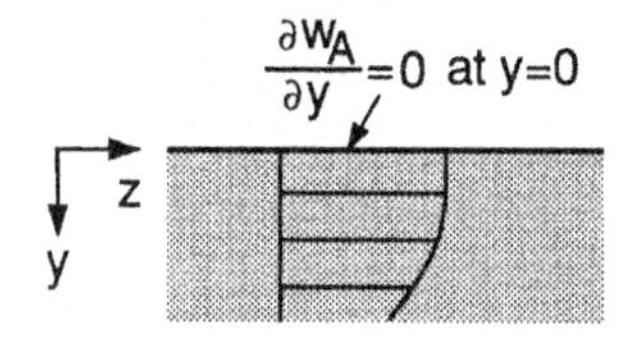

4. Convection exchange

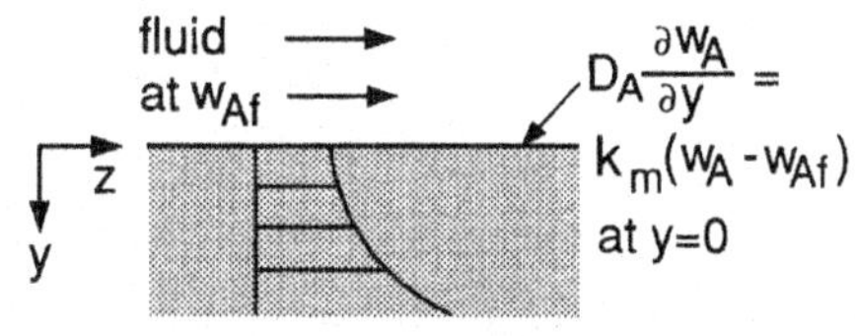

5.* Interface I/II: S/S,L/S,G/S,L/L,G/L

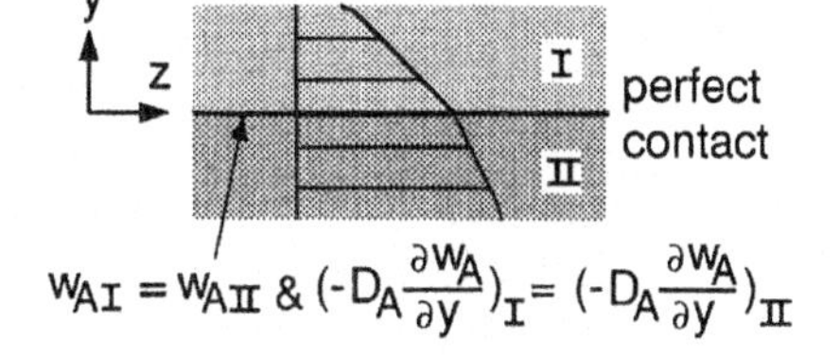

* When concentration distribution is to be found in both phases.

Fig. 3.3-1 Some mass transfer boundary conditions: rectangular coordinates.

for rectangular coordinates, as in case 4 of Fig. 3.3-1. For cylindrical coordinates, as in case 4 of Fig. 3.3-2, y in Eq. [3.3-33] is replaced by r.

5. For two phases in perfect contact with each other, the concentration and the diffusion flux are both continuous across the interface, that is, they are the same on both sides of the interface, as in case 5 of Figs. 3.3-1 and 3.3-2.

6. The boundary conditions at solid/solid and solid/fluid interfaces with phase transformations will be described in Chapter 5.

The boundary conditions for mass transfer in spheres are identical to those for mass transfer in cylinders, except that the axis of symmetry is now the point of symmetry.

Commonly Encountered Mass Transfer Boundary Conditions in Cylindrical Coordinates

1. Axis of symmetry

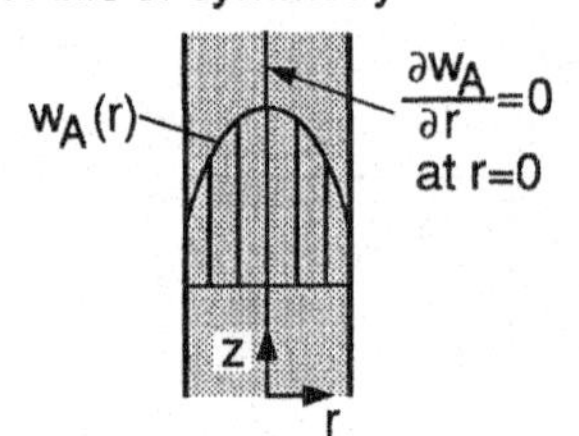

4. Convection exchange

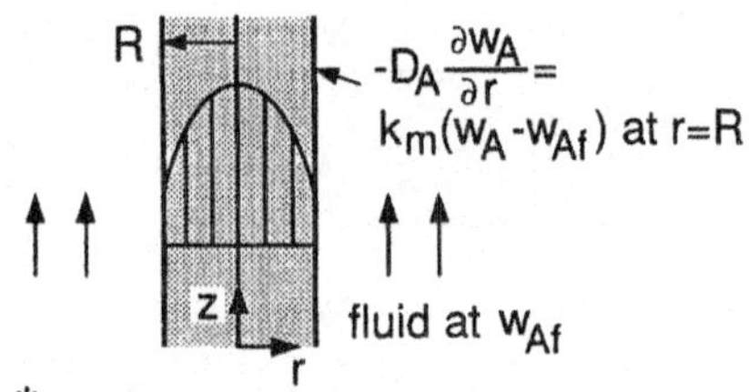

2. Constant surface concentration

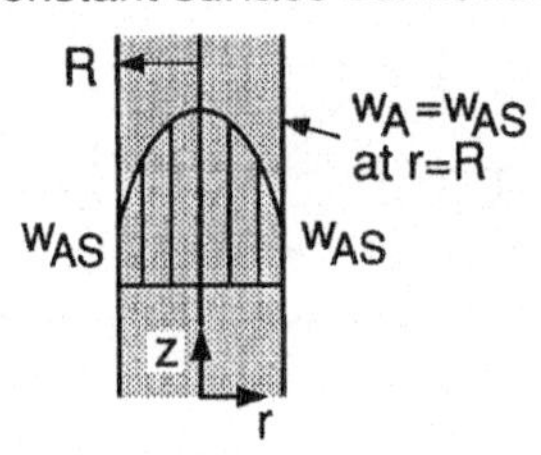

5.* Interface I/II : S/S, L/S, G/S

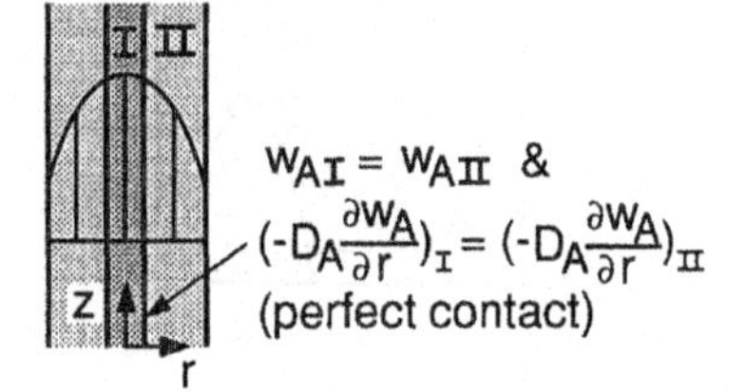

3. Sealed surface : no penetration, evaporation or reactions

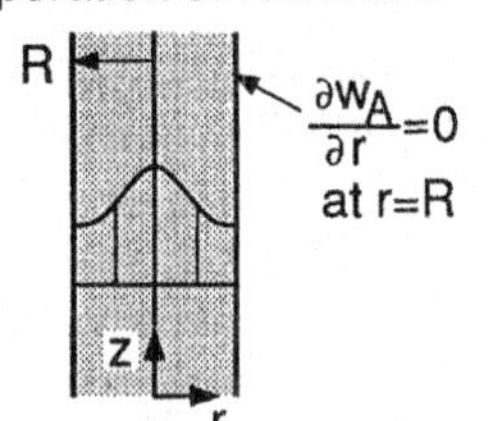

* When concentration distribution is to be found in both phases.

Fig. 3.3-2 Some mass transfer boundary conditions: cylindrical coordinates.

3.3.4. Solution Procedure

The purpose of the species continuity equation, as will be illustrated in the following examples, is to determine the concentration distribution. The step-by-step procedure for solving the species continuity equation, which is illustrated in Fig. 3.3-3 by the route represented by solid arrows, is as follows:

1. Choose a coordinate system that best describes the physical system geometrically.
2. Choose the species continuity equation from Table 3-3-1 or 3.3-2 for the coordinate system, if ρ and D_A are constant.

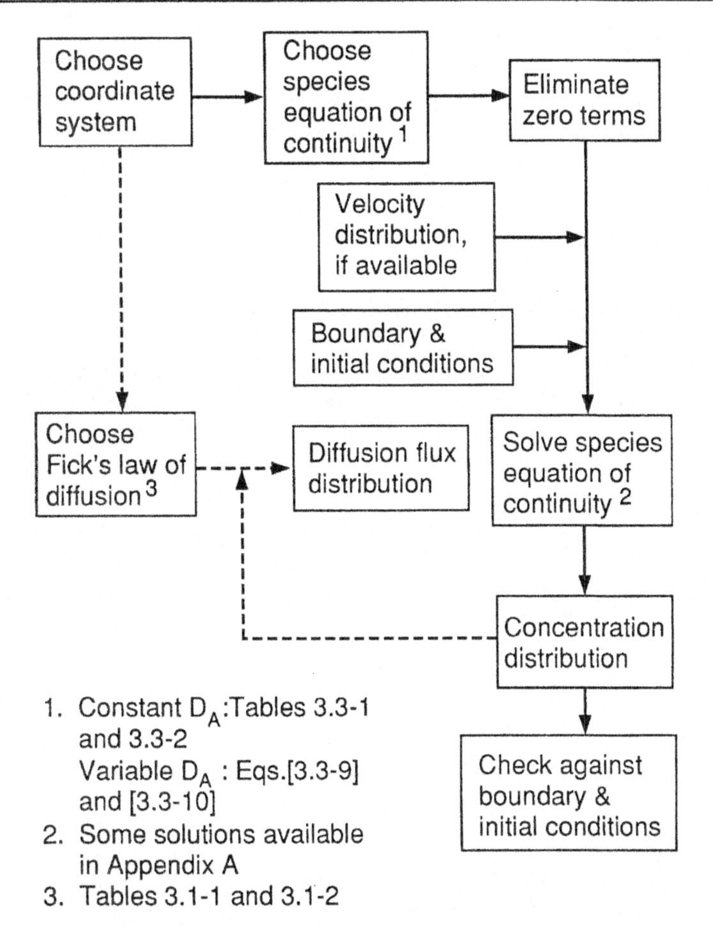

Fig. 3.3-3 Procedure for solving mass transfer problems by the species continuity equation.

3. Eliminate the zero terms from the species continuity equation.
4. Substitute the velocity distribution, if it is available and composition-independent, into the species continuity equation.
5. Set up the boundary and/or initial conditions.
6. Solve the species continuity equation, subject to the boundary and/or initial condition in step 5, for the concentration distribution. For convenience, some solutions are provided in Appendix A.
7. Check to see if the concentration distribution satisfies the boundary and/or initial conditions in step 5.

If the velocity and concentration distributions depend on each other, that is, if they are coupled, the equations of continuity and motion should also be considered. In other words, the equation of continuity, motion, and species continuity need to be solved simultaneously.

After the concentration distribution is determined, it can be substituted into Fick's law of diffusion to determine the diffusion flux distribution. This is illustrated in Fig. 3.3-3 by the route represented by broken arrows.

3.3.5. Examples

Example 3.3-1 Self-Diffusion of a Radioactive Isotope Tracer

Two solids of pure A are joined to each other with a thin film in between, as shown in Fig. 3.3-4*a*. The thin film is a solid solution of A and its radioactive isotope A^*. The quantity of A^* in moles per unit area of the joint is $\underline{M}$. At time $t = 0$, the system is annealed, that is, raised to an elevated temperature T and held there for a predetermined period of time, to allow diffusion of A^* into A. As A^* diffuses, its concentration at the joint continues to drop, as illustrated in Fig. 3.3-4*b*. Find the time-dependent concentration profile $c_{A^*}(x, t)$ and the diffusion flux $\underline{j}_{A^*x}$. Assume that the overall molar concentration c and the self-diffusion coefficient D_{A^*} are both constant.

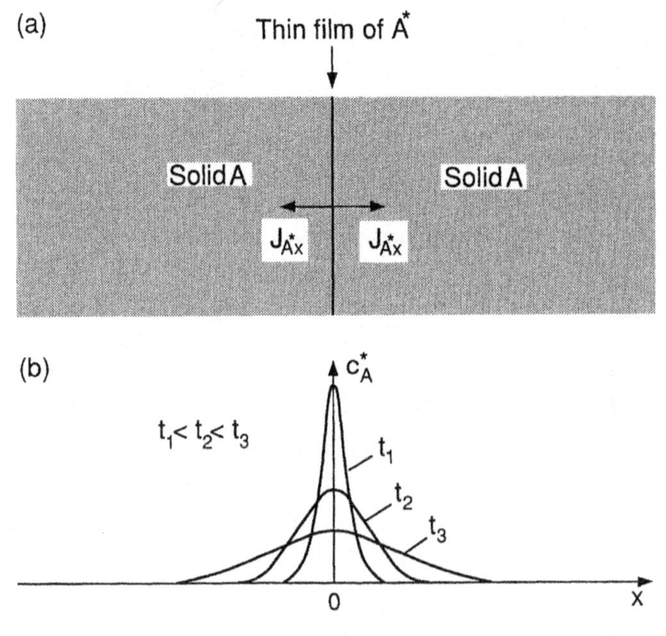

Fig. 3.3-4 Self-diffusion from a planar source between two solids.

Solution Apparently rectangular coordinates are most appropriate for the problem. Since c and D_{A*} are both constant, according to Eq. [A] of Table 3.3-2

$$\frac{\partial c_{A*}}{\partial t} + \left[v_x \frac{\partial c_{A*}}{\partial x} + v_y \frac{\partial c_{A*}}{\partial y} + v_z \frac{\partial c_{A*}}{\partial z} \right] = D_{A*} \left[\frac{\partial^2 c_{A*}}{\partial x^2} + \frac{\partial^2 c_{A*}}{\partial y^2} + \frac{\partial^2 c_{A*}}{\partial z^2} \right] + \underline{r}_{A*} \tag{3.3-34}$$

Since the solid is stationary, $v_x = v_y = v_z = 0$ and

$$\frac{\partial^2 c_{A*}}{\partial y^2} = \frac{\partial^2 c_{A*}}{\partial z^2} = 0$$

because concentration gradients exist in the x direction only. Since no chemical reactions are involved, $\underline{r}_{A*} = 0$. As such, Eq. [3.3-34] becomes

$$\frac{\partial c_{A*}}{\partial t} = D_{A*} \frac{\partial^2 c_{A*}}{\partial x^2} \tag{3.3-35}$$

The initial and boundary conditions are

$$c_{A*}(x, 0) = 0 \tag{3.3-36}$$

$$c_{A*}(\pm \infty, t) = 0 \tag{3.3-37}$$

and the mass conservation requirement is

$$\underline{M} = \int_{-\infty}^{\infty} c_{A*} dx \tag{3.3-38}$$

The solution, from Case P of Appendix A, is as follows:

$$\boxed{c_{A*} = \frac{\underline{M}}{2\sqrt{\pi_{A*} t}} \exp\left[\frac{-x^2}{4D_{A*}t}\right]} \tag{3.3-39}$$

and

$$\frac{\partial c_{A*}}{\partial x} = \frac{-\underline{M}x}{4\pi^{1/2}(D_{A*}t)^{3/2}} \exp\left[\frac{-x^2}{4D_{A*}t}\right] \tag{3.3-40}$$

From Eq. [A] of Table 3.1-2

$$\underline{j}_{A*x} = -D_{A*} \frac{\partial c_{A*}}{\partial x} \tag{3.3-41}$$

Substituting Eq. [3.3-40] into Eq. [3.3-41]

$$\underline{j}_{A^*x} = \frac{\underline{M}x}{4(\pi D_{A^*})^{1/2}t^{3/2}} \exp\left[\frac{-x^2}{4D_{A^*}t}\right] \qquad [3.3\text{-}42]$$

In a closely related case where only the solid on the right is present, Eqs, [3.3-37] and [3.3-38] are modified into

$$c_{A^*}(\infty, t) = 0 \qquad [3.3\text{-}43]$$

$$\underline{M} = \int_0^\infty c_{A^*}dx \qquad [3.3\text{-}44]$$

The solution, from Case Q of Appendix A, is the follows:

$$\boxed{c_{A^*} = \frac{\underline{M}}{\sqrt{\pi D_{A^*}t}} \exp\left[\frac{-x^2}{4D_{A^*}t}\right]} \qquad [3.3\text{-}45]$$

Equations [3.3-39] and [3.3-45] can both be used to determine D_{A^*} from diffusion experiments. The latter, which is preferred since it does not require joining two solids, can be converted into

$$\ln c_{A^*} = \ln\left[\frac{\underline{M}}{\sqrt{\pi D_{A^*}t}}\right] - \frac{x^2}{4D_{A^*}t} \qquad [3.3\text{-}46]$$

After annealing at the predetermined temperature and for the predetermined time period t, c_{A^*} can be measured at various x values with the help of the radioactive counter. The data can be plotted in the form of $\ln c_{A^*}$ versus x^2. According to Eq. [3.3-46], the slope of the straight line should be $-(4D_{A^*}t)^{-1}$. Since the annealing time t is known, D_{A^*} is readily obtained from the slope.

It is perhaps worth noting that it is not required to know $\underline{M}$, which happens to be difficult to measure because of the very small thickness of the film.

Figure 3.3-5 shows the expermental results for the self-diffusion of the ^{67}Cu isotope in single crystals of pure copper.[8] In Fig. 3.3-5*a* $\ln c_{A^*}$, in arbitrary radioactivity units, is plotted against x^2 for various annealing temperatures. In Fig. 3.3-5*b*, D_{A^*} is plotted against the reciprocal of the absolute temperature.

Equations [3.3-39] and [3.3-45] can also be applied to self-diffusion of A^* into an A–B alloy, either substitutional or interstitial. The thin film has the same chemical composition as the A–B alloy but with A^* replacing some of the A atoms; the mass of A^* per unit area of the joint is $\underline{M}$.

[8] S. J. Rothman and N. L. Peterson, *Phys. Stat.*, **35**, 305, 1969.

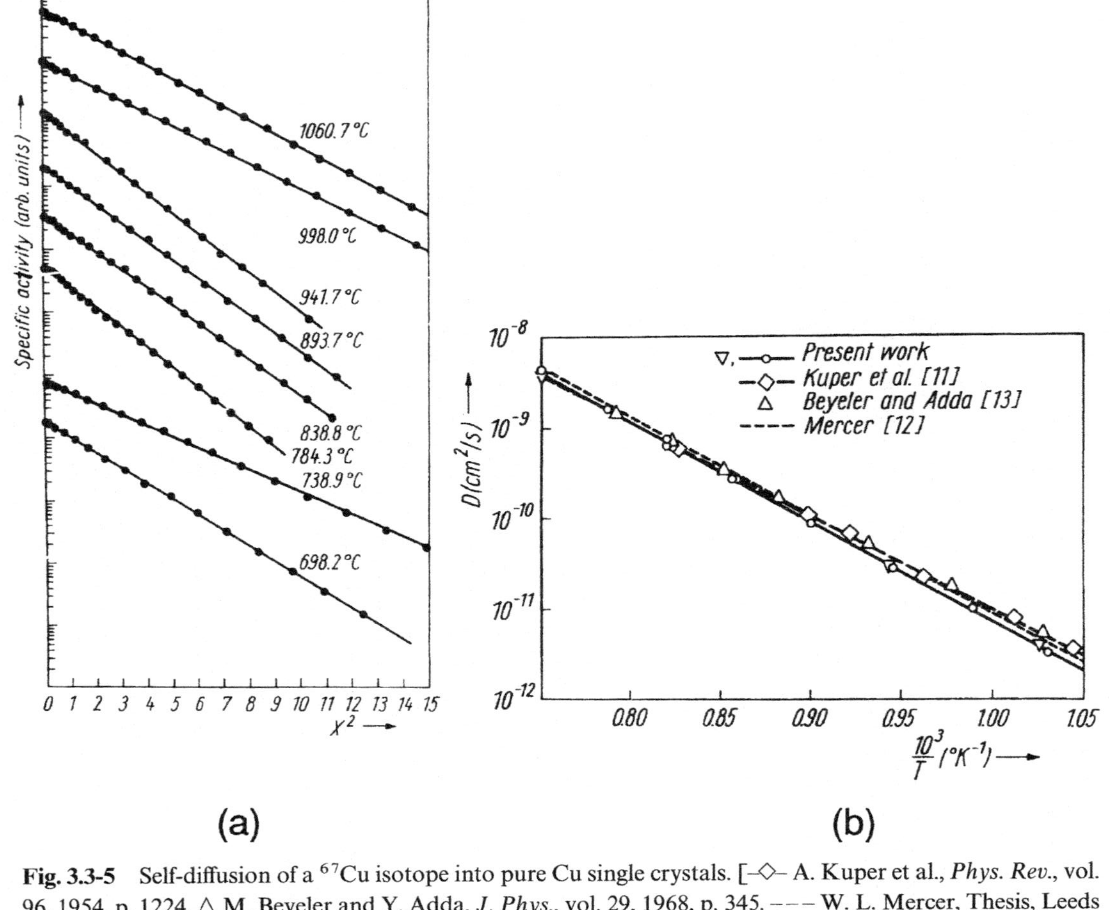

Fig. 3.3-5 Self-diffusion of a ^{67}Cu isotope into pure Cu single crystals. [–◇– A. Kuper et al., *Phys. Rev.*, vol. 96, 1954, p. 1224. △ M. Beyeler and Y. Adda, *J. Phys.*, vol. 29, 1968, p. 345. – – – W. L. Mercer, Thesis, Leeds University, 1995. (From S. J. Rothman and N. L. Peterson, *Physica Status Soilidi*, **35**, 1969, p. 305.)]

Example 3.3-2 Interstitial Diffusion Couple with Constant D_A

Two interstitial alloys of different initial compositions c_{A1} and c_{A2} are butt-joined together as illustrated in Fig. 3.3-6. At time $t = 0$, the couple is annealed at an elevated temperature T to allow diffusion of the interstitial atoms A across the joint. Assume that the overall molar density c and the intrinsic diffusion coefficient D_A are both constant. When c_{A1} and c_{A2} are both low in comparison to c, c is approximately constant. Find the time-dependent concentration profile $c_A(x, t)$.

Solution Since c and D_A are both constant, the concentration profile is symmetric with respect to the interface concentration c_{AS}: $c_{AS} = (c_{A1} + c_{A2})/2 =$ constant.

From Eq. [A] of Table 3.3-2 and the same reasons leading to Eq. [3.3-35]

$$\frac{\partial c_A}{\partial t} = D_A \frac{\partial^2 c_A}{\partial x^2} \qquad [3.3\text{-}47]$$

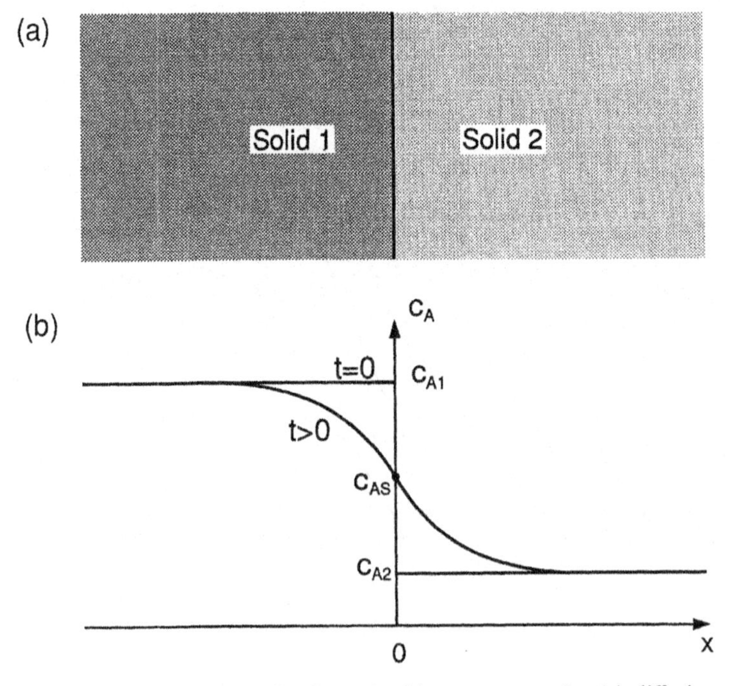

Fig. 3.3-6 Interstitial diffusion of solute A with a constant D_A: (*a*) diffusion couple; (*b*) concentration profile.

The initial and boundary conditions for solid 2 are

$$c_A(x, 0) = c_{A2} \tag{3.3-48}$$

$$c_A(0, t) = c_{AS} \tag{3.3-49}$$

$$c_A(\infty, t) = c_{A2} \tag{3.3-50}$$

The solution, from Case O of Appendix A, is as follows:

$$\boxed{\frac{c_A - c_{AS}}{c_{A2} - c_{AS}} = \operatorname{erf}\left[\frac{x}{\sqrt{4D_A t}}\right]} \tag{3.3-51}$$

The values of the error function can be found from Fig. 2.3.7.

From Eq. [2.3-51] we see that $\operatorname{erf}(-x) = -\operatorname{erf}(x)$. Therefore, Eq. [3.3-51] is also valid in solid 1. This equation can be used to determine the intrinsic diffusion coefficient D_A from experiments. The value of D_A that yields the best agreement with the measured concentration profile is selected. Since D_A can often be composition-dependent, Eq. [3.3-51] works better when c_{A1} is close to c_{A2}.

Example 3.3-3 Interstitial Diffusion Couple with Variable D_A

Consider a diffusion couple with the initial compositions of c_{A1} and c_{A2} in solids 1 and 2, respectively. The concentration profile after annealing is shown in Fig. 3.3-7. Determine the intrinsic diffusion coefficient D_A.

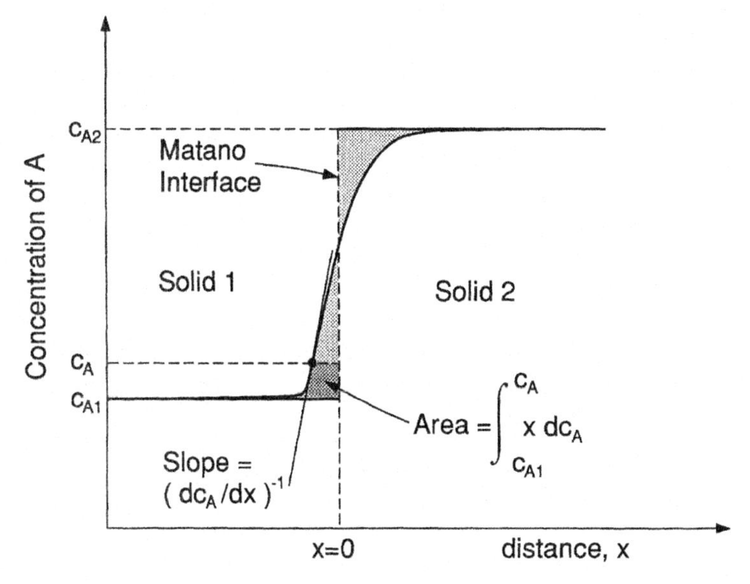

Fig. 3.3-7 Interstitial diffusion of solute A with a variable D_A.

Solution Assuming constant overall molar concentration c, Eq. [3.3-10] can be written as follows for one-dimensional diffusion in the x direction, with no bulk motion of material and chemical reactions:

$$\frac{\partial c_A}{\partial t} = \frac{\partial}{\partial x}\left(D_A \frac{\partial c_A}{\partial x}\right) \tag{3.3-52}$$

Let us combine x and t so that c_A can be expressed as a function of a single variable. This technique is called the **combination of variables**. Define

$$\eta = \frac{x}{t^{1/2}} \tag{3.3-53}$$

so that

$$\frac{\partial c_A}{\partial t} = \frac{\partial \eta}{\partial t}\frac{dc_A}{d\eta} = \frac{-1}{2}\frac{x}{t^{3/2}}\frac{dc_A}{d\eta} \tag{3.3-54}$$

and

$$\frac{\partial c_A}{\partial x} = \frac{\partial \eta}{\partial x}\frac{dc_A}{d\eta} = \frac{1}{t^{1/2}}\frac{dc_A}{d\eta} \tag{3.3-55}$$

Substituting Eqs. [3.3-53] through [3.3-55] into Eq. [3.3-52]

$$\frac{-1}{2}\frac{x}{t^{3/2}}\frac{dc_A}{d\eta} = \frac{1}{t^{1/2}}\frac{d}{d\eta}\left(D_A \frac{\partial c_A}{\partial x}\right) = \frac{1}{t}\frac{d}{d\eta}\left(D_A \frac{dc_A}{d\eta}\right) \tag{3.3-56}$$

Multiplying both sides by $t\,d\eta$

$$\frac{-1}{2}\eta\, dc_A = d\left(D_A \frac{dc_A}{d\eta}\right) \tag{3.3-57}$$

So, the partial differential equation, Eq. [3.3-52], is now transformed to an ordinary differential equation. The boundary conditions are

$$c_A = c_{A2} \quad \text{at} \quad x = \infty \quad \text{or} \quad \eta = \infty \tag{3.3-58}$$

and

$$c_A = c_{A1} \quad \text{at} \quad x = -\infty \quad \text{or} \quad \eta = -\infty \tag{3.3-59}$$

As such, Eq. [3.3-57] can be integrated as follows:

$$-\frac{1}{2}\int_{c_{A1}}^{c_A}\eta\, dc_A = \left(D_A\frac{dc_A}{d\eta}\right)\bigg|_{c_A} - \left(D_A\frac{dc_A}{d\eta}\right)\bigg|_{c_{A1}} = \left(D_A\frac{dc_A}{d\eta}\right)\bigg|_{c_A} \quad [3.3\text{-}60]$$

or

$$D_A|_{c_A} = \frac{-1}{2(dc_A/d\eta)|_{c_A}}\int_{c_{A1}}^{c_A}\eta\, dc_A \quad [3.3\text{-}61]$$

From Eq. [3.3-59] $dc_A/d\eta = dc_{A1}/d\eta = 0$ at $\eta = -\infty$, that is, $c_A = c_{A1}$. Similarly, $dc_A/d\eta = dc_{A2}/d\eta = 0$ at $\eta = \infty$, or, $c_A = c_{A2}$. As such, when $c_A = c_{A2}$, Eq. [3.3-60] becomes

$$\frac{-1}{2}\int_{c_{A1}}^{c_{A2}}\eta\, dc_A = 0 \quad [3.3\text{-}62]$$

Since concentration profiles are always measured at some fixed time t for diffusion, Eqs. [3.3-61] and [3.3-62] can be more conveniently expressed in terms of x rather than η. Substituting Eqs. [3.3-53] and [3.3.-55], these equations become

$$\boxed{D_A|_{c_A} = -\frac{1}{2t}\frac{1}{(dc_A/dx)|_{c_A}}\int_{c_{A1}}^{c_A} x dc_A} \quad [3.3\text{-}63]$$

and

$$\boxed{\int_{c_{A1}}^{c_{A2}} x\, dc_A = 0} \quad [3.3\text{-}64]$$

The transformation of the partial differential equation (Eq. [3.3-52]) into an ordinary equation (Eq. [3.3-57]) is due to Boltzmann.[9] Matano[10] first used Eqs. [3.3-63] and [3.3-64] to determine the diffusion coefficient from experiments, and this technique is known as the **Boltzmann–Matano method**. The origin of the special x coordinate satisfying Eq. [3.3-64] is called the **Matano interface**. Graphically, it is the line that makes the two shaded areas of Fig. 3.3-7 equal. After the interface location is decided, D_A (at c_A) can be determined from Eq. [3.3-63], from the darker shaded area and the reciprocal of the slope at c_A.

[9] L. Boltzmann, *Ann. Physik*, **53**, 960, 1894.

[10] C. Matano, *Jpn. Phys.*, **8**, 109, 1933.

Example 3.3-4 Substitutional Diffusion Couple with Constant $\tilde{D}$

In the diffusion couples of interstitial solid solutions only the diffusion of the interstitial species needs to be considered, as already shown in the previous two examples. In the diffusion couples of substitutional solid solutions, however, the interdiffusion of all species needs to be considered. Let the diffusion couple in Fig. 3.3-6*a* be two binary *A–B* substitutional alloys of different compositions. Assume that the overall molar density c and the interdiffusion coefficient $\tilde{D}$ are both constant. Find the concentration profiles $c_A(x, t)$ and $c_B(x, t)$.

Solution From Eq. [3.3-10] and the fact that there are no chemical reactions involved

$$\frac{\partial c_A}{\partial t} + \nabla \cdot (c_A \underline{v}) = \nabla \cdot (c D_A \nabla x_A) \qquad [3.3\text{-}65]$$

and

$$\frac{\partial c_B}{\partial t} + \nabla \cdot (c_B \underline{v}) = \nabla \cdot (c D_B \nabla x_B) \qquad [3.3\text{-}66]$$

Adding Eq. [3.3-66] to Eq. [3.3-65]

$$\frac{\partial}{\partial t}(c_A + c_B) + \nabla \cdot (c_A + c_B)\underline{v} = \nabla \cdot (c D_A \nabla x_A + c D_B \nabla x_B) \qquad [3.3\text{-}67]$$

Let us assume that the overall molar density c is constant:

$$c = c(x_A + x_B) = c_A + c_B = \text{constant} \qquad [3.3\text{-}68]$$

Substituting Eq. [3.3-68] into Eq. [3.3-67]

$$\nabla \cdot c\underline{v} = \nabla \cdot [c(D_A \nabla x_A + D_B \nabla x_B)] \qquad [3.3\text{-}69]$$

From Eq. [3.3-69] and the fact that $x_A + x_B = 1$

$$\boxed{\underline{v} = D_A \nabla x_A + D_B \nabla x_B = (D_A - D_B) \nabla x_A} \qquad [3.3\text{-}70]$$

Equation [3.3-70] suggests that if $D_A > D_B$ there will be a "convection" of solid material within the diffusion couple, keeping in mind that in the species continuity equation, Eq. [3.3-10], $\underline{v}$ is the fluid velocity. According to Eq. [3.3-70] this velocity is not uniform but varies with the local concentration gradient ∇x_A. Furthermore, it is in the direction of positive ∇x_A, that is, from where the concentration of A is low to where it is high. This convection of

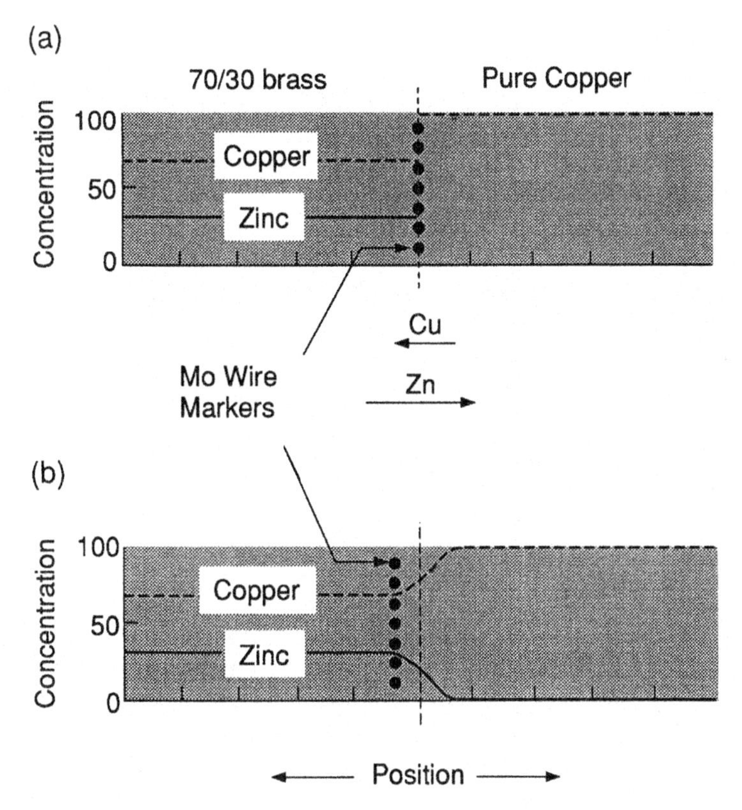

Fig. 3.3-8 Interdiffusion in a brass–copper diffusion couple: (*a*) original; (*b*) after diffusion.

material was observed by Smigelskas and Kirkendall in 1947[11] by placing inert Mo wire markers at an original copper–brass interface, as illustrated in Fig. 3.3-8*a*. Since $D_{Zn} > D_{Cu}$, more zinc atoms diffuse to the right into the copper than copper atoms diffuse to the left into the brass. The excess atoms at the right produce a convection of material including the markers to the left. Consequently, the final location of the markers is on the brass side of the original interface, as illustrated in Fig. 3.3-8*b*. For this reason $\underline{v}$ is often referred to as the **marker velocity**.

Substituting Eq. [3.3-70] into Eq. [3.3-65]

$$\frac{\partial c_A}{\partial t} + \nabla \cdot [c x_A (D_A - D_B) \nabla x_A] = \nabla \cdot (c D_A \nabla x_A) \qquad [3.3\text{-}71]$$

[11] A. D. Smigelskas and E. O. Kirkendall, *Trans. Met. Soc. AIME*, **171**, 130, 1947.

Since $x_A + x_B = 1$

$$\frac{\partial c_A}{\partial t} = \nabla \cdot [c(x_B D_A + x_A D_B)\nabla x_A] \qquad [3.3\text{-}72]$$

Let us define the **interdiffusion coefficient** $\tilde{D}$ as follows:

$$\boxed{\tilde{D} = x_B D_A + x_A D_B} \qquad [3.3\text{-}73]$$

and Eq. [3.3-72] becomes

$$\boxed{\frac{\partial c_A}{\partial t} = \nabla \cdot [c\tilde{D}\nabla x_A] = \nabla \cdot [\tilde{D}\nabla c_A]} \qquad [3.3\text{-}74]$$

Similarly, for species B

$$\frac{\partial c_B}{\partial t} = \nabla \cdot [c\tilde{D}\nabla x_B] = \nabla \cdot [\tilde{D}\nabla c_B] \qquad [3.3\text{-}75]$$

These equations are valid for situations were c is constant and where there are no chemical reactions involved. They are more convenient to use than Eqs. [3.3-65] and [3.3-66] since the velocity $\underline{\mathbf{v}}$ does not have to be dealt with directly.

Equations [3.3-70] through [3.3-75] are valid regardless of whether $\tilde{D}$ is constant or variable. In the example to follow, $\tilde{D}$ will be variable. In the present example, however, we shall consider the simpler case of a constant $\tilde{D}$.

For one-dimensional diffusion in the x-direction and assuming that $\tilde{D}$ is constant, Eqs. [3.3-74] and [3.3-75] reduce to

$$\frac{\partial c_A}{\partial t} = \tilde{D}\frac{\partial^2 c_A}{\partial x^2} \qquad [3.3\text{-}76]$$

and

$$\frac{\partial c_B}{\partial t} = \tilde{D}\frac{\partial^2 c_B}{\partial x^2} \qquad [3.3\text{-}77]$$

Equation [3.3-76] is identical to Eq. [3.3-47] for an interstitial diffusion couple, except that D_A is now replaced by $\tilde{D}$. The initial and boundary conditions are also identical to those given by Eqs. [3.3-48] through [3.3-50] for an interstitial diffusion couple:

$$c_A(x, 0) = c_{A2} \qquad [3.3\text{-}78]$$

$$c_A(0, t) = c_{AS} \qquad [3.3\text{-}79]$$

$$c_A(\infty, t) = c_{A2} \qquad [3.3\text{-}80]$$

where

$$c_{AS} = \tfrac{1}{2}(c_{A1} + c_{A2}) = \text{constant} \qquad [3.3\text{-}81]$$

The solution, from Case O of Appendix A, is as follows:

$$\boxed{\frac{c_A - c_{AS}}{c_{A2} - c_{AS}} = \text{erf}\left[\frac{x}{\sqrt{4\tilde{D}t}}\right]} \qquad [3.3\text{-}82]$$

and similarly

$$\boxed{\frac{c_B - c_{BS}}{c_{B2} - c_{BS}} = \text{erf}\left[\frac{x}{\sqrt{4\tilde{D}t}}\right]} \qquad [3.3\text{-}83]$$

Equation [3.3-82] can be used to determine the interdiffusion coefficient $\tilde{D}$. The value of $\tilde{D}$ that yields the best agreement with the measured concentration profile is selected.

In fact, once the interdiffusion coefficient $\tilde{D}$ is available, the intrinsic diffusion coefficients D_A and D_B can also be found, if the marker velocity is also available.

For one-dimensional diffusion in the x direction, Eq. [3.3-70] reduces to

$$\underline{v}_x = (D_A - D_B)\frac{\partial x_A}{\partial x} \qquad [3.3\text{-}84]$$

and from Eq. [3.3-73]

$$\tilde{D} = x_B D_A + x_A D_B \qquad [3.3\text{-}85]$$

These two equations were first derived by Darken[12] and are usually known as **Darken's equations**. They can be solved simultaneously to give

$$\boxed{D_A = \tilde{D} + x_A \underline{v}_x \left(\frac{\partial x_A}{\partial x}\right)^{-1}} \qquad [3.3\text{-}86]$$

and

$$\boxed{D_B = \tilde{D} - x_B \underline{v}_x \left(\frac{\partial x_A}{\partial x}\right)^{-1}} \qquad [3.3\text{-}87]$$

By using Eqs. [3.3-86] and [3.3-87], the intrinsic diffusion coefficients D_A and D_B corresponding to the x_A and x_B at the markers can be determined from $\tilde{D}$, and the $\underline{v}_x$ and $\partial x_A/\partial x$ at the markers.

[12] L. S. Darken, *Trans. Met. Soc. AIME*, **175**, 184, 1948.

Let x_m be the displacement of the markers. The marker velocity

$$\underline{v}_x = \frac{dx_m}{dt} \tag{3.3-88}$$

From Eq. [3.3-82] it is seen that for any concentration c_A, including that at the markers, the position x is proportional to $\sqrt{t}$. At such, we can expect the marker position

$$x_m = I\sqrt{t} \tag{3.3-89}$$

where I is a proportional constant.

Substituting Eq. [3.3-89] into Eq. [3.3-88]

$$\underline{v}_x = \frac{x_m}{2t} \tag{3.3-90}$$

As such, from the marker position change and the annealing time, the marker velocity can be found using Eq. [3.3-90].

Example 3.3-5 Substitutional Diffusion Couple with Variable $\tilde{D}$

Consider the diffusion couple of a binary A–B substitutional solid solutions, with the initial compositions of c_{A1} and c_{A2} in solids 1 and 2, respectively. The concentration profile after annealing is shown in Fig. 3.3-7. Determine the interdiffusion coefficient $\tilde{D}$. Assume that the molar density c is constant.

Solution From Eq. [3.3-74]

$$\frac{\partial c_A}{\partial t} = \frac{\partial}{\partial x}\left(\tilde{D}\frac{\partial c_A}{\partial x}\right) \tag{3.3-91}$$

The boundary conditions are

$$c_A = c_{A2} \quad \text{at} \quad x = \infty \tag{3.3-92}$$

and

$$c_A = c_{A1} \quad \text{at} \quad x = -\infty \tag{3.3-93}$$

These three equations are identical to those for an interstitial diffusion couple – Eqs. [3.3-52], [3.3-58], and [3.3-59] – except that D_A is replaced by $\tilde{D}$. As such

$$\tilde{D}|_{c_A} = -\frac{1}{2t}\frac{1}{(dc_A/dx)|_{c_A}}\int_{c_{A1}}^{c_A} x\, dc_A \tag{3.3-94}$$

and

$$\int_{c_{A1}}^{c_{A2}} x\,dc_A = 0 \qquad [3.3\text{-}95]$$

which are identical to Eqs. [3.3-63] and [3.3-64], respectively, for an interstitial diffusion couple, except that D_A is replaced by $\tilde{D}$. The origin of the special x coordinate satisfying Eq. [3.3-95] is the Matano interface. After the interface location is decided, $\tilde{D}$ (at c_A) can be determined from Eq. [3.3-94], from the darker shaded area in Fig. 3.3-7 and the reciprocal of the slope at c_A.

The intrinsic diffusion coefficients D_A and D_B at x_A can be calculated from Eq. [3.3-86] and [3.3-87], from the interdiffusion coefficient $\tilde{D}$ and the measured marker velocity $\underline{v}_x$ at x_A.

Figure 3.3-9 shows the experimental results from a diffusion couple of pure Pb and an In 36 at% (atom%) Pb alloy.[13] Figure 3.3-9*a* shows the concentration profiles – x_{In} versus x – from annealing at 175°C for 18 h (profile *A*1) and 118 h (profile *A*2). The x_{In} at the Matano interface (i.e., at $x = 0$) is 0.411. Figure 3.3-9*b* shows the measured $\tilde{D}$ against the reciprocal of temperature at the Matano interface.

Example 3.3-6 Diffusion and Reaction in a Porous Sphere[14]

As shown in Fig. 3.3-10, a chemical species *A* diffuses from a gas phase into a porous catalyst sphere of radius *R* in which it is converted into species *B*. The concentration of *A* at the surface of the sphere is c_{As}. The rate at which *A* is consumed per unit volume of the sphere is $\underline{r}_A = -k_1 a c_A$, where k_1 is the first-order reaction rate constant and *a* is the available catalytic surface area per unit volume of the sphere. Find the steady-state concentration distribution of *A* in the sphere. The effective diffusion coefficient of *A* in the sphere is D_A. Assume constant properties.

Solution The species continuity equation, Eq. [C] of Table 3.3-2,

$$\frac{\partial c_A}{\partial t} + v_r\frac{\partial c_A}{\partial r} + v_\theta\frac{1}{r}\frac{\partial c_A}{\partial \theta} + v_\phi\frac{1}{r\sin\theta}\frac{\partial c_A}{\partial \phi}$$

$$= D_A\left[\frac{1}{r^2}\frac{\partial}{\partial r}\left(r^2\frac{\partial c_A}{\partial r}\right) + \frac{1}{r^2\sin\theta}\frac{\partial}{\partial \theta}\left(\sin\theta\frac{\partial c_A}{\partial \theta}\right) + \frac{1}{r^2\sin^2\theta}\frac{\partial^2 c_A}{\partial \phi^2}\right] + \underline{r}_A \qquad [3.3\text{-}96]$$

[13] D. R. Campbell, K. N. Tu, and R. E. Robinson, *Acta Metallurgica*, **24**, 609, 1976.

[14] R. Aris, *Chem. Eng. Sci.* **6**, 265, 1957; E. W. Thiele, *Int. Eng. Chem.*, **31**, 916, 1957.

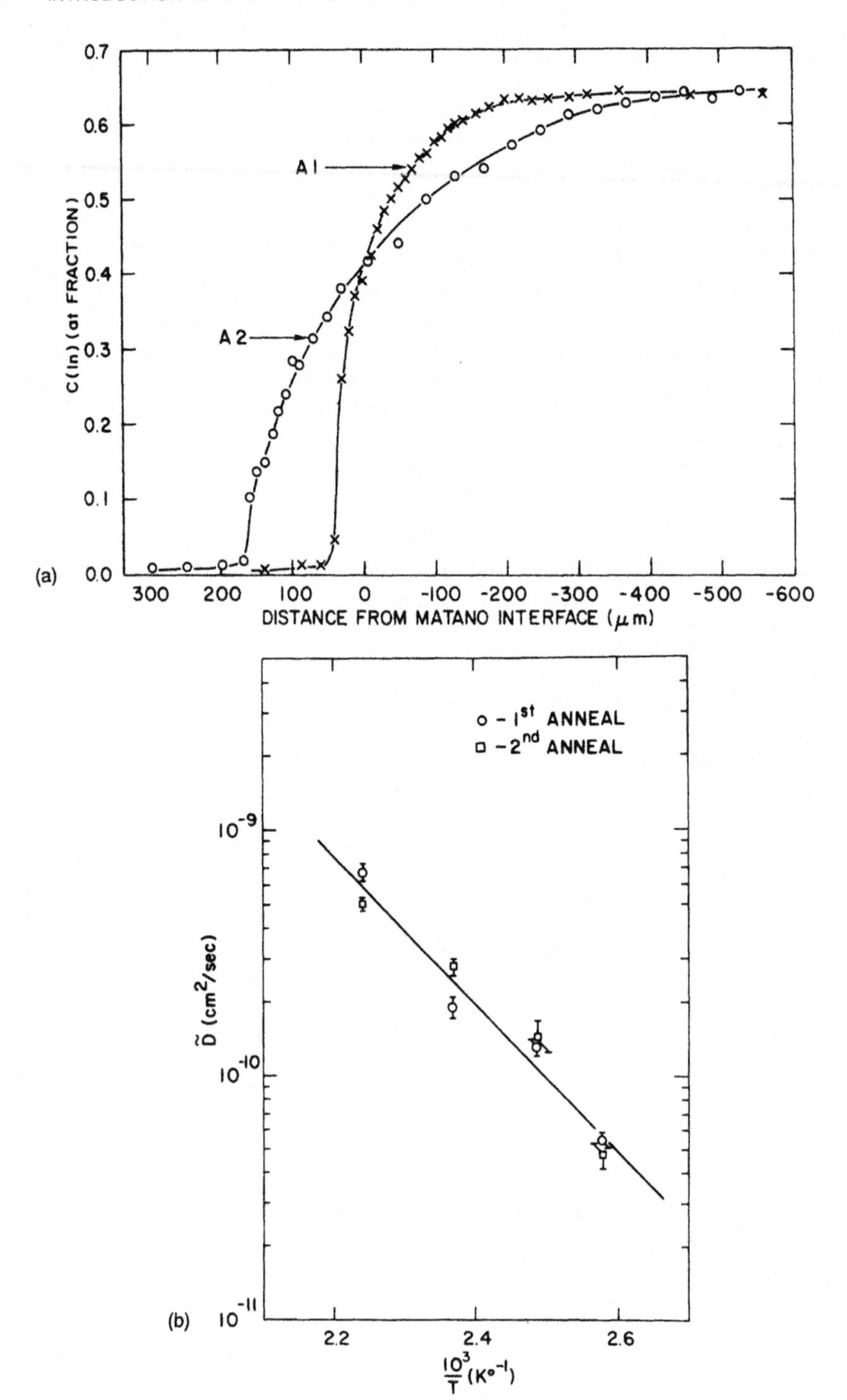

C(In) (at FRACTION)
A1
A2
(a)
DISTANCE FROM MATANO INTERFACE (μm)
○ - 1st ANNEAL
□ - 2nd ANNEAL
D̃ (cm²/sec)
(b)
10³/T (K°⁻¹)

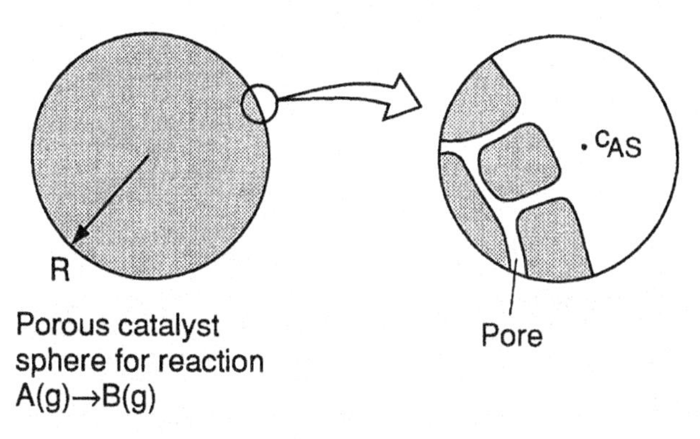

Fig. 3.3-10 Diffusion and chemical reaction in a porous catalyst sphere.

At the stady state $\partial c_A/\partial t$ is zero. Since there is no convection within the sphere, $v_r = v_\theta = v_\phi = 0$. With the absence of concentration gradients in the θ and ϕ directions, $\partial c_A/\partial\theta$ and $\partial^2 c_A/\partial\phi^2$ are both zero. As such, Eq. [3.3-96] reduces to

$$\frac{1}{r^2}\frac{d}{dr}\left(r^2\frac{dc_A}{dr}\right) = \frac{k_1 a}{D_A}c_A \tag{3.3-97}$$

The boundary conditions are

$$\frac{dc_A}{dr} = 0 \quad \text{at} \quad r = 0 \tag{3.3-98}$$

$$c_A = c_{As} \quad \text{at} \quad r = R \tag{3.3-99}$$

The solution, from Case N of Appendix A, is as follows

$$\boxed{\frac{c_A}{c_{As}} = \frac{R}{r}\left(\frac{e^{br} - e^{-br}}{e^{bR} - e^{-bR}}\right)} \tag{3.3-100}$$

where $b = \sqrt{k_1 a/D_A}$.

Fig. 3.3-9 Pb–In alloy diffusion couple: (*a*) concentration profile of In; (*b*) interdiffusion coefficient $\tilde{D}$. (Reprinted from D. R. Campbell, K. N. Tu, and R. E. Robinson, "Interdiffusion in a Bulk Couple of Pb and Pb–50% In Alloy," *Acta Metallurgica*, **24**, 1996, p. 609, with kind permission from Butterworth-Heinemann journals, Elsevier Science Ltd., The Boulevard, Langford Lane, Kidlington OX5 1GB, UK.)

The molar flow rate at the surface of the sphere is

$$W_A = 4\pi R^2 \left(-D_A \frac{dc_A}{dr} \right)\bigg|_{r=R} \qquad [3.3\text{-}101]$$

Substituting Eq. [3.3-100] into Eq. [3.3-101]

$$W_A = \frac{4\pi R c_{As}}{e^{bR} - e^{-bR}} \times [bR(e^{bR} + e^{-bR}) - (e^{bR} - e^{-bR})] \qquad [3.3\text{-}102]$$

Example 3.3-7 Diffusion into a Rising Film

As shown in Fig. 3.3-11, a liquid film of thickness L forms on a continuous belt that moves upward at a constant velocity V passing through a liquid bath of pure B. As the liquid film rises, it is exposed briefly to a gas containing species A, which has a small solubility of c_{A0} in B. Find the steady-state concentration distribution of A in the film. Assume constant properties and neglect end effects.

Solution The species continuity equation, Eq. [A] of Table 3.3-2, is as follows:

$$\frac{\partial c_A}{\partial t} + \left(v_x \frac{\partial c_A}{\partial x} + v_y \frac{\partial c_A}{\partial y} + v_z \frac{\partial c_A}{\partial z} \right) = D_A \left(\frac{\partial^2 c_A}{\partial x^2} + \frac{\partial^2 c_A}{\partial y^2} + \frac{\partial^2 c_A}{\partial z^2} \right) + \underline{r}_A \qquad [3.3\text{-}103]$$

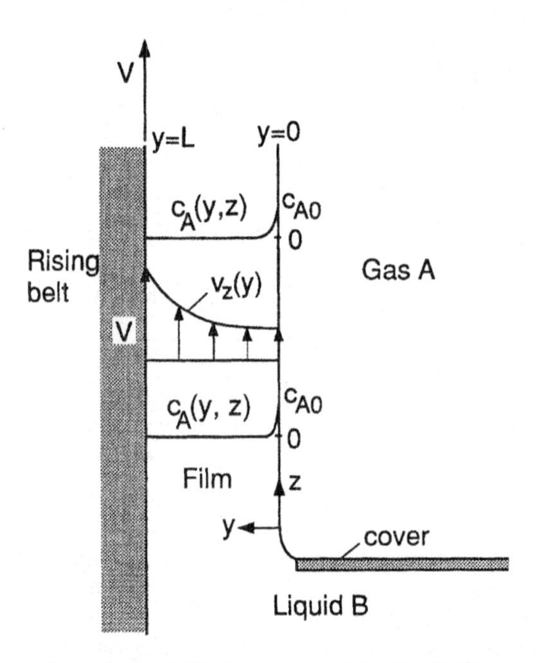

Fig. 3.3-11 Diffusion into a rising liquid film.

At the steady state $\partial c_A/\partial t$ is zero. Since the flow is in the z direction only $v_x = v_y = 0$. Since the belt width $\gg L$, $\partial c_A/\partial x = 0$. Mass transfer in the z direction is mainly by convection, and the diffusion term $D_A\, \partial^2 c_A/\partial z^2$ can be neglected. This is a good approximation since most liquids have a high Sc (Schmidt number). Finally, $\underline{r}_A = 0$ in the absence of chemical relations. As such, Eq. [3.3-103] reduces to

$$v_z \frac{\partial c_A}{\partial z} = D_A \frac{\partial^2 c_A}{\partial y^2} \qquad [3.3\text{-}104]$$

The velocity distribution $v_z(y)$ has already been solved in Example 1.5-3. From Eq. [1.5-68]

$$v_z = V + \frac{\rho g}{2\mu}(y^2 - L^2) \qquad [3.3\text{-}105]$$

Since the liquid film is exposed to the gas briefly, c_A rises significantly only near the free surface where $y \ll L$ and

$$v_z = v_{\min} = V - \frac{\rho g}{2\mu} L^2 \qquad [3.3\text{-}106]$$

Substituting into Eq. [3.3-104]

$$\frac{\partial c_A}{\partial z} = \frac{D_A}{v_{\min}} \frac{\partial^2 c_A}{\partial y^2} \qquad [3.3\text{-}107]$$

and the boundary conditions are

$$c_A(y, 0) = 0 \qquad [3.3\text{-}108]$$

$$c_A(0, z) = c_{A0} \qquad [3.3\text{-}109]$$

$$c_A(y, z) = 0 \quad \text{as} \quad y \to \infty \qquad [3.3\text{-}110]$$

The solution, from Case O of Appendix A, is as follows:

$$\frac{c_A - c_{A0}}{0 - c_{A0}} = \operatorname{erf}\left(\frac{y}{\sqrt{4D_A z/v_{\min}}}\right) \qquad [3.3\text{-}111]$$

From Eq. [A.7-25]

$$\frac{\partial c_A}{\partial y} = \frac{-c_{A0}}{\sqrt{\pi D_A z/v_{\min}}} e^{-(y^2/4D_A z/v_{\min})} \qquad [3.3\text{-}112]$$

The total molar transfer rate to a liquid film of height L and width w is

$$W_A = w \int_0^L -D_A \frac{\partial c_A}{\partial y}\bigg|_{y=0} dz \qquad [3.3\text{-}113]$$

Substituting Eq. [3.3-112] into Eq. [3.3-113]

$$W_A = \frac{w c_{A0} D_A}{\sqrt{\pi D_A/v_{\min}}} \int_0^L \frac{1}{\sqrt{z}} dz$$

$$= 2wLc_{A0}\sqrt{\frac{D_A v_{\min}}{\pi L}} \qquad [3.3\text{-}114]$$

Example 3.3-8 Mass Transfer in Laminar Flow over a Flat Plate

Consider the steady-state, laminar flow of an incompressible Newtonian fluid of uniform composition $w_{A\infty}$ over a wide flat plate of uniform surface composition w_{AS}, as illustrated in Fig. 3.3-12. Find the composition distribution in and the thickness of the diffusion boundary layer.

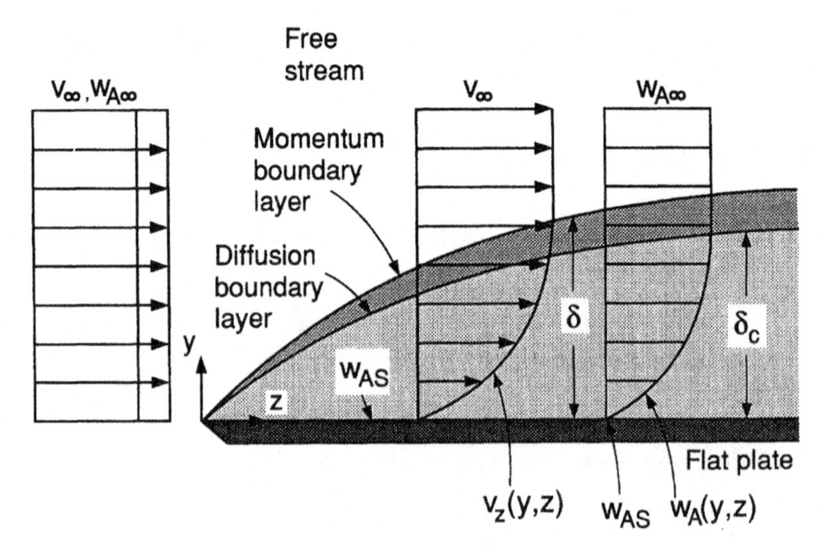

Fig. 3.3-12 Mass transfer in laminar flow over a flat plate.

Solution Assuming constant ρ, the species continuity equation, Eq. [A] of Table 3.3-1, can be divided by ρ to become

$$\frac{\partial w_A}{\partial t} + v_x \frac{\partial w_A}{\partial x} + v_y \frac{\partial w_A}{\partial y} + v_z \frac{\partial w_A}{\partial z} = D_A \left[\frac{\partial^2 w_A}{\partial x^2} + \frac{\partial^2 w_A}{\partial y^2} + \frac{\partial^2 w_A}{\partial z^2} \right] + \frac{r_A}{\rho} \quad [3.3\text{-}115]$$

At the steady state $\partial w_A/\partial t$ is zero. Since there is no fluid flow nor composition gradient in the x direction, v_x, $\partial w_A/\partial x$, and $\partial^2 w_A/\partial x^2$ are zero. Since mass transfer in the z direction is mainly by convection rather than by diffusion, $D_A \partial^2 w_A/\partial z^2$ is far smaller than $v_z\, \partial w_A/\partial z$ and can thus be neglected. Since there are no chemical reactions involved, r_A is zero.

The equations of continuity, motion, and species continuity are as follows:

Continuity:
$$\frac{\partial v_y}{\partial y} + \frac{\partial v_z}{\partial z} = 0 \quad [3.3\text{-}116]$$

Motion:
$$v_y \frac{\partial v_z}{\partial y} + v_z \frac{\partial v_z}{\partial z} = \nu \frac{\partial^2 v_z}{\partial y^2} \quad [3.3\text{-}117]$$

Species continuity:
$$v_y \frac{\partial w_A}{\partial y} + v_z \frac{\partial w_A}{\partial z} = D_A \frac{\partial^2 w_A}{\partial y^2} \quad [3.3\text{-}118]$$

From Eq. [3.3-116]

$$v_y = -\int_0^y \frac{\partial v_z}{\partial z}\, dy \quad [3.3\text{-}119]$$

which is substituted into Eq. [3.3-118] to give

$$-\left(\int_0^y \frac{\partial v_z}{\partial z}\, dy \right) \frac{\partial w_A}{\partial y} + v_z \frac{\partial w_A}{\partial z} = D_A \frac{\partial^2 w_A}{\partial y^2} \quad [3.3\text{-}120]$$

Since this equation is difficult to solve directly, the following approximate composition profile is assumed:

$$\boxed{\frac{w_A - w_{AS}}{w_{A\infty} - w_{AS}} = \frac{3}{2}\left(\frac{y}{\delta_c}\right) - \frac{1}{2}\left(\frac{y}{\delta_c}\right)^3} \quad [3.3\text{-}121]$$

where δ_c is the thickness of the diffusion boundary layer. This composition profile is reasonable since it satisfies the following boundary conditions:

$$w_A = w_{AS} \quad \text{at} \quad y = 0 \quad [3.3\text{-}122]$$

and

$$w_A = w_{A\infty} \quad \text{and} \quad \frac{\partial w_A}{\partial y} = 0 \quad \text{at} \quad y = \delta_c \qquad [3.3\text{-}123]$$

The concentration profile shown in Fig. 3.3-12 differs from that shown in Fig. 3.2-6a in that $w_{A\infty} > w_{AS}$ in the former while $w_{A\infty} < w_{AS}$ in the latter. This difference, however, will not affect the conclusion of the present example as either profile can be described by Eq. [3.3-121].

From Eq. [3.3-121]

$$\frac{\partial w_A}{\partial z} = \frac{(w_{A\infty} - w_{AS})}{2}\left(\frac{-3y}{\delta_c^2} + \frac{3y^3}{\delta_c^4}\right)\frac{d\delta_c}{dz} \qquad [3.3\text{-}124]$$

$$\frac{\partial w_A}{\partial y} = \frac{(w_{A\infty} - w_{AS})}{2}\left(\frac{3}{\delta_c} - \frac{3y^2}{\delta_c^3}\right) \qquad [3.3\text{-}125]$$

$$\frac{\partial^2 w_A}{\partial y^2} = \frac{-3(w_{A\infty} - w_{AS})y}{\delta_c^3} \qquad [3.3\text{-}126]$$

Substitute Eqs. [3.3-124] through [3.3-126], and Eq. [1.5-78] for v_z and Eq. [1.5-81] for $\partial v_z/\partial z$, into Eq. [3.3-120]. After rearranging and integrating with respect to y from 0 to δ_c, we get

$$\frac{8}{5}\delta\delta_c^3\left[-1 + \frac{1}{7}\left(\frac{\delta_c}{\delta}\right)^2\right]\frac{d\delta_c}{dz} + \frac{6}{5}\delta_c^4\left[\frac{2}{3} - \frac{1}{7}\left(\frac{\delta_c}{\delta}\right)^2\right]\frac{d\delta}{dz} = -\frac{8D_A}{v_\infty}\delta^2\delta_c \qquad [3.3\text{-}127]$$

This equation may be solved for δ_c as a function of δ, if z can be eliminated. From Eq. [1.5-87]

$$\delta = 4.64\sqrt{\frac{\nu z}{v_\infty}} \qquad [3.3\text{-}128]$$

which yields

$$2\delta\, d\delta = \frac{280}{13}\frac{\nu}{v_\infty}dz \qquad [3.3\text{-}129]$$

Let us define

$$\frac{\delta_c}{\delta} = a \qquad [3.3\text{-}130]$$

where a is a constant depending only on the physical properties of the fluid. Substituting Eqs. [3.3-128] through [3.3-130] into Eq. [3.3-127]

$$a^5 - 14a^3 + \frac{13}{\mathrm{Sc}} = 0 \qquad [3.3\text{-}131]$$

where the Schmidt number

$$\mathrm{Sc} = \frac{\nu}{D} = \frac{\mu}{\rho D} \qquad [3.3\text{-}132]$$

Let us now consider the case where $a < 1$, that is, where the diffusion boundary layer is thinner than the momentum boundary layer. Since $a^5 \ll 14a^3$, Eq. [3.3-131] reduces approximately to

$$a = \mathrm{Sc}^{-(1/3)} \qquad [3.3\text{-}133]$$

Substituting into Eq. [3.3-130]

$$\frac{\delta_c}{\delta} = \mathrm{Sc}^{-(1/3)} \qquad [3.3\text{-}134]$$

which is identical to Eq. [3.2-50], derived previously on the basis of integral species balance. Substituting Eq. [3.3-128]

$$\boxed{\delta_c = 4.64\,\mathrm{Sc}^{-(1/3)}\sqrt{\frac{\nu z}{v_\infty}}} \qquad [3.3\text{-}135]$$

Therefore, the thickness of the diffusion boundary layer increases with the square root of the distance from the leading edge of the plate. From Eqs. [3.3-121] and [3.3-135] the composition profile can be determined.

3.4. SPECIES DIFFERENTIAL MASS-BALANCE EQUATION IN STREAM FUNCTION

As described in the previous section, the differential governing equation for convective mass transfer is the species equation of continuity and is expressed in terms of the solute concentration and the velocity components involved. In the present section a different type of differential governing equation, expressed in concentration and the stream function will be introduced. This approach is often used in two-dimensional and axisymmetric problems, as will be seen in Chapters 8 and 9.

3.4.1. Two-Dimensional Problems

From the equation of continuity for species A (i.e., Eq. [A] of Table 3.3-2)

$$\frac{\partial c_A}{\partial t} + \left[v_x \frac{\partial c_A}{\partial x} + v_y \frac{\partial c_A}{\partial y} \right] = D_A \left[\frac{\partial^2 c_A}{\partial x^2} + \frac{\partial^2 c_A}{\partial y^2} \right] + \underline{r}_A \qquad [3.4\text{-}1]$$

From Eqs. [1.6-1] and [1.6-2]

$$v_x = \frac{\partial \psi}{\partial y} \qquad [3.4\text{-}2]$$

and

$$v_y = -\frac{\partial \psi}{\partial x} \qquad [3.4\text{-}3]$$

Substituting Eqs. [3.4-2] and [3.4-3] into Eq. [3.4-1]

$$\boxed{\frac{\partial c_A}{\partial t} + \left[\frac{\partial \psi}{\partial y}\frac{\partial c_A}{\partial x} - \frac{\partial \psi}{\partial x}\frac{\partial c_A}{\partial y} \right] = D_A \left[\frac{\partial^2 c_A}{\partial x^2} + \frac{\partial^2 c_A}{\partial y^2} \right] + \underline{r}_A} \qquad [3.4\text{-}4]$$

For two-dimensional convective mass transfer, this equation is solved simultaneously with Eqs. [1.6-11] and [1.6-12] for the concentration and velocity fields.

3.4.2. Axisymmetric Problems

From the equation of continuity for species A (i.e., Eq. [B] of Table 3.3-2)

$$\frac{\partial c_A}{\partial t} + \left[v_r \frac{\partial c_A}{\partial r} + v_z \frac{\partial c_A}{\partial z} \right] = D_A \left[\frac{1}{r}\frac{\partial}{\partial r}\left(r \frac{\partial c_A}{\partial r} \right) + \frac{\partial^2 c_A}{\partial z^2} \right] + \underline{r}_A \qquad [3.4\text{-}5]$$

From Eqs. [1.6-13] and [1.6-14]

$$v_r = \frac{1}{r}\frac{\partial \psi}{\partial z} \qquad [3.4\text{-}6]$$

and

$$v_z = \frac{-1}{r}\frac{\partial \psi}{\partial r} \qquad [3.4\text{-}7]$$

Substituting Eqs. [3.4-6] and [3.4-7] into Eq. [3.4-5]

$$\frac{\partial c_A}{\partial t} + \left[\frac{1}{r}\frac{\partial \psi}{\partial z}\frac{\partial c_A}{\partial r} - \frac{1}{r}\frac{\partial \psi}{\partial r}\frac{\partial c_A}{\partial z}\right] = D_A\left[\frac{1}{r}\frac{\partial}{\partial r}\left(r\frac{\partial c_A}{\partial r}\right) + \frac{\partial^2 c_A}{\partial z^2}\right] + \underline{r}_A \qquad [3.4\text{-}8]$$

For axisymmetric convective mass transfer, Eq. [3.4-8] is solved simultaneously with Eqs. [1.6-21] through [1.6-23] for the concentration and velocity fields.

3.5. TURBULENCE

3.5.1. Time-Smoothed Variables

Just like the local pressure and temperature, the local solute concentration is also affected by velocity fluctuations arising in turbulent flow.

As illustrated in Fig. 3.5-1, one may assume that the species concentration ρ_A at a fixed point in space over a given finite time interval t_0 can be resolved into the time-smoothed species concentration $\overline{\rho_A}$ and a fluctuation solute concentration term ρ_A' that accounts for the turbulent motion:

$$\rho_A = \overline{\rho_A} + \rho_A' \qquad [3.5\text{-}1]$$

where

$$\overline{\rho_A} = \frac{1}{t_0}\int_t^{t+t_0} \rho_A \, dt \qquad [3.5\text{-}2]$$

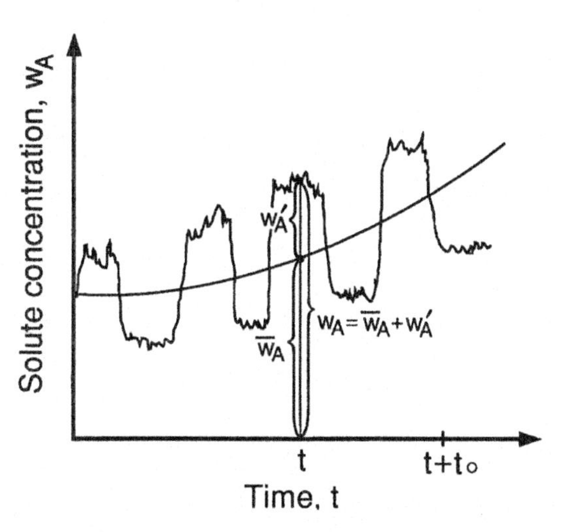

Fig. 3.5-1 Time-smoothed species concentration in turbulent flow.

The time interval t_0 is small with respect to the time over which $\overline{\rho_A}$ may vary in an unsteady state problem, but large with respect to the period of turbulent fluctuations. From the above two equations, it is obvious that

$$\overline{\rho'_A} = \frac{1}{t_0} \int_t^{t+t_0} \rho'_A \, dt = 0 \qquad [3.5\text{-}3]$$

3.5.2. Time-Smoothed Governing Equation

Substituting Eqs. [1.7-1] and [3.5-1] into Eq. [3.3-5] and then taking the time average according to Eqs. [1.7-2], [1.7-3], [3.5-2], and [3.5-3], we get the following **time-smoothed species continuity equation**:

$$\frac{\partial \bar{\rho}_A}{\partial t} + \nabla \cdot \bar{\rho}_A \, \bar{\mathbf{v}} = - \nabla \cdot \overline{\mathbf{j}_A} - \nabla \cdot \overline{\mathbf{j}'_A} + \overline{r_A} + \overline{r'_A} \qquad [3.5\text{-}4]$$

where $\overline{\mathbf{j}_A} = - D_A \nabla \bar{\rho}_A$ and the **turbulent mass flux** $\overline{\mathbf{j}'_A}$ is

$$\overline{\mathbf{j}'_A} = \overline{\rho'_A v'_x} + \overline{\rho'_A v'_y} + \overline{\rho'_A v'_z} \qquad [3.5\text{-}5]$$

Let us consider the reaction term r_A for the simple cases of the first- and the second-order reactions. If $r_A = - k_1 \rho_A$, where k_1 is the reaction constant for a first-order reaction, from Eq. [3.5-1]

$$\overline{r_A} = - k_1 \overline{\rho_A} \qquad [3.5\text{-}6]$$

and

$$\overline{r'_A} = 0 \qquad [3.5\text{-}7]$$

On the other hand, if $r_A = - k_2 \rho_A^2$, where k_2 is the reaction constant for a second-order reaction, from Eq. [3.5-1]

$$\overline{r_A} = - k_2 \overline{\rho_A^2} \qquad [3.5\text{-}8]$$

and

$$\overline{r'_A} = - k_2 \overline{\rho_A'^2} \qquad [3.5\text{-}9]$$

Equation [3.5-4] is the same as the species continuity equation for laminar flow, Eq. [3.3-5], except that the time-smoothed velocity and species concentration replace the instantaneous velocity and species concentration, and that two new terms $\nabla \cdot \overline{\mathbf{j}'_A}$ and $\overline{r'_A}$ arise.

3.5.3. Turbulent Mass Flux

Several semiempirical relations have been proposed for the turbulent mass flux $\overline{\mathbf{j}'_A}$ in order to solve Eq. [3.5-4] for solute distributions in turbulent flow.

3.5.3.1. *Eddy Diffusivity*

By analogy with Eq. [3.1-1], Fick's law of diffusion, one may write

$$\overline{j'_{Ay}} = -D'_A \frac{d\overline{\rho_A}}{dy} \qquad [3.5\text{-}10]$$

the coefficient D'_A is a turbulent or eddy viscosity and is usually position-dependent.

3.5.3.2. *Prandtl's Mixing Length*

By analogy with Eq. [1.7-13]

$$\overline{j'_{Ay}} = -\ell^2 \left|\frac{d\bar{v}_z}{dy}\right| \frac{d\bar{\rho}_A}{dy} \qquad [3.5\text{-}11]$$

and from Eq. [3.5-10]

$$D'_A = \ell^2 \left|\frac{d\bar{v}_z}{dy}\right| \qquad [3.5\text{-}12]$$

3.6. MASS TRANSFER CORRELATION

Convective mass transfer, described in Section 3.3, can be determined by solving the governing equations for fluid flow and mass transfer. Correlations that are derived theoretically but verified experimentally or are based on experimental data alone are useful for studying convective mass transfer. Some of these correlations will be presented in this section in dimensionless numbers.

3.6.1. External Flow

3.6.1.1. *Forced Convection over a Flat Plate*

Flow over a flat plate is laminar for local Reynolds number $\mathrm{Re}_z < 2 \times 10^5$. As described in Example 3.2-4, the following theoretical correlation has been suggested for laminar flow over a flat plate:

$$\mathrm{Sh}_z = 0.332\,\mathrm{Re}_z^{1/2}\,\mathrm{Sc}^{1/3} \qquad (0.6 < \mathrm{Sc} < 50) \qquad [3.6\text{-}1]$$

where

$$\text{Sh}_z \text{ (local Sherwood number)} = \frac{k_m z}{D_A} \quad [3.6\text{-}2]$$

$$\text{Re}_z \text{ (local Reynolds number)} = \frac{z v_\infty}{\nu} = \frac{z \rho v_\infty}{\mu} \quad [3.6\text{-}3]$$

$$\text{Sc (Schmidt number)} = \frac{\nu}{D_A} = \frac{\mu}{\rho D_A} \quad [3.6\text{-}4]$$

From these equations, the mass transfer coefficient averaged over a distance L from the leading edge of the plate is

$$k_L = \frac{1}{L} \int_0^L k_m \, dz \quad [3.6\text{-}5]$$

Substituting Eqs. [3.6-1] through [3.6-3] into Eq. [3.6-5]

$$k_L = 0.332 \frac{D_A}{L} \left(\frac{v_\infty}{\nu} \right)^{1/2} \text{Sc}^{1/3} \int_0^L z^{-1/2} \, dz \quad [3.6\text{-}6]$$

or

$$\boxed{\text{Sh}_L = 0.664 \, \text{Re}_L^{1/2} \, \text{Sc}^{1/3}} \qquad (0.6 < \text{Sc} < 50) \quad [3.6\text{-}7]$$

where

$$\text{Sh}_L \text{ (average Sherwood number)} = \frac{k_L L}{D_A} \quad [3.6\text{-}8]$$

$$\text{Re}_L \text{ (average Reynolds number)} = \frac{L v_\infty}{\nu} = \frac{L \rho v_\infty}{\mu} \quad [3.6\text{-}9]$$

The fluid properties in Eqs. [3.6-1] and [3.6-7] are evaluated at the so-called **film composition**:

$$w_{A_f} = \frac{w_{A\infty} + w_{As}}{2} \quad [3.6\text{-}10]$$

For turbulent flow over a flat plate, the following correlation has been suggested:

$$\boxed{\mathrm{Sh}_z = 0.0288\,\mathrm{Re}_z^{4/5}\,\mathrm{Sc}^{1/3}} \qquad (0.6 < \mathrm{Sc} < 60) \qquad [3.6\text{-}11]$$

and with help of Eq. [3.6-5] we get[15]

$$\boxed{\mathrm{Sh}_L = 0.036\,\mathrm{Re}_L^{4/5}\,\mathrm{Sc}^{1/3}} \qquad (0.6 < \mathrm{Sc} < 60) \qquad [3.6\text{-}12]$$

3.6.1.2. Forced Convection Past a Sphere

For a liquid flowing past a sphere of diameter D two empirical correlations have been suggested; the first[16] is

$$\boxed{\mathrm{Sh}_D = (4.0 + 1.21\,\mathrm{Re}_D^{2/3}\,\mathrm{Sc}^{2/3})^{1/2}} \qquad (\mathrm{Re}_D\,\mathrm{Sc} < 10{,}000) \qquad [3.6\text{-}13]$$

and the second

$$\boxed{\mathrm{Sh}_D = 1.01\,\mathrm{Re}_D^{1/3}\,\mathrm{Sc}^{1/3}} \qquad (\mathrm{Re}_D\,\mathrm{Sc} > 10{,}000) \qquad [3.6\text{-}14]$$

where

$$\mathrm{Sh}_D = \frac{k_L D}{D_A} \qquad [3.6\text{-}15]$$

$$\mathrm{Re}_D = \frac{D v_\infty}{\nu} = \frac{L \rho v_\infty}{\mu} \qquad [3.6\text{-}16]$$

For a gas flowing past a sphere of diameter D, the following empirical correlation has been suggested:[17]

$$\boxed{\mathrm{Sh}_D = 2.0 + 0.6\,\mathrm{Re}_D^{1/2}\,\mathrm{Sc}^{1/3}} \qquad (2 < \mathrm{Re}_D < 800;\ 0.6 < \mathrm{Sc} < 2.7) \qquad [3.6\text{-}17]$$

Note the analogy between Eqs. [3.6-17] and [2.6-20].

Equations [3.6-13], [3.6-14], and [3.6-17] are valid only when natural convection is negligible, that is when

$$\mathrm{Re}_D \geqslant 0.4\,\mathrm{Gr}_D^{1/2}\,\mathrm{Sc}^{-(1/6)} \qquad [3.6\text{-}18]$$

[15] J. R. Welty, C. E. Wicks, and R. E. Wilson, *Fundamentals of Momentum, Heat and Mass Transfer*, 3rd ed., Wiley, New York, 1984, p. 650.

[16] P. L. T. Brian and H. B. Hales, *AIChE J.*, **15**, 419, 1969.

[17] N. Froessling, *Gerlands Beitr. Geophys.*, **52**, 170, 1939.

3.6.2. Internal Flow

3.6.2.1. *Forced Convection inside a Circular Tube*

For turbulent flow in a tube of diameter D the following empirical correlation has been suggested:[18]

$$\boxed{\mathrm{Sh}_D = 0.023\,\mathrm{Re}_D^{0.83}\,\mathrm{Sc}^{1/3}} \qquad (2000 < \mathrm{Re}_D < 70{,}000;\ 1000 < \mathrm{Sc} < 2260) \qquad [3.6\text{-}19]$$

Notice the similarity between this equation and Eq. [2.6-23] for heat transfer.

3.6.3. Examples

Example 3.6-1 Turbulent flow over a Flat Plate

Consider the turbulent flow of air at 30 m/s, 1 atm, and 20°C over a flat plate 1 m long, with the air containing a low concentration of water vapor. Find the mass transfer coefficient h_L. Assume $\rho = 1.18 \times 10^{-3}$ g/cm^3, $\nu = 0.157$ cm^2/s, and $D_A = 0.220$ cm^2/s.

Solution The Reynolds and Schmidt numbers are as follows:

$$\mathrm{Re}_L = \frac{Lv_\infty}{\nu} = \frac{100\ \mathrm{cm} \times 3000\ \mathrm{cm/s}}{0.157\ \mathrm{cm^2/s}} = 1.91 \times 10^6 \qquad \text{(turbulent flow)} \qquad [3.6\text{-}20]$$

$$\mathrm{Sc} = \frac{\nu}{D_A} = \frac{0.157\ \mathrm{cm^2/s}}{0.220\ \mathrm{cm^2/s}} = 0.71 \qquad [3.6\text{-}21]$$

From Eq. [3.6-12]

$$\mathrm{Sh}_L = \frac{k_L L}{D_A} = 0.036\,\mathrm{Re}_L^{4/5}\,\mathrm{Sc}^{1/3} \qquad [3.6\text{-}22]$$

and so

$$k_L = 0.036 \times (1.91 \times 10^6)^{4/5} \times (0.71)^{1/3} \times \frac{0.220\ \mathrm{cm^2/s}}{100\ \mathrm{cm}} = 1.19 \times 10^{-4}\ \mathrm{cm/s} \qquad [3.6\text{-}23]$$

[18] W. H. Linton and T. K. Sherwood, *Chem. Eng. Progr.*, **46**, 258, 1950.

PROBLEMS[19]

3.1-1 Show that $\mathbf{j}_A = -\mathbf{j}_B$ for a binary system A–B.

3.1-2 Show that $\underline{\mathbf{j}}_A = -\underline{\mathbf{j}}_B$ for a binary system A–B.

3.2-1 A thin solid film with an initially uniform concentration of c_{Ai} is kept at a proper temperature to allow species A to be consumed by the reaction $\underline{r}_A = -k_1 c_A$. Find c_A as a function of time. Assume that there are no significant internal concentration gradients and no evaporation of A from the surface.

3.2-2 Consider the mixing tank in Fig. 3.2-3, assuming that there is a chemical reaction in the tank. The mass of A consumed per unit volume per unit time due to the reaction is given by $r_A = -k_1 \rho w_A$, where the minus sign indicates that A is being consumed.

3.2-3 Pure hydrogen gas is stored at 400°C and 9 atm in a steel tank with a wall thickness of 1 mm. At 400°C the diffusion coefficient of atomic hydrogen in steel is approximately 1×10^{-4} cm^2/s, and the solubility of atomic hydrogen in steel is about 3 ppm (by weight) under a hydrogen gas pressure of 1 atm.[20] Calculate the mass diffusion flux of hydrogen through the tank wall. The solubility of hydrogen in metals is described by Sievert's law:[21]

$$w_H = K_p (p_{H_2})^{1/2}$$

where K_p is the equilibrium constant and p_{H_2} is the partial pressure of hydrogen in equilibrium with the metal. The density of steel is about 7.7 g/cm^3.

3.2-4 Consider the diffusion of helium through the wall of a quartz cylinder of length L, inner diameter D, and wall thickness l ($\ll D$), as shown in Fig. P3.2-4 (see p. 264). The temperature is T. The initial pressure is P_{A0}. Assuming ideal gas, determine the pressure P_A as a function of time. The concentration of helium in quartz at the quartz/gas interface is

$$c_{AS} = K_p P_A$$

where K_p is the equilibrium constant for the reaction

$$A(\mathrm{g}) \rightarrow A \quad \text{(in solid solution)}$$

[19] The first two digits in problem numbers refer to section numbers in text.

[20] P. G. Shewmon, *Diffusion in Solids*, McGraw-Hill, New York, 1963, p. 37.

[21] L. S. Darken and R. W. Gurry, *Physical Chemistry of Metals*, McGraw-Hill, New York, 1953, p. 258.

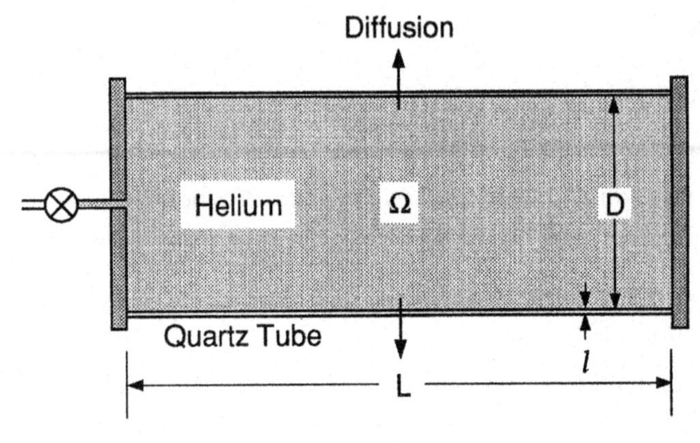

Fig. P3.2-4

3.3-1 Use the species continuity equation to determine the steady-state concentration distribution and diffusion flux in the sheet shown in Fig. 3.1-2. Assume constant properties and neglect end effects.

3.3-2 A fluid is held between two large horizontal plates separated by a gap of L. The concentration of the fluid at the lower plate is c_{A1}, while that at the upper one is c_{A2}. The lower plate moves at the constant velocity V while the upper one is at rest, as shown in Fig. P3.3-2. Find the steady-state concentration distribution in the fluid. Assume constant physical properties and neglect end effects.

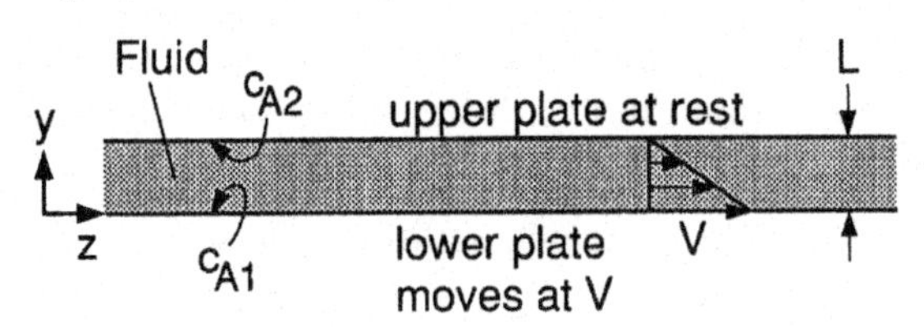

Fig. P3.3-2

3.3-3 As shown in Fig. P3.3-3, a chemical species A diffuses from a gas phase into a porous catalyst plate of thickness $2L$ in which it is converted into

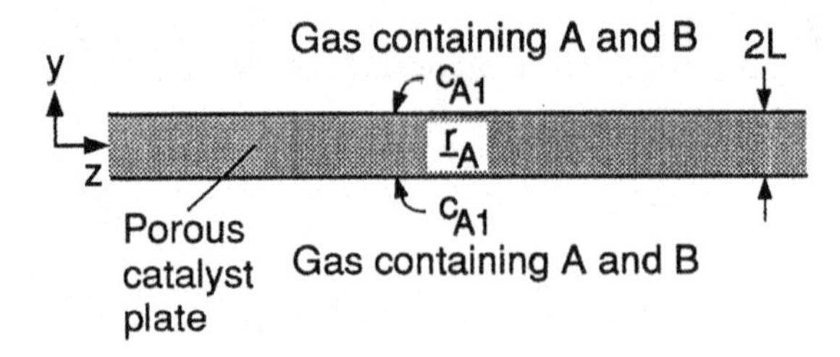

Fig. P3.3-3

species B. The concentration of A at the surface of the plate is c_{A1}. The rate at which A is consumed per unit volume of the plate is $\underline{r}_A = -k_1 a c_A$. Find the steady-state concentration distribution of A in the plate. The effective diffusion coefficient of A in the plates is D_A. Assume constant properties.

3.3-4 As shown in Fig. P3.3-4, a chemical species A diffuses from two gas phases into a porous catalyst plate of thickness L in which it is converted into species B. The concentrations of A at the bottom and top surfaces of the plate are c_{A1} and c_{A2}, respectively. The rate at which A is consumed per unit volume of the plate is $\underline{r}_A = -k_1 a c_A$. Find the steady-state concentration distribution of A in the plate. The effective diffusion coefficient of A in the plate is D_A. Assume constant properties.

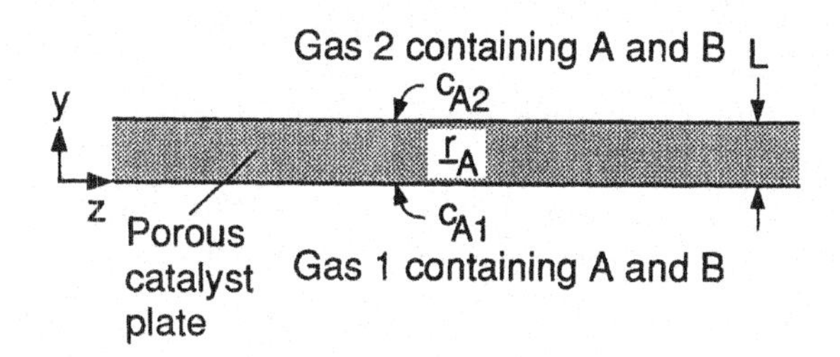

Fig. P3.3-4

3.3-5 As shown in Fig. P3.3-5, a chemical species A diffuses from a gas phase into a porous catalyst plate of thickness L from its bottom to be converted into species B. The concentration of A at the bottom surface of the plate is c_{A1}. The top surface of the plate is sealed. The rate at which A is consumed per unit volume of the plate is $\underline{r}_A = -k_1 a c_A$. Find the state-state concentration distribution of A in the plate. The effective diffusion coefficient of A in the plate is D_A. Assume constant properties.

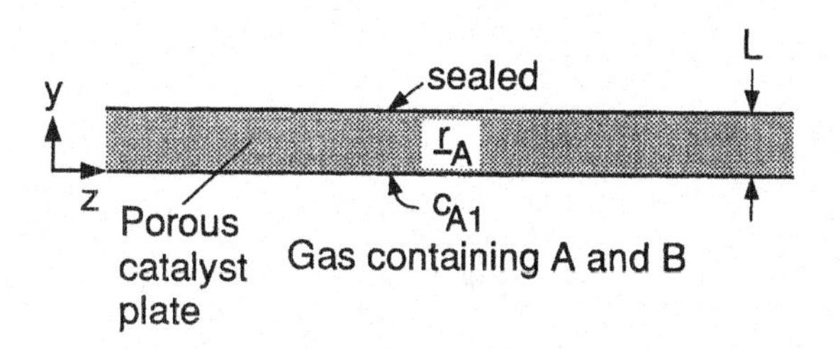

Fig. P3.3-5

3.3-6 A semiinfinite body occupying the space from $x = 0$ to $x = \infty$ is initially at a uniform concentration w_{Ai}. At time $t = 0$ the surface is suddenly raised to a concentration w_{As} by exposure to a proper gas and maintained at that concentration for $t > 0$, as illustrated in Fig. P3.3-6. Find the time-dependent concentration distribution $w_A(t, x)$ and the diffusion flux j_{Ax}. Assume constant physical properties.

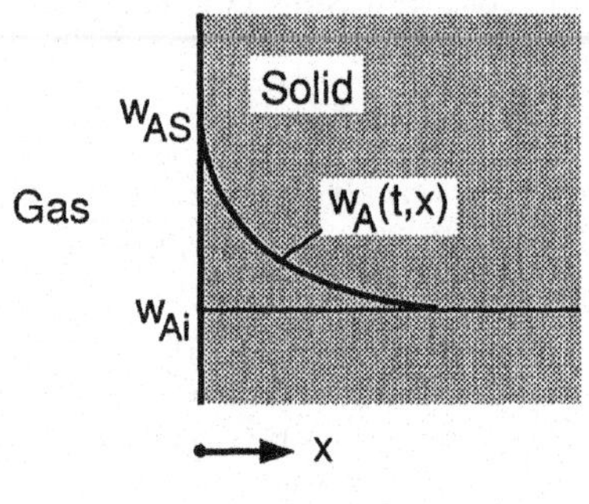

Fig. P3.3-6

3.3-7 A falling film of liquid B is exposed briefly to a gas containing A, which has a small solubility of c_{A0} in B, as shown in Fig. P3.3-7.[22] The thickness of the film is L. Find the steady-state concentration distribution of A in the film. Assume constant properties and neglect end effects.

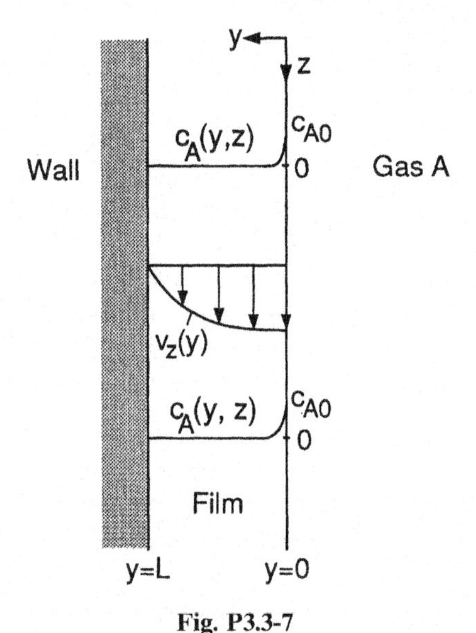

Fig. P3.3-7

3.3-8 A gas bubble of species A and radius R rises at the terminal velocity v_t through a quiescent liquid of pure B, as shown in Fig. P3.3-8[22] (see p. 267). Estimate the rate at which A is absorbed by B. The solubility of A in B is c_{A0}.

[22] R. B. Bird, W. E. Stewart, and E. N. Lightfood, *Transport Phenomena*, Wiley, New York, 1960.

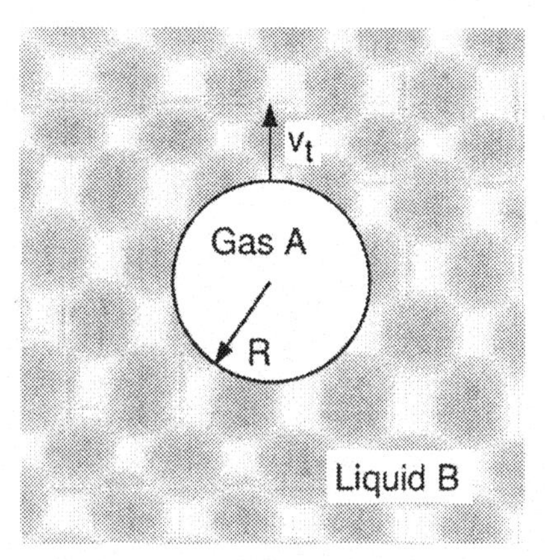

Fig. P3.3-8

3.3-9 A fluid enters the reaction zone of a tubular reactor at concentration c_{A0} and undergoes a reaction $\underline{r}_A = -k_1 a c_A$. Assume the so-called plug-flow situation, in which both v_z and c_A are independent of the radial position, as shown in Fig. P3.3-9. Assume also that mass transfer in the axial direction is due mainly to the convection rather than diffusion. Determine the steady-state axial concentration profile in the reactor.

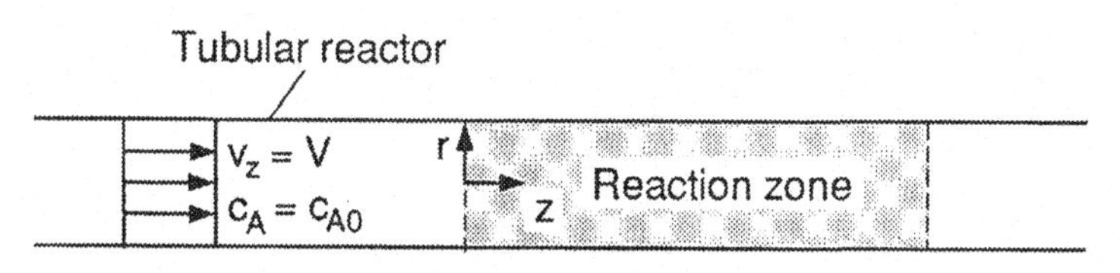

Fig. P3.3-9

3.3-10 A hollow sphere has a inner surface ($r = r_1$) maintained at c_{A1} and outer surface ($r = r_2$) at c_{A2}, respectively. Determine the steady-state concentration profile and the diffusion flux.

3.3-11 Gas A dissolves in, diffuses into and reacts with liquid B, and the concentration of A at the free surface is c_{A0}, as shown in Fig. P3.3-11 (see p. 268). The reaction rate per unit volume $\underline{r}_A = -k_1 c_A$. Find the steady-state concentration distribution of A in the liquid.[23] Assume constant properties, small concentration of A, and that the product of reaction does not interfere with the diffusion of A through B.

3.3-12 Consider the diffusion of the vapor of species A through a stagnant gas B in a tube,[23] as shown in Fig. P3.3-12 (see p. 268). Species B is insoluble in liquid A. In the gas phase the mole fractions of species A are x_{A0} at

[23] R. B. Bird, W. E. Stewart, and E. N. Lightfoot, ibid.

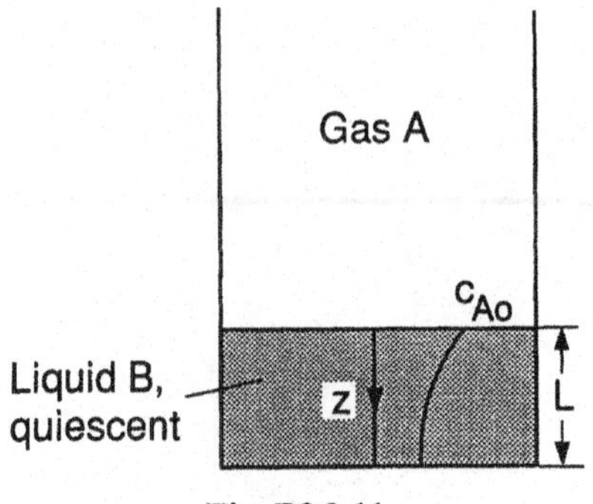

Fig. P3.3-11

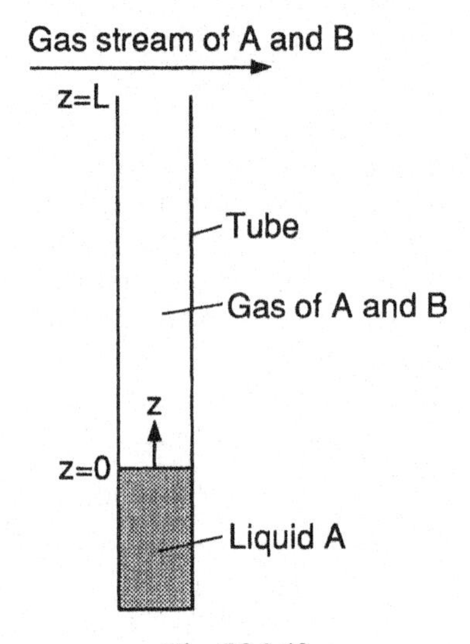

Fig. P3.3-12

the gas/liquid interface $z = 0$ and x_{AL} at the top of the tube $z = L$. Find the mole fraction of species A as a function of the distance from the interface at the steady state. Assume $\underline{\mathbf{n}}_B = 0$, constant properties, and no chemical reactions. (*Hint*: Use Eqs. [3.1-7], [3.1-16], and [3.3-6].)

3.3-13 Consider the flow over a flat plate shown in Fig. 3.2-6*a*. Prove that $\partial^2(w_A - w_{AS})/\partial y^2 = 0$ at $y = 0$. Does the concentration distribution given by Eq. [3.3-121] satisfy this condition?

CHAPTER 4

FLUID FLOW, HEAT TRANSFER, AND MASS TRANSFER: SIMILARITIES AND COUPLING

In the previous three chapters we discussed fluid flow, heat transfer, and mass transfer individually. In the present chapter we shall discuss them together, first their similarities and then their coupling. The latter will include buyoancy convection, solutal convection, and thermosolutal convection with or without magnetic damping. These types of convection, as will be shown later, in Chapters 8 and 9, are of great significance in materials processing.

4.1. SIMILARITIES AMONG DIFFERENT TYPES OF TRANSPORT

4.1.1. Basic Laws

The transfer of z momentum, heat, and species A occurs in the direction of decreasing v_z, T, and w_A, as summarized in Fig. 4.1-1. According to Eqs. [1.1-2], [2.1-1], and [3.1-1]

$$\tau_{yz} = -\mu \frac{dv_z}{dy} \quad \text{(Newton's law of viscosity)} \qquad [4.1\text{-}1]$$

$$q_y = -k \frac{dT}{dy} \quad \text{(Fourier's law of conduction)} \qquad [4.1\text{-}2]$$

$$j_{Ay} = -\rho D_A \frac{dw_A}{dy} \quad \text{(Fick's law of diffusion)} \qquad [4.1\text{-}3]$$

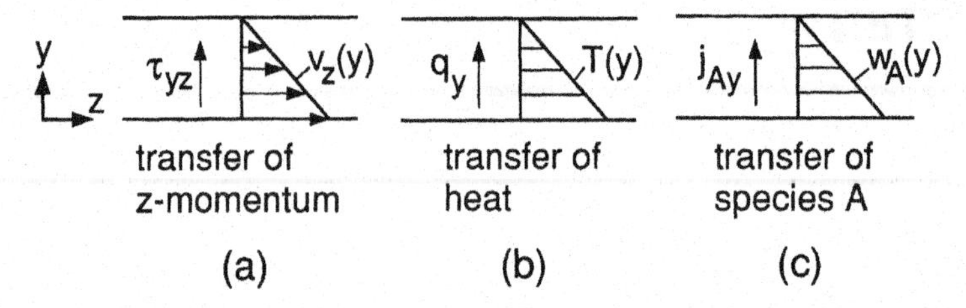

Fig. 4.1-1 Transport in the y direction: (a) z momentum; (b) heat; (c) species A. The vertical arrows indicate the direction of transport.

TABLE 4.1-1 Basic Laws

Phenomenological	Flux = − (Proportional constant) × (Driving force in gradient)	[A]
General	$\mathbf{j}_\phi = -\Gamma_\phi \nabla \phi$	[B]
Momentum[a]	$\boldsymbol{\tau} = -\mu(\nabla \mathbf{v} + \nabla \mathbf{v}^T)$	[C]
Energy	$\mathbf{q} = -k\nabla T$	[D]
Species	$\mathbf{j}_A = -\rho D_A \nabla w_A$	[E]

[a]For incompressible Newtonian fluids.

The three basic laws share the same form as follows:

$$\begin{pmatrix}\text{Flux of}\\ \text{transport}\\ \text{property}\end{pmatrix} = -\begin{pmatrix}\text{proportional}\\ \text{constant}\end{pmatrix} \times \begin{pmatrix}\text{gradient of}\\ \text{transport}\\ \text{property}\end{pmatrix} \qquad [4.1\text{-}4]$$

or

$$j_{\phi y} = -\Gamma_\phi \frac{d\phi}{dy} \qquad [4.1\text{-}5]$$

The three-dimensional forms of these basic laws are summarized in Table 4.1-1.

For constant physical properties, Eqs. [4.1-1] through [4.1-3] can be written as follows:

$$\tau_{yz} = -\nu \frac{d}{dy}(\rho v_z) \qquad [4.1\text{-}6]$$

$$q_y = -\alpha \frac{d}{dy}(\rho C_v T) \qquad [4.1\text{-}7]$$

$$j_{Ay} = -D_A \frac{d}{dy}(\rho_A) \qquad [4.1\text{-}8]$$

These equations share the same form as follows:

$$\begin{pmatrix}\text{Flux of}\\ \text{transport}\\ \text{property}\end{pmatrix} = -\begin{pmatrix}\text{diffusivity}\\ \text{of transport}\\ \text{property}\end{pmatrix}\times\begin{pmatrix}\text{gradient of}\\ \text{transport property}\\ \text{concentration}\end{pmatrix} \qquad [4.1\text{-}9]$$

In other words, ν, α, and D_A are the diffusivities of momentum, heat, and mass, respectively, and ρv_z, $\rho C_v T$, and ρ_A are the concentrations of z momentum, thermal energy, and species mass, respectively. In the cgs unit, the diffusivities have the same unit of cm^2/s.

4.1.2. Coefficients of Transfer

Figure 4.1-2 shows the transfer of z momentum, heat, and species A from an interface, where they are more abundant, to an adjacent fluid (upper row), and from an adjacent fluid, where they are more abundant, to an interface (lower row). The coefficients of transfer, according to Eqs. [1.1-35], [2.1-14], and [3.1-21], are defined as follows:

$$C_f' = \left|\frac{\tau_{yz}|_{y=0}}{v_{z0} - v_{z\infty}}\right| = \left|\frac{-\mu(\partial v_z/\partial y)|_{y=0}}{0 - v_\infty}\right|$$

(momentum transfer coefficient) [4.1-10]

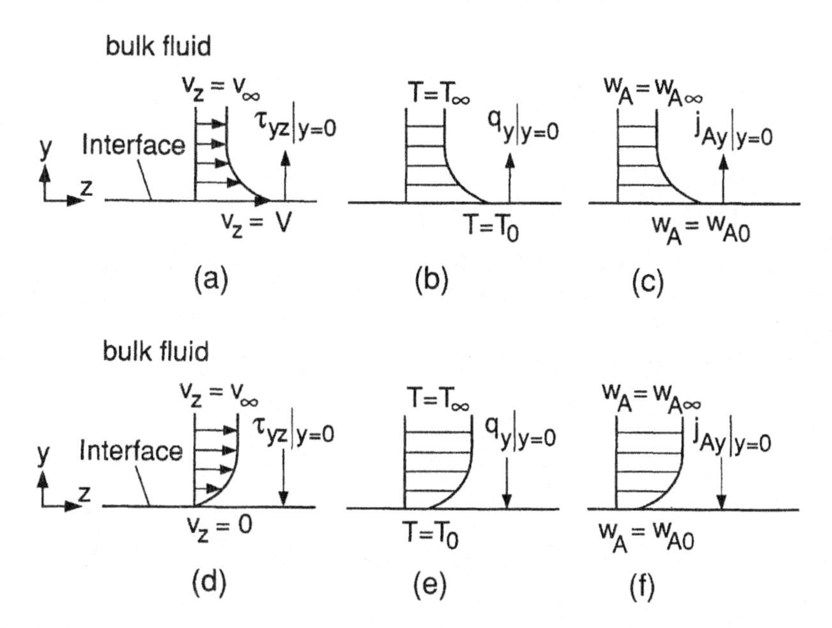

Fig. 4.1-2 Transport between a fluid and a solid boundary: (*a*, *d*) z momentum; (*b*, *e*) heat; (*c*, *f*) species A. The vertical arrows indicate the directions of transport.

$$h = \left| \frac{q_y|_{y=0}}{T_0 - T_\infty} \right| = \left| \frac{-k(\partial T/\partial y)|_{y=0}}{T_0 - T_\infty} \right| \quad \text{(heat transfer coefficient)} \qquad [4.1\text{-}11]$$

$$k_m = \left| \frac{j_{Ay}|_{y=0}}{\rho w_{A0} - \rho w_{A\infty}} \right| = \left| \frac{-D_A(\partial w_A/\partial y)|_{y=0}}{w_{A0} - w_{A\infty}} \right| \quad \text{(mass transfer coefficient)} \qquad [4.1\text{-}12]$$

As mentioned in Section 3.1.6, Eq. [4.1-12] is for low solubility of species A in the fluid. These coefficients share the same form as follows:

$$\begin{pmatrix} \text{Coefficient} \\ \text{of transfer} \end{pmatrix} = \left(\frac{\text{flux at the interface}}{\text{difference in transport property}} \right) \qquad [4.1\text{-}13]$$

or

$$k_\phi = \left| \frac{j_\phi|_{y=0}}{\phi_0 - \phi_\infty} \right| = \left| \frac{-\Gamma_\phi(d\phi/dy)|_{y=0}}{\phi_0 - \phi_\infty} \right| \qquad [4.1\text{-}14]$$

It is common to divide C_f' by $\rho v_\infty/2$ to make the denominator appear in the form of the kinetic energy $\rho v_\infty^2/2$. As shown in Eq. [1.1-36], the so-called friction coefficient is defined by

$$C_f = \frac{C_f'}{\frac{1}{2}\rho v_\infty} = \left| \frac{\tau_{yz}|_{y=0}}{\frac{1}{2}\rho v_\infty^2} \right| \qquad [4.1\text{-}15]$$

4.1.3. The Chilton–Colburn Analogy

The analogous behavior of momentum, heat, and mass transfer is apparent from Examples 1.4-6, 2.2-5, and 3.2-4, where laminar flow over a flat plate was considered. From Eqs. [1.4-67], [2.2-71], and [3.2-56], at a distance z from the leading edge of the plate,

$$\frac{C_{fz}}{2} = 0.323\,\mathrm{Re}_z^{-(1/2)} \qquad [4.1\text{-}16]$$

$$\frac{hz}{k} = 0.323\,\mathrm{Pr}^{1/3}\mathrm{Re}_z^{1/2} \qquad [4.1\text{-}17]$$

$$\frac{k_m z}{D_A} = 0.323\,\mathrm{Sc}^{1/3}\mathrm{Re}_z^{1/2} \qquad [4.1\text{-}18]$$

where

$$\text{Re}_z = \frac{z\rho v_\infty}{\mu} \qquad \text{(local Reynolds number)} \qquad [4.1\text{-}19]$$

$$\text{Pr} = \frac{\nu}{\alpha} = \frac{C_p\mu}{k} \qquad \text{(Prandtl number)} \qquad [4.1\text{-}20]$$

$$\text{Sc} = \frac{\nu}{D_A} = \frac{\mu}{\rho D_A} \qquad \text{(Schmidt number)} \qquad [4.1\text{-}21]$$

and v_∞ is the velocity of the fluid approaching the flat plate.

Equations [4.1-16] through [4.1-18] can be rearranged as follows

$$\frac{C_{fz}}{2} = 0.323\,\text{Re}^{-(1/2)} \qquad [4.1\text{-}22]$$

$$\frac{hz}{k}\frac{1}{\text{Pr}\,\text{Re}_z}\text{Pr}^{2/3} = 0.323\,\text{Re}_z^{-(1/2)} \qquad [4.1\text{-}23]$$

$$\frac{k_m z}{D_A}\frac{1}{\text{Sc}\,\text{Re}_z}\text{Sc}^{2/3} = 0.323\,\text{Re}_z^{-(1/2)} \qquad [4.1\text{-}24]$$

Since these equations have the same RHS, we see

$$\frac{C_{fz}}{2} = \frac{hz}{k}\frac{1}{\text{Pr}\,\text{Re}_z}\text{Pr}^{2/3} = \frac{k_m z}{D_A}\frac{1}{\text{Sc}\,\text{Re}_z}\text{Sc}^{2/3} \qquad [4.1\text{-}25]$$

Substituting Eqs. [4.1-19] through [4.1-21] into Eq. [4.1-25], we obtain

$$\boxed{\frac{C_{fz}}{2} = \frac{h}{v_\infty \rho C_p}\text{Pr}^{2/3} = \frac{k_m}{v_\infty}\text{Sc}^{2/3}} \qquad [4.1\text{-}26]$$

This equation, known as the **Chilton–Colburn analogy**,[1] is often written as follows

$$\boxed{\frac{C_{fz}}{2} = j_{\text{H}} = j_{\text{D}}} \qquad [4.1\text{-}27]$$

[1] A. P. Colburn, *Trans. A.I.Ch.E.*, **29**, 174, 1933; T. H. Chilton and A. P. Colburn, *Ind. Eng. Chem.*, **26**, 1183, 1934.

where the **j factor for heat transfer**

$$j_{\mathrm{H}} = \frac{h}{v_{\infty}\rho C_p}\mathrm{Pr}^{2/3} \tag{4.1-28}$$

and the **j factor for mass transfer**

$$j_{\mathrm{D}} = \frac{k_m}{v_{\infty}}\mathrm{Sc}^{2/3} \tag{4.1-29}$$

The Chilton–Colburn analogy for momentum, heat and mass transfer has been derived here on the basis of laminar flow over a flat plate. However, it has been observed to be a reasonable approximation in laminar and turbulent flow in systems of other geometries provided no form drag is present. Form drag, which has no counterpart in heat and mass transfer, makes $C_f/2$ greater than j_{H} and j_{D}, for example, in flow around (normal to) cylinders. However, when form drag is present, the Chilton–Colburn analogy between heat and mass transfer can still be valid, that is,

$$j_{\mathrm{H}} = j_{\mathrm{D}} \tag{4.1-30}$$

or

$$\frac{h}{v_{\infty}\rho C_p}\mathrm{Pr}^{2/3} = \frac{k_m}{v_{\infty}}\mathrm{Sc}^{2/3} \tag{4.1-31}$$

These equations are considered valid for liquids and gases within the ranges $0.6 < \mathrm{Sc} < 2500$ and $0.6 < \mathrm{Pr} < 100$. They have been observed to be a reasonable approximation for various geometries, such as flow over flat plates, flow around cylinders, and flow in pipes.

The Chilton–Colburn analogy is useful in that it allows one unknown transfer coefficient to be evaluated from another transfer coefficient which is known or measured in the same geometry. For example, by using Eq. [4.1-26] the mass transfer coefficient k_m (for low solubility of species A in the fluid) can be estimated from a heat transfer coefficient h already measured for the same geometry.

It is worth mentioning that for the limiting case of $\mathrm{Pr} = 1$, we see that from Eq. [4.1-26]

$$\frac{C_{fz}}{2} = \frac{h}{v_{\infty}\rho C_p} \tag{4.1-32}$$

which is known as the **Reynolds analogy**, in honor of Reynolds' first recognition of the analogous behavior of momentum and heat transfer in 1874. Modifications of the Reynolds analogy, other than the Chilton–Colburn analogy, have also been proposed, including the Prandtl analogy and the von Karman analogy.

4.1.4. Integral-Balance Equations

The integral-balance equations governing momentum, heat, and species transfer, according to Eqs. [1.4-3], [2.2-6], and [3.2-4], respectively, are as follows

$$\frac{\partial}{\partial t}\iiint_{\Omega} \rho \mathbf{v}\, d\Omega = -\iint_{A} (\rho \mathbf{v}\mathbf{v}\cdot\mathbf{n})\, dA - \iint_{A} \boldsymbol{\tau}\cdot\mathbf{n}\, dA + \iiint_{\Omega} (\mathbf{f}_b - \nabla p)\, d\Omega$$

(momentum transfer) [4.1-33]

$$\frac{\partial}{\partial t}\iiint_{\Omega} \rho C_v T\, d\Omega = -\iint_{A} (\rho \mathbf{v} C_v T)\cdot\mathbf{n}\, dA - \iint_{A} \mathbf{q}\cdot\mathbf{n}\, dA + \iiint_{\Omega} s\, d\Omega$$

(heat transfer) [4.1-34]

$$\frac{\partial}{\partial t}\iiint_{\Omega} \rho w_A\, d\Omega = -\iint_{A} (\rho \mathbf{v} w_A)\cdot\mathbf{n}\, dA - \iint_{A} \mathbf{j}_A\cdot\mathbf{n}\, dA + \iiint_{\Omega} r_A\, d\Omega$$

(species transfer) [4.1-35]

In Eq. [4.1-33] the pressure term has been converted from a surface integral to a volume integral using a Gauss divergence type theorem (i.e., Eq. [A.4-2]). Furthermore, the body force $\mathbf{f}_b$ and pressure gradient ∇p can be considered as the rate of momentum generation due to these forces. In Eq. [4.1-34] the kinetic and potential energy, and the pressure, viscous, and shaft work are not included since they are either negligible or irrelevant in most materials processing problems. In Eq. [4.1-35] $\rho w_A = \rho_A$.

These integral balance equations share the same form as follows:

$$\begin{pmatrix}\text{Rate of}\\ \text{accumulation}\end{pmatrix} = \begin{pmatrix}\text{rate of net}\\ \text{inflow by}\\ \text{convection}\end{pmatrix} + \begin{pmatrix}\text{rate of}\\ \text{other}\\ \text{net inflow}\end{pmatrix} + \begin{pmatrix}\text{rate of}\\ \text{generation}\end{pmatrix} \qquad [4.1\text{-}36]$$

or

$$\frac{\partial}{\partial t}\iiint_{\Omega} \rho\phi\, d\Omega = -\iint_{A} (\rho \mathbf{v}\phi)\cdot\mathbf{n}\, dA - \iint_{A} \mathbf{j}_\phi\cdot\mathbf{n}\, dA + \iiint_{\Omega} s_\phi\, d\Omega \qquad [4.1\text{-}37]$$

These equations are summarized in Table 4.1-2. The following integral mass-balance equation, Eq. [1.2-4], is also included in the table:

$$\frac{\partial}{\partial t}\iiint_{\Omega} \rho \, d\Omega = -\iint_{A} (\rho \mathbf{v}) \cdot \mathbf{n} \, dA \qquad [4.1\text{-}38]$$

4.1.5. Overall Balance Equations

The overall balance equations for momentum, heat, and species transfer according to Eqs. [1.4-9], [2.2-8], and [3.2-7], respectively, are as follows

$$\frac{d\mathbf{P}}{dt} = (m\mathbf{v})_{\text{in}} - (m\mathbf{v})_{\text{out}} + \mathbf{F}_v + (\mathbf{F}_p + \mathbf{F}_b) \qquad \text{(momentum transfer)} \qquad [4.1\text{-}39]$$

$$\frac{dE_T}{dt} = (mC_v T)_{\text{in}} - (mC_v T)_{\text{out}} + Q + S \qquad \text{(heat transfer)} \qquad [4.1\text{-}40]$$

$$\frac{dM_{A\Omega}}{dt} = (mw_A)_{\text{in}} - (mw_A)_{\text{out}} + J_A + R_A \qquad \text{(species transfer)} \qquad [4.1\text{-}41]$$

These overall balance equations share the same form as follows

$$\begin{pmatrix}\text{Rate of}\\ \text{accumulation}\end{pmatrix} = \begin{pmatrix}\text{rate of inflow}\\ \text{by convection}\end{pmatrix} - \begin{pmatrix}\text{rate of outflow}\\ \text{by convection}\end{pmatrix} + \begin{pmatrix}\text{rate of other net inflow}\\ \text{from surroundings}\end{pmatrix} + \begin{pmatrix}\text{rate of}\\ \text{generation}\end{pmatrix} \qquad [4.1\text{-}42]$$

or

$$\frac{d\Phi}{dt} = (m\phi)_{\text{in}} - (m\phi)_{\text{out}} + J_\phi + S_\phi \qquad [4.1\text{-}43]$$

where the total momentum, thermal energy, or species A in the control volume Ω is

$$\Phi = \iiint_{\Omega} \rho\phi \, d\Omega \qquad [4.1\text{-}44]$$

In Eq. [4.1-39] the viscous force $\mathbf{F}_v$ at the wall can be considered as the rate of momentum transfer through the wall by molecular diffusion. The pressure force $\mathbf{F}_p$ and the body force $\mathbf{F}_b$, on the other hand, can be considered as the rate of

TABLE 4.1-2 Integral Balance Equations (Variable Physical Properties)

Conservation Law	Rate of Accumulation	= Rate of Net Inflow by Convection	+ Rate of Other Net Inflow	+ Rate of Generation	[A]
General	$\frac{\partial}{\partial t}\iiint_{\Omega} \rho\phi \, d\Omega$	$= -\iint_{A} (\rho\mathbf{v}\phi)\cdot\mathbf{n}\, dA$	$-\iint_{A} \mathbf{j}_{\phi}\cdot\mathbf{n}\, dA$	$+\iiint_{\Omega} s_{\phi}\, d\Omega$	[B]
Mass $\phi = 1,\ \mathbf{j}_{\phi} = s_{\phi} = 0$	$\frac{\partial}{\partial t}\iiint_{\Omega} \rho \, d\Omega$	$= -\iint_{A} (\rho\mathbf{v})\cdot\mathbf{n}\, dA$			[C]
Momentum $\phi = \mathbf{v},\ \mathbf{j}_{\phi} = \boldsymbol{\tau}$ $s_{\phi} = \mathbf{f}_b - \nabla p$	$\frac{\partial}{\partial t}\iiint_{\Omega} \rho\mathbf{v} \, d\Omega$	$= -\iint_{A} (\rho\mathbf{v}\mathbf{v})\cdot\mathbf{n}\, dA$	$-\iint_{A} \boldsymbol{\tau}\cdot\mathbf{n}\, dA$	$+\iiint_{\Omega} (\mathbf{f}_b - \nabla p)\, d\Omega$	[D]
Energy[a] $\phi = C_v T,\ \mathbf{j}_{\phi} = \mathbf{q}$ $s_{\phi} = s$	$(\frac{\partial}{\partial t}\iiint_{\Omega} \rho C_v T \, d\Omega$	$= -\iint_{A} (\rho\mathbf{v}C_v T)\cdot\mathbf{n}\, dA$	$-\iint_{A} \mathbf{q}\cdot\mathbf{n}\, dA$	$+\iiint_{\Omega} s\, d\Omega$	[E]
Species $\phi = w_A,\ \mathbf{j}_{\phi} = \mathbf{j}_A$ $s_{\phi} = r_A$	$\frac{\partial}{\partial t}\iiint_{\Omega} \rho w_A \, d\Omega$	$= -\iint_{A} (\rho\mathbf{v}w_A)\cdot\mathbf{n}\, dA$	$-\iint_{A} \mathbf{j}_A\cdot\mathbf{n}\, dA$	$+\iiint_{\Omega} r_A\, d\Omega$	[F]

[a] $\phi = C_v T$ or $C_p T$ or H; kinetic and potential energy, and pressure, viscous, and shaft work not included.

momentum generation due to the action of these forces. In Eq. [4.1-40] Q is by conduction, which is similar to diffusion.

The above equations are summarized in Table 4.1-3. The following overall mass balance equation (i.e. Eq. [1.2-6]), is also included in the table

$$\frac{dM}{dt} = (m)_{\text{in}} - (m)_{\text{out}} \quad [4.1\text{-}45]$$

4.1.6. Differential Balance Equations

The differential balance equations governing momentum, heat, and species transfer, according to Eqs. [1.5-6], [2.3-5] and [3.3-5], respectively, are as follows:

$$\frac{\partial}{\partial t}(\rho \mathbf{v}) = -\nabla\cdot(\rho\mathbf{v}\mathbf{v}) - \nabla\cdot\boldsymbol{\tau} + (\mathbf{f}_b - \nabla p) \quad \text{(momentum transfer)} \quad [4.1\text{-}46]$$

$$\frac{\partial}{\partial t}(\rho C_v T) = -\nabla\cdot(\rho\mathbf{v}C_v T) - \nabla\cdot\mathbf{q} + s \quad \text{(heat transfer)} \quad [4.1\text{-}47]$$

$$\frac{\partial}{\partial t}(\rho w_A) = -\nabla\cdot(\rho\mathbf{v}w_A) - \nabla\cdot\mathbf{j}_A + r_A \quad \text{(species transfer)} \quad [4.1\text{-}48]$$

In Eq. [4.1-47] the viscous dissipation is neglected and in Eq. [4.1-48] $\rho w_A = \rho_A$.

These differential balance equations share the same form as follows:

$$\begin{pmatrix}\text{Rate of}\\ \text{accumulation}\end{pmatrix} = \begin{pmatrix}\text{rate of}\\ \text{net inflow}\\ \text{by convection}\end{pmatrix} + \begin{pmatrix}\text{rate of}\\ \text{other}\\ \text{net inflow}\end{pmatrix} + \begin{pmatrix}\text{rate of}\\ \text{generation}\end{pmatrix} \quad [4.1\text{-}49]$$

or

$$\frac{\partial}{\partial t}(\rho\phi) = -\nabla\cdot(\rho\mathbf{v}\phi) - \nabla\cdot\mathbf{j}_\phi + s_\phi \quad [4.1\text{-}50]$$

These equations are summarized in Table 4.1-4. The following equation of continuity, Eq. [1.3-4], is also included in the table:

$$\frac{\partial}{\partial t}(\rho) = -\nabla\cdot(\rho\mathbf{v}) \quad [4.1\text{-}51]$$

Table 4.1-5 summarizes these equations for incompressible fluids.

TABLE 4.1-3 Overall Balance Equations

Conservation Law	Rate of Accumulation	= Rate of Inflow by Convection	− Rate of Outflow by Convection	+ Rate of Other Transfer from Surroundings	+ Rate of Generation	[A]
General	$\frac{d\Phi}{dt} = \frac{d}{dt}\iiint_\Omega \rho\phi\, d\Omega$	$= (m\phi)_{in}$	$- (m\phi)_{out}$	$+ J_\phi$	$+ S_\phi$	[B]
Mass $\phi = 1$, $J_\phi = S_\phi = 0$	$\frac{dM}{dt} = \frac{d}{dt}\iiint_\Omega \rho\, d\Omega$	$= (m)_{in}$	$- (m)_{out}$			[C]
Momentum $\phi = \mathbf{v}$, $J_\phi = \mathbf{F}_v$ $S_\phi = \mathbf{F}_p + \mathbf{F}_b$	$\frac{d\mathbf{P}}{\mathrm{dt}} = \frac{d}{dt}\iiint_\Omega \rho\mathbf{v}\, d\Omega$	$= (m\mathbf{v})_{in}$	$- (m\mathbf{v})_{out}$	$+ \mathbf{F}_v$	$+ (\mathbf{F}_p + \mathbf{F}_b)$	[D]
Energy[a] $\phi = C_v T$, $J_\phi = Q$, $S_\phi = S$	$\frac{dE_T}{dt} = \frac{d}{dt}\iiint_\Omega \rho C_v T\, d\Omega$	$= (mC_v T)_{in}$	$- (mC_v T)_{out}$	$+ Q$	$+ S$	[E]
Species $\phi = w_A$, $J_\phi = J_A$, $S_\phi = R_A$	$\frac{dM_{A\Omega}}{dt} = \frac{d}{dt}\iiint_\Omega \rho w_A\, d\Omega$	$= (mw_A)_{in}$	$- (mw_A)_{out}$	$+ J_A$	$+ R_A$	[F]

[a] $\phi = C_v T$ of $C_p T$ of H; kinetic and potential energy, and pressure, viscous, and shaft work not included.

TABLE 4.1-4 Differential Balance Equations (Variable Physical Properties)

Conservation Law	Rate of Accumulation	= Rate of Net Inflow by Convection	+ Rate of Other Net Input	+ Rate of Generation	[A]
General	$\frac{\partial}{\partial t}(\rho\phi)$	$= -\nabla\cdot(\rho\mathbf{v}\phi)$	$-\nabla\cdot\mathbf{j}_\phi$	$+s_\phi$	[B]
Mass $\phi = 1,\ \mathbf{j}_\phi = s_\phi = 0$	$\frac{\partial}{\partial t}(\rho)$	$= -\nabla\cdot(\rho\mathbf{v})$			[C]
Momentum $\phi = \mathbf{v},\ \mathbf{j}_\phi = \boldsymbol{\tau}$ $s_\phi = \mathbf{f}_b - \nabla p$	$\frac{\partial}{\partial t}(\rho\mathbf{v})$	$= -\nabla\cdot(\rho\mathbf{v}\mathbf{v})$	$-\nabla\cdot\boldsymbol{\tau}$	$+(\mathbf{f}_b - \nabla p)$	[D]
Energy[a] $\phi = C_vT,\ \mathbf{j}_\phi = \mathbf{q}$ $s_\phi = s$	$\frac{\partial}{\partial t}(\rho C_vT)$	$= -\nabla\cdot(\rho\mathbf{v}C_vT)$	$-\nabla\cdot\mathbf{q}$	$+s$	[E]
Species $\phi = w_A,\ \mathbf{j}_\phi = \mathbf{j}_A$ $s_\phi = r_A$	$\frac{\partial}{\partial t}(\rho w_A)$	$= -\nabla\cdot(\rho\mathbf{v}w_A)$	$-\nabla\cdot\mathbf{j}_A$	$+r_A$	[F]

[a] $\phi = C_vT$ or C_pT or H; viscous disspation neglected.

TABLE 4.1-5 Differential Balance Equations for Incompressible Fluids

	Rate of Accumulation	= Rate of Net Inflow by Convection	+ Rate of Other Net Input	+ Rate of Generation	[A]
General	$\rho \frac{\partial \phi}{\partial t}$	$= -\rho \mathbf{v} \cdot \nabla \phi$	$- \nabla \cdot \mathbf{j}_\phi$	$+ s_\phi$	[B]
Momentum $\phi = \mathbf{v}$, $\mathbf{j}_\phi = \boldsymbol{\tau}$ $s_\phi = \mathbf{f}_b - \nabla p$	$\rho \frac{\partial \mathbf{v}}{\partial t}$	$= -\rho \mathbf{v} \cdot \nabla \mathbf{v}$	$- \nabla \cdot \boldsymbol{\tau}$	$+ (\mathbf{f}_b - \nabla p)$	[C]
Energy[a] $\phi = C_v T$, $\mathbf{j}_\phi = \mathbf{q}$ $s_\phi = s$	$\rho \frac{\partial}{\partial t}(C_v T)$	$= -\rho \mathbf{v} \cdot \nabla (C_v T)$	$- \nabla \cdot \mathbf{q}$	$+ s$	[D]
Species $\phi = w_A$, $\mathbf{j}_\phi = \mathbf{j}_A$ $s_\phi = r_A$	$\rho \frac{\partial w_A}{\partial t}$	$= -\rho \mathbf{v} \cdot \nabla w_A$	$- \nabla \cdot \mathbf{j}_A$	$+ r_A$	[E]

[a] $\phi = C_v T$ or $C_p T$ or H; viscous disspation neglected.

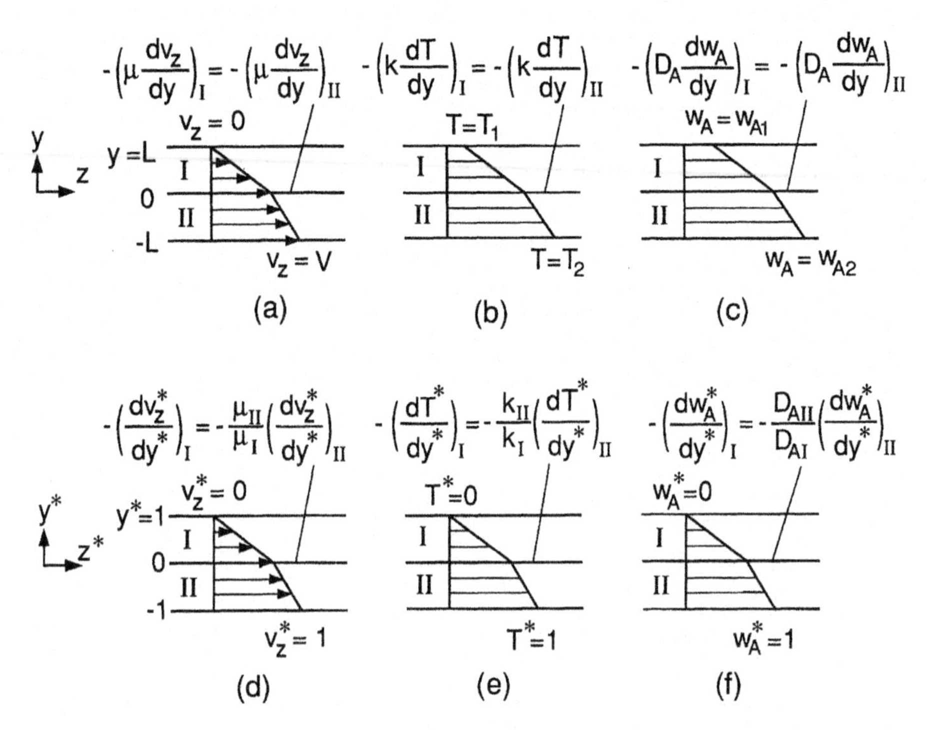

Fig. 4.1-3 Transport through two layers of materials: (*a*, *d*) *z* momentum; (*b*, *e*) heat; (*c*, *f*) species *A*.

4.1.7. Examples

Example 4.1-1 Transport through Two Layers of Materials

Two immiscible liquids (I and II) of different viscosities are held between two horizontal plates, the lower plate moving at a constant velocity V, as shown in Fig. 4.1-3*a*. As shown in Fig. 4.1-3*b*, two plates of different thermal conductivities are in perfect contact, the top surface kept at a constant temperature T_1 and the bottom, at T_2. As shown in Fig. 4.1-3*c*, two plates containing species A and of different diffusivities are in perfect contact, the top surface kept at a constant concentration w_{A1} and the bottom, at w_{A2}. Show the similarities among the three different cases. Find the velocity, temperature, and concentration distributions in the liquids or plates.

Solution The governing equations and boundary conditions are summarized in Table 4.1-6. The three cases are similar in the following ways: (1) the governing equations all have only one term, the diffusion term; (2) at the interface ($y = 0$), continuity is observed in both the property and the flux; and (3) at both the top surface ($y = L$) and the bottom surface ($y = -L$), the properties are specified.

TABLE 4.1-6 Governing Equations and Boundary Conditions for Example 4.1-1

	Fluid Flow	Heat Transfer	Mass Transfer	Location
Governing equations	$\frac{d^2 v_z}{dy^2} = 0$	$\frac{d^2 T}{dy^2} = 0$	$\frac{d^2 w_A}{dy^2} = 0$	I and II
Boundary conditions	$v_z = 0$	$T = T_1$	$w_A = w_{AI}$	$y = L$
	$(v_z)_I = (v_z)_{II}$	$(T)_I = (T)_{II}$	$(w_A)_I = (w_A)_{II}$	$y = 0$
	$-\left(\mu \frac{dv_z}{dy}\right)_I = -\left(\mu \frac{dv_z}{dy}\right)_{II}$	$-\left(k \frac{dT}{dy}\right)_I = -\left(k \frac{dT}{dy}\right)_{II}$	$-\left(D_A \frac{dw_A}{dy}\right)_I = -\left(D_A \frac{dw_A}{dy}\right)_{II}$	$y = 0$
	$(v_z)_{II} = V$	$(T)_{II} = T_2$	$(w_A)_{II} = w_{A2}$	$y = -L$

Let us define the following dimensionless variables

$$y^* = \frac{y}{L} \qquad \text{(dimensionless coordinate)} \qquad [4.1\text{-}52]$$

$$v_z^* = \frac{v_z - 0}{V - 0} = \frac{v_z}{V} \qquad \text{(dimensionless velocity)} \qquad [4.1\text{-}53]$$

$$T^* = \frac{T - T_1}{T_2 - T_1} \qquad \text{(dimensionless temperature)} \qquad [4.1\text{-}54]$$

$$w_A^* = \frac{w_A - w_{A1}}{w_{A2} - w_{A1}} \qquad \text{(dimensionless concentration)} \qquad [4.1\text{-}55]$$

The dimensionless governing equations, boundary conditions, and solutions are summarized in Table 4.1-7. Since the governing equations and boundary conditions are identical in form, the dimensionless solutions should also be identical, as shown in the table and Figs. 4.1-3*d*–*f*. For the case of fluid flow, the solution has already been given, in Example 1.5-5.

Example 4.1-2 Transport through a Film or Slab

A liquid overflows from the top of a vertical wall and forms a falling film due to the gravity force ρg, as shown in Fig. 4.1-4*a*. A slab is resistance-heated at the rate of s per unit volume by passing an electric current through it. Its right surface is thermally insulated while its left surface is kept at a constant temperature T_1, as shown in Fig. 4.1-4*b*. A porous slab of species B is kept at a uniform temperature at which B is converted slowly to a gaseous species A at a uniform and constant rate $\underline{r}_A$ per unit volume of the slab, as shown in Fig. 4.1-4*c*. The right surface of the slab is sealed while the left surface is kept at a constant concentration c_{A1} by being exposed to a gas of a controlled partial pressure of A. Show the similarities among the three different cases. Find the velocity, temperature, and concentration distributions in the film or slabs. Neglect end effects.

Solution The governing equations and boundary conditions are summarized in Table 4.1-8. The three cases are similar in three respects: (1) the governing equations all have a diffusion term and a source term, (2) at the left surface ($y = L$) the transport property is specified, and (3) at the right surface ($y = 0$) the flux is zero.

Let us define the following dimensionless variables:

$$y^* = \frac{y}{L} \qquad \text{(dimensionless coordinate)} \qquad [4.1\text{-}56]$$

TABLE 4.1-7 Dimensionless Governing Equations and Boundary Conditions, and Solutions for Example 4.1-1

	Fluid Flow	Heat Transfer	Mass Transfer	Location
Governing equations	$\frac{d^2v_z^*}{dy^{*2}} = 0$	$\frac{d^2T^*}{dy^{*2}} = 0$	$\frac{d^2w_A^*}{dy^{*2}} = 0$	I and II
Boundary conditions	$v_z^* = 0$	$T^* = 0$	$w_A^* = 0$	$y^* = 1$
	$(v_z^*)_{\mathrm{I}} = (v_z^*)_{\mathrm{II}}$	$(T^*)_{\mathrm{I}} = (T^*)_{\mathrm{II}}$	$(w_A^*)_{\mathrm{I}} = (w_A^*)_{\mathrm{II}}$	$y^* = 0$
	$\left(\frac{dv_z^*}{dy^*}\right)_{\mathrm{I}} = \frac{\mu_{\mathrm{II}}}{\mu_{\mathrm{I}}}\left(\frac{dv_z^*}{dy^*}\right)_{\mathrm{II}}$	$\left(\frac{dT^*}{dy^*}\right)_{\mathrm{I}} = \frac{k_{\mathrm{II}}}{k_{\mathrm{I}}}\left(\frac{dT^*}{dy^*}\right)_{\mathrm{II}}$	$\left(\frac{dw_A^*}{dy^*}\right)_{\mathrm{I}} = \frac{D_{A\mathrm{II}}}{D_{A\mathrm{I}}}\left(\frac{dw_A^*}{dy^*}\right)_{\mathrm{II}}$	$y^* = 0$
	$(v_z^*)_{\mathrm{II}} = 1$	$(T^*)_{\mathrm{II}} = 1$	$(w_A^*)_{\mathrm{II}} = 1$	$y^* = -1$
Solutions	$(v_z^*)_{\mathrm{I}} = \frac{\mu_{\mathrm{II}}}{\mu_{\mathrm{I}} + \mu_{\mathrm{II}}}(1 - y^*)$	$(T^*)_{\mathrm{I}} = \frac{k_{\mathrm{II}}}{k_{\mathrm{I}} + k_{\mathrm{II}}}(1 - y^*)$	$(w_A^*)_{\mathrm{I}} = \frac{D_{A\mathrm{II}}}{D_{A\mathrm{I}} + D_{A\mathrm{II}}}(1 - y^*)$	I
	$(v_z^*)_{\mathrm{II}} = 1 - \frac{\mu_{\mathrm{I}}}{\mu_{\mathrm{I}} + \mu_{\mathrm{II}}}(1 + y^*)$	$(T^*)_{\mathrm{II}} = 1 - \frac{k_{\mathrm{I}}}{k_{\mathrm{I}} + k_{\mathrm{II}}}(1 + y^*)$	$(w_A^*)_{\mathrm{II}} = 1 - \frac{D_{A\mathrm{I}}}{D_{A\mathrm{I}} + D_{A\mathrm{II}}}(1 + y^*)$	II

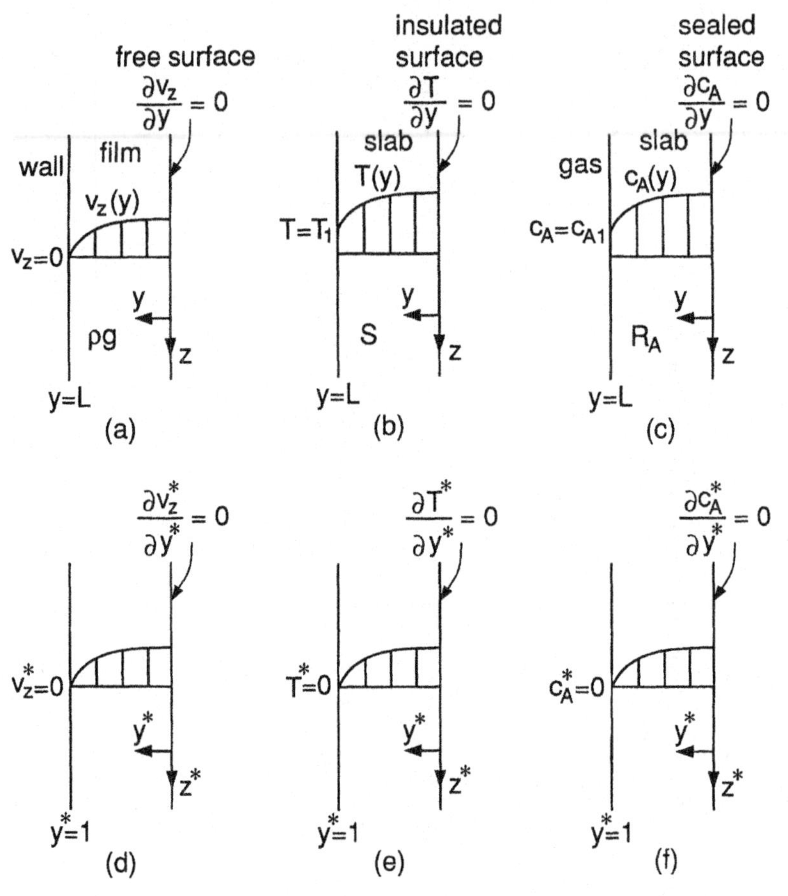

Fig. 4.1-4 Transport through a film or slab: (*a*, *d*) *z* momentum in a falling film; (*b*, *e*) heat in a slab; (*c*, *f*) species *A* in slab.

$$v_z^* = \frac{v_z - 0}{L^2(\rho g/\mu)} \quad \text{(dimensionless velocity)} \tag{4.1-57}$$

$$T^* = \frac{T - T_1}{L^2(s/k)} \quad \text{(dimensionless temperature)} \tag{4.1-58}$$

$$c_A^* = \frac{c_A - c_{A1}}{L^2(\underline{r}_A/D_A)} \quad \text{(dimensionless construction)} \tag{4.1-59}$$

The dimensionless governing equations, boundary conditions, and solutions are summarized in Table 4.1-9. Since the dimensionless governing equations and boundary conditions are identical, the dimensionless solutions should also be identical.

TABLE 4.1-8 Governing Equations and Boundary Conditions for Example 4.1-2

	Fluid Flow	Heat Transfer	Mass Transfer	Location
Governing equations	$\frac{d^2v_z}{dy^2} = \frac{-\rho g}{\mu}$	$\frac{d^2T}{dy^2} = \frac{-s}{k}$	$\frac{d^2c_A}{dy^2} = \frac{-\underline{r}_A}{D_A}$	0–L
Boundary conditions	$v_z = 0$	$T = T_1$	$c_A = c_{AI}$	$y = L$
	$\frac{dv_z}{dy} = 0$	$\frac{dT}{dy} = 0$	$\frac{dc_A}{dy} = 0$	$y = 0$

TABLE 4.1-9 Dimensionless Governing Equations, Boundary Conditions, and Solutions for Example 4.1-2

	Fluid Flow	Heat Transfer	Mass Transfer	Location
Governing equations	$\frac{d^2v_z^*}{dy^{*2}} = -1$	$\frac{d^2T^*}{dy^{*2}} = -1$	$\frac{d^2c_A^*}{dy^{*2}} = -1$	$y^* = 0$–1
Boundary conditions	$v_z^* = 0$	$T^* = 0$	$c_A^* = 0$	$y^* = 1$
	$\frac{dv_z^*}{dy^*} = 0$	$\frac{dT^*}{dy^*} = 0$	$\frac{dc_A^*}{dy^*} = 0$	$y^* = 0$
Solutions	$v_z^* = -\frac{1}{2}(y^{*2} - 1)$	$T^* = -\frac{1}{2}(y^{*2} - 1)$	$c_A^* = -\frac{1}{2}(y^{*2} - 1)$	$y^* = 0$–1

The dimensionless governing equations and boundary conditions are of the following general form:

$$\frac{d^2\phi}{du^2} = 1 \quad [4.1\text{-}60]$$

$$\phi = 0 \quad \text{at} \quad u = 1 \quad [4.1\text{-}61]$$

$$\frac{d\phi}{du} = 0 \quad \text{at} \quad u = 0 \quad [4.1\text{-}62]$$

The solution, from Case D of Appendix A, is as follows:

$$\phi = -\tfrac{1}{2}(u^2 - 1) \quad [4.1\text{-}63]$$

It should be pointed out that for most chemical reactions $\underline{r}_A$ is not a constant but concentration-dependent. For example, for a first-order reaction $A \rightarrow B$,

$\underline{r}_A = -k_1 c_A$ and the similarities no longer exist between mass transfer and fluid flow and heat transfer.

4.2. COUPLING OF DIFFERENT TYPES OF TRANSPORT

For simplicity of discussion we shall assume that the physical properties (ρ, μ, C_v, k, and D_A) are constant. From Eqs. [1.3-5], [1.5-8], [2.3-8], and [3.3-12], the governing equations are as follows:

$$\text{Continuity:} \quad \nabla \cdot \mathbf{v} = 0 \qquad [4.2\text{-}1]$$

$$\text{Motion:} \quad \rho \frac{\partial \mathbf{v}}{\partial t} + \rho \mathbf{v} \cdot \nabla \mathbf{v} = -\nabla p + \mu \nabla^2 \mathbf{v} + \mathbf{f}_b \qquad [4.2\text{-}2]$$

$$\text{Energy:} \quad \rho C_v \left[\frac{\partial T}{\partial t} + \mathbf{v} \cdot \nabla T \right] = k \nabla^2 T + s \qquad [4.2\text{-}3]$$

$$\text{Species:} \quad \frac{\partial c_A}{\partial t} + \mathbf{v} \cdot \nabla c_A = D_A \nabla^2 c_A + \underline{r}_A \qquad [4.2\text{-}4]$$

It is clear from Eqs. [4.2-3] and [4.2-4] that fluid flow can affect heat and mass transfer significantly. In fact, heat and mass transfer can also affect fluid flow significantly, even though this is not immediately clear from Eq. [4.2-2]. Fluid flow induced by temperature and/or concentration variations in the fluid will be discussed in this section. Since it is often desirable to damp this type of fluid flow by magnetic fields, the effect of the Lorentz force on fluid flow will also be discussed.

As mentioned in Section 1.4.1, the body force acting on the fluid can be the gravity or Lorentz force. Convection in a fluid can be induced by the gravity force if its density is not uniform, as in casting and crystal growth. If the density variations are due to temperature variations, the resultant convection is called **buoyancy convection**. On the other hand, if the density variations are due to variations in the solute concentration, the resultant convection is called **solutal convection**. If variations in temperature and the solute concentration both contribute to the density variations, the resultant convection is called **thermosolutal convection**.

Convection in a fluid can also be induced by a nonuniform Lorentz force, as in the case of arc welding. On the other hand, in casting and crystal growth an external magnetic field can be used to create a uniform Lorentz force to damp buoyancy or thermosolutal convection.

4.2.1. Buoyancy Convection

The so-called **thermal expansion coefficient** is defined by

$$\beta_T = -\frac{1}{\rho}\left(\frac{\partial \rho}{\partial T}\right)_{P,C} \qquad [4.2\text{-}5]$$

This physical property of the fluid provides a measure of the fraction change in the density due to a temperature change under constant pressure and composition. The minus sign is included since the fluid density decreases with increasing temperature. As an approximation

$$\beta_T = -\frac{1}{\rho}\left(\frac{\rho - \rho_0}{T - T_0}\right) \qquad [4.2\text{-}6]$$

where T_0 is a reference temperature and ρ_0 is the density at T_0. As such

$$\rho = \rho_0 + (\rho - \rho_0) = \rho_0 - \rho\beta_T(T - T_0) \qquad [4.2\text{-}7]$$

If the gravity force $\rho\mathbf{g}$ alone is involved, Eq. [4.2-2] becomes

$$\rho\frac{\partial \mathbf{v}}{\partial t} + \rho\mathbf{v}\cdot\nabla\mathbf{v} = -\nabla p + \mu\nabla^2\mathbf{v} + \rho\mathbf{g} \qquad [4.2\text{-}8]$$

Substituting Eq. [4.2-7] into Eq. [4.2-8]

$$\rho\frac{\partial \mathbf{v}}{\partial t} + \rho\mathbf{v}\cdot\nabla\mathbf{v} = -\nabla p + \mu\nabla^2\mathbf{v} + \rho_0\mathbf{g} - \rho\beta_T\mathbf{g}(T - T_0) \qquad [4.2\text{-}9]$$

Through the last term of the equation, fluid flow is coupled with heat transfer.

Since the gravitational acceleration is derivable from a potential

$$\mathbf{g} = -\nabla\phi \qquad [4.2\text{-}10]$$

Substituting Eq. [4.2-10] into Eq. [4.2-9]

$$\boxed{\rho\frac{\partial \mathbf{v}}{\partial t} + \rho\mathbf{v}\cdot\nabla\mathbf{v} = -\nabla P + \mu\nabla^2\mathbf{v} - \rho\beta_T\mathbf{g}(T - T_0)} \qquad [4.2\text{-}11]$$

where

$$P = p + \rho_0\phi \qquad [4.2\text{-}12]$$

In this equation $\rho_0\phi$ is the hydrostatic pressure. If the z direction is taken vertically upward, $\phi = gz$ and $\rho_0\phi = \rho_0 gz$.

In Eq. [4.2-11] the density ρ is considered constant everywhere except in β_T. This is known as the **Boussinesq approximation**. If $T = T_0$, Eq. [4.2-11] is identical to Eq. [4.2-8] except P has replaced p. This change makes no difference as long as the pressure does not appear explicitly in the boundary conditions.[2] If the actual pressure is required, one needs only to correct for the hydrostatic pressure $\rho_0\phi$.

4.2.2. Solutal Convection

If the fluid is a solution containing a significant amount of solute A, its density can be affected by the concentration of A, namely, c_A. A **solutal expansion coefficient** similar to β_T can be defined by

$$\beta_S = -\frac{1}{\rho}\left(\frac{\partial \rho}{\partial c_A}\right)_{P,T} \qquad [4.2\text{-}13]$$

This physical property of the fluid provides a measure of the fraction change in the density due to a chanage in the solute concentration under constant pressure and temperature. The minus sign is included merely for consistency with the definition of β_T. It is not implied that the fluid density always has to decrease with increasing concentration of A, unless A is indeed a lighter component. As an approximation

$$\beta_S = -\frac{1}{\rho}\left(\frac{\rho - \rho_0}{c_A - c_{A0}}\right) \qquad [4.2\text{-}14]$$

where c_{A0} is a reference concentration and ρ_0 is the density at c_{A0}. As such

$$\rho = \rho_0 + (\rho - \rho_0) = \rho_0 - \rho\beta_S(c_A - c_{A0}) \qquad [4.2\text{-}15]$$

Substituting Eqs. [4.2-10] and [4.2-15] into Eq. [4.2-8]

$$\boxed{\rho\frac{\partial \mathbf{v}}{\partial t} + \rho\mathbf{v}\cdot\nabla\mathbf{v} = -\nabla P + \mu\nabla^2\mathbf{v} - \rho\beta_S\mathbf{g}(c_A - c_{A0})} \qquad [4.2\text{-}16]$$

Through the last term of the equation, fluid flow is coupled with mass transfer.

Note that the fluid density is considered constant except in β_S; this, as already mentioned previously, is the Boussinesq approximation.

When convection is associated with the density dependence on both temperature and the solute concentration, or **thermosolutal convection**, Eq. [4.2-16] can

[2] D. J. Tritton, *Physical Fluid Dynamics*, Van Nostrand Reinhold, New York, 1977, p. 128.

TABLE 4.2-1 The Equation of Motion in Rectangular Coordinates (x, y, z) (for Thermosolutal Convection of a Newtonian Fluid)

x component

$$\rho\left[\frac{\partial v_x}{\partial t}+v_x\frac{\partial v_x}{\partial x}+v_y\frac{\partial v_x}{\partial y}+v_z\frac{\partial v_x}{\partial z}\right]$$

$$=-\frac{\partial P}{\partial x}+\mu\left[\frac{\partial^2 v_x}{\partial x^2}+\frac{\partial^2 v_x}{\partial y^2}+\frac{\partial^2 v_x}{\partial z^2}\right]-\rho\beta_T g_x(T-T_0)-\rho\beta_S g_x(c_A-c_{A0}) \quad \text{[A]}$$

y component

$$\rho\left[\frac{\partial v_y}{\partial t}+v_x\frac{\partial v_y}{\partial x}+v_y\frac{\partial v_y}{\partial y}+v_z\frac{\partial v_y}{\partial z}\right]$$

$$=-\frac{\partial P}{\partial y}+\mu\left[\frac{\partial^2 v_y}{\partial x^2}+\frac{\partial^2 v_y}{\partial y^2}+\frac{\partial^2 v_y}{\partial z^2}\right]-\rho\beta_T g_y(T-T_0)-\rho\beta_S g_y(c_A-c_{A0}) \quad \text{[B]}$$

z component

$$\rho\left[\frac{\partial v_z}{\partial t}+v_x\frac{\partial v_z}{\partial x}+v_y\frac{\partial v_z}{\partial y}+v_z\frac{\partial v_z}{\partial z}\right]$$

$$=-\frac{\partial P}{\partial z}+\mu\left[\frac{\partial^2 v_z}{\partial x^2}+\frac{\partial^2 v_z}{\partial y^2}+\frac{\partial^2 v_z}{\partial z^2}\right]-\rho\beta_T g_z(T-T_0)-\rho\beta_S g_z(c_A-c_{A0}) \quad \text{[C]}$$

be modified as follows:

$$\boxed{\rho\frac{\partial \mathbf{v}}{\partial t}+\rho\mathbf{v}\cdot\nabla\mathbf{v}=-\nabla P+\mu\nabla^2\mathbf{v}-\rho\beta_T\mathbf{g}(T-T_0)-\rho\beta_S\mathbf{g}(c_A-c_{A0})} \quad \text{[4.2-17]}$$

In Tables 4.2-1 through 4.2-3, Eq. [4.2-17] is given in the expanded form for rectangular, cylindrical, and spherical coordinates.

4.2.3. Convection in the Presence of the Lorentz Force

Like the gravity force, the Lorentz or electromagnetic force is a body force. It is defined by

$$\mathbf{f}_L=\mathbf{J}\times\mathbf{B} \quad \text{[4.2-18]}$$

where $\mathbf{J}$ is the current density vector and $\mathbf{B}$ the magnetic flux vector.

In arc welding the diverging $\mathbf{J}$ interacts with the $\mathbf{B}$ it induces to form a nonuniform $\mathbf{f}_L$ field, causing stirring in the weld pool and the arc. This stirring helps weld penetration and uniform mixing of the filler metal with the melted base metal. In crystal growth and continuous casting, on the other hand, it is

TABLE 4.2-2 The Equation of Motion in Cylindrical Coordinates (r, θ, z) (for Thermosolutal Convection of a Newtonian Fluid)

r component

$$\rho\left[\frac{\partial v_r}{\partial t} + v_r\frac{\partial v_r}{\partial r} + \frac{v_\theta}{r}\frac{\partial v_r}{\partial \theta} - \frac{v_\theta^2}{r} + v_z\frac{\partial v_r}{\partial z}\right]$$
$$= -\frac{\partial P}{\partial r} + \mu\left[\frac{\partial}{\partial r}\left(\frac{1}{r}\frac{\partial}{\partial r}(rv_r)\right) + \frac{1}{r^2}\frac{\partial^2 v_r}{\partial \theta^2} - \frac{2}{r^2}\frac{\partial v_\theta}{\partial \theta} + \frac{\partial^2 v_r}{\partial z^2}\right]$$
$$- \rho\beta_T g_r(T - T_0) - \rho\beta_S g_r(c_A - c_{A0}) \qquad \text{[A]}$$

θ component

$$\rho\left[\frac{\partial v_\theta}{\partial t} + v_r\frac{\partial v_\theta}{\partial r} + \frac{v_\theta}{r}\frac{\partial v_\theta}{\partial \theta} + \frac{v_r v_\theta}{r} + v_z\frac{\partial v_\theta}{\partial z}\right]$$
$$= -\frac{1}{r}\frac{\partial P}{\partial \theta} + \mu\left[\frac{\partial}{\partial r}\left(\frac{1}{r}\frac{\partial}{\partial r}(rv_\theta)\right) + \frac{1}{r^2}\frac{\partial^2 v_\theta}{\partial \theta^2} + \frac{2}{r^2}\frac{\partial v_r}{\partial \theta} + \frac{\partial^2 v_\theta}{\partial z^2}\right]$$
$$- \rho\beta_T g_\theta(T - T_0) - \rho\beta_S g_\theta(c_A - c_{A0}) \qquad \text{[B]}$$

z component

$$\rho\left[\frac{\partial v_z}{\partial t} + v_r\frac{\partial v_z}{\partial r} + \frac{v_\theta}{r}\frac{\partial v_z}{\partial \theta} + v_z\frac{\partial v_z}{\partial z}\right]$$
$$= -\frac{\partial P}{\partial z} + \mu\left[\frac{1}{r}\frac{\partial}{\partial r}\left(r\frac{\partial v_z}{\partial r}\right) + \frac{1}{r^2}\frac{\partial^2 v_z}{\partial \theta^2} + \frac{\partial^2 v_z}{\partial z^2}\right]$$
$$- \rho\beta_T g_z(T - T_0) - \rho\beta_S g_z(c_A - c_{A0}) \qquad \text{[C]}$$

often desirable to reduce convection. An external magnet can be used to produce a uniform magnetic field. It interacts with the electric current induced by the convection of the melt and damps it.

Substituting Eq. [4.2-18] into Eq. [4.2-2]

$$\rho\frac{\partial \mathbf{v}}{\partial t} + \rho\mathbf{v}\cdot\nabla\mathbf{v} = -\nabla p + \mu\nabla^2\mathbf{v} + \rho\mathbf{g} + \mathbf{J}\times\mathbf{B} \qquad \text{[4.2-19]}$$

for forced convection in the presence of the Lorentz force. From Eqs. [4.2-17] and [4.2-18]

$$\rho\frac{\partial \mathbf{v}}{\partial t} + \rho\mathbf{v}\cdot\nabla\mathbf{v} = -\nabla P + \mu\nabla^2\mathbf{v} - \rho\beta_T\mathbf{g}(T - T_0) - \rho\beta_S\mathbf{g}(c_A - c_{A0}) + \mathbf{J}\times\mathbf{B} \qquad \text{[4.2-20]}$$

for thermosolutal convection in the presence of the Lorentz force.

TABLE 4.2-3 Equation of Motion in Spherical Coordinates (r, θ, ϕ) (for Thermosolutal Convection of a Newtonian Fluid)[a]

r component

$$\rho\left[\frac{\partial v_r}{\partial t} + v_r\frac{\partial v_r}{\partial r} + \frac{v_\theta}{r}\frac{\partial v_r}{\partial \theta} + \frac{v_\phi}{r\sin\theta}\frac{\partial v_r}{\partial \phi} - \frac{v_\theta^2 + v_\phi^2}{r}\right]$$

$$= -\frac{\partial P}{\partial r} + \mu\left[\nabla^2 v_r - \frac{2}{r^2}v_r - \frac{2}{r^2}\frac{\partial v_\theta}{\partial \theta} - \frac{2}{r^2}v_\theta\cot\theta - \frac{2}{r^2\sin\theta}\frac{\partial v_\phi}{\partial \phi}\right]$$

$$- \rho\beta_T g_r(T - T_0) - \rho\beta_S g_r(c_A - c_{A0}) \qquad [A]$$

θ component

$$\rho\left[\frac{\partial v_\theta}{\partial t} + v_r\frac{\partial v_\theta}{\partial r} + \frac{v_\theta}{r}\frac{\partial v_\theta}{\partial \theta} + \frac{v_\phi}{r\sin\theta}\frac{\partial v_\theta}{\partial \phi} + \frac{v_r v_\theta}{r} - \frac{v_\phi^2\cot\theta}{r}\right]$$

$$= -\frac{1}{r}\frac{\partial P}{\partial \theta} + \mu\left[\nabla^2 v_\theta + \frac{2}{r^2}\frac{\partial v_r}{\partial \theta} - \frac{v_\theta}{r^2\sin^2\theta} - \frac{2\cos\theta}{r^2\sin^2\theta}\frac{\partial v_\phi}{\partial \phi}\right]$$

$$- \rho\beta_T g_\theta(T - T_0) - \rho\beta_S g_\theta(c_A - c_{A0}) \qquad [B]$$

ϕ component

$$\rho\left[\frac{\partial v_\phi}{\partial t} + v_r\frac{\partial v_\phi}{\partial r} + \frac{v_\theta}{r}\frac{\partial v_\phi}{\partial \theta} + \frac{v_\phi}{r\sin\theta}\frac{\partial v_\phi}{\partial \phi} + \frac{v_\phi v_r}{r} + \frac{v_\theta v_\phi}{r}\cot\theta\right]$$

$$= -\frac{1}{r\sin\theta}\frac{\partial P}{\partial \phi} + \mu\left[\nabla^2 v_\phi - \frac{v_\phi}{r^2\sin^2\theta} + \frac{2}{r^2\sin\theta}\frac{\partial v_r}{\partial \phi} + \frac{2\cos\theta}{r^2\sin^2\theta}\frac{\partial v_\theta}{\partial \phi}\right]$$

$$- \rho\beta_T g_\phi(T - T_0) - \rho\beta_S g_\phi(c_A - c_{A0}) \qquad [C]$$

[a] In these equations:

$$\nabla^2 = \frac{1}{r^2}\frac{\partial}{\partial r}\left(r^2\frac{\partial}{\partial r}\right) + \frac{1}{r^2\sin\theta}\frac{\partial}{\partial \theta}\left(\sin\theta\frac{\partial}{\partial \theta}\right) + \frac{1}{r^2\sin^2\theta}\left(\frac{\partial^2}{\partial \phi^2}\right)$$

Let us now consider the case where a uniform magnetic field is applied externally to damp thermosolutal convection, as in the crystal growth of semiconductors. The magnetic flux vector

$$\mathbf{B} = B_0\mathbf{e}_B \qquad [4.2\text{-}21]$$

where $\mathbf{e}_B$ is the unit vector in the direction $\mathbf{B}$ is applied. The current density vector induced by fluid motion under the magnetic field

$$\mathbf{J} = \sigma\mathbf{v} \times \mathbf{B} \qquad [4.2\text{-}22]$$

where σ is the electrical conductivity of the fluid.

Substituting Eqs. [4.2-21] and [4.2-22] into Eq. [4.2-20], we get the following equation of motion for thermosolutal convection damped by a uniform magnetic field:

$$\rho\frac{\partial \mathbf{v}}{\partial t} + \rho\mathbf{v}\cdot\nabla\mathbf{v}\left(= \rho\frac{D\mathbf{v}}{Dt}\right) = -\nabla P + \mu\nabla^2\mathbf{v} - \rho\beta_T\mathbf{g}(T - T_0)$$

$$\qquad\qquad\qquad (1) \qquad\qquad (2) \qquad\qquad (3) \qquad\qquad (4)$$

$$- \rho\beta_S\mathbf{g}(c_A - c_{A0}) + \sigma B_0^2\mathbf{v}\times\mathbf{e}_B\times\mathbf{e}_B$$

$$(5) \qquad\qquad (6) \qquad [4.2\text{-}23]$$

The Lorentz force in this equation is an idealization valid when the melt is electrically conductive enough that convection of charge by the flow is insignificant. From Eq. [1.5-10] term (1) is the inertia force per unit volume. Terms (2), (3), (4), (5), and (6) are the pressure force, viscous force, thermal buoyancy force, solutal buoyancy force, and Lorentz force per unit volume, respectively.

4.2.4. Stream Function and Vorticity

4.2.4.1. Two-Dimensional Problems

The stream function ψ and vorticity ω, as in Section 1.6.1, are defined as follows:

$$v_x = \frac{\partial \psi}{\partial y} \qquad [4.2\text{-}24]$$

$$v_y = -\frac{\partial \psi}{\partial x} \qquad [4.2\text{-}25]$$

and

$$\omega = \frac{\partial v_y}{\partial x} - \frac{\partial v_x}{\partial y} \qquad [4.2\text{-}26]$$

From Eqs. [A] and [B] of Table 4.2-1

$$\rho\left[\frac{\partial v_x}{\partial t} + v_x\frac{\partial v_x}{\partial x} + v_y\frac{\partial v_x}{\partial y}\right]$$

$$= -\frac{\partial P}{\partial x} + \mu\left[\frac{\partial^2 v_x}{\partial x^2} + \frac{\partial^2 v_x}{\partial y^2}\right] - \rho\beta_T g_x(T - T_0) - \rho\beta_S g_x(c_A - c_{A0}) \qquad [4.2\text{-}27]$$

and

$$\rho\left[\frac{\partial v_y}{\partial t}+v_x\frac{\partial v_y}{\partial x}+v_y\frac{\partial v_y}{\partial y}\right]$$
$$=-\frac{\partial P}{\partial y}+\mu\left[\frac{\partial^2 v_y}{\partial x^2}+\frac{\partial^2 v_y}{\partial y^2}\right]-\rho\beta_T g_y(T-T_0)-\rho\beta_S g_y(c_A-c_{A0}) \quad [4.2\text{-}28]$$

By differentiating Eq. [4.2-27] with respect to y and Eq. [4.2-28] with respect to x and subtracting, we can eliminate the pressure term. Substituting Eqs. [4.2-24] through [4.2-26] into the resulting equation, we can obtain the following vorticity equation:

$$\rho\left(\frac{\partial\omega}{\partial t}+\frac{\partial\psi}{\partial y}\frac{\partial\omega}{\partial x}-\frac{\partial\psi}{\partial x}\frac{\partial\omega}{\partial y}\right)=\mu\left(\frac{\partial^2\omega}{\partial x^2}+\frac{\partial^2\omega}{\partial y^2}\right)-\rho\beta_T\left(g_x\frac{\partial T}{\partial y}-g_y\frac{\partial T}{\partial x}\right)$$
$$-\rho\beta_S\left(g_x\frac{\partial c_A}{\partial y}-g_y\frac{\partial c_A}{\partial x}\right) \quad [4.2\text{-}29]$$

The stream-function equation remains identical to that for forced convection (Eq. [1.6-12]):

$$\left(\frac{\partial^2\psi}{\partial x^2}+\frac{\partial^2\psi}{\partial y}\right)=-\omega \quad [4.2\text{-}30]$$

4.2.4.2. Axisymmetric Problems

The stream function ψ and vorticity ω, as in Section 1.6.2, are defined as follows:

$$v_r=\frac{1}{r}\frac{\partial\psi}{\partial z} \quad [4.2\text{-}31]$$

$$v_z=\frac{-1}{r}\frac{\partial\psi}{\partial r} \quad [4.2\text{-}32]$$

and

$$\omega=\frac{\partial v_r}{\partial z}-\frac{\partial v_z}{\partial r} \quad [4.2\text{-}33]$$

Normally the z axis is in the vertical direction and so $g_r=g_\theta=0$. From Eqs. [A] and [C] of Table 4.2-2

$$\rho\left[\frac{\partial v_r}{\partial t}+v_r\frac{\partial v_r}{\partial r}-\frac{v_\theta^2}{r}+v_z\frac{\partial v_r}{\partial z}\right]=-\frac{\partial P}{\partial r}+\mu\left(\frac{\partial}{\partial r}\left[\frac{1}{r}\frac{\partial}{\partial r}(rv_r)\right]+\frac{\partial^2 v_r}{\partial z^2}\right) \quad [4.2\text{-}34]$$

and

$$\rho\left[\frac{\partial v_z}{\partial t}+v_r\frac{\partial v_z}{\partial r}+v_z\frac{\partial v_z}{\partial z}\right]=-\frac{\partial P}{\partial z}+\mu\left(\left[\frac{1}{r}\frac{\partial}{\partial r}\left(r\frac{\partial v_z}{\partial r}\right)\right]+\frac{\partial^2 v_z}{\partial z^2}\right)$$
$$-\rho\beta_T g_z(T-T_0)-\rho\beta_S g_z(c_A-c_{A0}) \qquad [4.2\text{-}35]$$

By differentiating Eq. [4.2-34] with respect to z and Eq. [4.2-35] with respect to r and subtracting, we can eliminate the pressure term. Substituting Eqs. [4.2-31] through [4.2-33] into the resulting equation, we can obtain the following vorticity equation:

$$\rho\left[\frac{\partial\omega}{\partial t}-\frac{\partial\psi}{\partial z}\frac{\partial}{\partial r}\left[\frac{\omega}{r}\right]+\frac{\partial\psi}{\partial r}\frac{\partial}{\partial z}\left[\frac{\omega}{r}\right]\right]$$
$$=\mu\left(\frac{\partial}{\partial r}\left[\frac{1}{r}\frac{\partial(r\omega)}{\partial r}\right]+\frac{\partial}{\partial z}\left[\frac{1}{r}\frac{\partial(r\omega)}{\partial z}\right]\right)+\rho\frac{\partial}{\partial z}\left(\frac{v_\theta^2}{r}\right)+\rho\beta_T g_z\frac{\partial T}{\partial r}$$
$$+\rho\beta_S g_z\frac{\partial c_A}{\partial r} \qquad [4.2\text{-}36]$$

The stream-function and circulation equations remain identical to those for forced convection, Eqs. [1.6-22] and [1.6-23]:

$$\frac{\partial}{\partial r}\left(\frac{1}{r}\frac{\partial\psi}{\partial r}\right)+\frac{\partial}{\partial z}\left(\frac{1}{r}\frac{\partial\psi}{\partial z}\right)=\omega \qquad [4.2\text{-}37]$$

and

$$\rho\left[\frac{\partial v_\theta}{\partial t}+\frac{1}{r^2}\frac{\partial\psi}{\partial z}\frac{\partial}{\partial r}(rv_\theta)-\frac{1}{r}\frac{\partial\psi}{\partial r}\frac{\partial v_\theta}{\partial z}\right]=\mu\left(\frac{\partial}{\partial r}\left[\frac{1}{r}\frac{\partial}{\partial r}(rv_\theta)\right]+\frac{\partial^2 v_\theta}{\partial z^2}\right) \qquad [4.2\text{-}38]$$

4.2.5. Dimensionless Form

Let us define the following dimensionless quantities for thermosolutal convection

$$x^*, y^*, z^* = \frac{x}{L}, \frac{y}{L}, \frac{z}{L} \qquad \text{(dimensionless coordinates)} \qquad [4.2\text{-}39]$$

$$\nabla^* = L\nabla = \mathbf{e}_x\frac{\partial}{\partial x^*}+\mathbf{e}_y\frac{\partial}{\partial y^*}+\mathbf{e}_z\frac{\partial}{\partial z^*} \qquad \text{(dimensionless operator)} \qquad [4.2\text{-}40]$$

$$\nabla^{*2} = L^2\nabla^2 = \frac{\partial^2}{\partial x^{*2}} + \frac{\partial^2}{\partial y^{*2}} + \frac{\partial^2}{\partial z^{*2}} \quad \text{(dimensionless operator)} \qquad [4.2\text{-}41]$$

$$\mathbf{v}^* = \frac{\mathbf{v}L}{\nu} \quad \text{(dimensionless velocity)} \qquad [4.2\text{-}42]$$

$$t^* = \frac{t\nu}{L^2} \quad \text{(dimensionless time)} \qquad [4.2\text{-}43]$$

$$P^* = \frac{(P - P_0)L^2}{\mu\nu} \quad \text{(dimensionless pressure)} \qquad [4.2\text{-}44]$$

As mentioned in Section 1.5-2, in forced convection a characteristic velocity V is often readily available for use as a reference velocity. This, however, is not the case in buoyancy, solutal, or thermosolutal convection. One way to overcome this problem is to define the **characteristic velocity** as ν/L, where ν is the kinematic viscosity and L the **characteristic length**. This is because ν/L has the same unit as velocity. The **characteristic time** L/V now becomes $L/(\nu/L)$ or L^2/ν. By substituting ν/L for V in Eqs. [1.5-12] through [1.5-14], Eqs. [4.2-42] through [4.2-44] can be obtained. It should be mentioned that characteristic velocities other than ν/L have also been used.

From Eq. [4.2-23], with characteristic velocity ν/L, characteristic length L, and characteristic time L^2/ν, the forces per unit volume in thermosolutal convection can be expressed qualitatively as follows:

Inertia force: $$\frac{\rho Dv}{Dt} = \rho\frac{\nu/L}{L^2/\nu} = \frac{\rho\nu^2}{L^3} \qquad [4.2\text{-}45]$$

Pressure force: $$\nabla P = \frac{P}{L} \qquad [4.2\text{-}46]$$

Viscous force: $$\mu\nabla^2 v = \mu\frac{\nu/L}{L^2} = \frac{\mu\nu}{L^3} \qquad [4.2\text{-}47]$$

Thermal buoyancy force: $$\rho\beta_T g(T - T_0) = \rho\beta_T g(T_1 - T_0) \qquad [4.2\text{-}48]$$

Solutal buoyancy force: $$\rho\beta_S g(c_A - c_{A0}) = \rho\beta_S g(c_{A1} - c_{A0}) \qquad [4.2\text{-}49]$$

Lorentz force: $$\sigma B_0^2 v = \sigma B_0^2\frac{\nu}{L} = \frac{\sigma B_0^2\nu}{L} \qquad [4.2\text{-}50]$$

Substituting the dimensionless variables in Eqs. [4.2-39] through [4.2-44] into Eqs. [4.2-1], [4.2-3], (4.2-4], and [4.2-23], the following equations can be obtained for thermosolutal convection with magnetic damping:

Continuity: $$\nabla^* \cdot \mathbf{v}^* = 0 \qquad [4.2\text{-}51]$$

Motion: $$\frac{\partial \mathbf{v}^*}{\partial t^*} + \mathbf{v}^* \cdot \nabla^* \mathbf{v}^* = -\nabla^* P^* + \nabla^{*2} \mathbf{v}^* - \frac{1}{\text{Pr}}(\text{Ra}_T T^* + \text{Ra}_S c_A^*)\mathbf{e}_g + \text{Ha}^2 \mathbf{v}^* \times \mathbf{e}_B \times \mathbf{e}_B \qquad [4.2\text{-}52]$$

Energy: $$\frac{\partial T^*}{\partial t^*} + \mathbf{v}^* \cdot \nabla^* T^* = \frac{1}{\text{Pr}} \nabla^{*2} T^* \qquad [4.2\text{-}53]$$

Species: $$\frac{\partial c_A^*}{\partial t^*} + \mathbf{v}^* \cdot \nabla^* c_A^* = \frac{1}{\text{Sc}} \nabla^{*2} c_A^* \qquad [4.2\text{-}54]$$

where

$$T^* = \frac{T - T_0}{T_1 - T_0} \qquad \text{(dimensionless temperature)} \qquad [4.2\text{-}55]$$

$$c_A^* = \frac{c_A - c_{A0}}{c_{A1} - c_{A0}} \qquad \text{(dimensionless concentration)} \qquad [4.2\text{-}56]$$

$$\text{Pr} = \frac{\nu}{\alpha} \qquad \left(\text{Prandtl number} = \frac{\text{viscous diffusivity } \nu}{\text{thermal diffusivity } \alpha}\right) \qquad [4.2\text{-}57]$$

$$\text{Sc} = \frac{\nu}{D_A} \qquad \left(\text{Schmidt number} = \frac{\text{viscous diffusivity } \nu}{\text{species diffusivity } D_A}\right) \qquad [4.2\text{-}58]$$

$$\text{Ra}_T = \frac{g\beta_T(T_1 - T_0)L^3}{\nu\alpha}$$

$$\left(\text{thermal Rayleigh number} = \frac{\text{thermal buoyancy force } \rho\beta_T g(T_1 - T_0)}{\text{viscous force } \mu\nu/L^3} \times \frac{\nu}{\alpha} = \text{Gr}_T\text{Pr}\right) \qquad [4.2\text{-}59]$$

$$\text{Ra}_S = \frac{g\beta_S(c_{A1} - c_{A0})L^3}{\nu D_A}$$

$$\left(\text{solutal Rayleigh number} = \frac{\text{solutal buoyance force } \rho\beta_S g(c_{A1} - c_{A0})}{\text{viscous force } \mu\nu/L^3} \times \frac{\nu}{D_A} = \text{Gr}_S\text{Sc}\right) \qquad [4.2\text{-}60]$$

$$\mathrm{Ha} = B_0 L \left(\frac{\sigma}{\mu}\right)^{1/2} \qquad \left(\text{Hartmann number} = \sqrt{\frac{\text{Lorentz force } \sigma B_0^2 \nu / L}{\text{viscous force } \mu\nu / L^3}}\right)$$

[4.2-61]

$$\mathrm{Gr}_T = \frac{\beta_T g (T_1 - T_0) L^3}{\nu^2}$$

$$\left(\text{thermal Grashof number} = \frac{\text{thermal buoyancy force } \rho\beta_T g (T_1 - T_0)}{\text{viscous force } \mu\nu / L^3}\right)$$

[4.2-62]

$$\mathrm{Gr}_S = \frac{\beta_S g (c_{A1} - c_{A0}) L^3}{\nu^2}$$

$$\left(\text{solutal Grashof number} = \frac{\text{solutal buoyancy force } \rho\beta_S g (c_{A1} - c_{A0})}{\text{viscous force } \mu\nu / L^3}\right)$$

[4.2-63]

4.2.6. Correlations for Buoyancy Convection

4.2.6.1. Natural Convection on a Vertical Plate

For a fluid of bulk temperature T_∞ in contact with a vertical plate of a different temperature T_0, natural convection occurs in a boundary layer on the plate. Transition from laminar to turbulent flow has been determined to occur near[3]

$$\mathrm{Ra}_{T,x} = \mathrm{Gr}_{T,x}\mathrm{Pr} = 10^9 \qquad [4.2\text{-}64]$$

where

$$\mathrm{Ra}_{T,x} \text{ (local thermal Rayleigh number)} = \frac{g\beta_T (T_0 - T_\infty) x^3}{\nu\alpha} \qquad [4.2\text{-}65]$$

$$\mathrm{Gr}_{T,x} \text{ (local thermal Grashof number)} = \frac{g\beta_T (T_0 - T_\infty) x^3}{\nu^2} \qquad [4.2\text{-}66]$$

$$\mathrm{Pr} \text{ (Prandtl number)} = \frac{\nu}{\alpha} \qquad [4.2\text{-}67]$$

[3] F. P. Incropera and D. P. DeWitt, *Introduction to Heat Transfer*, 2nd ed., Wiley, New York, 1990, p. 497.

The following empirical correlation for a vertical plate of length L has been proposed for all levels of $Ra_{T,L}$:[4]

$$\boxed{Nu_L = \left\{0.825 + \frac{0.387Ra_{T,L}^{1/6}}{[1 + (0.492/Pr)^{9/16}]^{8/27}}\right\}^2} \qquad [4.2\text{-}68]$$

where

$$Nu_L \text{ (Nusselt number)} = \frac{hL}{k} \qquad [4.2\text{-}69]$$

$$Ra_{T,L} \text{ (thermal Rayleigh number)} = \frac{g\beta_T(T_0 - T_\infty)L^3}{\nu\alpha} \qquad [4.2\text{-}70]$$

4.2.6.2. Natural Convection on a Vertical Cylinder

Equation [4.2-68] for a vertical plate can also be used, if the boundary layer thickness is small as compared to the cylinder diameter D. This criterion is as follows:[5]

$$\frac{D}{L} \geq \frac{35}{Gr_{T,L}^{1/4}} \qquad [4.2\text{-}71]$$

where L is the cylinder height and

$$Gr_{T,L} \text{ (thermal Grashof number)} = \frac{g\beta_T(T_0 - T_\infty)L^3}{\nu^2} \qquad [4.2\text{-}72]$$

4.2.6.3. Natural Convection on a Horizontal Plate

It is apparent that natural convection will be much different for a hot surface facing up than down. For a plate with a charateristic length L, specifically, the ratio of the plate surface area to perimeter, the following empirical correlations[6] have suggested for a hot surface facing up or a cold surface facing down:

$$\boxed{Nu_L = 0.54Ra_{T,L}^{1/4}} \quad (10^5 < Ra_{T,L} < 2 \times 10^7) \qquad [4.2\text{-}73]$$

where

$$\boxed{Nu_L = 0.14Ra_{T,L}^{1/3}} \quad (2 \times 10^7 < Ra_{T,L} < 3 \times 10^{10}) \qquad [4.2\text{-}74]$$

[4] S. W. Churchill and H. H. S. Chu, *Int. J. Heat Mass Transfer*, **18**, 1323, 1975.

[5] E. M. Sparrow and J. L. Gregg, *Trans. AIME*, **78**, p. 1823, 1956.

[6] W. H. McAdams, *Heat Transmission*, 3rd ed., McGraw-Hill, New York, 1957.

TABLE 4.2-4 Constants in Eq. [4.2-77] for Natural Convection on a Horizontal Cylinder

Ra_D	a	b
10^{-10}–10^{-2}	0.675	0.058
10^{-2}–10^{2}	1.02	0.148
10^{2}–10^{4}	0.850	0.188
10^{4}–10^{7}	0.480	0.250
10^{7}–10^{12}	0.125	0.333

Source: V.T. Morgan, *Advances in Heat Transfer*, vol. II, T. F. Irvine and J. P. Hartnett, eds., Academic Press, New York, 1975, pp. 199–264.

and for a hot surface facing down or a cold surface facing up,

$$\boxed{Nu_L = 0.27 Ra_{T,L}^{1/4}} \qquad (3 \times 10^5 < Ra_{T,L} < 10^{10}) \qquad [4.2\text{-}75]$$

In these equations the fluid properties should be evaluated at the film temperature

$$T_f = \frac{T_0 + T_\infty}{2} \qquad [4.2\text{-}76]$$

4.2.6.4. *Natural Convection on a Horizontal Cylinder*

For an isothermal cylinder of diameter D and sufficient length that end effects are negligible, two empirical correlations have been suggested. The first correlation is[7]

$$\boxed{Nu_D = a Ra_{T,D}^{b}} \qquad (10^{-5} < Ra_{T,D} < 10^{12}) \qquad [4.2\text{-}77]$$

where

$$Nu_D \text{ (Nusselt number)} = \frac{hD}{k} \qquad [4.2\text{-}78]$$

$$Ra_{T,D} \text{ (thermal Rayleigh number)} = \frac{g\beta_T(T_0 - T_\infty)D^3}{\nu\alpha} \qquad [4.2\text{-}79]$$

The constants a and b are given in Table 4.2-4.

[7] V. T. Morgan, *Advances in Heat Transfer*, Vol. II, T. F. Irvine and J. P. Hartnett, eds., Academic Press, New York, 1975, pp. 199–264.

The second correlation is[8]

$$\mathrm{Nu}_D = \left\{0.60 + \frac{0.387\mathrm{Ra}_{T,D}^{1/6}}{[1 + (0.559/\mathrm{Pr})^{9/16}]^{8/27}}\right\}^2 \qquad (10^{-5} < \mathrm{Ra}_{T,D} < 10^{12}) \tag{4.2-80}$$

4.2.6.5. Natural Convection on a Sphere

For an isothermal sphere of diameter D the following empirical correlation has been suggested:[9]

$$\mathrm{Nu}_D = 2 + \frac{0.589\mathrm{Ra}_{T,D}^{1/4}}{[1 + (0.469/\mathrm{Pr})^{9/16}]^{4/9}} \qquad (\mathrm{Ra}_{T,D} < 10^{11},\ \mathrm{Pr} > 0.7) \tag{4.2-81}$$

4.2.7. Examples

Example 4.2-1 Buoyancy Convection between Two Vertical Isothermal Walls

A fluid is held in a rectangular container that has two large vertical walls separated by a small distance $2w$, as shown in part in Fig. 4.2-1. The left wall is heated and maintained at a constant temperature T_h, while the right wall is cooled and maintained at a constant temperature, T_c. Let T_{av} be the average temperature: $(T_h + T_c)/2$. According to Eqs. [4.2-5] and [4.2-59], the thermal expansion coefficient of the fluid β_T and the thermal Rayleigh number of the system Ra_T are defined as follows:

$$\beta_T = -\frac{1}{\rho}\left[\frac{\partial \rho}{\partial T}\right]_{p,c} \tag{4.2-82}$$

$$\mathrm{Ra}_T = \frac{g\beta_T w^3(T_h - T_{av})}{\nu\alpha} = \frac{g\beta_T w^3(T_h - T_c)}{2\nu\alpha} \tag{4.2-83}$$

where ν and α are the kinematic viscosity and the thermal diffusivity of the fluid, respectively. Notice that the characteristic length L and the reference temperatures T_1 and T_0 in Eq. [4.2-59] have been replaced by w, T_h, and T_{av}, respectively. Find the steady-state velocity distribution in the fluid as a function of Ra_T.

[8] S. W. Churchill and H. H. S. Chu, *Int. J. Heat Mass Transfer*, **18**, 1049, 1975.

[9] S. W. Churchill, *Heat Exchanger Design Handbook*, E. U. S. Schundler, Ed.-in-Chief, Hemisphere, New York, 1983.

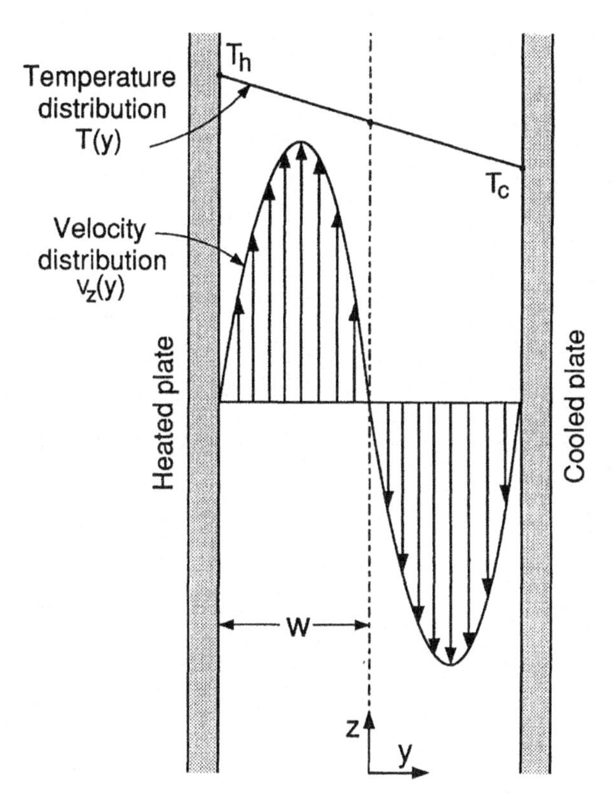

Fig. 4.2-1 Buoyancy convection between two vertical isothermal walls.

Solution The dimensionless equation of energy according to Eq. [4.2-53] is as follows:

$$\frac{\partial T^*}{\partial t^*} + \left[v_x^* \frac{\partial T^*}{\partial x^*} + v_y^* \frac{\partial T^*}{\partial y^*} + v_z^* \frac{\partial T^*}{\partial z^*} \right] = \frac{1}{\mathrm{Pr}} \left[\frac{\partial^2 T^*}{\partial x^{*2}} + \frac{\partial^2 T^*}{\partial y^{*2}} + \frac{\partial^2 T^*}{\partial z^{*2}} \right] \tag{4.2-84}$$

The time-dependent term is zero at the steady state. The convection terms are zero since $v_x = v_y = 0$ and $\partial T/\partial z = 0$. Regarding the conduction terms, there are no temperature variations in the x and z directions. As such, Eq. [4.2-84] reduces to

$$\frac{d^2 T^*}{dy^{*2}} = 0 \tag{4.2-85}$$

where according to Eq. [4.2-55]

$$T^* = \frac{T - T_{\mathrm{av}}}{T_h - T_{\mathrm{av}}} \tag{4.2-86}$$

and according to Eqn. [4.2-39]

$$y^* = \frac{y}{w} \qquad [4.2\text{-}87]$$

The boundary conditions are

$$T^* = 1 \quad \text{at} \quad y^* = -1 \qquad [4.2\text{-}88]$$

$$T^* = -1 \quad \text{at} \quad y^* = 1 \qquad [4.2\text{-}89]$$

The solution, from Case C of Appendix A, is as follows:

$$\boxed{T^* = -y^*} \qquad [4.2\text{-}90]$$

Equation [4.2-90] shows that heat transfer is totally unaffected by convection. This is because the convection in the present case happens to be in the direction perpendicular to heat transfer. This linear temperature distribution is shown, in the dimensional form, in Fig. 4.2-1.

The z component of the dimensionless equation of motion, Eq. [4.2-52], is as follows:

$$\frac{\partial v_z^*}{\partial t^*} + \left[v_x^* \frac{\partial v_z^*}{\partial x^*} + v_y^* \frac{\partial v_z^*}{\partial y^*} + v_z^* \frac{\partial v_z^*}{\partial z^*} \right]$$
$$= -\frac{\partial P^*}{\partial z^*} + \left[\frac{\partial^2 v_z^*}{\partial x^{*2}} + \frac{\partial^2 v_z^*}{\partial y^{*2}} + \frac{\partial^2 v_z^*}{\partial z^{*2}} \right] + \frac{1}{\text{Pr}} \text{Ra}_T T^* \qquad [4.2\text{-}91]$$

Note that since the z direction is taken vertically upward, the z component of $\mathbf{e}_g$ is -1. Regarding Eqn. [4.2-91], the time-dependent term is zero at the steady state. The convection terms are also zero since $v_x = v_y = 0$ and $\partial v_z / \partial z = 0$ from continuity. Regarding the viscous-force terms, there are no velocity variations in the x and z directions. As such, Eq. [4.2-91] reduces to

$$\frac{d^2 v_z^*}{dy^{*2}} = \frac{dP^*}{dz^*} - \frac{1}{\text{Pr}} \text{Ra}_T T^* \qquad [4.2\text{-}92]$$

If the fluid were at a uniform temperature T_{av} (i.e., $T^* = 0$), there would be no convection at all and Eq. [4.2-92] would reduce to

$$\frac{dP^*}{dz^*} = 0 \qquad [4.2\text{-}93]$$

If the buoyancy convection were weak, Eq. [4.2-93] could be a reasonable approximation and Eq. [4.2-92] would reduce to

$$\frac{d^2v_z^*}{dy^{*2}} = \frac{-1}{\text{Pr}}\text{Ra}_T T^* \quad [4.2\text{-}94]$$

Substituting Eq. [4.2-90] into the equation

$$\frac{d^2v_z^*}{dy^{*2}} = \frac{1}{\text{Pr}}\text{Ra}_T y^* \quad [4.2\text{-}95]$$

The boundary conditions are

$$v_z^* = 0 \quad \text{at} \quad y^* = -1 \quad [4.2\text{-}96]$$

$$v_z^* = 0 \quad \text{at} \quad y^* = 1 \quad [4.2\text{-}97]$$

The solution, from Case F of Appendix A, is as follows

$$\boxed{v_z^* = \frac{1}{6}\frac{1}{\text{Pr}}\text{Ra}_T(y^{*3} - y^*)} \quad [4.2\text{-}98]$$

This velocity distribution is shown, in the dimensional form, in Fig. 4.2-1. As shown, the hotter and hence lighter fluid on the left rises, while the cooler and hence heavier fluid on the right sinks.

Example 4.2-2 Buoyancy Convection between Two Horizontal Insulating Walls[10]

A fluid is held in a rectangular container that has two large horizontal walls separated by a small distance $2L$; both walls are insulating. The left vertical wall is kept somewhat hotter than the right one to induce a steady buoyancy convection in the fluid, as shown in Fig. 4.2-2. It has been observed that except near the sidewalls the temperature drop from the top wall to the bottom ΔT is constant. Find the velocity and temperature distributions in the fluid. Neglect end effects.

Solution Except near the sidewalls the fluid velocity v_z ($v_x = v_y = 0$) is expected to vary with y only. The fluid temperature T, however, is expected to vary with y as well as z. This is because hotter fluid is lighter and tends to stay on the top. A schematic sketch of the velocity and temperature distributions in the fluid is shown in Fig. 4.2-2.

[10] A. G. Kirdyashkin, *Int. J. Heat Mass Transfer*, **27**, 1205, 1984.

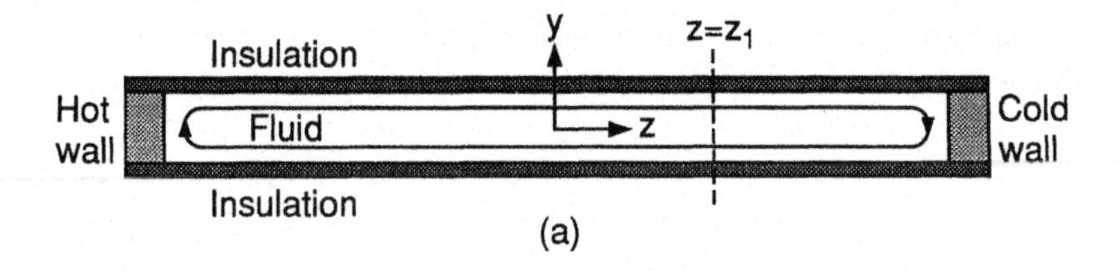

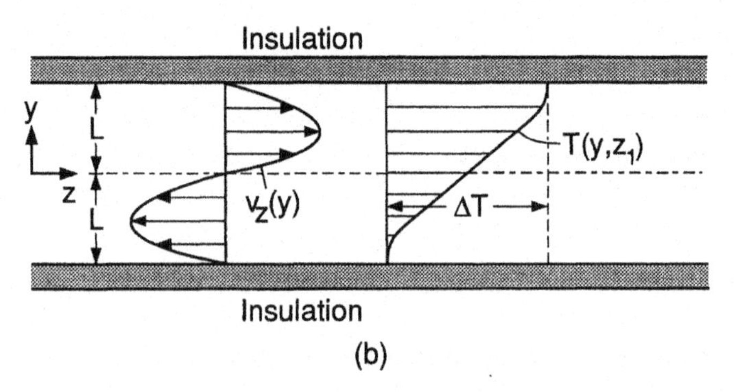

Fig. 4.2-2 A liquid held between two horizontal insulating walls: (*a*) convection pattern; (*b*) velocity and temperature profiles.

From Eqs. [B] and [C] of Table 4.2-1 and Eq. [A] of Table 2.3-1

$$\rho\left[\frac{\partial v_y}{\partial t} + v_x\frac{\partial v_y}{\partial x} + v_y\frac{\partial v_y}{\partial y} + v_z\frac{\partial v_y}{\partial z}\right]$$

$$= -\frac{\partial P}{\partial y} + \mu\left[\frac{\partial^2 v_y}{\partial x^2} + \frac{\partial^2 v_y}{\partial y^2} + \frac{\partial^2 v_y}{\partial z^2}\right] + \rho\beta_T g(T - T_0) \qquad (y \text{ momentum}) \qquad [4.2\text{-}99]$$

$$\rho\left[\frac{\partial v_z}{\partial t} + v_x\frac{\partial v_z}{\partial x} + v_y\frac{\partial v_z}{\partial y} + v_z\frac{\partial v_z}{\partial z}\right] = -\frac{\partial P}{\partial z} + \mu\left[\frac{\partial^2 v_z}{\partial x^2} + \frac{\partial^2 v_z}{\partial y^2} + \frac{\partial^2 v_z}{\partial z^2}\right] \qquad (z \text{ momentum}) \qquad [4.2\text{-}100]$$

$$\rho C_v\left[\frac{\partial T}{\partial t} + v_x\frac{\partial T}{\partial x} + v_y\frac{\partial T}{\partial y} + v_z\frac{\partial T}{\partial z}\right] = k\left[\frac{\partial^2 T}{\partial x^2} + \frac{\partial^2 T}{\partial y^2} + \frac{\partial^2 T}{\partial z^2}\right] \qquad (\text{energy}) \qquad [4.2\text{-}101]$$

Note that $g_y = -g$, $g_z = 0$, and $s = 0$. At the steady state these equations reduce to

$$\frac{1}{\rho}\frac{\partial P}{\partial y} = \beta_T g(T - T_0) \qquad (y\text{-momentum}) \qquad [4.2\text{-}102]$$

$$\frac{1}{\rho}\frac{\partial P}{\partial z} = \nu\frac{\partial^2 v_z}{\partial y^2} \qquad (z\text{-momentum}) \qquad [4.2\text{-}103]$$

$$v_z\frac{\partial T}{\partial z} = \alpha\left(\frac{\partial^2 T}{\partial y^2} + \frac{\partial^2 T}{\partial z^2}\right) \qquad (\text{energy}) \qquad [4.2\text{-}104]$$

Differentiating Eq. [4.2-102] with respect to z and Eq. [4.2-103] with respect to y, and substracting to get rid of P

$$\nu\frac{\partial^3 v_z}{\partial y^3} = \beta_T g\frac{\partial T}{\partial z} \qquad (\text{motion}) \qquad [4.2\text{-}105]$$

The LHS of the equation is a function of y or constant while the RHS is a function of z or constant. Therefore, both should be equal to a constant:

$$\frac{\partial T}{\partial z} = A \text{ (constant)} \qquad [4.2\text{-}106]$$

and

$$\frac{\partial^3 v_z}{\partial y^3} = \frac{\beta_T g}{\nu} A \text{ (constant)} \qquad [4.2\text{-}107]$$

Substituting Eq. [4.2-106] into Eq. [4.2-104]

$$\frac{\partial^2 T}{\partial y^2} = \frac{A}{\alpha} v_z \qquad [4.2\text{-}108]$$

Integrating Eq. [4.2-106]

$$T = T(y) + Az \qquad [4.2\text{-}109]$$

where $T(y)$ is an integration constant depending on y alone. At $y = L$ and $-L$

$$T_L = T(L) + Az \equiv T_1 + Az \qquad [4.2\text{-}110]$$

and

$$T_{-L} = T(-L) + Az \equiv T_2 + Az \qquad [4.2\text{-}111]$$

Therefore, the temperature drop from the top wall to the bottom is constant:

$$\Delta T = T_L - T_{-L} = T_1 - T_2 \quad \text{(constant)} \qquad [4.2\text{-}112]$$

Let us define the following dimensionless quantities:

$$\mathrm{Ra}_T = \frac{\beta_T g \, \Delta T L^3}{\alpha \nu} \qquad [4.2\text{-}113]$$

$$v_z^* = \frac{v_z}{\mathrm{Ra}_T \, \alpha / L} = \frac{\nu v_z}{\beta_T g \, \Delta T L^2} \qquad [4.2\text{-}114]$$

$$T^* = \frac{T(y) - T_2}{\Delta T} \equiv \frac{T - Az - T_2}{\Delta T} \qquad [4.2\text{-}115]$$

$$A^* = \frac{AL}{\Delta T} \qquad [4.2\text{-}116]$$

$$y^* = \frac{y}{L} \qquad [4.2\text{-}117]$$

$$z^* = \frac{z}{L} \qquad [4.2\text{-}118]$$

Substituting these dimensionless quantities into Eqs. [4.2-107] and [4.2-108]

$$\frac{\partial^3 v_z^*}{\partial y^{*3}} = A^* \qquad [4.2\text{-}119]$$

and

$$\frac{\partial^2 T^*}{\partial y^{*2}} = A^* \, \mathrm{Ra}_T v_z^* \qquad [4.2\text{-}120]$$

Integrating Eq. [4.2-119]

$$(A^*)^{-1} v_z^* = \frac{y^{*3}}{3!} + c_1 \frac{y^{*2}}{2!} + c_2 y^* + c_3 \qquad [4.2\text{-}121]$$

Substituting Eq. [4.2-121] into Eq. [4.2-120] and integrating

$$(A^{*2}\mathrm{Ra}_T)^{-1}T^* = \frac{y^{*5}}{5!} + c_1\frac{y^{*4}}{4!} + c_2\frac{y^{*3}}{3!} + c_3\frac{y^{*2}}{2!} + c_4 y^* + c_5 \quad [4.2\text{-}122]$$

The velocity boundary conditions are as follows:

$$v_z^* = 0 \quad \text{at} \quad y^* = 1 \quad [4.2\text{-}123]$$

$$v_z^* = 0 \quad \text{at} \quad y^* = -1 \quad [4.2\text{-}124]$$

and

$$\int_{-1}^{1} v_z^*\, dy^* = 0 \quad [4.2\text{-}125]$$

The first two equations are the no-slip conditions at the top and bottom walls. The last equation states that there is no net flow of fluid across any vertical cross section. It is a more flexible boundary condition than $v_z^* = 0$ at $y^* = 0$ since symmetry is not required.

The thermal boundary conditions are as follows:

$$T^* = 1 \quad \text{and} \quad \frac{\partial T^*}{\partial y^*} = 0 \quad \text{at} \quad y^* = 1 \quad [4.2\text{-}126]$$

and

$$T^* = 0 \quad \text{and} \quad \frac{\partial T^*}{\partial y^*} = 0 \quad \text{at} \quad y^* = -1 \quad [4.2\text{-}127]$$

The last two equations are based on Eqs. [4.2-110] and [4.2-111], and the insulating top and bottom walls.

Substituting these boundary conditions into Eq. [4.2-121] and [4.2-122]

$$A^* = -\left(\frac{45}{2\mathrm{Ra}_T}\right)^{1/2} \quad [4.2\text{-}128]$$

$$\boxed{v_z^* = \left(\frac{5}{8\mathrm{Ra}_T}\right)^{1/2}(y^* - y^{*3})} \quad [4.2\text{-}129]$$

$$\boxed{T^* = \tfrac{1}{16}(3y^{*5} - 10y^{*3} + 15y^* + 8)} \quad [4.2\text{-}130]$$

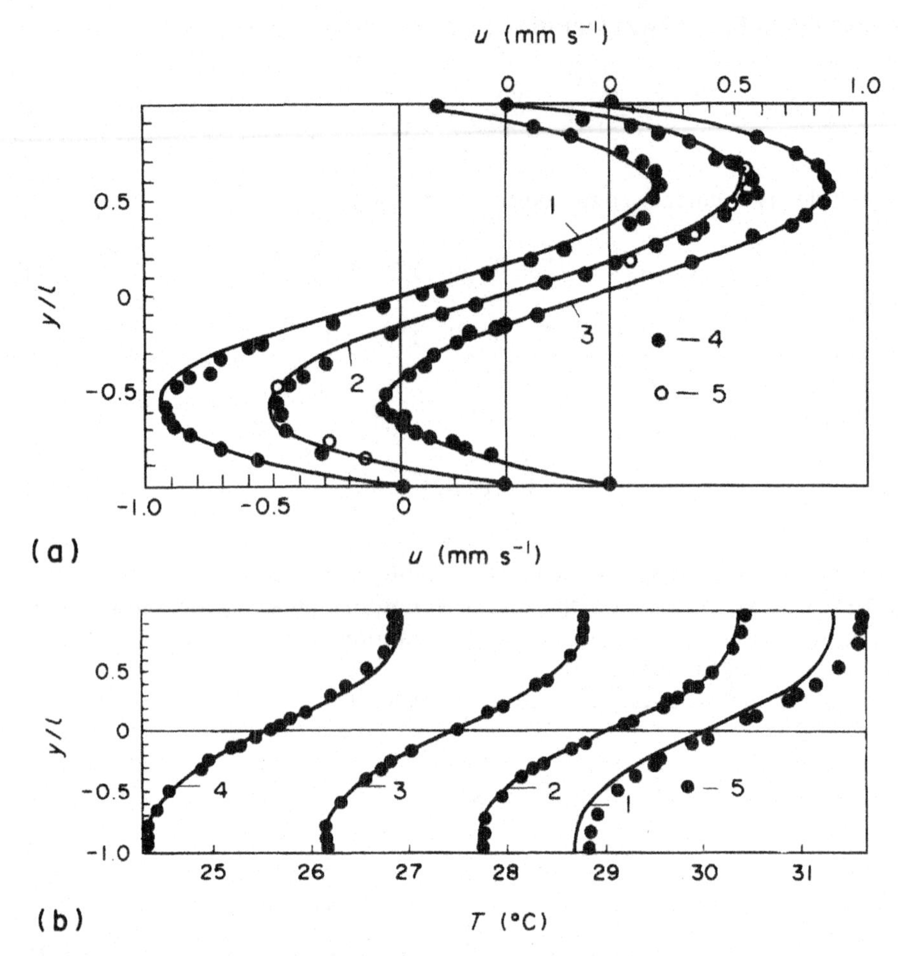

Fig. 4.2-3 Buoyancy convection in water between two horizontal insulating walls: (*a*) velocity profiles; (*b*) temperature profiles. (Reprinted from A.G. Kirdyashkin, "Thermogravitational and Thermocapillary Flows in a Horizontal Liquid Layer under the Conditions of Horizontal Temperature Gradient," *International Journal of Heat and Mass Transfer*, **27**, 1984, p. 1205, with kind permission from Elsevier Science Ltd., The Boulevard, Langford Lane, Kidlington OX 1GB, UK.)

Figure 4.2-3 shows the measured and calculated velocities (u is v_z) and the water temperatures at various cross sections.[11] The conditions are as follows: thickness $(2L) = 14.45$ mm, width $= 810$ mm, $\mathrm{Ra}_T = 2.2 \times 10^4$, hot wall at 26.5°C, and cold wall at 18.5°C. As shown, the water in the upper half flows

[11] A. G. Kirdyashkin, ibid.

toward the cold wall, while that in the lower half flows toward the hot wall. Also, the water is hotter near the top and the hot wall.

It should be mentioned that when Rayleigh number is increased to beyond a certain critical value, flow oscillation occurs and the fluid fluctuates in both velocity and temperature. For example, periodic temperature fluctuations have been observed in mercury enclosed in a shallow rectangular box.[12] The critical Ra_T for flow oscillation depends on the properties and dimensions of the fluid.

PROBLEMS[13]

4.1-1 Verify the similarities (identical dimensionless solutions) among the three types of transport shown in Fig. P4.1-1.

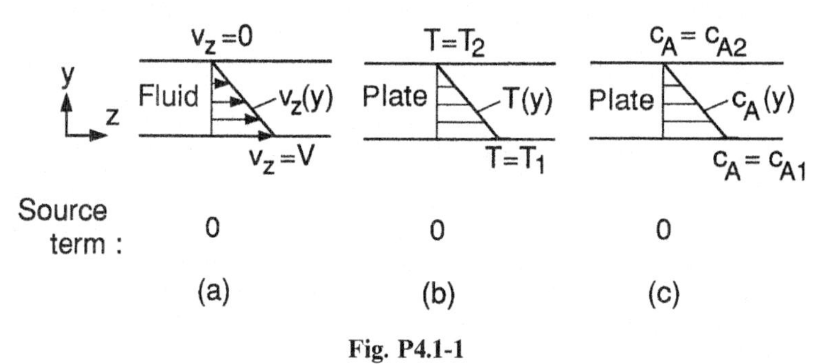

Fig. P4.1-1

4.1-2 Verify the similarities among the three types of transport shown in Fig. P4.1-2.

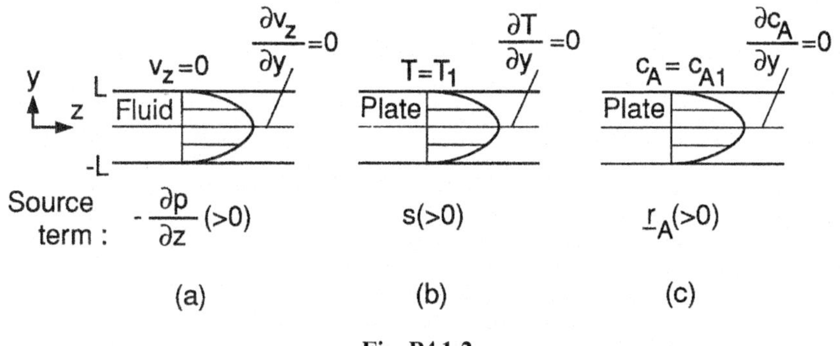

Fig. P4.1-2

4.1-3 Verify the similarities among the three types of transport shown in Fig. P4.1-3 (see p. 312).

[12] J. M. Pratte and J. E. Hart, *J. Crystal Growth*, **102**, 54, 1990.

[13] The first two digits in problem numbers refer to section numbers in text.

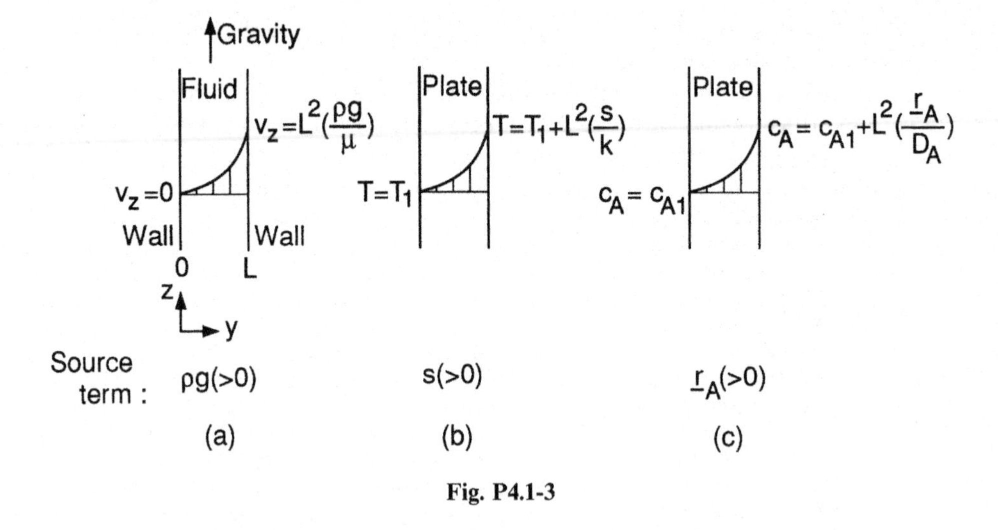

Fig. P4.1-3

4.1-4 Verify the similarities among the three types of transport shown in Fig. P4.1-4.

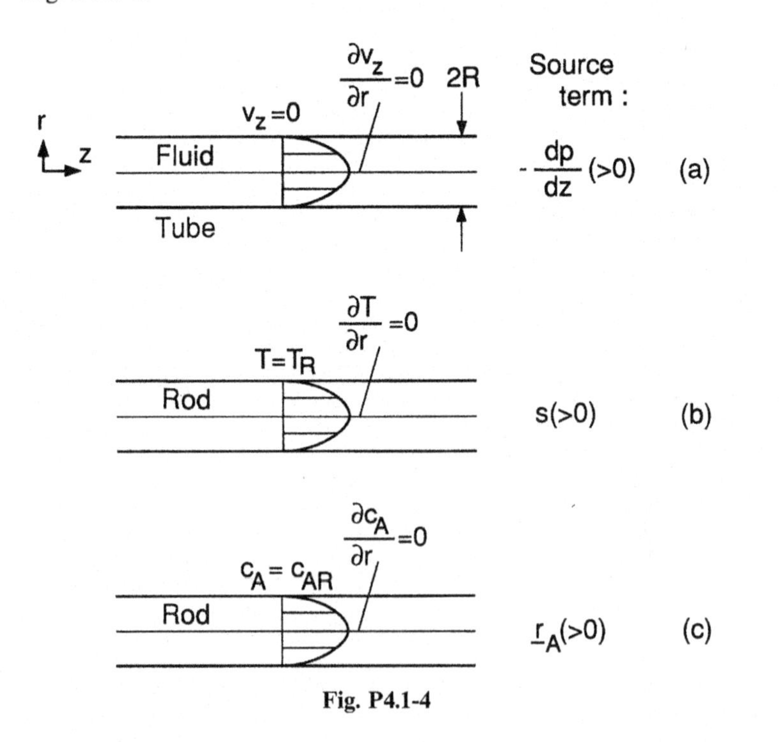

Fig. P4.1-4

4.1-5 Verify the similarities among the three types of transport shown in Fig. P4.1-5 (see p. 313).

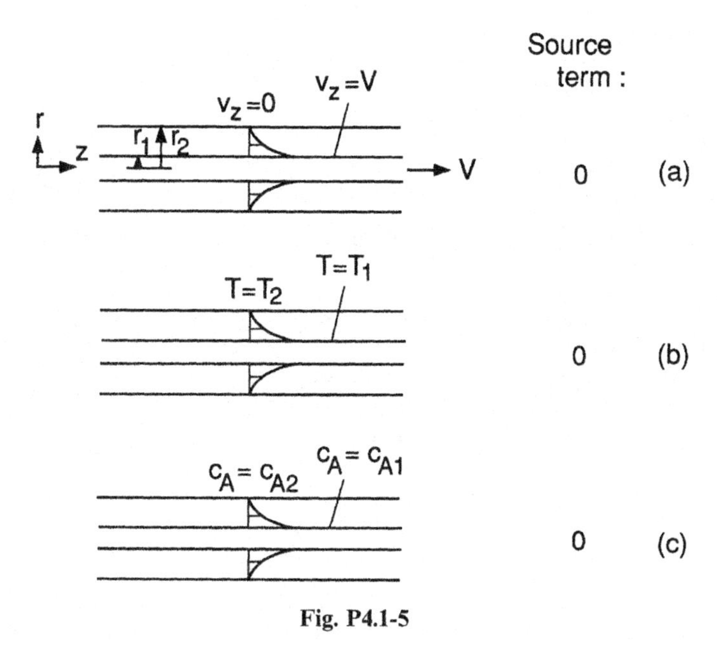

Fig. P4.1-5

4.1-6 Verify the similarities among the three types of transport shown in Fig. P4.1-6.

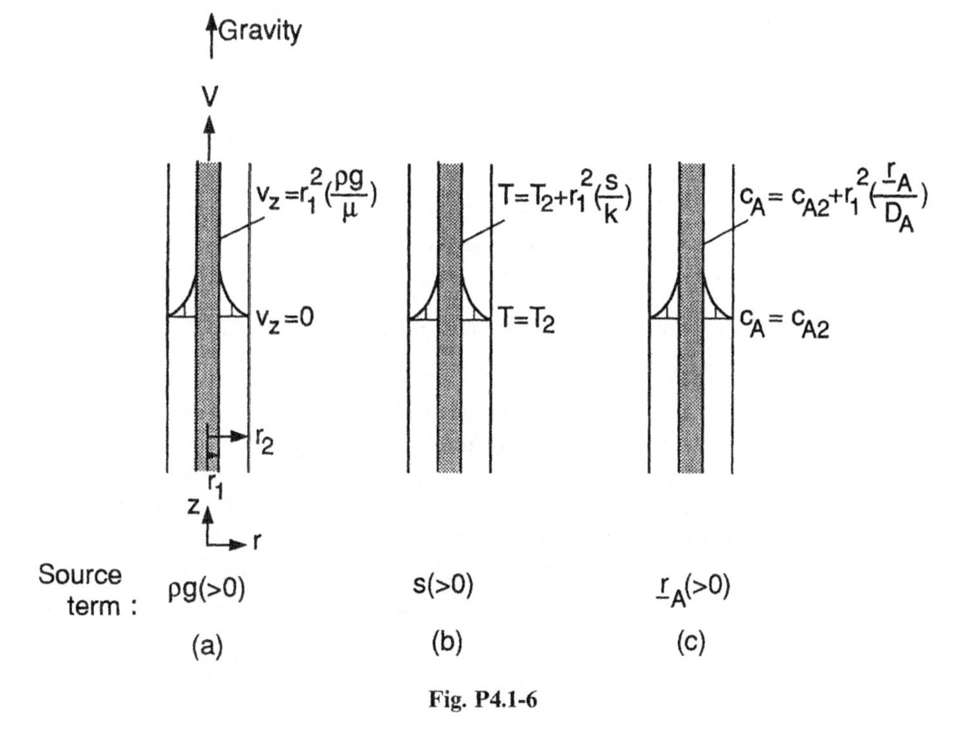

Fig. P4.1-6

4.1-7 Verify the similarities among the three types of transport shown in Fig. P4.1-7.

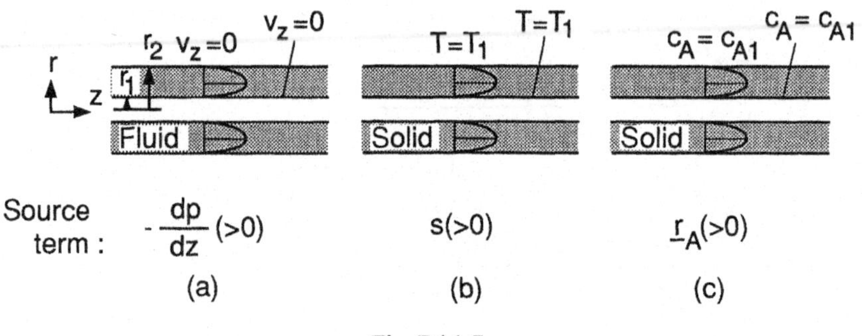

Fig. P4.1-7

4.1-8 Verify the similarities among the three types of transient transport in semiinfinite materials shown in Fig. P4.1-8.

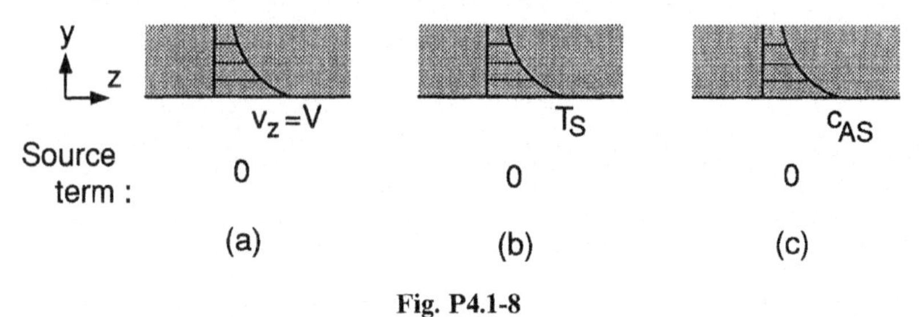

Fig. P4.1-8

4.1-9 For turbulent flow in a pipe of diameter D, the following equation has been proposed by Dittus and Boelter

$$\mathrm{Nu}_D = \frac{hD}{k} = 0.023\ \mathrm{Re}_D^{0.8}\,\mathrm{Pr}^{1/3}$$

Show the corresponding equation for the mass transfer coefficient k_m and compare it against Eq. [3.6-19].

4.2-1 A solution containing a light solute A is held in a rectangular container that has two vertical walls separated by a small distance $2w$, as shown in Fig. P4.2-1 (see p. 315). The coating on the left wall has a higher concentration of solute A than that on the right wall. At the steady state the concentrations of the solution are c_{A1} at the left wall and c_{A2} at the right one ($c_{A1} > c_{A2}$). Find the steady-state velocity distribution in the fluid as a function of the solutal Rayleigh number Ra_S where

$$\mathrm{Ra}_S = \frac{g\beta_S w^3(c_{A1} - c_{A,\mathrm{av}})}{\nu D_A} = \frac{g\beta_S w^3(c_{A1} - c_{A2})}{2\nu D_A}$$

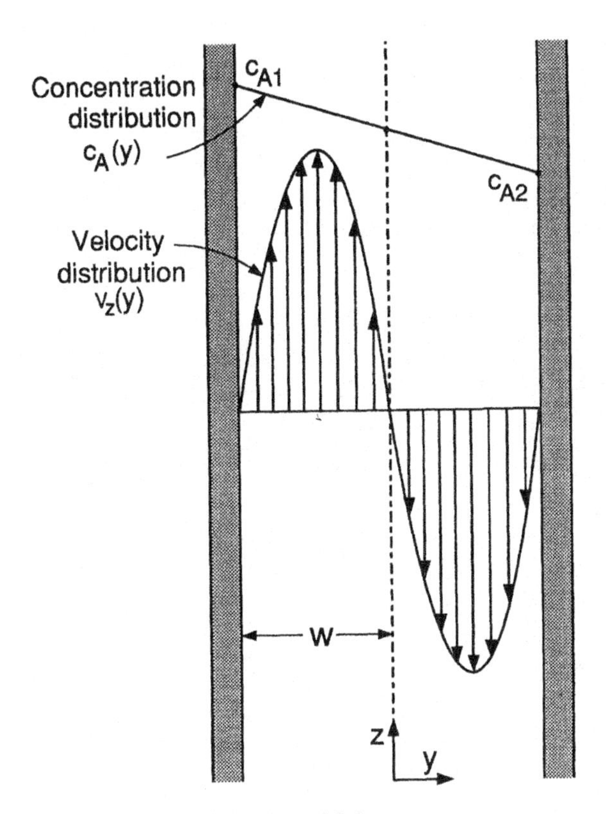

Fig. P4.2-1

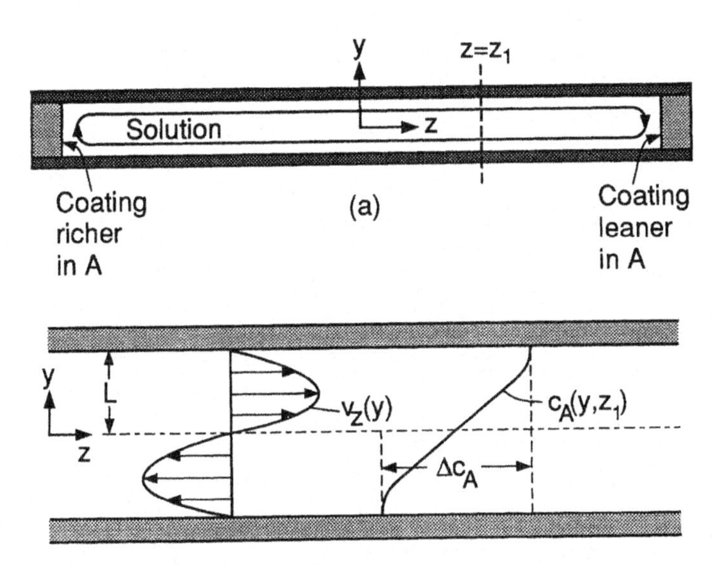

Fig. P4.2-2

4.2-2 A solution containing a light solute A is held in a rectangular container that has two large horizontal walls separated by a small distance $2L$. The two sidewalls are coated. The coating on the left wall has a higher concentration of solute A than that on the right one, as shown in Fig. P4.2-2 (see p. 315). Except near the sidewalls the concentration drop of the solution from the top wall to the bottom Δc_A is constant. Find the velocity and concentration distributions in the solution. Neglect end effects.

CHAPTER 5

BOUNDARY CONDITIONS AT INTERFACES

Materials processing often involves transformation or conversion of materials from one phase to another. The shape and velocity of the interface between two different phases during processing can have a significant effect on the quality of the resultant materials. During materials processing, interfaces and transport phenomena interact with each other, that is, become coupled, through the boundary conditions at the interfaces.

In the present chapter, boundary conditions at the interfaces during materials processing are described.

5.1. SOLID/LIQUID INTERFACE

5.1.1. Fluid Flow

Consider the transformation of a melt (L) into a solid (S) in a mold or crucible, as illustrated in Fig. 5.1-1*a*. The coordinates (r, z) or (x, y, z) are stationary with respect to the mold. The unit vector $\mathbf{n}$ is normal to the S/L interface pointing into the melt. As heat is being dissipated, the S/L interface advances normal to itself into the melt at a local growth rate $\mathbf{R}$. Note that the melt at the interface is not pushed at the velocity $\mathbf{R}$. Rather, it merely transforms into solid at the rate of $\mathbf{R}$.

Most metals (and oxides) shrink on solidification, that is, the density of the solid ρ_S is greater than that of the liquid ρ_L. As such, the melt at the S/L interface is slightly sucked toward the interface to feed the shrinkage while it transforms into solid.

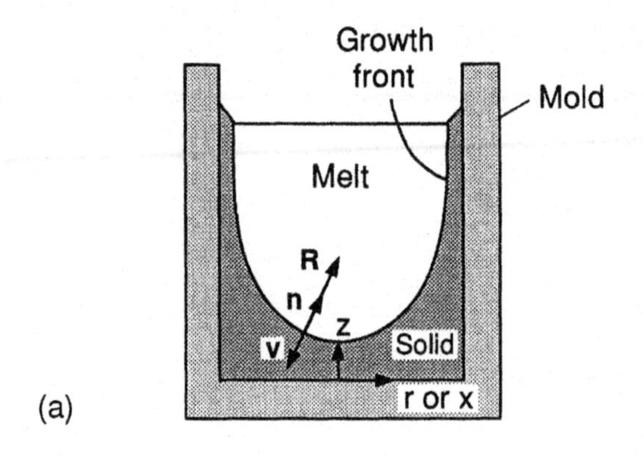

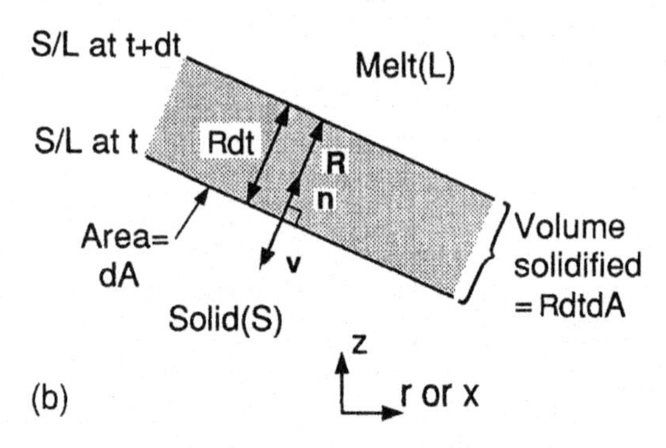

Fig. 5.1-1 Solid/liquid (S/L) interface during casting in a stationary mold: (*a*) overall view; (*b*) enlarged view.

Figure 5.1-1*b* shows a differential surface element dA of the S/L interface, advancing into the melt from time t to time $t + dt$ at the local growth rate of R. As shown, the volume of the solid formed in dt is $R\,dt\,dA$. Assuming $\rho_S > \rho_L$ for the purpose of discussion, the amount of the melt needed to feed the shrinkage is

$$(\rho_S - \rho_L)\mathbf{R}\,dt\,dA = \rho_L(-\mathbf{v})\,dt\,dA \qquad [5.1\text{-}1]$$

where $\mathbf{v}$ is the melt velocity at the S/L interface. Since $\mathbf{v}$ is in the opposite direction of $\mathbf{R}$, a minus sign is needed. Dividing Eq. [5.1-1] by $(-\rho_L\,dt\,dA)$.

$$\boxed{\mathbf{v} = -\left(\frac{\rho_S - \rho_L}{\rho_L}\right)\mathbf{R}} \qquad [5.1\text{-}2]$$

If the velocity due to the density difference is neglected as an approximation

$$\mathbf{v} = 0 \qquad [5.1\text{-}3]$$

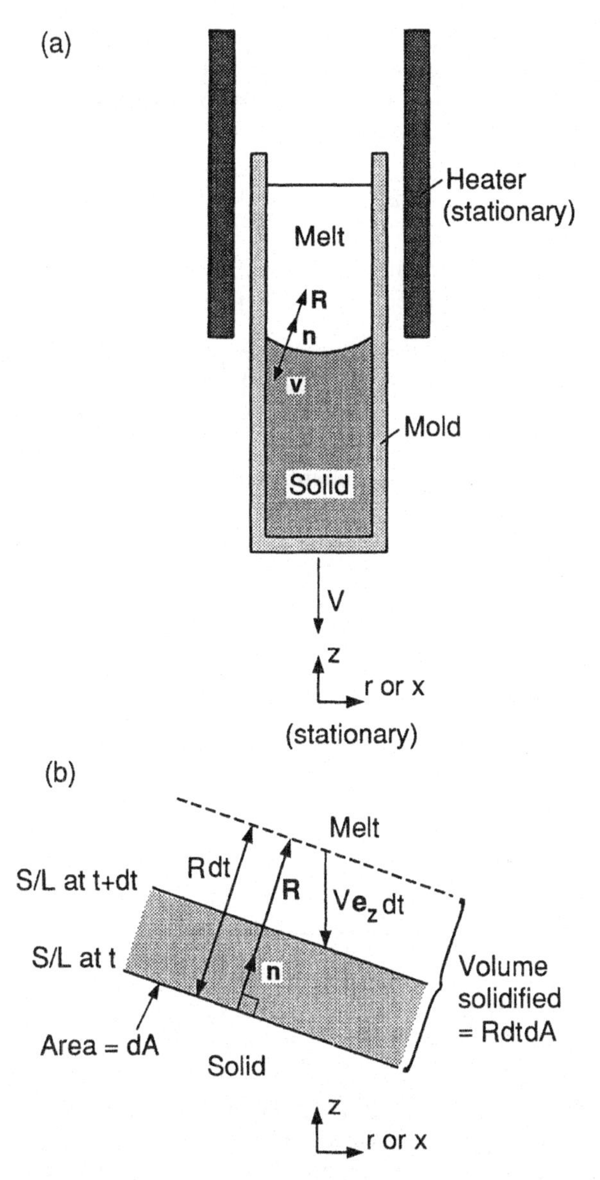

Fig. 5.1-2 Solid/liquid (S/L) interface during casting in a moving mold: (*a*) overall view; (*b*) enlarged view.

Let us now consider the case where the solid is withdrawn at a constant speed V in the negative z direction of stationary coordinates (r, z) or (x, y, z), as illustrated in Fig. 5.1-2*a*. Since it is in the negative z direction, $V < 0$. The unit vector $\mathbf{n}$ is again normal to the S/L interface pointing into the melt. The heater is stationary with respect to the coordinates.

As shown in Fig. 5.1-2*b*, the volume of the solid formed in dt is $R\,dt\,dA$. The advancement of the S/L interface is not $R\,dt$ but shorter because of withdrawal of the solid. Assuming $\rho_S > \rho_L$ for the purpose of discussion, the amount of the melt needed to feed the shrinkage is

$$(\rho_S - \rho_L)\mathbf{R}\,dt\,dA = \rho_L[-(\mathbf{v} - V\mathbf{e}_z)]\,dt\,dA \qquad [5.1\text{-}4]$$

Notice that $V\mathbf{e}_z$, which is due to the withdrawal of the solid rather than shrinkage feeding, is subtracted from $\mathbf{v}$. Dividing by $(-\rho_L\,dt\,dA)$

$$\boxed{\mathbf{v} = -\left(\frac{\rho_S - \rho_L}{\rho_L}\right)\mathbf{R} + V\mathbf{e}_z} \qquad [5.1\text{-}5]$$

If the velocity due to the density difference is neglected as an approximation

$$\mathbf{v} = V\mathbf{e}_z \qquad [5.1\text{-}6]$$

which is just the no-slip condition at the interface.

5.1.2. Heat Transfer

The boundary conditions for the S/L interface shown in Fig. 5.1-1 are as follows:

$$T = T_m \qquad [5.1\text{-}7]$$

and

$$\boxed{\left(k\frac{\partial T}{\partial n}\right)_L + \rho_S H_f\,\mathbf{n}\cdot\mathbf{R} = \left(k\frac{\partial T}{\partial n}\right)_S} \qquad [5.1\text{-}8]$$

(1) (2) (3)

where T_m is the melting point and H_f is the heat of fusion ($H_f > 0$). The second condition, is called the **Stefan condition** (see Fig. 5.1-3), states that the heat conducted from the melt to the interface (1) and the heat of fusion released at the interface (2) must be conducted away from the interface into the solid (3) in order to satisfy the conservation of energy at the interface. Term (1) is proportional to the slope of the tangent to the melt temperature profile at the interface. Likewise, term (3) is proportional to the slope of the tangent to the solid temperature profile at the interface. The temperature gradient $\partial T/\partial n$ at the interface is positive in both phases since the temperature increases in the order of the solid, interface, and melt. Also, $\mathbf{n}\cdot\mathbf{R} > 0$ since $\mathbf{n}$ and $\mathbf{R}$ are in the same direction. As such, all three terms in Eq. [5.1-8] are positive in value.

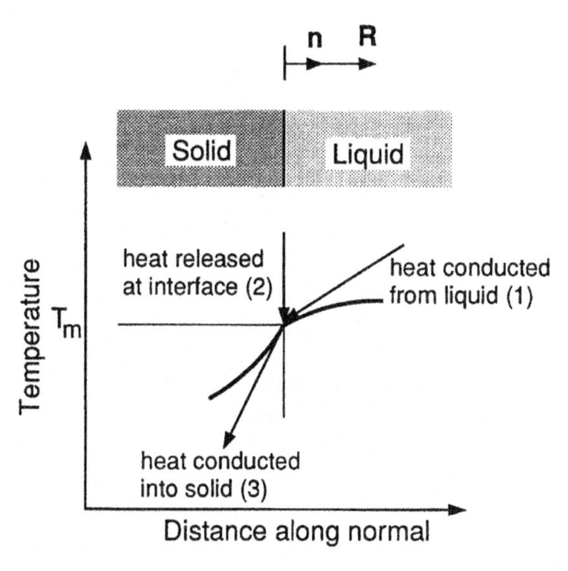

Fig. 5.1-3 Temperature profile near a solid/liquid interface during solidification.

Since $\partial T/\partial n = \mathbf{n}\cdot\nabla T$, from Eq. [5.1-8]

$$(k\mathbf{n}\cdot\nabla T)_L + \rho_S H_f \mathbf{n}\cdot\mathbf{R} = (k\mathbf{n}\cdot\nabla T)_S \qquad [5.1\text{-}9]$$

The same boundary conditions apply to the S/L interface shown in Fig. 5.1-2.

5.1.3. Mass Transfer

Figure 5.1-4*a* shows a portion of a binary phase diagram, with the concentration in the mass fraction of the solute (or dopant) *A*. At the S/L interface the temperature is T^*, the solute concentration of the melt w_{AL}^*, and the solute concentration of the solid w_{AS}^*. The **equilibrium segregation coefficient** k_0 is defined as the ratio of w_{AS}^*/w_{AL}^*. As such, the solute concentration of the solid at the interface $w_{AS}^* = k_0 w_{AL}^*$. For $k_0 < 1$, as in the present case, $k_0 w_{AL}^* < w_{AL}^*$ and the solute is rejected by the solid into the melt at a rate that is proportional to $(w_{AL}^* - k_0 w_{AL}^*)$. This results in a solute-rich layer in the melt at the interface, as illustrated in Fig. 5.1-5.

The boundary condition for the S/L interface shown in in Fig. 5.1-1 is the following conservation of the solute:

$$\boxed{\rho_S(w_{AL}^* - k_0 w_{AL}^*)\mathbf{n}\cdot\mathbf{R} = \left(-\rho D_A \frac{\partial w_A}{\partial n}\right)_L + \left(\rho D_A \frac{\partial w_A}{\partial n}\right)_S} \qquad [5.1\text{-}10]$$

(1) (2) (3)

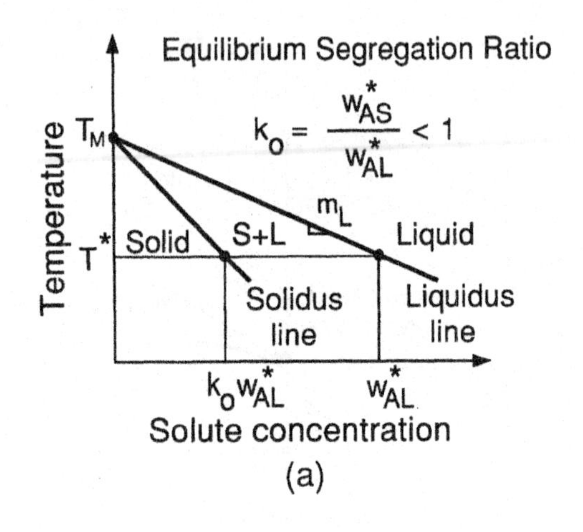

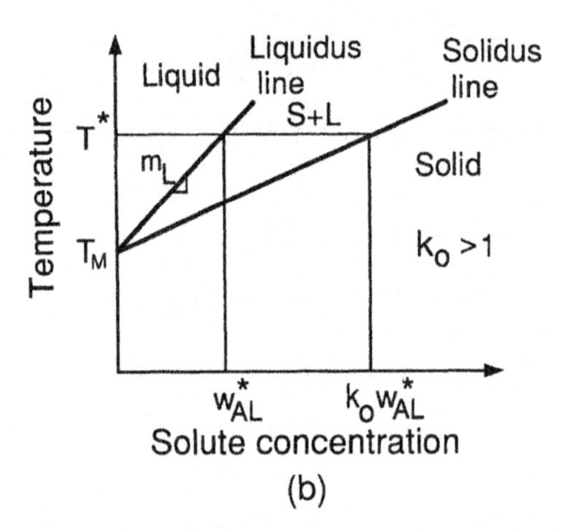

Fig. 5.1-4 Phase diagram and equilibrium segregation ratio k_0: (*a*) $k_0 < 1$; (*b*) $k_0 > 1$.

where D_A is the diffusion coefficient of the solute. As illustrated in Fig. 5.1-5, this equation states that the solute released at the interface (1) must diffuse away into the liquid (2) and the solid (3). Term (2) is proportional to the slope of the tangent to the melt composition profile at the interface. Likewise, term (3) is proportional to the slope of the tangent to the solid composition profile at the interface. A minus sign is required in term (2) since $(\partial w_A/\partial n)_L$ is negative. As such, all three terms in Eq. [5.1-10] are positive in value. It has been assumed in Eq. [5.1-10] that the density is uniform and is the same in the solid and the melt.

Since the diffusion coefficient of a solid is usually about three orders of magnitude smaller than that of its melt, solid-state diffusion, specifically, term (3) in Eq. [5.1-10], can often be neglected.

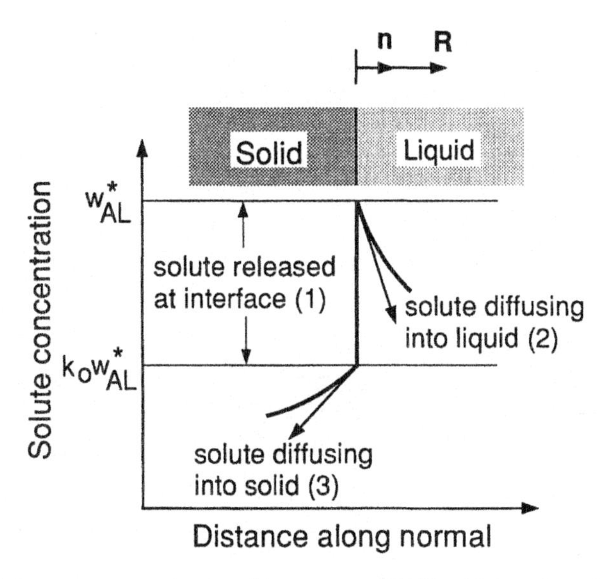

Fig. 5.1-5 Solute concentration profile near a solid/liquid interface during solidification.

Equation [5.1-10] can be written in the form

$$(\rho D_A \mathbf{n} \cdot \nabla w_A)_L + \rho_S (w^*_{AL} - k_0 w^*_{AL}) \mathbf{n} \cdot \mathbf{R} = (\rho D_A \mathbf{n} \cdot \nabla w_A)_S \qquad [5.1\text{-}11]$$

which is analogous to the thermal boundary condition at the S/L interface given by Eq. [5.1-9]. The concentration difference between the melt and the solid $(w_{AL} - k_0 w_{AL})$ in Eq. [5.1-11] is analogous to the energy difference between the melt and the solid $\rho_S H_f$ in Eq. [5.1-9].

Boundary condition similar to Eq. [5.1-11] can be written for the case where $k_0 > 1$, which is the case shown in Fig. 5.1-4*b*. In this case, the solute is absorbed into the solid at the interface.

Since the material is no longer pure, Eq. [5.1-7] should be modified to

$$T = T_L \qquad [5.1\text{-}12]$$

where T_L, as shown in Fig. 5.1-4, is the liquidus temperature corresponding to the composition w_A. If the liquidus line of the phase diagram is assumed to be a straight line, Eq. [5.1-12] becomes

$$T = T_m + m_L w_{AL} \qquad [5.1\text{-}13]$$

where m_L is the slope of the liquidus line.

The same boundary conditions apply to the S/L interface shown in Fig. 5.1-2.

5.1.4. Examples

Example 5.1-1 Zone Melting in a Vertical Tube

As shown in Fig. 5.1-6*a*, a solid feed rod contained in a vertical tube passes through a stationary ring heater at a constant rate in the negative z direction. Assume that the feed rod is sufficiently long that a steady-state condition is reached, that is, the molten zone remains constant in shape and position.

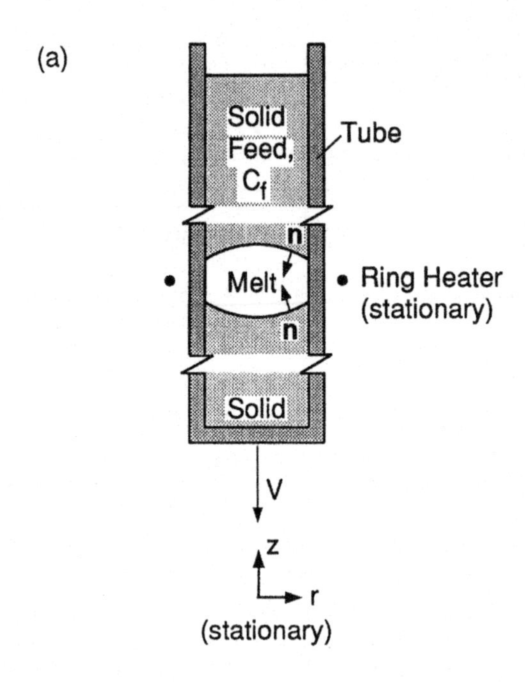

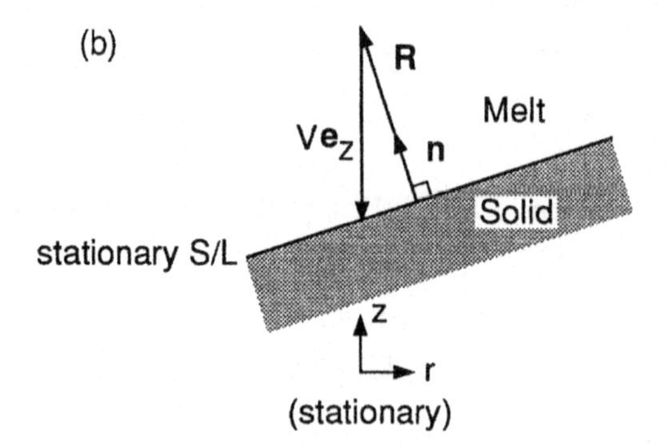

Fig. 5.1-6 Zone melting in a vertical tube.

Neglect the density difference between the solid and the melt. Give the boundary conditions for the molten zone.

Solution The fluid flow boundary conditions are as follows:

1. At the two S/L interfaces, from Eq. [5.1-6]

$$v_z = V \qquad [5.1\text{-}14]$$

$$v_r = 0 \qquad [5.1\text{-}15]$$

Note that $V < 0$ since it is in negative z direction.

2. At the axis ($r = 0$), assuming axisymmetry

$$\frac{\partial v_z}{\partial r} = 0 \qquad [5.1\text{-}16]$$

$$v_r = 0 \qquad [5.1\text{-}17]$$

3. At the tube inner wall, from the no-slip condition

$$v_z = V \qquad [5.1\text{-}18]$$

$$v_r = 0 \qquad [5.1\text{-}19]$$

The heat transfer boundary conditions are as follows:

1. At the lower S/L interface, that is, the growth front, from Eq. [5.1-8]

$$\underset{(1)}{\left(k\frac{\partial T}{\partial n}\right)_L} + \underset{(2)}{\rho_S H_f \mathbf{n}\cdot\mathbf{R}} = \underset{(3)}{\left(k\frac{\partial T}{\partial n}\right)_S} \qquad [5.1\text{-}20]$$

In other words, the heat conducted from the melt to the interface (1) and the heat released at the interface (2) are conducted from the interface to the solid (3). At this interface $\partial T/\partial n > 0$ and $\mathbf{n}\cdot\mathbf{R} > 0$, that is, all three terms are positive in value. As shown in Fig. 5.1-6*b*, the lower S/L interface is stationary at the steady state and so

$$\mathbf{n}\cdot\mathbf{R} = -V\mathbf{n}\cdot\mathbf{e}_z \qquad \text{(steady state)} \qquad [5.1\text{-}21]$$

Substituting into Eq. [5.1-20]

$$\left(k\frac{\partial T}{\partial n}\right)_L + \rho_S H_f(-V\mathbf{n}\cdot\mathbf{e}_z) = \left(k\frac{\partial T}{\partial n}\right)_S \qquad \text{(steady state)} \qquad [5.1\text{-}22]$$

2. At the upper S/L interface, or, the melting front,

$$\boxed{\left(k\frac{\partial T}{\partial n}\right)_L = \rho_S H_f(-\mathbf{n}\cdot\mathbf{R}') + \left(k\frac{\partial T}{\partial n}\right)_S} \qquad [5.1\text{-}23]$$

(1) (2) (3)

where the normal $\mathbf{n}$ points into the melt and the local melting rate $\mathbf{R}'$ points in the opposite direction of $\mathbf{n}$. In other words, the heat conducted from the melt to the interface (1) is absorbed by fusion at the interface (2) and conducted from the interface into the solid (3). At this interface $\partial T/\partial n > 0$ and $\mathbf{n}\cdot\mathbf{R}' < 0$, that is, all three terms are positive in value. Since the upper S/L interface is stationary at the steady state

$$\mathbf{n}\cdot\mathbf{R}' = -V\mathbf{n}\cdot\mathbf{e}_z \qquad \text{(steady state)} \qquad [5.1\text{-}24]$$

Substituting into Eq. [5.1-23]

$$\left(k\frac{\partial T}{\partial n}\right)_L = \rho_S H_f V\mathbf{n}\cdot\mathbf{e}_z + \left(k\frac{\partial T}{\partial n}\right)_S \qquad \text{(steady state)} \qquad [5.1\text{-}25]$$

which is identical to Eq. [5.1-22].

At both S/L interfaces, from Eq. [5.1-13],

$$T = T_m + m_L w_{AL} \qquad [5.1\text{-}26]$$

3. At the axis, assuming axisymmetry

$$\frac{\partial T}{\partial r} = 0 \qquad [5.1\text{-}27]$$

4. At the tube inner wall

$$\left(-k\frac{\partial T}{\partial r}\right)_L = \left(-k\frac{\partial T}{\partial r}\right)_{\text{wall}} \quad \text{and} \quad T_L = T_{\text{wall}} \qquad [5.1\text{-}28]$$

The mass transfer boundary conditions, assuming negligible solid-state diffusion, are as follows:

1. At the lower S/L interface (the growth front), from Eq. [5.1-10] and neglecting solid-state diffusion

$$(w_{AL}^* - k_0 w_{AL}^*)\mathbf{n}\cdot\mathbf{R} = \left(-D_A \frac{\partial w_A}{\partial n}\right)_L \qquad [5.1\text{-}29]$$

(1) (2)

In other words, the solute released at the interface (1), assuming $k_0 < 1$, must diffuse away into the liquid (2). At this interface $(\partial w_{AL}/\partial n)_L < 0$ and both terms are positive in value. Substituting Eq. [5.1-21]

$$(w_{AL}^* - k_0 w_{AL}^*)(-V\mathbf{n}\cdot\mathbf{e}_z) = \left(-D_{AL}\frac{\partial w_{AL}}{\partial n}\right)_L \quad \text{(steady state)} \qquad [5.1\text{-}30]$$

2. At the upper S/L interface (the feed front), neglecting solid-state diffusion,

$$\boxed{(w_{AL} - w_{Af})(-\mathbf{n}\cdot\mathbf{R}') = \left(D_A\frac{\partial w_A}{\partial n}\right)_L} \qquad [5.1\text{-}31]$$

(1) (2)

In other words, the solute absorbed at the interface (1) equals the solute diffusing from the melt to the interface (2). At the interface $\partial w_A/\partial n > 0$ and both terms are positive in value. Substituting Eq. [5.1-24]

$$(w_{AL} - w_{Af})V\mathbf{n}\cdot\mathbf{e}_z = \left(D_A\frac{\partial w_A}{\partial n}\right)_L \quad \text{(steady state)} \qquad [5.1\text{-}32]$$

3. At the axis, assuming axisymmetry,

$$\frac{\partial w_{AL}}{\partial r} = 0 \qquad [5.1\text{-}33]$$

4. At the tube inner wall, assuming no diffusion through the wall

$$\frac{\partial w_{AL}}{\partial r} = 0 \qquad [5.1\text{-}34]$$

5.2. LIQUID/GAS INTERFACE

5.2.1. Fluid Flow

In this section we consider fluid flow boundary conditions at the interface between liquid (L) and gas (G). These boundary conditions have a significant effect on the fluid flow near the interface and on the shape of the interface itself.

TABLE 5.2-1 Surface Tensions of Some Molten Metals at their Melting Points

Metal	γ (dyn/cm)	$d\gamma/dT$ (dyn $cm^{-1}\,°C^{-1}$)
Ag	966	− 0.19
Al	914	− 0.35
Au	1169	− 0.25
Bi	378	− 0.07
Ca	361	− 0.10
Cd	570	− 0.26
Co	1873	− 0.49
Cu	1303	− 0.23
Fe	1872	− 0.49
Ga	718	− 0.10
Hg	498	− 0.20
In	556	− 0.09
K	115	− 0.08
Li	398	− 0.14
Mg	559	− 0.35
Na	191	− 0.10
Ni	1778	− 0.38
Pb	458	− 0.13
Sb	367	− 0.05
Sn	560	− 0.09
Zn	782	− 0.17

Source: Selected from T. Iida and R. I. L. Guthrie, *The Physical Properties of Liquid Metals*, Oxford University Press, Oxford, 1988.

The surface tension γ of some liquid materials are listed in Tables 5.2-1 through 5.2-3. For most pure materials, γ tends to decrease with increasing temperature, as shown in the tables. In fact, γ also tends to vary significantly with the concentration of impurities,[1] as shown in Fig. 5.2-1 for liquid aluminium.[2]

Significant surface tension gradients are often present along a L/G interface, due to the presence of temperature and/or composition gradients along the interface. These surface tension gradients can induce significant shear stresses and hence fluid flow along the interface: **thermocapillary** or **Marangoni convection.**

[1] F. D. Richardson, *Physical Chemistry of Melts in Metallurgy*, Vol. 2, Academic Press, London, 1974, p. 430.

[2] A. M. Korol'kov, *Casting Properties of Metals and Alloys*, Consultants Bureau, New York, 1960, p. 37.

TABLE 5.2-2 Surface Tensions of Some Semiconductor Melts

Materials	γ (dyn/cm)	$d\gamma/dT$ (dyn cm^{-1} °C^{-1})
Ge	621	-0.26
Si	874	-0.28
GaAs	401	-0.18
GaSb	440	-0.15

Source: Selected from R. Rupp and G. Muller, *J. Crystal Growth*, **113**, 131, 1991.

TABLE 5.2-3 Surface Tensions of Some Transparent Fluids

Materials	Surface Tension (dyn cm^{-1} °C^{-1})
Silicone oil (5cs)[a]	$\gamma = 18.7 + (T - 25\,°\text{C}) \times (-0.060)$
$NaNO_3$[b]	$\gamma = 120 + (T - 307\,°\text{C}) \times (-0.056)$
Ethanol[c]	$\gamma = 25.5 + (T - 0\,°\text{C}) \times (-0.09)$

[a] C. V. Burkersroda, A. Prakash, and J. N. Koster, *Microgravity Quarterly*, **4**, 93, 1994.
[b] F. Preisser, D. Schwabe, and A. Scharmann, *J. Fluid Mechanics*, **126**, 545, 1983.
[c] J. Metzger and D. Schwabe , *Physicochem. Hydrodynamics*, **10**, 263, 1988.

Thermocapillary flow has been a subject of active research in fluid dynamics in recent years, notably by the experimental studies of Ostrach and Komotani in the United States and Chun and Schwabe in Germany.[3] Computational studies have also been conducted by numerous investigators. Microgravity provides the ideal condition for experiments on thermocapillary convection since the effect of buoyancy convection can be eliminated.

The simplest case of thermocapillary convection in a rectangular container is illustrated in Fig. 5.2-2*a*. The left wall is kept at temperature T_L and the right wall at a lower temperature T_R. The bottom wall is insulated. The boundary condition at the L/G interface (i.e., the free surface) is as follows:

$$\tau_{yx} = -\mu \frac{\partial v_x}{\partial y} = \underbrace{\frac{\partial \gamma}{\partial T}\frac{\partial T}{\partial x}}_{(1)} + \underbrace{\frac{\partial \gamma}{\partial w_A}\frac{\partial w_A}{\partial x}}_{(2)} \qquad [5.2\text{-}1]$$

[3] See references listed in Y. Tao, R. Sakidja, and S. Kou, *Int. J. Heat Mass Transfer*, **38**, 503, 1995.

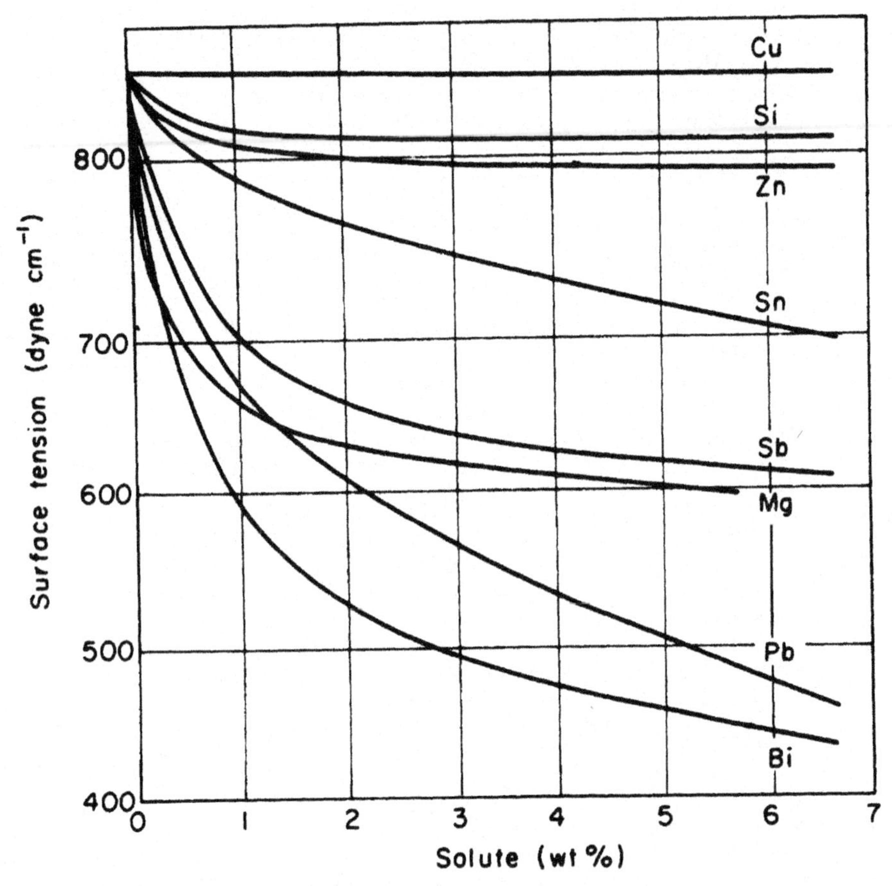

Fig. 5.2-1 Effect of alloying elements on the surface tension of molten aluminium. (F. D. Richardson, *Physical Chemistry of Melts in Metallurgy*, vol. 2, Academic Press, London, 1974, p. 430.)

Terms (1) and (2) represent thermally and compositionally induced shear stresses at the free surface, respectively. For a pure material in the container $\tau_{yx} > 0$ since $\partial\gamma/\partial T < 0$ and $\partial T/\partial x < 0$. In other words, the liquid is pulled along the free surface from higher temperature (left) to lower temperature (right). This creates a clockwise flow loop, as illustrated in Fig. 5.2-2*a*.

Figure 5.2-2*b* shows the flow pattern in an $NaNO_3$ melt in a 1-cm-wide rectangular container.[4] The left and the right walls are at 340 and 332°C, respectively. The flow is dominated by thermocapillary convection, and is clockwise in direction. The melt immediately adjacent to the left wall floats nearly vertically upward because of the buoyancy effect.

[4] Y. J. Kim and S. Kou, unpublished work, University of Wisconsin, Madison, WI, 1991.

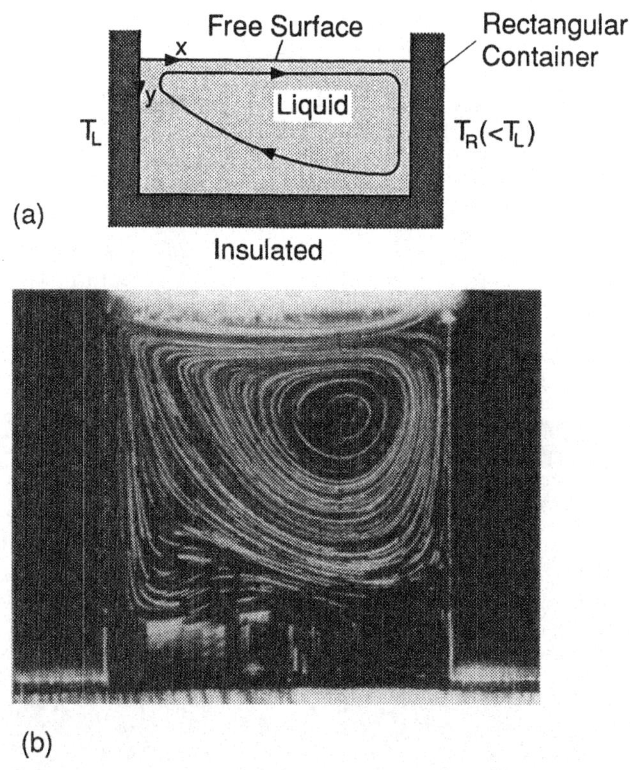

Fig. 5.2-2 Thermocapillary convection in a fluid in a rectangular container: (*a*) schematic sketch; (*b*) $NaNO_3$ melt. (Y. J. Kim and S. Kou, unpublished work at the University of Wisconsin, Madison, WI.)

The case of thermocapillary convection in a cylindrical container is illustrated in Fig. 5.2-3*a*. The center of the free surface is heated by a heat source, such as a laser beam. The boundary condition at the free surface is as follows:

$$\tau_{zr} = -\mu\frac{\partial v_r}{\partial z} = \underset{(1)}{\frac{\partial\gamma}{\partial T}\frac{\partial T}{\partial r}} + \underset{(2)}{\frac{\partial\gamma}{\partial w_A}\frac{\partial w_A}{\partial r}} \qquad [5.2\text{-}2]$$

Again, terms (1) and (2) represent thermally and compositionally induced shear stresses at the free surface, respectively. For a pure material in the container $\tau_{zr} > 0$ since $\partial\gamma/\partial T < 0$ and $\partial T/\partial r < 0$. As such, the liquid is pulled radially outward along the free surface. The resultant flow pattern shows two flow loops on the meridian plane, which is the longitudinal cross section at the axis. The loop on the right is clockwise whereas that on the left is counterclockwise. Figure 5.2-3*b* shows the flow pattern in a silicone oil in a Plexiglas container.[5]

[5] M. P. Wernet and A. Pline, *Experiments in Fluids*, **15**, 295, 1993.

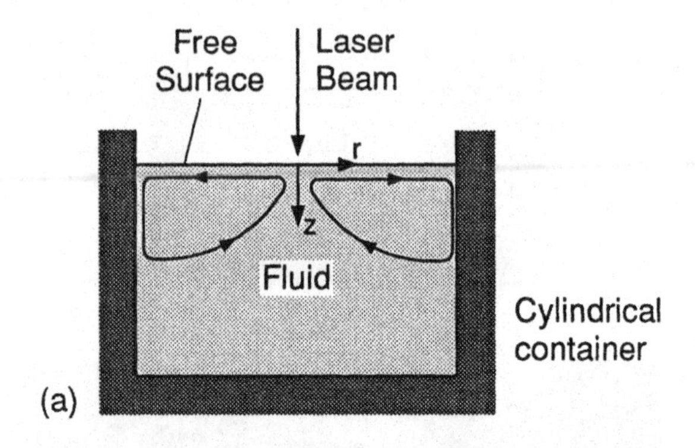

Fig. 5.2-3 Thermocapillary convection in a fluid in a cylindrical container: (*a*) schematic sketch; (*b*) silicone oil. (From M. P. Wernet and A Pline, "Particle Displacement Tracking Technique and Cramer–Rao Low Bound Error in Centroid Estimates From CCD Imagery." *Experiments in Fluids*, **15**, 1993, p. 295)

The loop on the right is clockwise in direction, whereas that on the left is counterclockwise, consistent with those illustrated in Fig. 5.2-3*a*.

The case of thermocapillary convection in a vertical liquid bridge is illustrated in Fig. 5.2-4*a*. The lower rod is kept at temperature T_L and the upper rod, at a higher temperature T_U. The boundary condition is as follows:

$$\tau_{ns} = -\mu \frac{\partial v_s}{\partial n} = \frac{\partial \gamma}{\partial T}\frac{\partial T}{\partial s} + \frac{\partial \gamma}{\partial w_A}\frac{\partial w_A}{\partial s} \qquad [5.2\text{-}3]$$

where the subscripts n and s denote directions normal and tangent to the free surface, respectively. For a pure material $\tau_{ns} > 0$ since $\partial\gamma/\partial T < 0$ and $\partial T/\partial s < 0$. As such, the liquid is pulled downward along the free surface. The resultant flow

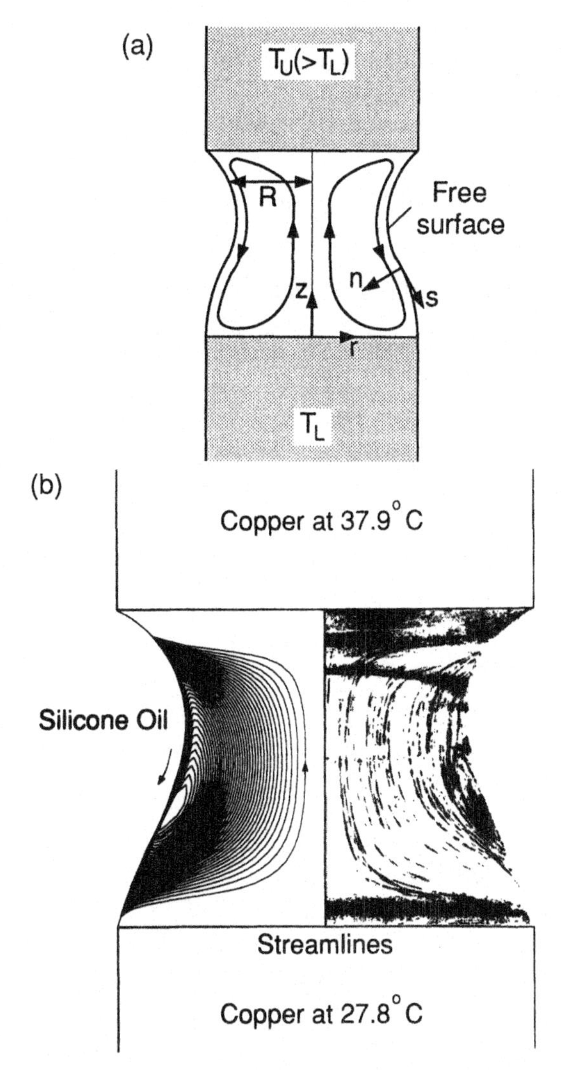

Fig. 5.2-4 Thermocapillary convection in a liquid bridge: (*a*) schematic sketch; (*b*) silicone oil. (Y. Tao, R. Sakidja, and S. Kou, *International Journal of Heat and Mass Transfer*, **38**, 1995, p. 503.)

pattern shows two loops on the meridian plane, a clockwise loop on the right, and a counterclockwise one on the left. Figure 5.2-4*b* shows the calculated (left) and observed (right) flow patterns in a silicone oil bridge held between two copper rods of different temperatures.[6]

[6] Y. Tao, R. Sakidja, and S. Kou, *Int. J. Heat Mass Transfer*, **38**, p. 503, 1995.

It should be pointed out that as the temperature difference between the copper rods is increased to beyond a critical value, flow oscillation occurs and convection becomes neither steady nor axisymmetric. The critical value depends on the properties and dimensions of the fluid.

Equation [5.2-3] can be written in terms of vectors and tensors as follows:[7]

$$\mathbf{ns}:\boldsymbol{\tau} = \frac{\partial \gamma}{\partial T}(\mathbf{s}\cdot\nabla T) + \frac{\partial \gamma}{\partial w_A}(\mathbf{s}\cdot\nabla w_A) \qquad [5.2\text{-}4]$$

Under the steady-state condition no fluid crosses the free surface and therefore

$$v_n = 0 \qquad [5.2\text{-}5]$$

or

$$\mathbf{v}\cdot\mathbf{n} = 0 \qquad [5.2\text{-}6]$$

The shape of the free surface is governed by the following normal-stress balance:[7]

$$\boxed{\tau_{nn} = -\mu\frac{\partial v_n}{\partial n} = \gamma\left(\frac{1}{R_1} + \frac{1}{R_2}\right) - P_a} \qquad [5.2\text{-}7]$$

where R_1 and R_2 respectively are the radii of curvature and P_a the ambient pressure. It can be shown that

$$\frac{1}{R_1} + \frac{1}{R_2} = \frac{R(d^2R/dz^2) - (dR/dz)^2 - 1}{R[1 + (dR/dz)^2]^{3/2}} \qquad [5.2\text{-}8]$$

where R is the radius of the free surface (Fig. 5.2-4*a*).

Equations [5.2-7] can also be written in terms of vectors and tensors as follows:[7]

$$\mathbf{nn}:\boldsymbol{\tau} = \gamma\left(\frac{1}{R_1} + \frac{1}{R_2}\right) - P_a \qquad [5.2\text{-}9]$$

5.2.2. Heat Transfer

The boundary conditions at a L/G interface are similar to those already described in Chapter 2. For example, in case 5 of Fig. 2.3-1, the boundary can be a L/G interface. A radiation heat flux term q_{rad} should be added

[7] C. W. Lan and S. Kou, *J. Crystal Growth*, **108**, p. 351, 1991.

to the convective heat flux term $h(T - T_f)$ if radiation to the interface through the gas phase is significant.

5.2.3. Mass Transfer

The boundary conditions at a L/G interface are similar to those already described in Chapter 3. For example, in case 4 of Fig. 3.3-1, the boundary can be a L/G interface.

5.2.4. Examples

Example 5.2-1 Zone Melting without a Container

A solid feed rod passes through a stationary ring heater at a constant speed V, as illustrated in Fig. 5.2-5*a*. The molten zone is self-supported by the surface tension of the melt, the so-called floating zone. Give the boundary conditions for the molten zone.

Solution At the steady state the boundary conditions at the solid/melt interfaces and the axis are identical to those described in Example 5.1-1 for zone melting in a vertical tube. The boundary conditions at the free surface are as follows:

Fluid Flow:

$$\tau_{ns} = -\mu \frac{\partial v_s}{\partial n} = \frac{\partial \gamma}{\partial T}\frac{\partial T}{\partial s} + \frac{\partial \gamma}{\partial w_A}\frac{\partial w_A}{\partial s} \qquad [5.2\text{-}10]$$

$$\tau_{nn} = -\mu \frac{\partial v_n}{\partial n} = \gamma\left(\frac{1}{R_1} + \frac{1}{R_2}\right) - P_a \qquad [5.2\text{-}11]$$

Heat Transfer:

$$\left(-k \frac{\partial T}{\partial n}\right)_L = h(T - T_a) + q_{\text{rad}} \qquad [5.2\text{-}12]$$

where h is the heat transfer coefficient, T_a the ambient temperature, and q_{rad} the radiation heat flux from the ring heater.

Mass Transfer:

$$\frac{\partial w_A}{\partial n} = 0 \qquad [5.2\text{-}13]$$

assuming that there is no evaporation of material from the free surface.

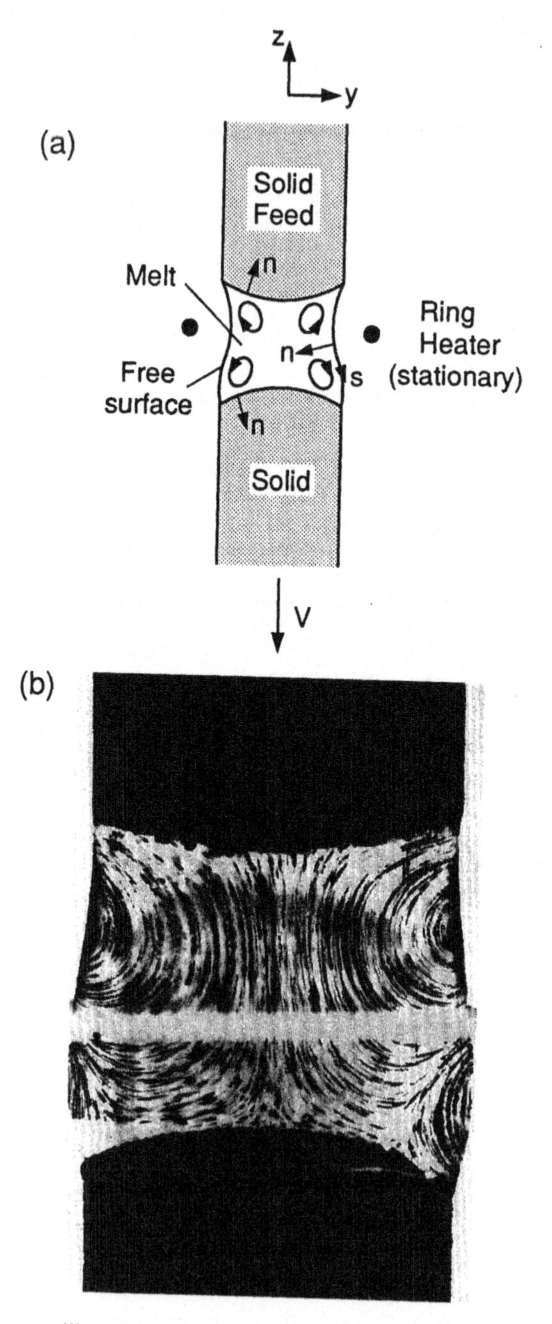

Fig. 5.2-5 Thermocapillary convection in a molten zone: (*a*) schematic sketch; (*b*) $NaNO_3$. (Y. Tao and S. Kou, unpublished work at the University of Wisconsin Madison, WI.)

The thermocapillary convection pattern in the melt is illustrated in Fig. 5.2-5*a*. The surface temperature gradient $\partial T/\partial s < 0$ near the lower end of the free surface but > 0 near the upper end. For the pure material $\partial \gamma/\partial T < 0$. As such, from Eq. [5.2-10] $\tau_{ns} > 0$ near the lower end of the free surface and the melt flows downward along the free surface. Near the upper end of the free surface, on the other hand, $\tau_{ns} < 0$, and the melt flows upward along the free surface.

Figure 5.2-5*b* shows the flow pattern in the floating zone in a stationary ($V = 0$), 4-mm-diameter rod of $NaNO_3$ ($\partial \gamma/\partial T < 0$).[8] The upper right loop is counterclockwise and the lower right loop is clockwise. On the other hand, the upper left loop is clockwise and the lower left loop counterclockwise.

Example 5.2-2 Convection in a Shallow Pool under a Horizontal Temperature Gradient[9]

A fluid in a rectangular container has a free surface on the top and a depth of $2L$. The top and bottom walls are insulators. The left wall is kept somewhat hotter than the right one to induce a steady buoyancy/thermocapillary convection, as shown in Fig. 5.2-6. It has been observed that, except near the sidewalls, the temperature drop from the surface to the bottom of the fluid ΔT is constant. Find the velocity and temperature distributions in the fluid. Neglect end effects and heat losses from the free surface.

Solution This problem is identical to that in Example 4.2-2, except that here the top surface of the fluid is no longer rigid but free. In other words, thermocapillary convection as well as buoyancy convection can be induced by the horizontal temperature gradient across the fluid.

At the free surface $y = L$

$$\mu \frac{\partial v_z}{\partial y} = \frac{\partial \gamma}{\partial T} \frac{\partial T}{\partial z} \tag{5.2-14}$$

Since both sides of the equation are positive, there is no need for a minus sign on the LHS.

Let us define the **Marangoni number** as follows:

$$\text{Ma} = \frac{(\partial \gamma/\partial T) L \Delta T}{\mu \alpha} \tag{5.2-15}$$

which can be considered as a measure of the driving force for thermocapillary convection.

[8]Y. Tao and S. Kou, unpublished research, University of Wisconsin, Madison, WI, 1992.

[9]A. G. Kirdyashkin, *Int. J. Heat Mass Transfer*, **27**, 1205, 1984.

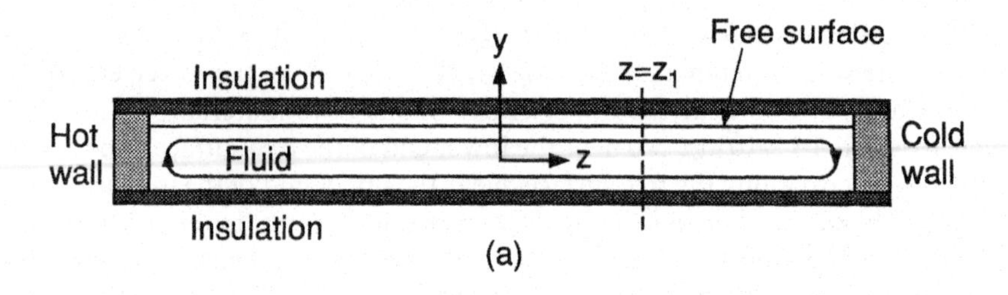

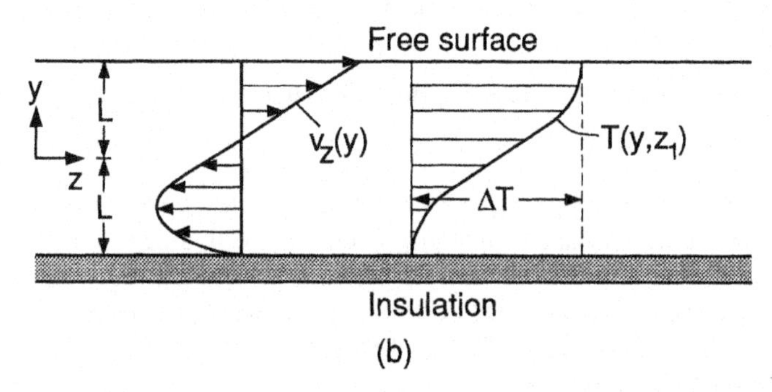

Fig. 5.2-6 A shallow pool under a horizontal temperature gradients: (*a*) convection pattern; (*b*) velocity and temperature profiles.

From Eq. [5.2-15] and other dimensionless quantities already defined, in Example 4.2-2, Eq. [5.2-14] becomes

$$\frac{\partial v_z^*}{\partial y^*} = A^* \frac{\text{Ma}}{\text{Ra}_T} \equiv A^* a \quad \text{at} \quad y^* = 1 \qquad [5.2\text{-}16]$$

The governing equations are identical to those in Example 4.2-2 and hence are not repeated here. The boundary conditions are also identical, except that at the top surface of the fluid ($y^* = 1$) the velocity boundary condition $v_z^* = 0$ is now replaced by Eq. [5.2-16].

The solutions are as follows:

$$A^* = -\left(\frac{30}{(3 - 5a)\text{Ra}_T}\right)^{1/2} \qquad [5.2\text{-}17]$$

$$\boxed{v_z^* = A^* \left[\frac{y^{*3}}{6} + \left(\frac{3a - 1}{8}\right) y^{*2} + \frac{a - 1}{4} y^* - \frac{3a - 1}{24}\right]} \qquad [5.2\text{-}18]$$

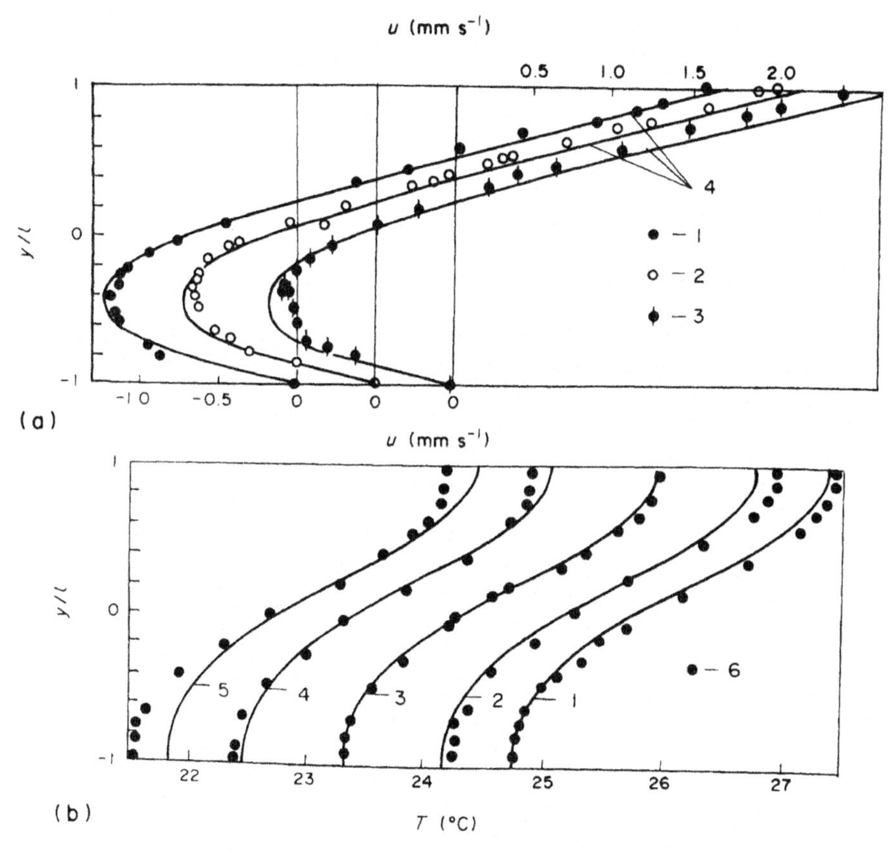

Fig. 5.2-7 Combined thermocapillary/buoyancy convection in a shallow pool of alcohol: (*a*) velocity profiles; (*b*) temperature profiles. (Reprinted from A. G. Kirdyashkin, "Thermogravitational and Thermocapillary Flows in a Horizontal Liquid Layer under the conditions of Horizontal Temperature Gradient," *International Journal of Heat and Mass Transfer*, **27**, 1984, p. 1205, with kind permission from Elsevier Science Ltd., The Boulevard, Langford Lane, Kidlington OX 1GB, UK.)

$$T^* = \left(\frac{5}{12 - 20a}\right)\left[\frac{y^{*5}}{5} + \left(\frac{3a-1}{4}\right)y^{*4} + (a-1)y^{*3} + \frac{1-3a}{2}y^{*2} + (2-3a)y^* - \frac{5a}{4} + \frac{19}{20}\right] \qquad [5.2\text{-}19]$$

where $a = \mathrm{Ma}/\mathrm{Ra}_T$.

Figure 5.2-7 shows the measured and calculated velocities (u is v_z) and temperatures in alcohol at various cross sections.[10] The conditions are as

[10] A. G. Kirdyashkin, ibid.

follows: thickness ($2L$) = 9.4 mm, width = 412 mm, Ma = -1.1×10^4, $a = 0.5$, hot wall at 32.5°C, and cold wall at 18.1°C. As shown, the fluid near the free surface flows toward the cold wall while that below it flows toward the hot wall. Also, the fluid is hotter at the top.

5.3. SOLID/GAS INTERFACE

5.3.1. Fluid Flow

The boundary conditions at a S/G interface are similar to those described in Chapter 1. For example, in cases 1 and 2 in Figs. 1.5-1 and 1.5-2 the boundaries can be a S/G interface.

5.3.2. Heat Transfer

The boundary conditions at a S/G interface are similar to those described in Chapter 2. For example, in cases 4 and 5 in Figs. 2.3-1 and 2.3-2 the boundaries can be a S/G interface. A radiation heat flux term q_{rad} should be added to the convective heat flux term $h(T - T_f)$ if radiation to the interface through the gas phase is significant.

If the temperature field in the solid as well as that in the gas is to be determined, then the following boundary conditions can be used:[11]

$$(-k\mathbf{n}\cdot\nabla T)_s = (-k\mathbf{n}\cdot\nabla T)_G + q_{\mathrm{rad}} \qquad [5.3\text{-}1]$$

where q_{rad} is the radiation heat flux. Since, as in chemical vapor deposition, the concentration of the reactant gas in the inert carrier gas is usually very low, the heat of reaction is often negligible.

5.3.3. Mass Transfer

Consider the case, as in chemical vapor deposition, where a chemical species A in a mixture gas (G) reaches a solid (S) whereupon it is consumed by a chemical reaction at the rate of $\underline{r}_A$. From the conservation of A at the S/G interface ($\mathbf{n}$ points into G), the diffusion flux of A from the gas mixture equals $\underline{r}_A$:

$$\boxed{-D_A\mathbf{n}\cdot\nabla c_A = \underline{r}_A} \qquad [5.3\text{-}2]$$

where c_A is the molar concentration of species A in the gas mixture and D_A, the diffusion coefficient of A in the gas mixture. For a first-order reaction,

[11] D. I. Fotiadis, S. Kieda, and K. F. Jenson, *J. Crystal Growth*, **102**, 441, 1990.

namely $\underline{r}_A = -k_1 c_A$,

$$-D_A \mathbf{n}\cdot\nabla c_A = -k_1 c_A \qquad [5.3\text{-}3]$$

where k_1 is the first-order reaction reaction rate constant.

If thermal diffusion (Eq. [3.1-10]) in the gas is significant, Eq. [5.3-2] should be modified as follows:

$$-D_A\left\{\mathbf{n}\cdot\left[\nabla c_A + \frac{1}{\rho}\underline{\alpha}_T c_A c_{cg}\nabla(\ln T)\right]\right\} = \underline{r}_A \qquad [5.3\text{-}4]$$

where $\underline{\alpha}_T$ is the molar thermal diffusion factor, c_{cg} the molar concentration of the carrier gas, and T the temperature.

5.4. SOLID/SOLID INTERFACE

5.4.1. Heat Transfer

At the interface between solid I and solid II, which are in perfect contact with each other,

$$\boxed{T_{\text{I}} = T_{\text{II}}} \qquad [5.4\text{-}1]$$

and

$$\boxed{(-k\mathbf{n}\cdot\nabla T)_{\text{I}} = (-k\mathbf{n}\cdot\nabla T)_{\text{II}}} \qquad [5.4\text{-}2]$$

where $\mathbf{n}$ is a unit vector normal to the S/S interface pointing into solid I. These two boundary conditions were shown in case 6 of Fig. 2.3-1.

Since the heat of transformation from one solid phase to the other is usually negligible, Eq. [5.4-2] states that heat conducted from solid I to the interface is conducted into solid II.

5.4.2. Mass Transfer

Let us consider the solid-state transformation from solid I to solid II. Figure 5.4-1 shows a portion of a binary phase diagram; the concentration is in the mass fraction of the solute. Solid I and solid II are equivalent to the liquid and the solid in Fig. 5.1-4*a*, respectively. As such, Eq. [5.1-10] can be rewritten for the S/S interface as follows:

$$\boxed{\rho_{\text{I}}(w_{A\text{I}}^* - k_0 w_{A\text{I}}^*)\mathbf{n}\cdot\mathbf{R} = \left(-\rho D_A\frac{\partial w_A}{\partial n}\right)_{\text{I}} + \left(\rho D_A\frac{\partial w_A}{\partial n}\right)_{\text{II}}} \qquad [5.4\text{-}3]$$

(1) (2) (3)

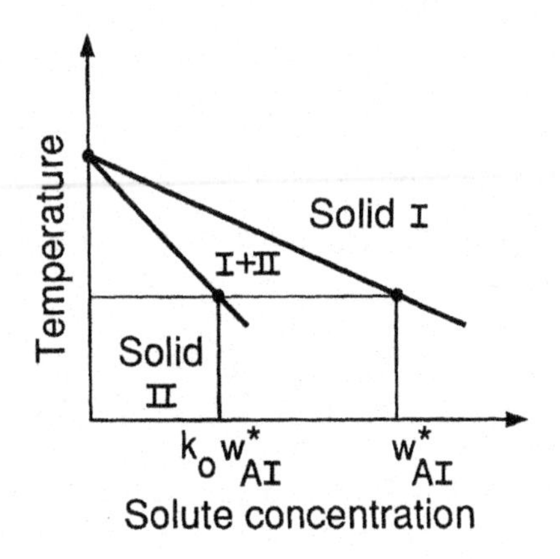

Fig. 5.4-1 A portion of a phase diagram relevant to solid-state phase transformation.

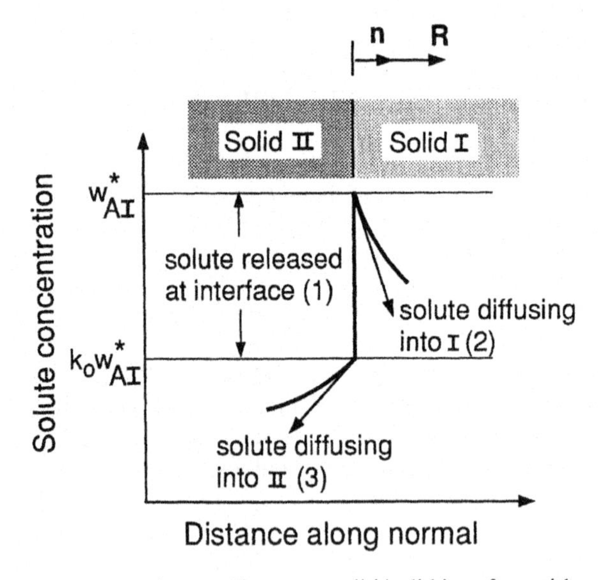

Fig. 5.4-2 Solute concentration profile near a solid/solid interface with solute diffusion from the interface to solid I.

where **R** is the growth rate of solid II into solid I in the direction normal to the interface.

This boundary condition describes the conservation of solute at the interface. As illustrated in Fig. 5.4-2, the solute released at the interface (1) must diffuse away into solid I (2) and solid II (3). Term (2) is proportional to the slope of the tangent to the composition profile of solid I at the interface. Likewise, term (3) is proportional to the slope of the tangent to the composition profile of solid II at

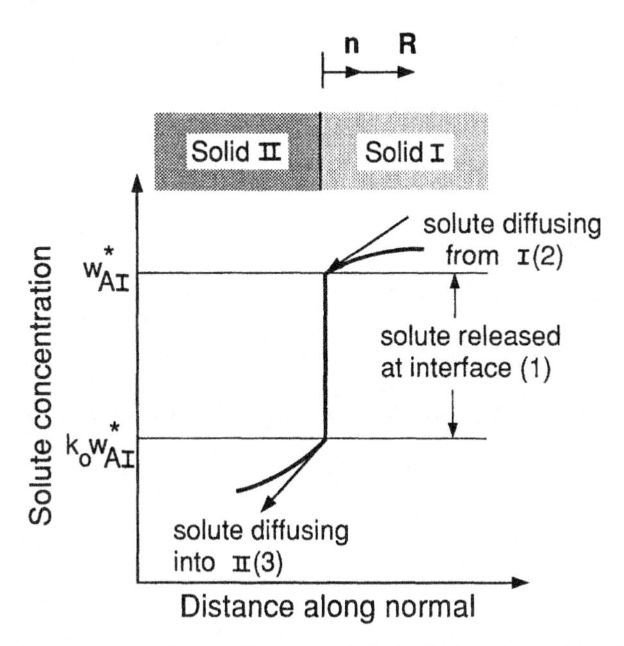

Fig. 5.4-3 Solute concentration profile near a solid/solid interface with solute diffusion from solid I to the interface.

the interface. Notice that $(\partial w_A/\partial n)_\text{I} < 0$ and $(\partial w_A/\partial n)_\text{II} > 0$. As such, all three terms in Eq. [5.4-3] are positive in value.

In some cases, the transformation of solid I to solid II is caused by continuously pumping the solute from the initial surface of solid I, as in the γ to α transformation of carbon steel by decarburization. As shown in Fig. 5.4-3, $(\partial w_A/\partial n)_\text{I} > 0$ and $(\partial w_A/\partial n)_\text{II} > 0$. In other words, $\partial w_A/\partial n > 0$ on both sides of the interface. As such, the solute released at the interface (I) and the solute that diffuses from solid I to the interface (2) must diffuse away from the interface into solid II (3):

$$\underset{(1)}{\rho_\text{I}(w^*_{A\text{I}} - k_0 w^*_{A\text{I}})\mathbf{n}\cdot\mathbf{R}} + \underset{(2)}{\left(\rho D_A \frac{\partial w_A}{\partial n}\right)_\text{I}} = \underset{(3)}{\left(\rho D_A \frac{\partial w_A}{\partial n}\right)_\text{II}} \qquad [5.4\text{-}4]$$

which is identical in Eq. [5.4-3].

5.4.3. Examples

Example 5.4-1 γ/α **Transformation in Steel**

A carbon steel initially containing 0.10 wt% C is kept at a uniform temperature of 875°C in the γ phase. Its surface is exposed to a decarburizing gas and kept at

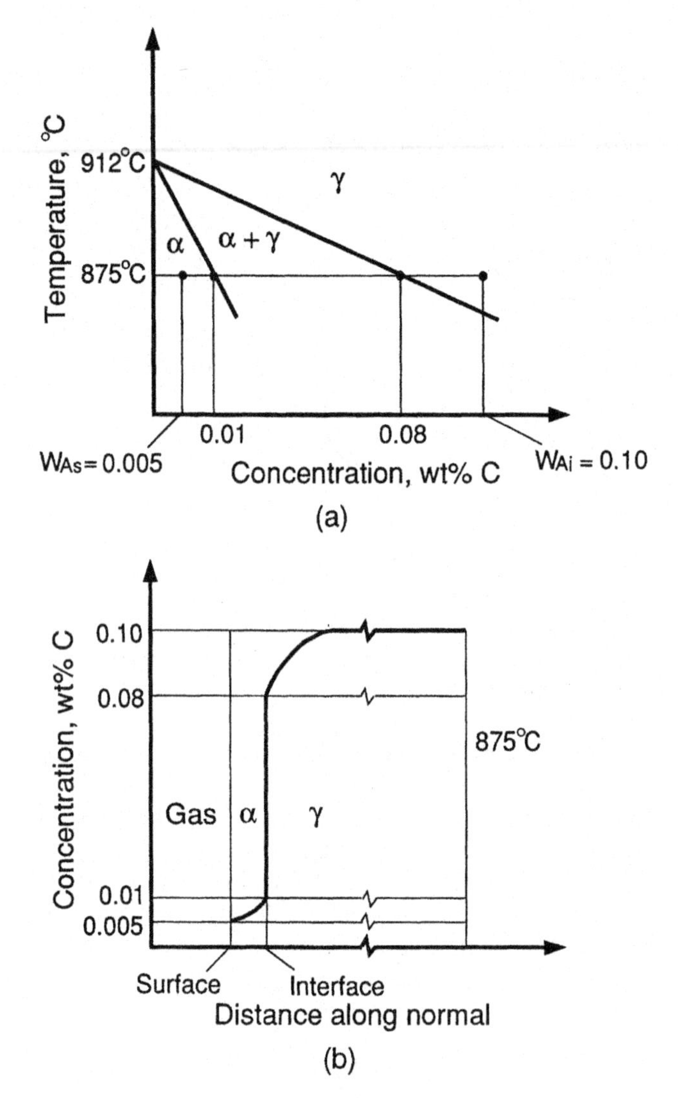

Fig. 5.4-4 γ/α phase transformation in carbon steel: (*a*) phase diagram; (*b*) carbon concentration profile.

a constant carbon content of 0.005 wt%. Carbon continues to diffuse to the surface and reacts with the gas, thus allowing the α-phase to grow into the γ phase. The phase diagram is shown in Fig. 5.4-4. Describe the mass transfer boundary conditions. Neglect the density difference between α and γ.

Solution Since the steel is kept at a constant and uniform temperature, heat transfer is not a relevant issue here. As illustrated in Fig. 5.4-4, the mass transfer

boundary conditions are as follows:
At the surface of the steel

$$w_A = w_{As} = 0.005 \text{ wt\%} \quad [5.4\text{-}5]$$

At the γ/α interface, from Eq. [5.4-4]

$$(0.08 \text{ wt\%} - 0.01 \text{ wt\%})\mathbf{n}\cdot\mathbf{R} + \left(D_A \frac{\partial w_A}{\partial n}\right)_\gamma = \left(D_A \frac{\partial w_A}{\partial n}\right)_\alpha \quad [5.4\text{-}6]$$

In addition, the following initial condition is also needed:

$$w_A = w_{Ai} = 0.10 \text{ wt\%} \quad [5.4\text{-}7]$$

PROBLEMS[12]

5.1-1 A solid material of width a, height b, and depth c is insulated everywhere except for its left surface, as shown in Figure P5.1-1. The depth c is sufficiently large to ensure two dimensional heat transfer and fluid flow. The solid is initially at a uniform temperature T_i. At time zero its left surface is raised to a constant temperature T_{hot} to allow melting to occur. Give the boundary conditions for heat transfer and fluid flow.

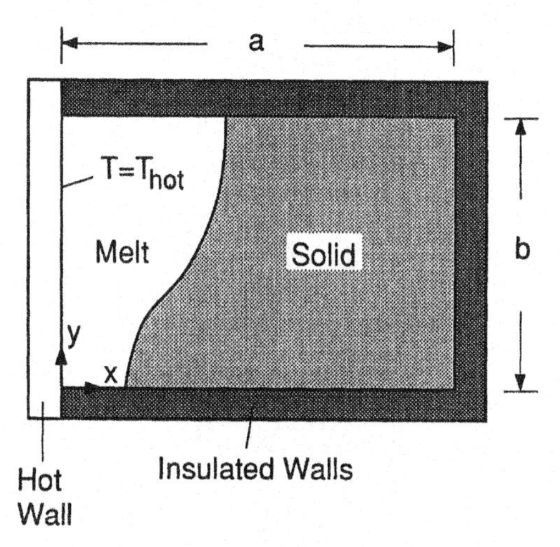

Fig. P5.1-1

5.1-2 A melt is contained in a rectangular crucible of length a, height b, and depth c, as shown in Fig. P5.1-2 (see p. 346). Depth c is sufficiently large

[12]The first two digits in problem numbers refer to section numbers in text.

that convection in the melt can be considered two-dimensional. The temperature at the right wall T_R and that at the left wall T_L are both lowered gradually, with $T_R < T_L$ to allow the melt to solidify from the right. Give the boundary conditions for heat transfer and fluid flow.

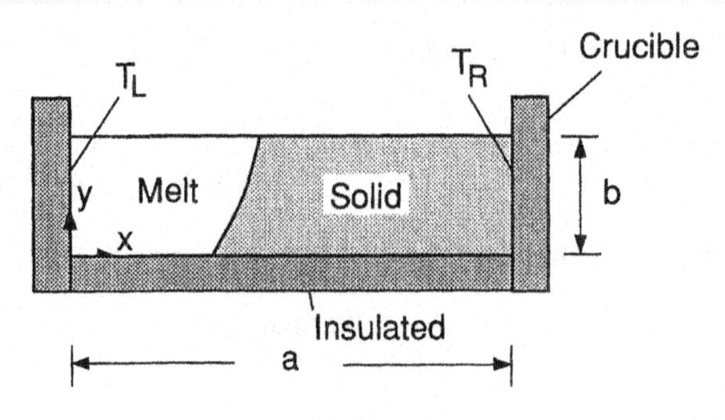

Fig. P5.1-2

5.1-3 The top of a long rod of $NaNO_3$ is in contact with and melted by the bottom end of a graphite rod of the same diameter, as shown in Fig. P5.1-3. The bottom temperature is kept constant at T_h. The melt is short so that its free surface is essentially cylindrical. Give the boundary conditions for fluid flow and heat transfer.[13]

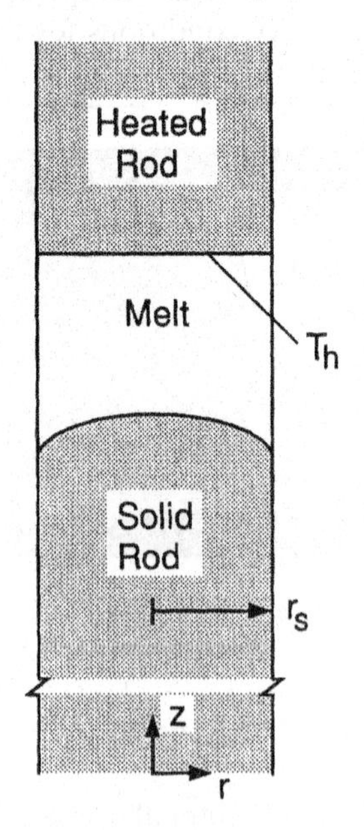

Fig. P5.1-3

[13] C. W. Lan, Y. J. Kim, and S. Kou, *Int. J. Num. Meth. Fluids*, **12**, 59, 1991.

5.2-1 A silicone oil is held in the rectangular container shown in Fig. 5.2-2*a*. Give the boundary conditions for fluid flow and heat transfer.

5.2-2 A silicone oil is held in the cylindrical container shown in Fig. 5.2-3*a*. Give the boundary conditions for fluid flow and heat transfer. Let a and q be the radius and heat flux of the laser beam, respectively.

5.2-3 Give the boundary conditions for fluid flow and heat transfer for the silicone oil bridge shown in Fig. 5.2-4.

5.2-4 Give the boundary conditions for fluid flow and heat transfer for the stationary molten zone in an $NaNO_3$ rod of 4 mm diameter, such as that shown in Fig. 5.2-5. Assume steady-state, axisymmetric convection.

5.2-5 An $NaNO_3$ melt is contained in a rectangular cavity and heated with a horizontal Pt wire heater slightly above and parallel to the centerline of its free surface, as shown in Fig. P5.2-5*a*. The melt is 2 cm wide, 1 cm high, and 2 cm deep.[14] The current passing through the wire heater is I. The flow pattern, shown in Fig. P5.2-5*b*, is for $I = 0$ A. The directions of flow are indicated by the arrows. Sketch the distributions of the temperature and surface tension along the free surface of the melt, and explain the flow pattern.

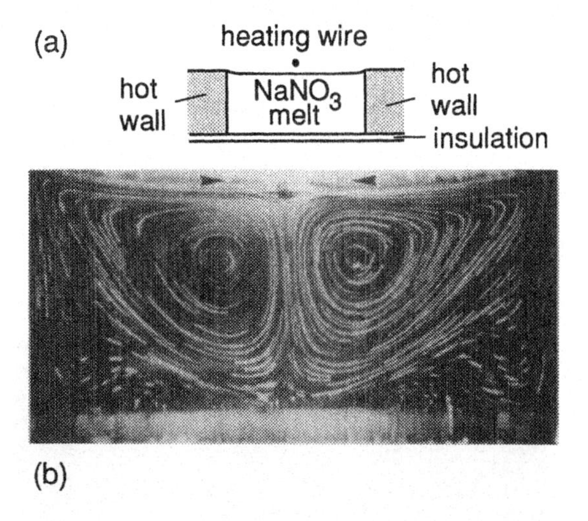

Fig. P5.2-5

5.2-6 Repeat Problem 5.2-5 for the flow pattern shown in Fig. P5.2-6 (see p. 348) for $I = 20$ A.

5.2-7 Repeat Problem 5.2-5 for the flow pattern shown in Fig. P5.2-7 (see p. 348) for $I = 25$ A.

[14]Y. J. Kim ad S. Kou, unpublished research, University of Wisconsin, Madison, WI 1989.

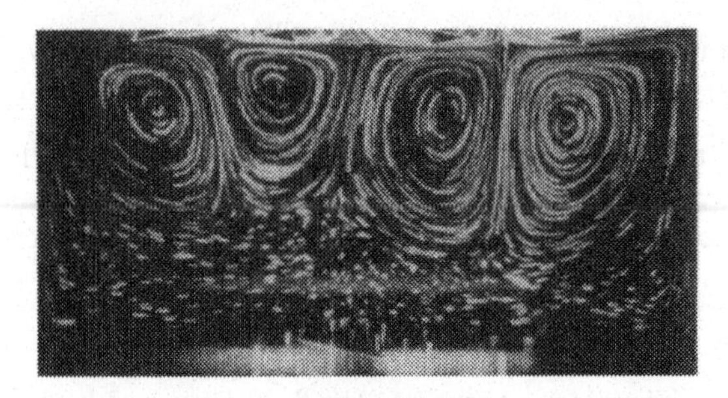

Fig. P5.2-6

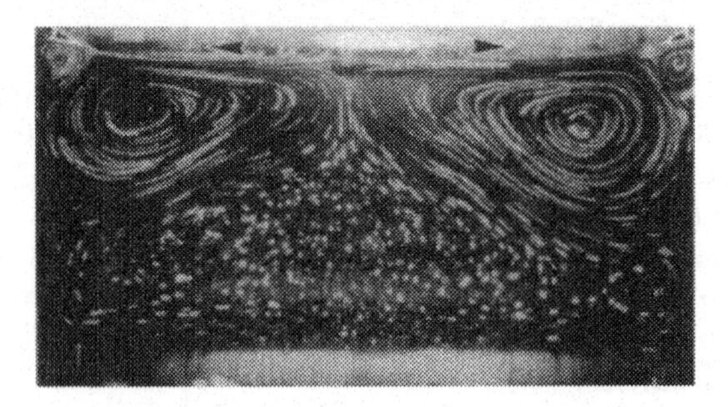

Fig. P5.2-7

5.2-8 Repeat Problem 5.2-5 for the case where the wire heater is shifted horizontally 5 mm to the left of the centerline, as shown in Fig. P5.2-8*a*. The flow pattern is shown in Fig. P5.2-8*b* for $I = 20\,\text{A}$.

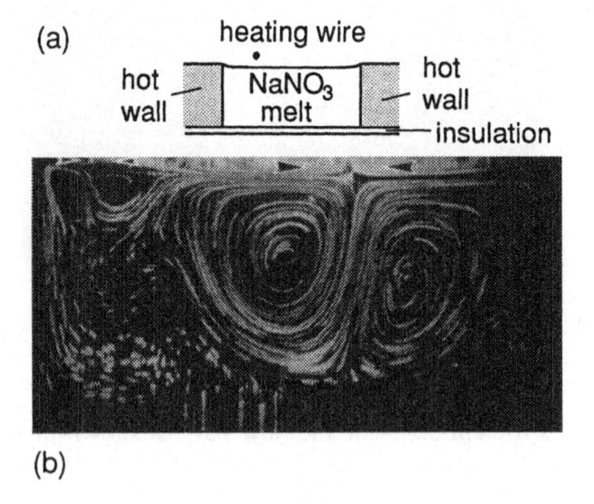

Fig. P5.2-8

5.2-9 Repeat Problem 5.2-8 for the flow pattern shown in Fig. P5.2-9 (see p. 349) for $I = 27\,\text{A}$.

Fig. P5.2-9

5.2-10 A silicone oil is contained between two parallel horizontal plates; the top plate is hotter than the bottom. A gas bubble is injected with a syringe through the top plate, as shown in Fig. P5.2-10.[15] Describe the flow pattern around the bubble.

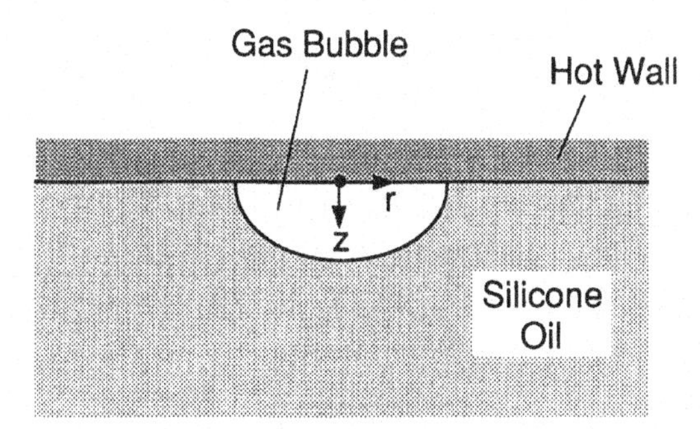

Fig. P5.2-10

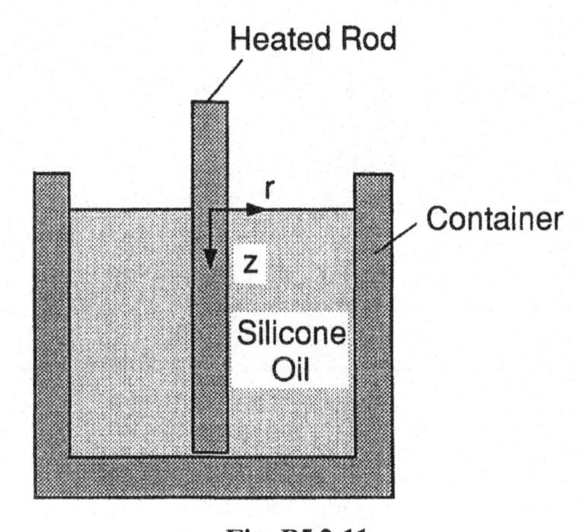

Fig. P5.2-11

[15] K. Wozniak and G. Worniak, *Exp. Fluids*, **10**, 12, 1990.

5.2-11 A heated vertical rod is immersed along the axis of a silicone oil in a cylindrical container under microgravity,[16] as shown in Fig. P5.2-11 (see p. 349). Describe the flow pattern (assume steady and axisymmetric).

5.2-12 The bottom end of a small (e.g., 2-mm diameter) vertical rod is melted by heating from below, as shown in Fig. P5.2-12. Describe the flow pattern in the pendant drop.

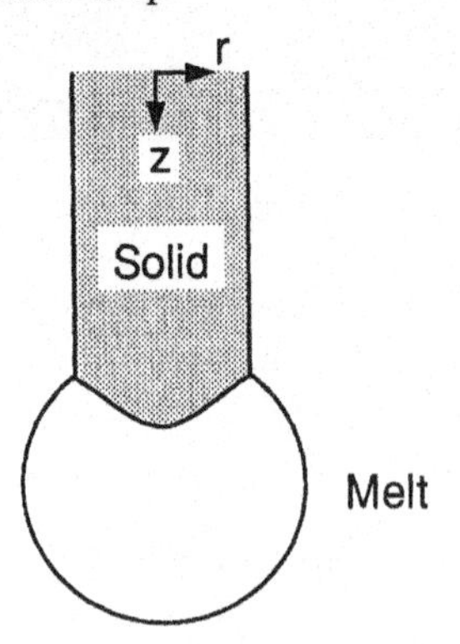

Fig. P5.2-12

5.2-13 Describe the convection in the melt in Problem 5.1-2 under microgravity. What about under normal gravity?

5.2-14 Explain the convection pattern in the melt shown in Fig. P5.2-14, which is produced by heating the top of a $NaNO_3$ rod with a graphite-rod heater.[17]

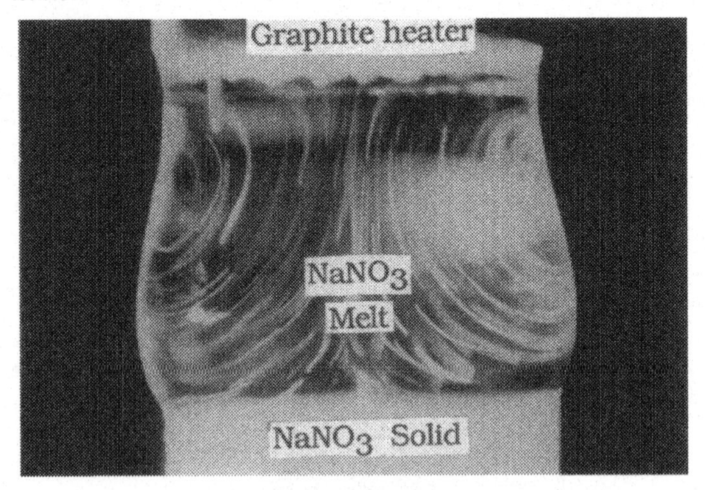

Fig. P5.2-14

[16] S. Ostrach, Y. Kamotani and J. Lee, *Adv. Space Research*, **13** (7), 97, 1993.

[17] C. W. Lan, Y. J. Kim, and S. Kou, *J. Crystal Growth*, **104**, 801, 1990.

PART II

APPLICATIONS OF TRANSPORT PHENOMENA IN MATERIALS PROCESSING

CHAPTER 6

SELECTED MATERIALS PROCESSING TECHNOLOGIES

Transport phenomena in materials processing often involve phase transformations or conversions from liquid to solid, solid to solid, or gas to solid. Transport phenomena significantly affect the way these transformations or conversions take place during materials processing and therefore the quality of the resultant products. In the present chapter, representative materials processing technologies involving liquid-to-solid, solid-to-solid, and gas-to-solid processing are described briefly.

6.1. LIQUID-TO-SOLID PROCESSING

6.1.1. Crystal Growth

Most solid materials are polycrystalline, consisting of numerous small grains. Although atoms in a single grain are arranged in good order, each grain is oriented differently and the atoms become highly disordered along grain boundaries. This causes electronic or optical signals to scatter as they pass through a polycrystalline material. In order to have the atoms arranged in as perfectly good order as possible, the whole piece of material must be a single grain, specifically, a **single crystal**.

Single crystals are most often grown from the melt, although they can also be grown from the vapor, from aqueous solution, or even in the solid state. The most widely used melt growth methods are the Czochralski, Bridgman, and floating-zone processes.

6.1.1.1. Czochralski Crystal Growth

The Czochralski process is the most widely used method for crystal growth of electronic and optical materials. As illustrated in Fig. 6.1-1*a*, the signal crystal is pulled from the melt contained in a heated crucible. To initiate the process, a single crystal seed is dipped into the melt to cause its tip to melt. The melt temperature is then reduced to allow the seed crystal to grow as it is pulled from the melt. The seed is rotated during pulling to help the crystal grow round. Crystal rotation, sometimes coupled with crucible rotation, can also be used to help control convection in the melt. This, in turn, helps control the shape of the crystal/melt interface during growth and the radial compositional uniformity in

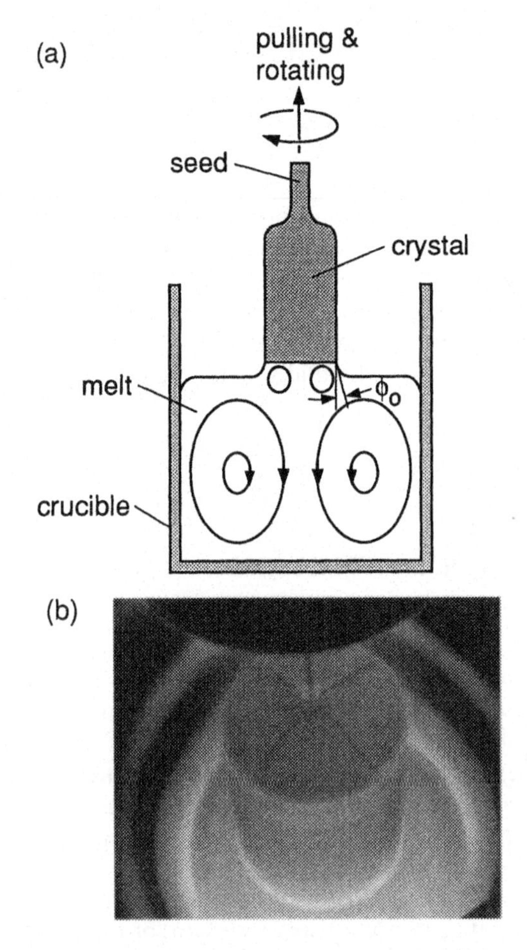

Fig. 6.1-1 Czochralski crystal growth: (*a*) process; (*b*) growth of a Si crystal. (Courtesy of Leybold Technologies, Inc.)

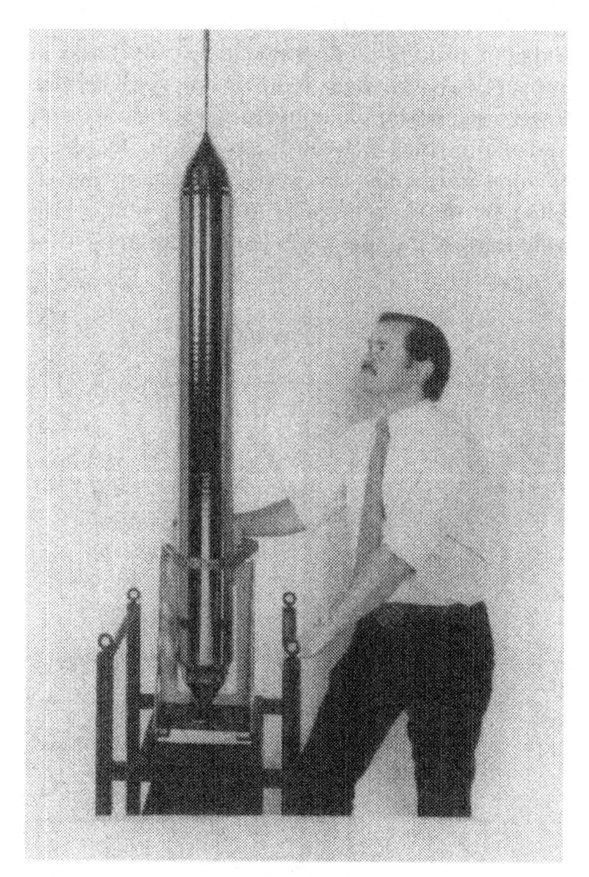

Fig. 6.1-2 A 15-cm (6-in)-diameter Czochralski Si crystal grown in a Kayex Hamco furnace. (Courtesy of G. S. Kayex, Inc.)

the resultant crystal. Figure 6.1-1*b* shows the Czochralski pulling of a Si single crystal[1]. A Czochralski grown single crystal of Si is shown in Fig. 6.1-2.[2]

The equilibrium growth angle ϕ_0 is the angle between the surface of a crystal growing with a constant cross-sectional area or diameter and the tangent to the free surface of the melt at the crystal/melt/gas trijunction.[3] It is a physical property of the crystal material, for example, 11° for Si and 8° for Ge. To keep the crystal cross-section or diameter uniform, the growth angle ϕ must be equal to ϕ_0.

[1] Leybold Technologies, Inc., Alzenau, Germany.

[2] G. S. Kayex, Inc., Rochester, New York.

[3] T. Surek and B. Chalmers, *J. Crystal Growth*, **29**, 1, 1975.

6.1.1.2. Bridgman Crytal Growth

The vertical Bridgman process is illustrated in Fig. 6.1-3*a*. As shown, the furnace consists of an upper zone where the temperature is above the melting point of the crystal, a lower zone where the temperature is below the melting point, and sometimes an adiabatic zone between the two. Initially, the ampoule is raised into the upper zone until only the lower portion of the single crystal seed remains unmelted in the lower zone. After the temperature stabilizes, the ampoule is lowered slowly into the lower zone to initiate crystal growth from the

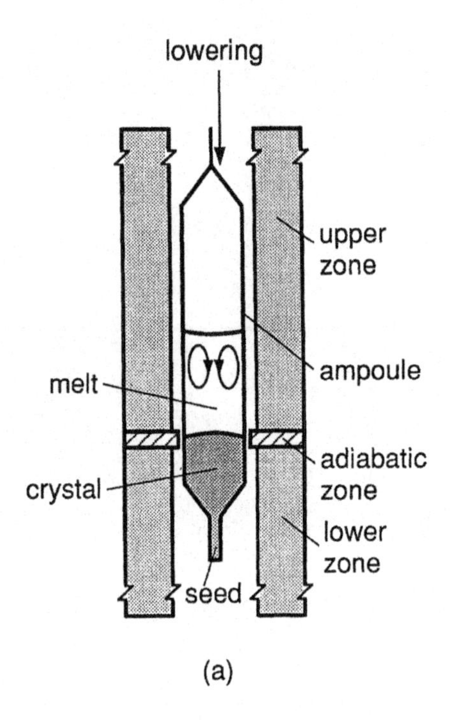

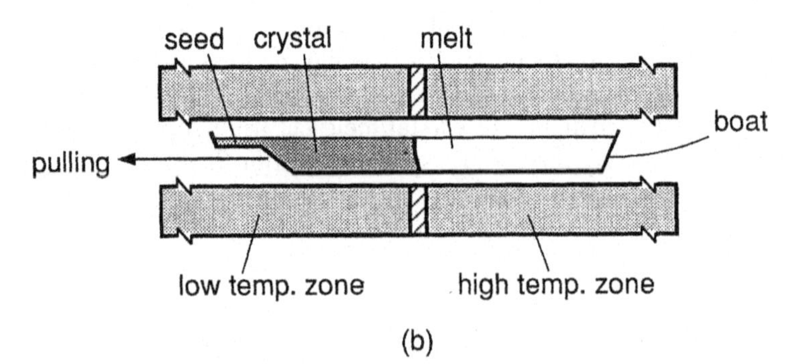

Fig. 6.1-3 Bridgman crystal growth: (*a*) vertical; (*b*) horizontal.

Fig. 6.1-4 A vertical Bridgman GaP crystal. (From E. Monberg, in *Handbook of Crystal Growth*, vol. 2a, D. T. J. Hurle, ed., North Holland, New York, p. 51. Copyright 1994, Elsevier Science.)

seed. As the ampoule is being lowered, the single crystal grows at the expense of the melt. After crystal growth is over, the temperature of the lower zone is reduced gradually to allow the crystal to cool down.

Bridgman growth can also be carried out horizontally, called the **horizontal Bridgman process**, as illustrated in Fig. 6.1-3*b*. A boat-shaped crucible is used in this case. Figure 6.1-4 shows a GaP single crystal grown by vertical Bridgman.[4]

6.1.1.3. Zone Melting

Zone melting was first used by Pfann[5] as a purification technique. It, however, can also be used for crystal growth. Figure 6.1-5*a* shows zone-melting crystal growth in a horizontal boat. The heater, which surrounds a small portion of the boat, produces a short molten zone in the sample. Before crystal growth, the entire boat is filled with the feed. To initiate the process the heater is positioned near the seed so that a portion of the seed can be melted. After the temperature in the sample has stabilized, the heater is moved at a constant velocity away from the seed to cause the crystal to grow from upon the unmelted portion of the seed.

Zone melting crystal growth can also be conducted without a crucible, as in the floating-zone process shown in Fig. 6.1-5*b*. The molten zone is sustained by the surface tension of the melt, and also by electromagnetic levitation if an induction heater is used. The advantage of the floating-zone process is that it is free from contamination by the crucible material, which is particularly significant when growing single crystals from high-melting-point or reactive materials. The disadvantage, however, is that the molten zone has a tendency to collapse under gravity. Figure 6.1-6*a* shows a 6-mm-diameter single crystal of $NaNO_3$ and a Pt ring heater.[6] Figure 6.1-6*b* shows a 50-mm-diameter single crystal of

[4] E. Monberg, in *Handbook of Crystal Growth*, vol. 2a, D. T. J. Hurle, ed., North Holland, Amsterdam, 1994, p. 51.

[5] W. G. Pfann, *Zone Melting*, 3rd ed., Wiley, New York, 1966.

[6] C. W. Lan and S. Kou, *J. Crystal Growth*, **119**, 281, 1992.

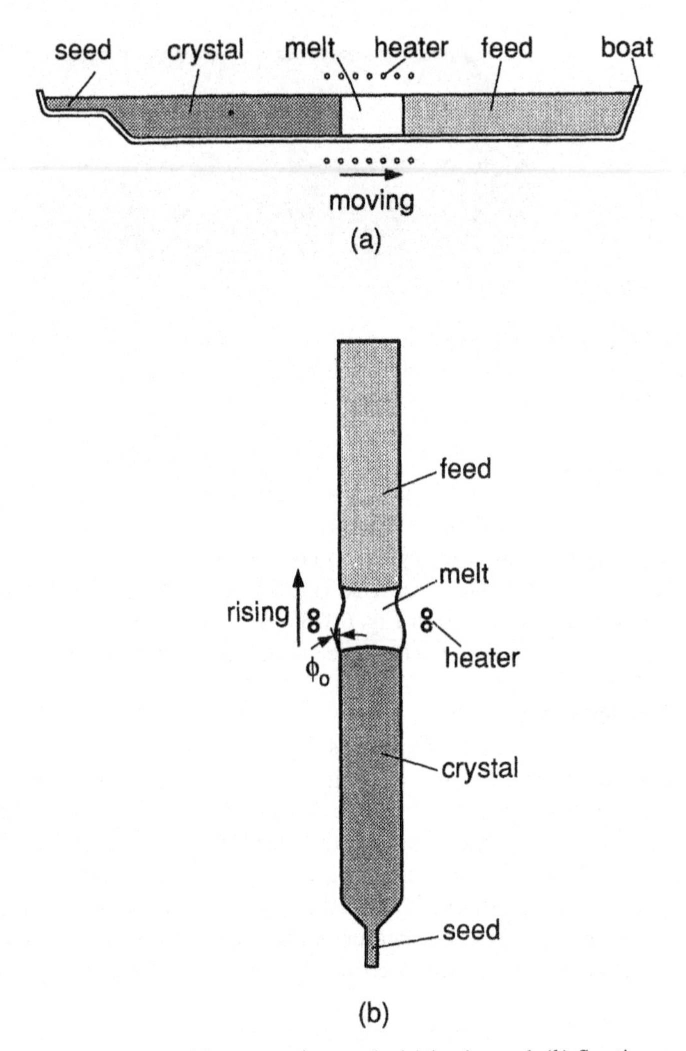

Fig. 6.1-5 Zone melting crystal growth: (*a*) horizontal; (*b*) floating zone.

silicon and a disk-shape induction heater, which has a hole (called the needle eye) to let the melt pass through.[7] The high surface tension and low density of the silicon melt make the needle-eye technique feasible.

6.1.1.4. Macrosegregation

Single crystals are often doped with a very small amount of a foreign material, called **dopant**. For one example, single crystals of Si doped with B or P are used

[7] T. Ciszek, National Renewable Energy Laboratory, Golden, CO.

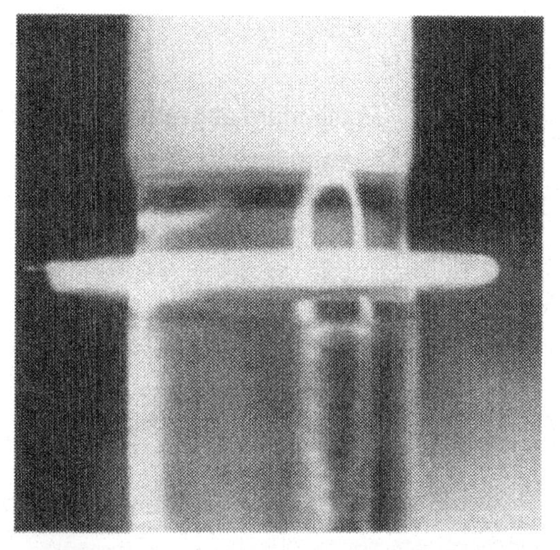

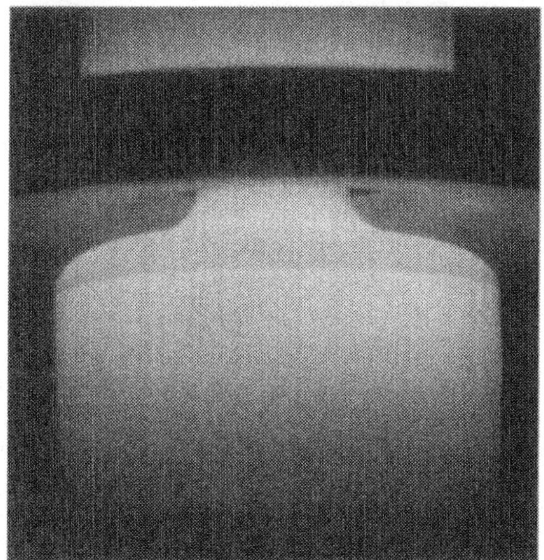

Fig. 6.1-6 Floating-zone crystals: (*a*) 6-mm-diameter $NaNO_3$ (from C. W. Lan and S. Kou, *Journal of Crystal Growth*, **119**, 1992, p. 281); (*b*) 50-mm-diameter Si (Courtesy of T. Ciszek, National Renewable Energy Laboratory, Golden, CO.)

as a p- or n-type semiconductor, respectively. For another example, single crystals of $Y_3Al_5O_{12}$ (YAG, i.e., yttrium aluminum garnet) doped with Nd are a widely used solid-state laser. Segregation of the dopant in a single crystal, however, can result in significant spatial variations of physical properties in the crystal. As such, **dopant segregation** is an important issue in crystal growth.

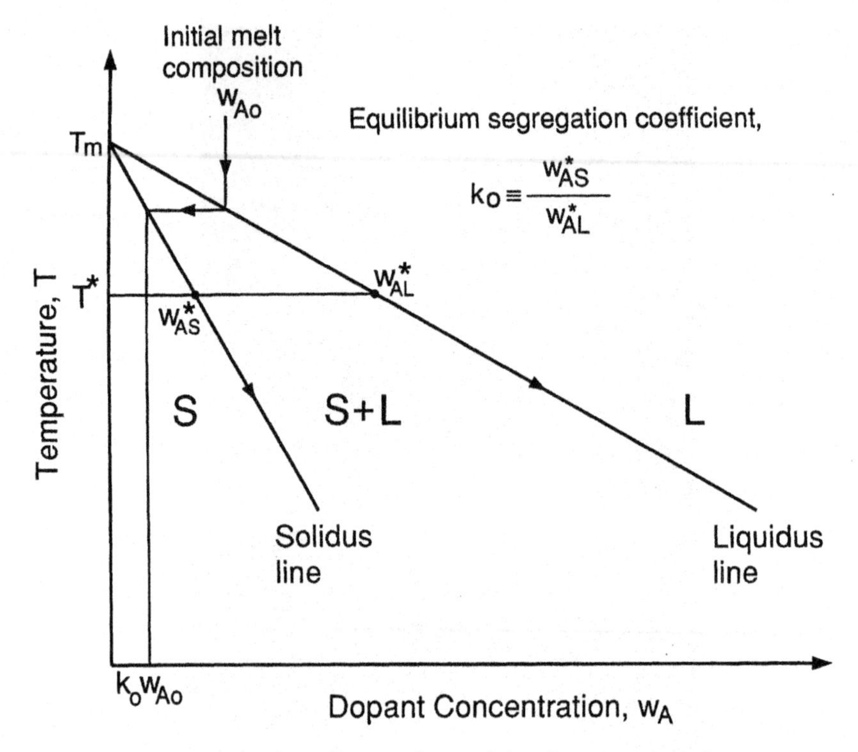

Fig. 6.1-7 A phase diagram for studying dopant segregation.

Let us consider dopant segregation in light of the phase diagram shown in Fig. 6.1-7. A phase diagram is a temperature/composition diagram showing the locations and boundaries of various phases in the material system, such as Si–B or Sn–Pb. In the present phase diagram A denotes the dopant, S and L denote the crystal (solid) and the melt (liquid), respectively, and the asterisk (*) denotes the melt/crystal interface (growth front). The composition of the crystal at the growth front w^*_{AS} is defined by the solidus line, whereas the composition of the melt at the growth front w^*_{AL} by the liquidus line. These two compositions are related to each other through the **equilibrium segregation coefficient** defined as follows:

$$k_0 = \frac{w^*_{AS}}{w^*_{AL}} \tag{6.1-1}$$

Consider crystal growth from a melt of an initial dopant concentration w_{A0}. As shown in Fig. 6.1-7, the first solid to appear should have the composition $k_0 w_{A0}$, that is $w^*_{AS} = k_0 w^*_{AL} = k_0 w_{A0}$ at the starting end of the crystal.

A k_0 less than unity, as in the present case, means that the crystal cannot accommodate as much dopant as the melt, and the dopant is thus rejected by the crystal into the melt. This causes w_{AL}^* to increase during crystal growth, which, in turn, causes w_{AS}^* to increase since $w_{AS}^* = k_0 w_{AL}^*$. As indicated by the arrows along the liquidus and solidus lines in Fig. 6.1-7, w_{AL}^* and w_{AS}^* increase during crystal growth. For a k_0 greater than unity, the opposite is true. In either case, the composition of the crystal w_{AS}^* is nonuniform. Figures 6.1-8*a* and *b* show dopant segregation in a Czochralski-grown Cd-doped InSb crystal[8] ($k_0 = 0.274$) and a Bridgman-grown Li-doped $NaNO_3$ crystal[9] ($k_0 = 0.06$), respectively.

6.1.1.5. Dopant Striations

Dopant segregation can occur over the macroscopic scale of the crystal length, as described above. It can also occur over a microscopic scale. Thermal fluctuations due to unsteady convection in the melt often result in fluctuations in the growth rate and hence dopant concentration. This type of dopant segregation is called **dopant striations.**

Unsteady buoyancy convection can often occur in horizontal Bridgman and Czochralski crystal growth of semiconductors. Figure 6.1-9 shows what appears to be the first application of a steady magnetic field to damp unsteady convection in crystal growth. As shown, oscillatory thermal fluctuations and dopant striations are reduced in the horizontal Bridgman growth of a Te-doped InSb crystal.[10] Dopant striations have also been observed in Czochralski growth of doped Si crystals.[11] In floating-zone crystal growth, thermocapillary convection (Section 5.2) can easily become oscillatory and result in dopant striations, as in the P-doped silicon crystal shown in Fig. 6.1-10.[12] Magnetic damping also helps reduce dopant striations in floating-zone crystals.

6.1.1.6. Constitutional Supercooling

Severe microsegregation can occur in the resultant crystal if the growth front is not planar but cellular or even dendritic. To ensure a planar growth front, the G/R ratio has to be sufficiently high, where G is the temperature gradient in the melt at the growth front and R the growth rate. An insufficient G/R ratio can cause a planar growth front to break down. Figure 6.1-11 is a sequence of photographs showing a planar growth front breaking down to a cellular one, during the crystal growth of a pivalic acid doped with 0.32 mol% ethanol.[13] The growth direction is vertically upward. The dopant ethanol segregates heavily

[8] M. H. Lin and S. Kou, *J. Crystal Growth*, **152**, 256, 1995.

[9] Y. Tao and S. Kou, *J. Crystal Growth*, in press.

[10] H. P. Utech and M. C. Flemings, *J. Appl. Phys.*, **37**, 2021, 1996.

[11] K. Hoshi, N. Isawa, T. Suzuki, and Y. Ohkubo, *J. Electrochem. Soc.* **132**, 693, 1985.

[12] A. Croll, W. Mueller, and R. Nitsche, *Proc. 6th European Symposium on Material Sciences under Microgravity Conditions*, Bordeaux, France, Dec. 2–5, 1986, ESA SP-256 (Feb. 1987), p. 87

[13] L. X. Liu and J. S. Kirkaldy, *J. Crystal Growth*, **144**, 335, 1994.

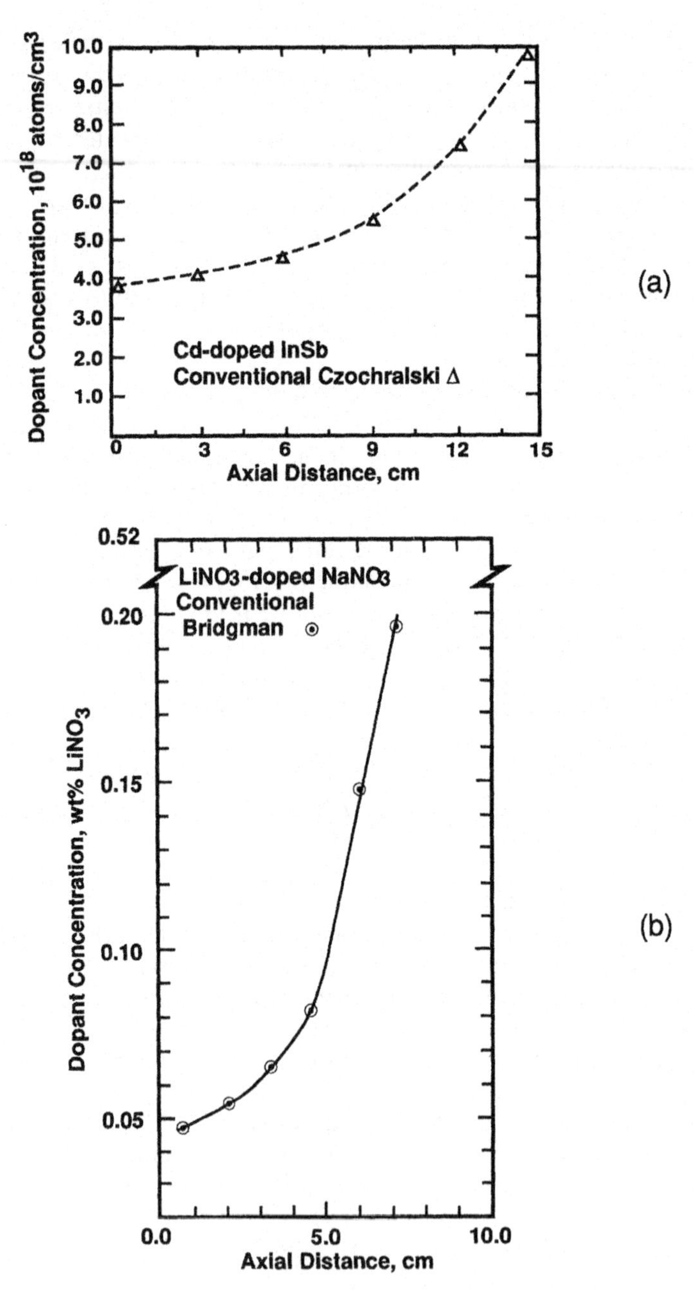

Fig. 6.1-8 Dopant segregation along crystals: (*a*) Czochralski InSb (from M. H. Lin and S. Kou, *Journal of Crystal Growth*, **152**, 256, 1995.); (*b*) Bridgman $NaNO_3$ (from Y. Tao and S. Kou, *Journal of Crystal of Growth*, in press.)

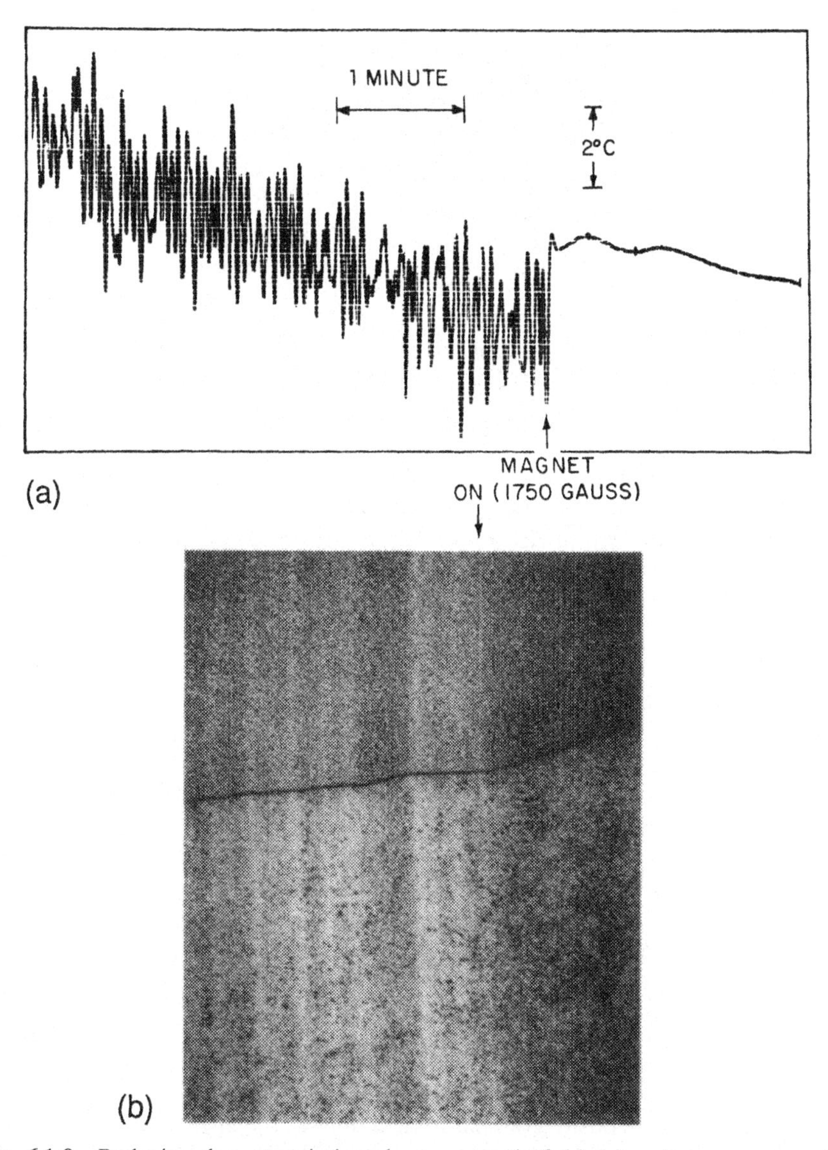

Fig. 6.1-9 Reducing dopant striations by a magnetic field: (*a*) melt temperature; (*b*) dopant striations in a Te-doped InSb bicrystal. (From H. P. Utech and M. C. Flemings, *Journal of Applied Physics*, **37**, 1966, p. 2021.)

along the cell boundaries, making the crystals highly nonuniform in composition. Figure 6.1-12 shows a dendritic growth front during the crystal growth of a carbon tetrabromide doped with several percent impurity.[14] The growth

[14] K. A. Jackson, in *Solidification*, American Society for Metals, Metals Park, OH, 1971, p. 134.

Fig. 6.1-10 Dopant striations in a floating-zone Si crystal. (From A. Croll, W. Mueller-Sebert, and R. Nitsche, *Proceedings of 7th European Symposium on Materials and Fluid Sciences in Microgravity*, ESA SP-259, 1990, p. 263.)

direction is essentially vertically upward. The impurity, lighter in color, segregates heavily in between the dendrite arms.

In the crystal growth of electronic and optical materials the growth front is almost always planar, while in the casting and welding of alloys the growth front is almost always dendritic.

Let us consider the criterion for a planar growth front to remain stable. As illustrated in Fig. 6.1-13*a*, a solute-rich layer forms in the liquid (melt) at the growth front, by the solute rejected by the crystal ($k_0 < 1$). The lowest temperature for the solute-rich layer to remain stable as liquid, can be obtained with the help of the liquidus line $T_\ell(w_{AL})$ in the phase diagram in Fig. 6.1-13*b*. This temperature is shown in Fig. 6.1-13*c* as a function of the distance from the interface $T_\ell(n)$, where n is the distance along the normal to the growth front.

In order for a planar growth front to remain stable, the solute-rich layer must be a stable liquid phase, that is, without formation of any solid ahead of the growth front. This requires that the actual temperature in the liquid be above $T_\ell(n)$:

$$G = \left.\frac{\partial T}{\partial n}\right|_{n=0} \geq \left.\frac{\partial T_\ell}{\partial n}\right|_{n=0} \qquad [6.1\text{-}2]$$

where G is the temperature gradient in the liquid at the growth front. As shown in Fig. 6.1-13*d*, a planar growth front breaks down to a cellular, columnar dendritic, and eventually equiaxed dendritic one, as G is reduced more and more.

Since $\partial T_\ell/\partial n = (\partial T_\ell/\partial w_{AL})(\partial w_{AL}/\partial n)$, Eq. [6.1-2] can be written as

$$G \geq \left.\frac{\partial T_\ell}{\partial w_{AL}}\frac{\partial w_{AL}}{\partial n}\right|_{n=0} = m_L \left.\frac{\partial w_{AL}}{\partial n}\right|_{n=0} \qquad [6.1\text{-}3]$$

where m_L is the slope of the liquidus line (< 0), as shown in Fig. 6.1-13*b*.

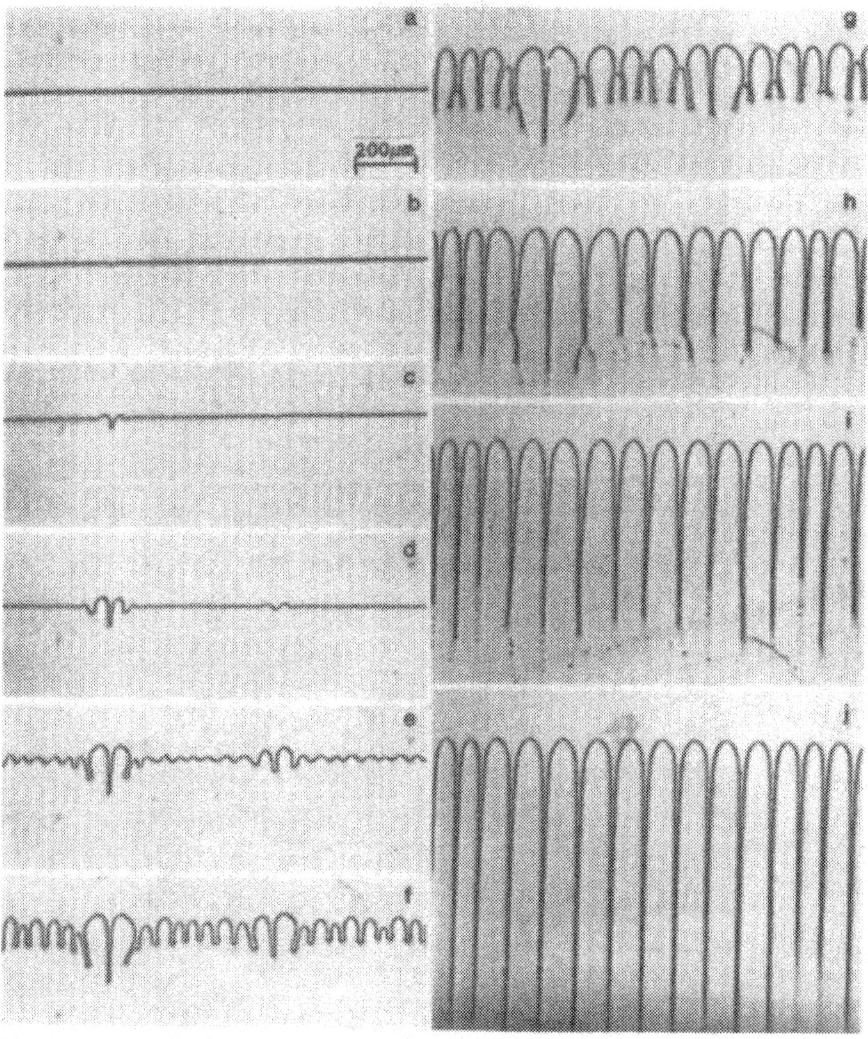

Fig. 6.1-11 Breakdown of a planar solid/liquid interface to a cellular one. (From L. X. Liu and J. S. Kirkaldy, "The Nature of the Planar Instability and Pattern Selection Mechanisms Attending Directional Thin Film Alloy Solidification," *Journal of Crystal Growth*, **144**, 1994, p. 335.)

From Eq. [5.1-10] and assuming $\rho_S = \rho_L$

$$w_{AL}^*(1 - k_0)R = -D_{AL} \left. \frac{\partial w_{AL}}{\partial n} \right|_{n=0} \qquad [6.1\text{-}4]$$

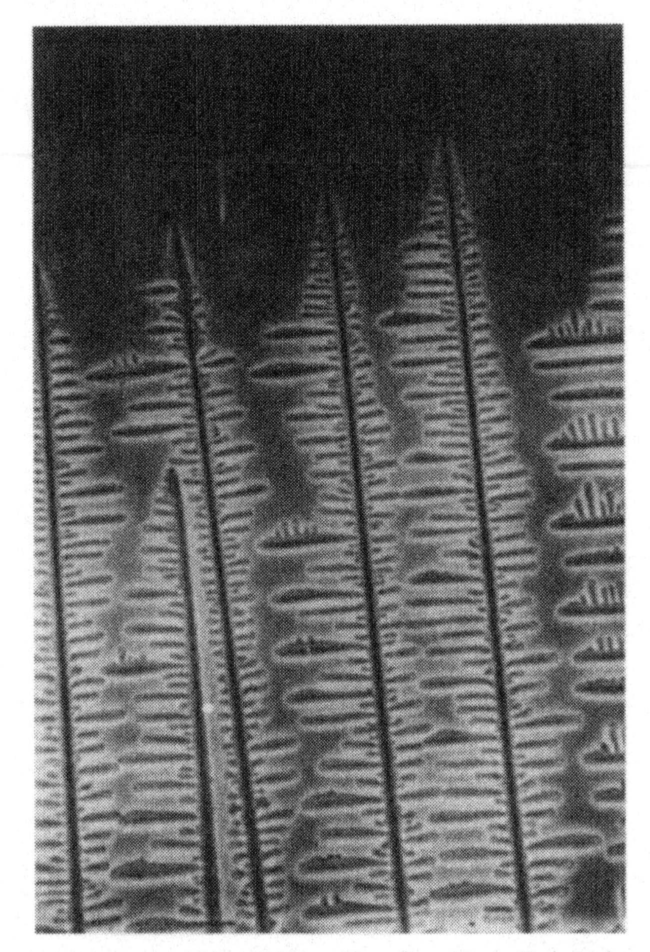

Fig. 6.1-12 A columnar dendritic solid/liquid interface. (From K. A. Jackson, in *Solidification*, ASM, Metals Park, OH, 1971, p. 134.)

where the growth rate $R = \mathbf{n} \cdot \mathbf{R}$ since $\mathbf{n}$ and $\mathbf{R}$ are in the same direction. Diffusion in the solid is neglected. Substituting Eq. [6.1-4] into Eq. [6.1-3].

$$\boxed{\frac{G}{R} \geq -\frac{m_L w_{AL}^*(1 - k_0)}{D_{AL}}} \qquad [6.1\text{-}5]$$

which is the criterion for a planar growth front to be stable; this is the **constitutional supercooling theory** developed by Chalmers and his associates.[15]

More detailed information on crystal growth can be found elsewhere.[16]

[15] J. W. Rutter and B. Chalmers, *Can. J. Phys.*, **31**, 15, 1953; W. A. Tiller, K. A. Jackson, J. W. Rutter, and B. Chalmers, *Acta Metallurgica*, **1**, 428, 1953.

[16] D. T. J. Hurle, ed., *Handbook of Crystal Growth*, Elsevier, Amsterdam, 1994.

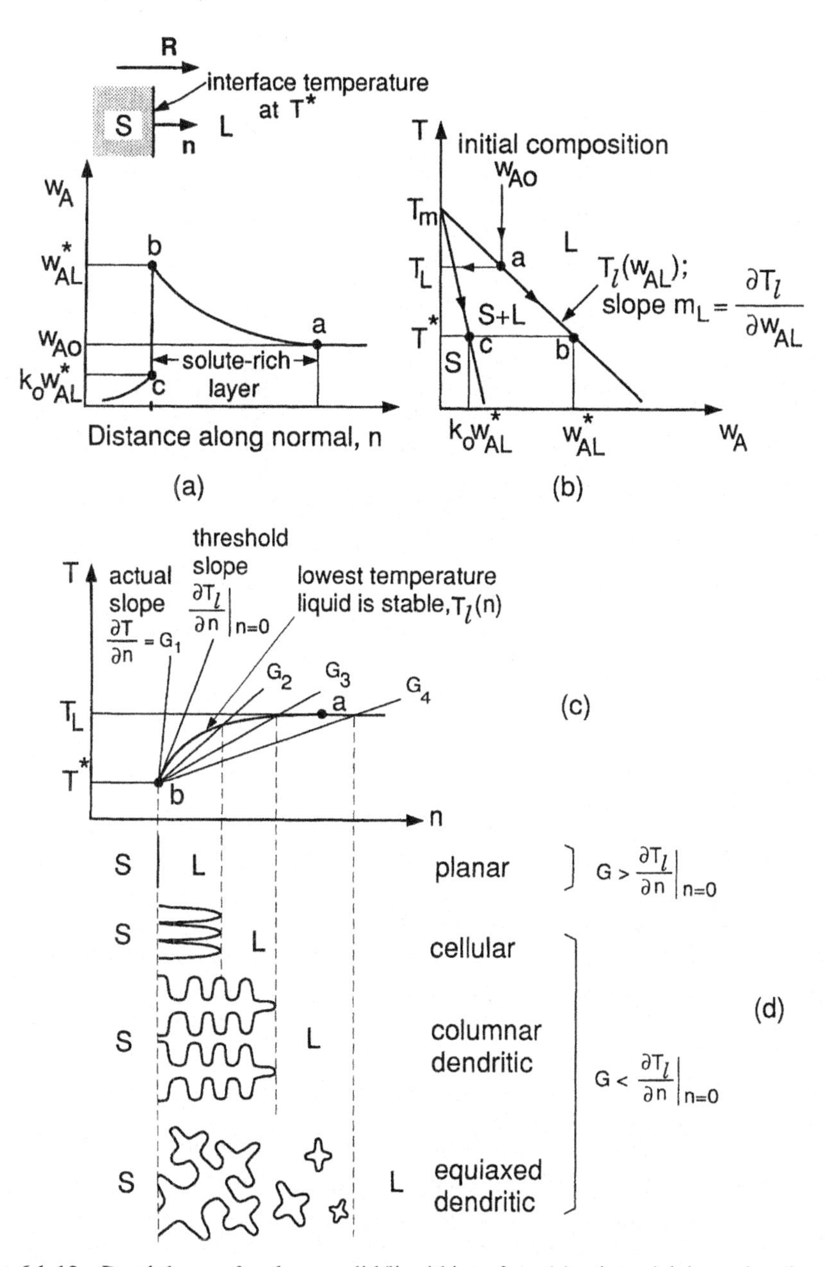

Fig. 6.1-13 Breakdown of a planar solid/liquid interface. (*a*) solute-rich boundary layer; (*b*) phase diagram; (*c*) temperature gradients; (*d*) interface morphology.

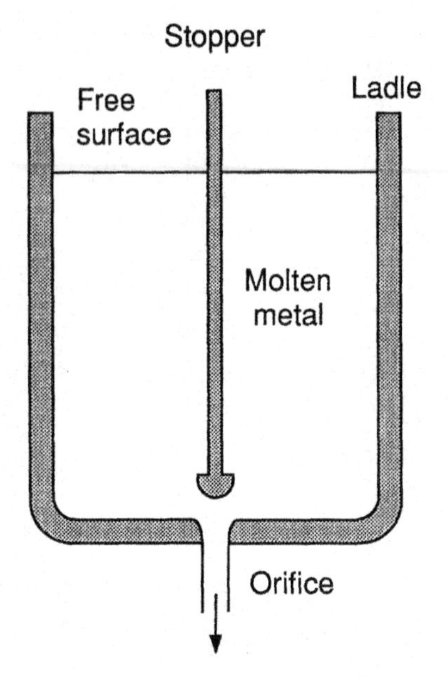

Fig. 6.1-14 A ladle for metal casting.

6.1.2. Casting

In metal casting the molten metal is discharged from a container into a mold where it solidifies into a polycrystalline solid as heat is dissipated through the mold wall. In mass production where a relatively large amount of molten metal is required, the container is often in the form of a ladle. As illustrated in Fig. 6.1-14, a **ladle** consists of a steel shell on the outside with a thick lining of high-temperature insulator to help keep the metal molten for an extended period of time. A stopper activates or stops the flow of molten metal from the bottom orifice of the ladle.

6.1.2.1. Continuous Casting

In continuous casting the mold is open at both ends, as illustrated in Fig. 6.1-15. The molten metal is introduced from one end of the mold and the solidifying metal is withdrawn from the opposite end in a continuous fashion. The mold, which is water-cooled, is made of high-thermal-conductivity material, such as copper and aluminum in the continuous casting of steel and aluminum, respectively. The transverse cross section of the solidified metal is defined by that of the mold interior.

Figure 6.1-16 illustrates the continuous casting of steel. The intermediate reservoir between the ladle and the mold, called the **tundish**, helps regulate the flow of the molten metal into the mold. During casting the level of the molten

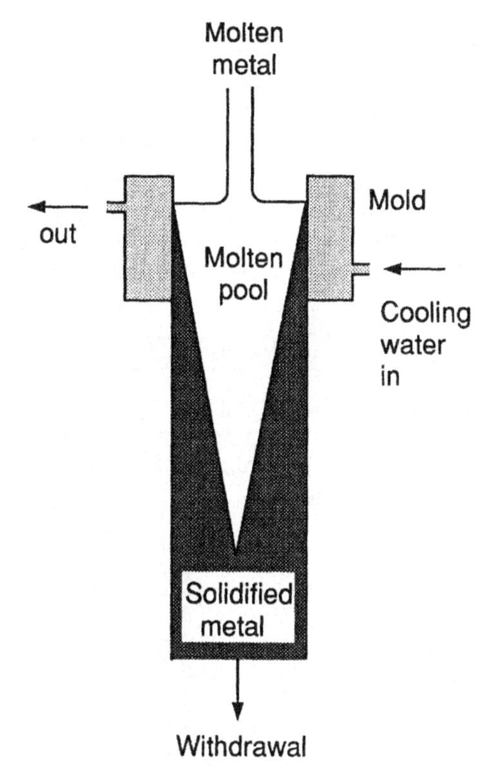

Fig. 6.1-15 Continuous metal casting.

metal continues to drop in the ladle but remains constant in the tundish. This helps maintain a steady flow of the molten metal into the mold.

To initiate continuous casting, a dummy of the same material and transverse cross section as the desired solidified metal is inserted a short distance into the mold from its exit. After the molten metal is introduced into the mold, the dummy can be withdrawn from the mold to initiate casting.

6.1.2.2. Near-Net-Shape Casting

In the production of sheet steel or aluminum, a thick slab is produced by continuous casting first. It is then hot rolled to reduced its thickness step by step to thin sheets. The new trend, however, is to eliminate hot rolling by near-net-shape casting, that is, by continuous casting of a thin sheet. This technology, called **strip casting**, can be done by using a pair of water-cooled rotating steel belts or rolls as the mold. Figure 6.1-17 is an example of **twin-belt casting** of thin slab steel.

In **ingot** or **shape casting**, on the other hand, a closed mold is used, that is, there is no exit through which the solidifying metal can be wirhdrawn during casting. Figure 6.1-18 is a simple illustration showing an ingot solidifying in such

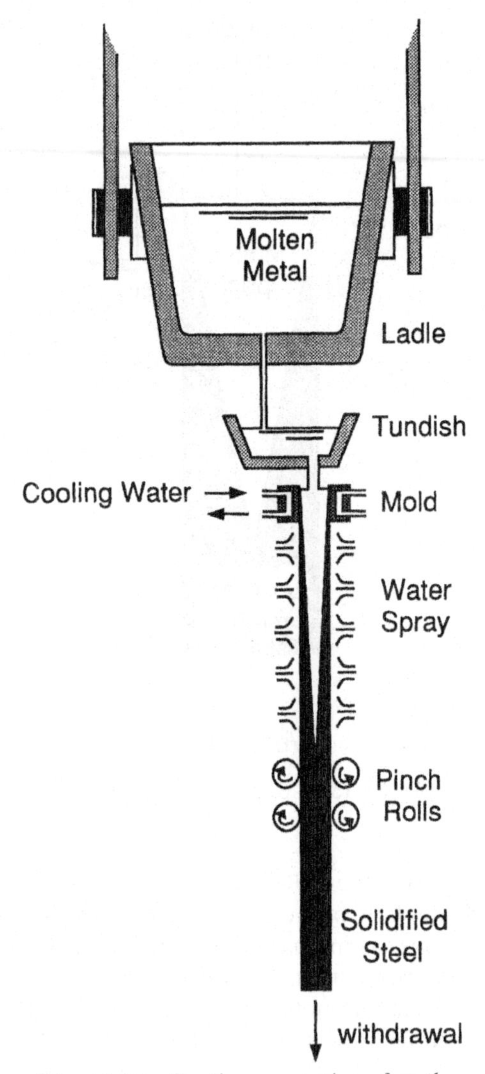

Fig. 6.1-16 Continuous casting of steel.

a mold. Various types of closed molds can be used for preparing shape casting, including sand molds, metal molds, and ceramic shell molds.

6.1.2.3. Sand Casting

In sand casting, the molten metal is poured into the cavity of a sand mold; the cavity is identical in shape to the desired casting. Figure 6.1-19 illustrates a typical sand mold after the pattern halves have been withdrawn.[17] The pattern,

[17] *Casting Design Handbook*, American Society for Metals, Metals Park, OH, 1962, p. 1.

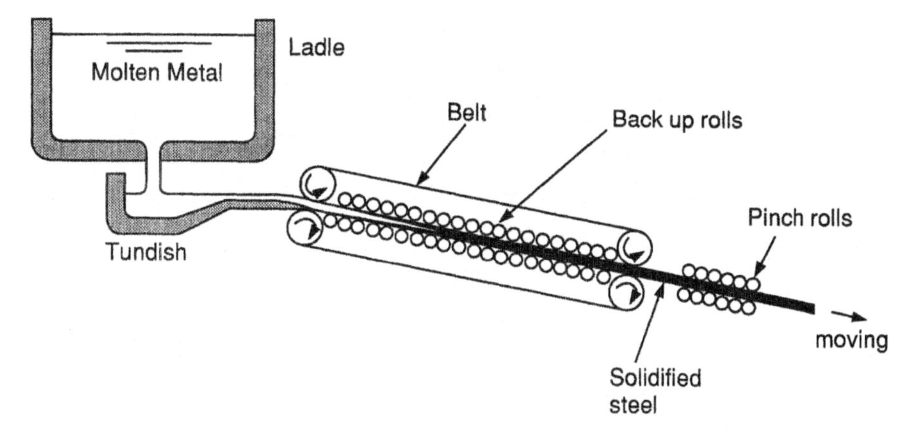

Fig. 6.1-17 Twin-belt continuous casting.

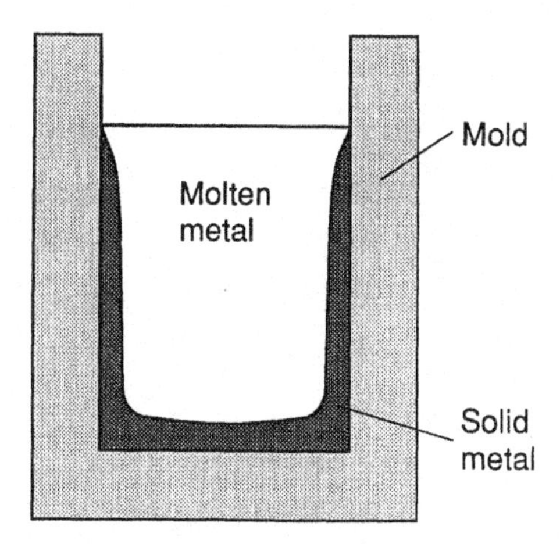

Fig. 6.1-18 A solidifying ingot.

usually made of wood, is used to make the proper impression in the sand mold —the mold cavity—by ramming the sand in place over the pattern. The top half of the sand mold is called the **cope** and the bottom half, the **drag**. The sand contains small amounts of water, clay, and binding agents to develop enough strength for the mold cavity to be retained before and during pouring of the molten metal. The core is placed in the mold cavity to produce a hole through the casting. The core is usually made of baked oil or resin-bond sand to achieve greater strength and to reduce the amount of volatile components.

The molten metal is introduced into the mold cavity through a gating system, which in the present case includes a vertical sprue and a short horizontal gate connecting the sprue to the mold cavity. The riser is designed to absorb the

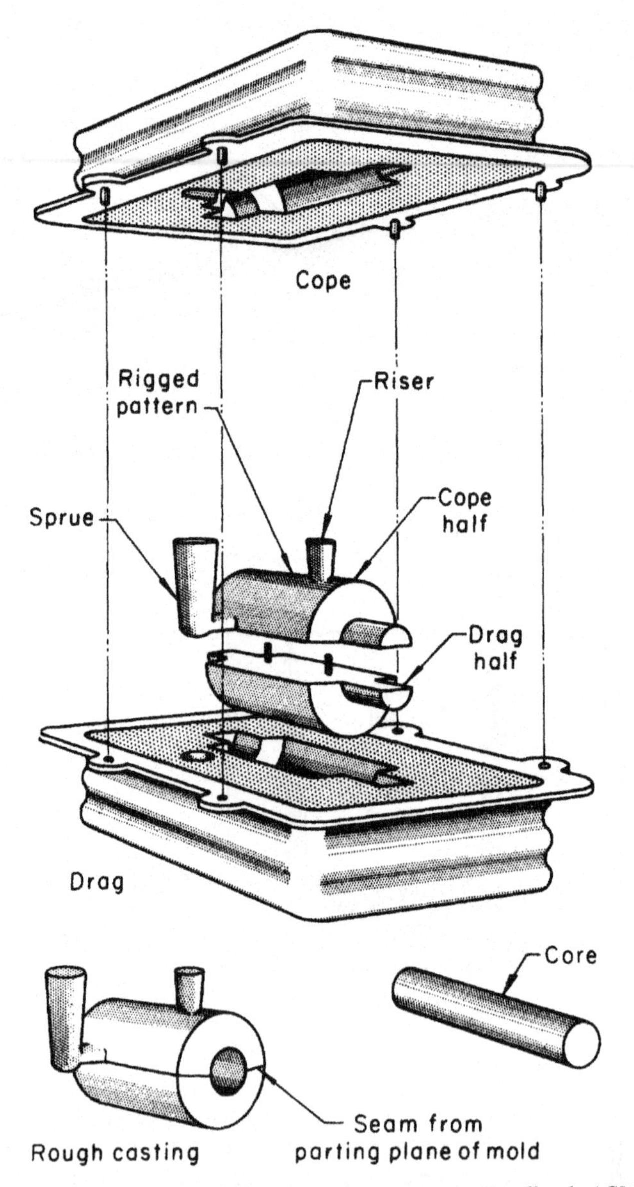

Fig. 6.1-19 A sand casting mold. (From *Casting Design Handbook*, ASM, Metals Park, OH, 1962, p. 1.)

volume decrease of the molten metal during solidification, in order to prevent the formation of shrinkage cavity inside the casting. It is well known that most metals shrink when they solidify. Aluminum, for example, shrinks about 7% when it solidifies.

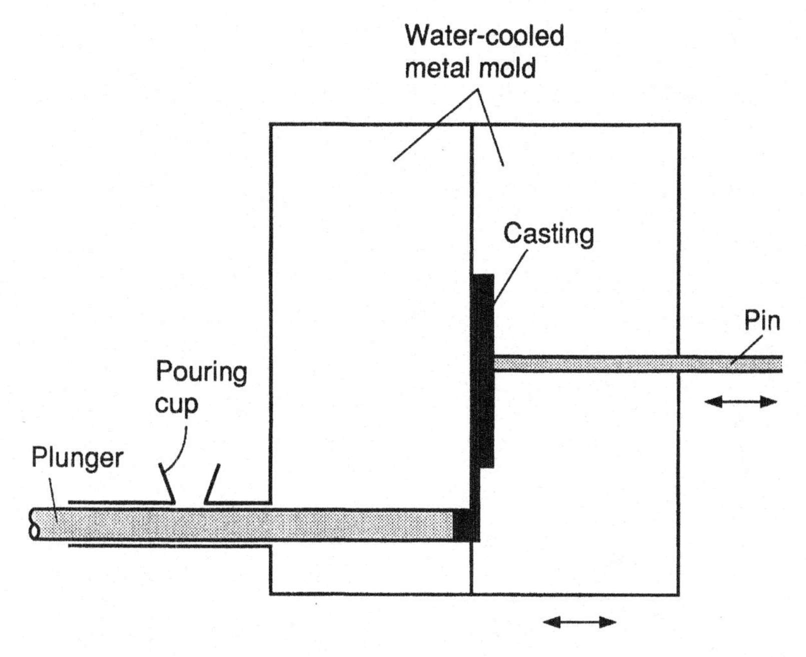

Fig. 6.1-20 Die casting.

In production, a large number of molds line up on a conveyor belt to receive, one by one, the molten metal from a ladle.

6.1.2.4. Die Casting

In die casting, as illustrated in Fig. 6.1-20, the molten metal is forced by a plunger into the cavity of a water-cooled metal mold, specifically, the die. After the metal solidifies, the die is opened and the casting is knocked out by the pin. Die casting is widely used for production of large quantities of simple nonferrous parts, such as aluminum, magnesium, zinc, and brass. Since the metal mold is water-cooled and the cross section of the casting thin, solidification is very fast. Consequently, high production rates such as several hundred shots per hour are common. In highly mechanized foundries, automatic metal ladles or pumps are used.

More detailed information on casting is available elsewhere.[18]

6.1.2.5. Microsegregation

In practical casting and welding the materials, which are mostly alloys, almost always solidify dendritically, as already mentioned in the previous section.

[18] M. C. Flemings, *Solidification Processing*, McGraw-Hill, New York, 1974.

As a result of dendritic solidification, the solute material tends to segregate across the dendrite arms. Since the dendrite arms are microscopic in size, for example, 100 μm, this type of solute segregation is called **microsegregation**.

Let us consider the casting of a binary alloy of a solute concentration w_{A0}; species A is the solute. An eutectic-type phase diagram is shown in Fig. 6.1-21*a* for the purpose of discussion, and the resultant dendritic structure is shown in Fig. 6.1-21*b*. Point *a* at the centerline of the dendrite arm represents the first material to solidify, which occurs at the liquidus temperature T_L. According to the phase diagram and assuming negligible solid-state diffusion, the composition at point *a* is $k_0 w_{A0}$, where k_0 is the equilibrium segregation coefficient.

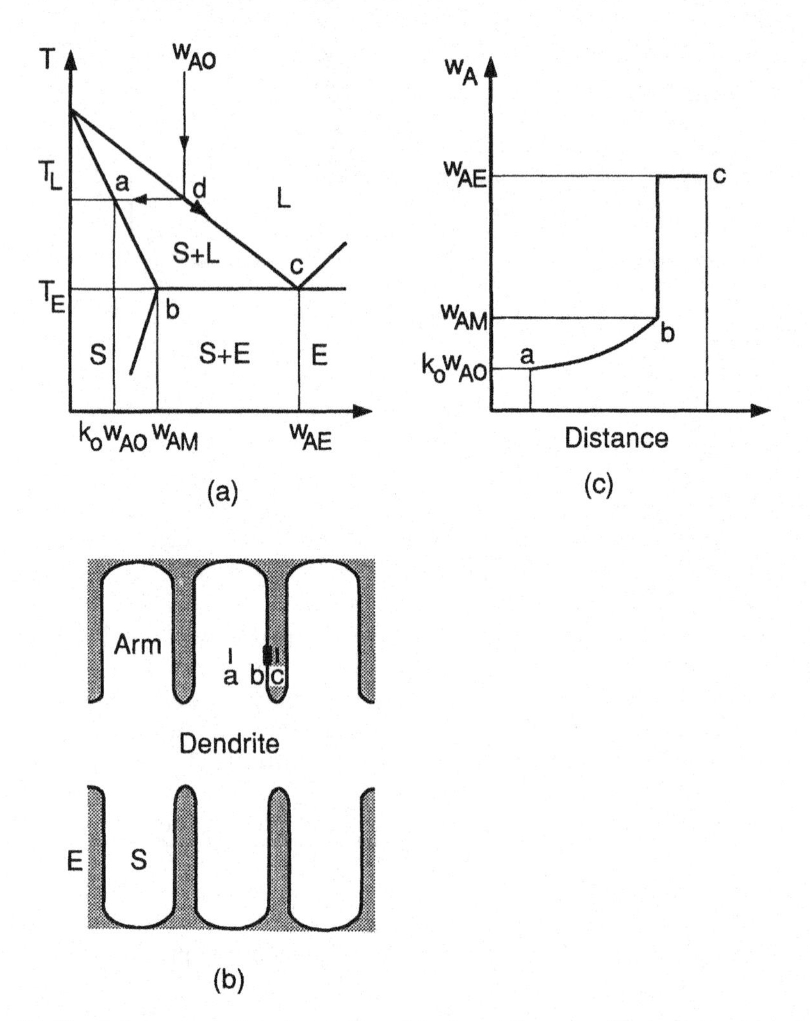

Fig. 6.1-21 Microsegregation in casting: (*a*) phase diagram; (*b*) dendrite; (*c*) solute concentration profile.

As solidification continues, that is, as temperature drops from the liquidus temperature T_L to the eutectic temperature T_E, the dendrite arm grows from point *a* to point *b*. At the same time, the compositions of the solid and the liquid at the arm/liquid interface increase from $k_0 w_{A0}$ to w_{AM} along the solidus line $\overline{ab}$ and from w_{A0} to w_{AE} along the liquidus line $\overline{dc}$, respectively. At T_E, the interdendritic liquid of composition w_{AE} solidifies into solid eutectic *E* without any further changes in composition. The resultant microsegregation is shown in Fig. 6.1-21*c*. The quantitative description of microsegregation will be given in Section 8.1.5.

6.1.2.6. Macrosegregation

Besides microsegregation, macrosegregation can also develop in casting, that is, solute segregation on the scale of the casting or ingots. One important

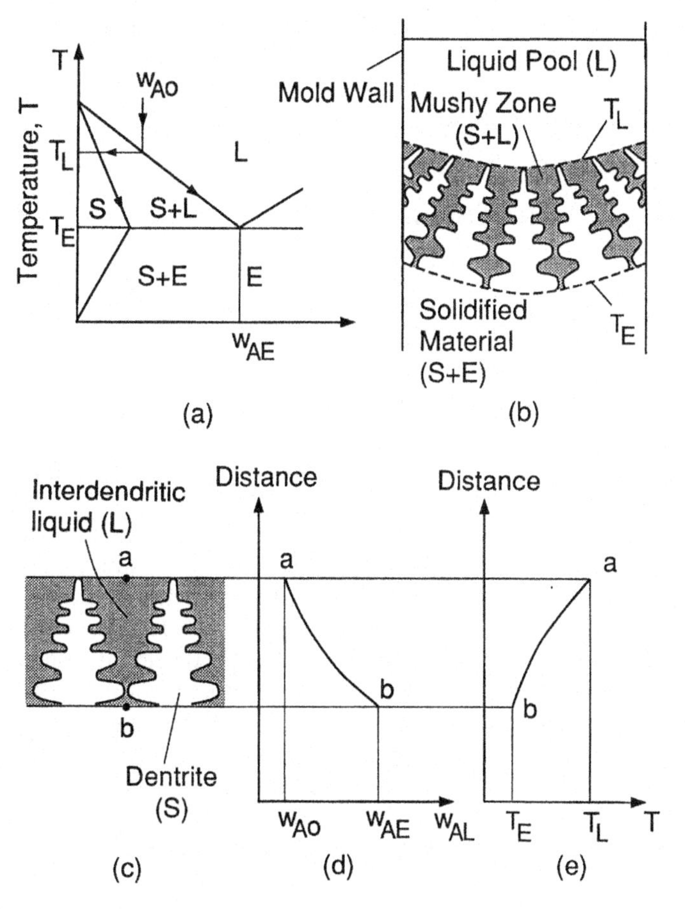

Fig. 6.1-22 Macrosegregation in casting: (*a*) phase diagram; (*b*) solidification of an ingot; (*c*) the mushy zone; (*d*) solute concentration of interdendritic liquid; (*e*) temperature distribution in the mushy zone.

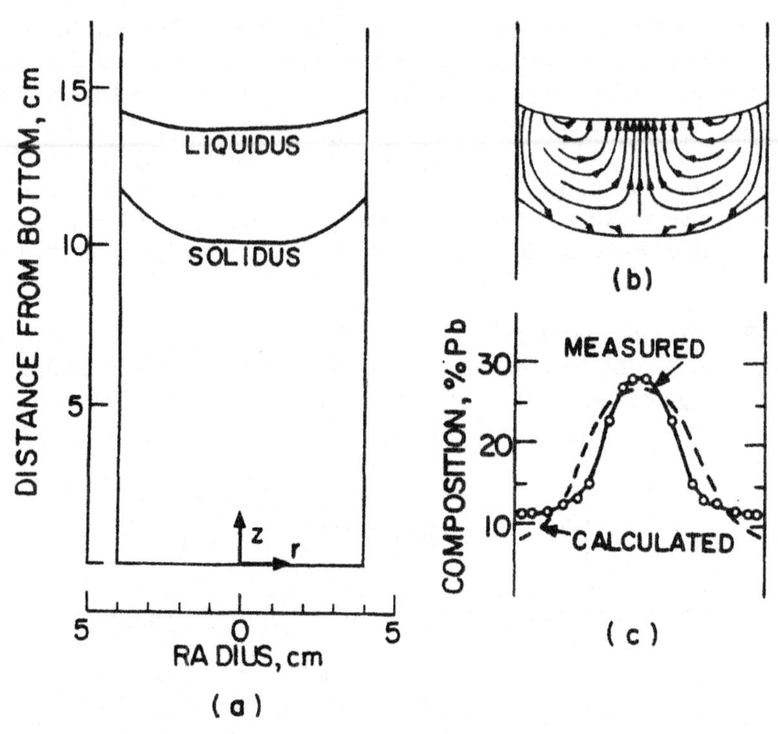

Fig. 6.1-23 Macrosegregation of Pb in a Sn–15% Pb ingot. (From S. Kou, D. R. Poirier, and M. C. Flemings, *Electric Furnance Proceedings*, **35**, 1977, p. 221.)

mechanism for macrosegregation in casting is convection of the interdendritic liquid in the mushy zone,[19] which is the region of a solid/liquid mixture during solidification.

A binary alloy of composition w_{A0} (Fig. 6.1-22*a*) is being cast continuously into a cylindrical ingot (Fig. 6.1-22*b*). The interdendritic liquid (Fig. 6.1-22*c*) is uniform neither in composition (Fig. 6.1-22*d*) nor in temperature (Fig. 6.1-22*e*). Let us consider the case where the solute is significantly heavier than the solvent, such as Pb in a Sn–15% Pb alloy. From Fig. 6.1-22*d* the liquid along the bottom of the mushy zone, which is represented by the isotherm T_E in Fig. 6.1-22*b*, has a maximum solute concentration of w_{AE} and hence a maximum density. As such, this heavy solute-rich liquid tends to sink, flowing along the bottom of the mushy zone toward the axis and resulting in solute segregation toward the axis of the ingot. Figure 6.1-23 shows the interdendritic convection in the mushy zone of a Sn–15% Pb ingot and the resultant macrosegregation.[20] As shown, Pb

[19] M. C. Fleming, ibid.

[20] S. Kou, D. R. Poirier, and M. C. Flemings, *Electric Furnace Proc.*, **35**, 221, 1977.

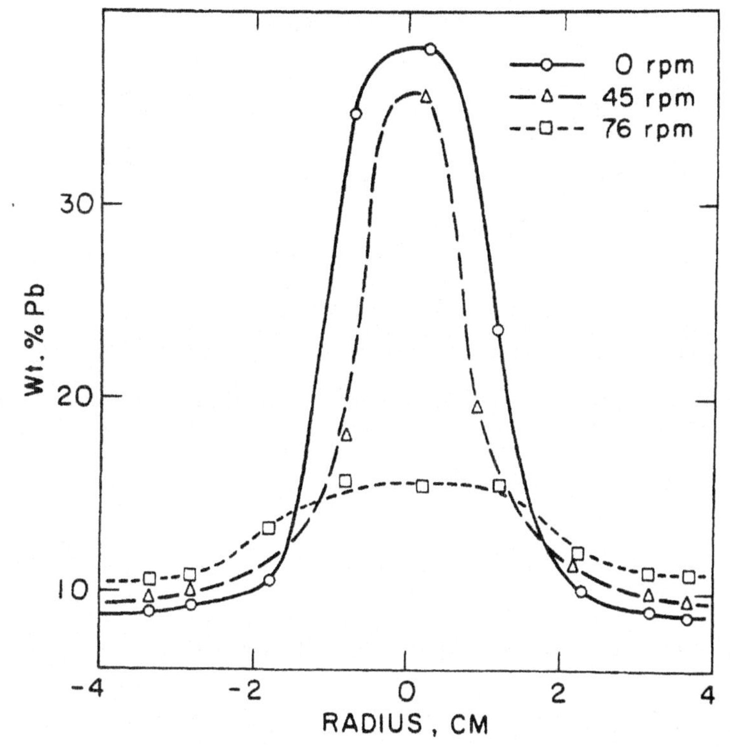

Fig. 6.1-24 Effect of rotation on macrosegregation in a Sn–12% Pb ingot. (From S. Kou, D. R. Poirier, and M. C. Flemings, *Metallurgical Transactions*, **9B**, 1978, p. 711.)

segregates toward the axis of the ingot. Figure 6.1-24 shows the effect of rotation observed in a Sn–12% Pb ingot.[21] The centrifugal force reduces the tendency of the heavy Pb-rich liquid to flow and segregate toward the axis.

6.1.3. Welding

Fusion welding is a joining process in which the coalescence of metal is accomplished by fusion. Fusion welding processes can be classified into three categories: gas welding, arc welding, and high-energy-beam welding, among which arc welding is most widely used.

6.1.3.1. Gas Tungsten Arc Welding

Gas tungsten arc welding (GTAW), as illustrated in Fig. 6.1-25,[22] is a process in which coalescence of metals is produced by heating with an arc between a

[21] S. Kou, D. R. Poirier, and M. C. Flemings, *Metallurgical Trans. B*, **9B**, 711, 1978.

[22] S. Kou, *Welding Metallurgy*, Wiley, New York, 1987.

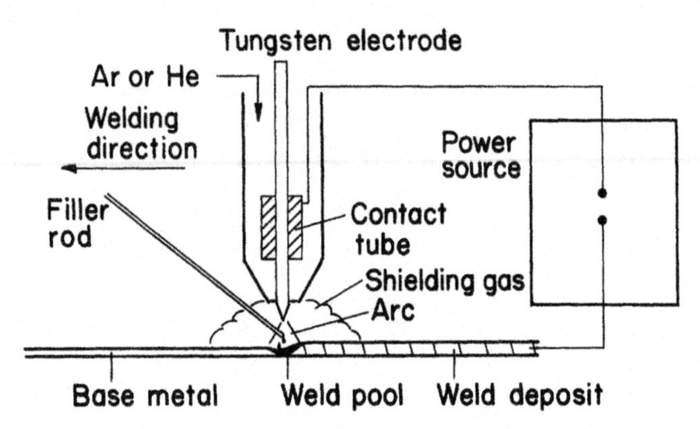

Fig. 6.1-25 Gas tungsten arc welding. (From S. Kou, *Welding Metallurgy*, Copyright, 1987, John Wiley and Sons, New York, Reprinted by permission of John Wiley and Sons, Inc.)

nonconsumable tungsten electrode and the workpiece. Since shielding of the electrode and the weld pool is usually by inert gases such as argon and helium, the process is often called the **tungsten inert gas** (TIG) welding process. The contact tube inside the welding torch conducts the current from the power source to the electrode. It is usually water-cooled in order to prevent the electrode from becoming overheated. For butt joints of thin sheets, filler wires are seldom required.

The GTAW process is suitable for joining thin sections because of its limited allowable heat inputs. Since GTAW is a very clean welding process, it can be used to weld reactive metals, such as titanium, or metals that form refractory oxides, such as aluminum and magnesium. The disadvantage of GTAW, however, is its relatively, low power and hence welding rate. This is because excessive welding currents can cause the tungsten electrode to melt and results in brittle tungsten inclusions in the weld.

6.1.3.2. Gas Metal Arc Welding

Gas metal arc welding (GMAW), as illustrated in Fig. 6.1-26, is a process which produces coalescence of metals by heating them with an arc established between a continuous filler-metal electrode (consumable) and the workpiece. Since shielding of the electrode and the weld pool is usually by argon or helium, the process is often called the **metal inert gas** (MIG) welding process. Since melting of electrode is intended, the welding current and hence the power and the welding rate can be significantly higher than those in GTAW.

6.1.3.3. Laser and Electron Beam Welding

There are two major types of high-energy-beam welding processes: the electron-beam welding (EBW) process and the laser-beam welding (LBW) process. In

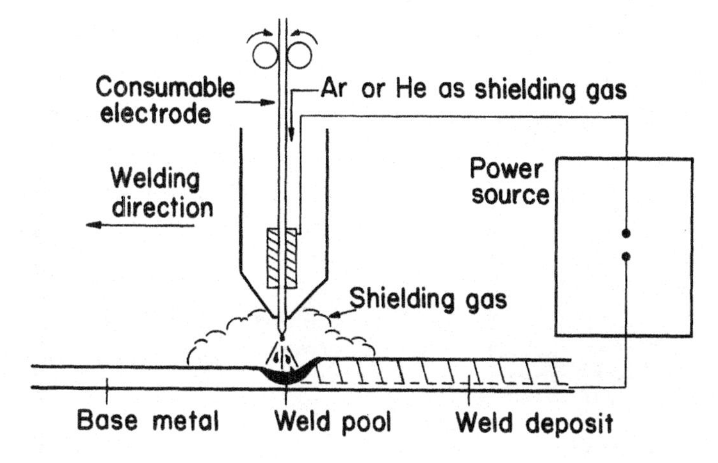

Fig. 6.1-26 Gas metal arc welding. From S. Kou, ibid.

both processes a high intensity beam is focused onto the workpiece (specimen), as illustrated in Figs. 6.1-27*a* and 6.1-27*b*.[23] For example, the electron beam can be focussed to a diameter ranging from 0.3 to 0.8 mm and the resultant power density can be as high as 10^{10} W/m^2.

Due to the very high power density, the metal immediately adjacent to the beam is vaporized to form a deep hole called the **keyhole**, as illustrated in Fig. 6.1-27*c*. As a result, high-energy-beam welds are much deeper and narrower than arc welds. In fact, keyholing is often possible even with relatively thick workpieces. This not only allows the use of very high welding speeds but also permits single-pass welding of joints that would normally require multipass arc welding.

Electron-beam welding requires a vacuum chamber to keep the beam from being scattered by air and to contain the X ray generated by the electron beam during welding. Laser-beam welding does not require vacuum, but the penetration power is significantly lower than that of an electron beam.

6.1.3.4. Filler Wires

Filler wires are not used in EBW and LBW, are optional in GTAW, and are mandatory in GMAW. Welding without the use of filler wires is called **autogenous welding**. Filler wires normally differ from the workpiece in composition. As expected, the weld metal composition depends not only on the compositions of the filler wire and the workpiece but also on their relative amounts in the weld. These relative amounts are shown in Fig. 6.1-28 for two commonly

[23] Y. Arata, *Development of Ultra High Energy Density Heat Source and Its Application to Heat Processing*, Okada Memorial Japan, Society, Tokyo, 1985.

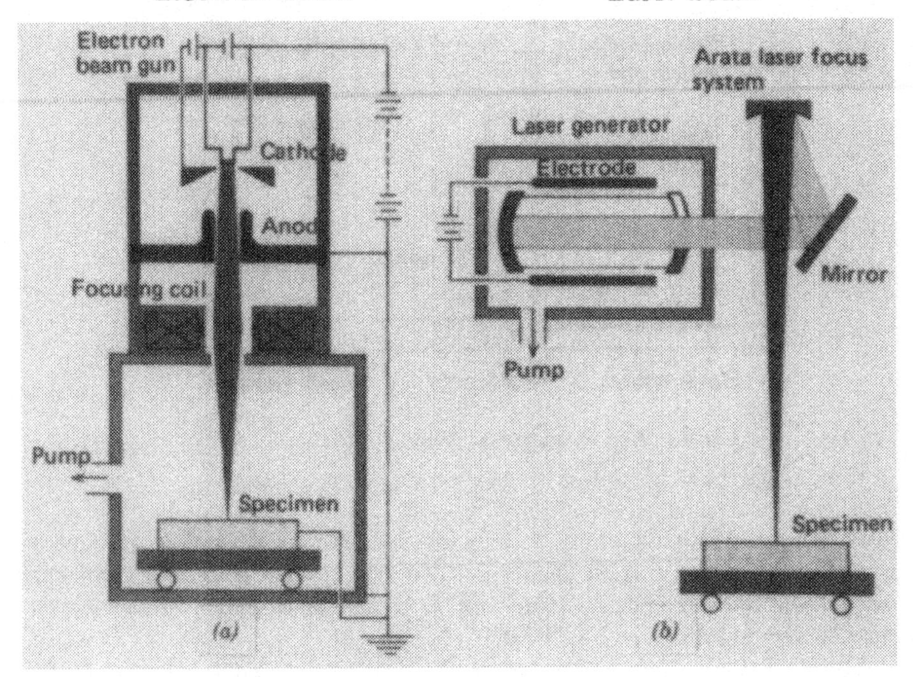

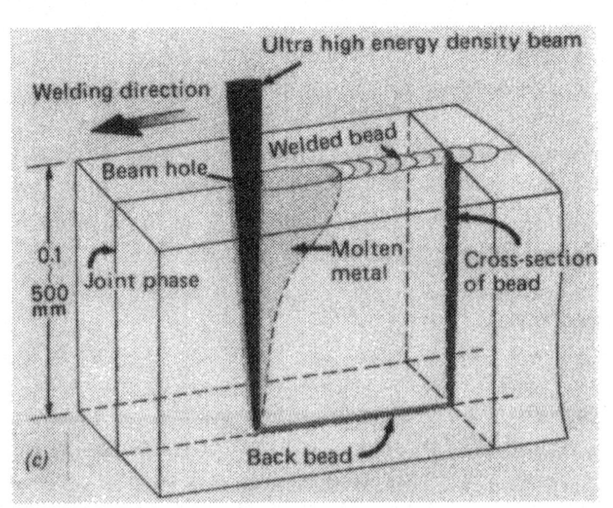

Fig. 6.1-27 High-energy-beam welding: (*a*) electron-beam welding; (*b*) laser-beam welding; (*c*) keyhole. (From Y. Arata, *Development of Ultra High Energy Density Heat Source and Its Application to Heat Processing*, Okada Memorial Japan Society, Tokyo, 1985.)

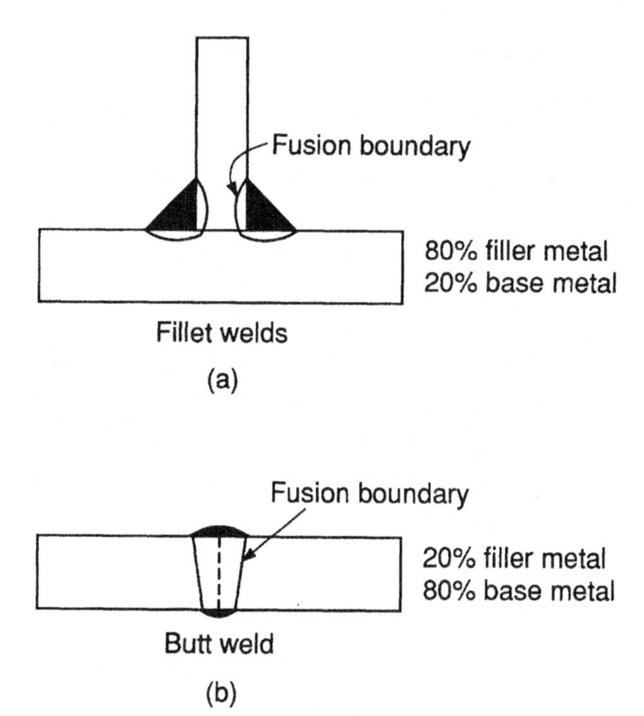

Fig. 6.1-28 Contributions of filler and base metals to weld metal: (*a*) fillet welds; (*b*) butt weld.

encountered joint designs. The dark areas only represent the amounts of filler wires deposited: they do not imply no mixing between the filler wire and the melted base metal.

6.1.3.5. Driving Forces for Convection in Welding

Let us consider the driving forces for convection in the weld pool in gas tungsten arc welding without a filler wire.

The first driving force is the buoyancy force, as shown in Fig. 6.1-29*a*. For all liquids the density ρ decreases with increasing temperature T. The melt at point a should be hotter and hence lighter than the melt at point b, which is closer to the pool boundary and hence the freezing temperature. As such, the former should rise while the latter sink, resulting in a flow pattern similar to that shown in Fig. 6.1-29*b*.

The second driving force is the Lorentz force, as shown in Fig. 6.1-29*c*. The density of the electric current is much higher near the center of the pool surface than near the pool boundary. As such, a Lorentz force $\mathbf{F}$ is created by the nonuniform electric current field and the magnetic field it induces. This force pushes the melt near the axis of the pool inward and downward, resulting in a flow pattern similar to that shown in Fig. 6.1-29*d*.

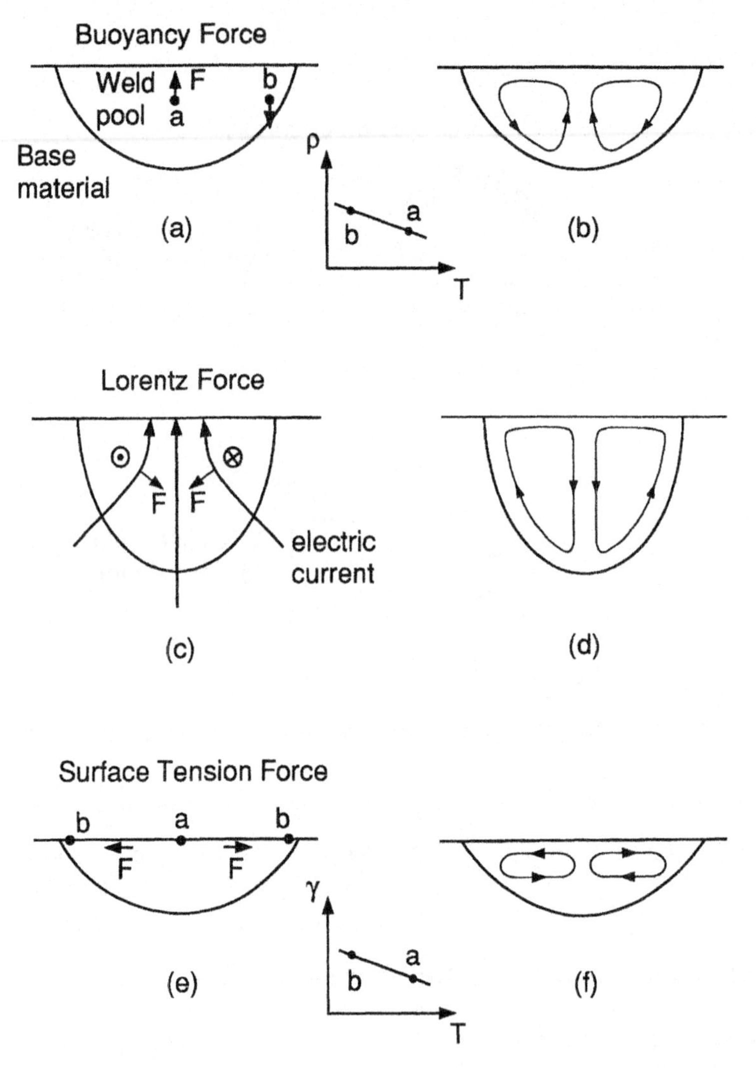

Fig. 6.1-29 Driving forces for weld pool convection: (*a*, *b*) buoyancy force; (*c*, *d*) Lorentz force; (*e*, *f*) surface tension force.

The third driving force is the surface tension force, as shown in Fig. 6.1-29*e*. For most liquids the surface tension γ decreases with increasing temperature T. Therefore, the surface tension is lower at point *a*, where the temperature is higher, and higher at point *b*, where the temperature is lower. As such, the melt at the pool surface is driven radially outward, resulting in a flow pattern similar to that shown in Fig. 6.1-29*f*.

Let us now consider the driving force for convection in a gas tungsten arc. As shown in Fig. 6.1-30*a*, the density of the electric current is higher near the tip of

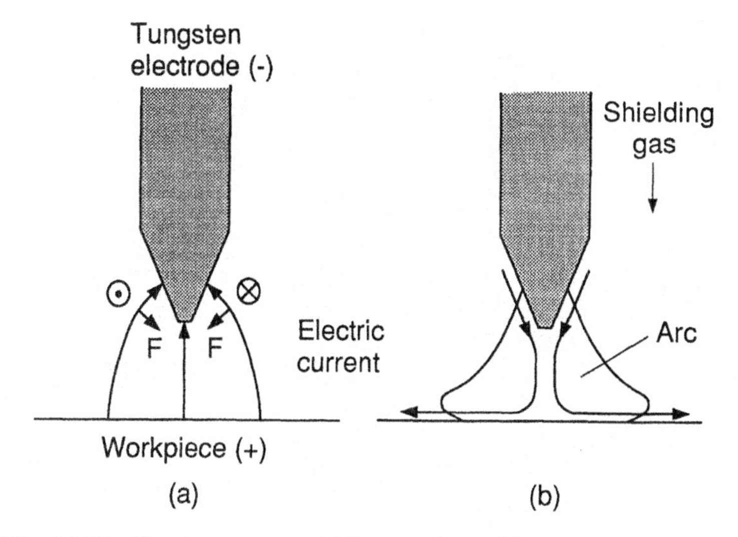

Fig. 6.1-30 Gas tungsten arc: (*a*) Lorentz force; (*b*) convection in the arc.

the electrode and lower below it. As such, a Lorentz force **F** is created by the nonuniform electric-current field and the magnetic field it induces. The force pushes the ionic gas (e.g., Ar^+) in the arc inward and downward, resulting in a flow pattern similar to that shown in Fig. 6.1-30*b*.

More detailed information on fusion welding is available elsewhere.[24]

6.1.4. Powder Processing

A molten metal can be broken up into fine droplets by being subjected to a high-velocity gas jet or a high centrifugal force. The former is often called **gas atomization** and the latter, **centrifugal atomization**. In either case, the mechanical energy is converted into the surface energy, to create new surface for these droplets to form. The droplets cool rapidly into fine powder, due to their small mass and their high-speed travel through the inert-gas atmosphere in the atomization chamber. Such rapidly solidified materials, which often have compositional uniformity and mechanical properties superior to their conventionally cast counterparts, can be consolidated into useful parts under high pressure and temperature.

6.1.4.1. Gas Atomization

Figures 6.1-31*a* and 6.1-31*b* are examples of gas atomization and the powder so produced.[25] High-velocity (Mach $\geqslant 2.5$) pulses of an inert gas impinge on the

[24] S. Kou, *Welding Metallurgy*, Wiley, New York, 1987.

[25] S. J. Savage and F. H. Froes, *J. Metals*, p. 20, April 1984.

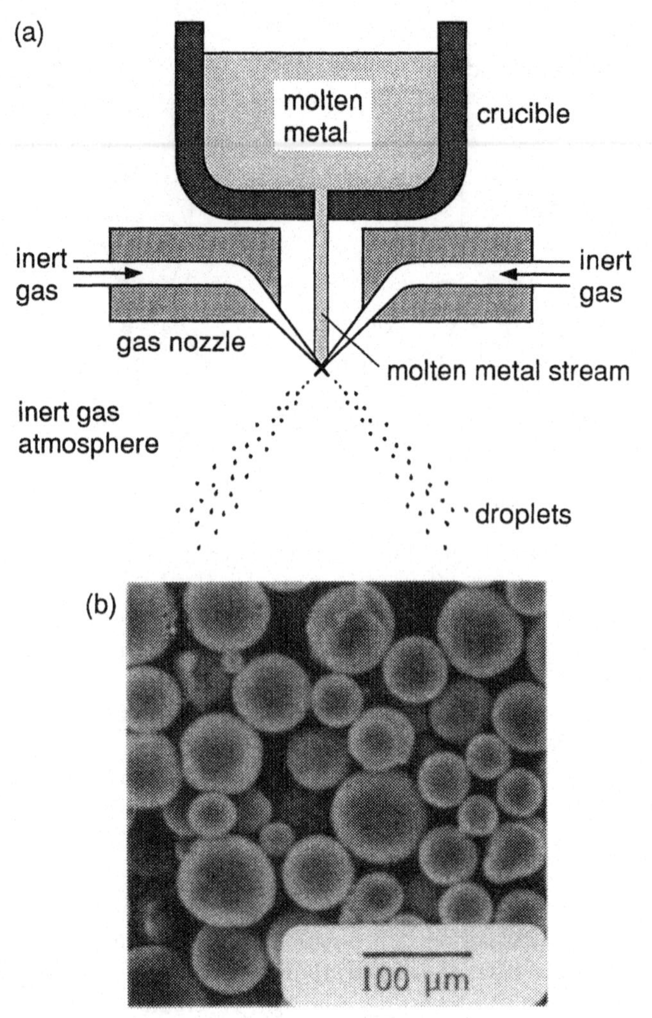

Fig. 6.1-31 Gas atomization: (*a*) process; (*b*) Co alloy powder. (From S. J. Savage and F. H. Froes, *Journal of Metals*, April 1984, p. 20.)

molten metal stream and shear it into fine droplets. Average particle sizes of < 20 μm and cooling rates in excess of 10^6 °C/s have been reported.

6.1.4.2. Centrifugal Atomization

Figures 6.1-32*a* and 6.1-32*b* are examples of centrifugal atomization and the powder so produced. An arc, laser beam, or electron beam can be used as the heat source. As compared to the gas atomization process, the cooling rates are lower ($\sim 10^3$ °C/s), but the melt is free from crucible contamination.

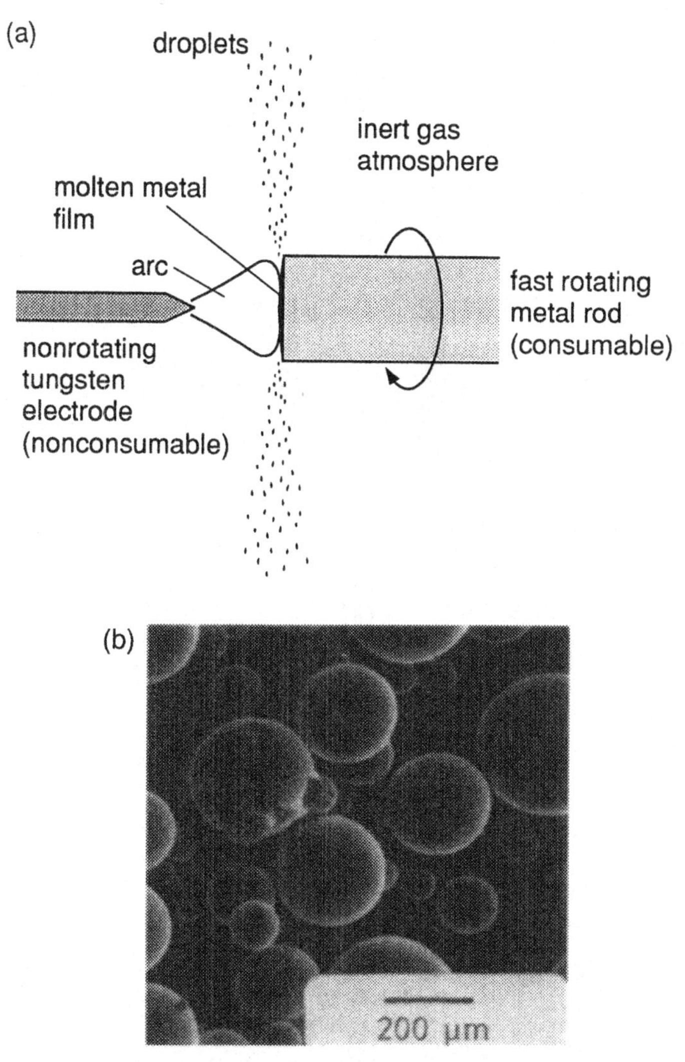

Fig. 6.1-32 Centrifugal atomization: (*a*) process; (*b*) Ti alloy powder. (From S. J. Savage and F. H. Froes, *Journal of Metals*, April 1984, p. 20.)

6.1.5. Fiber Processing

6.1.5.1. Optical Fiber Drawing

The process of drawing optical fiber is illustrated in Fig. 6.1-33. A glass rod, (i.e., the preform) is heated and melted in a cylindrical furnace and is drawn into a fiber, say, 150 μm in diameter at 2 m/s. The high viscosity of the molten glass makes fiber drawing possible. Coating the fiber with an organic substance helps preserve the purity of the fiber and increase its strength. The fiber has to be

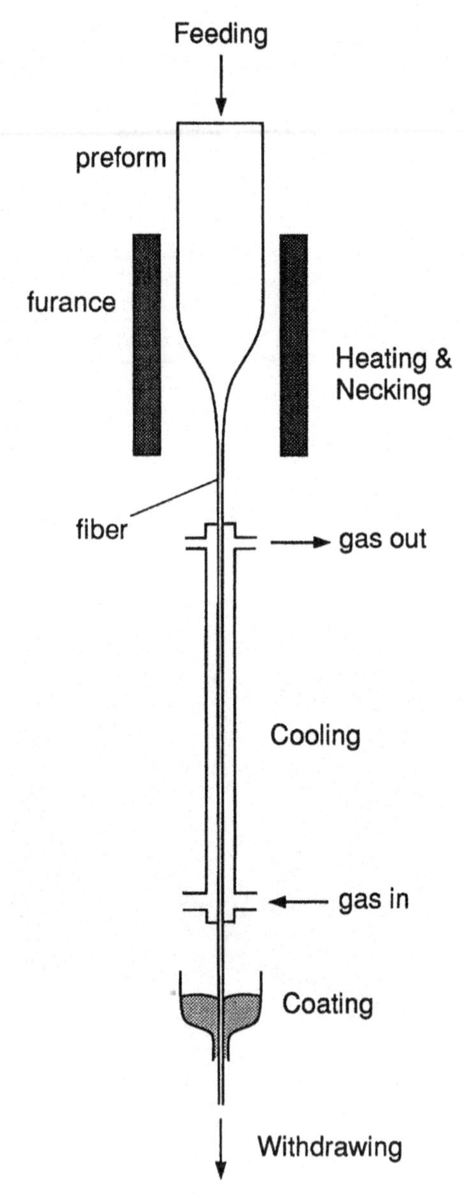

Fig. 6.1-33 Optical fiber drawing.

cooled from $\sim 1600\,^\circ C$ at the exit of the furnace to $< 200\,^\circ C$ to allow coating to be applied. Natural cooling in air is insufficient, if high production rates, that is, high fiber drawing speeds, are desired. Accelerated cooling has been achieved by surrounding the fiber with a channel through which a cooling gas (e.g., He) is

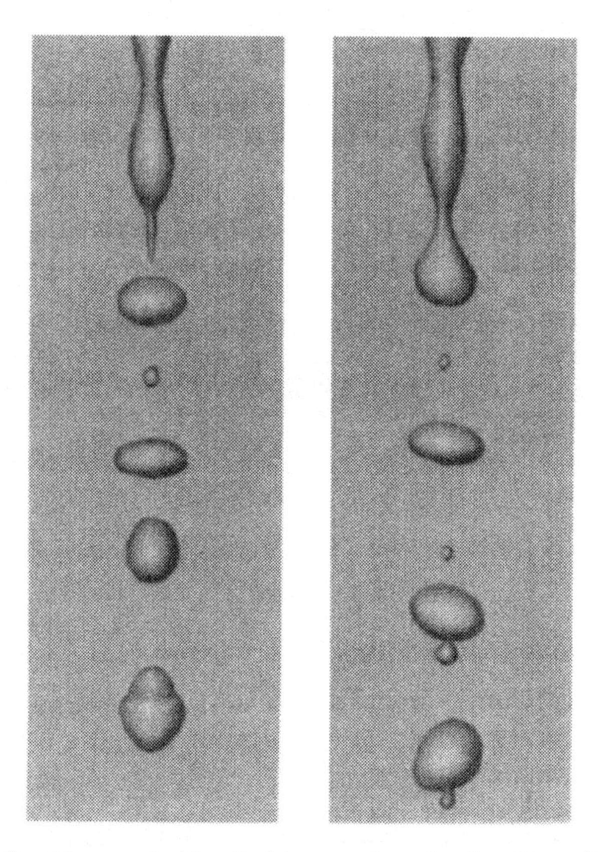

Fig. 6.1-34 Breaking up of a fine liquid stream into droplets (From W. R. Marshall, *Atomization and Spray Drying*, Chem. Engr. Progr. Monograph Series, No. 2, 1954, AIChE, New York.)

passed. The cooling channel can, for example, be a water-cooled aluminum tube of $\sim$5 mm inner diameter.

6.1.5.2. Inviscid Melt Spinning

Drawing of a glass fiber from a highly viscous melt, as described in the previous section, is a routine practice in production today. Similar production of a metal or ceramic fiber from its low-viscosity melt, however, is more difficult. This is because a very fine stream of an inviscid fluid is hydrodynamically unstable. The stream tends to form Rayleigh waves and break up into droplets, due to the influence of the surface tension, as illustrated in Fig. 6.1-34.[26]

[26] W. R. Marshall, *Atomization and Spray Drying*, Chem. Engr. Progr. Monograph Series, No. 2, 1954, AIChE, New York.

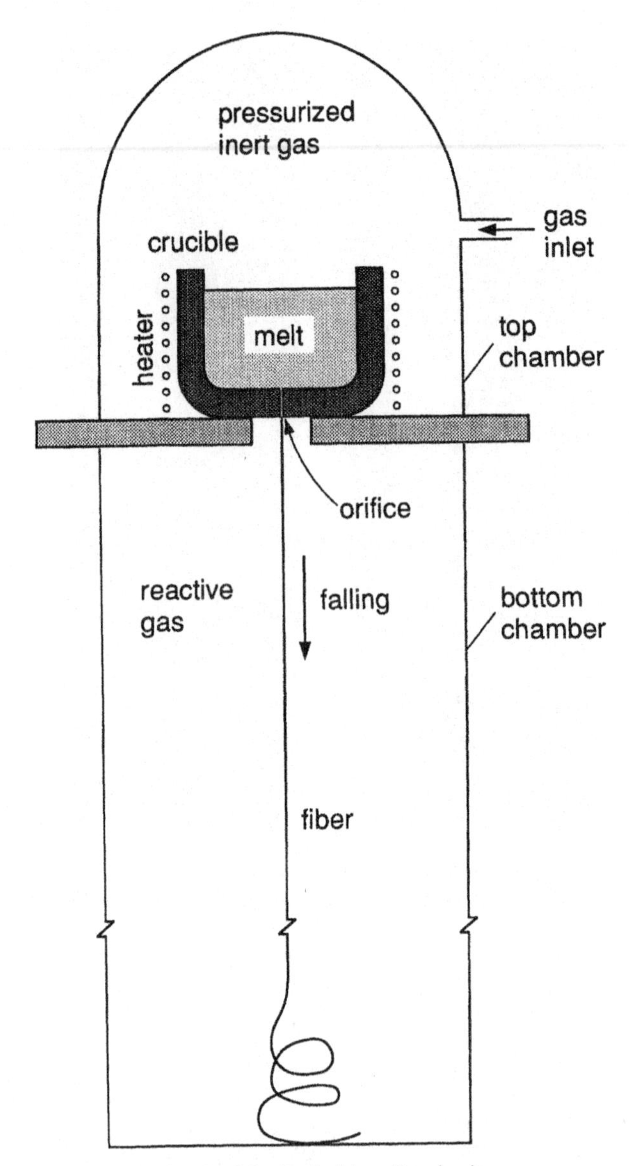

Fig. 6.1-35 Inviscid melt spinning.

To overcome this problem, the inviscid melt spinning process has been developed.[27] By exposing the melt stream to a reactive gas, a very thin sheath can form on the surface and, therefore, stabilizes the stream. For example, oxygen can stabilize an aluminum stream by forming an Al_2O_3 sheath. Similarly, propane can stabilize an Al_2O_3 stream by depositing a carbon sheath.

[27] F. T. Wallenberger, N. E. Weston, and S. A. Dunn, *J. Materials Research*, **5**, 2682, 1990.

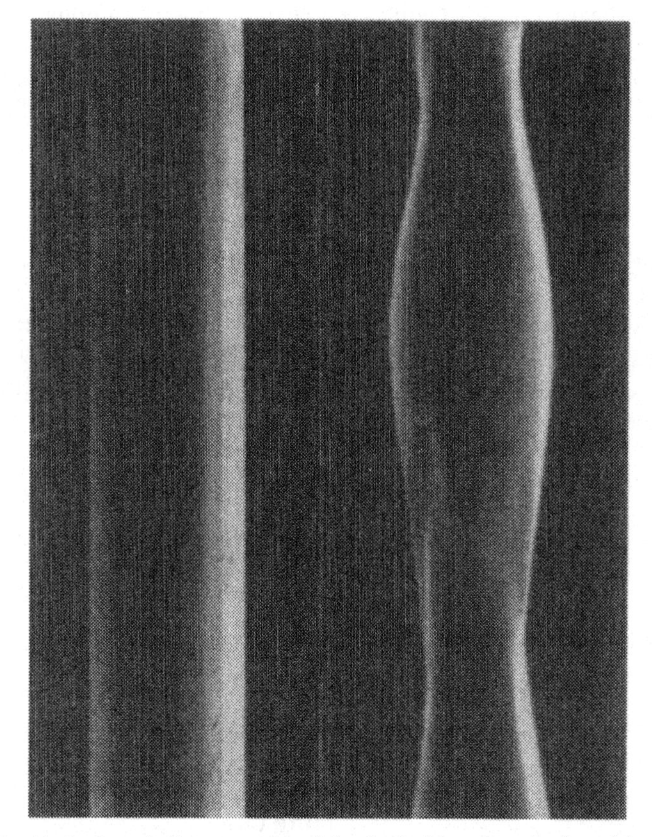

Fig. 6.1-36 Al_2O_3-based fibers produced by inviscid melt spinning: (*a*) with sheath stabilization; (*b*) without sheath stabilization. (From F. T. Wallenberger, N. E. Weston, and S. A. Dunn, "Inviscid Melt Spinning: As Spun Crystalline Alumina Fibers," *Journal of Materials Research*, **5**, 1990, p. 2682.)

As illustrated in Fig. 6.1-35, the melt in the top chamber is forced through a small orifice at the bottom of the crucible by a pressurized inert gas. On exposure to the reactive gas in the bottom chamber, the melt stream exiting from the orifice is stabilized to solidify into a continuous fiber. Figures 6.1-36*a* and 6.1-36*b* show Al_2O_3 fibers produced with and without proper sheath stabilization, respectively.

6.2. SOLID-TO-SOLID PROCESSING

Many steels, aluminum alloys, and nickel-base superalloys are heat-treatable, that is, they can be heat-treated to develop the desired mechanical properties.

Although the heat-treating schedules can be quite different among these materials, they usually involve heat treating at an elevated temperature and subsequent cooling at a predetermined rate to the room temperature. Reheating at a relatively low temperature may or may not be required, depending on the material involved.

6.2.1. Bulk Heat Treating

For applications where strength is critical, the material is usually heat-treated all the way through, that is, bulk heat treating. In such heat treating, the material to be heat-treated is often heated up in a resistance or induction furnace. After reaching the desired heat-treating temperature, it is removed from the furnace and quenched in water (e.g., steels and aluminum) or just cooled in air (e.g., Ni-base superalloys and some stainless steels).

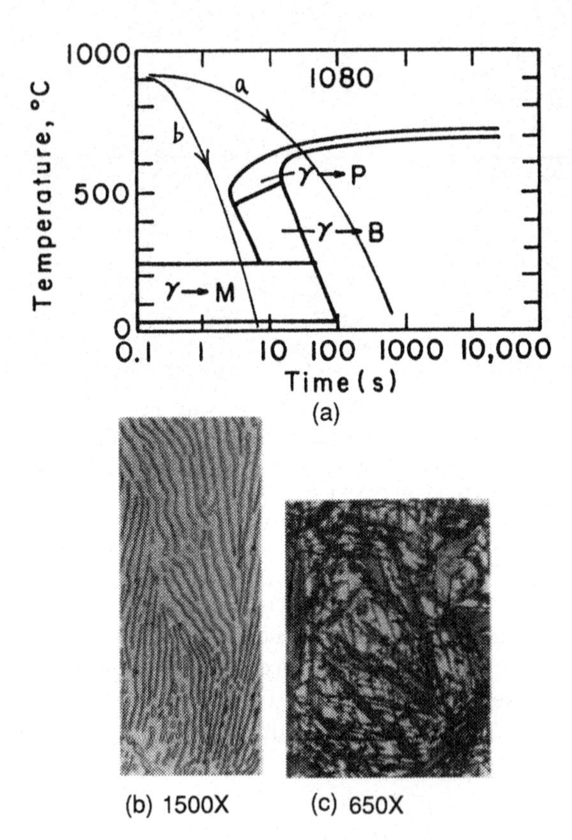

Fig. 6.2-1 Heat treating of steel: (*a*) continuous cooling transformation diagram of 1080 steel. (From C. R. Brooks, *Heat Treatment of Ferrous Alloys*, Hemisphere Publishing, Washington, D.C., 1979, p. 44. Reproduced with permission.) (*b*) Pearlite; (*c*) martensite. (From *Metals Handbooks*, vol. 8, 8th ed., ASM, Metals Park, OH, 1973, p. 65. Reprinted by permission.)

The desired cooling rate from the elevated temperature is often determined from the **continuous cooling transformation diagram**. Let us use 1080 steel (essentially Fe–0.80 wt% carbon) as an example, as shown in Fig. 6.2-1. First, the material is kept at the heat-treating temperature of 900 °C to form the austenite phase (γ). On cooling the austenite phase can transform to lower-temperature phases such as the equilibrium phase pearlite (P), and the nonequilibrium phases martensite (M) and bainite (B). The continuous cooling transformation diagram of 1080 steel is shown in Fig. 6.2-1*a*.[28] With cooling path *a*, the steel transforms slowly from austenite to pearlite during cooling and the resultant steel exhibits a pearlitic microstructure. An example of pearlite is shown in Fig. 6.2-1*b*.[29] With cooling path *b*, however, the steel transforms quickly from austenite to martensite and the resultant steel exhibits a martensitic microstructure. An example of martensite is shown in Fig. 6.2-1*c*.[29] Since martensite is much harder than pearlite, the fast-cooled steel exhibits a higher strength and a lower ductility than the slowly cooled one. This example shows that the rate of cooling from the heat-treating temperature can be used to control the microstructure and properties of alloys.

6.2.2. Surface Heat Treating

For applications where wear or erosion resistance is critical, the surface of a steel can be hardened by heat treating, that is, surface heat treating. In so doing the surface of the material is heated up rapidly. Rapid heating is necessary; otherwise heat will be dissipated into the bulk material rather than utilized for surface hardening. Rapid surface heating is possible with an induction coil, a laser beam, or an electron beam. The laser or electron beam is usually defocused or oscillated to prevent undesirable surface melting.

The use of laser or electron beam for surface hardening has increased significantly recently. It is very attractive since the beam heats up the surface instantaneously without heating up the bulk material. In this way, when the beam is turned off, or more commonly, when the surface is scanned under the beam, the cool bulk material acts as a perfect heat sink to rapidly self-quench the surface to form martensite. Figure 6.2-2*a* is an illustration of laser surface heat treating. A laser-hardened surface layer of an alloy steel is shown in Fig. 6.2-2*b*.[30]

[28] C. R. Brooks, *Heat Treatment of Ferrous Alloys*, Hemisphere Publishing, Washington, DC, 1979, p. 44.

[29] *Metals Handbooks*, vol. 8, 8th ed., ASM, Metals Park, OH, 1973, p. 65.

[30] Y. Arata, *Development of Ultra High Density Heat Source and Its Application to Heat Processing*, Okada Memorial Japan Society, 1985, p. 512.

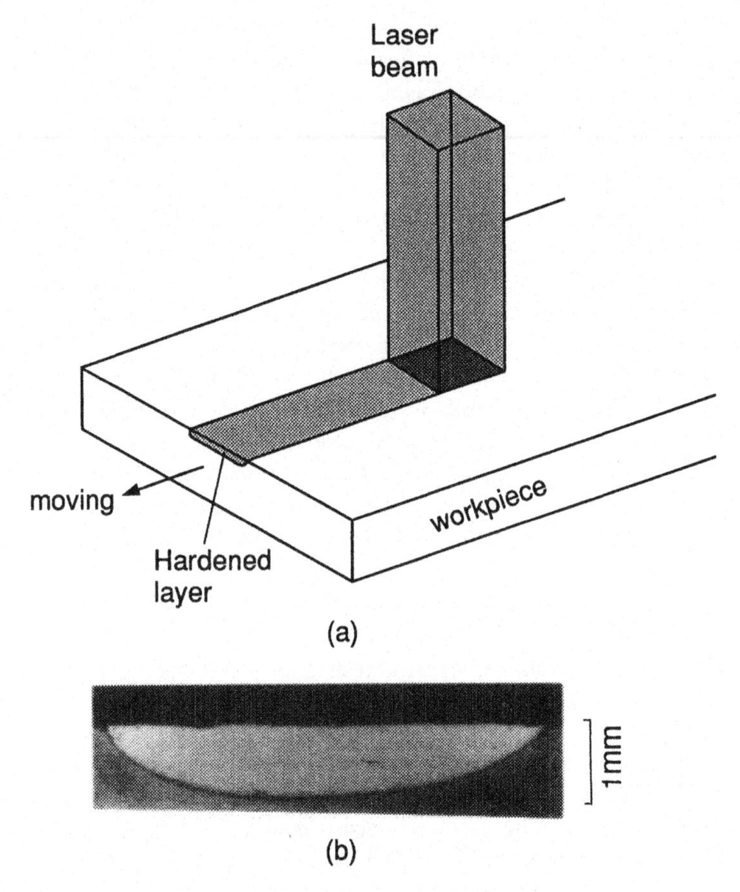

Fig. 6.2-2 Laser transformation hardening: (*a*) process; (*b*) hardened layer in an alloy steel. (From Y. Arata, *Development of Ultra High Energy Density Heat Source and Its Application to Heat Processing*, Okada Memorial Japan Society, Tokyo, 1985.)

6.3. GAS-TO-SOLID PROCESSING

6.3.1. Surface Heat Treating

Carburizing is a surface heat treating process in which the carbon content of the surface of a steel is increased, usually to between 0.8 and 1 wt%, by exposure to a gas atmosphere at an elevated temperature, often between 850 and 950 °C.[31] Subsequent rapid cooling allows the high-carbon surface layer to transform to martensite, thus producing a hardened surface layer for wear resistance, as shown in the gear in Fig. 6.3-1.[32]

[31] G. Krauss, *Principles of Heat Treatment of Steel*, American Society for Metals, Metals Park, OH, 1980.

[32] *Metals Handbooks*, vol. 9, 9th ed, American Society for Metals, Metals Park, OH, 1985, p. 226.

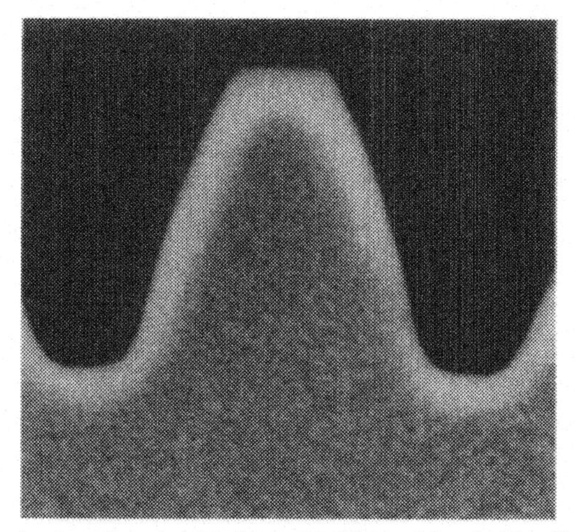

Fig. 6.3-1 Carburized surface layer of a gear. (From *Metals Handbook*, vol. 9, 9th ed., ASM, Metals Park, OH, 1985, p. 226.)

As illustrated in Fig. 6.3-2, the gas atmosphere can be a mixture of CO and CO_2, with or without an inert gas such as N_2, to cause carburization by the following reaction:

$$2CO(g) \rightarrow CO_2(g) + C(s) \tag{6.3-1}$$

For a given gas composition and temperature, the surface carbon content is fixed and it remains constant as carbon diffuses into steel. The equilibrium constant for the reaction is as follows

$$K_P = \frac{P_{CO_2}}{P_{CO}^2} a_C \tag{6.3-2}$$

where P_{CO_2} and P_{CO} are the partial pressures of CO_2 and CO in the gas mixture, respectively. The activity of carbon a_C is a function of the carbon concentration w_C as follows:

$$a_C = f_C w_C \tag{6.3-3}$$

where f_C is the activity coefficient. The equilibrium constant K_P has been determined to be a function of temperature T as follows:[33]

$$\log K_P = \frac{-8918}{T\ (K)} + 9.1148 \tag{6.3-4}$$

[33] G. Krauss, *Principles of Heat Treatment of Steel*, American Society for Metals, Metals Park, OH, 1980, p. 253.

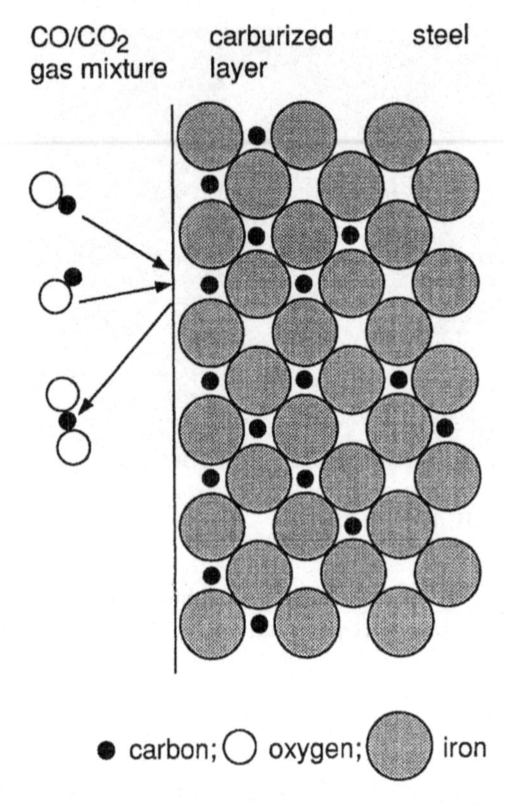

Fig. 6.3-2 Carburizing of a steel.

From Eqs. [6.3-2] through [6.3-4], it is seen that the surface carbon concentration w_C depends on both temperature T and a parameter K defined by

$$K = \frac{P_{CO}^2}{P_{CO_2}} \quad [6.3\text{-}5]$$

Figure 6.3-3 can be used to find w_C from T and K; the total pressure of the gas mixture is 1 atm.[34] Similar information is also available for carburization by the reaction

$$CH_4(g) \rightarrow 2H_2(g) + C(s) \quad [6.3\text{-}6]$$

[34] J. B. Austin and M. J. Day, in *Controlled Atmospheres*, American Society for Metals, Metals Park, OH, 1942.

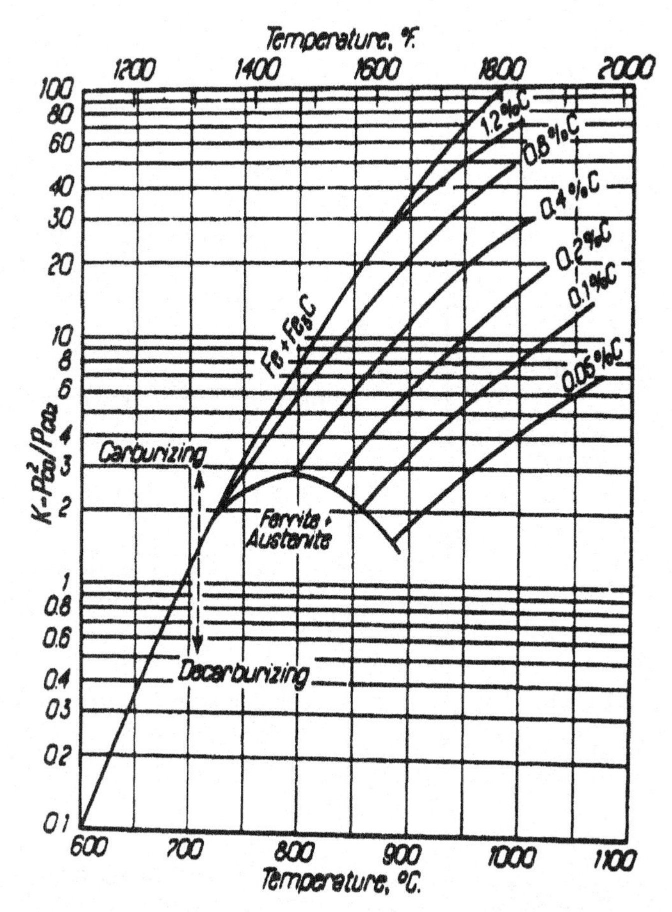

Fig. 6.3-3 Composition of carburizing gas (1 atm) in equilibrium with various carbon-content irons. (From J. B. Austin and M. J. Day, in *Controlled Atmospheres*, ASM, Metals Park, OH, 1942.)

6.3.2. Semiconductor Device Fabrication

The fabrication of silicon devices is illustrated in Fig. 6.3-4.[35] A heavily doped n-type Si wafer is used as the starting substrate (Fig. 6.3-4*a*). A thin film of n-type Si is deposited on the substrate (Fig. 6.3-4*b*), for example, by chemical vapor deposition. A SiO_2 layer is then formed by thermal oxidation (Fig. 6.3-4*c*). This step is followed by the photolithography process (Figs. 6.3-4*d* and 6.3-4*e*). The oxide is first coated with a thin film of organic polymer called the photoresist and then covered with a mask (Fig. 6.3-4*d*), which is transparent to ultraviolet

[35] E. S. Yang, *Fundamentals of Semiconductor Devices*, McGraw-Hill, New York, 1987.

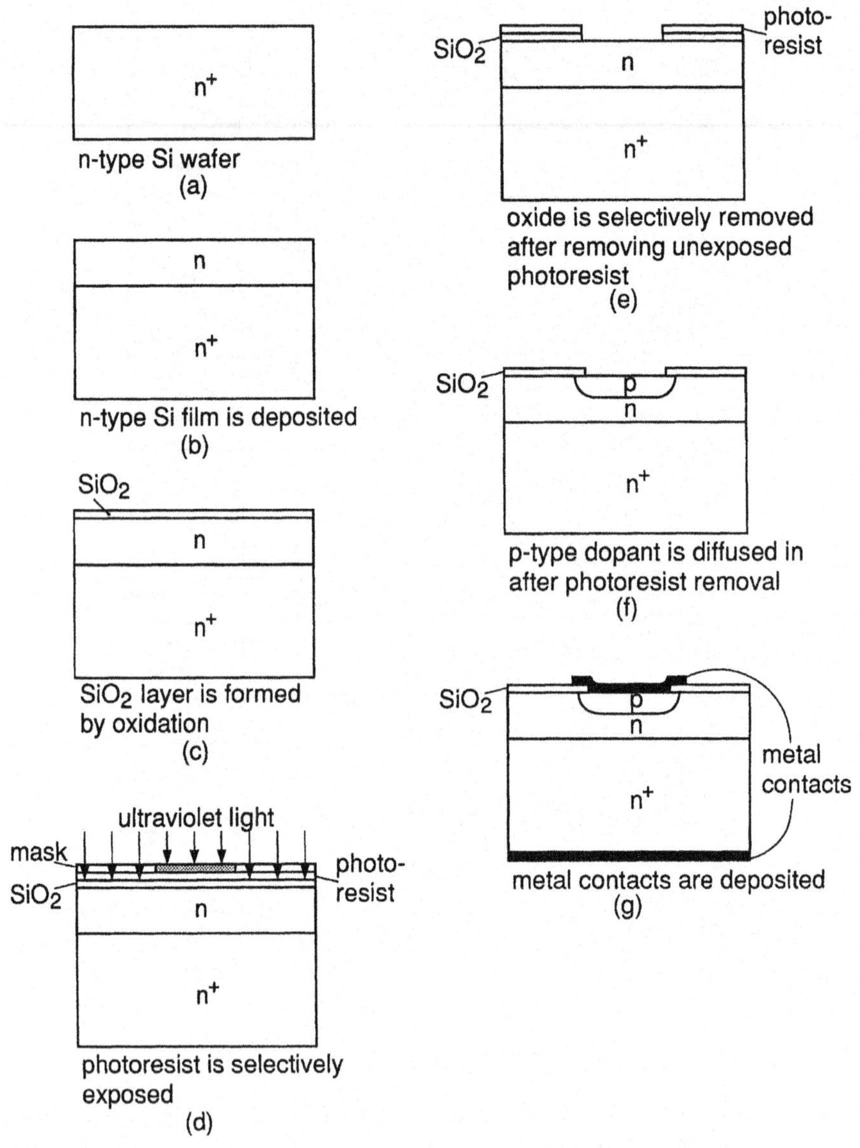

Fig. 6.3-4 Semiconductor device fabrication. (Redrawn from E. S. Yang, *Fundamentals of Semiconductor Devices*, Copyright 1987, McGraw-Hill, New York, Reproduced with permission of McGraw-Hill.)

light except in the opaque region denoted by the shaded area. On exposure to ultraviolet light, the photoresist is polymerized. In the area covered by the opaque region of the mask, however, the unexposed photoresist remains soluble in a special solvent and is so removed. The naked area of the SiO_2 layer is then removed by etching (Fig. 6.3-4*e*). Subsequently, a p-type dopant is diffused through the oxide opening to form a p–n junction (Fig. 6.3-4*f*). Finally, metal films are deposited and etched to form the metal contacts (Fig. 6.3-4*g*), where connecting wires can be attached. The steps of chemical vapor deposition (Fig. 6.3-4*b*), thermal oxidation (Fig. 6.3-4*c*), and thermal diffusion (Fig. 6.3-4*f*) will be discussed in greater detail below.

6.3.2.1. Chemical Vapor Deposition

Chemical vapor deposition is a widely used process for growing thin films on Si and other substrates. One simple version of this process is illustrated in Fig. 6.3-5. The Si wafer, placed on a rotatable graphite susceptor in a quartz tube reactor, is heated by heating the graphite susceptor to typically above 1000 °C with an induction heater. The vapor does not deposit on the quartz tube as quartz cannot be induction-heated.

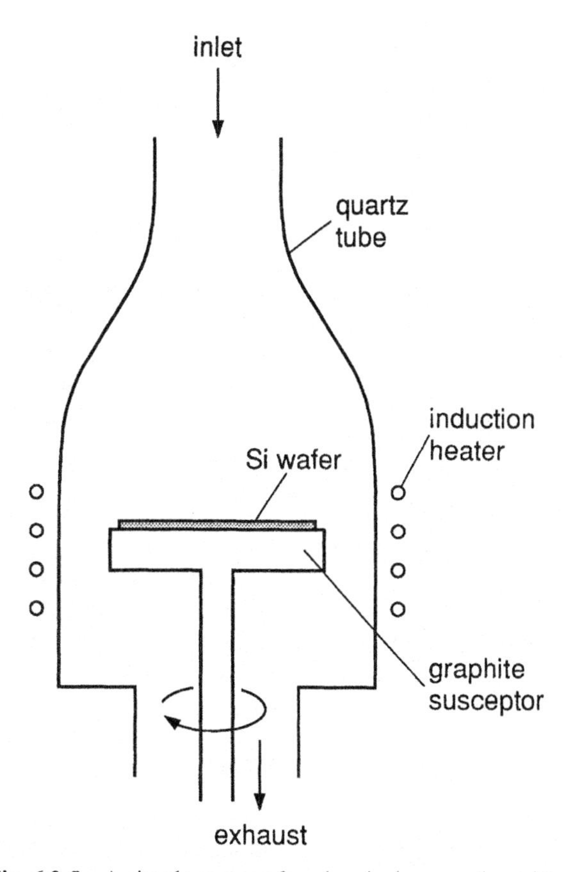

Fig. 6.3-5 A simple reactor for chemical vapor deposition.

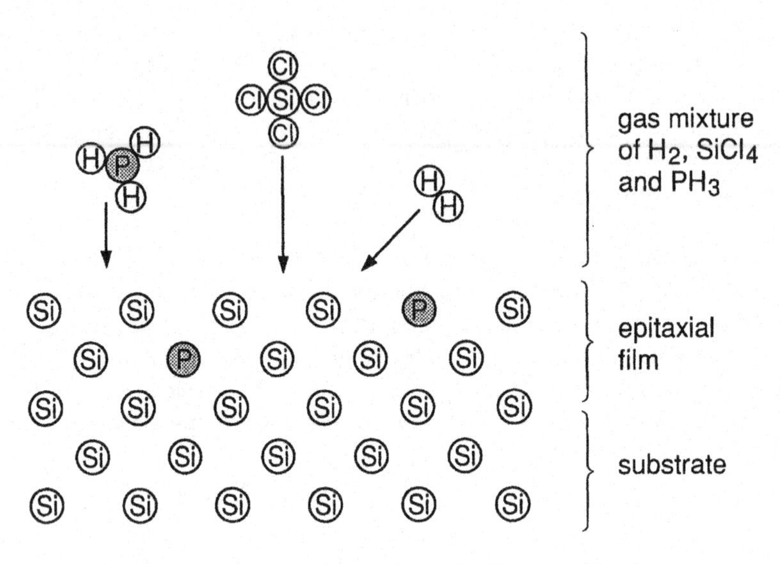

Fig. 6.3-6 Deposition of a p-doped Si film on a Si substrate.

The inlet gas is hydrogen containing a controlled concentration of silicon tetrachloride. The basic reaction is

$$SiCl_4(g) + 2H_2(g) \rightleftharpoons Si(s) + 4HCl(g) \qquad [6.3\text{-}7]$$

The Si single-crystal thin film grows on the substrate with the same lattice structure and orientation as the substrate; this is, **epitaxial growth**.

Chemicals containing the atoms to be doped in the thin film are introduced in the inlet gas; examples are phosphine (PH_3) for n-type doping and diborane (B_2H_6) for p-type doping.[36] Figure 6.3-6 illustrates the formation of a P-doped Si film.

6.3.2.2. Thermal Oxidation

Thermal oxidation in Si device fabrication is to form a SiO_2 layer (Fig. 6.3-4*c*) that can protect the device surface and/or provide a mask for selective diffusion (Fig. 6.3-4*f*). Either a dry or a steam oxidation process can be used, as shown by the following chemical reactions:

$$\text{Dry oxidation:} \quad Si(s) + O_2(g) \rightarrow SiO_2(s) \qquad [6.3\text{-}8]$$

$$\text{Steam oxidation:} \quad Si(s) + 2H_2O(g) \rightarrow SiO_2(s) + 2H_2(g) \qquad [6.3\text{-}9]$$

[36] R. M. Warner, Jr. and J. N. Fordemwalt, *Integrated Circuits*, McGraw-Hill, New York, 1965, p. 295.

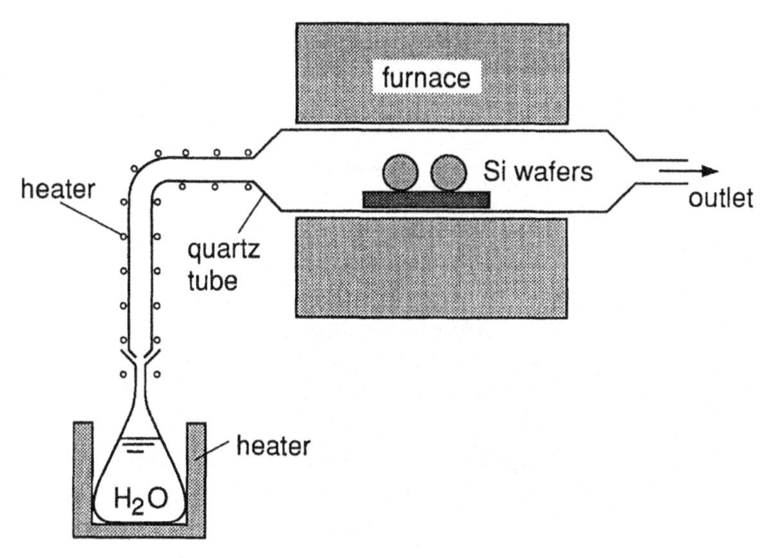

Fig. 6.3-7 Steam oxidation of Si wafers. (Redrawn from E. S. Yang, *Fundamentals of Semiconductor Devices*, Copyright 1987, McGraw-Hill, New York. Reproduced with permission of McGraw-Hill.)

The steam oxidation process is illustrated in Fig. 6.3-7.[37] As shown in Fig. 6.3-8, the growth mechanism of the SiO_2 layer is such that the oxidant, either O_2(g) or H_2O(g), diffuses through the layer to the SiO_2/Si interface and react with Si to form SiO_2. The concentration of the oxidant at the SiO_2 surface depends on the temperature, and the gas flow rate and its equivalent solubility in SiO_2. At 1000 °C it is 5×10^{16} molecules/cm^3 for dry oxygen and 3×10^{19} molecules/cm^3 for water vapor at the atmospheric pressure.[37]

6.3.2.3. Thermal Diffusion

Thermal diffusion in semiconductor device fabrication consists of two steps: predeposition and drive-in diffusion.

In **predeposition** the wafer is exposed briefly to a dopant-containing gas atmosphere at an elevated temperature so that its surface is saturated with the dopant, as illustrated in Fig. 6.3-9*a*. A typical diffusion system is shown schematically in Fig. 6.3-10; the furnace temperature ranges from 800 to 1200 °C.[37] The liquid dopant source can be boron tribromide BBr_3 for boron diffusion

[37] E. S. Yang, *Fundamentals of Semiconductor Devices*, McGraw-Hill, New York, 1978.

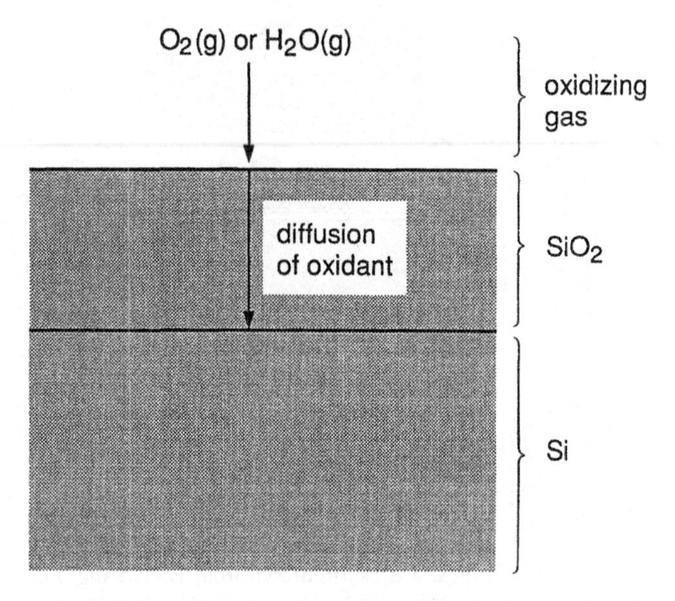

Fig. 6.3-8 Diffusion of oxidant through an SiO_2 layer on a Si substrate.

in silicon. The BBr_3 vapor, which is produced by bubbling an inert carrier gas (e.g., N_2) through the liquid source, is allowed to react with oxygen according to the following reaction:

$$4BBr_3(g) + 3O_2(g) \rightarrow 2B_2O_3(g) + 6Br_2(g) \qquad [6.3\text{-}10]$$

The gaseous B_2O_3 then reacts with silicon as follows

$$2B_2O_3(g) + 3Si(s) \rightarrow 4B(s) + 3SiO_2(g) \qquad [6.3\text{-}11]$$

The boron so produced is incorporated into silicon, whereas the SiO_2 forms a thin layer on the surface. The concentration of the dopant at the surface of the wafer is nominally equal to the solubility of the dopant in silicon, which is given in Fig. 6.3-11 as a function of temperature for several dopants in silicon.[38]

After predeposition, extended thermal diffusion can be applied to reduce the surface dopant concentration and push the dopant deeper into the bulk of the substrate. This step, called **drive-in diffusion**, is illustrated in Figs. 6.3-9*b*.

[38] S. M. Sze, *Semiconductor Devices, Physics and Technology*, Wiley, New York, 1985, p. 319.

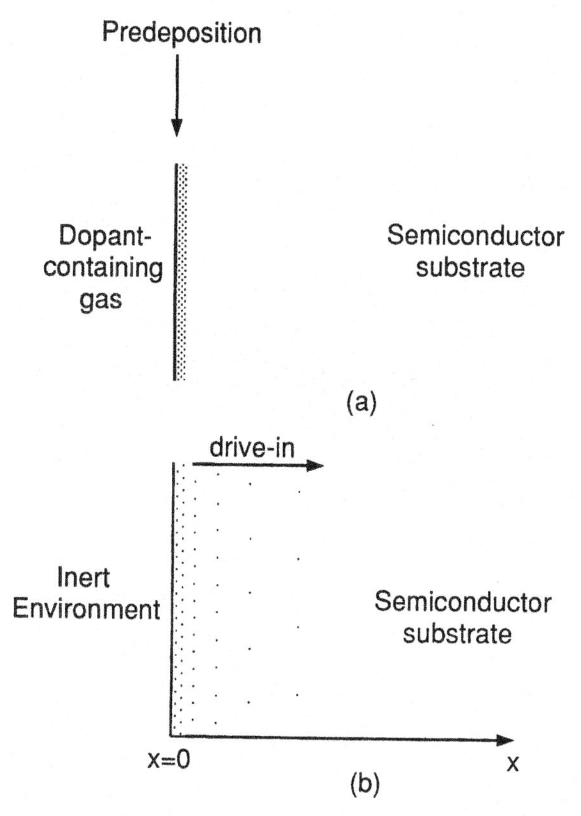

Fig. 6.3-9 Thermal diffusion in a semiconductor substrate: (*a*) predeposition; (*b*) drive-in.

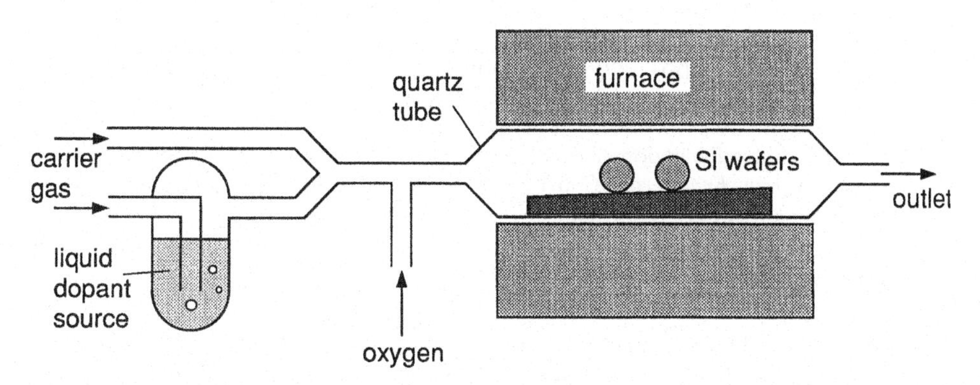

Fig. 6.3-10 Thermal diffusion system for Si wafers. (Redrawn from E. S. Yang, *Fundamentals of Semiconductor Devices*, Copyright 1987, McGraw-Hill, New York. Reproduced with permission of McGraw-Hill.)

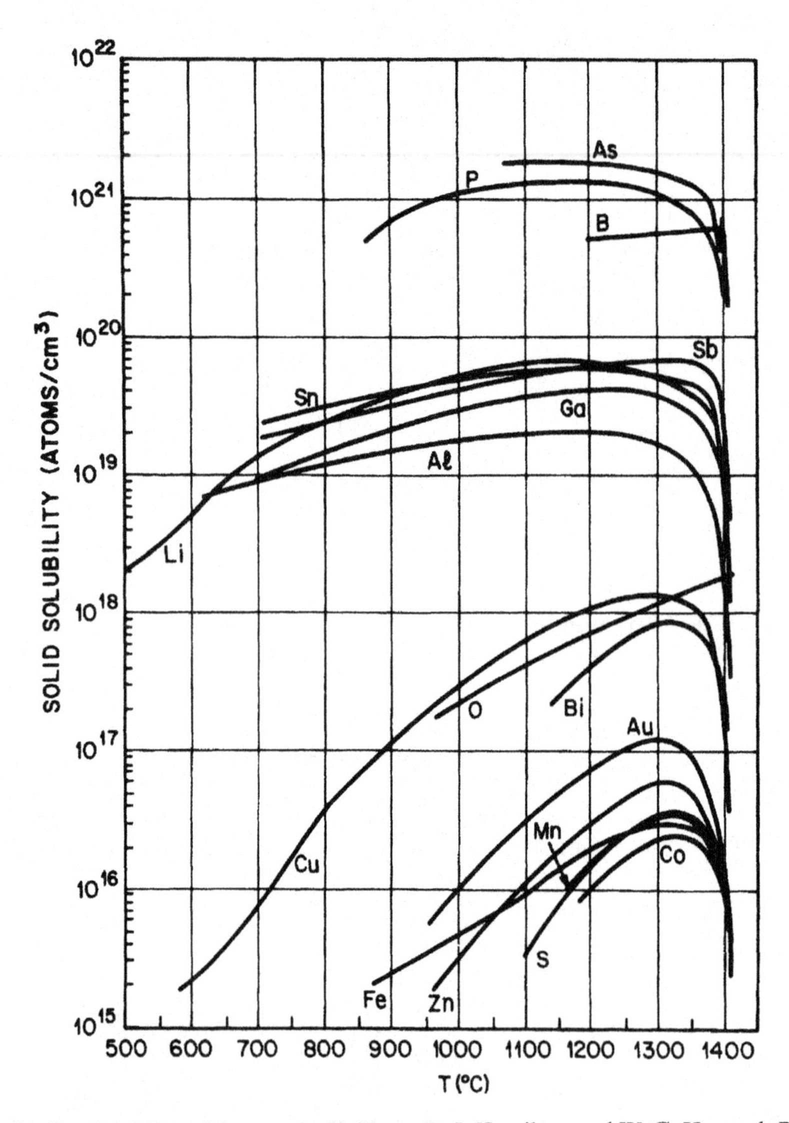

Fig. 6.3-11 Solubility of dopants in Si. (From D. J. Hamilton and W. G. Howard, *Basic Integrated Circuit Engineering*, Copyright 1975, McGraw-Hill, New York. Reproduced with permission of McGraw-Hill.)

PROBLEMS[39]

6.1-1 Consider the Czochralski pulling of a single crystal. How would the crystal diameter and the growth front shape be affected if the pulling rate were increased?

[39] The first two digits in problem numbers refer to section numbers in text.

6.1-2 Consider the Czochralski pulling of a single crystal. How would the crystal diameter and the growth front shape be affected if the crucible temperature were increased?

6.1-3 Is Czochralski pulling feasible in microgravity, that is, in space? Explain.

6.1-4 Rate the following three versions of Bridgman crystal growth in the decreasing order of the extent of buoyancy convection: (a) crystal growing from the bottom, (b) crystal growing horizontally, and (c) crystal growing from the top.

6.1-5 What problems are expected if the melt wets the Bridgman crucible?

6.1-6 How is the tendency for a floating zone to collapse under normal gravity related to the density and the surface tension of the melt?

6.1-7 Can an induction heater be used to produce and even help levitate a floating zone in a semiconductor, metallic, or oxide material?

6.1-8 Can magnetic damping of convection be expected to work in the crystal growth of a semiconductor, metallic or oxide material?

6.1-9 Why is it more difficult to maintain a planar growth front for an equilibrium segregation coefficient $k_0 = 0.01$ than 0.2?

6.1-10 Why does a planar growth front tend to break down more easily if the initial dopant concentration w_{A0} is higher?

6.1-11 To maintain a planar growth front the G/R ratio must be sufficiently high. For metals a high R can be used by keeping G very high. Why is this practice not recommended for semiconductors?

6.1-12 In continuous casting a high concavity of the growth front toward the melt can often reduce the quality of the solidified metal significantly. How is the concavity affected by: (a) the inlet melt temperature, (b) the casting speed, and (c) the transverse cross-sectional area of the ingot?

6.1-13 Higher cooling rates often result in better-quality castings. Why are cooling rates lower in sand casting than in die casting?

6.1-14 Sand casting often allows more flexibility in designing a complex shape casting than permanent mold casting (e.g., die casting). Explain why.

6.1-15 Show that the direction of the Lorentz force in the weld pool is not reversed when the polarity – that is, the direction of current flow – is reversed.

6.1-16 The efficiency of a welding heat source is defined as the ratio of the power actually delivered to the workpiece over the power available for welding. Why is the efficiency higher with GMAW than with GTAW?

6.1-17 A metal surface, especially copper and aluminum, can have a high reflectivity to a laser beam, especially a high-power CO_2 laser. How does this affect the efficiency of the heat source? How can the efficiency be improved?

6.1-18 In electron-beam welding the voltage and current are on the levels of kilovolts and milliamperes, respectively. How significant can the Lorentz force be expected in electron-beam welding?

6.1-19 What is the force that keeps the keyhole in electron- or laser-beam welding from closing?

6.1-20 Consider to the effect of rotation on radial macrosegregation shown in Fig. 6.1-24. What would radial segregation be like if the rotation speed were increased much further, say, to 120 rpm?[40]

6.1-21 Consider the solidification of the ingot shown in Fig. 6.1-22. How is radial macrosegregation affected by the shape of the mushy zone and the solidification rate[41] (i.e., how fast the mushy zone travels upward)?

[40] S. Kou, D. R. Poirier, and M. C. Flemings, *Metallurgical Trans. B*, **9B**, 711, 1978.

[41] S. Kou, D. R. Poirier, and M. C. Flemings, *Electric Furnace Proc.*, **53**, 221, 1977.

CHAPTER 7

FLUID FLOW IN MATERIALS PROCESSING

In our discussion on fluid flow in Chapter 1, we derived the overall mass-balance equation (Eq. [1.2-6]). Bernoulli's equation (Eq. [1.9-13]), and the equation of motion (Eq. [1.5-9]). In present chapter these equations will be applied to materials processing, the first two in Section 7.1 and the third one, throughout the rest of the chapter.

7.1. OVERALL BALANCE

7.1.1. Crystal Growth: Czochralski

In the Czochralski process (Section 6.1.1) the single crystal is pulled from the melt contained in a heated crucible, as illustrated in Fig. 7.1-1. The cross-sectional areas of the crystal and the crucible are A_S and A_L, respectively. The crystal is pulled at a constant speed V with respect to the crystal growth laboratory. This speed, however, is not the actual production speed of the crystal V_p.

As the crystal is being pulled from the melt, the melt level drops. Let V_L (taken positive) be the speed at which the melt level drops. Since the crystal rises at the speed $(V + V_L)$ relative to the melt level, the production rate of the crystal is

$$V_p = V + V_L \qquad [7.1\text{-}1]$$

Let the melt be the control volume Ω, the density of the crystal ρ_S, and the density of the melt ρ_L. From Eq. [1.2-6]

$$\frac{d}{dt}(\rho_L \Omega) = 0 - \rho_S(V + V_L)A_S \qquad [7.1\text{-}2]$$

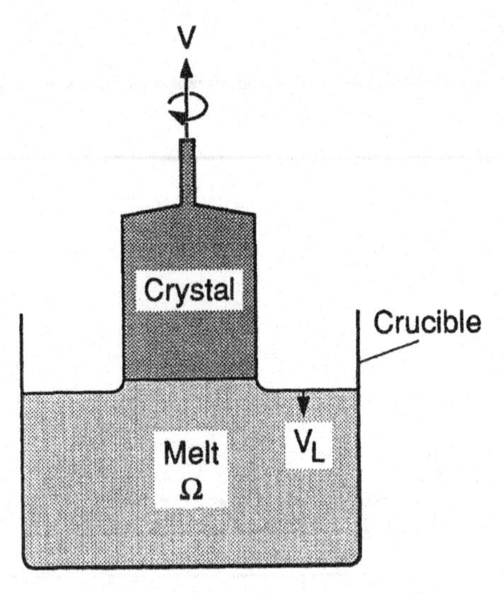

Fig. 7.1-1 Czochralski crystal pulling.

or

$$-\rho_L V_L A_L = -\rho_S (V + V_L) A_S \qquad [7.1\text{-}3]$$

Therefore

$$V_L = \frac{\rho_S A_S V}{\rho_L A_L - \rho_S A_S} \qquad [7.1\text{-}4]$$

Substituting Eq. [7.1-4] into Eq. [7.1-1] yields

$$\boxed{\frac{V_p}{V} = \frac{1}{1 - (\rho_S A_S / \rho_L A_L)}} \qquad [7.1\text{-}5]$$

This equation is identical to that derived by Brandle.[1] In practice the crystal: crucible diameter ratio can be up to a maximum of 0.5, corresponding to an A_S/A_L of 0.25. If $\rho_S = \rho_L$ is assumed as an approximation, $V_p/V = 1.33$, which is significantly greater than 1.

7.1.2. Crystal Growth: Floating-Crucible Czochralski

A floating crucible is often used to help grow a single crystal of uniform composition, as will be described later in Chapter 9. As shown in Fig. 7.1-2, the

[1] C. D. Brandle, in *Crystal Growth*, 2nd ed., B. R. Pamplin, ed., Pergamon Press, Oxford, UK, 1980, p. 275.

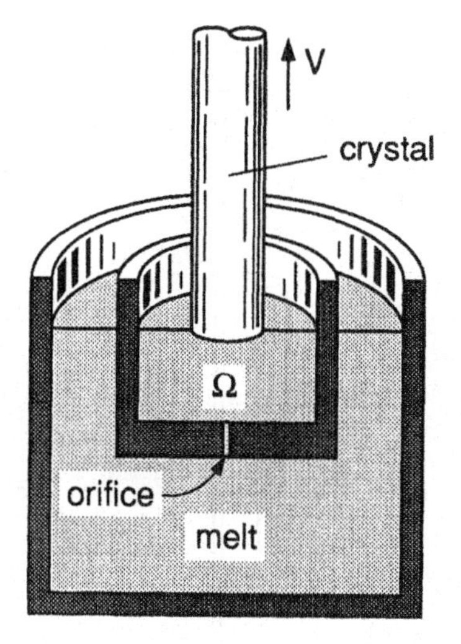

Fig. 7.1-2 Floating-crucible Czochralski crystal pulling.

crystal is pulled from the melt in a floating crucible having an orifice at the bottom. To ensure a uniform crystal composition the average velocity of the melt going through the orifice V_0 must be sufficiently high, to override dopant backdiffusion through the orifice. This will be explained further in Chapter 9.

To determine V_0, let the melt in the floating crucible be the control volume Ω. Since the crucible is floating, Ω is constant. From Eq. [1.2-6]

$$0 = \rho_L V_0 A_0 - \rho_S A_S V_p \qquad [7.1\text{-}6]$$

and so

$$V_0 = \frac{\rho_S A_S V_p}{\rho_L A_0} \qquad [7.1\text{-}7]$$

where V_p the crystal production speed.

Substituting Eq. [7.1-5] into Eq. [7.1-7]

$$\boxed{\frac{V_0}{V} = \frac{\rho_S A_S A_L}{A_0(\rho_L A_L - \rho_S A_S)}} \qquad [7.1\text{-}8]$$

This equation will become useful later in Chapter 9.

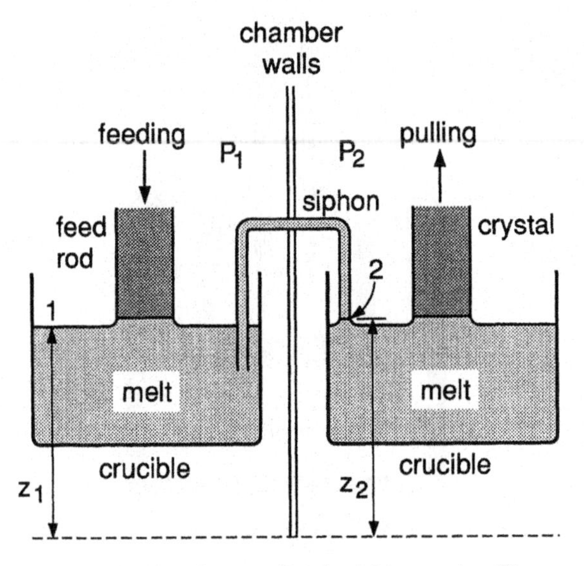

Fig. 7.1-3 Continuous Czochralski crystal pulling.

7.1.3. Crystal Growth: Continuous Czochralski

Czochralski crystal pulling (Section 6.1.1.1) is a batch process. In order to improve the crystal quality and yield, continuous pulling has been investigated,[2] one earlier version being shown in Fig. 7.1-3. The melt is supplied from chamaber 1 (left) of pressure p_1, through a siphon tube of inner cross-sectional area A_0, to chamber 2 (right) of pressure p_2. A feed rod is lowered into the melt in chamber 1 as the crystal is pulled from the melt in chamber 2. The heights of the left melt level (1) and the siphon tube exit (2) are kept constant at z_1 and z_2, respectively.

Let us consider the question of how fast should crystal be pulled and the rod be fed.

From the Bernoulli equation (i.e. Eq. [1.9-13])

$$0 + \rho g z_1 + p_1 = \tfrac{1}{2}\rho v_2^2 + \rho g z_2 + p_2 \qquad [7.1\text{-}9]$$

Therefore

$$v_2 = \sqrt{2\left[g(z_1 - z_2) + \frac{(p_1 - p_2)}{\rho}\right]} \qquad [7.1\text{-}10]$$

[2] G. Fiegl, *Solid State Technol.*, p. 121, Aug. 1983.

and since the mass flow rate $m = A_0 \rho v_2$

$$m = A_0 \rho \sqrt{2\left[g(z_1 - z_2) + \frac{(p_1 - p_2)}{\rho}\right]} \qquad [7.1\text{-}11]$$

This is the rate, in terms of mass per unit time, at which the crystal should be pulled and at which the rod should be fed.

7.1.4. Casting: Shrinkage Pipes

For most metals the density of the solid is about 3–6% greater than that of the liquid ρ_L. At such, a shrinkage pipe is often formed in the resultant casting, as illustrated in Fig. 7.1-4.[3] In order to overcome this problem a riser can be used, as already mentioned in Section 6.1.2. The riser, which is often surrounded by an insulation or even exothermic material, is also called a "hot top." The purpose is to ensure that there will be molten metal available to feed the shrinkage of the casting, as illustrated in Fig. 7.1-5*a*.

The optimum design of the riser, as illustrated in Fig. 7.1-5*b*, is such that the shrinkage pipe extend just to, but not into, the casting.

Before proceeding to determine the optimum riser design, let us consider the volume of the solid in the casting Ω_{SC} at time t after the molten metal is introduced. According to the Chvorinov rule, which will be derived in Section 8.3

$$\Omega_{SC} = C_c A_c \sqrt{t} \qquad [7.1\text{-}12]$$

where C_c is a proportional constant depending on the physical properties of the mold and the casting, and A_c is the casting/mold interface. Similarly

$$\Omega_{SR} = C_R A_R \sqrt{t} \qquad [7.1\text{-}13]$$

where R denotes the riser.

Let $d\Omega_{SC}/dt$ be the rate of increase in the volume of the solid in the casting. The rate of increase in the mass of the casting is $(\rho_S - \rho_L)\, d\Omega_{SC}/dt$, which comes from the liquid in the riser. Let the metal in the riser (i.e., $\rho_S \Omega_{SR} + \rho_L \Omega_{LR}$) be the control volume. From Eq. [1.2-6]

$$\frac{d}{dt}(\rho_S \Omega_{SR} + \rho_L \Omega_{LR}) = 0 - (\rho_S - \rho_L)\frac{d\Omega_{SC}}{dt} \qquad [7.1\text{-}14]$$

[3] A. Ohno, *The Solidification of Metals*, Chijin Shokan, Tokyo, 1976, p. 127.

Fig. 7.1-4 Shrinkage pipe in an aluminum ingot. (From A. Ohno, *The Solidification of Metals*, Chijin Shokan, Tokyo, 1976, p. 127.)

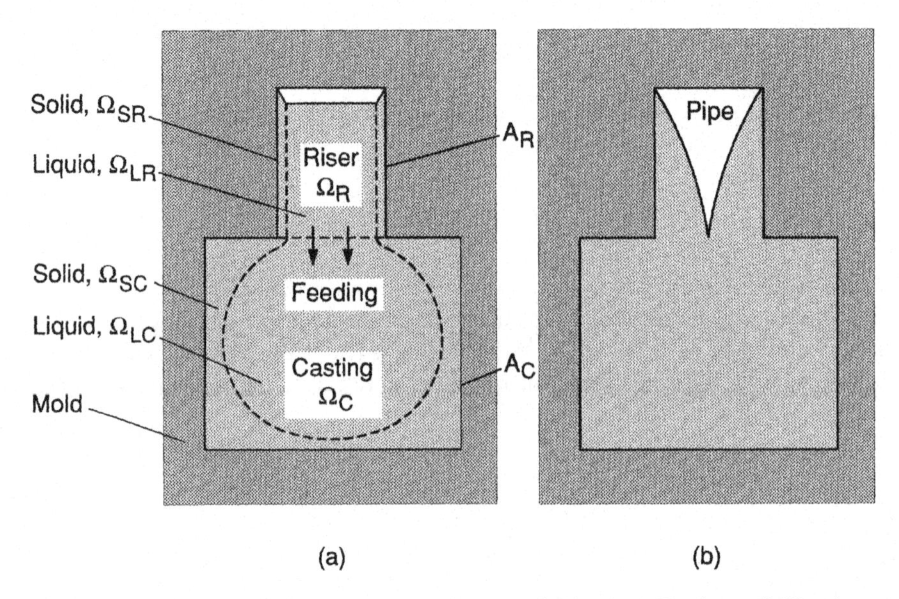

Fig. 7.1-5 Metal casting with a riser: (*a*) during solidification; (*b*) after solidification.

Substituting Eqs. [7.1-12] and [7.1-13] into Eq. [7.1-14]

$$\rho_S C_R A_R \frac{1}{2\sqrt{t}} + \rho_L \frac{d\Omega_{LR}}{dt} = -(\rho_S - \rho_L) C_C A_C \frac{1}{2\sqrt{t}} \qquad [7.1\text{-}15]$$

On rearranging and integrating

$$\int_{\Omega_R}^{0} \rho_L d\Omega_{LR} = -[\rho_S C_R A_R + (\rho_S - \rho_L) C_C A_C] \int_0^{t_f} \frac{1}{2\sqrt{t}} dt \qquad [7.1\text{-}16]$$

where Ω_R is the volume of the riser cavity. Note that the optimum riser design is such that the liquid metal in the riser, Ω_{LR}, just runs out at the time solidification in the casting finishes, t_f. As such

$$-\rho_L \Omega_R = -[\rho_S C_R A_R + (\rho_S - \rho_L) C_C A_C] \sqrt{t_f} \qquad [7.1\text{-}17]$$

From Eqs. [7.1-12]

$$\Omega_C = C_C A_C \sqrt{t_f} \qquad [7.1\text{-}18]$$

where Ω_C is the volume of the casting.

Substituting Eqs. [7.1-18] into Eq. [7.1-17] and dividing by $(-\rho_S \Omega_C)$

$$\boxed{\frac{\rho_L \Omega_R}{\rho_S \Omega_C} = \frac{C_R A_R}{C_C A_C} + \frac{\rho_S - \rho_L}{\rho_S}} \qquad [7.1\text{-}19]$$

This equation is identical to that derived by Adams and Taylor.[4]

7.1.5. Casting: Ladle Discharging

In metal casting (Section 6.1.2), such as continuous, ingot and sand casting, a ladle is used to hold the molten metal. Let us consider the discharge of the molten metal from the ladle shown in Fig. 7.1-6. For simplicity, the inside diameter (ID) D of the ladle is assumed constant. Since the orifice diameter d is much smaller than the ladle diameter D, the velocity at which the free surface (1) drops is negligible as compared to the velocity of the metal discharging from the orifice (2). From the Bernoulli equation, (Eq. [1.9-13]):

$$0 + 0 + 1 \text{ atm} = \tfrac{1}{2}\rho v_2^2 + \rho g(-H) + 1 \text{ atm} \qquad [7.1\text{-}20]$$

[4] C. M. Adams, Jr. and H. F. Taylor, *Trans. Am. Foundryman Soc.*, **61**, 686, 1953.

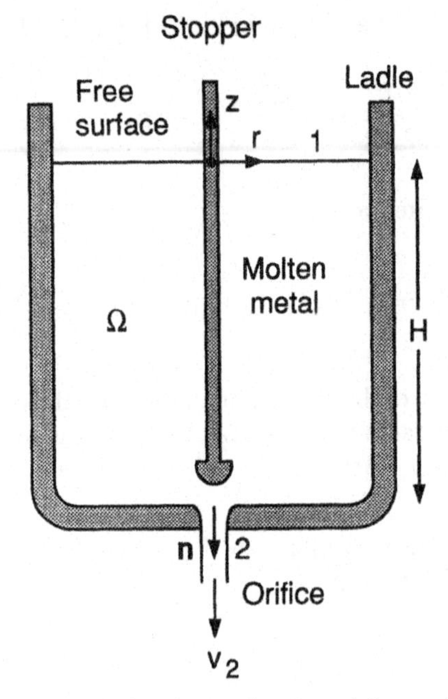

Fig. 7.1-6 Discharging molten metal from a ladel.

and hence

$$v_2 = \sqrt{2gH} \qquad [7.1\text{-}21]$$

where v_2 is the average velocity of the molten metal discharging from the orifice and H is the depth of the molten metal in the ladle.

Let us now determine the pouring time required for making a casting. Let Ω be the volume of the molten metal in the ladle. Equation [1.2-6] reduces to

$$\frac{dM}{dt} = 0 - \rho v_2 \left(\frac{\pi}{4} d^2 \right) \qquad [7.1\text{-}22]$$

where $\pi d^2/4$ is the cross-sectional area of the orifice. Substituting Eq. [7.1-21] into Eq. [7.1-22], we obtain

$$-\frac{dM}{dt} = \frac{\pi}{4} \rho d^2 \sqrt{2gH} \qquad [7.1\text{-}23]$$

Neglecting the volume of the stopper in the ladle

$$M = \rho \left(\frac{\pi}{4} D^2 H \right) \qquad [7.1\text{-}24]$$

From Eqs. [7.1-23] and [7.1-24], we can eliminate H and get

$$\frac{-1}{2}\frac{1}{\sqrt{M}}\,dM = \sqrt{\frac{\pi\rho g}{2}}\frac{d^2}{2D}\,dt \qquad [7.1\text{-}25]$$

Let M_i be the initial mass of the molten metal in the ladle and m_c be the mass required for making the casting. Integrating from $M = M_i$ at $t = 0$ to $M = M_i - m_c$ at $t = t$, we have

$$\sqrt{M_i} - \sqrt{M_i - m_c} = \sqrt{\frac{\pi\rho g}{8}}\frac{d^2 t}{D} \qquad [7.1\text{-}26]$$

Therefore, the required discharging time is

$$\boxed{t = \sqrt{\frac{8}{\pi\rho g}}\frac{D}{d^2}\left(\sqrt{M_i} - \sqrt{M_i - m_c}\right)} \qquad [7.1\text{-}27]$$

7.1.6. Casting: Design of Sprues

Let us consider the flow of the molten metal in the sprue in sand casting. The molten metal is seldom poured directly into the sprue; rather, it is poured into a pouring basin at the top of the mold. Furthermore, the sprue is usually tapered. As illustrated in Fig. 7.1-7,[5] the danger of turbulence and vortexing, with subsequent air entrapment at the sprue entrance, is thereby reduced.

As an approximation, let us consider the pouring basin in Fig. 7.1-8 as a ladle. Let point 1 be at the sprue entrance and point 2 any other location in the sprue. From Eq. [7.1-21]

$$v_1 = \sqrt{2gZ_1} \qquad [7.1\text{-}28]$$

and

$$v_2 = \sqrt{2gZ_2} \qquad [7.1\text{-}29]$$

From mass balance (i.e. Eq. [1.2-8])

$$v_1 A_1 = v_2 A_2 \qquad [7.1\text{-}30]$$

[5] H. F. Taylor, M. C. Flemings, and J. Wulff, *Foundary Engineering*, Wiley, 1959, pp. 186, 199.

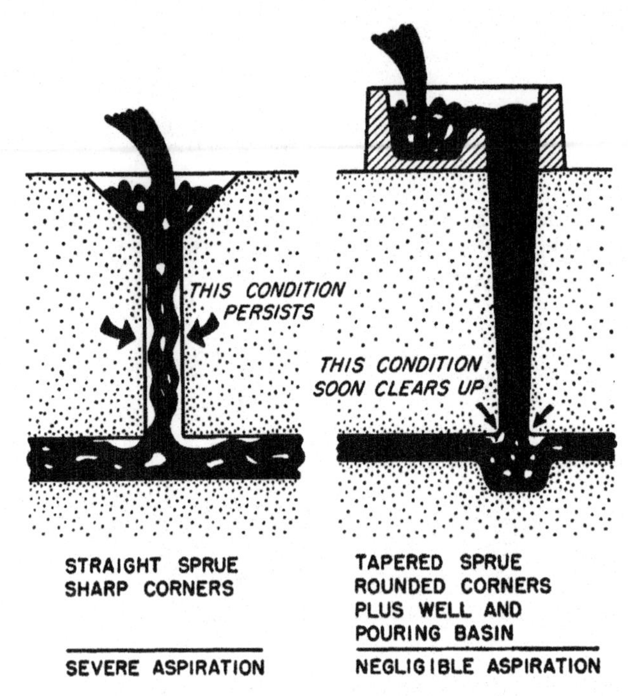

Fig. 7.1-7 Effect of sprue on aspiration during metal casting. (From H. F. Taylor, M. C. Flemings, and J. Wulff, *Foundry Engineering*, John Wiley, and Sons, New York, 1959.)

where A denotes the cross-sectional area of the molten metal stream. Therefore, from Eqs. [7.1-28] through [7.1-30]

$$\boxed{\frac{A_2}{A_1} = \sqrt{\frac{z_1}{z_2}}} \quad [7.1\text{-}31]$$

This equation suggests that the ideal sprue should have a parabolic taper. In practice, however, a straight taper is usually sufficient.

7.2. ONE-DIMENSIONAL FLUID FLOW

7.2.1. Fiber Processing: Inviscid Melt Spinning of Fibers

The inviscid melt spinning process (Section 6.1.5) is used for the preparation of fibers from low-viscosity melts such as molten ceramics and metals. As shown in Fig. 7.2-1, let p_t and p_b be the gas pressures at the top and bottom chambers, respectively; H the melt depth in the crucible; and L and R the length and radius of the orifice, respectively. The pressures at the inlet and outlet of the orifice,

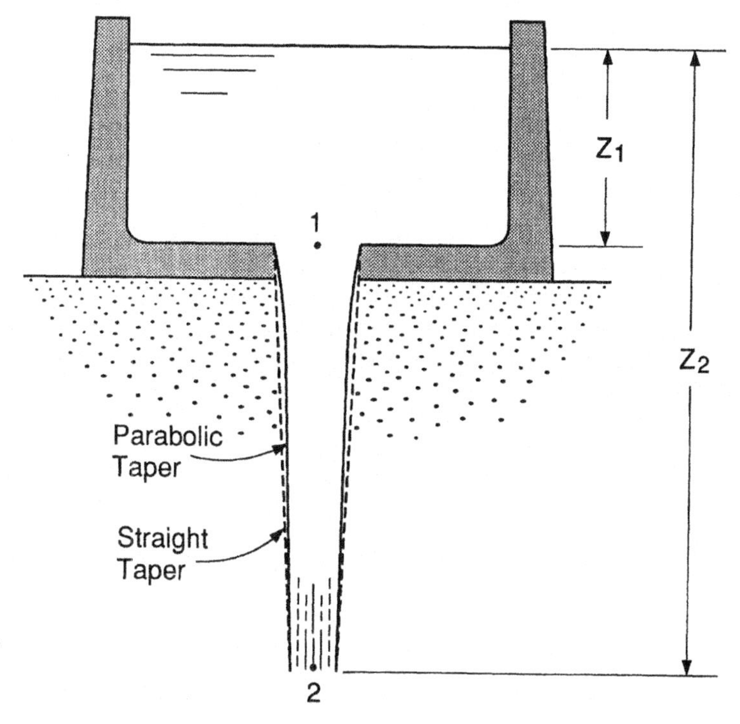

Fig. 7.1-8 Tapering of a sprue in metal casting. (From H. F. Taylor, M. C. Flemings, and J. Wulff, *Foundry Engineering*, John Wiley, and Sons, New York, 1959.)

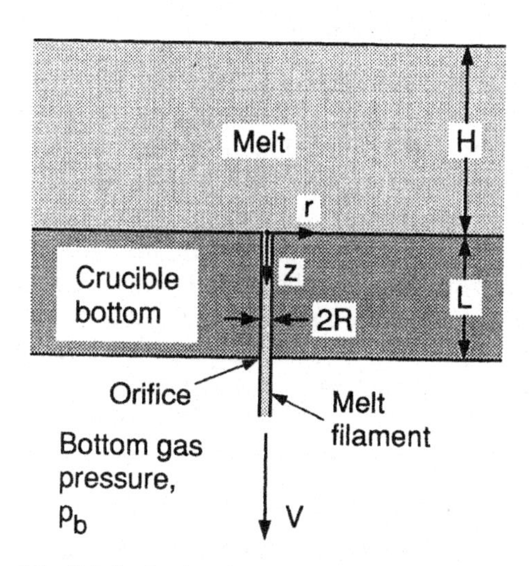

Fig. 7.2-1 Inviscid melt spinning of a fiber.

p_0 and p_L, respectively, are as follows

$$p_0 = p_t + \rho g H \qquad [7.2\text{-}1]$$

and

$$p_L = p_b \qquad [7.2\text{-}2]$$

Molten metals and ceramics are Newtonian and the flow is laminar in the small orifice. Since the radius of the orifice R is much smaller than the length L, the end effects are negligible. Substituting Eqs. [7.2-1] and [7.2-2] into the Hagen–Poisseulle law (i.e., Eq. [1.5-52]), the volume flow rate of the melt through the orifice is

$$Q = \frac{\pi[(p_t + \rho g H - p_b) + \rho g L]R^4}{8\mu L} \qquad [7.2\text{-}3]$$

Substituting the fiber production speed

$$V = \frac{Q}{\pi R^2} \qquad [7.2\text{-}4]$$

into Eq. [7.2-3], we get

$$\boxed{V = \frac{[(p_t - p_b) + \rho g(H + L)]R^2}{8\mu L}} \qquad [7.2\text{-}5]$$

7.2.2. Casting: Twin-Belt Casting

Let us consider the segregation of oxide inclusions in the twin-belt continuous casting (Section 6.1.2) of thin slabs. For simplicity the melt flow is assumed to be one-dimensional in the casting direction at the casting speed V, in view of the high casting speed and small thickness involved.

Oxide inclusions of SiO_2 or Al_2O_3 are formed in steelmaking when a deoxidant Si or Al is added to the molten steel to react with and hence consume the oxygen dissolved in the molten steel. Excessive oxygen in the molten steel can react with carbon to form CO(g) during solidification to cause gas porosity in the resultant castings and, therefore, should be removed before casting. Figure 7.2-2*a* is an example of such SiO_2 inclusion particles[6] (5000 times magnification). Figure 7.2-2*b* shows the segregation of oxide inclusions in a 4-cm-thick, twin-belt cast slab of carbon steel.[7] Too high an inclusion content

[6] M. Meyers and M. C. Flemings, *Metallurgical Trans.*, **3**, 2225, 1972.

[7] Y. Yoshihara, in *Casting of Neat Net Shape Products*, Y. Sahai, R. S., Carbonara, and C. E. Mobley, eds., The Metallurgical Society, Warrendale, PA, 1988, p. 99.

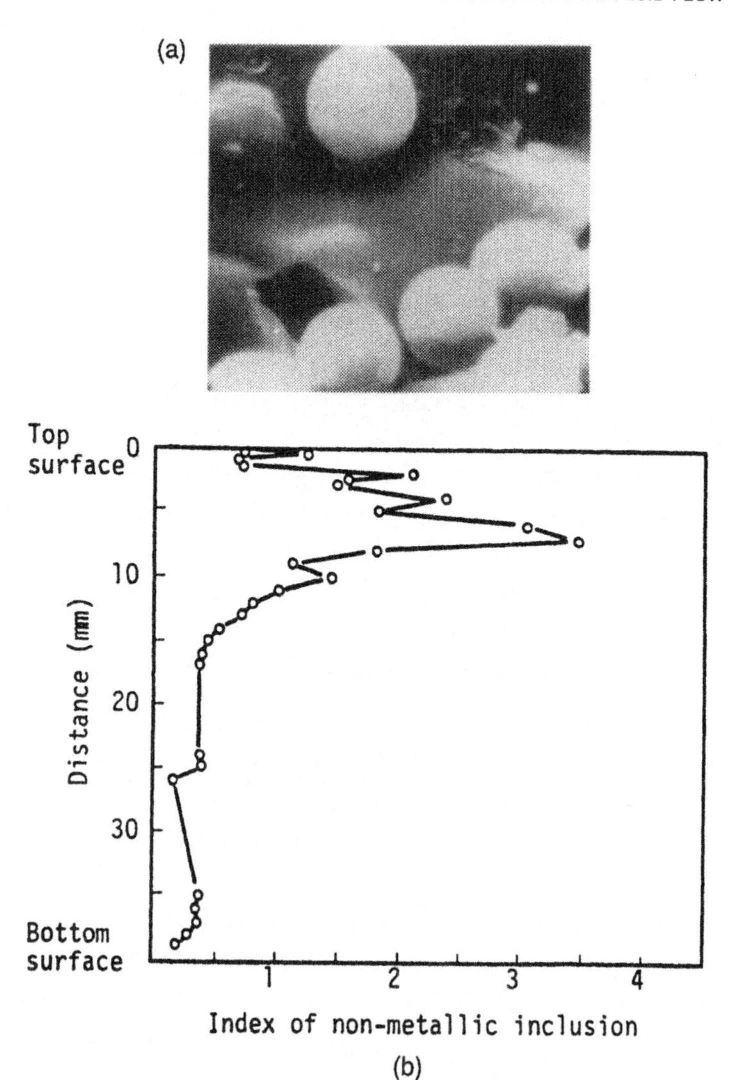

Fig. 7.2-2 Inclusions in steels: (*a*) SiO_2 particles, 5000× (M. Meyers and M. C. Flemings, *Metallurgical Transactions*, **3**, 1972, p. 2225); (*b*) inclusion segregation in a thin slab (Y. Yoshihara, in *Casting of Near Net Shape Products*, Y. Sahai et al., eds., The Metallurgical Society, Warrendale, PA, 1988, p. 99.)

or severe segregation of inclusions can cause deterioration of the mechanical properties of the slab.

The inclusion particles are significantly lighter than the molten steel. The densities of SiO_2, Al_2O_3, and the molten steel are about 2.2, 2.4, and 7.3 g cm^3, respectively. As such, the inclusion particles tend to float up during casting.

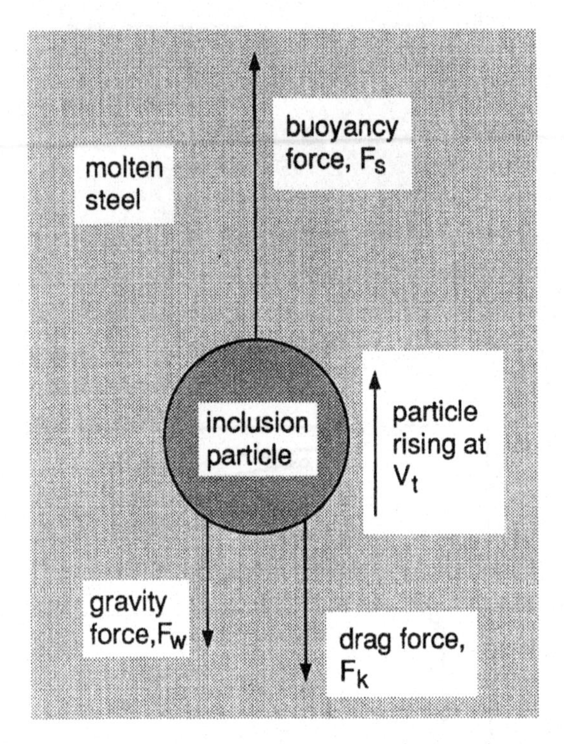

Fig. 7.2-3 Floating of an inclusion particle in molten steel.

Since the particles are rather small, the Reynolds number is very small and the Stokes law (Eq. [1.4-46]) can be applied.

As illustrated in Fig. 7.2-3, the buoyancy force (F_s) acts in the vertically upward direction, while the gravity force (F_w) and the drag force (F_k) act in the opposite direction. After a brief acceleration, the inclusion particles reach a constant, that is, terminal, velocity V_t. When this state is reached, the sum of all the forces acting on the particles must be zero:

$$F_s = F_w + F_k \tag{7.2-6}$$

Therefore, according to Eqs. [1.4-45] and [1.4-46]

$$\tfrac{4}{3}\pi R^3 \rho_m g = \tfrac{4}{3}\pi R^3 \rho_i g + 6\pi\mu R V_t \tag{7.2-7}$$

where ρ_m and ρ_i are the densities of the molten metal and the inclusion particles, respectively. From Eq. [7.2-7]

$$\boxed{V_t = \frac{2g(\rho_m - \rho_i)R^2}{9\mu}} \tag{7.2-8}$$

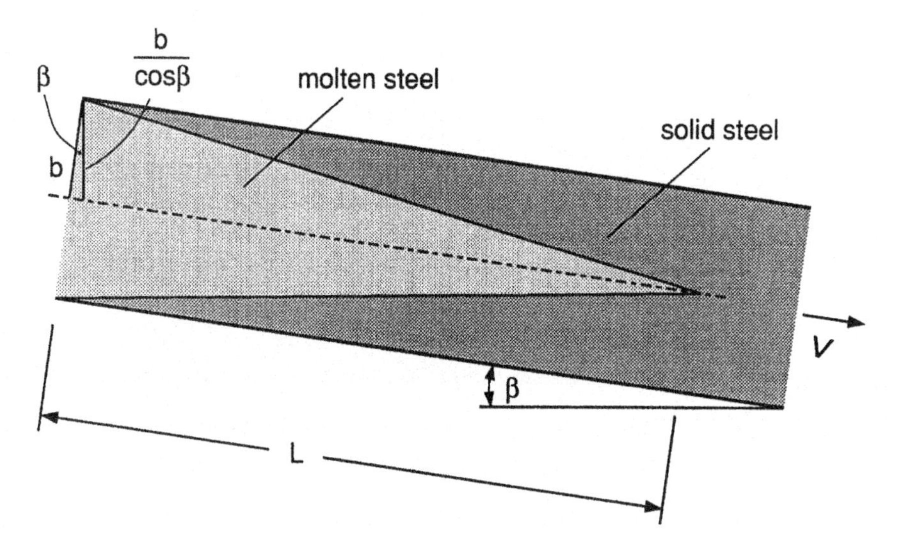

Fig. 7.2-4 Continuous casting of a steel slab.

As shown in Fig. 7.2-4, let L be the length of the molten pool in the mold and β the inclination angle of the mold. Let us now consider an inclusion particle on the midthickness plane of the mold. If ρ_i were equal to ρ_m, the particle would travel along the midthickness plane for a time period of L/V before being caught by the solidification front, where V is the casting speed. Therefore, L/V can be considered as a characteristic time. This characteristic time corresponds to a distance of $V_t(L/V)$ that the particle travels vertically upward since in reality $\rho_i < \rho_m$. The vertical distance between the midthickness plane and the roof of the mold is $b/\cos\beta$, where b is the half-thickness of the slab. As a rough estimate, inclusion segregation can occur if

$$\boxed{\frac{V_t L}{V} \geq \frac{b}{\cos\beta}} \qquad [7.2\text{-}9]$$

7.3. MULTIDIMENSIONAL FLUID FLOW

7.3.1. Semiconductor Device Fabrication: Spin Coating

The spin-coating process is used to coat a thin film of photoresist on the wafer (Section 6.3.2). A viscous solution containing the photoresist material is made to form a thin (on the order of 1 μm) uniform film on the wafer by spinning the wafer, as illustrated in Fig. 7.3-1.

Let us assume that the film is an incompressible Newtonian fluid. Let us consider the equation of motion according to Eq. [A] of Table 1.5-2. Since the

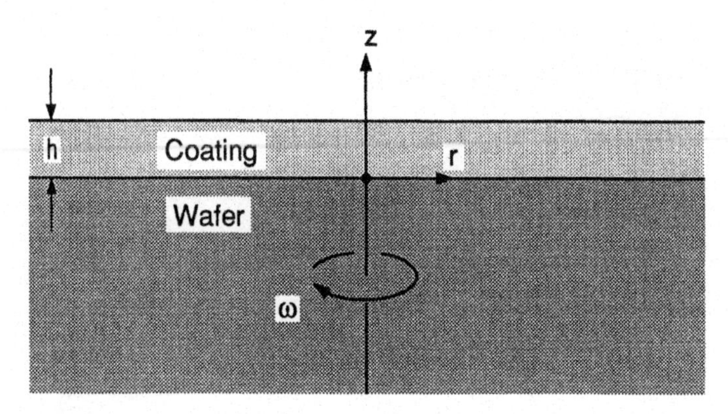

Fig. 7.3-1 Spin coating a photoresist film on a wafer.

film is very thin and viscous, the flow can be considered creeping flow (Section 1.5.1) with respect to the wafer; thus, all terms on the LHS of the equation can be neglected except for $-\rho v_\theta^2/r$. There is no radial pressure gradient (i.e., $\partial p/\partial r = 0$), and the flow is axisymmetric. Since the film is very thin, the viscous force due to the variation of v_r in the r direction is negligible as compared to that due to the variation of v_r in the z direction. As such, the equation of motion reduces to

$$\frac{-\rho v_\theta^2}{r} = \mu \frac{d^2 v_r}{dz^2} \qquad [7.3\text{-}1]$$

Since the film is viscous and very thin, it moves with the wafer as a rigid body in the θ direction:

$$v_\theta = r\omega \qquad [7.3\text{-}2]$$

Substituting Eq. [7.3-2] into Eq. [7.3-1]

$$\frac{d^2 v_r}{dz^2} = -\frac{1}{\nu} r\omega^2 \qquad [7.3\text{-}3]$$

The boundary conditions are

$$\frac{dv_r}{dz} = 0 \quad \text{at} \quad z = h \qquad [7.3\text{-}4]$$

and

$$v_r = 0 \quad \text{at} \quad z = 0 \qquad [7.3\text{-}5]$$

The solution, from Case D of Appendix A, is

$$v_r = \frac{r\omega^2}{2\nu}(2zh - z^2) \qquad [7.3\text{-}6]$$

The equation of continuity according to Eq. [B] of Table 1.3-1 reduces to

$$\frac{1}{r}\frac{\partial}{\partial r}(rv_r) + \frac{\partial v_z}{\partial z} = 0 \qquad [7.3\text{-}7]$$

Substituting Eq. [7.3-6] into Eq. [7.3-7]

$$\frac{dv_z}{dz} = -\frac{\omega^2}{2\nu}(2zh - z^2)\frac{1}{r}\frac{d}{dr}(r^2) \qquad [7.3\text{-}8]$$

or

$$\frac{dv_z}{dz} = -\frac{\omega^2}{\nu}(2zh - z^2) \qquad [7.3\text{-}9]$$

On integration from $v_z = 0$ at $z = 0$

$$v_z = -\frac{\omega^2}{\nu}\left(hz^2 - \frac{1}{3}z^3\right) \qquad [7.3\text{-}10]$$

Since

$$v_z|_h = \frac{dh}{dt} \qquad [7.3\text{-}11]$$

Eq. [7.3-10] becomes

$$\frac{dh}{dt} = -\frac{\omega^2}{\nu}\left(h^3 - \frac{1}{3}h^3\right) = -\frac{2\omega^2 h^3}{3\nu} \qquad [7.3\text{-}12]$$

On integration

$$\int_{h_0}^{h}\frac{1}{h^3}\,dh = -\int_0^t \frac{2\omega^2}{3\nu}\,dt \qquad [7.3\text{-}13]$$

or

$$\frac{1}{2}\left(\frac{1}{h^2} - \frac{1}{h_0^2}\right) = \frac{2\omega^2 t}{3\nu} \qquad [7.3\text{-}14]$$

For very thin films (i.e., $h \ll h_0$)

$$h = \left(\frac{3\nu}{4\omega^2 t}\right)^{1/2} \qquad [7.3\text{-}15]$$

Equation [7.3-15] has been derived by Middleman and Hochberg.[8] Notice that h is independent of the initial thickness h_0.

7.3.2. Casting: Continuous Casting

In continuous casting (Section 6.1.2) the molten steel is often introduced into the mold through a submerged entry nozzle, as shown in Fig. 7.3-2 for slab casting.[9]

The governing equations are as follows (z direction downward):

$$\frac{\partial v_x}{\partial x} + \frac{\partial v_y}{\partial y} + \frac{\partial v_z}{\partial z} = 0 \qquad \text{(continuity)} \qquad [7.3\text{-}16]$$

$$\rho\left(v_x \frac{\partial v_x}{\partial x} + v_y \frac{\partial v_x}{\partial y} + v_z \frac{\partial v_x}{\partial z}\right) = -\frac{\partial p}{\partial x} + \frac{\partial}{\partial x}\left(2\mu_{\text{eff}} \frac{\partial v_x}{\partial x}\right) + \frac{\partial}{\partial y}\left(\mu_{\text{eff}}\left[\frac{\partial v_x}{\partial y} + \frac{\partial v_y}{\partial x}\right]\right) + \frac{\partial}{\partial z}\left(\mu_{\text{eff}}\left[\frac{\partial v_x}{\partial z} + \frac{\partial v_z}{\partial x}\right]\right) \qquad (x \text{ momentum}) \qquad [7.3\text{-}17]$$

$$\rho\left(v_x \frac{\partial v_y}{\partial x} + v_y \frac{\partial v_y}{\partial y} + v_z \frac{\partial v_y}{\partial z}\right) = -\frac{\partial p}{\partial y} + \frac{\partial}{\partial x}\left(\mu_{\text{eff}}\left[\frac{\partial v_y}{\partial x} + \frac{\partial v_x}{\partial y}\right]\right) + \frac{\partial}{\partial y}\left(2\mu_{\text{eff}}\left[\frac{\partial v_y}{\partial y}\right]\right) + \frac{\partial}{\partial z}\left(\mu_{\text{eff}}\left[\frac{\partial v_y}{\partial z} + \frac{\partial v_z}{\partial y}\right]\right) \qquad (y \text{ momentum}) \qquad [7.3\text{-}18]$$

$$\rho\left(v_x \frac{\partial v_z}{\partial x} + v_y \frac{\partial v_z}{\partial y} + v_z \frac{\partial v_z}{\partial z}\right) = -\frac{\partial p}{\partial z} + \frac{\partial}{\partial x}\left(\mu_{\text{eff}}\left[\frac{\partial v_z}{\partial x} + \frac{\partial v_x}{\partial z}\right]\right) + \frac{\partial}{\partial y}\left(\mu_{\text{eff}}\left[\frac{\partial v_z}{\partial y} + \frac{\partial v_y}{\partial z}\right]\right) + \frac{\partial}{\partial z}\left(2\mu_{\text{eff}} \frac{\partial v_z}{\partial z}\right) + \rho g_z \qquad (z \text{ momentum}) \qquad [7.3\text{-}19]$$

As shown, the fluid is assumed incompressible, steady, and turbulent. The last three equations can be obtained by substituting Eq. [1.1-11], with μ_{eff} replacing μ, into Eq. [C] of Table 4.1-5.

[8] S. Middleman and A. K. Hochberg, *Process Engineering Analysis in Semiconductor Device Fabrication*, McGraw-Hill, New York, 1993, p. 314.

[9] B. G. Thomas, L. J. Mika, and F. M. Najjar, *Metallurgical Trans. B*, **21B**, 387, 1990.

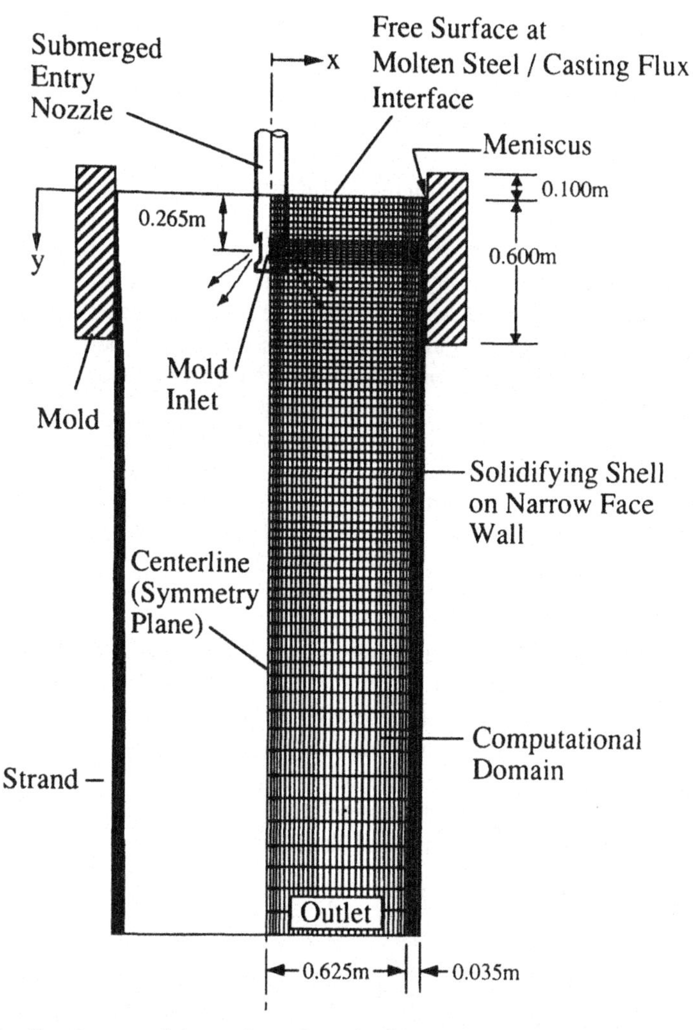

Fig. 7.3-2 Continuous slab casting of steel with a submerged entry nozzle. (B. G. Thomas, L. J. Mika, and F. M. Najjar, *Metallurgical Transactions*, **21B**, 1990, p. 387.)

The effective viscosity is the sum of the laminar (or molecular) and turbulent (or eddy) viscosity components:

$$\mu_{eff} = \mu + \mu' \qquad [7.3\text{-}20]$$

and, on the basis of the K-ε model (Section 1.7)

$$\mu' = C_\mu \rho \frac{K^2}{\varepsilon} \qquad [7.3\text{-}21]$$

where C_μ is a proportional constant, K the turbulent kinetic energy, and ε is its rate of its dissipation. Two additional equations, one for K and the other for ε, are needed to solve for the turbulent viscosity μ'.

The boundary conditions for the velocity components include the no-slip condition at the solid surfaces, zero gradients at and no flow across the symmetry planes, and zero gradients at the top surface. Additional boundary conditions are needed for K and ε.[10]

Figure 7.3-3 shows the flow of molten steel in the top 3-m portion of the strand during the casting of a 1.32×0.22-m slab at 1 m/min. As shown, the molten metal leaves the nozzle as a jet, called the **inlet jet**. This jet impinges on the shell solidifying on the mold wall and separates the molten metal pool into two recirculating regions, the lower one penetrating deep into the pool. The superheat of the incoming molten steel tends to be directed by the jet to the impingement point, thus eroding the local solidifying shell. Dangerous breakout of the molten steel can occur if the impingement point is near or below the mold exit.

The effect of fluid flow on continuous casting will be discussed further in the exercise problems, for example, surface smoothness and inclusions.

7.3.3. Casting: Mold Filling

In shape casting (Section 6.12) the molten metal is introduced into the mold. The way the molten metal enters and fills the mold cavity can often have a significant effect on the quality of the resultant casting.

Let us consider the three-dimensional fluid flow during mold filling. The governing equations are identical to Eqs. [7.3-16] through [7.3-19]. In order to locate the free surface of the molten metal, the mold cavity is divided into many small stationary volume elements and the following equation is applied:

$$\frac{\partial F}{\partial t} + v_x \frac{\partial F}{\partial x} + v_y \frac{\partial F}{\partial y} + v_z \frac{\partial F}{\partial z} = 0 \qquad [7.3\text{-}22]$$

where F is the volume fraction of the molten metal in the element. The value of F is 0 for empty elements, 1 for completely filled elements, and between 0 and 1 for elements containing the free surface. Equation [7.3-22] is identical to the equation of continuity except that the density of the fluid is replaced by the volume fraction of the fluid here. This approach of solving transient flow problems was developed in a computational fluid dynamic technique called SOLA-VOF[11] and was first applied to mold filling in metal casting by Hwang and Stoehr[12] in 1983. The boundary conditions include no-slip conditions at the mold wall. Since then, numerous computational studies based on this type

[10] X. Huang and B. G. Thomas, *Metallurgical Trans. B*, **24B**, 379, 1993.

[11] B. D. Nichols, C. W. Hurt, and R. C. Hotchkiss: "SOLA-VOF: A Solution Algorithm for Transient Flow with Multiple Free Boundaries," Report LA-8355, Los Alamos Scientific Lab, Los Alamos, NM, 1980.

[12] W. S. Hwang and R. A. Stoehr, *J. Metals*. p. 22, Oct. 1983.

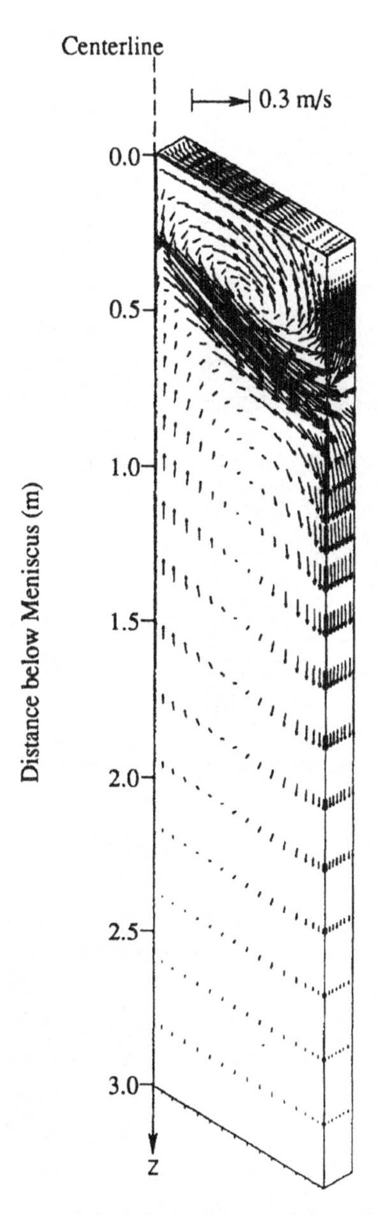

Fig. 7.3-3 Flow of molten steel during continuous slab casting. (X. Huang and B. G. Thomas, *Metallurgical Transactions B*, **24B**, 1993, p. 379.)

of approach have been conducted on mold filling in metal casting.[13] Heat transfer has also been considered in order to study solidification.

[13] For example, see T. S. Piwonka, V. Voller, and L. Katgerman, eds. *Modeling of Casting, Welding and Advanced Solidification Processes—VI*, The Minerals, Metals and Materials Society, Warrendale, PA, 1993.

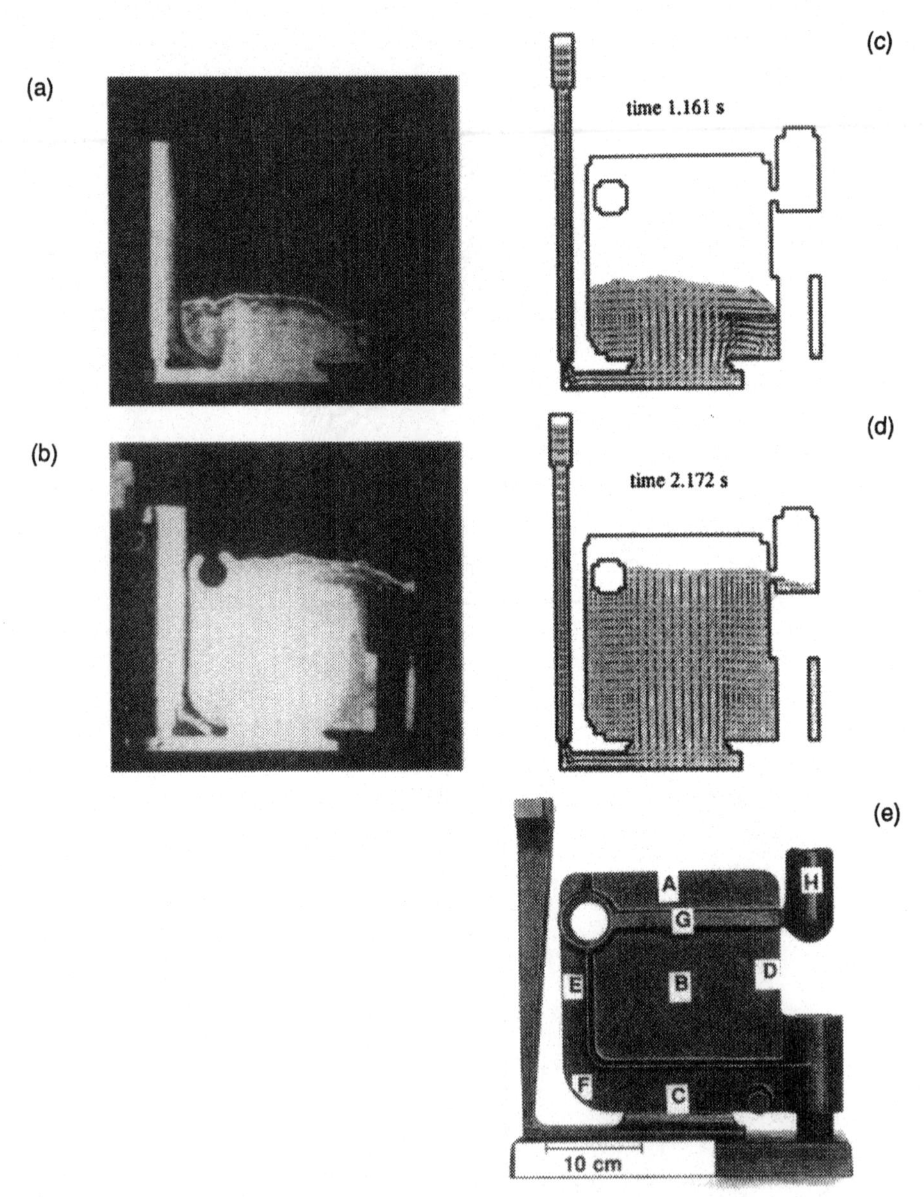

Fig. 7.3-4 Mold filling of a nodule iron casting: (*a*, *b*) flow visualization; (*c*, *d*) computer simulation; (*e*) resultant casting. (Z. A. Xu and F. Mampaey, in *Modeling of Casting, Welding and Advanced Solidification Processes—VI*, T. S. Piwonka, V. Voller, and L. Katgerman, eds., The Metallurgical Society, Warrendale, PA, 1993, p. 485.)

The mold filling of a nodule iron casting is shown in Fig. 7.3-4.[14] Computer simulation is compared with flow visualization through a heat-resistant glass.

PROBLEMS[15]

7.1-1 A single crystal of cross-sectional area A_S is Czochralski-pulled at a constant speed V with respect to the crystal growth laboratory. The crucible, of inner cross-sectional area A_L, is raised at an appropriate speed V_c so that the melt level remains stationary with respect to the crystal growth laboratory. What should V_c be and what is the actual crystal production speed V_p?

7.1-2 Consider the semicontinuous Czochralski process shown in Fig. P7.1-2.[16] A single crystal of cross-sectional area A_S is being pulled at

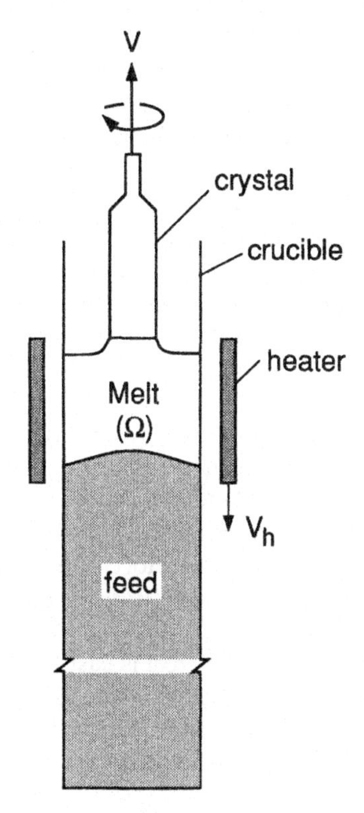

Fig. P7.1-2

[14] Z. A. Xu and F. Mampaey, in T. S. Piwonka, V. Voller, and L. Katgerman, eds., *Modeling of Casting, Welding and Advanced Solidification Processes—VI*. The Minerals, Metals and Materials Society, Warrendale, PA, 1993, p. 485.

[15] First two digits in problem numbers refer to section numbers in text.

[16] M. H. Lin and S. Kou, *J. Crystal Growth*, **142**, 392, 1994.

a constant speed V and the heater lowered at a constant speed V_h, both with respect to the crystal growth laboratory. As the heater is lowered, the melt is replenished by the feed below it. The inner cross-sectional area of the crucible is A_L. At the steady state, the melt remains constant in volume and its lowering speed remains constant at V_h. What is the actual crystal production speed V_p?

7.1-3 Another way of conducting the semicontinuous Czochralski process shown in Fig. P7.1-2 is to keep the heater stationary and raise the crucibe at V_c so as to keep the melt level constant with respect to the crystal growth laboratory. What should V_c be and what is the actual crystal production speed V_p?

7.1-4 Referring to the continuous casting process shown in Fig. 6.1-16, what is the velocity of the molten steel arriving at the top of the mold, given that the level of the molten steel in the tundish is 20 cm above the mold? What would it be if no tundish were used and if the level of the molten steel in the ladle were about 300 cm above the mold? From these velocities, what can be said about the benefit of the tundish?

7.2-1 In the zone refining of organic compounds shown in Fig. P7.2-1 (see p. 429), the ring heater and hence the molten zone inside the glass tube move upward at the speed V. Tiny gas bubbles often nucleate in the melt near the growth front (i.e., the lower melt/solid interface), where gas-forming impurities accumulate. What is the terminal velocity of the bubbles?

7.2-2 Suppose the density and viscosity of the organic fluid in Problem 7.2-1 are 0.8 g cm^{-3} and 0.008 $\text{g cm}^{-1}\text{s}^{-1}$, respectively. What is the terminal velocity of the bubbles if their average radius is 0.005 cm? Natural convection in the melt zone is negligible if the glass tube is small in diameter and the molten zone is short. Would these bubbles be trapped in the refined material if the zone travel speed V were 1 cm/min?

7.2-3 A 4-cm-thick slab of a carbon steel is cast by the twin-belt process. The casting condition is as follows: $\beta = 10°$, $L = 300$ cm, and $V = 5$ cm/s. The viscosity of the molten steel is about 0.05 $\text{g s}^{-1}\text{cm}^{-1}$. Estimate how large the inclusion particles should be, if they are expected to segregate significantly.

7.3-1 Consider the flow of molten steel shown in Fig. 7.3-3 for the continuous casting of a steel slab and the angle of the nozzle ports shown in Fig. 7.3-2. How is the angle of the inlet jet affected by the nominal angle of the nozzle ports? Would a significanlty downward port angle, say, 30° down, increase the chance of molten steel breakout?

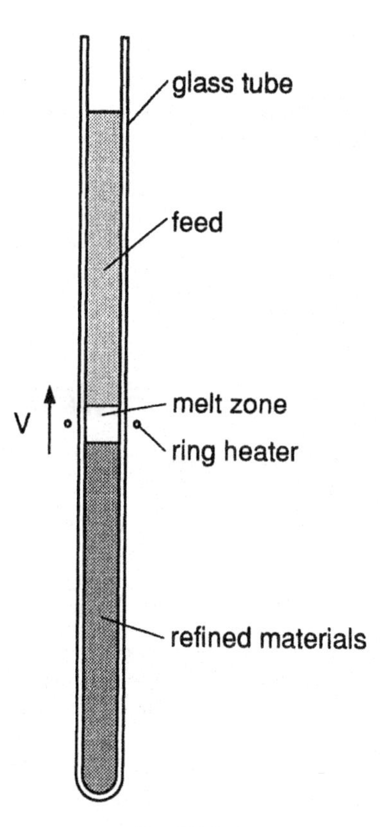

Fig. P7.2-1

7.3-2 Referring to Problem 7.3-1, it has been shown that the inlet jet cannot be made to point upward even when the nozzle port angle is significantly upward, say, 30° up. Explain why.

7.3-3 Consider the flow of molten steel shown in Fig. 7.3-3 for the continuous casting of a steel slab. How is the intensity of flow affected if the casting speed were increased, For example, from 0.5 to 1.5 mm/min? How is the thickness of the solidifying shell affected (keep in mind that the jet is hotter than the rest of the molten pool)? How is the chance of the molten steel breakout affected?

7.3-4 Consider the flow of molten steel shown in Fig. 7.3-3 for the continuous casting of a steel slab. How is the point of impingement affected if the width of the mold is increased, for instance, from 0.9 to 1.8 m? How is the chance of molten steel breakout affected?

7.3-5 Consider the flow of molten steel shown in Fig. 7.3-3 for the continuous casting of a steel slab. How is the point of impingement affected if the nozzle submergence depth is increased, say, from 0.17 to 0.37 m? How is the chance of molten steel breakout affected?

7.3-6 Consider the magnetic damping of convection during continuous casting, as illustrated in Fig. P7.3-6.[17] The molten steel to be cast carries inclusion particles (Section 7.2-2) with it into the mold and the pool. Explain how magnetic damping can help produce cleaner steels, that is, with fewer inclusion particles.

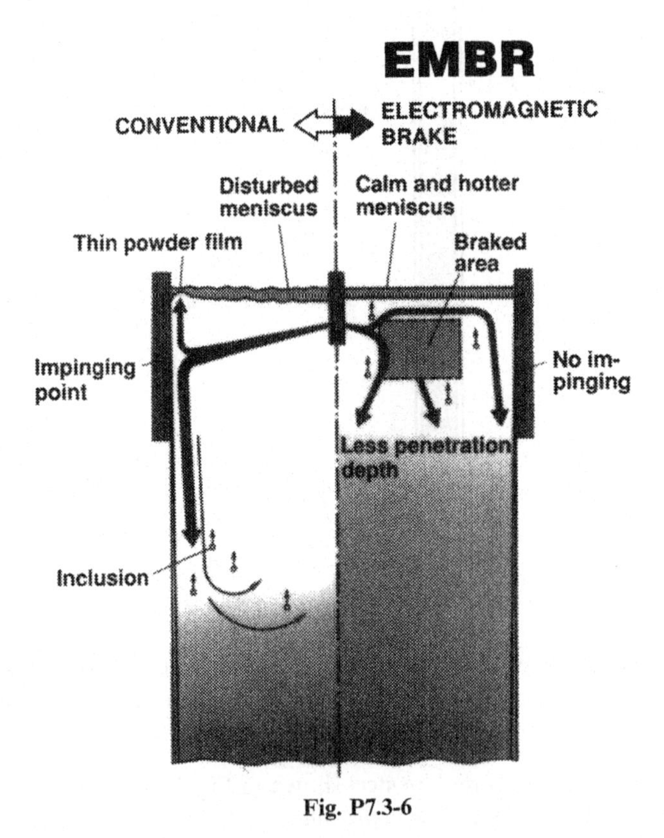

Fig. P7.3-6

7.3-7 Consider the magnetic damping of convection during continuous casting, as illustrated in Fig. P7.3-6. Explain why magnetic damping can help produce smoother slab surfaces by magnetic damping. Mold powder (for

[17] Courtesy of ABB Metallurgy, Vasteras, Sweden.

lubrication) entrapment by the molten steel at the mold wall can lead to poor surface smoothness and even surface cracking.

7.3-8 Consider the magnetic damping of convection during continuous casting, as illustrated in Fig. P7.3-6. Explain why magnetic damping reduces the risk of molten steel breakout.

CHAPTER 8

HEAT TRANSFER IN MATERIALS PROCESSING

In our discussion of heat transfer in Chapter 2, we derived the overall energy-balance equation (Eq. [2.2-8]) and the differential energy-balance equation (Eq. [2.3-8]). In the present chapter these equations will be applied to materials processing, the former in Section 8.1 and the latter throughout the rest of the chapter.

8.1. OVERALL BALANCE

8.1.1. Bulk Heat Treating: Thin Sheets

Let us consider the heat treating (Section 6.2.1) of a thin sheet. A thin sheet of mass M, initially held at the heat treating temperature T_i, is cooled or quenched in a liquid of temperature T_f. Let us assume that the temperature gradients within the sheet are negligible. In other words, the temperature within the sheet is uniform and this temperature is a function of time only. This condition of cooling is often referred to as **Newtonian cooling**.

From Newton's law of cooling the heat loss from the sheet is

$$q = h(T - T_f) \qquad [8.1\text{-}1]$$

where h is the heat transfer coefficient, T the sheet temperature and T_f the liquid temperature.

Let us consider the sheet as the control volume Ω. The overall energy-balance equation, (Eq. [2.2-8]) reduces to

$$MC_v \frac{dT}{dt} = Q = -qA \qquad [8.1\text{-}2]$$

where the surface area A includes both sides of the sheet. Substituting Eq. [8.1-1] into Eq. [8.1-2], we obtain

$$MC_v \frac{dT}{dt} = -h(T - T_f)A \qquad [8.1\text{-}3]$$

The initial condition is

$$T = T_i \quad \text{at} \quad t = 0 \qquad [8.1\text{-}4]$$

The solution, from Case A of Appendix A, is as follows:

$$\boxed{\frac{T - T_f}{T_i - T_f} = \exp\left[-\frac{hAt}{MC_v}\right]} \qquad [8.1\text{-}5]$$

This equation also applies to gas cooling if radiation is negligible.

8.1.2. Powder Processing: Cooling of Droplets

Let us consider the cooling of a droplet of radius R such as that in gas or centrifugal atomization (Section 6.1.4). In view of its small size, the droplet cools with negligible internal temperature gradients; this is Newtonian cooling. The heat flux at the droplet surface, due to convection and radiation, is as follows:

$$q = \underset{\text{(convection)}}{h(T - T_\infty)} + \underset{\text{(radiation)}}{\sigma\varepsilon(T^4 - T_\infty^4)} \qquad [8.1\text{-}6]$$

wher T_∞ is the ambient temperature, σ the Stefan-Boltzmann constant, and ε the emissivity of the droplet.

Let us consider the droplet as the control volume Ω. From Eq. [2.2-8]

$$\frac{d}{dt}\left(\frac{4}{3}\pi R^3 \rho C_v T\right) = Q = -q(4\pi R^2) \qquad [8.1\text{-}7]$$

Substituting Eq. [8.1-6]

$$\frac{4}{3}\pi R^3 \rho C_v \frac{dT}{dt} = -[h(T - T_\infty) + \sigma\varepsilon(T^4 - T_\infty^4)](4\pi R^2) \qquad [8.1\text{-}8]$$

or

$$\frac{dT}{h(T - T_\infty) + \sigma\varepsilon(T^4 - T_\infty^4)} = -\frac{3}{R\rho C_v}\,dt \qquad [8.1\text{-}9]$$

which can be integrated from the initial condition of $T = T_i$ at $t = 0$ to yield

$$\int_{T_i}^{T} \frac{dT}{h(T - T_\infty) + \sigma\varepsilon(T^4 - T_\infty^4)} = -\frac{3t}{R\rho C_v} \qquad [8.1\text{-}10]$$

where T_i is the initial temperature of the droplet. The integration needs to be done numerically. For low-temperature materials, radiation may be neglected as an approximation and Eq. [8.1-9] reduces to

$$\frac{dT}{dt} = -\frac{3h}{R\rho C_v}(T - T_\infty) \qquad [8.1\text{-}11]$$

As already mentioned, the initial condition is

$$T = T_i \quad \text{at} \quad t = 0 \qquad [8.1\text{-}12]$$

The solution, from Case A of Appendix A, is as follows

$$\boxed{\frac{T - T_\infty}{T_i - T_\infty} = \exp\left[-\frac{3ht}{R\rho C_v}\right]} \qquad [8.1\text{-}13]$$

From Eq. [8.1-13] the time for the droplet to reach the nucleation temperature T_n is

$$t_n = -\frac{R\rho C_v}{3h}\ln\left[\frac{T_n - T_\infty}{T_i - T_\infty}\right] \qquad [8.1\text{-}14]$$

The nucleation temperature is the temperature at which the solid phase starts to form in the liquid. Depending on the size, cooling rate, and material of the droplet, the nucleation temperature can be substantially below the normal freezing temperature.

8.1.3. Casting: Die Casting

In die casting (Section 6.1.2) the molten metal is injected into a water-cooled metal mold (i.e., the die), which has a cavity with the shape of the desired casting. Unlike in sand casting, the resistance to heat flow is not in the mold. This is because the die is made of metal and is water-cooled. Since die castings are usually rather thin, the resistance to heat flow is not in the castings, either. As illustrated in Fig. 8.1-1, it is primarily at the mold/metal interface, where temperature drops significantly.

As a result of thermal contraction and solidification shrinkage, there can be a very small gap between the mold and the solid metal that grows from the mold

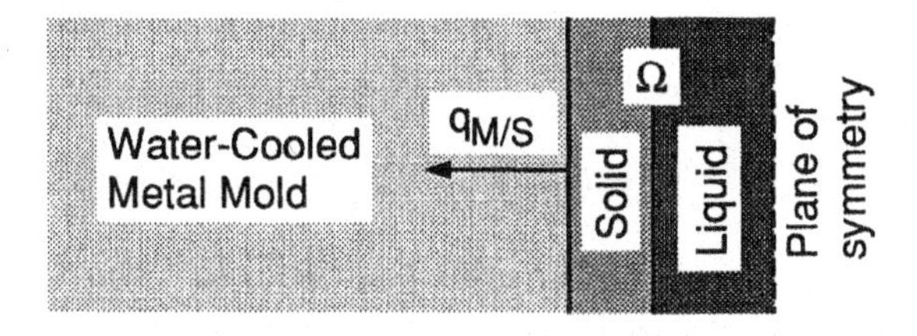

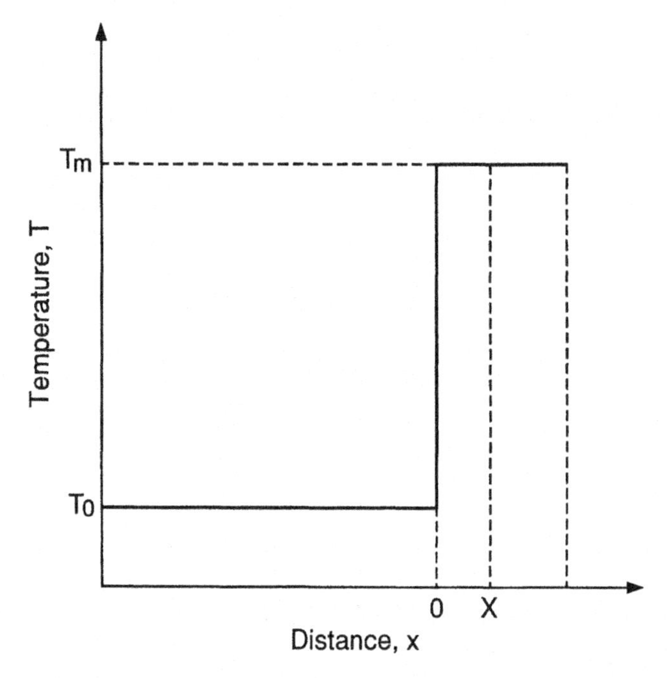

Fig. 8.1-1 Heat flow and temperature distribution in die casting.

wall. In die casting, the thermal resistance at the gap dominates since that in the mold and the metal is negligible.

In the following heat flow analysis we shall assume: (1) the mold remains at its original temperature T_0, (2) the molten metal is poured with a superheating that is negligible as compared to its melting point, and (3) the temperature drop is primarily across the mold/metal (M/S) interface. From Case 7 of Fig. 2.3-1, the heat flow across this interface is given by

$$q_{M/S} = h(T_m - T_0) \tag{8.1-15}$$

where h is the heat transfer coefficient, T_m the melting point of the metal, and T_0 the temperature of the water-cooled mold.

Let us consider the metal, from the mold to the midplane, as the control volume Ω. Since $T = T_m$ in Ω, Eq. [2.2-8] reduces to

$$0 = Q + S = -q_{M/S}A_{M/S} + S \quad [8.1\text{-}16]$$

where $A_{M/S}$ is the area of the mold/solid interface. Let X be the thickness of the solidified metal. Since the mass solidifying per unit time is $\rho_s A_{S/L}\, dX/dt$, the rate of heat generated $S = \rho_S H_f A_{S/L}\, dX/dt$, where ρ_S is the density of the solid metal, H_f the heat of fusion per unit mass of the metal, and $A_{S/L}$ the area of the solid/liquid interface. Substituting S and Eq. [8.1-15] into Eq. [8.1-16]

$$\rho_S H_f A_{S/L} \frac{dX}{dt} = h(T_m - T_0)A_{M/S} \quad [8.1\text{-}17]$$

For castings with a thin cross section of any shapes, $A_{S/L} \cong A_{M/S}$ and

$$\frac{dX}{dt} = \frac{h(T_m - T_0)}{\rho_s H_f} \quad [8.1\text{-}18]$$

Therefore, the solidification rate dX/dt is constant and independent of the thermal properties of the mold. The above equation is integrated from the initial condition of $X = 0$ at $t = 0$ to become

$$\boxed{X = \frac{h(T_m - T_0)\, t}{\rho_s H_f}} \quad [8.1\text{-}19]$$

This equation, which has been derived by Flemings,[1] indicates that the thickness of the solid metal is a linear function of time.

8.1.4. Casting: Strip Casting

In strip casting (Section 6.1.2) the shape of the liquid pool is of practical interest. For simplicity, assumptions (1) through (3) in Section 8.1.3 are adopted here again.

Let us consider the overall energy-balance equation, Eq. [2.2-8]. The control volume, shown as the shaded areas in Fig. 8.1-2, has an inlet at x, where the fraction of the solid is f_S, and an outlet at $x + dx$, where the fraction of the solid is $f_S + df_S$. Let w and a be the width and thickness of the strip, respectively, and V the casting speed. Since all surfaces of the control volume are at T_m and since there is no heat flow across the plane of symmetry, at the steady state Eq. [2.2-8]

[1] M. C. Flemings, *Solidification Processing*, McGraw-Hill, New York, 1974.

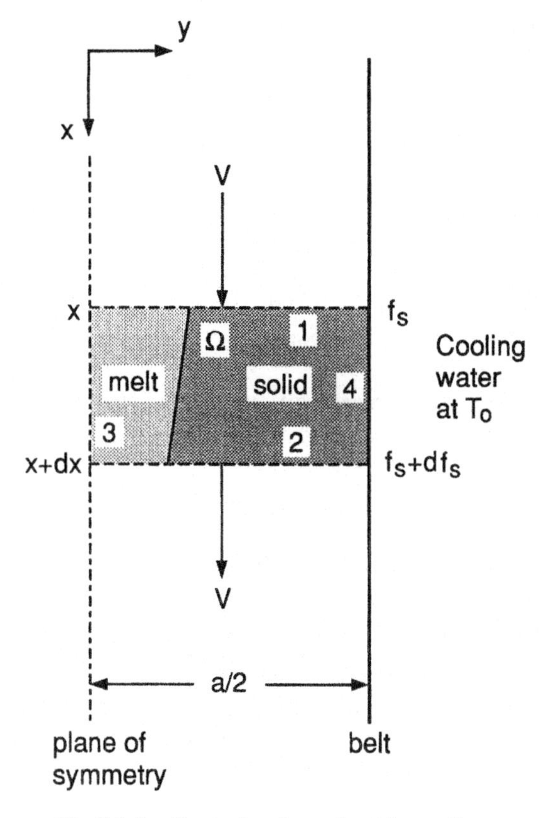

Fig. 8.1-2 Control volume in strip casting.

reduces to

$$0 = Q + S = -h(T_m - T_0)w\,dx + S \qquad [8.1\text{-}20]$$

Since the mass going through Ω per unit time is $(a/2)wV\rho$ and its fraction solid increases by df_S, the rate of heat generation in Ω is $(a/2)wV\rho H_f\,df_S$:

$$S = \frac{a}{2}wV\rho H_f\,df_S \qquad [8.1\text{-}21]$$

Substituting into Eq. [8.1-20] and dividing it by w

$$\frac{a}{2}V\rho H_f\,df_S = h(T_m - T_0)\,dx \qquad [8.1\text{-}22]$$

which can be integrated from the initial condition of $f_S = 0$ at $x = 0$ to yield

$$\boxed{\frac{f_S a}{2} = \frac{h(T_m - T_0)x}{V\rho H_f}} \qquad [8.1\text{-}23]$$

Therefore, the thickness of the solid metal (i.e., $f_S a/2$) increases linearly with the distance along the casting direction. Since $f_S = 1$ at the end of the pool, the length of the pool is

$$L = \frac{aV\rho H_f}{2h(T_m - T_0)} \qquad [8.1\text{-}24]$$

Therefore, the pool length increases with increasing casting speed. It is interesting to note that Eq. [8.1-24] can also be obtained simply by substituting $f_S a/2$ for X and x/V for t in Eq. [8.1-19].

8.2. ONE-DIMENSIONAL CONDUCTION

8.2.1. Bulk Heat Treating: Uniform Surface Quenching

A semiinfinite solid, initially at a uniform heat-treating temperature T_i, is quenched uniformly over its bounding surface with a fluid at temperature T_f, as illustrated in Fig. 8.2-1*a*. Such a problem is often encountered in heat treating of steels. In the **Jominy end-quench test**, for example, a steel bar is removed from a heat-treating furnace and immediately quenched with water at its lower end, as illustrated in Fig. 8.2-1*b* (side-surface heat losses neglected). A solid of finite thickness can be considered semiinfinite in extent, if it is much thicker than the surface layer to be transformed. This is because the surface layer is already quenched before cooling can be felt at the opposite surface of the solid.

The energy equation, Eq. [A] of Table 2.3-1, reduces to

$$\frac{\partial T}{\partial t} = \alpha \frac{\partial^2 T}{\partial x^2} \qquad [8.2\text{-}1]$$

where $\alpha = k/(\rho C_v)$. The initial and boundary conditions are

$$T(x, 0) = T_i \qquad [8.2\text{-}2]$$

$$k\frac{\partial T(0, t)}{\partial x} = h[T(0, t) - T_f] \qquad [8.2\text{-}3]$$

$$T(\infty, t) = T_i \qquad [8.2\text{-}4]$$

where h is the heat transfer coefficient. Note that the coordinate x is such that $k(\partial T/\partial x)$ is positive in Eq. [8.2-3], and a minus sign is thus not needed.

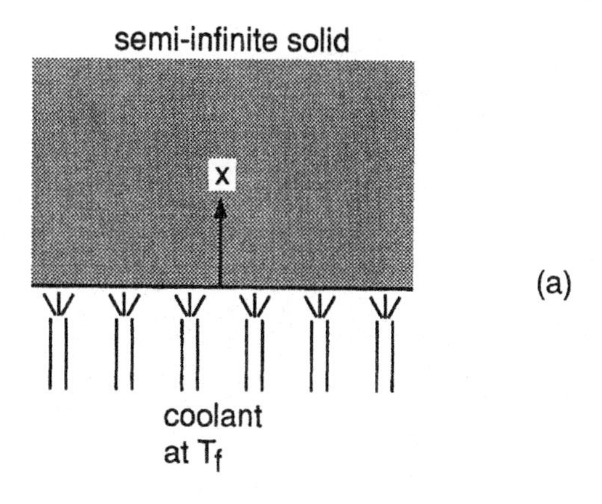

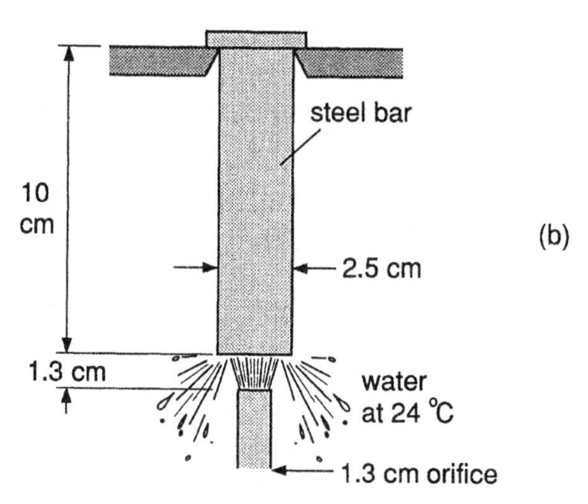

Fig. 8.2-1 Surface quenching after heat treating: (*a*) a semiinfinite solid; (*b*) Jominy end-quench test.

The solution, from Case S of Appendix A, is as follows:

$$\boxed{\theta = \frac{T - T_f}{T_i - T_f} = \operatorname{erf}\left(\frac{x}{\sqrt{4\alpha t}}\right) + \exp\left(\frac{hx}{k} + \frac{h^2 \alpha t}{k^2}\right)\left[1 - \operatorname{erf}\left(\frac{x}{\sqrt{4\alpha t}} + \frac{h}{k}\sqrt{\alpha t}\right)\right]} \quad [8.2\text{-}5]$$

Equation [8.2-5] is plotted in Fig. 8.2-2[2] for various values of $x/\sqrt{4\alpha t}$.

[2] J. C. Slattery, *Momentum, Energy and Mass Transfer in Continua*, 2nd ed., McGraw-Hill, New York, 1981, p. 316.

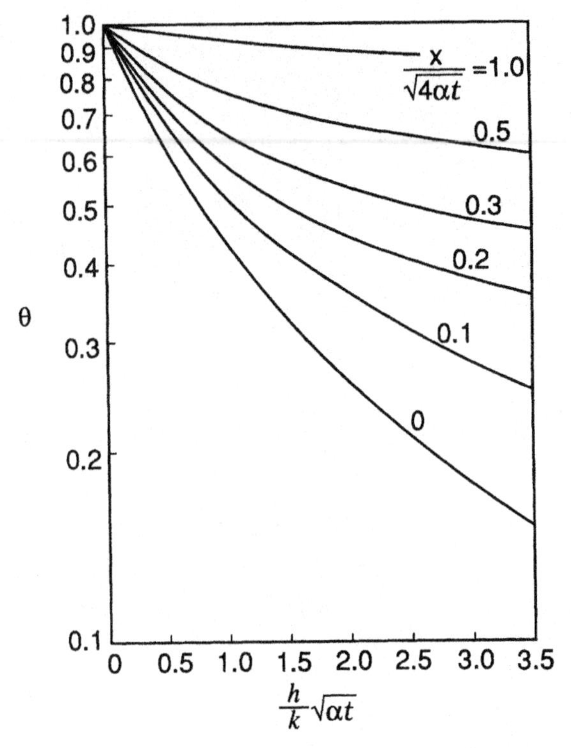

Fig. 8.2-2 Dimensionless temperature in a surface-quenched semiinfinite solid.

8.2.2. Surface Heat Treating: Uniform Surface Heating

Figure 8.2-3 illustrates the surface heat treating of a semiinfinite workpiece with a defocused laser beam. The beam is uniform in power density and much larger in diameter than the thickness of the surface layer to be heat-treated. In most of the area under the beam, heat flow can be approximated as one-dimensional in the z direction. Furthermore, a workpiece of finite thickness can often be considered semiinfinite since the thin surface layer is often already transformed before heating can be felt at the bottom of the workpiece.

It is worth mentioning at this point that special mirrors are available for making the power density of a high-power laser beam either uniform or Gaussian. For example, the power a 15-kW continuous-wave CO_2 laser can be distributed uniformly over a 1.27×1.27-cm square. In addition, high-speed oscillators have been used to rapidly scan a laser or electron beam over a substantial surface area, say, 1.27×1.27 cm square, producing a uniform or Gaussian power density distribution.[3] The sizes of these defocused beams are

[3] D. S. Gnanamuthu, in *Applications of Lasers in Materials Processing*, E. A. Metzbower, ed., American Society for Metals, Metal Park, OH, 1979, p. 177.

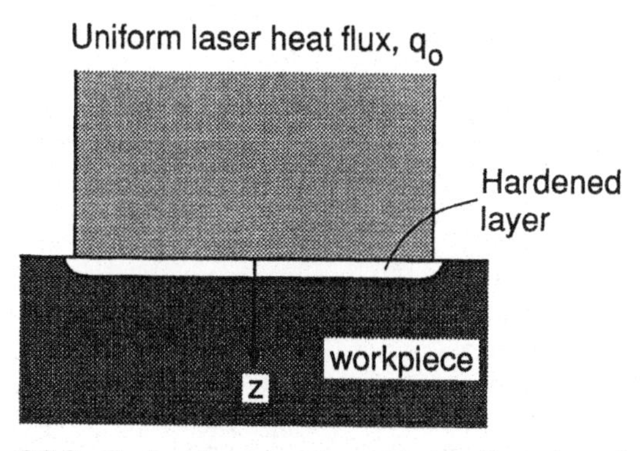

Fig. 8.2-3 Surface heat treating with a uniform heat flux.

usually significantly larger than the thickness of the heat treated layer (e.g., ≤ 1 mm).

As in the previous section

$$\frac{\partial T}{\partial t} = \alpha \frac{\partial^2 T}{\partial z^2} \qquad [8.2\text{-}6]$$

The initial and boundary conditions are

$$T(z, 0) = T_i \qquad [8.2\text{-}7]$$

$$-k \frac{\partial T(0, t)}{\partial z} = q_0 \qquad [8.2\text{-}8]$$

$$T(\infty, t) = T_i \qquad [8.2\text{-}9]$$

where q_0 is the uniform heat flux at the surface of the specimen.

The solution, from Case R of Appendix A, is as follows:

$$T - T_i = \frac{q_0}{k} \sqrt{4\alpha t} \, i\text{erf}c \left[\frac{z}{\sqrt{4\alpha t}} \right] \qquad [8.2\text{-}10]$$

where

$$i\text{erf}c(\xi) \equiv \int_{\xi}^{\infty} [1 - \text{erf}(\xi)] \, d\xi = \frac{1}{\sqrt{\pi}} e^{-\xi^2} - \xi [1 - \text{erf}(\xi)] \qquad [8.2\text{-}11]$$

TABLE 8.2-1 The Function $i\,\mathrm{erfc}(x)$

x	$2\,i\,\mathrm{erfc}\,x$	x	$2\,i\,\mathrm{erfc}\,x$
0	1.1284	0.95	0.1173
0.05	1.0312	1.0	0.1005
0.1	0.9396	1.1	0.0729
0.15	0.8537	1.2	0.0521
0.2	0.7732	1.3	0.0366
0.25	0.6982	1.4	0.0253
0.3	0.6284	1.5	0.0172
0.35	0.5639	1.6	0.0115
0.4	0.5043	1.7	0.0076
0.45	0.4495	1.8	0.0049
0.5	0.3993	1.9	0.0031
0.55	0.3535	2.0	0.0020
0.6	0.3119	2.1	0.0012
0.65	0.2742	2.2	0.0007
0.7	0.2402	2.3	0.0004
0.75	0.2097	2.4	0.0002
0.8	0.1823	2.5	0.0001
0.85	0.1580	2.6	0.0001
0.9	0.1364		

Equation [8.2-10] describes the temperature distribution in the workpiece after the heat flux q_0 has been turned on for a time period of t. The function $i\,\mathrm{erfc}$ is given in Table 8.2-1. Surface heat losses are neglected.

Suppose at time $t = \tau$ the heat flux q_0 is turned off to allow the surface layer of the workpiece to cool off. An equation similar to Eq. [8.2-10] is given below:[4]

$$T - T_i = \frac{2q_0\sqrt{\alpha}}{k}\left\{\sqrt{t}\; i\,\mathrm{erfc}\left[\frac{z}{\sqrt{4\alpha t}}\right] - \sqrt{(t-\tau)}\; i\,\mathrm{erfc}\left[\frac{z}{\sqrt{4\alpha(t-\tau)}}\right]\right\} \tag{8.2-12}$$

Note that the first term on the RHS of this equation is identical to that on the RHS of Eq. [8.2-10]. Equation [8.2-12] describes the temperature distribution in the workpiece after the heat flux q_0 has been turned off. The constant τ is the duration of the heat flux.

[4] H. S. Carslaw and J. C. Jaeger, *Conduction of Heat in Solids*, 2nd ed., Oxford University Press, London, 1959, p. 76.

Equations [8.2-10] and [8.2-12] can be combined as follows:

$$T - T_i = \left(\frac{2q_0\sqrt{\alpha}}{k}\right)\left\{\sqrt{t}\, i\,\mathrm{erfc}\left[\frac{z}{\sqrt{4\alpha t}}\right] - K\sqrt{(t-\tau)}\, i\,\mathrm{erfc}\left[\frac{z}{\sqrt{4\alpha(t-\tau)}}\right]\right\}$$

[8.2-13]

where

$$K = \begin{cases} 0 & \text{for} \quad t \le \tau \\ 1 & \text{for} \quad t > \tau \end{cases}$$

8.2.3 Bulk Heat Treating: Plates, Cylinders, and Spheres

In Section 8.1.1 the temperature gradients within the thin sheet were neglected in the heat flow analysis. For a flat plate that is thick and/or has a low thermal conductivity, however, internal temperature gradients cannot be neglected. In other words, the temperature within the plate is a function of position as well as time, as illustrated in Fig. 8.2-4. This condition is called **non-Newtonian cooling**.

The energy equation, Eq. [A] of Table 2.3-1, reduces to

$$\frac{\partial T}{\partial t} = \alpha \frac{\partial^2 T}{\partial x^2} \qquad [8.2\text{-}14]$$

The initial and boundary conditions are

$$T(x, 0) = T_i \qquad [8.2\text{-}15]$$

$$\frac{\partial T(0, t)}{\partial x} = 0 \qquad [8.2\text{-}16]$$

$$-k\frac{\partial T(L, t)}{\partial x} = h[T(L, t) - T_f] \qquad [8.2\text{-}17]$$

where T_i is the initial temperature of the plate, L the half-thickness of the plate, h the heat transfer coefficient, and T_f the temperature of the bulk fluid (coolant). The solution, from Case T of Appendix A, is as follows:

$$\frac{T - T_f}{T_i - T_f} = 2\sum_{n=1}^{\infty} \frac{\sin(\lambda_n L)}{(\lambda_n L) + \sin(\lambda_n L)\cos(\lambda_n L)} \times \exp\left[-(\lambda_n L)^2\left(\frac{\alpha t}{L^2}\right)\right]\cos\left[(\lambda_n L)\left(\frac{x}{L}\right)\right] \qquad [8.2\text{-}18]$$

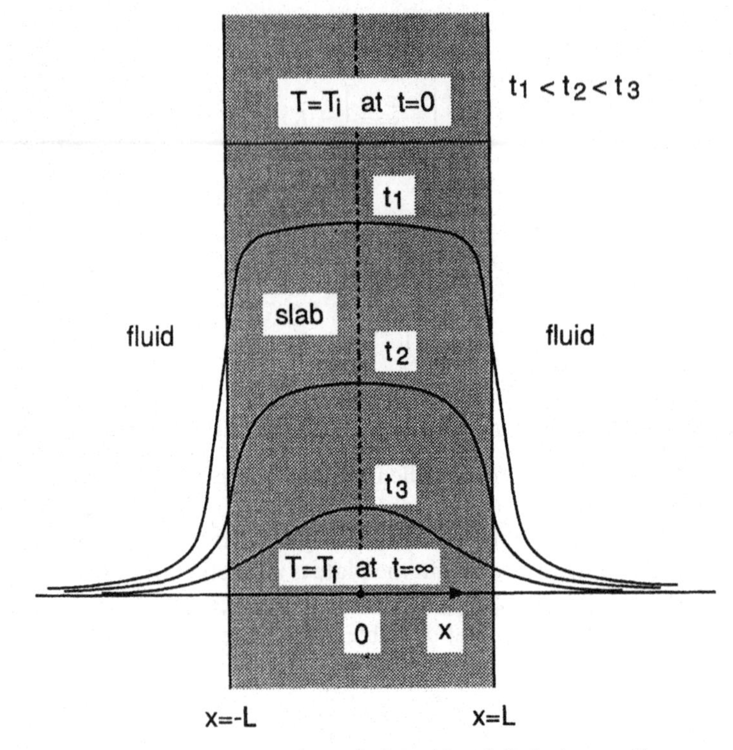

Fig. 8.2-4 Temperature variations in a slab during cooling.

From Eq. [A.7-106]

$$(\lambda L)\tan(\lambda L) = \text{Bi} \qquad [8.2\text{-}19]$$

where

$$\text{Bi} = \frac{hL}{k} \qquad [8.2\text{-}20]$$

As such, $\lambda_n L$ is a function of the **Biot number** Bi. According to Eq. [8.2-18], the dimensionless temperature $(T - T_f)/(T_i - T_f)$ is a function of Bi, $\alpha t/L^2$, and x/L. The variable $\alpha t/L^2$ is the **Fourier number** Fo and x/L is the dimensionless coordinate. For convenience, the results calculated from Eqs. [8.2-18] and [8.2-19] are summarized in Fig. 8.2-5, as a function of the dimensionless groups Bi, Fo, and x/L. Similarly, results for cylinders and spheres are also provided in Figs. 8.2-6 and 8.2-7, respectively.

The procedure for determining the temperature at a given location and time from Figs. 8.2-5 through 8.2-7 is as follows: (1) choose the proper figure for the desired geometry (e.g., Fig. 8.2-5 for flat plates), (2) choose the proper graph for the desired location (e.g., Fig. 8.2-5*a* for the midplane), (3) choose the proper curve for the desired Biot number (e.g., the rightmost curve for Bi = 0.1), and (4) choose the proper point on the curve for the desired time and get the dimensionless temperature $(T - T_f)/(T_i - T_f)$ (e.g., ~0.4 for Fo = 10).

8.2.4. Welding: Nonconsumable Electrodes

A gas tungsten arc welding electrode (Section 6.1.3) is shown in Fig. 8.2-8. Savage et al.[5] measured the electrode temperature between the water-cooled contact tube and the tungsten electrode tip. Figure 8.2-9 shows the electrode temperature as a function of the distance from the electrode tip, specifically, $L - z$, where L and z are as shown in Fig. 8.2-8. They observed that for a given electrode diameter and tip geometry, the tip temperature is essentially independent of the welding current. By "tip geometry" we mean the angle of the conical electrode tip and the amount of material truncated off from the end of the conical tip. This surprising phenomenon was explained based on the concept of the **workfunction**, which is the energy required to remove one electron from the electrode tip. It was suggested that the higher the welding current, the faster the electrons are emitted from the electrode tip and hence the more the electrode tip is cooled by pumping off energy in the form of the workfunction. Savage et al. also observed that when the welding current is too high for a given electrode size, the electrode tip does not melt and become a pendent drop. Instead, the electrode melts in half, at a location between the tip and the contact tube.

Suppose the electrode temperature is T_0 at the lower end of the contact tube and T_t at the tip of the electrode. Let us calculate the steady-state temperature distribution along the electrode for a given welding current I and electrode extension L. For simplicity, we will assume that heat losses from the electrode surface and the radial temperature gradients in the electrode are negligible. Since the electrode is stationary and since the temperature gradients are mainly in the z direction, the energy equation, Eq. [B] of Table 2.3-1, reduces to

$$0 = k\frac{\partial^2 T}{\partial z^2} + s \qquad [8.2\text{-}21]$$

or

$$\frac{d^2 T}{dz^2} = -\frac{s}{k} \qquad [8.2\text{-}22]$$

[5] W. F. Savage, S. S. Strunk, and Y. Ishikawa, *Welding J.*, **44**, 489s, 1965.

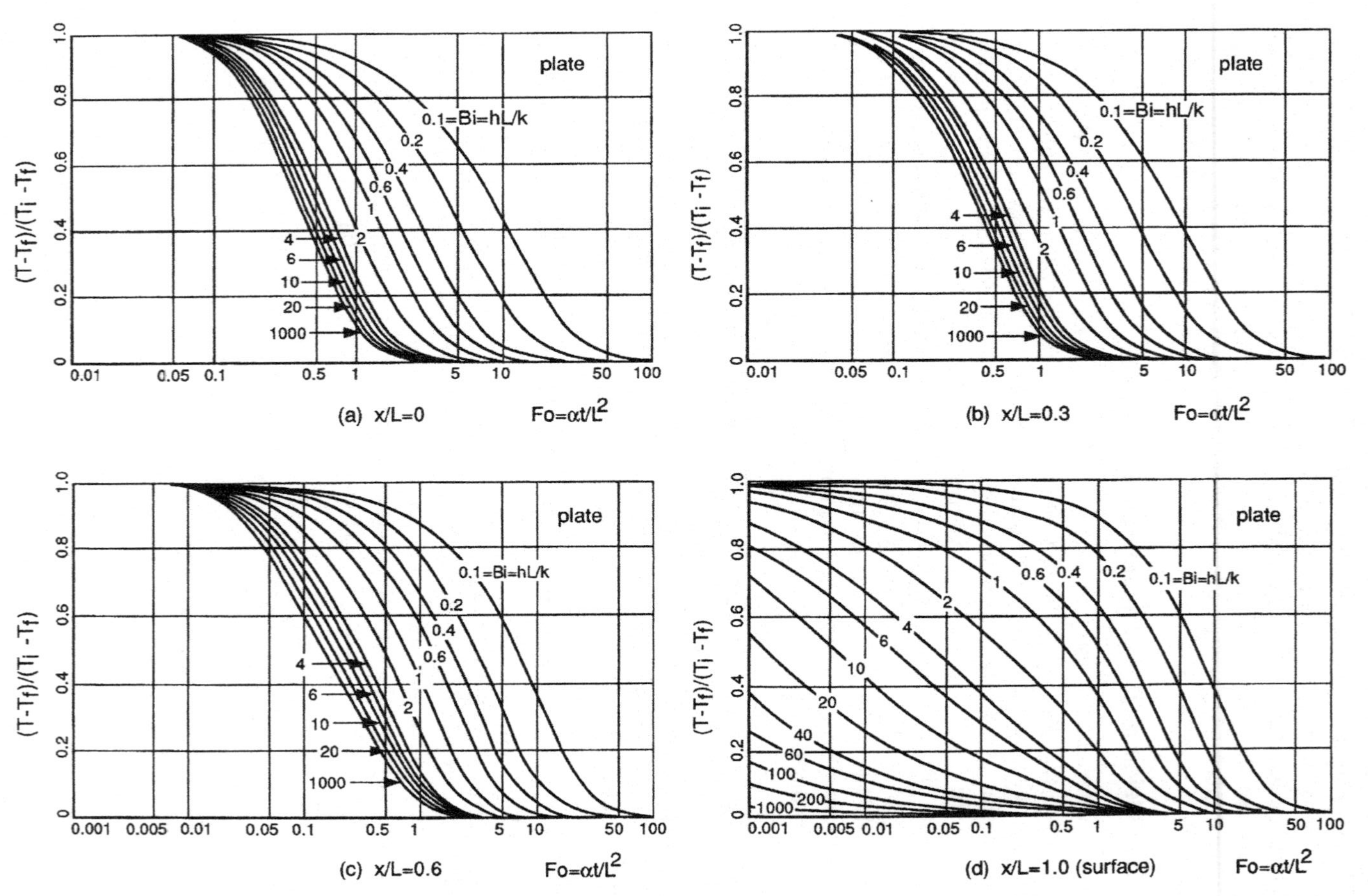

Fig. 8.2-5 Dimensionless temperature in a wide slab: (*a*)–(*d*) from midplane to surface.

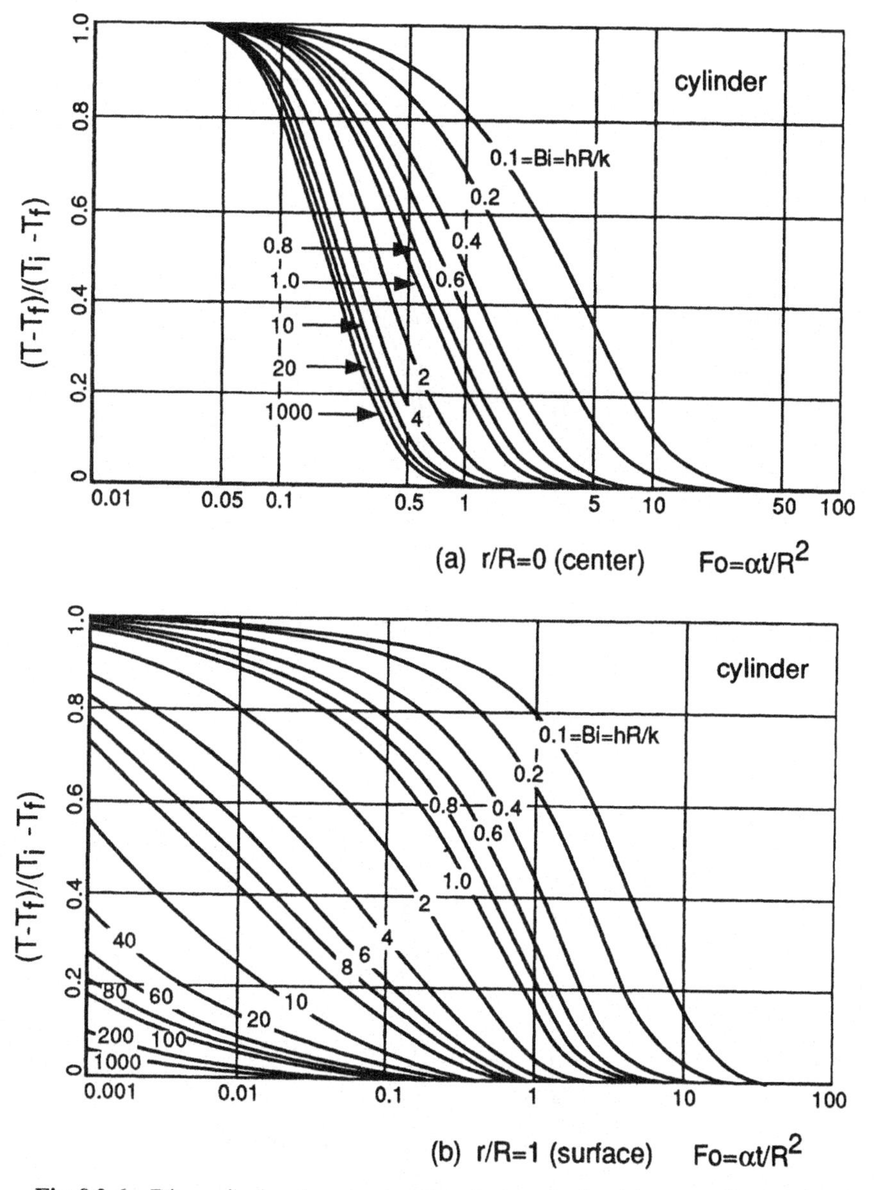

Fig. 8.2-6 Dimensionless temperature in a long cylinder: (*a*) center; (*b*) surface.

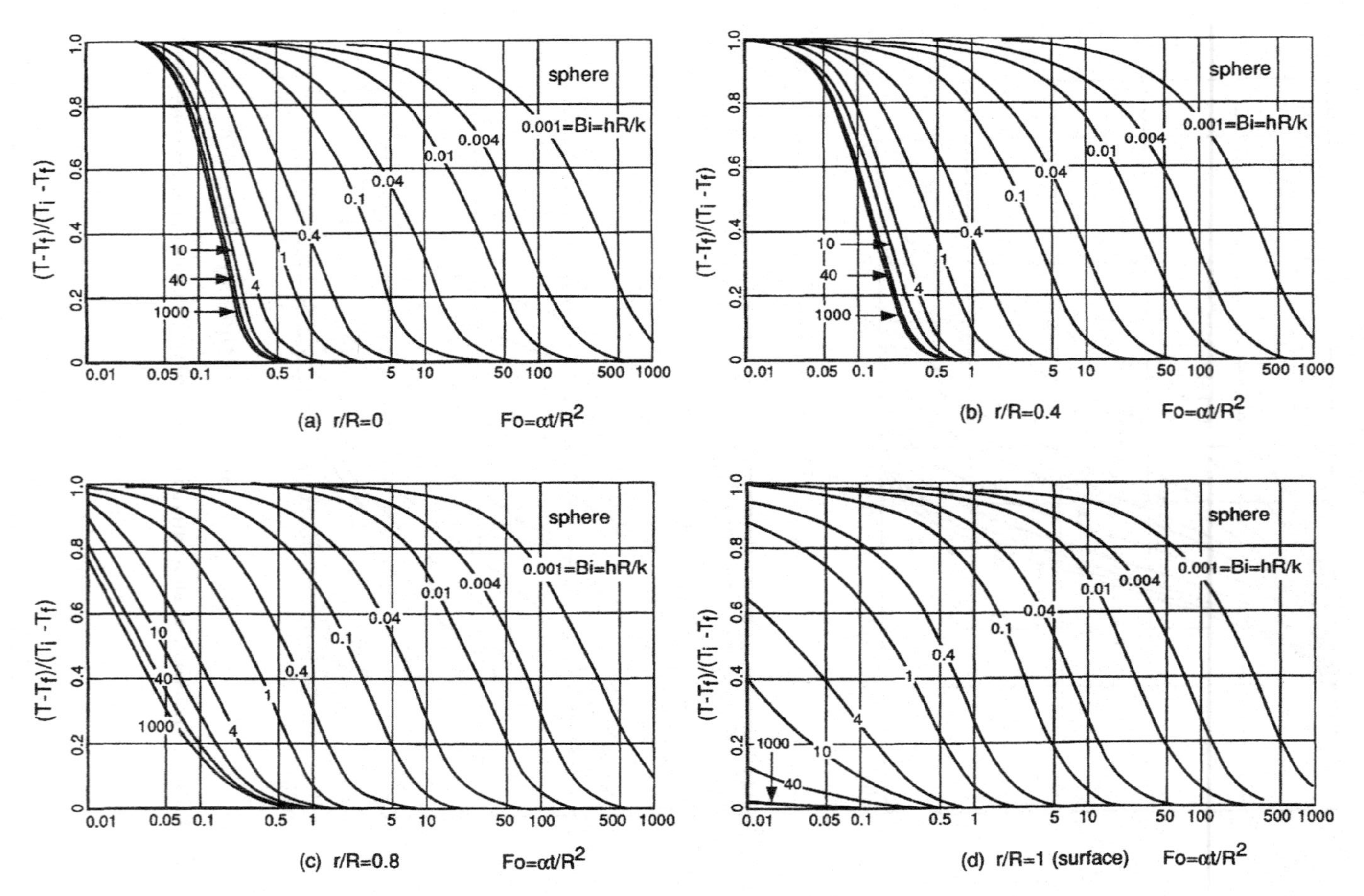

Fig. 8.2-7 Dimensionless temperature in a sphere: (*a*)–(*d*) from center to surface.

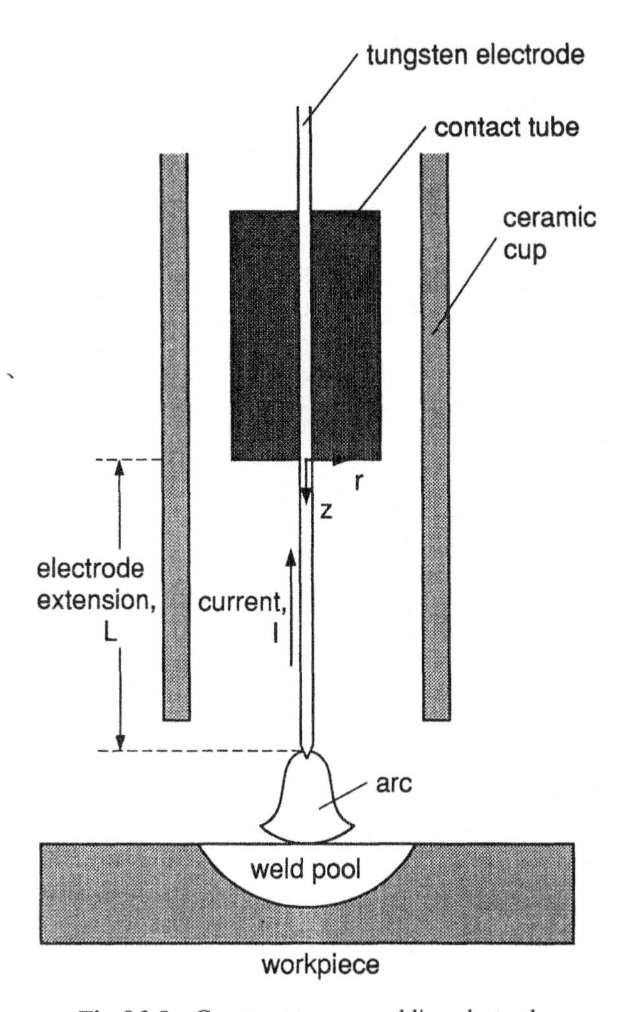

Fig. 8.2-8 Gas tungsten arc welding electrode.

As in Example 2.3-1, the rate of heat generation per unit volume, s, is given by

$$s = \frac{1}{k_e}\left(\frac{I}{\pi R^2}\right)^2 \qquad [8.2\text{-}23]$$

where k_e and R are the electrical conductivity and radius of the electrode, respectively.

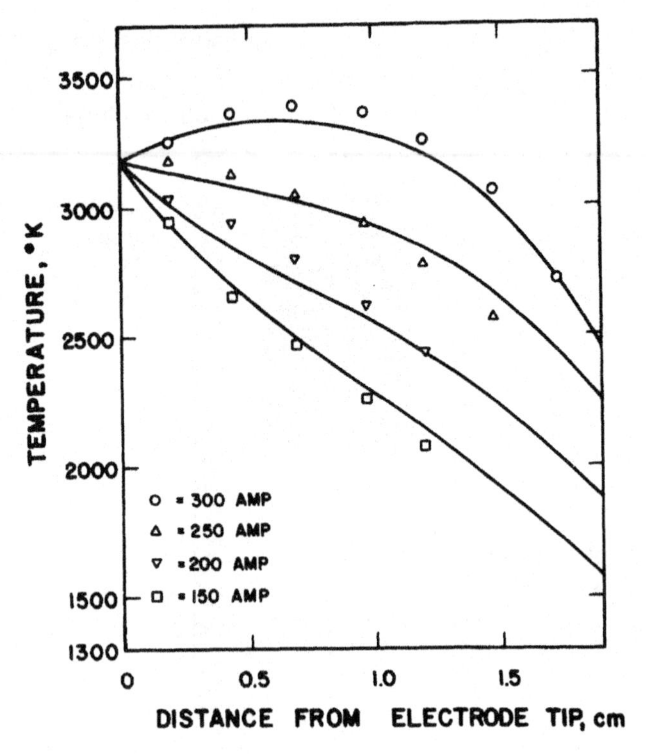

Fig. 8.2-9 Temperature distributions in a gas tungsten arc welding electrode. (From S. Kou and M. C. Tsai, *Welding Journal*, **64**, 1985, p. 226s.)

The boundary conditions are as follows:

$$T = T_0 \quad \text{at} \quad z = 0 \tag{8.2-24}$$

$$T = T_t \quad \text{at} \quad z = L \tag{8.2-25}$$

The solution, from Case C of Appendix A, is as follows:

$$\boxed{T = \frac{s}{2k}(Lz - z^2) + \left(\frac{T_t - T_0}{L}\right)z + T_0} \tag{8.2-26}$$

This equation shows that for a small current I, the first term on the LHS is small and T is thus a linear function of z. However, as I increases, this term becomes more significant and T is no longer a linear function of T. This characteristic of Eq. [8.2-26] is consistent with the data shown in Fig. 8.2-9. The solid curves shown in the figure are based on more elaborate calculations considering temperature-dependent physical properties of tungsten and surface heat losses.[6]

From Eq. [8.2-26]

$$\frac{dT}{dz} = \frac{sL}{2k} - \frac{s}{k}z + \frac{T_t - T_0}{L} \qquad [8.2\text{-}27]$$

Therefore, the location of the maximum temperature, specifically, $dT/dz = 0$, is

$$z|_{T=T_{max}} = \frac{k(T_t - T_0)}{Ls} + \frac{L}{2} \qquad [8.2\text{-}28]$$

This explains why the electrode melts in half rather than at its tip.

8.2.5. Welding: Consumable Electrodes

A gas metal arc welding electrode (Section 6.1.3) is shown in Fig. 8.2-10. Let us calculate the steady-state temperature distribution in the filler wire. For simplicity, we assume that resistance heating is negligible. According to Eq. [8.2-23], this is true when the electrical conductivity of the wire is high (e.g., aluminum), or when the current density in the wire is small. We also assume that heat losses from the wire surface and the radial temperature gradients in the wire are both negligible, and that the physical properties of the wire are constant.

The energy equation, Eq. [B] of Table 2.3-1, reduces to

$$\rho C_v v_z \frac{\partial T}{\partial z} = k \frac{\partial^2 T}{\partial z^2} \qquad [8.2\text{-}29]$$

and with $v_z = V$ to

$$\frac{d^2 T}{dz^2} = \frac{V}{\alpha}\frac{dT}{dz} \qquad [8.2\text{-}30]$$

where

$$\alpha = \frac{k}{\rho C_v} \qquad [8.2\text{-}31]$$

The boundary conditions are

$$T = T_0 \quad \text{at} \quad z = 0 \qquad [8.2\text{-}32]$$

$$T = T_m \quad \text{at} \quad z = L \qquad [8.2\text{-}33]$$

where T_m is the melting point of the filler wire.

[6] S. Kou and M. C. Tsai, *Welding J.*, **64**, 226s, 1985.

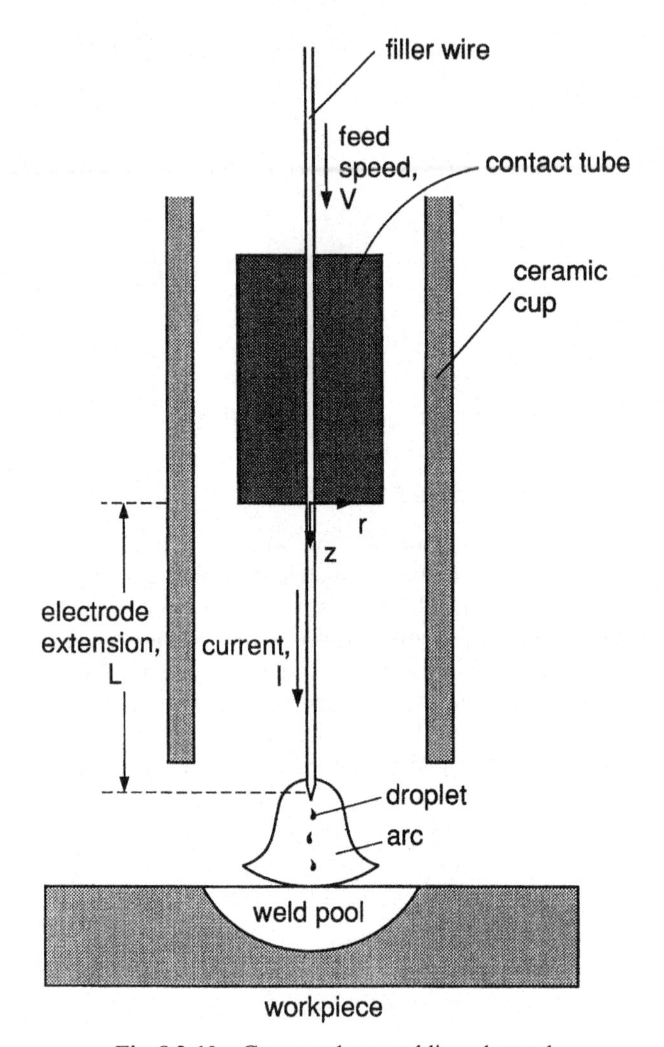

Fig. 8.2-10 Gas metal arc welding electrode.

The solution, from Case E of Appendix A, is as follows:

$$\frac{T - T_0}{T_m - T_0} = \frac{e^{Vz/\alpha} - 1}{e^{VL/\alpha} - 1} \qquad [8.2\text{-}34]$$

or

$$\frac{T_m - T}{T_m - T_0} = \frac{1 - \exp\{-(VL/\alpha)[1 - (z/L)]\}}{1 - e^{-(VL/\alpha)}} \qquad [8.2\text{-}35]$$

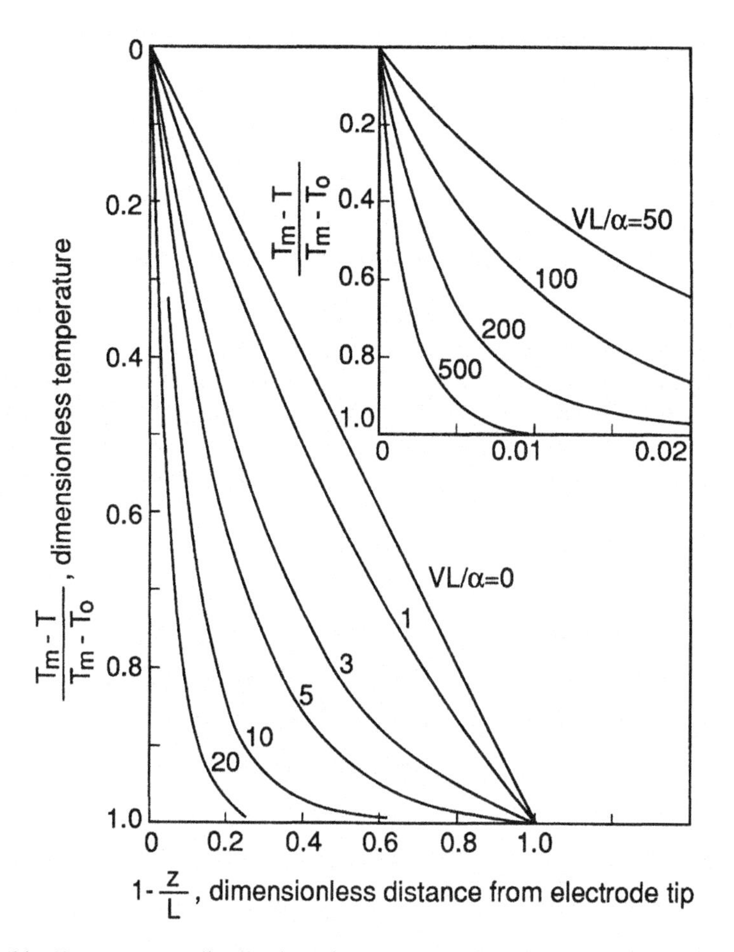

Fig. 8.2-11 Temperature distributions in a gas metal arc welding electrode. (From P. S. Meyers, O. A. Uyehara, and G. L. Borman, "Fundamentals of Heat Flow in Welding," Welding Research Council, Bulletin 123, New York, 1967.)

In Fig. 8.2-11[7] the dimensionless temperature is plotted as a function of the dimensionless distance from the electrode tip for various levels of the dimensionless feed rate. As shown, at high feed rates the temperature distribution in the feed wire becomes highly nonlinear, and the temperature gradient near the wire tip becomes rather steep.

[7] P. S. Meyers, O. A. Uyehara, and G. L. Borman, Welding Research Council Bulletin 123, Welding Research Council, New York, 1967.

8.3. ONE-DIMENSIONAL CONDUCTION WITH SOLIDIFICATION

8.3.1. Casting: Sand Casting

In sand casting (Section 6.1.2) the molten metal is poured into a sand mold, which has a cavity with the shape of the desired casting. Since the thermal conductivity of the metal casting is much larger than that of the sand mold, the resistance to heat flow and hence the temperature drop along the path of heat flow are primarily in the same mold, as illustrated in Fig. 8.3-1. Furthermore, since the sand mold is usually much thicker than the casting, the mold can be considered semiinfinite as an approximation.

For simplicity, we assume that heat flow is one-dimensional. For a casting that has a cross-sectional area much smaller than its surface area, this appears to be a reasonable approximation. This is also true in the early stage of solidification in most ingots or castings. Referring to the casting of the aluminum ingot

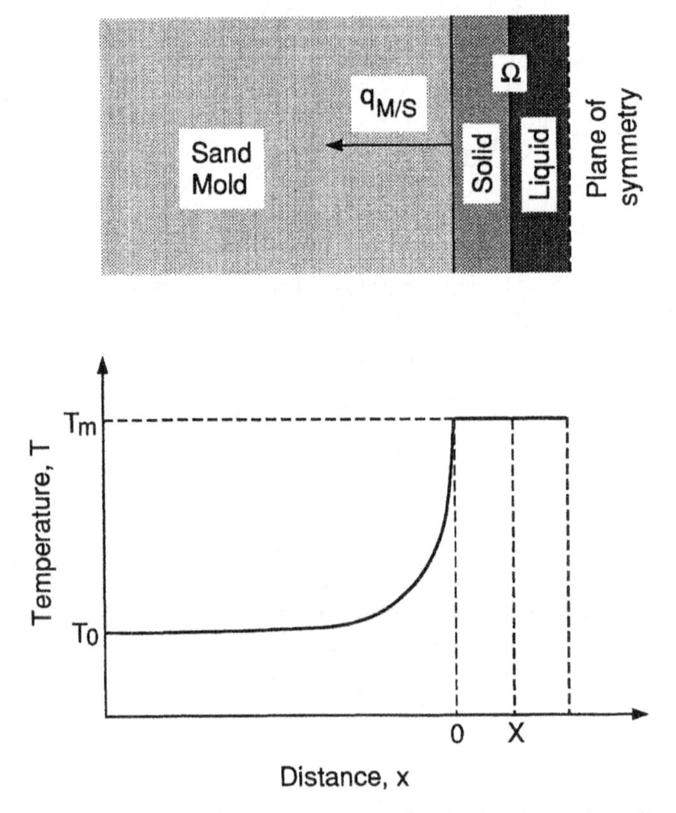

Fig. 8.3-1 Heat flow and temperature distribution in sand casting.

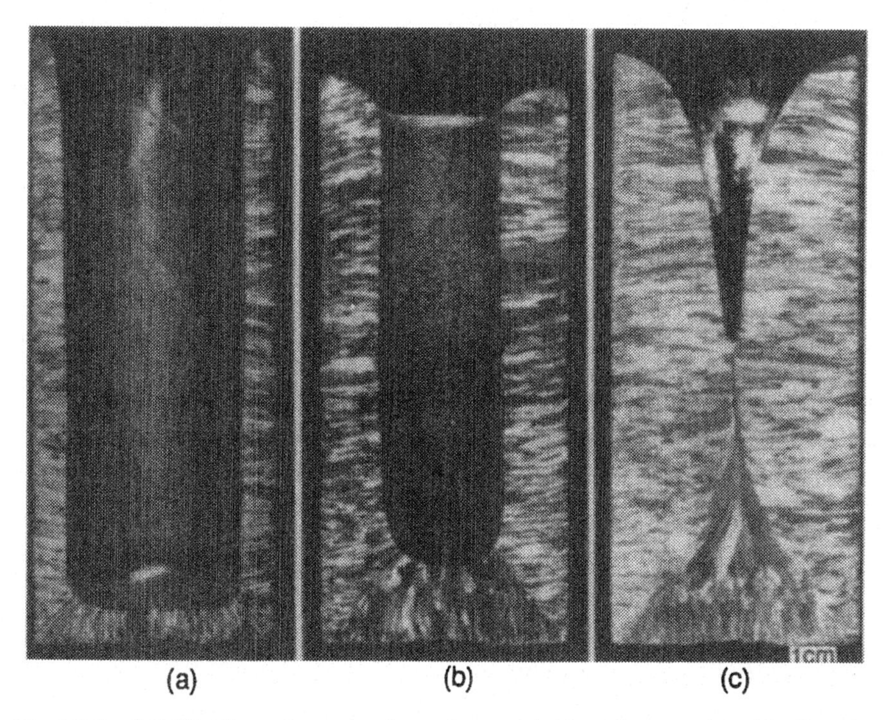

Fig. 8.3-2 Solidification of an aluminum ingot: (*a*) 20 s after pouring; (*b*) 40 s after pouring; (*c*) solidification completed. (From A. Ohno, *The Solidification of Metals*, Chijin Shokan, Tokyo, 1976, p. 127.)

shown in Fig. 8.3-2 as an example,[8] decanting the liquid metal 20 s after it was poured into the mold (Fig. 8.3-2*a*) revealed the one-dimensional nature of heat flow at this early stage of solidification. We also assume that the thermal properties of the mold and the casting are constant, and that the metal is poured with a superheating that is negligible as compared to its melting point.

The energy equation, Eq. [A] of Table 2.3-1, reduces to

$$\frac{\partial T}{\partial t} = \alpha \frac{\partial^2 T}{\partial x^2} \qquad [8.3\text{-}1]$$

where α is the thermal diffusivity of the sand mold. Note that the origin of the x coordinate coincides with the mold/metal interface and that the positive x direction points into the metal. The initial and boundary conditions are

$$T(x, 0) = T_0 \qquad [8.3\text{-}2]$$

[8] A. Ohno, *The Solidification of Metals*, Chijin Shokan Co. Ltd., Tokyo, 1976, p. 127.

$$T(0, t) = T_m \qquad [8.3\text{-}3]$$

$$T(-\infty, t) = T_0 \qquad [8.3\text{-}4]$$

where T_0 is the initial mold temperature and T_m the melting point of the metal. Regarding Eq. [8.3-3], the mold/metal interface is instantaneously raised to and stays at the melting point of the metal, since the metal is poured with negligible superheating. The solution, from Case O of Appendix A, is as follows:

$$\frac{T - T_m}{T_0 - T_m} = \text{erf}\left[\frac{-x}{\sqrt{4\alpha t}}\right] \quad \text{(mold)} \qquad [8.3\text{-}5]$$

This equation describes the temperature distribution in the sand mold at time t after the metal is poured into the mold. From Eq. [A.7-25]

$$\frac{\partial T}{\partial x} = \frac{-1}{\sqrt{\pi\alpha t}}(T_0 - T_m)e^{-(x^2/4\alpha t)} \qquad [8.3\text{-}6]$$

Since the heat flux is taken positive and since $\partial T/\partial x > 0$, Fourier's law of conduction can be written as follows

$$q = k\frac{\partial T}{\partial x} \qquad [8.3\text{-}7]$$

where k is the thermal conductivity of the mold. Substituting Eq. [8.3-6] into Eq. [8.3-7]

$$q_{M/S} = \frac{-k}{\sqrt{\pi\alpha t}}(T_0 - T_m)e^{-(x^2/4\alpha t)}\bigg|_{x=0} = \sqrt{\frac{k\rho C_v}{\pi t}}(T_m - T_0) \qquad [8.3\text{-}8]$$

where the subscript M/S denotes the interface between the mold and the solid metal, and k is the thermal conductivity of the mold.

Let us now consider the metal, from the mold to the midplane, as the control volume Ω. Since T is uniform and constant at T_m in Ω, the overall energy-balance equation, Eq. [2.2-8], reduces to

$$0 = Q + S = -q_{M/S}A_{M/S} + S \qquad [8.3\text{-}9]$$

Let X be the thickness of the solid metal and H_f the heat of fusion per unit mass of the metal. Since the mass solidifying per unit time is $\rho_S A_{S/L}\,dX/dt$, the rate of heat generation $S = \rho_S H_f A_{S/L}\,dX/dt$, where the subscript S/L denotes the interface between the solid metal and the liquid metal. Substituting Eq. [8.3-8] and S into Eq. [8.3-9]

$$\rho_S H_f A_{S/L} \frac{dX}{dt} = \sqrt{\frac{k\rho C_v}{\pi t}} (T_m - T_0) A_{M/S} \qquad [8.3\text{-}10]$$

For a casting with a cross-sectional area much smaller than its surface area (e.g., a thin, wide casting), $A_{S/L} \approx A_{M/S}$ and Eq. [8.3-10] becomes

$$\frac{dX}{dt} = \underbrace{\left(\frac{T_m - T_0}{\rho_S H_f}\right)}_{\text{(metal)}} \underbrace{\sqrt{\frac{k\rho C_v}{\pi}}}_{\text{(mold)}} \frac{1}{\sqrt{t}} \qquad [8.3\text{-}11]$$

Therefore, the solidification rate is affected by the thermal properties of both the metal and the mold. The above equation is integrated from the initial condition of $X = 0$ at $t = 0$ to become

$$\boxed{X = \frac{2}{\sqrt{\pi}} \left(\frac{T_m - T_0}{\rho_S H_f}\right) \sqrt{k\rho C_v} \sqrt{t}} \qquad [8.3\text{-}12]$$

This equation has been derived by Flemings.[9] It states that the thickness of the solid metal is a parabolic function of time.

In this one-dimensional heat flow analysis, the thickness of the metal solidified X can be related to the volume of the metal solidified Ω_S and the area of the mold/solid interface $A_{M/S}$ as follows:

$$X = \frac{\Omega_S}{A_{M/S}} \qquad [8.3\text{-}13]$$

If we extend this simple relationship to castings of any shapes as an approximation, from Eq. [8.3-12]

$$\frac{\Omega_S}{A_{M/S}} = C\sqrt{t} \qquad [8.3\text{-}14]$$

where C is a proportional constant related to the physical properties of the mold and the metal. This equation, called the **Chvorinov rule**, was verified by Chvorinov's experimental results[10] shown in Fig. 8.3-3.[9]

[9] M. C. Flemings, *Solidification Processing*, McGraw-Hill, New York, 1974.
[10] N. Chvorinov, *Giesserei*, **27**, 177, 1940.

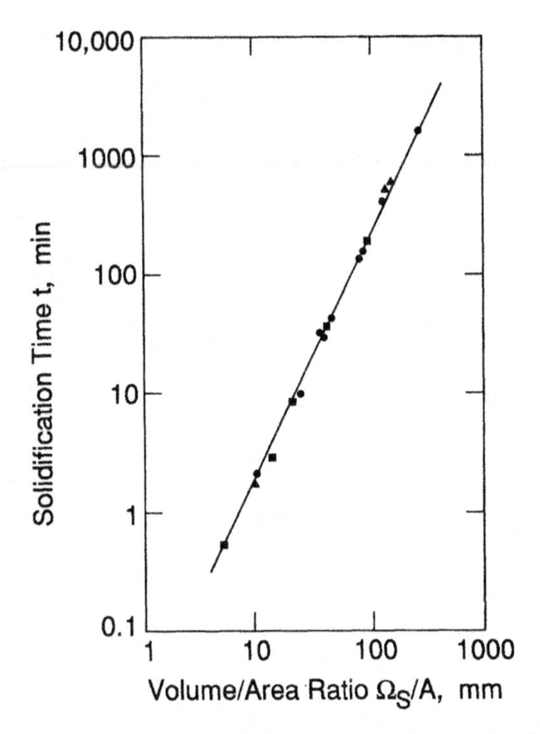

Fig. 8.3-3 Solidification time as a function of the volume/area ratio of the casting. (From M. C. Flemings, *Solidification Processing*, copyright 1974, McGraw-Hill, New York. Reproduced with permission of McGraw-Hill.)

8.3.2. Fiber Processing: Inviscid Melt Spinning

Inviscid melt spinning (Section 6.1.5) can be used to produce fine metal or ceramic fibers. As illustrated in Fig. 8.3-4, the melt leaves the orifice in the form of a filament and solidifies at a short distance L below the orifice outlet. The diameter of the fiber is determined by that of the orifice and the velocity V, by the gas pressure.

Let us set the origin of the z coordinate at the outlet of the orifice. Let us assume (1) that heat flow is steady and is one-dimensional in the z direction, (2) that thermal properties are consant and same for the solid and the melt, (3) negligible surface heat losses, (4) that the temperature of the melt at the orifice exit is essentially the same as that of the bulk melt, T_M, and (5) that the melt solidifies with negligible undercooling, that is, at the melting point T_m. In view of the small diameter of the fine wire or fiber, the assumption of one-dimensional heat flow is reasonable.

The energy equation, Eq. [B] of Table 2.3-1, reduces to

$$\rho C_v v_z \frac{\partial T}{\partial z} = k \frac{\partial^2 T}{\partial z^2} \qquad [8.3\text{-}15]$$

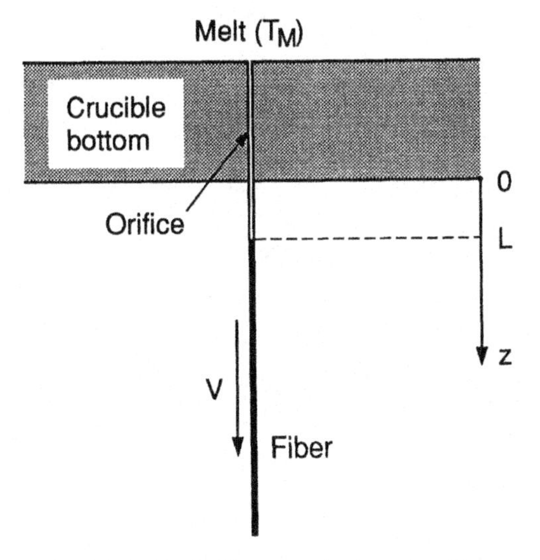

Fig. 8.3-4 Inviscid melt spinning of a fiber.

and with $v_z = V$ to

$$\frac{d^2T}{dz^2} = \frac{V}{\alpha}\frac{dT}{dz} \qquad [8.3\text{-}16]$$

where the thermal diffusivity

$$\alpha = \frac{k}{\rho C_v} \qquad [8.3\text{-}17]$$

The boundary conditions for the liquid are

$$T = T_M \quad \text{at} \quad z = 0 \qquad [8.3\text{-}18]$$

$$T = T_m \quad \text{at} \quad z = L \qquad [8.3\text{-}19]$$

The solution, from Case E of Appendix A, is as follows

$$\frac{T - T_M}{T_m - T_M} = \frac{e^{Vz/\alpha} - 1}{e^{VL/\alpha} - 1} \quad \text{(liquid)} \qquad [8.3\text{-}20]$$

Since surface heat losses are neglected, the temperature is uniform at T_m in the solid:

$$T = T_m \quad \text{(solid)} \qquad [8.3\text{-}21]$$

The length L can be determined from the Stefan condition given by Eq. [5.1-8]:

$$-k\frac{dT}{dz}\bigg|_{L^-} + \rho H_f V = -k\frac{dT}{dz}\bigg|_{L^+} \quad \text{at} \quad z = L \qquad [8.3\text{-}22]$$

where H_f is the heat of fusion (> 0), L^- denotes the melt side of the melt/solid interface, and L^+ denotes the solid (fiber) side of the same interface. Substituting Eqs. [8.3-20] and [8.3-21] into Eq. [8.3-22]

$$-k\frac{V(T_m - T_M)}{\alpha(e^{VL/\alpha} - 1)}e^{VL/\alpha} + \rho H_f V = 0 \qquad [8.3\text{-}23]$$

From this equation

$$e^{VL/\alpha} = \frac{\rho H_f V\alpha}{\rho H_f V\alpha - kV(T_m - T_M)} \qquad [8.3\text{-}24]$$

and

$$\boxed{L = \frac{\alpha}{V}\ln\left(\frac{\rho H_f V\alpha}{\rho H_f V\alpha - kV(T_m - T_M)}\right)} \qquad [8.3\text{-}25]$$

Substituting Eq. [8.3-24] into Eq. [8.3-20]

$$\frac{T - T_M}{T_m - T_M} = (e^{Vz/\alpha} - 1)\left[\frac{\rho H_f \alpha}{k(T_m - T_M)} - 1\right] \quad \text{(liquid)} \qquad [8.3\text{-}26]$$

Subsituting Eq. [8.3-17] into this equation

$$\boxed{\frac{T - T_M}{T_m - T_M} = (e^{Vz/\alpha} - 1)\left[\frac{H_f}{C_v(T_m - T_M)} - 1\right]} \quad \text{(liquid)} \qquad [8.3\text{-}27]$$

8.4. MULTIDIMENSIONAL CONDUCTION

8.4.1. Surface Heat Treating: Spot Laser Heating

In Section 8.2.2 we used a one-dimensional heat flow approximation for spot laser surface heat treating (Fig. 8.2-3). While this approximation is reasonable in certain cases, we shall deal with three-dimensional heat flow in the present section.

Consider a semiinfinite workpiece subjected to a heat source q of radius a, as illustrated in Fig. 8.4-1. Let us consider the surface heat source as consisting of individual point heat sources. For example, let us consider the point heat source

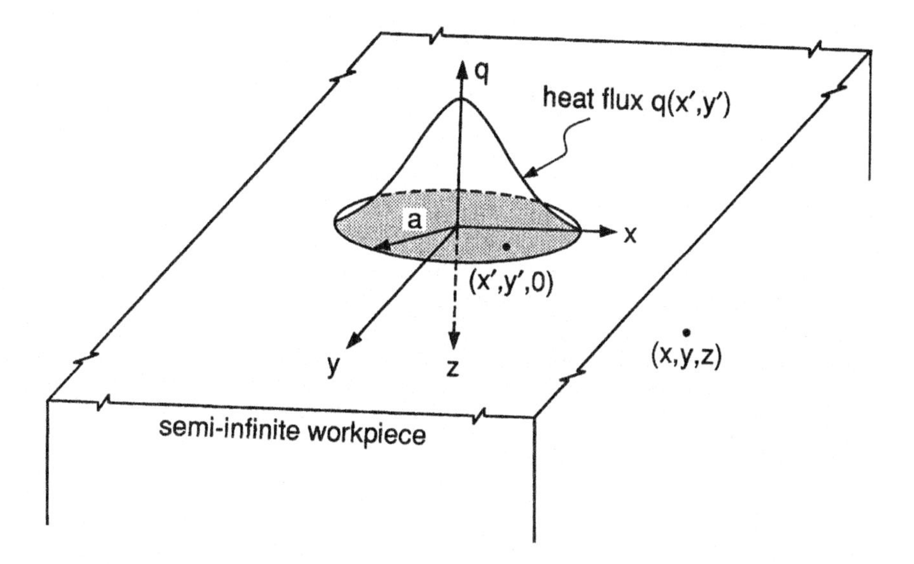

Fig. 8.4-1 Surface heat treating of a stationary workpiece.

located at point $(x', y', 0)$. At time $t = 0$ this point heat source releases instantaneously an amount of heat Q'. Let us consider the response of the workpiece, initially at a uniform temperature T_i, to this point heat source.

From the equation of energy, Eq. [A] of Table 2.3-1

$$\frac{\partial T}{\partial t} = \alpha\left[\frac{\partial^2 T}{\partial x^2} + \frac{\partial^2 T}{\partial y^2} + \frac{\partial^2 T}{\partial x^2}\right] \qquad [8.4\text{-}1]$$

or

$$\frac{\partial(T - T_i)}{\partial t} = \alpha\left[\frac{\partial^2 (T - T_i)}{\partial x^2} + \frac{\partial^2 (T - T_i)}{\partial y^2} + \frac{\partial^2 (T - T_i)}{\partial z^2}\right] \qquad [8.4\text{-}2]$$

where α and T_i are the thermal diffusivity and initial temperature of the workpiece, respectively. The initial and boundary conditions are

$$T(x, y, z, 0) - T_i = 0 \qquad [\text{except at } (x', y', 0)] \qquad [8.4\text{-}3]$$

$$T(\pm\infty, \pm\infty, \infty, t) - T_i = 0 \qquad [8.4\text{-}4]$$

and the energy conservation requirement is

$$\rho C_v \int_0^{\infty}\int_{-\infty}^{\infty}\int_{-\infty}^{\infty} (T - T_i)\, dx\, dy\, dz = Q' \qquad [8.4\text{-}5]$$

Equation [8.4-5] states that the thermal-energy rise in the workpiece is due to the heat input Q'. Heat losses from the workpiece are neglected. The solution, from Case U of Appendix A, is as follows:

$$T - T_i = \frac{2Q'}{\rho C_v (4\pi\alpha t)^{3/2}} \exp\left\{\frac{-[(x-x')^2 + (y-y')^2 + z^2]}{4\alpha t}\right\} \qquad [8.4\text{-}6]$$

Equation [8.4-6] is for a quantity of heat Q' given instantaneously at $t = 0$. If heat is given continuously at a constant rate starting from time $t = 0$, then Eq. [8.4-6] should be rewritten as

$$T - T_i = \frac{2Q}{\rho C_v (4\pi\alpha)^{3/2}} \int_0^t t^{-(3/2)} \exp\left\{\frac{-[(x-x')^2 + (y-y')^2 + z^2]}{4\alpha t}\right\} dt \qquad [8.4\text{-}7]$$

where Q is the amount of heat given by the point source per unit time. For a distributed heat source, the above equation can be further modified to

$$T - T_i = \frac{2}{\rho C_v (4\pi\alpha)^{3/2}} \int_0^t \int_{-\infty}^{\infty} \int_{-\infty}^{\infty} q(x', y')\, t^{-(3/2)} \times \exp\left\{\frac{-[(x-x')^2 + (y-y')^2 + z^2]}{4\alpha t}\right\} dx'\, dy'\, dt \qquad [8.4\text{-}8]$$

where $q(x', y')$ is the power-density distribution of the heat flux on the workpiece surface.

Let us consider the typical case of a **Gaussian heat source**:

$$q(x', y') = \frac{Q}{\pi a^2} \exp\left\{\frac{-(x'^2 + y'^2)}{a^2}\right\} \qquad [8.4\text{-}9]$$

where a can be considered as an effective radius of the heat source. Equation [8.4-9] can be substituted into Eq. [8.4-8] and the surface integral:

$$\int_{-\infty}^{\infty} \int_{-\infty}^{\infty} q(x', y')\, t^{-(3/2)} \exp\left\{\frac{-[(x-x')^2 + (y-y')^2 + z^2]}{4\alpha t}\right\} dx'\, dy'$$
$$= \frac{Q}{\pi a^2 t^{3/2}} \exp\left(-\frac{z^2}{4\alpha t}\right) \int_{-\infty}^{\infty} \exp\left\{-\left[\frac{x'^2}{a^2} + \frac{(x-x')^2}{4\alpha t}\right]\right\} dx'$$
$$\times \int_{-\infty}^{\infty} \exp\left\{-\left[\frac{y'^2}{a^2} + \frac{(y-y')^2}{4\alpha t}\right]\right\} dy' \qquad [8.4\text{-}10]$$

To help carry out the integration, the following formula can be used:

$$\int_{-\infty}^{\infty} \exp\left\{-\left[\frac{\xi^2}{a_1^2}+\frac{(a_2-\xi)^2}{a_3}\right]\right\}d\xi = \left[\frac{\pi a_3 a_1^2}{a_3+a_1^2}\right]^{1/2} \exp\left\{-\frac{a_2^2}{(a_3+a_1^2)}\right\} \tag{8.4-11}$$

The RHS of Eq. [8.4-10] now becomes

$$\frac{Q}{\pi a^2 t^{3/2}} \exp\left(-\frac{z^2}{4\alpha t}\right)\cdot\left[\frac{\pi 4\alpha t a^2}{4\alpha t+a^2}\right]^{1/2} \exp\left[-\frac{x^2}{(4\alpha t+a^2)}\right]\cdot\left[\frac{\pi 4\alpha t a^2}{(4\alpha t+a^2)}\right]^{1/2}$$
$$\times \exp\left[-\frac{y^2}{(4\alpha t+a^2)}\right] = 4\alpha Q t^{-(1/2)} \frac{1}{4\alpha t+a^2} \exp\left[-\left(\frac{x^2+y^2}{4\alpha t+a^2}+\frac{z^2}{4\alpha t}\right)\right] \tag{8.4-12}$$

Substituting this quantity for the surface integral in Eq. [8.4-8]

$$\boxed{T-T_i = \frac{Q}{\rho C_v \pi^{3/2}\alpha^{1/2}} \int_0^t \frac{1}{(4\alpha t+a^2)t^{1/2}} \exp\left[-\left(\frac{x^2+y^2}{4\alpha t+a^2}+\frac{z^2}{4\alpha t}\right)\right]dt} \tag{8.4-13}$$

This equation describes the temperature distribution in a semiinfinite workpiece due to a Gaussian heat source of radius a and constant power Q at time t after the heat source is turned on.

8.4.2. Surface Heat Treating: Scanning Laser Heating

In the previous section a stationary workpiece is heat treated with a stationary laser beam. When a large surface area is to be heat-treated, it is much more effective if the workpiece is traversed under the laser beam.[11]

As shown in Fig. 8.4-2, both the heat flux and the coordinates (x, y, z) are stationary, the origin of the coordinates coinciding with the center of the heat flux q. The workpiece moves at a constant velocity V in the positive x direction. Since a point in the workpiece located at (x, y, z) at time t was located at $(x - Vt, y, z)$ at time zero, Eq. [8.4-7] should now be modified as

$$T-T_i = \frac{2Q}{\rho C_v (4\pi\alpha)^{3/2}} \int_0^t t^{-(3/2)} \exp\left\{\frac{-[(x-Vt-x')^2+(y-y')^2+z^2]}{4\alpha t}\right\} dt \tag{8.4-14}$$

[11] D. S. Gnanamuthu, in *Applications of Lasers in Materials Processing*, E. A. Metzbower, ed., American Society for Materials Park, OH, 1977, p. 177

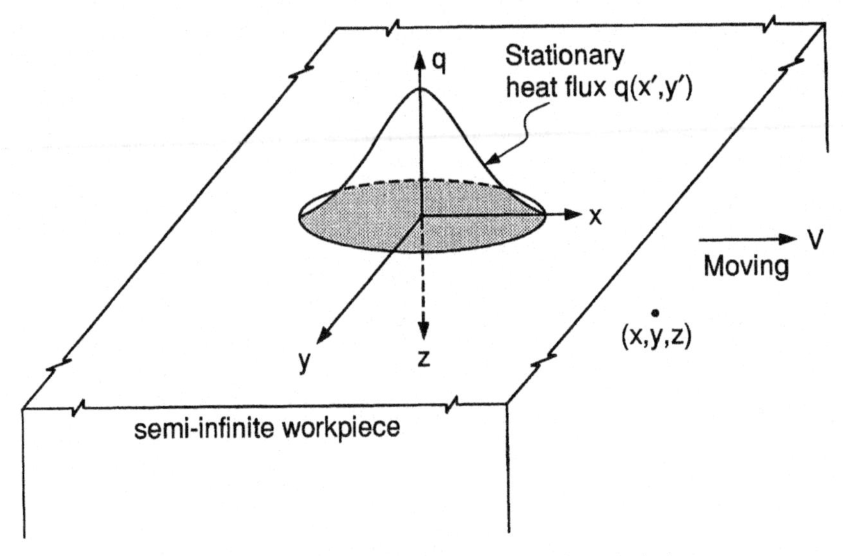

Fig. 8.4-2 Surface heat treating of a moving workpiece.

Similarly, for a distributed heat source, Eq. [8.4-8] should be modified as

$$T - T_i = \frac{2}{\rho C_v (4\pi\alpha)^{3/2}} \int_0^t \int_{-\infty}^{\infty} \int_{-\infty}^{\infty} q(x', y') t^{-(3/2)}$$

$$\times \exp\left\{\frac{-[(x - Vt - x')^2 + (y - y')^2 + z^2]}{4\alpha t}\right\} dx'\, dy'\, dt \qquad [8.4\text{-}15]$$

Following an integration procedure similar to that from Eqs. [8.4-10] through [8.4-12], Eq. [8.4-15] becomes as follows for a Gaussian heat source:

$$T - T_i = \frac{Q}{\rho C_v \pi^{3/2} \alpha^{1/2}} \int_0^t \frac{1}{(4\alpha t + a^2) t^{1/2}} \exp\left\{-\left[\frac{(x - Vt)^2 + y^2}{4\alpha t + a^2} + \frac{z^2}{4\alpha t}\right]\right\} dt$$

[8.4-16]

After the heat source has been turned on for a long time ($t \to \infty$), the temperature field becomes steady with respect to the coordinates (x, y, z). This steady-state temperature field is described by

$$\boxed{T - T_i = \frac{Q}{\rho C_v \pi^{3/2} \alpha^{1/2}} \int_0^{\infty} \frac{1}{(4\alpha t + a^2) t^{1/2}} \exp\left\{-\left[\frac{(x - Vt)^2 + y^2}{4\alpha t + a^2} + \frac{z^2}{4\alpha t}\right]\right\} dt}$$

[8.4-17]

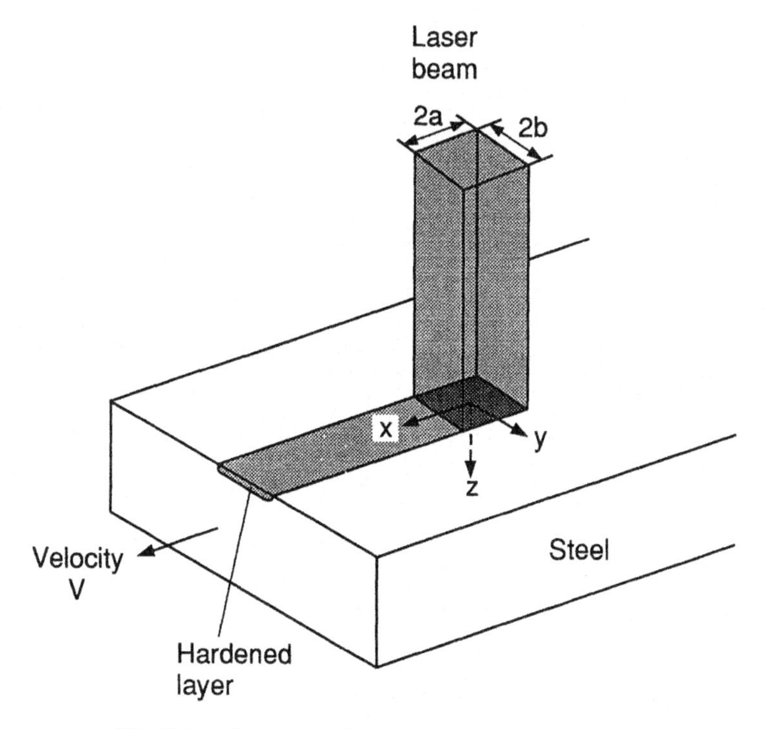

Fig. 8.4-3 Laser transformation hardening of steel.

Equation [8.4-17] has been derived by Cline et al.[12] and Arata et al.[13]

Similar equations for a rectangular Gaussian and rectangular uniform heat source have also be derived.[13] These equations involve integration of the error function from zero to infinity and numerical computations are required. An alternative is to solve the equation of energy numerically. Kou et al. have developed computer models for the laser transformation hardening of flat plates[14] and cylindrical bodies.[15] The former will be described briefly below and the latter will be left as exercise problems.

Consider the transformation hardening of steel with a uniform rectangular laser beam, as shown in Fig. 8.4-3. Since the thermal conductivity and specific heat of steel are often temperature-dependent, an energy equation for variable

[12] H. E. Cline, and T. R. Anthony, *J. Appl. Phys.*, **48**, 3895, 1977.

[13] Y. Arata, H. Maruo, and I. Miyamoto, International Welding Institute Document IV-241-78, International Welding Institute, 1978.

[14] S. Kou, D. K. Sun, and Y. P. Le, *Metallurgical Trans.*, **14A**, 643, 1983.

[15] S. Kou and D. K. Sun, *Metallurgical Trans.*, **14A**, 1859, 1983.

properties is preferred. From Eq. [E] of Table 4.1-4

$$\frac{\partial}{\partial t}(\rho C_v T) + V\frac{\partial}{\partial x}(\rho C_v T) = \nabla\cdot(k\nabla T) \qquad [8.4\text{-}18]$$

noting that $v_x = V$ and $v_y = v_z = 0$.

The boundary conditions are as follows:

1. $\dfrac{\partial T}{\partial y} = 0 \quad \text{at} \quad y = 0$ [8.4-19]

2. $T = T_0 \quad \text{as} \quad (x^2 + y^2 + z^2)^{1/2} \to \infty$ [8.4-20]

3. $-k\dfrac{\partial T}{\partial z} = \dfrac{\eta Q_0}{4ab} \quad \text{for} \quad z = 0, |x| \leq a, |y| \leq b$ [8.4-21]

4. $-k\dfrac{\partial T}{\partial z} = h(T - T_a) + \varepsilon\sigma(T^4 - T_a^4) \quad \text{for} \quad z = 0, |x| > a, |y| > b$ [8.4-22]

where T_0 is the initial temperature, Q_0 the nominal beam power, and η the absorptivity of the beam by the workpiece surface. The heat losses on the RHS of the last equation turn out to be negligible ($< 1\%$ of the heat input).

The governing equation (and the boundary conditions) can be expressed in the dimensionless form when the physical properties are assumed constant:

$$\frac{\partial T^*}{\partial t^*} = -\frac{\partial T^*}{\partial x^*} + \nabla^{*2} T^* \qquad [8.4\text{-}23]$$

where

$$T^* = \frac{k\alpha(T - T_0)}{\eta Q_0 V} \qquad [8.4\text{-}24]$$

$$t^* = \frac{tV^2}{\alpha} \qquad [8.4\text{-}25]$$

$$x^*, y^*, z^* = \frac{xV}{\alpha}, \frac{yV}{\alpha}, \frac{zV}{\alpha} \qquad [8.4\text{-}26]$$

$$\nabla^*, \nabla^{*2} = \frac{\alpha}{V}\nabla, \left(\frac{\alpha}{V}\right)^2 \nabla^2 \qquad [8.4\text{-}27]$$

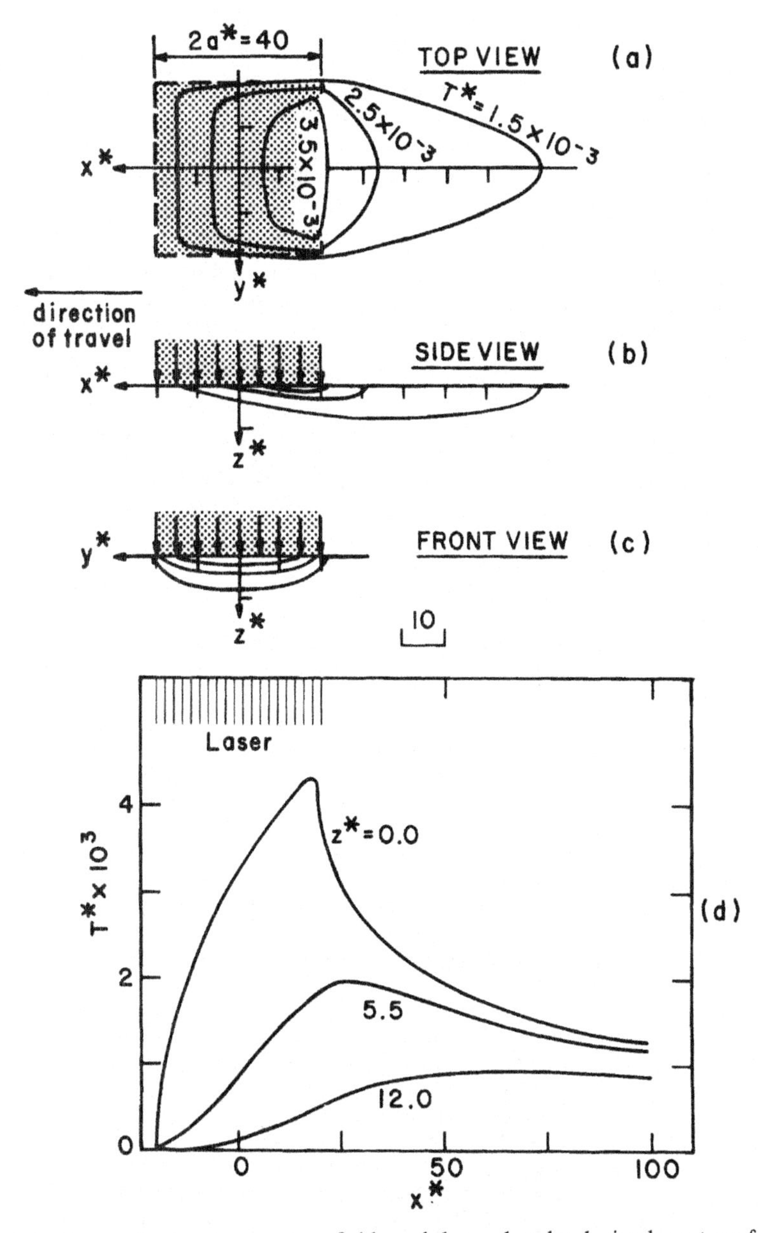

Fig. 8.4-4 Dimensionless temperature fields and thermal cycles during laser transformation hardening: (*a*) top view; (*b*) side view; (c) front view; (*d*) thermal cycles at $y = 0$. (From S. Kou, D. K. Sun, and Y. P. Le, *Metallurgical Transactions*, **14A**, 1983, p. 643.)

A dimensionless temperature distribution for a uniform square laser beam with $a^* = 20$ is shown in Fig. 8.4-4, where $a^* = aV/\alpha$. As shown, the maximum temperature at the workpiece surface is near the trailing edge of the laser beam, rather than near the center. Figure 8.4-4*c* can be used to determine the shape of the hardened layer if the actual phase transformation temperature is known.

The dimensionless peak temperature at $y = 0$ is shown in Fig. 8.4-5 as a function of the dimensionless depth and half-beam size. From this the maximum dimensionless temperature at the workpiece surface ($y = z = 0$) and the onset of surface melting can be found, as shown in Fig. 8.4-6. As shown in Fig. 8.4-6*b*

$$T^*_{max} = 0.293a^{*-1.4} \qquad (5 \le a^* \le 50) \qquad [8.4\text{-}28]$$

The one-dimensional heat flow equation in Fig. 8.4-6*a* can be obtained from Eq. [8.2-10], and this will be left as an exercise problem.

Since melting is undesirable in transformation hardening, Eq. [8.4-28] can be used as a guide to ensure that the maximum surface temperature stays below the

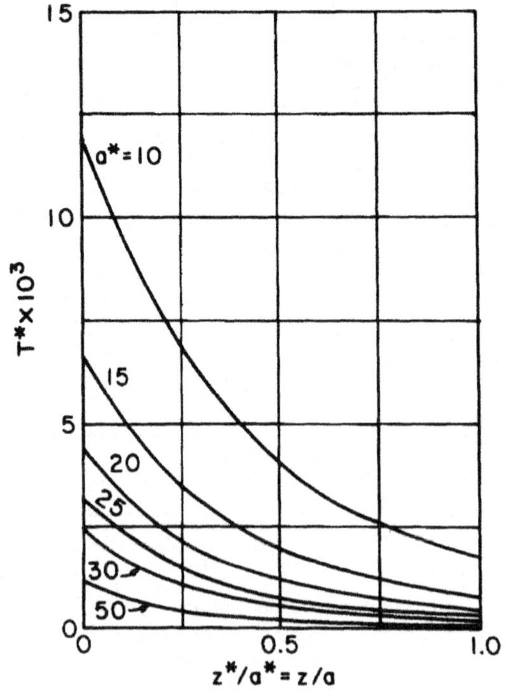

Fig. 8.4-5 Dimensionless peak temperature at $y = 0$ as a function of the dimensionless depth and half-beam size. (From S. Kou, D. K. Sun and Y. P. Le, *Metallurgical Transaction*, **14A**, 1983, p. 643.)

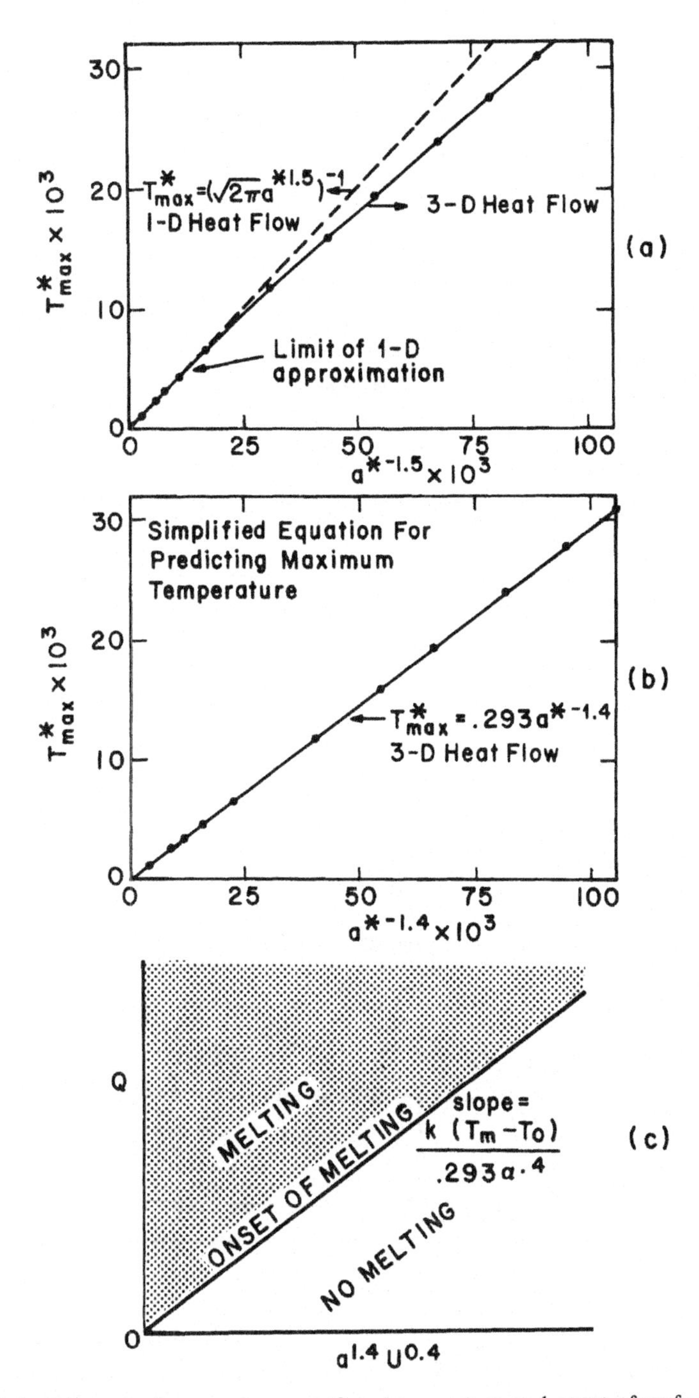

Fig. 8.4-6 Dimensionless maximum surface temperature and onset of surface melting: (*a*) 1-D and 3-D solutions; (*b*) a simple correlation; (*c*) onset of surface melting. (From S. Kou, D. K. Sun, and Y. P. Le, *Metallurgical Transactions*, **14A**, 1983, p. 643.)

melting point T_m. In other words

$$T_m^* = \frac{k\alpha(T_m - T_0)}{\eta Q_0 V} = 0.293\left(\frac{aV}{\alpha}\right)^{-1.4} \quad [8.4\text{-}29]$$

On rearranging

$$\frac{\eta Q_0}{a^{1.4}V^{0.4}} = c_m \quad [8.4\text{-}30]$$

where

$$c_m = \frac{k(T_m - T_0)}{0.293\alpha^{0.4}} \quad [8.4\text{-}31]$$

For a given material c_m is a constant, as shown in Fig. 8.4-6*c*.

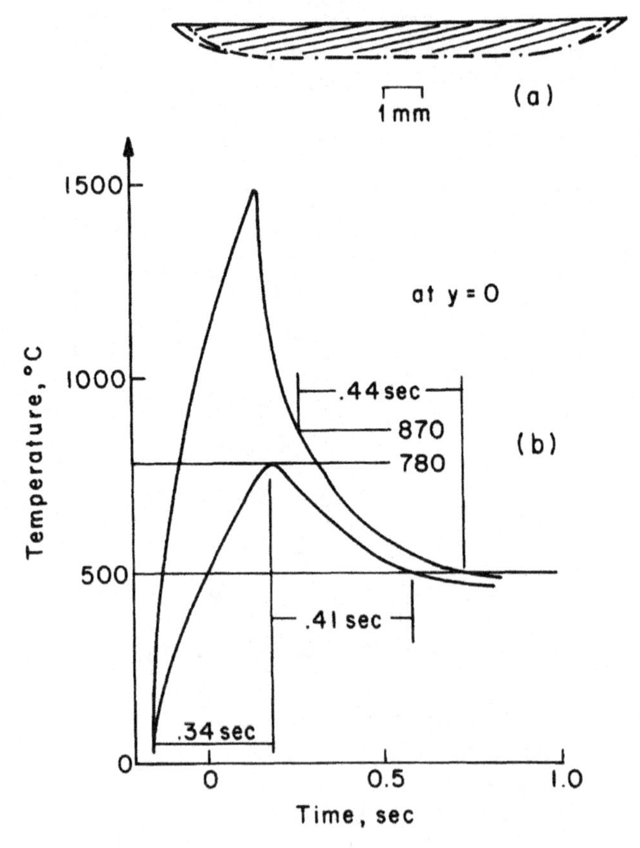

Fig. 8.4-7 Laser transformation hardening of 1018 steel with a 12 × 12-mm uniform square beam at 5.7 kW and 38 mm/s: (*a*) transformed zone; (*b*) thermal cycles at top and bottom of transformed zone. (From S. Kou, D. K. Sun, and Y. P. Le, *Metallurgical Transactions*, **14A**, 1983, p. 643.)

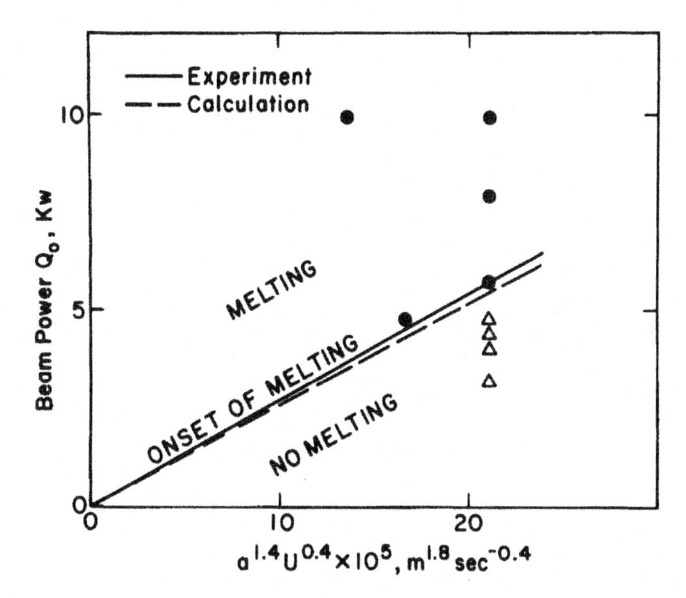

Fig. 8.4-8 Onset of surface melting in laser transformation hardening of 1018 steel with a 12 × 12-mm uniform square beam. (From S. Kou, D. K. Sun, and Y. P. Le, *Metallurgical Transactions*, **14A**, 1983, p. 643.)

Figure 8.4-7*a* shows the calculated transformed zone in a 1018 steel heat-treated by a uniform 12 × 12-mm square laser beam at the nominal beam power $Q_0 = 5.70$ kW and the travel speed $V = 38.1$ mm/s. The workpiece surface was sprayed with Krylon flat black to boost the beam absorptivity η to around 85%. The thermal responses at the top and bottom of the transformed zone are shown in Fig. 8.4-7*b*. According to the continuous cooling transformation diagram for 1018 steel, a 0.44-s cooling time from 870 to 550°C should result in the formation of martensite, which is consistent with the microstructure observed at the workpiece surface. The prediction of the onset of surface melting by Eq. [8.4-30] is compared with the experimental results in Fig. 8.4-8.

8.4.3. Welding: Thick Plates

Figure 8.4-9 illustrates schematically the **bead-on-plate welding** of a thick workpiece. This type of welding is often used by welding engineers and metallurgists as a simple means for studying the response of the workpiece material on welding, including, for instance, the microstructures in the fusion zone (i.e., the bead) and the heat-affected zone. Because of the limited heat input, the depth and width of the bead are much smaller than the thickness and width of the workpiece. Therefore, the workpiece can often be considered semiinfinite in the heat transfer analysis.

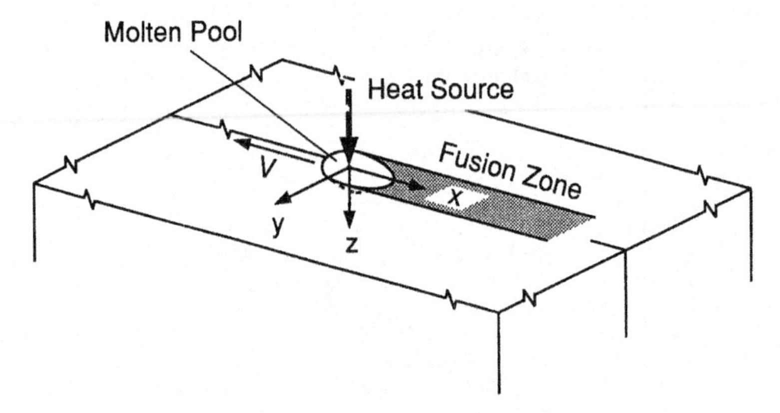

Fig. 8.4-9 Welding of thick plates.

Heat transfer in a thick workpiece during bead-on-plate welding was first considered by Rosenthal,[16] assuming a point heat source. As illustrated in Fig. 8.4-9, the heat source travels at a constant velocity V in the negative x direction. Mathematically, this is equivalent to the case where the heat source remains stationary while the workpiece travels at a constant velocity in the positive x direction. In fact, in laser-beam and electron-beam welding the beam normally remains stationary while the workpiece is traversed under the beam. Therefore, in our analysis here we will consider the heat source as stationary and the workpiece moving. The origin of the stationary coordinates coincides with the point heat source.

According to Eq. [8.4-14] the quantity

$$T - T_i = \frac{2Q}{\rho C_v (4\pi\alpha)^{3/2}} \int_0^t t^{-(3/2)} \exp\left\{\frac{-[(x - Vt - x')^2 + (y - y')^2 + z^2]}{4\alpha t}\right\} dt \qquad [8.4\text{-}32]$$

represents the temperature rise $(T - T_i)$ at time t and point (x, y, z) in the moving semiinfinite workpiece, due to a stationary point heat source Q located at $(x', y', 0)$ on the workpiece surface. Since the origin of the coordinate system coincides with the point heat source, $x' = y' = 0$. After heat has been emitted for a long time, a steady-state temperature distribution, with respect to the stationary coordinate system, is established:

$$T - T_i = \frac{2Q}{\rho C_v (4\pi\alpha)^{3/2}} \int_0^\infty t^{-(3/2)} \exp\left\{\frac{-[(x - Vt)^2 + y^2 + z^2]}{4\alpha t}\right\} dt \qquad [8.4\text{-}33]$$

[16] D. Rosenthal, *Transaction ASME*, p. 849, Nov. 1946.

Let us define radius R as

$$R = (x^2 + y^2 + z)^{1/2} \tag{8.4-34}$$

Equation [8.4-33] can be rearranged as

$$T - T_i = \frac{2Q}{\rho C_v (4\pi\alpha)^{3/2}} \exp\left(\frac{Vx}{2\alpha}\right) \int_0^{\infty} t^{-(3/2)} \exp\left\{-\left[\frac{R^2}{4\alpha t} + \frac{V^2 t}{4\alpha}\right]\right\} dt \tag{8.4-35}$$

or

$$T - T_i = \frac{2Q}{\rho C_v (4\pi\alpha)^{3/2}} \exp\left(\frac{Vx}{2\alpha}\right)\left[\frac{-4\alpha^{1/2}}{R}\right] \times \int_0^{\infty} \exp\left\{-\left[\frac{R^2}{4\alpha t} + \frac{V^2 R^2}{16\alpha^2}\frac{4\alpha t}{R^2}\right]\right\} d\left[\frac{R}{\sqrt{4\alpha t}}\right] \tag{8.4-36}$$

In order to carry out the integration, the following formula can be used:[17]

$$\int_{\infty}^{0} \exp\left(-ax^2 - \frac{b}{x^2}\right) dx = -\frac{1}{2}\sqrt{\frac{\pi}{a}} \exp(-2\sqrt{ab}) \tag{8.4-37}$$

Substituting Eq. [8.4-37] into Eq. [8.4-36], we get

$$\boxed{T - T_i = \frac{Q}{2\pi k R} \exp\left[\frac{V(x - R)}{2\alpha}\right]} \tag{8.4-38}$$

where k is the thermal conductivity of the workpiece.

Equation [8.4-38] is Rosenthal's equation for three-dimensional heat flow in a semiinfinite workpiece during welding. It is based on the following simplifying assumptions: (1) a point heat source, (2) constant thermal properties, (3) no melting nor solidification, and (4) no heat losses from the workpiece surface. Despite the restrictions imposed by these simplifying assumptions, this equation has been widely used in the welding community mainly because of its simplicity.

In addition to the temperature distribution in the workpiece, the equation can also be used to calculate the thermal history experienced by any location in the workpiece: the **thermal cycle**. To do this, the temperature distribution along the welding direction (i.e., T vs. x) can be readily converted into a temperature/time plot, simply the setting time $t = (x - 0)/V$. Figures 8.4-10 and 8.4-11 show the results calculated for the welding of 1018 steel at two different levels of heat

[17] I. S. Gradshteyn and I. M. Ryzhik, *Table of Integrals, Series, and Products*, Academic Press, New York, 1980, p. 307.

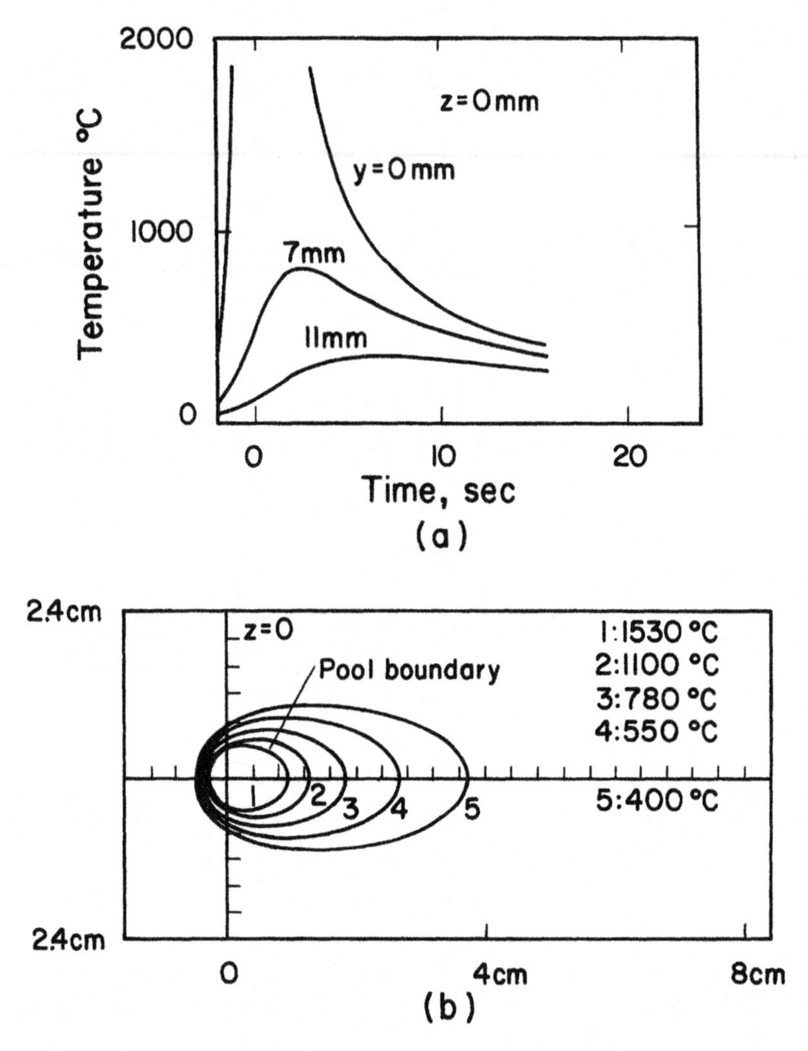

Fig. 8.4-10 Calculated results from Rosenthal's 3-D equation: (*a*) thermal cycles; (*b*) isotherms. $V = 2.4$ mm/s, $Q = 3200$ W, and material = 1018 steel. (From S. Kou, *Welding Metallurgy*, copyright 1987, John Wiley and Sons, New York. Reprinted by permission of John Wiley and Sons, Inc.)

input and welding speed.[18] As shown in the thermal cycles in Figs. 8.4-10*a* and 8.4-11*a*, the temperature at the position of the heat source approaches infinity. This is due to the singularity problem associated with Rosenthal's point heat source assumption. Table 8.4-1 lists the physical properties of some workpiece materials.

[18] S. Kou, *Welding Metallurgy*, Wiley, New York, 1987, pp. 39–41

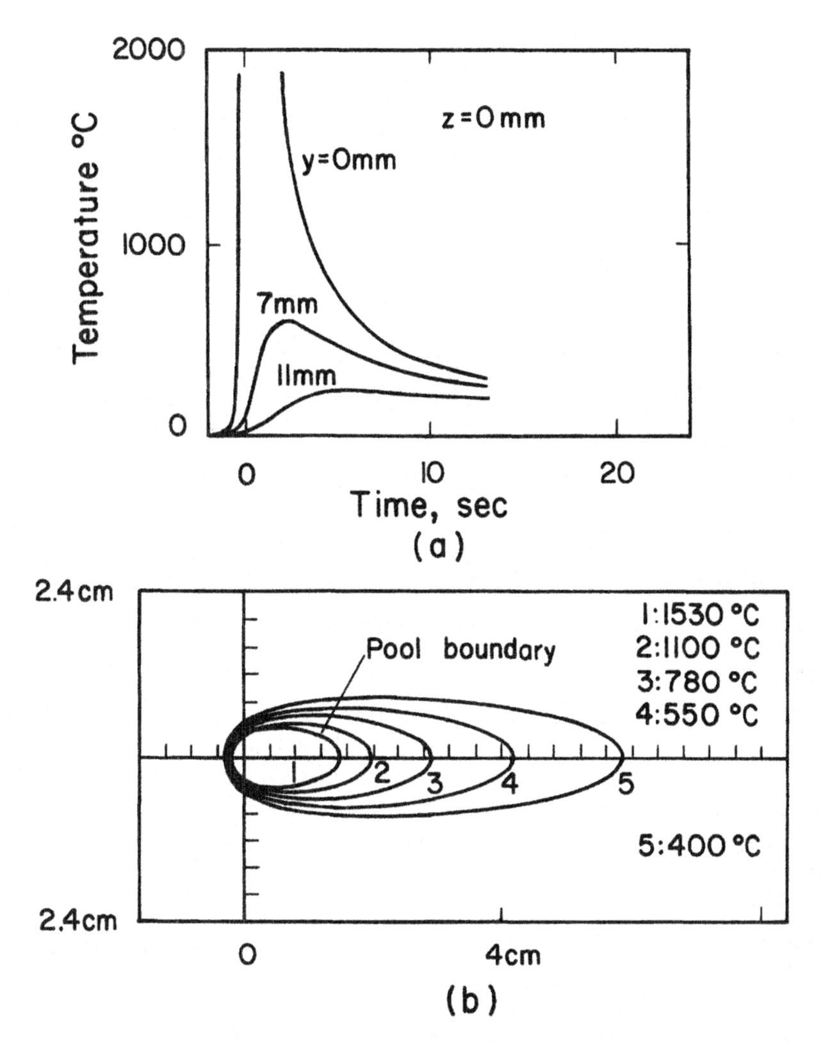

Fig. 8.4-11 Same as Fig. 8.4-10 but with $V = 6.2$ mm/s and $Q = 5000$ W. The weld pool is more elongated and the cooling rates are higher because of the higher welding speed. (From S. Kou, *Welding Metallurgy*, copyright 1987, John Wiley and Sons, New York. Reprinted by permission of John Wiley and Sons, Inc.)

Kou and Le[19] have relaxed Rosenthal's assumptions and calculated three-dimensional heat conduction during GTA (gas tungsten arc) welding. As already mentioned the heat source and the coordinate system can be considered stationary while the workpiece moves in the x direction at a constant velocity V. From

[19] S. Kou and Y. Le, *Metallurgical Trans. A*, **14A**, 2245, 1983.

TABLE 8.4-1 Thermal Properties for Several Materials

Material	Thermal Diffusivity α (m^2/s)	Volume Thermal Capacity ρC_s (J m^{-3} K^{-1})	Thermal Conductivity k (J m^{-1} s^{-1} K^{-1})	Melting Point (K)
Aluminum	8.5×10^{-5}	2.7×10^6	229	933
Carbon steel	9.1×10^{-6}	4.5×10^6	41.0	1800
9% Ni steel	1.1×10^{-5}	3.2×10^6	35.2	1673
Stainless steel	5.3×10^{-6}	4.7×10^6	24.9	1773
Inconel 600	4.7×10^{-6}	3.9×10^6	18.3	1673
Ti alloy	9.0×10^{-6}	3.0×10^6	27.0	1923
Copper	9.6×10^{-5}	4.0×10^6	384.0	1336
Monel 400	8.0×10^{-6}	4.4×10^6	35.2	1573

the energy equation Eq. [2.3-7]) and with $H = C_v T$ and $s = 0$

$$\frac{\partial}{\partial t}(\rho H) + V\frac{\partial}{\partial x}(\rho H) = \nabla \cdot (k \nabla T) \qquad [8.4\text{-}39]$$

Note that in order to consider the heat of fusion, the enthalpy per unit mass H rather than $C_v T$ is used. Except during the initial and final transients of welding, the temperature field in a workpiece of sufficient length is steady with respect to the heat source and the coordinate system. As such, the time-dependent term in energy equation can be dropped. The boundary conditions are as follows:

1. $$\frac{\partial T}{\partial y} = 0 \quad \text{at} \quad y = 0 \qquad [8.4\text{-}40]$$

2. $$q = \frac{3Q}{\pi a^2}\exp\left(\frac{x^2 + y^2}{-a^2/3}\right) \quad \text{for} \quad z = 0 \quad \text{and} \quad (x^2 + y^2)^{1/2} \le a \qquad [8.4\text{-}41]$$

3. $$q = -[h(T - T_a) + \varepsilon\sigma(T^4 - T_a^4)] \quad \text{for} \quad z = 0$$

$$\text{and} \quad (x^2 + y^2)^{1/2} > a \quad \text{and for} \quad z = b \qquad [8.4\text{-}42]$$

4. $$q = -[h(T - T_a) + \varepsilon\sigma(T^4 - T_a^4)] \quad \text{at} \quad y = \pm\frac{w}{2} \qquad [8.4\text{-}43]$$

5. $$T = T_0 \quad \text{as} \quad x \to -\infty \qquad [8.4\text{-}44]$$

6. $$\frac{\partial T}{\partial x} = 0 \quad \text{as} \quad x \to \infty \qquad [8.4\text{-}45]$$

where Q is the power delivered into the workpiece, a the effective radius of the Gaussian heat source at the workpiece surface, b the workpiece thickness, and w the workpiece width. In boundary conditions of Eqs. [3 and 4], heat losses due to convection and radiation are considered. The last two boundary conditions state that far ahead of the heat source the workpiece is at its initial temperature T_0, while far behind it the temperature gradient levels off.

Figure 8.4-12 shows the calculated results for a GTA weld in a 3.2-mm thick 6061 aluminum alloy. The fusion boundary is represented by the liquidus temperature $T_L = 652°C$. The conditions are as follows: arc voltage 10 V, DC current (electrode negative) 110 A, welding speed 5.5 mm/s, and arc radius 3 mm. An arc efficiency of $\eta = 78\%$ (i.e., $Q = 0.78 \times 110\text{ A} \times 10\text{ V}$) and an effective liquid thermal conductivity 1.5 times the value of the solid thermal conductivity give the best agreement with the experiment. An effective liquid thermal conductivity is used to take into account the effect of convection since convective heat transfer is not calculated.

8.4.4. Welding: Thin Sheets

Figure 8.4-13 illustrates schematically the welding of thin sheets. Because of the small thickness of the workpiece, temperature variations in the thickness direction can be neglected and heat flow in the workpiece can be considered two-dimensional. A simple way to verify this is to examine the transverse cross section of the weld. If the weld penetrates vertically through the workpiece and the fusion boundaries of the weld are nearly parallel to each other, heat flow during welding should be essentially two-dimensional. It should be pointed out that the workpiece does not always have to be physically very thin for heat flow to become essentially two-dimensional. An obvious example of this is electron-beam welding, where the high power-density beam can readily penetrate through a physically thick workpiece and produce a weld with nearly parallel fusion boundaries. Figure 8.4-14*a* shows an electron-beam weld in 12.7-mm-thick aluminum.[20] Notice that the fusion boundaries are nearly parallel to each other, thus suggesting two-dimensional heat flow during welding. On the contrary, the gas tungsten arc weld in a 3.2-mm-thick 6061 aluminum alloy shown in Fig. 8.4-14*b*, in spite of the smaller thickness, clearly suggests three-dimensional heat flow during welding.[21]

Let us assume that the heat source is a line source and that it coincides with the z axis. If q represents the heat input per unit thickness, the heat input due to a differential element of length dz' of the heat source located at a distance z' below the surface is $q\,dz'$. The differential temperature increase $d(T - T_i)$ according to Eq. [8.4-38] is as follows:

$$d(T - T_i) = \frac{q\,dz'}{4\pi k R'} \exp\left[\frac{V(x - R')}{2\alpha}\right] \qquad [8.4\text{-}46]$$

[20] W. J. Farrell, ASTME Paper CP 63-208, 1962–1963.

[21] S. Kou, *Welding Metallurgy*, Wiley, New York, 1987, p. 38.

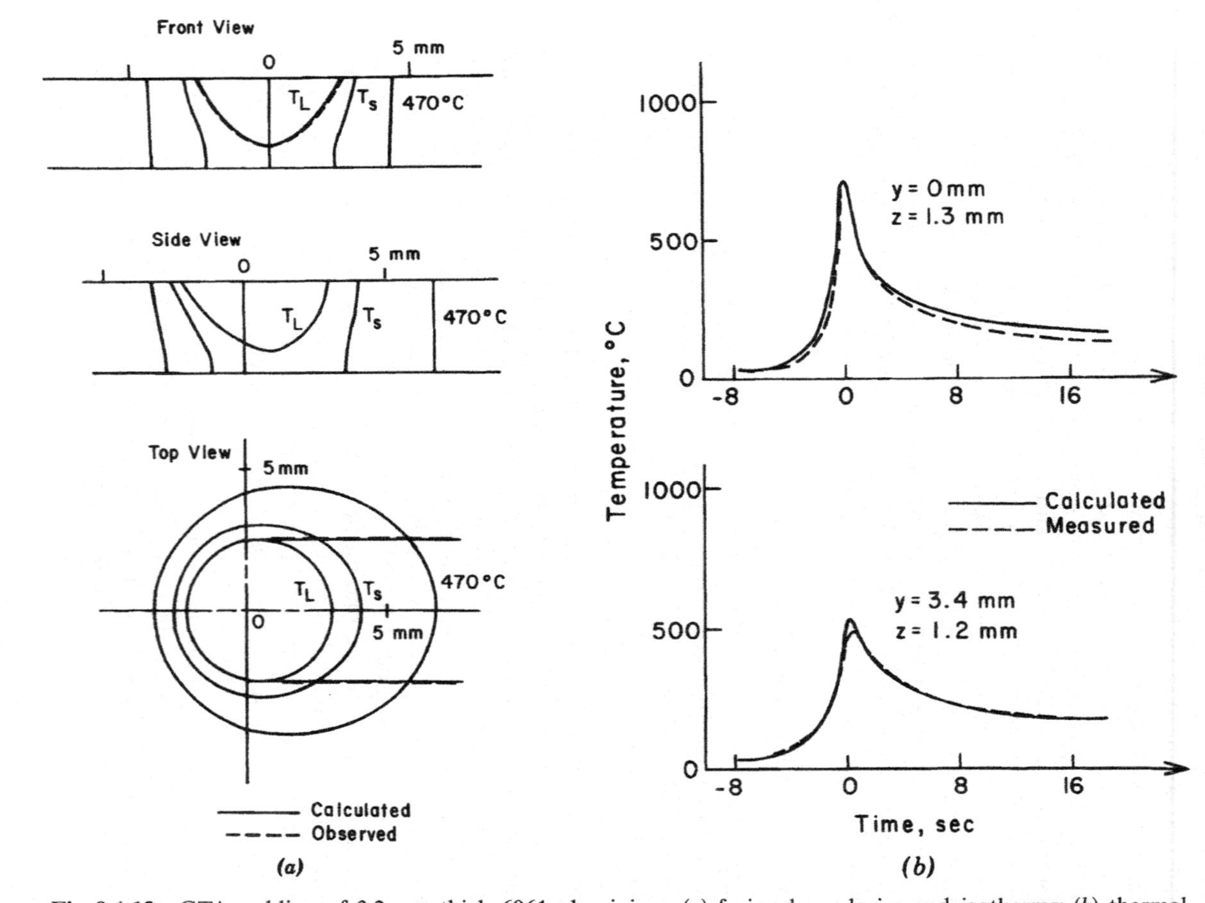

Fig. 8.4-12 GTA welding of 3.2-mm-thick 6061 aluminium: (*a*) fusion boundaries and isotherms; (*b*) thermal cycles. (From S. Kou and Y. Le, *Metallurgical Transactions*, **14A**, 1983, p. 2245.)

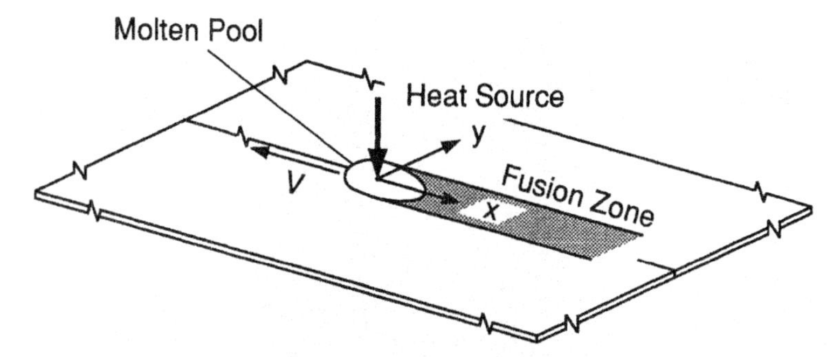

Fig. 8.4-13 Welding of thin sheets.

where $R' = [x^2 + y^2 + (z - z')^2]^{1/2}$, that is, the distance between any point (x, y, z) and the heat source at $(0, 0, z')$. This temperature increase is only half of that according to Eq. [8.4-38] since the heat source is now in the workpiece rather than at its surface. In other words, there is now material above as well as below the heat source to dissipate heat. As such, the total temperature increase due to the line source is as follows:

$$T - T_i = \int_{-\infty}^{\infty} \frac{q}{4\pi k R'} \exp\left[\frac{V(x - R')}{2\alpha}\right] dz' \qquad [8.4\text{-}47]$$

For heat flow to be two-dimensional, the line source must be uniform: $q = Q/L$, where L is the length of the heat source, that is, the thickness of the workpiece. As such, Eq. [8.4-47] becomes

$$T - T_i = \frac{Q}{4\pi k L} \exp\left(\frac{Vx}{2\alpha}\right) \int_{-\infty}^{\infty} \frac{1}{R'} \exp\left(\frac{-VR'}{2\alpha}\right) dz' \qquad [8.4\text{-}48]$$

The integral in this equation turns out to be

$$2K_0\left[\frac{V(x^2 + y^2)^{1/2}}{2\alpha}\right]$$

where $K_0(\xi)$ is the modified Bessel function of the second kind of order zero,[22] and is plotted in Fig. 8.4-15.[23] It has been shown that $K_0(\xi)$ approaches $-\ln \xi$

[22] D. Rosenthal, *Welding J.*, **20**, 220s, 1941.

[23] M. Abramowitz and I. A. Stegun, eds., *Handbook of Mathematical Functions*, National Bureau of Standards, Washington, DC, 1964.

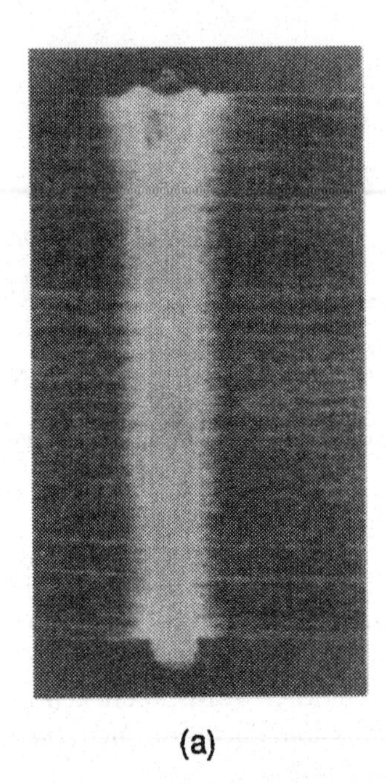

(a)

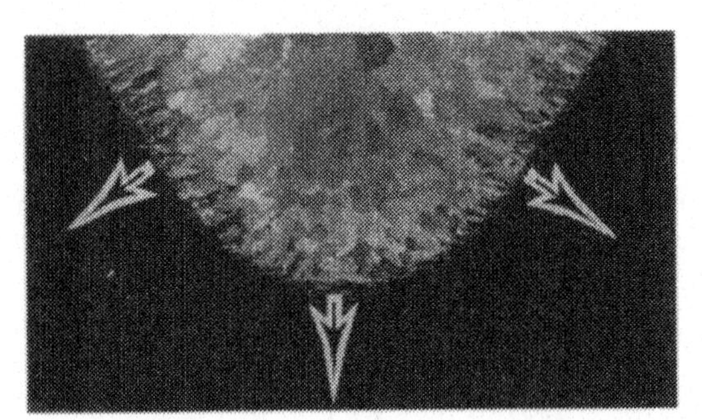

(b)

Fig. 8.4-14 Transverse cross sections of aluminium welds: (*a*) electron-beam welding; (*b*) gas tungsten arc welding. (From S. Kou, *Welding Metallurgy*, copyright 1987, John Wiley and Sons, New York. Reprinted by permission of John Wiley and Sons, Inc.)

as ξ approaches zero. Equation [8.4-48] now becomes

$$\boxed{T - T_i = \frac{Q}{2\pi k L} \exp\left(\frac{Vx}{2\alpha}\right) K_0 \left(\frac{Vr}{2\alpha}\right)} \qquad [8.4\text{-}49]$$

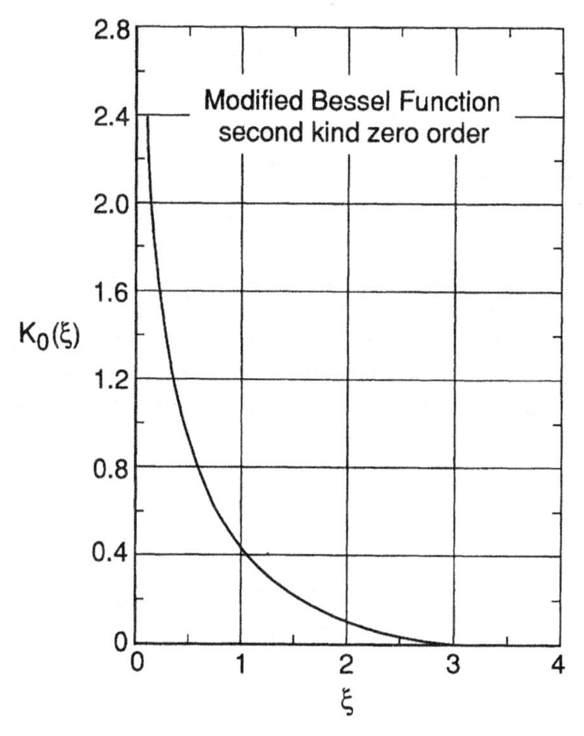

Fig. 8.4-15 Modified Bessel function of the second kind and zero order, K_0.

where the radius

$$r = (x^2 + y^2)^{1/2} \qquad [8.4\text{-}50]$$

Equation [8.4-49] is Rosenthal's equation for two-dimensional heat flow in a wide and thin workpiece during welding. As already pointed out, for a fully penetrating heat source such as the electron beam, the workpiece can, in fact, be physically thick while still maintaining two-dimensional heat flow. Similar to Eq. [8.4-38] for a thick workpiece, this equation is based on the same four simplifying assumptions of Rosenthal's three-dimensional heat flow equation. Consequently, it also suffers from the singularity problem caused by the line source assumption, specifically, that the temperature reaches infinity at the location of the heat source.

Kou[24] has relaxed Rosenthal's assumptions and calculated two-dimensional eat condition during GTA welding of thin sheets. Because of the two-dimensional nature of the problem, the heat input and heat losses no longer appear as

[24] S. Kou, *Metallurgical Trans.*, **12A**, 2025, 1981.

boundary conditions, but as the internal heat source and heat sink, respectively. The energy equation and boundary conditions are left as an exercise problem.

8.4.5. Powder Processing: Cooling of Droplets

In powder processing (Section 6.1.4) fine melt droplets are often supercooled to below the melting point T_M. Nucleation of the solid phase occurs at a temperature T_N, which depends on the material and the cooling rate involved. At this point the heat of fusion H_f is evolved and the droplet temperature is reversed briefly, which is called **recalescence**. The enthalpy (H) versus temperature relationship is illustrated in Fig. 8.4-16, where f_s is the weight fraction of the solid phase during recalescence and H_{SM} is the enthalpy per unit mass of the solid phase at the melting point. The specific heat in the solid-puls-liquid region $(0 < f_s < 1)$ is the weighted average $(1 - f_s)C_L + f_s C_S$, where C_L and C_S are the specific heats of liquid and solid, respectively. As shown[25], at T

$$H - H_{SM} = (1 - f_s)H_f - (T_M - T)[(1 - f_s)C_L + f_s C_S] \quad [8.4\text{-}51]$$

With $\mathbf{v} = s = 0$ and $\partial(C_v T) = \partial H$, the energy equation Eq. [2.3-7] becomes

$$\rho \frac{\partial H}{\partial t} = \nabla \cdot (k \nabla T) \quad [8.4\text{-}52]$$

The boundary condition is

$$-k \frac{\partial T}{\partial r} = h(T - T_\infty) + \sigma\varepsilon(T^4 - T_\infty^4) \quad [8.4\text{-}53]$$

where h is the heat transfer coefficient, T_∞ the ambient temperature, σ the Stefan–Boltzmann constant, and ε the emissivity of the droplet. The initial condition is

$$T = T_i \quad \text{at} \quad t = 0 \quad [8.4\text{-}54]$$

where T_i is the initial droplet temperature.

Levi[25] et al. have solved this problem numerically by using the superimposed bispherical (rotational bipolar) coordinate system shown in Fig. 8.4-17. Single nucleation point and negligible radiation were assumed. Figure 8.4-18 shows the

[25] C. G. Levi and R. Mehrabian, *Metallurgical Trans.*, **13A**, 221, 1982.

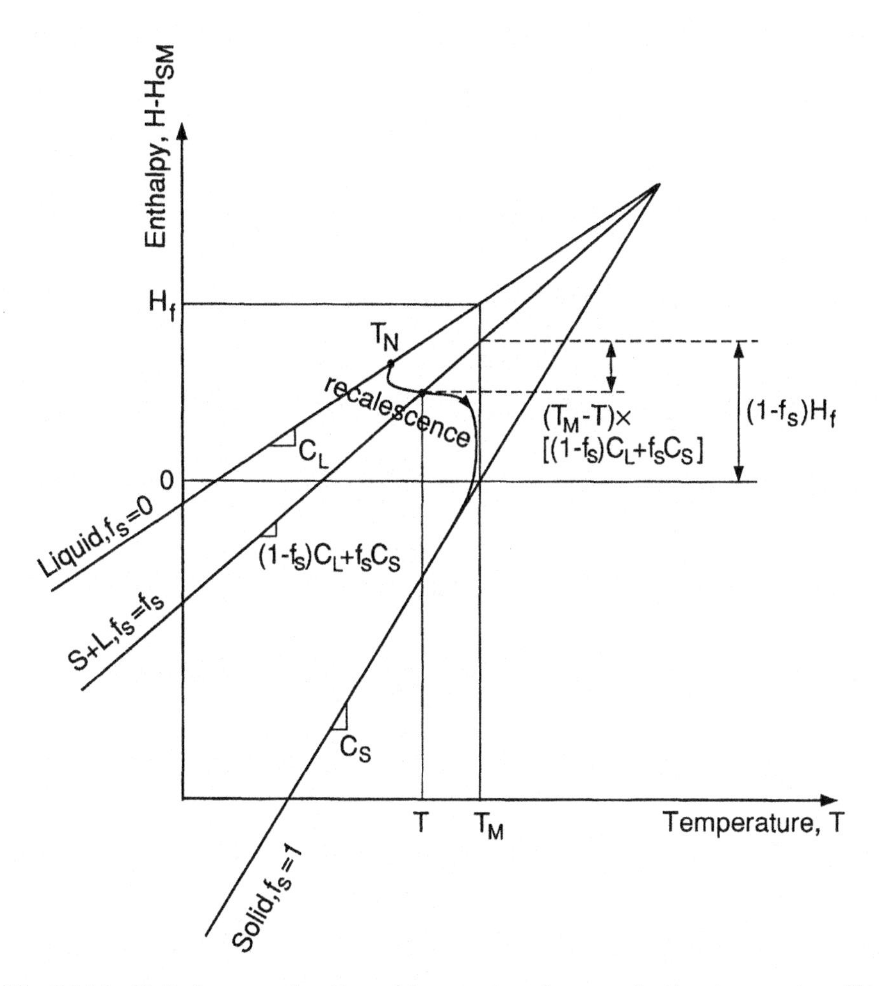

Fig. 8.4-16 Enthalpy as a function of temperature for a nucleation temperature T_N significantly below the melting point T_M.

calculated shapes and velocities of the advancing solid/liquid interface in 1, 10, and 100 μm diameter aluminum droplets. The interface velocity was assumed to be a function of the extent of undercooling, that is, $T_M - T_N$. For the Newtonian cooling case, from Eq. [8.1-8]

$$\frac{R\rho}{3}\frac{dH}{dt} = -h(T - T_\infty) \qquad [8.4\text{-}55]$$

As shown in the figure, Newtonian cooling (dotted line, $f_s = 0.5$) is a reasonable approximation only for small droplets, for example, 1 μm.

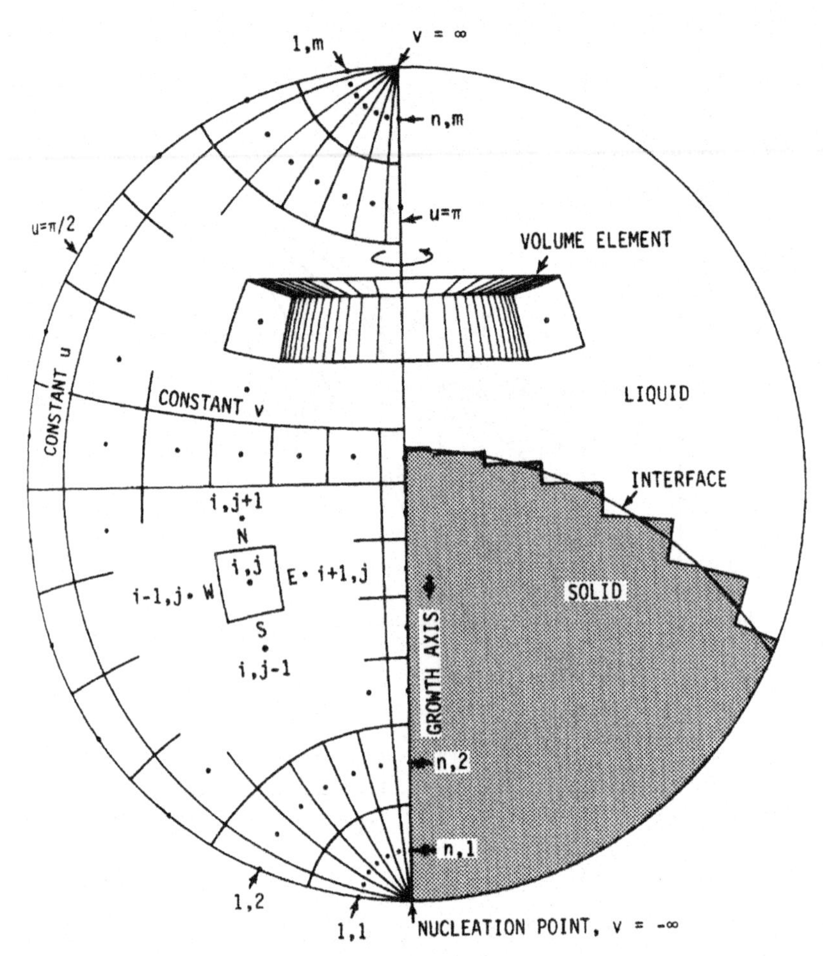

Fig. 8.4-17 Grid mesh in bispherical coordinates for computing solidification of an undercooled melt droplet. (From C. G. Levi and R. Mehrabian, *Metallurgical Transactions*, **13A**, 1982, p. 221.)

8.5. CONVECTIVE HEAT TRANSFER

8.5.1. Crystal Growth: Czochralski

The shape of the crystal/melt interface during crystal growth and the extent of dopant segregation in the resultant crystal can affect the crystal quality significantly. A crystal grown with a nearly flat crystal/melt interface and with minimal dopant segregation often tends to have less defects and more uniform physical properties. The interface shape and dopant segregation are significantly affected by convection in the melt during crystal growth.

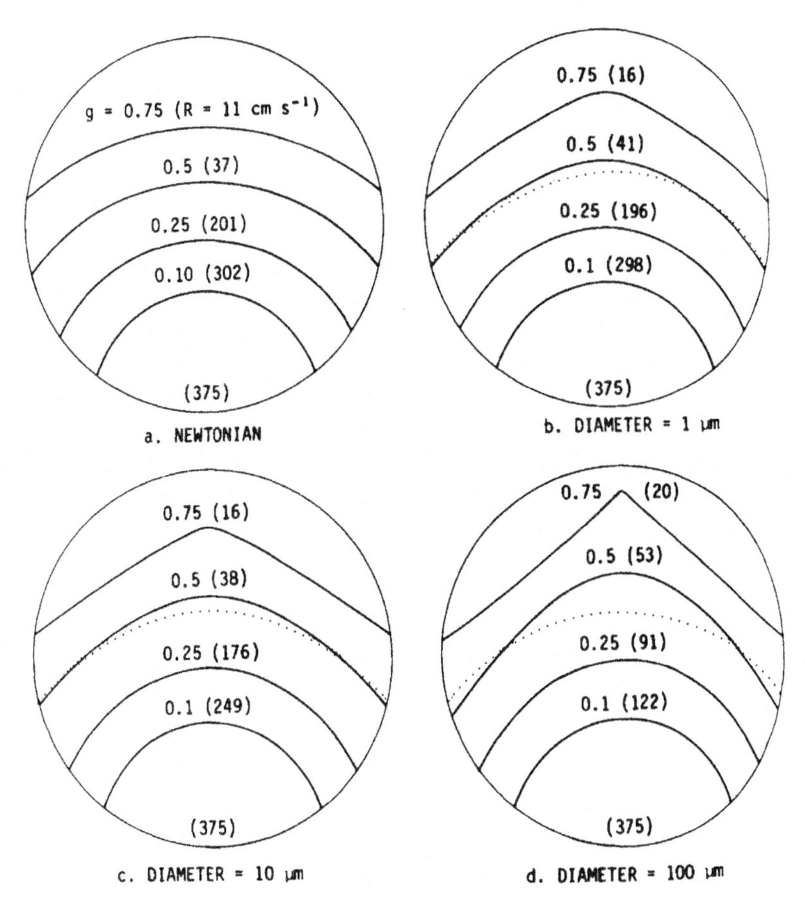

Fig. 8.4-18 Calculated interface shapes and velocities during undercooled solidification of aluminum droplets: (*a*) Newtonian cooling; (*b*)–(*d*) non-Newtonian cooling. (From C. G. Levi and R. Mehrabian, *Metallurgical Transactions*, **13A**, 1982, p. 221.)

Computation of simultaneous fluid flow and heat transfer in Czochralski crystal growth has recently become a very active area of research. Extensive work has been conducted by numerous investigators, notably Brown and Derby and their associates. Some results selected from their work on Czochralski crystal growth will be presented here. Dupret and Van Der Bogaert have reviewed the numerical computation of Czochralski and Bridgman crystal growth.[26]

The governing equations, assuming laminar flow, are as follows:

Continuity: $\nabla \cdot \mathbf{v} = 0$ [8.5-1]

[26] F. Dupret and N. Van Den Bogaert, in *Handbook of Crystal Growth*, Vol. 2b, D. T. J. Hurle, ed., Elsevier Science, Amsterdam, 1994, p. 875.

Motion: $$\rho\frac{\partial \mathbf{v}}{\partial t} + \rho\mathbf{v}\cdot\nabla\mathbf{v} = -\nabla P + \mu\nabla^2\mathbf{v} - \rho\beta_T\mathbf{g}(T - T_0) + \sigma B_0^2\mathbf{v}\times\mathbf{e}_B\times\mathbf{e}_B \quad [8.5\text{-}2]$$

Energy: $$\frac{\partial T}{\partial t} + \mathbf{v}\cdot\nabla T = \alpha\nabla^2 T \quad [8.5\text{-}3]$$

In Eq. [8.5-2], which is Eq. [4.2-23] without solutal convention, β_T is the thermal expansion coefficient and B_0 a uniform magnetic field applied in the direction $\mathbf{e}_B$. In the following discussion the magnetic field, if it is applied, is assumed to be in the z direction, specifically, $\mathbf{e}_B = \mathbf{e}_z$, where $\mathbf{e}_z$ is the unit vector in the z-direction. The z direction is taken vertically upward.

Let us consider the Czochralski configuration shown in Fig. 8.5-1.[27] The quasi-steady-state assumption[28] will be made, namely, that the melt volume decrease during growth is handled by performing a sequence of steady-state calculations at decreasing values of the melt volume. For simplicity, convection in the melt is assumed to be axisymmetric and laminar and the crystal diameter constant. It should be pointed out that in reality convection can be oscillatory, asymmetric, and even turbulent.

The boundary conditions are too numerous to describe here. The boundary conditions for the melt, including those at the crystal/melt and melt/gas interfaces, are as follows:

Crystal/Melt Interface (∂D_0):

$$v_r = v_z = 0 \quad [8.5\text{-}4]$$

$$v_\theta = \omega_s r \quad [8.5\text{-}5]$$

$$T = T_m \quad [8.5\text{-}6]$$

$$(-k\mathbf{n}_0\cdot\nabla T)_L + \rho H_f(-\mathbf{n}_0\cdot\mathbf{R}) = (-k\mathbf{n}_0\cdot\nabla T)_s \quad [8.5\text{-}7]$$

where ω_s is the crystal rotation speed, k the thermal conductivity, T_m the melting point, H_f the heat of fusion, and R the local growth rate. The first two equations are the no-slip condition, while the last equation is the Stefan condition. The last equation is the same as Eq. [5.1-9], except that $\mathbf{n}_0 = -\mathbf{n}$.

Centerline ($r = 0$):

$$\frac{\partial v_z}{\partial r} = \frac{\partial T}{\partial r} = 0 \quad [8.5\text{-}8]$$

$$v_r = v_\theta = 0 \quad [8.5\text{-}9]$$

These two equations follow from the axisymmetry assumption.

[27] P. A. Sackinger, R. A. Brown, and J. J. Derby, *Int. J. Num. Meth. Fluids*, **9**, 453, 1989.

[28] J. J. Derby and R. A. Brown, *J. Crystal Growth*, **87**, 251, 1987.

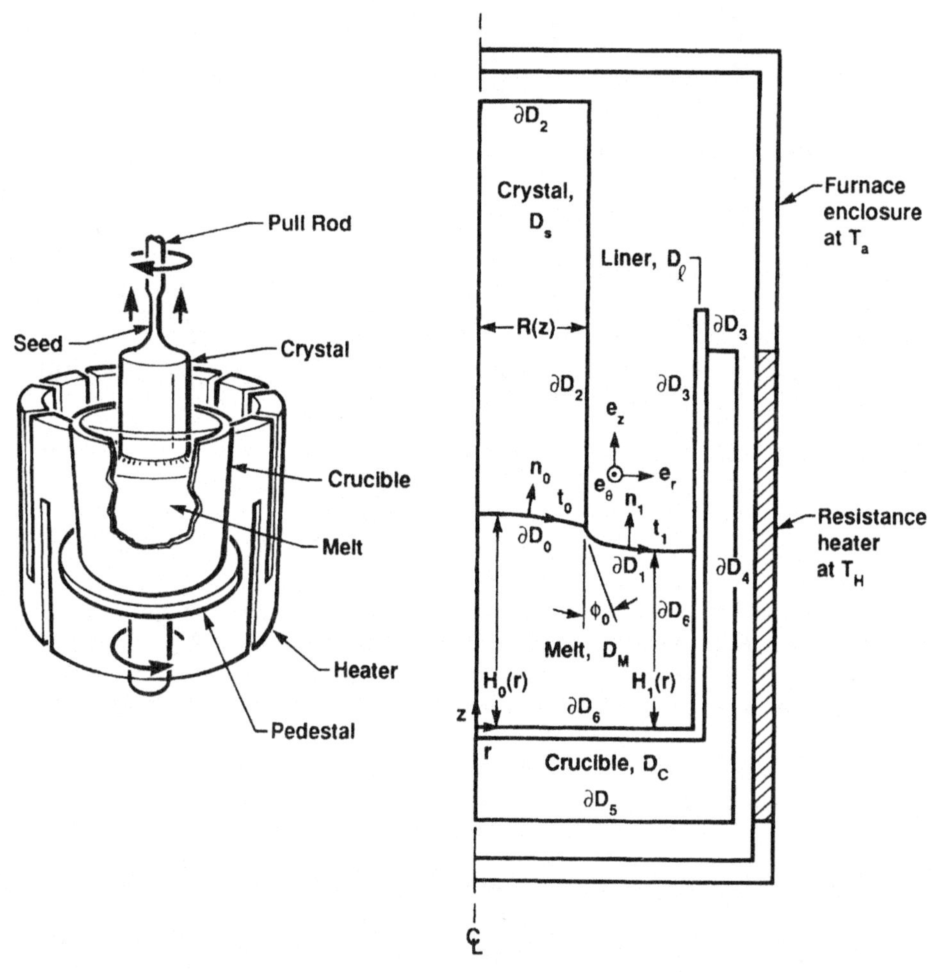

Fig. 8.5-1 Czochralski crystal growth. (From P. A. Sackinger, R. A. Brown, and J. J. Derby, "A Finite Element Method for Analysis of Fluid Flow, Heat Transfer, and Free Interfaces in Czochralski Crystal Growth," *International Journal of Numerical Methods in Fluids*, vol. 9, 1989, p. 453, John Wiley and Sons, Ltd. Reprinted by permission of John Wiley and Sons, Ltd.)

Crucible Wall (∂D_6):

$$v_r = v_z = 0 \qquad [8.5\text{-}10]$$

$$v_\theta = \omega_c r \qquad [8.5\text{-}11]$$

$$(T)_L = (T)_c \qquad [8.5\text{-}12]$$

$$(k\mathbf{n}_6 \cdot \nabla T)_L = (k\mathbf{n}_6 \cdot \nabla T)_c \qquad [8.5\text{-}13]$$

The first two equations are the no-slip condition at the inner surface of the crucible (called "liner" in Fig. 8.5-1), where ω_c is the crucible rotation speed. The last two equations describe the continuity in temperature and the heat flux at the interface (case 6, Fig. 2.3-2).

Free Surface (∂D_1):

$$\mathbf{n}_1 \cdot \mathbf{v} = 0 \qquad [8.5\text{-}14]$$

$$\mathbf{t}_1 \mathbf{n}_1 \colon \boldsymbol{\tau} = \frac{\partial \gamma}{\partial T}(\mathbf{t}_1 \cdot \nabla T) \qquad [8.5\text{-}15]$$

$$\mathbf{e}_\theta \mathbf{n}_1 \colon \boldsymbol{\tau} = 0 \qquad [8.5\text{-}16]$$

$$\mathbf{n}_1 \mathbf{n}_1 \colon \boldsymbol{\tau} = \gamma\left(\frac{1}{R_1} + \frac{1}{R_2}\right) - P_a \qquad [8.5\text{-}17]$$

$$(k\mathbf{n}_1 \cdot \nabla T)_L = h_1(T - T_a) + \varepsilon_1 \sigma (T^4 - T_a^4) \qquad [8.5\text{-}18]$$

where γ is the surface tension of the melt, R_1 and R_2 the radii of curvature of the free surface, P_a a reference pressure, h_1 the heat transfer coefficient, ε_1 the emissivity, σ the Stefan–Boltzmann constant, and T_a the effective ambient temperature. The first equation states that no fluid crosses a steady free surface, Eq. [5.2-6]. The second equation describes the shear stress due to surface tension gradients along the free surface, Eq. [5.2-4]; **t** and **n** are the tangential and normal unit vectors, respectively. The third equation is similar to the second one, in which the surface tension gradients are zero in the θ direction as a result of the axisymmetry requirement. The fourth equation describes the normal stress balance at the free surface, Eq. [5.2-9]. The last equation describes the heat losses from the free surface due to convection and radiation.

The finite-element method and Newton's method were used to solve the governing equations.[29] The flow fields were presented in terms of streamlines, with

$$v_r = -\frac{1}{r}\frac{\partial \psi}{\partial z} \qquad [8.5\text{-}19]$$

$$v_z = \frac{1}{r}\frac{\partial \psi}{\partial r} \qquad [8.5\text{-}20]$$

where ψ is the stream function. This definition of the stream function is identical to that given by Eqs. [1.6-13] and [1.6-14] except for the location of the minus sign.

[29] P. A. Sackinger, R. A. Brown, and J. J. Derby, *Int. J. Num. Meth. Fluids*, **9**, 453, 1989.

Figure 8.5-2 shows the evolution of the temperature and flow fields and the interface shapes with decreasing melt volume.[30] Because of the sidewall heating of the crucible the melt temperature is higher at the sidewall than the centerline. This causes a toroidal flow with melt rising up the sidewall and falling along the centerline.

As the melt volume decreases, the toroidal flow is flattened by the decreasing aspect ratio of the melt and even splits into nested vortices. Meanwhile, the crystal/melt interface convexity and the crucible sidewall temperature both increase since the effectiveness of convective heat flow from the heated crucible sidewall to the centerline is reduced.

Figure 8.5-3 shows the evolution of the temperature and flow fields with increasing strength of an applied axial magnetic field. As the magnetic field B is increased, convection in the bulk melt is damped as indicated by the decreasing absolute values of the minimum and maximum stream functions. Meanwhile, the center of the flow vortex moves toward the crucible sidewall, forming the so-called **Hartmann layer**. Since heat flow becomes more conduction-dominated, the isotherms become less distorted and the crucible sidewall temperature increases. The Hartmann number, Eq. [4.2-61], is defined by

$$\mathrm{Ha} = B_0 R_{\mathrm{crucible}} \left(\frac{\sigma}{\mu}\right)^{1/2} \qquad [8.5\text{-}21]$$

where B_0 is the magnetic flux and σ is the electrical conductivity of the melt.

The results described above are typical of low-Pr melts such as semiconductors (0.01–0.1). For high-Pr melts such as oxides ($1 \sim 10$), however, the results can be quite different.

Figure 8.5-4 shows the temperature and flow fields in the growth of an oxide crystal.[31] Strong buoyancy convection causes the streamlines near the crucible bottom to undulate, as the warm fluid plunging down the centerline is retarded by the colder fluid near the crucible bottom.[32] Besides streamline undulation, a high Pr also causes the isotherms to be severely distorted by convection, and the crystal/melt interface shape to be very sensitive to convection.

Figure 8.5-5 shows two distinctly different flow patterns in Czochralski melts. The flow pattern in Si ($\mathrm{Pr} \approx 0.01$)[33] shown in Fig. 8.5-5*a* appears similar to that calculated in Fig. 8.5-2 (upper left) for low-Pr materials. It is revealed by using X ray and a cleverly designed tracer particle. The flow pattern in $NaNO_3$ ($\mathrm{Pr} \approx 9$)[34] shown in Fig. 8.5-5*b* reveals clear flow undulation near the crucible bottom, similar to that predicted in Fig 8.5-4 for high-Pr materials.

[30] R. A. Brown, T. A. Kinney, P. A. Sackinger, and D. E. Bornside, *J. Crystal Growth*, **97**, 99, 1989.

[31] Q. Xiao and J. J. Derby, *J. Crystal Growth*, **128**, 188, 1993.

[32] P. A. Sackinger, R. A. Brown, and J. J. Derby, *Int. J. Num. Meth. Fluids*, **9**, 453, 1989.

[33] M. Watanabe, M. Eguchi, K. Kakimoto, Y. Baros, and T. Hibiya, *J. Crystal Growth*, **128**, 288, 1993.

[34] Y. Tao and S. Kou, unpublished work at University of Wisconsin, Madison, WI, 1993.

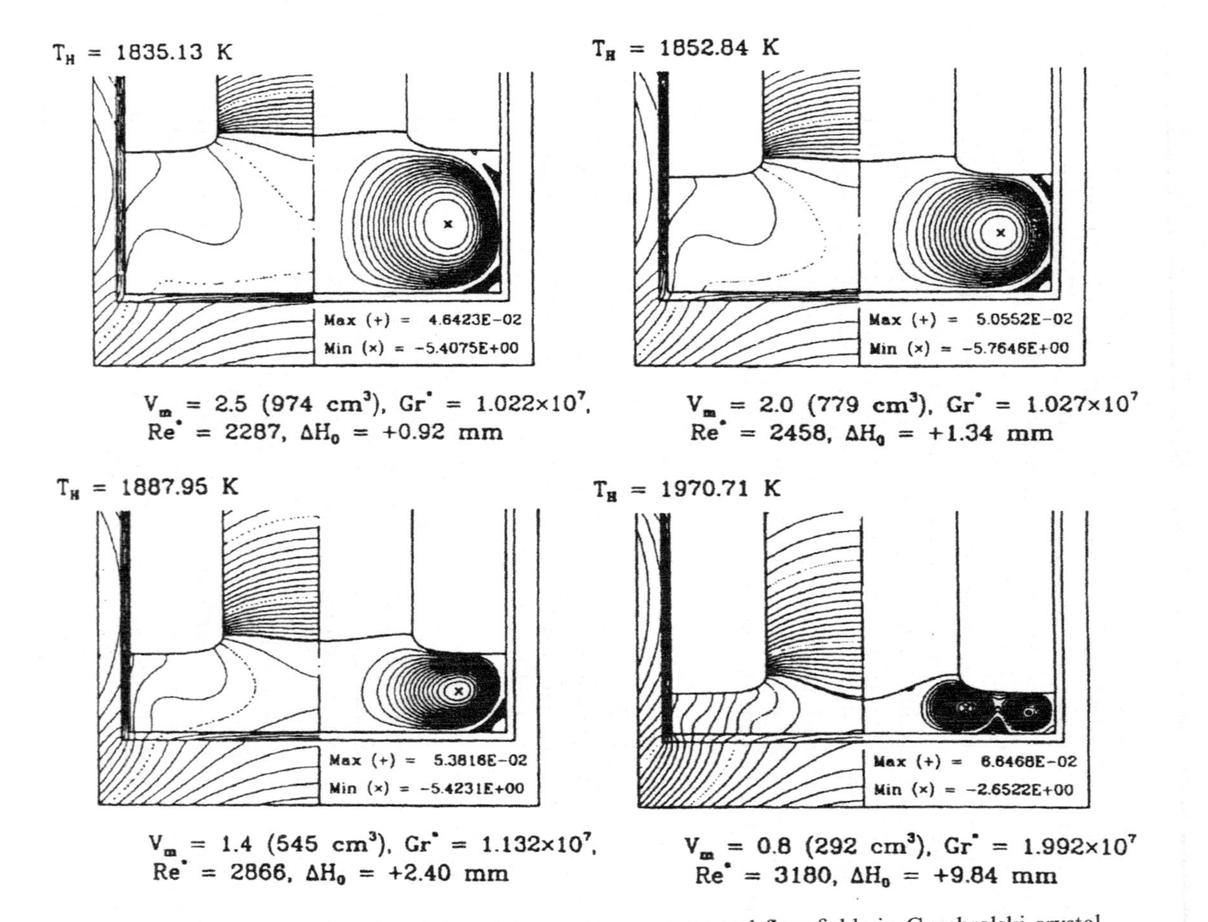

Fig. 8.5-2 Effect of decreasing melt volume on temperature and flow fields in Czochralski crystal growth. (R. A. Brown, T. A. Kinney, P. A. Sackinger, and D. E. Bornside, "Toward an Integrated Analysis of Czochralski Growth," *Journal of Crystal Growth*, **97**, 1989, p. 99.)

In oxide crystal growth internal radiation can be significant. As an approximation. Xiao and Derby[35] have considered the limiting case of an optically thin crystal in which only surfaces interact. For instance, at the crystal/melt interface the Stefan condition, Eq. [8.5-7], is modified to become

$$\underset{(1)}{(-k\mathbf{n}_0 \cdot \nabla T)_L} + \underset{(2)}{\rho_S H_f(-\mathbf{n}_0 \cdot \mathbf{R})} = \underset{(3)}{(-k\mathbf{n}_0 \nabla T)_S} + \underset{(4)}{n^2 \varepsilon_{Sm} \sigma(T^4 - T^4_{\text{eff,in}})} \qquad [8.5\text{-}22]$$

In other words, the heat conducted from the melt to the growth front (1) and the heat of fusion released at the growth front (2) escape into the crystal by conduction (3) and radiation (4). In Eq. [8.5-22] n is the refractive index of the crystal, ε_{Sm} the emissivity of the crystal/melt interface, and $T_{\text{eff,in}}$ the effective radiation temperature calculated for the enclosure defined by the crystal interior. The extra heat loss by internal radiation results in a deep or convex interface, as often observed in the growth of YAG (yttrium aluminum garnet; an oxide) crystals for solid state lasers.

Three-dimensional (asymmetric) unsteady convection has been observed in Czocharlski growth of Si[36] and $NaNO_3$.[37] Attempts have been made in numerical computation to consider three-dimensional (3-D) unsteady convection (e.g., by Xiao and Derby[38]), turbulence (e.g., by Kinney and Brown[39]), crucible partition (e.g., Jafri et al.[40]), gas convection in high-pressure Czochralski (e.g., by Zhang et al.[41]), magnetic fields (e.g., by Mihelcic et al.[42] and Williams et al.[43]), and mass transfer (e.g., Toh et al.[44]).

8.5.2. Crystal Growth: Floating Zone

Let us consider the floating-zone crystal growth configuration shown schematically in Fig. 8.5-6. The molten zone is produced by a ring heater, which is not shown. The ring heater and the (r, z) coordinates are both stationary while the

[35] Q. Xiao and J. J. Derby, *J. Crystal Growth*, **128**, 188, 1993.

[36] K. Kakimoto, *Prog. Crystal Growth Charact.*, **30**, 191, 1995.

[37] Y. Tao and S. Kou, unpublished work at University of Wisconsin, Madison, WI, 1993.

[38] Q. Xiao and J. J. Derby, *J. Crystal Growth*, **152**, 169, 1995.

[39] T. A. Kinney and R. A. Brown, *J. Crystal. Growth*, **132**, 551, 1993.

[40] I. H. Jafri, V. Prasad, A. P. Anselmo, and K. P. Gupta, *J. Crystal. Growth*, **154**, 280, 1995.

[41] T. Zhang, and V. Prasad, *J. Crystal Growth*, **155**, 47, 1995.

[42] M. Mihelcic and K. Wingerath, *J. Crystal Growth*, **82**, 318, 1987.

[43] M. G. Williams, J. S. Walker, and W. E. Langlois, *J. Crystal Growth*, **100**, 233, 1990.

[44] K. Toh and H. Ozoe, *J. Crystal Growth*, **130**, 645, 1993.

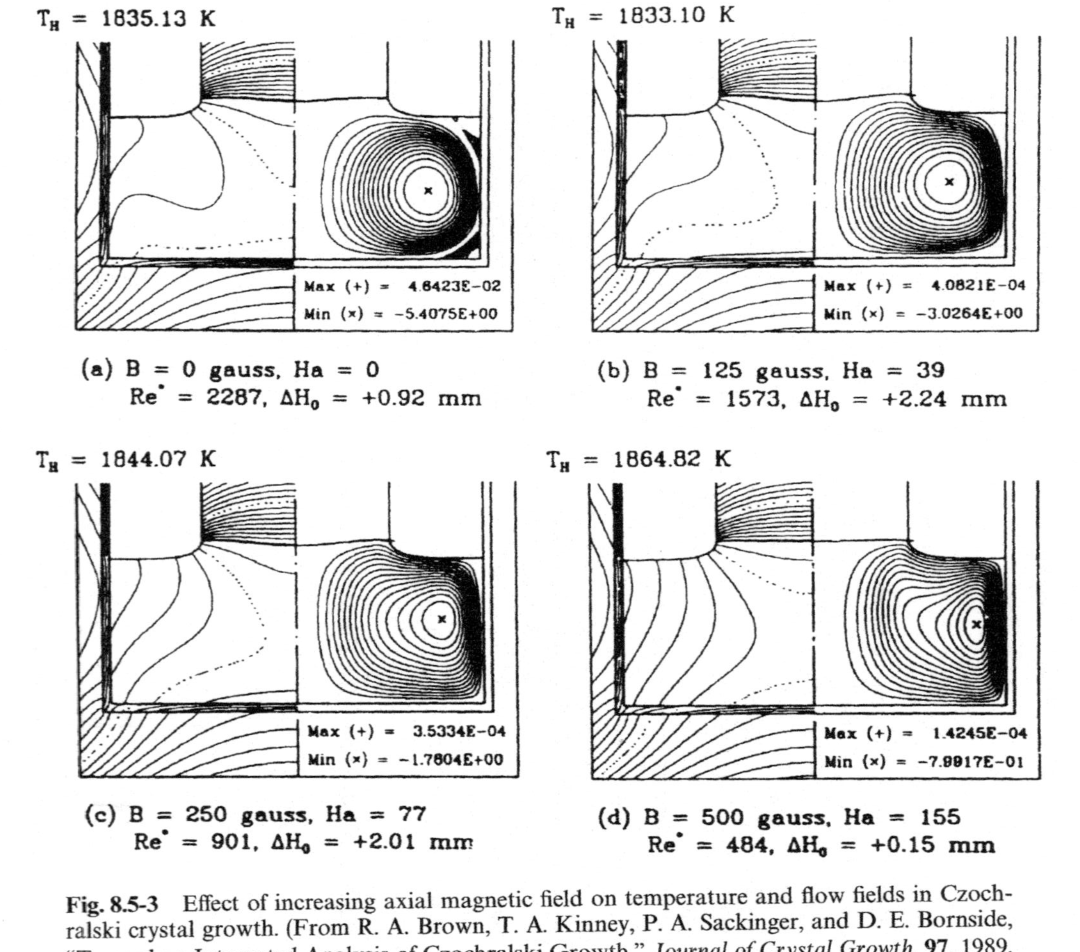

Fig. 8.5-3 Effect of increasing axial magnetic field on temperature and flow fields in Czochralski crystal growth. (From R. A. Brown, T. A. Kinney, P. A. Sackinger, and D. E. Bornside, "Toward an Integrated Analysis of Czochralski Growth," *Journal of Crystal Growth*, **97**, 1989, p. 99.)

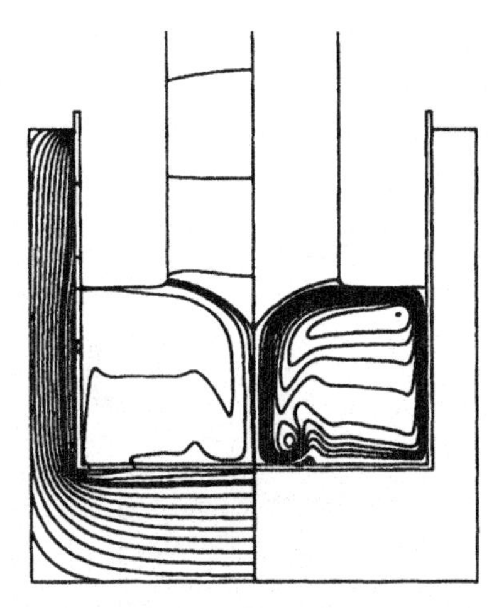

Fig. 8.5-4 Temperature (left) and flow (right) fields in Czochralski growth of an oxide. (From Q. Xiao and J. J. Derby, "The Role of Internal Radiation and Melt Convection in Czochralski Oxide Growth: Deep Interfaces, Interface Inversion, and Spiraling," *Journal of Crystal Growth*, **128**, 1993, p. 188.)

crystal and the feed rod move downward. The ambient-temperature distribution due to the ring heater is shown. For simplicity, convection in the molten zone is assumed axisymmetric and laminar. In reality, however, convection can be oscillatory and asymmetric, as, in Si. The governing equations are as follows:[45]

Vorticity:
$$\frac{\partial \omega}{\partial t} - \frac{\partial \psi}{\partial z}\frac{\partial}{\partial r}\left(\frac{\omega}{r}\right) + \frac{\partial \psi}{\partial r}\frac{\partial}{\partial z}\left(\frac{\omega}{r}\right)$$
$$= \nu\left(\frac{\partial}{\partial r}\left[\frac{1}{r}\frac{\partial (r\omega)}{\partial r}\right] + \frac{\partial}{\partial z}\left[\frac{1}{r}\frac{\partial (r\omega)}{\partial z}\right]\right) + \frac{\partial}{\partial z}\left(\frac{v_\theta^2}{r}\right) - \beta_T g\frac{\partial T}{\partial r} - \beta_S g\frac{\partial c_A}{\partial r} \tag{8.5-23}$$

Circulation:
$$\frac{\partial v_\theta}{\partial t} + \frac{1}{r^2}\frac{\partial \psi}{\partial z}\frac{\partial}{\partial z}(rv_\theta) - \frac{1}{r}\frac{\partial \psi}{\partial r}\frac{\partial v_\theta}{\partial z} = \nu\left(\frac{\partial}{\partial r}\left[\frac{1}{r}\frac{\partial}{\partial r}(rv_\theta)\right] + \frac{\partial^2 v_\theta}{\partial z^2}\right) \tag{8.5-24}$$

Stream function:
$$\frac{\partial}{\partial r}\left(\frac{1}{r}\frac{\partial \psi}{\partial r}\right) + \frac{\partial}{\partial z}\left(\frac{1}{r}\frac{\partial \psi}{\partial z}\right) = \omega \tag{8.5-25}$$

[45] C. W. Lan and S. Kou, *J. Crystal. Growth*, **114**, 517, 1991.

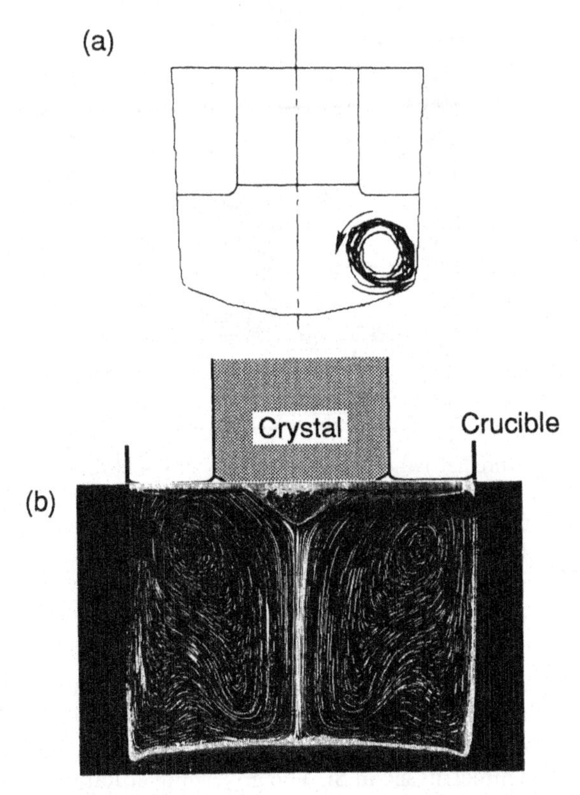

Fig. 8.5-5 Flow patterns in Czochralski melts: (*a*) Si in a 7.5-cm-ID crucible. (M. Watanabe, M. Eguchi, K. Kakimoto, Y. Baros, and T. Hibiya, *Journal of Crystal Growth*, **128**, 1993, p. 288.); (*b*) $NaNO_3$ in a 3.5-cm-ID crucible. (Y. Tao and S. Kou, unpublished work at University of Wisconsin, Madison, WI.)

Energy: $$\frac{\partial T}{\partial t} + \frac{1}{r}\frac{\partial \psi}{\partial z}\frac{\partial T}{\partial r} - \frac{1}{r}\frac{\partial \psi}{\partial r}\frac{\partial T}{\partial z} = \alpha\left[\frac{1}{r}\frac{\partial}{\partial r}\left(r\frac{\partial T}{\partial r}\right) + \frac{\partial^2 T}{\partial z^2}\right] \qquad [8.5\text{-}26]$$

The first three equations are from Eqs. [4.2-36] through [4.2-38] for thermosolutal convection in terms of the stream function ψ and vorticity ω, with $g_z = -g$. The last equation is Eq. [2.4-8] without heat generation s. In the following discussion the feed rod and the crystal will be assumed to be sufficiently long so that the steady-state condition is achieved; thus, the molten zone remains unchanged in its shape, location, temperature field, and velocity field.

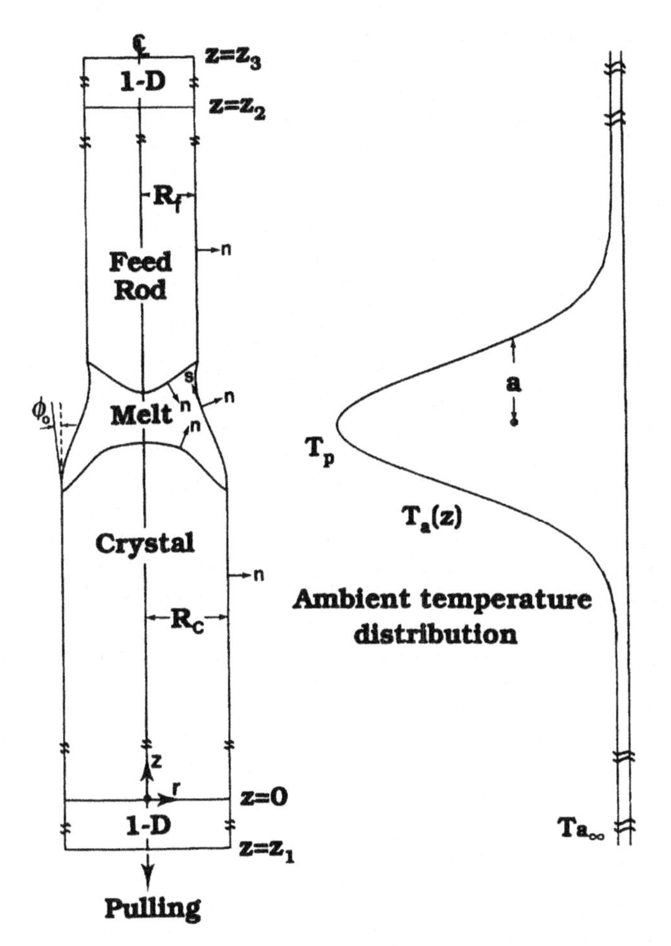

Fig. 8.5-6 Floating-zone crystal growth. (From C. W. Lan and S. Kou, *Journal of Crystal Growth*, **108**, 1991, p. 351.)

The stream function and vorticity, according to Eq. [1.6-13] through [1.6-15], are defined as follows:

$$v_r = \frac{1}{r}\frac{\partial \psi}{\partial z} \qquad [8.5\text{-}27]$$

$$v_z = -\frac{1}{r}\frac{\partial \psi}{\partial r} \qquad [8.5\text{-}28]$$

$$\omega = \frac{\partial v_r}{\partial z} - \frac{\partial v_z}{\partial r} \qquad [8.5\text{-}29]$$

The boundary conditions[46] are too numerous to describe here. The boundary conditions for fluid flow and heat transfer in the molten zone are as follows:

Crystal/Melt Interface:

$$\psi = -\tfrac{1}{2}V_c r^2 \tag{8.5-30}$$

$$\omega = \frac{\partial v_r}{\partial z} - \frac{\partial v_z}{\partial r} \quad \text{with} \quad v_r = 0,\ v_z = V_c \tag{8.5-31}$$

$$v_\theta = r\Omega_c \tag{8.5-32}$$

$$T = T_m \tag{8.5-33}$$

$$(k\mathbf{n}\cdot\nabla T)_L + \rho_S H_f(-V_c)\mathbf{n}\cdot\mathbf{e}_z = (k\mathbf{n}\cdot\nabla T)_c \tag{8.5-34}$$

where V_c is the crystal pulling velocity (< 0), Ω_c the crystal rotation speed and $\mathbf{e}_z$ the unit vector in the z direction. The first equation is obtained by integrating Eq. [8.5-28] from the centerline, with $v_z = V_c$. As already shown in Example 5.1-1, at the steady state $v_z = V_c$ and $v_r = 0$. The last equation is the Stefan condition, Eq. [5.1-22].

Feed/Melt Interface:

$$\psi = -\tfrac{1}{2}V_f r^2 \tag{8.5-35}$$

$$\omega = \frac{\partial v_r}{\partial z} - \frac{\partial v_z}{\partial r} \quad \text{with} \quad v_r = 0,\ v_z = V_f \tag{8.5-36}$$

$$v_\theta = r\Omega_f \tag{8.5-37}$$

$$T = T_m \tag{8.5-38}$$

$$(k\mathbf{n}\cdot\nabla T)_L = \rho_S H_f V_f \mathbf{n}\cdot\mathbf{e}_z + (k\mathbf{n}\cdot\nabla T)_f \tag{8.5-39}$$

where V_f and Ω_f are the feed pushing velocity (< 0) and the feed rotation speed, respectively. Equation [8.5-39] is the Stefan condition, Eq. [5.1-25].

Centerline ($r = 0$):

$$\psi = 0 \tag{8.5-40}$$

$$\omega = 0 \tag{8.5-41}$$

[46] C. W. Lan ad S. Kou, ibid.

$$v_\theta = 0 \quad [8.5\text{-}42]$$

$$\frac{\partial T}{\partial r} = 0 \quad [8.5\text{-}43]$$

The stream function is set to zero at the centerline as a reference. The zero vorticity is the result of $\partial v_r/\partial z = \partial v_z/\partial r = 0$ at the centerline due to axisymmetry. The zero temperature gradient is also due to axisymmetry.

Free Surface:

$$\psi = -\tfrac{1}{2} V_f R_f^2 \quad [8.5\text{-}44]$$

$$\omega = \frac{\partial v_r}{\partial z} - \frac{\partial v_z}{\partial r} \quad \text{with} \quad [8.5\text{-}45]$$

$$\mathbf{ns}\text{: } \tau = \frac{\partial \gamma}{\partial T}(s \cdot \nabla T) \quad \text{and} \quad [8.5\text{-}46]$$

$$\mathbf{nn}\text{: } \tau = \gamma\left(\frac{1}{R_1} + \frac{1}{R_2}\right) - P_a \quad [8.5\text{-}47]$$

$$\mathbf{ne}_\theta\text{: } \tau = 0 \quad [8.5\text{-}48]$$

$$(-k\mathbf{n} \cdot \nabla T)_L = h(T - T_a) + \varepsilon\sigma(T^4 - T_a^4) \quad [8.5\text{-}49]$$

where R_f is the feed radius, h the heat transfer coefficient, T_a the effective ambient temperature, ε the emissivity of the free surface, and σ the Stefan–Boltzmann constant. For simplicity, the ambient temperature distribution is assumed Gaussian. The last four equations are similar to Eqs. [8.5-15] through [8.5-18], which have already been described.

Because the unknown crystal/melt interface, the feed/melt interface, and the free surface are all curved, the Stefan condition and the vorticity boundary conditions cannot be implemented properly. In view of this, Lan and Kou[47] have transformed the cylindrical coordinates into body-fitted general (nonorthogonal) curvilinear coordinates and used the control-volume finite-different method to obtain the solutions.

Figure 8.5-7 shows the results for the floating-zone growth of $NaNO_3$ crystals of 4 mm diameter at two different growth rates.[47] Flow visualization experiments have confirmed that convection in the $NaNO_3$ molten zone during crystal

[47] C. W. Lan and S. Kou, *J. Crystal Growth*, **108**, 351, 1991.

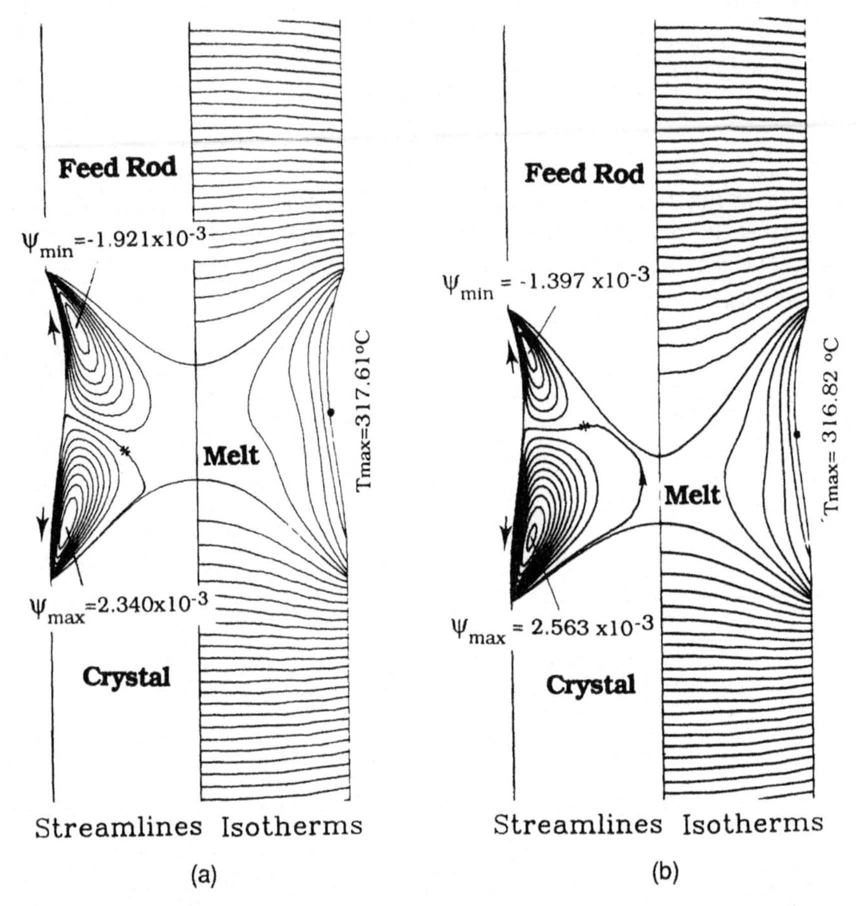

Fig. 8.5-7 Effect of growth rate on flow and temperature fields in 4-mm-diameter $NaNO_3$: (*a*) 1.46×10^{-4} cm/s; (*b*) 7.29×10^{-4} cm/s. (From C. W. Lan and S. Kou, *Journal of Crystal Growth*, **108**, 1991, p. 351.)

growth is steady and axisymmetric.[48] The surface tension of the $NaNO_3$ melt decreases with increasing temperature; $\partial\gamma/\partial T$ is -0.056 dyn cm^{-1} °C^{-1}. In either case, there exist two flow cells near the free surface, the upper one clockwise in direction and the lower one counterclockwise. These two cells are dominated by thermocapillary convection, as was explained in Section 5.2. The stream function in Figs. 8.5-7 through 8.5-9 is defined differently from Eqs. [8.5-27] and [8.5-28]. It equals the ψ defined by these two equations multiplied by $-\rho_L$ (i.e., -1.9 g cm^{-3}).

The ring heater surrounding the molten zone produces a maximum melt temperature and hence a minimum surface tension ($\partial\gamma/\partial T < 0$) near the middle

[48] Y. Tao and S. Kou, *J. Crystal Growth*, **137**, 72, 1994.

of the free surface. As such, the melt is pulled toward the feed rod and the crystal along the free surface, resulting in the two flow cells. Since this convection favors heat transfer from the heater to the surfaces of the feed rod and the crystal, the crystal/melt and feed/melt interfaces become very convex toward the melt. As the growth rate is increased, the crystal/melt interface becomes less convex while the crystal/feed interface becomes even more convex.

Figure 8.5-8 shows the effect of crystal/feed counterrotation on the flow and temperature fields in the floating-zone growth of 6-mm-diameter $NaNO_3$ crystals.[49] The counterrotation produces a centrifugal force near the crystal/melt and feed/melt interfaces, which pushes the melt near the interfaces radially outward. As the rotation speed increases, the forced-convection cells grow at the expense of the thermocapillary-convection cells. This helps deliver heat from the ring heater to the centerline of the molten zone, causing the interfaces to switch from convex to concave. The high sensitivity of the shapes of the interfaces to convection and the severe distortions of the isotherms by convection are typical of a high-Pr material such as $NaNO_3$ ($Pr = 9$).

These results suggest two benefits of counterrotation in high-Pr materials: (1) the shape of the crystal/melt interface can be effectively controlled and (2) the zone length can be reduced to prevent collapse of the molten zone due to gravity, without freezing the melt near the centerline.[50]

Figure 8.5-9 shows the calculated and observed convection patterns during the floating-zone growth of a 6-mm-diameter $NaNO_3$ crystal.[51] The lens effect of the molten zone needs to be taken into account in order to compare the convection patterns and even the shapes of the melt/solid interfaces. Equations based on optical analysis have been derived for this purpose.[52] Further discussion on the effect of process variables on floating-zone crystal growth is left as exercise problems.

Before leaving the subject of floating-zone crystal growth, it is worth mentioning that Muller and Rupp[53] have calculated asymmetric oscillatory thermocapillary convection in a cylindrical liquid bridge, that is, a liquid column between two vertical rods.

8.5.3. Welding: The Weld Pool

Convection can significantly affect the shape of the weld pool, especially the depth of penetration, which is of great practical interest in welding. The work of Szekely and his associates[54] represents the most significant breakthrough in the

[49] C. W. Lan and S. Kou, *J. Crystal Growth*, **114**, 517, 1991.

[50] C. W. Lan and S. Kou, *J. Crystal Growth*, **119**, 281, 1992.

[51] C. W. Lan and S. Kou, unpublished work at the University of Wisconsin, Madison, WI, 1992.

[52] C. W. Lan and S. Kou, *J. Crystal Growth*, **132**, 471, 1993.

[53] G. Muller and R. Rupp, in *Crystal Properties and Preparation*, Vol. 35 (*Trans. Tech. Switzerland*, 1991), p. 138.

[54] G. M. Oreper, T. W. Wagar, and J. Szekely, *Welding J.*, **62**, 307s, 1983.

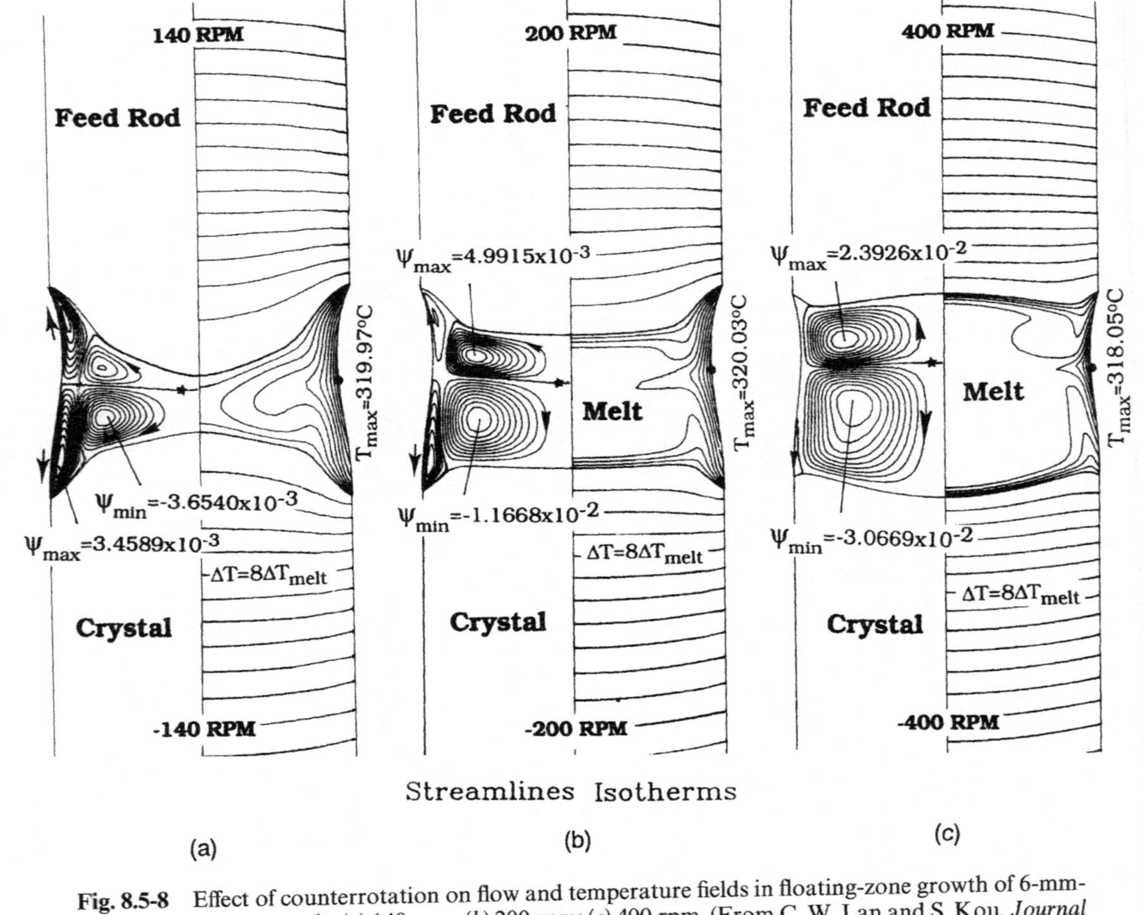

Fig. 8.5-8 Effect of counterrotation on flow and temperature fields in floating-zone growth of 6-mm-diameter $NaNO_3$ crystals: (*a*) 140 rpm; (*b*) 200 rpm; (*c*) 400 rpm. (From C. W. Lan and S. Kou, *Journal of Crystal Growth*, **114**, 1991, p. 517.)

computer simulation of weld pool convection. Numerous studies that followed, including many of Szekely's, further refined the computation of weld pool convection. Recently, Zacharia et al. have extended the study of weld pool convection to the welding of stainless steel,[55] and DebRoy and coworkers to the weld metal composition change.[56]

The driving forces for weld pool convection and the convection patterns they produce have been described in Section 6.1.3.5.

Let us consider a pool produced by GTA welding (Section 6.1.3), as illustrated in Fig. 8.5-10. For simplicity, autogenous welding will be considered; no filler metals are used, and convection is assumed laminar and incompressible.

The governing equations are as follows:

Continuity: $$\nabla \cdot \mathbf{v} = 0 \qquad [8.5\text{-}50]$$

Motion: $$\rho \frac{\partial \mathbf{v}}{\partial t} + \rho(\mathbf{v} \cdot \nabla)\mathbf{v} = -\nabla P + \mu \nabla^2 \mathbf{v} - \rho \beta_T \mathbf{g}(T - T_0) + \mathbf{J} \times \mathbf{B} \qquad [8.5\text{-}51]$$

Energy: $$\rho \frac{\partial}{\partial t}(C_v T) + \rho \mathbf{v} \cdot \nabla(C_v T) = \nabla \cdot (k \nabla T) \qquad [8.5\text{-}52]$$

Equation [8.5-51] is Eq. [4.2-20] without solutal convection. Assuming axisymmetric current distribution in the workpiece, the Lorentz force

$$\mathbf{J} \times \mathbf{B} = B_\theta (J_r \mathbf{e}_z - J_z \mathbf{e}_r) \qquad [8.5\text{-}53]$$

At the weld pool surface the current density distribution is assumed as follows:

$$J_z = \frac{3I}{\pi b^2} \exp\left(\frac{-3r^2}{b^2}\right) \quad \text{at} \quad z = 0 \qquad [8.5\text{-}54]$$

where I is the welding current and b the effective radius of the current distribution at the pool surface. This Gaussian current density distribution is an

[55] T. Zacharia, S. A. David, J. M. Vitek, and T. DebRoy, in *Recent Trends in Welding Science and Technology*, S. A. David and J. M. Vitek, eds., ASM International, Materials Park, OH, 1990, p. 25; T. Zacharia, S. A. David, and J. M. Vitek, in *International Trends in Welding Science and Technology*, S. A. David and J. M. Vitek, eds., ASM International, Materials Park, OH, 1993, p. 29.

[56] K. Mundra and T. DebRoy, in *International Trends in Welding Science and Technology*, S. David and J. M. Vitek, eds., ASM International, Materials Park, OH, 1993, p. 75.

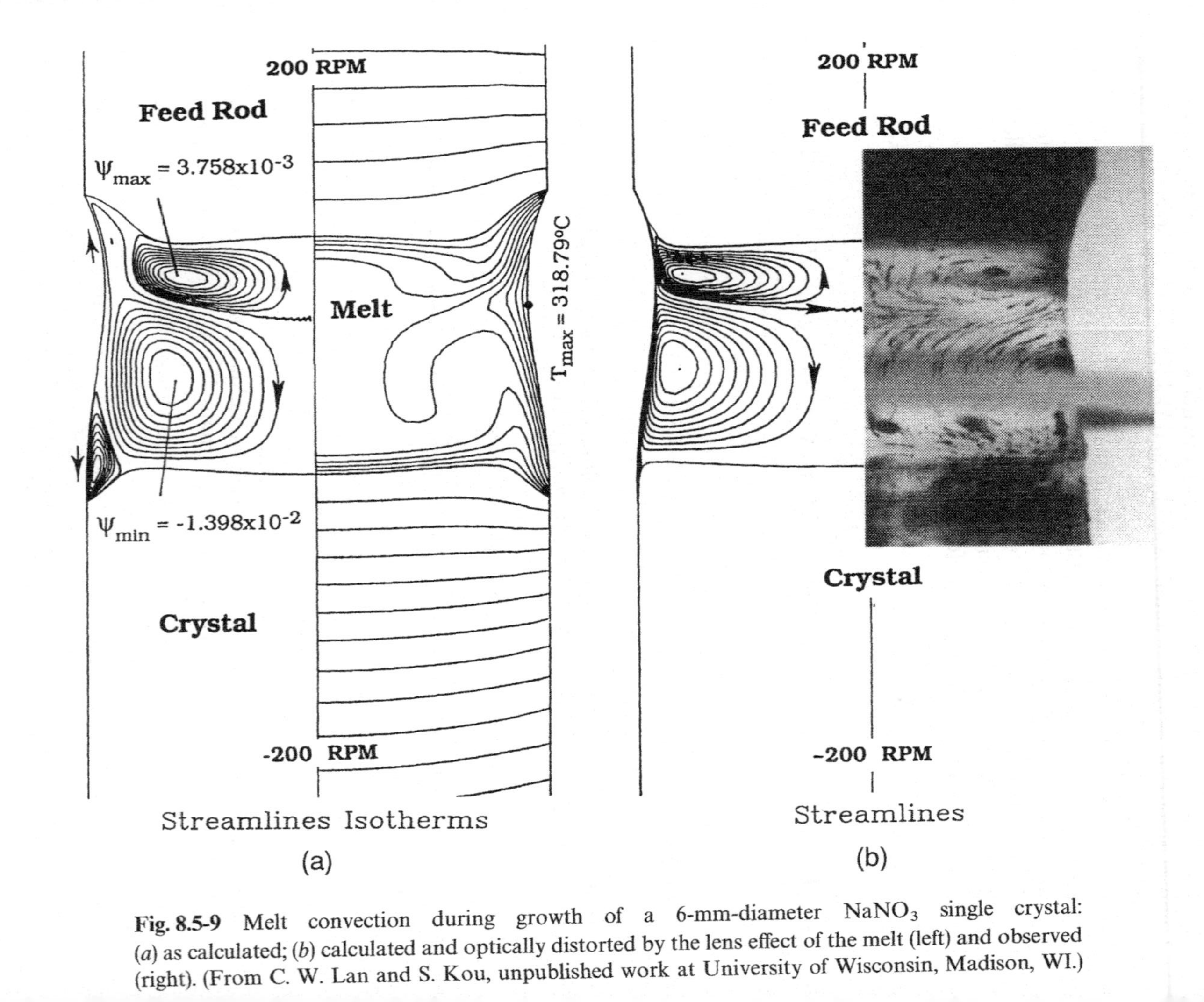

Fig. 8.5-9 Melt convection during growth of a 6-mm-diameter $NaNO_3$ single crystal: (*a*) as calculated; (*b*) calculated and optically distorted by the lens effect of the melt (left) and observed (right). (From C. W. Lan and S. Kou, unpublished work at University of Wisconsin, Madison, WI.)

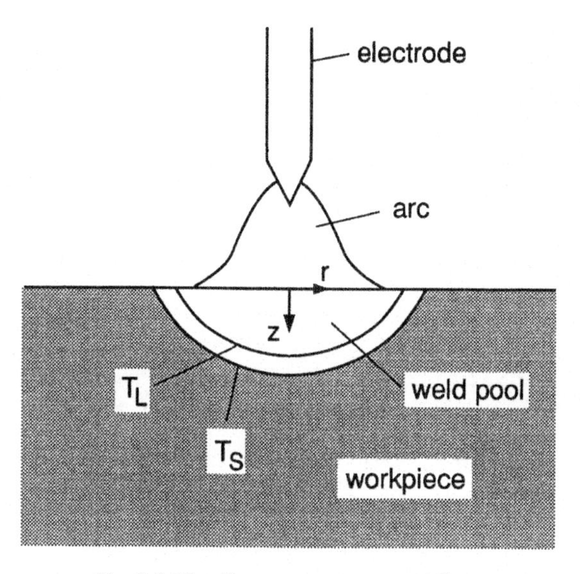

Fig. 8.5-10 Gas tungsten arc welding.

approximation, as shown in Fig. 8.5-11*a*.[57] From Maxwell's equations it can be shown that for a thick plate[58]

$$J_z = \frac{I}{2\pi}\int_0^\infty \lambda J_0(\lambda r)\exp\left(\frac{-\lambda z - \lambda^2 b^2}{12}\right) d\lambda \qquad [8.5\text{-}55]$$

$$J_r = \frac{I}{2\pi}\int_0^\infty \lambda J_1(\lambda r)\exp\left(\frac{-\lambda z - \lambda^2 b^2}{12}\right) d\lambda \qquad [8.5\text{-}56]$$

$$B_\theta = \frac{\mu_m I}{2\pi}\int_0^\infty J_1(\lambda r)\exp\left(\frac{-\lambda z - \lambda^2 b^2}{12}\right) d\lambda \qquad [8.5\text{-}57]$$

where J_0 and J_1 are the Bessel functions of the first kind and of the zero and first orders, respectively, and μ_m is the magnetic permeability. Equations [8.5-55] through [8.5-57] can be substituted into Eq. [8.5-53] to find the Lorentz force in Eq. [8.5-51].

In welding and casting of most practical materials the solid/melt transformation occurs over a finite temperature range (T_S to T_L) rather than at a discrete temperature (T_m). Since the Stefan condition (Section 5.1.2) no longer applies, it

[57] M. Lu and S. Kou, *Welding J.*, **67**, 29s, 1988.

[58] S. Kou and D. K. Sun, *Metallurgical Trans. A*, **16A**, 203, 1985.

is desirable to replace C_vT by the enthalpy per unit mass H in Eq. [8.5-52] so that the heat of fusion is automatically covered, as illustrated in Fig. 8.5-12. Other techniques have also been used, including boosting C_v by $H_f/(T_L - T_S)$, where H_f is the heat of fusion.

For a stationary weld pool with a flat free surface and axisymmetric heat and fluid flow, the boundary conditions for fluid flow are as follows:

Free Surface ($z = 0$):

$$v_z = 0 \tag{8.5-58}$$

$$-\mu \frac{\partial v_r}{\partial z} = \frac{\partial T}{\partial r}\frac{\partial \gamma}{\partial T} + \frac{\partial w_A}{\partial r}\frac{\partial \gamma}{\partial w_A} \tag{8.5-59}$$

The first equation is the result of the flat free surface. The second equation is Eq. [5.2-2].

Centerline ($r = 0$):

$$v_r = 0 \tag{8.5-60}$$

$$\frac{\partial v_z}{\partial r} = 0 \tag{8.5-61}$$

which are the result of the axisymmetry assumption.

Pool Boundary ($T = T_L$):

$$v_r = v_z = 0 \tag{8.5-62}$$

which is the no-slip condition.

The boundary conditions for heat flow in the weld pool are as follows:

Top Surface ($z = 0$):

$$-k\frac{\partial T}{\partial z} = \frac{3Q}{\pi a^2}\exp\left(\frac{-3r^2}{a^2}\right), \qquad r \leq a \tag{8.5-63}$$

$$-k\frac{\partial T}{\partial z} = h(T - T_a) + \sigma\varepsilon(T^4 - T_a^4), \qquad r > a \tag{8.5-64}$$

where Q is the power delivered into the workpiece from the heat source, a the effective radius of the heat source, and T_a the ambient temperature. The

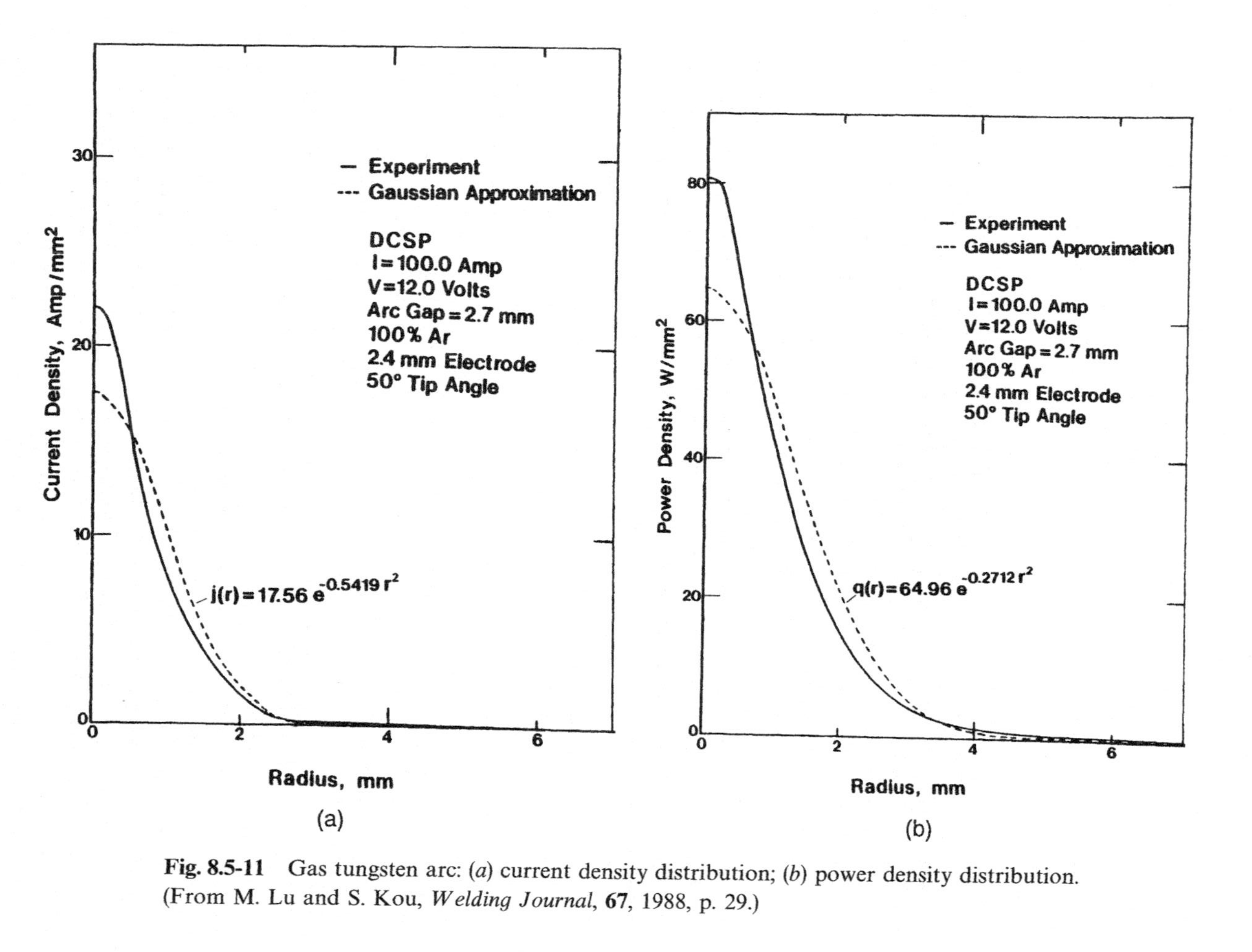

Fig. 8.5-11 Gas tungsten arc: (*a*) current density distribution; (*b*) power density distribution. (From M. Lu and S. Kou, *Welding Journal*, **67**, 1988, p. 29.)

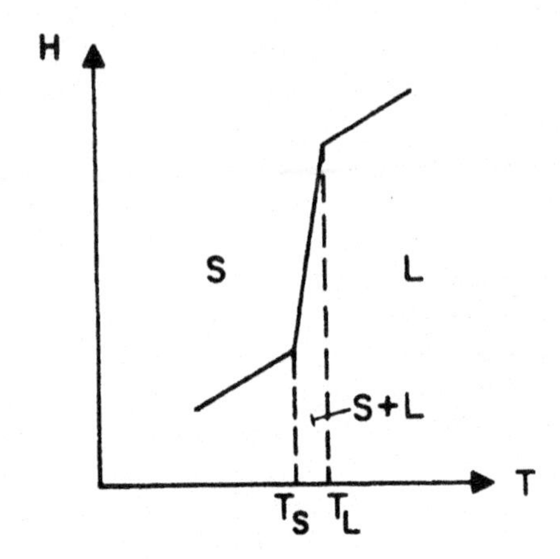

Fig. 8.5-12 Enthalpy as a function of temperature for an alloy with a freezing temperature range of T_S to T_L.

Gaussian power density distribution in Eq. [8.5-63] is only an approximation, as shown in Fig. 8-5-11*b*. In the second equation the first and second RHS terms represent heat losses from the workpiece surface due to convection and radiation, respectively.

Centerline ($r = 0$):

$$\frac{\partial T}{\partial r} = 0 \qquad [8.5\text{-}65]$$

Far Below Top Surface ($z \to \infty$):

$$T = T_\infty \qquad [8.5\text{-}66]$$

Far from Centerline ($r \to \infty$):

$$T = T_\infty \qquad [8.5\text{-}67]$$

Figure 8.5-13 shows the steady-state velocity and temperature fields in stationary weld pools of 6061 aluminum alloy. With the buoyancy force alone (Figs. 8.5-13*a* and 8.5-13*b*), the melt rises along the centerline and sinks along the pool boundary. The maximum velocity is on the order of 1 cm/s. With the Lorentz force alone (Figs. 8.5-13*c* and 8.5-13*d*), however, the melt sinks along the centerline and rises along the pool boundary, due to the inward and downward Lorentz force induced by the diverging current density field in the pool. The maximum velocity is on the order of 10 cm/s. Since this flow pattern favors convective heat transfer from the arc to the pool bottom,

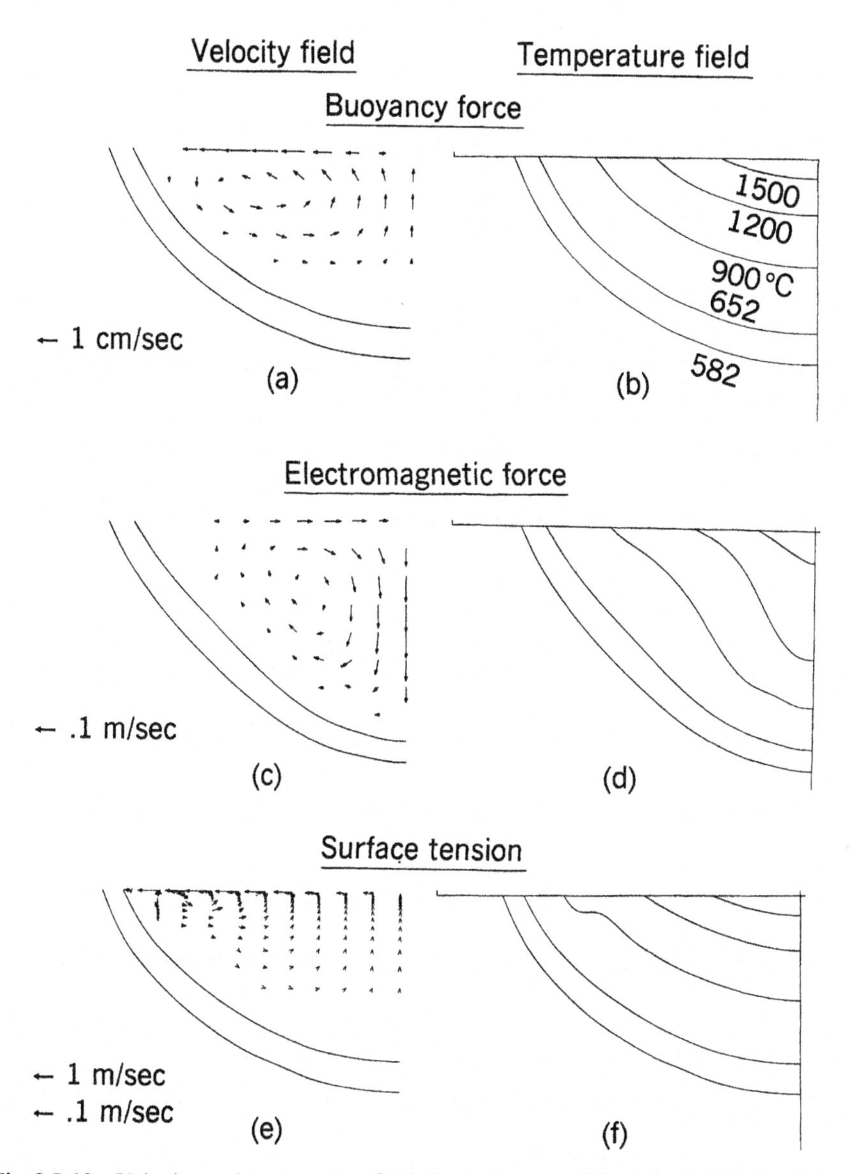

Fig. 8.5-13 Velocity and temperature fields in stationary weld pools of 6061 aluminium alloy: (*a*, *b*) buoyancy force; (*c*, *d*) Lorentz force; (*e*, *f*) surface tension force. (From S. Kou and D. K. Sun, *Metallurgical Transactions*, **16A**, 1985, p. 203.)

the pool becomes significantly deeper. With the surface-tension force along (Figs. 8.5-13*e* and 8.5-13*f*), the melt flows radially outward along the pool surface; the maximum velocity is on the order of 1 m/s. Since $\partial\gamma/\partial T = -0.35$ dyn cm^{-1} °C^{-1} (from pure aluminum; $\partial\gamma/\partial w_A$ not available), the hotter and hence lower surface-tension melt near the center of the pool surface is pulled radially outward.

The effect of the Lorentz force in increasing the penetration depth is illustrated in Fig. 8.5-14,[59] with the help of Woods metal (a low-melting-point material). With the tip of a heated copper rod in contact with the top surface, a shallow pool is produced, as shown in Fig. 8.5-14*a*. When a 75-A current is passed through the rod into the pool, the penetration depth increases dramatically, as shown in Fig. 8.5-14*b*.

Figure 8.5-15 shows the thermocapillary (or Marangoni) convection in a hemispherical silicone oil pool ($\partial\gamma/\partial T = -0.06$ dyn cm^{-1} °C^{-1}) induced by a hot wire in contact with the center of the pool surface.[60] The flow along the pool surface is radially outward.

With the three driving forces in Fig. 8.5-13 acting simultaneously, the flow field consists of two cells, an upper outer cell dominated by the surface tension force and a lower inner cell dominated by the Lorentz force. These cells can be shown more explicitly with streamlines than velocity vectors. As shown in Fig. 8.5-16*a*[61] for a stainless-steel arc weld, the thermocapillary cell is clockwise while the Lorentz force cell is counterclockwise. As shown in Fig. 8.5-16*b*, the thermocapillary cell is reversed in direction to join the Lorentz force cell when $\partial\gamma/\partial T$ is switched from negative to positive. Since convective heat transfer from the arc to the pool bottom is favored, the pool becomes much deeper.

Figure 8.5-17 shows the surface tension as a function of temperature for two molten steels; the lower curve for the steel has approximately 160 ppm more sulfur than the other.[62] As shown, the presence of a surface active element such as sulfur (or oxygen, selenium or tellurium) can switch the $\partial\gamma/\partial T$ of a molten steel from negative to positive and increase the depth : width ratio of the weld pool. With $\partial\gamma/\partial T > 0$ the radially inward surface flow causes the melt to go downward along the centerline, promoting convective heat transfer from the heat source to the pool bottom and producing a deeper pool.

Figure 8.5-18 shows two laser weld pools in 6061 aluminum.[63] The free surfaces, calculated with the help of Eq. [5.2-9], are allowed to deform. With $\partial\gamma/\partial T = -0.35$ dyn cm^{-1} °C^{-1} the radially outward surface flow causes the free surface to rise near the pool boundary. With $\partial\gamma/\partial T = 0.1$ dyn cm^{-1} °C^{-1}, however, the radially inward surface flow causes a small hump near the center of the pool surface.

In fact, the free-surface shape of the weld pool can also be affected by the thermal expansion of the molten metal and, especially, the volume increase of

[59] S. Kou and D. K. Sun, ibid.

[60] Y. Tao and S. Kou, unpublished work at the University of Wisconsin, Madison, WI, 1993.

[61] M. Kanouff and R. Greif, *Int. J. Heat Mass Transfer*, **35**, 967, 1992.

[62] C. R. Heiple and P. Burgardt, in *Weldability of Materials*, R. A. Patterson and K. W. Mahin, eds., ASM International, Materials Park, OH, 1990, p. 73.

[63] M. C. Tsai and S. Kou, *Int. J. Num. Meth. in Fluids*, **9**, 1503, 1989.

Buoyancy force

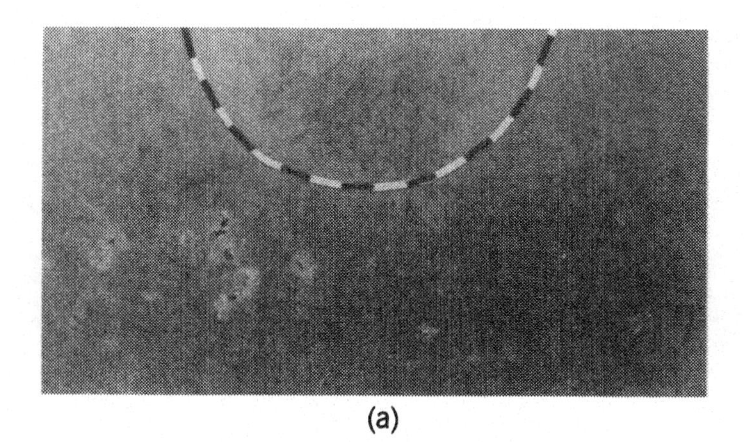

(a)

Electromagnetic force

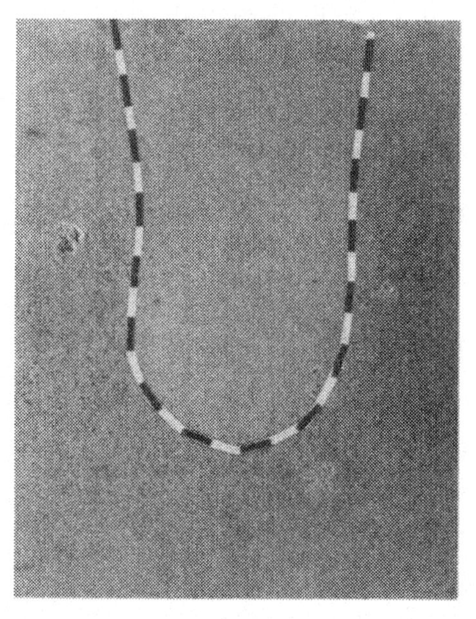

(b)

Fig. 8.5-14 Weld penetration in Woods metal: (*a*) buoyancy force; (*b*) Lorentz force. (From S. Kou and D. K. Sun, *Metallurgical Transactions*, **16A**, 1985, p. 203.)

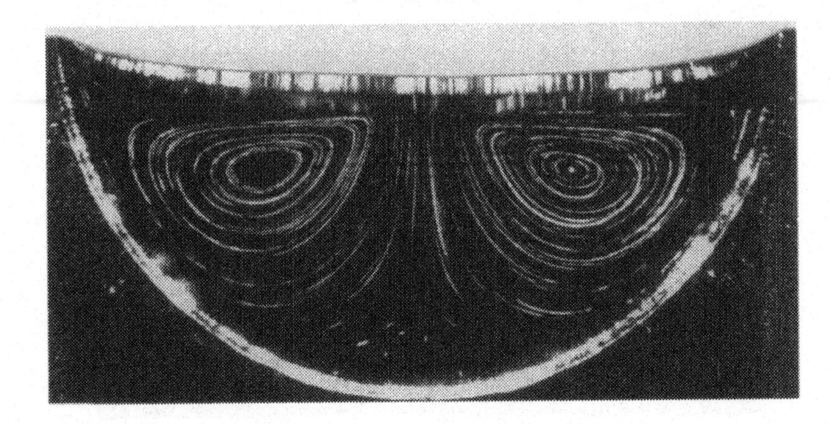

Fig. 8.5-15 Flow pattern in a simulated weld pool of silicone oil. (Y. Tao and S. Kou, unpublished work at University of Wisconsin, Madison, WI.)

the metal on melting ($\rho_L < \rho_S$). To illustrate this, let us consider the simple case of convection due to the buoyancy force alone. As shown in Fig. 8.5-19, the calculated pool surface bulges significantly,[64] mainly due to the about 6% volume expansion of aluminum on melting. As shown in Fig. 8.5-20, the assumption of a nondeformable flat pool surface can lead to too deep a weld pool.

Three-dimensional convection in moving weld pools has been calculated by Kou and Wang[65] and by many subsequent investigators. The boundary conditions are left as an exercise problem.

8.5.4. Welding: The Arc

Convective heat transfer in arc plasmas has been studied by many investigators, notably Pfender,[66] Szekely,[67] and their associates.

Tsai and Kou[68] have studied gas tungsten welding arcs (Section 6.1.3), considering the electrode-tip geometry, the arc length realistic for welding, and the presence of the shielding gas nozzle, as shown in Fig. 8.5-21. Since the physical properties of the shielding gas (Ar) are temperature-dependent, the governing equations in the form of variable properties are used (see Table 4.1-4).

[64] M. C. Tsai and S. Kou, *Num. Heat Transfer*, Part A, **17**, 73, 1990.

[65] S. Kou and Y. H. Wang, *Metallurgical Trans. A.*, **17A**, 2265, 2271, 1986.

[66] K. C. Hsu, K. Etemadi, and E. Pfender, *J. Appl. Phys.*, **54**, 1293; 1983; K. C. Hsu and E. Pfender, *J. Appl. Phys.*, **54**, 4359, 1983.

[67] J. McKelliget and J. Szekely, *Metallurgical Trans.*, **17A**, 1139, 1986.

[68] M. C. Tsai and S. Kou, *Int. J. Heat Mass Transfer*, **33**, 2089, 1990.

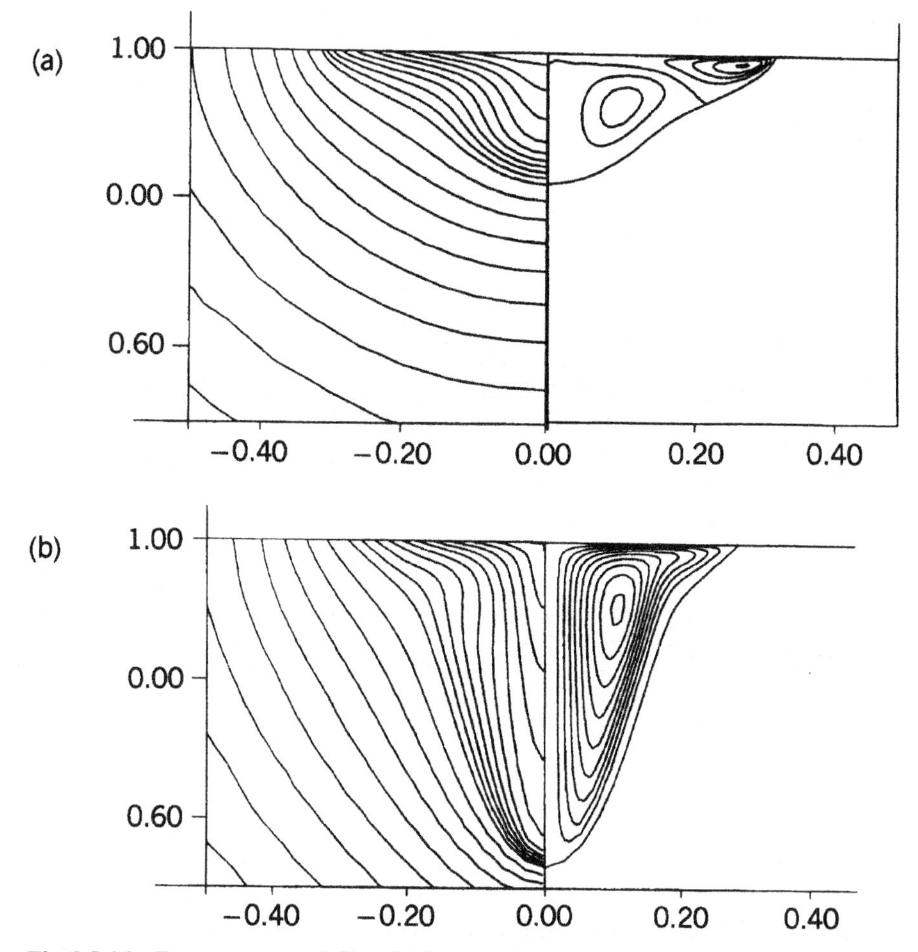

Fig. 8.5-16 Temperature and flow fields in stationary gas tungsten arc weld pools of stainless steel: (*a*) $\partial\gamma/\partial T = -0.01$ dyn cm^{-1} °C^{-1}; (*b*) $\partial\gamma/\partial T = 0.01$ dyn cm^{-1} °C^{-1}. (Reprinted from M. Kanouf and R. Greif, "The Unsteady Development of a GTA Weld Pool," *International Journal of Heat and Mass Transfer*, **35**, 1992, p. 967, with kind permission from Elsevier Science Ltd., The Boulevard Langford Lane, Kidlington OX 1GB, UK.)

Assuming steady-state, laminar flow, the governing equations are as follows:

Continuity: $$0 = -\nabla\cdot(\rho\mathbf{v}) \quad [8.5\text{-}68]$$

Momentum: $$0 = -\nabla\cdot(\rho\mathbf{v}\mathbf{v}) - \nabla\cdot\boldsymbol{\tau} + (\mathbf{f}_b - \nabla p) \quad [8.5\text{-}69]$$

Energy: $$0 = -\nabla\cdot(\rho\mathbf{v}H) = -\nabla\cdot\mathbf{q} + s \quad [8.5\text{-}70]$$

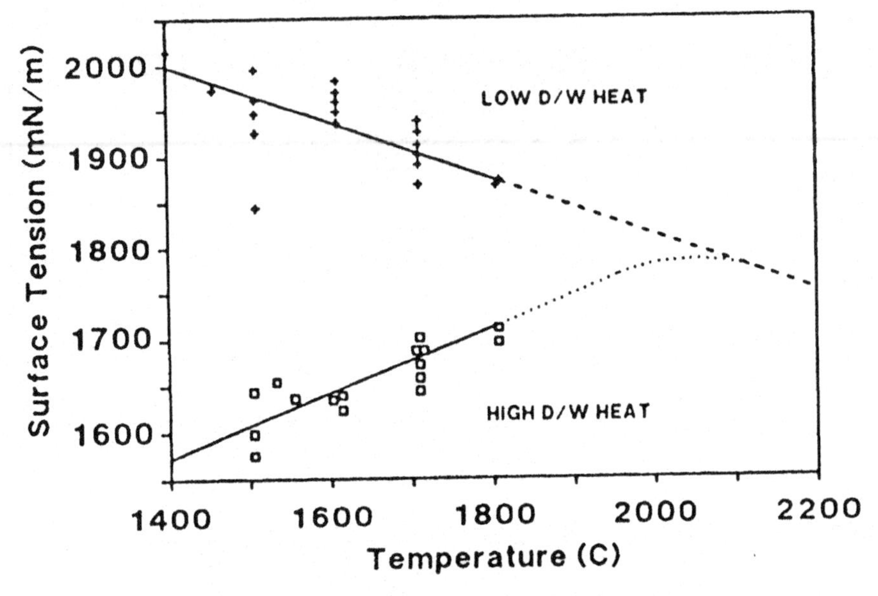

Fig. 8.5-17 Negative $\partial\gamma/\partial T$ and low weld depth: width ratio in one steel (top curve), and positive $\partial\gamma/\partial T$ and high weld depth: width ratio in another steel with 160 ppm more sulfur (bottom). (From C. R. Heiple and P. Burgardt, in *Weldability of Materials*, R. A. Patterson and K. W. Mahin, eds., ASM International, 1990, p. 73.)

In Eq. [8.5-69], the body force

$$\mathbf{f}_b = \mathbf{J} \times \mathbf{B} \qquad [8.5\text{-}71]$$

where **J** is the current density vector and **B** the self-induced magnetic-field vector, and they are solved with the help of the Maxwell equations. The gravity force is negligible. In Eq. [8.5-70], H is the enthalpy per unit mass. The source term

$$s = s_J - s_R - s_e \qquad [8.5\text{-}72]$$

where s_J, s_R, and s_e are the Joule heating caused by the arc resistance; s_R, the optically thin radiation loss per unit volume; and s_e, the transport of electron enthalpy due to the fact that the electron velocity is much higher than the heavy-particle velocity. The details of these terms are available elsewhere.[69]

The boundary conditions[69] include the no-slip condition at all gas/solid interfaces, the symmetry condition at the axis ($r = 0$), and zero heat flux at the

[69] M. C. Tsai and S. Kou, ibid.

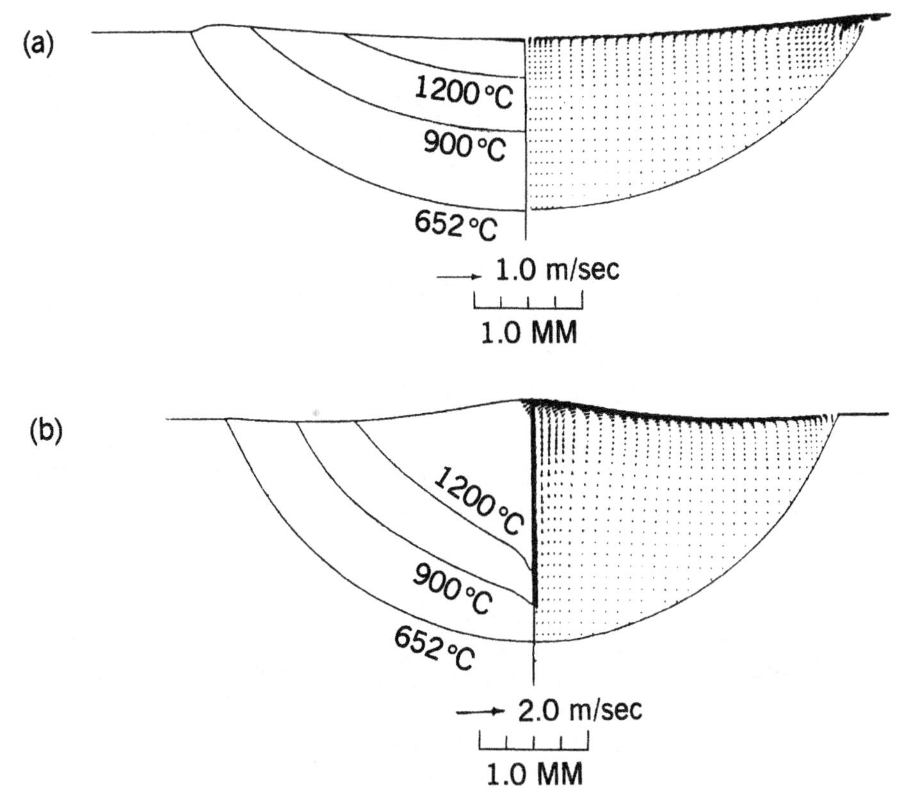

Fig. 8.5-18 Effect of $\partial\gamma/\partial T$ on laser-beam welds: (*a*) $\partial\gamma/\partial T = -0.35$ dyn cm^{-1} °C; (*b*) $\partial\gamma/\partial T = 0.1$ dyn cm^{-1} °C^{-1}. (From M. C. Tsai and S. Kou, *International Journal of Numerical Methods in Fluids*, **9**, 1989, p. 1503.)

gas/nozzle interface. The measured electrode temperature distribution and plasma temperature distribution immediately adjacent to the workpiece surface were also used.

Figures 8.5-22 shows the calculated velocity and temperature fields in a 2-mm-long Ar arc produced by a 3.2-mm-diameter electrode at 200 A. As described in Section 6.1.3.5, the Lorentz force near the electrode tip pushes the fluid downward and inward. This force produces a jet near the electrode tip as shown in Fig. 8.5-22*a*.

The calculated arc temperature distribution in Fig. 8.5-22*b* is close to the measured one of Haddad and Farmer[70] from 21,000 down to 13,000 K. The calculated 11,000-K isotherm (the outermost one) deviates from the measured one significantly. At temperatures around and below 10,000 K, the local electron

[70] G. N. Haddad and A. J. D. Farmer, *Welding J.*, **64**, 339, 1985.

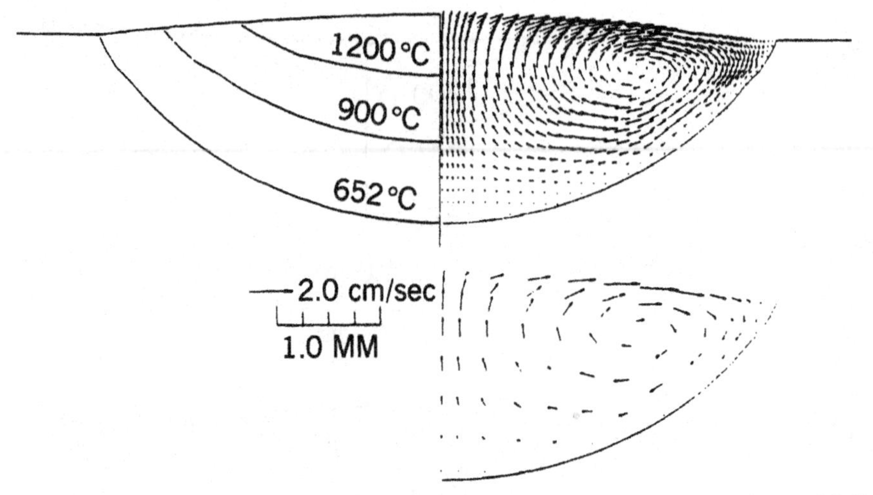

Fig. 8.5-19 Bulging of the weld pool surface (buoyancy convection alone). (From M. C. Tsai and S. Kou, *Numerical Heat Transfer*, **17A**, 1990, p. 73.)

and fluid temperatures may no longer be the same.[71] As such, a one-temperature (or -enthalpy) energy equation such as Eq. [8.5-70] may no longer be valid.

It is worth noting that the extent of the ionization of the argon shielding gas and hence the electrical conductivity and Joule heating in the arc are affected by the arc temperature distribution. Since the current density and hence the Lorentz force are affected by the electrical conductivity, heat transfer and fluid flow in the arc are coupled.

8.5.5. Fiber Processing: Optical Fiber

Accelerated cooling in optical fiber drawing (Section 6.1.5) has been studied recently by several investigators, including, Jaluria,[72] Polymeropoulos,[73] and their coworkers.

Figure 8.5-23 shows a section of a cooling channel of inner radius r_w. A fiber of radius r_f is being drawn at the constant speed of V_f in the negative z direction ($V_f < 0$). The cooling gas can be introduced from the bottom of the cooling channel (opposing flow), from the top (aiding flow), or through the channel wall near its bottom (peripheral flow).

Since the physical properties of the cooling gas may vary with temperature significantly, it is desirable to use the governing equations in the form of variable

[71] K. C. Hsu and E. Pfender, *J. Appl. Phys.*, **53**, 4359, 1983.

[72] S. R. Choudhury and Y. Jaluria, in *Advanced Computations in Materials Processing*, V. Prasad and R. V. Arimilli, ASME Publication HTD-Vol. 241, New York, 1993, p. 57.

[73] T. Vaskopulos, C. Polymeropoulos, and A. Zebid, in *Heat and Mass Transfer in Materials Processing and Manufacturing*, D. A. Zumbrunner et al., eds., ASME Publication HTD-Vol. 261, New York, 1993, p. 21.

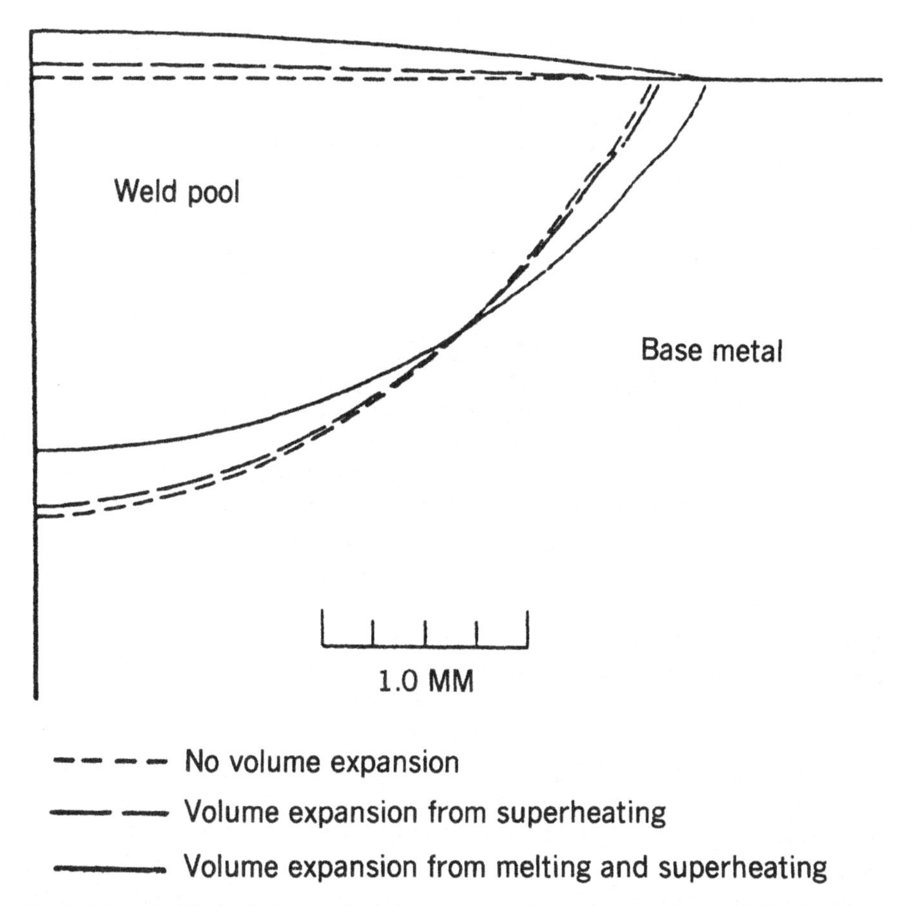

Fig. 8.5-20 Bulging of the weld pool surface and its effect on the pool depth. (From M. C. Tsai and S. Kou, *Numerical Heat Transfer*, **17A**, 1990, p. 73.)

properties (see Table 4.1-4). Assuming steady-state, laminar flow

Continuity: $$0 = -\nabla \cdot (\rho \mathbf{v}) \qquad [8.5\text{-}73]$$

Momentum: $$0 = -\nabla \cdot (\rho \mathbf{v}\mathbf{v}) - \nabla \cdot \boldsymbol{\tau} - (\rho \mathbf{g} - \nabla p) \qquad [8.5\text{-}74]$$

Energy: $$0 = -\nabla \cdot (\rho \mathbf{v} C_v T) - \nabla \cdot \mathbf{q} \qquad [8.5\text{-}75]$$

The ideal-gas law $\rho = pM/RT$ can be used to calculate the gas density, where M is the atomic or molecular weight of the gas.

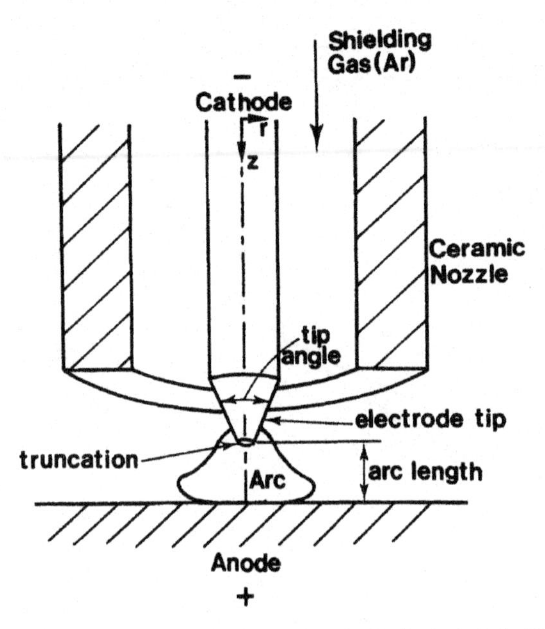

Fig. 8.5-21 Gas tungsten welding arc. (From M. C. Tsai and S. Kou, *International Journal of Heat and Mass Transfer*, **33**, 1990, p. 2089.)

For the fiber, only the energy equation is needed. Since the fiber diameter is very small, radial temperature gradients can be neglected. In view of the high drawing speed and low fiber thermal conductivity, axial heat transfer due to conduction can be expected to be negligible as compared to that due to motion. Since heat flow is one-dimensional, the surface heat loss can be considered a heat sink (see Example 2.3-2). As such

$$\text{Energy:} \quad 0 = -\rho_f C_f V_f \frac{\partial T_f}{\partial z} + s_f \qquad [8.5\text{-}76]$$

where the subscript f denotes the fiber. The heat sink s_f (< 0) includes heat losses from the fiber to the cooling gas by convection and to the channel wall by radiation (q_R):

$$s_f = \left[\left(k \frac{\partial T}{\partial r} \right) \bigg|_{r_f} - q_R \right] \frac{2}{r_f} \qquad [8.5\text{-}77]$$

where k is the thermal conductivity of the cooling gas.

The boundary conditions for the cooling gas in the case of opposing flow are as follows:

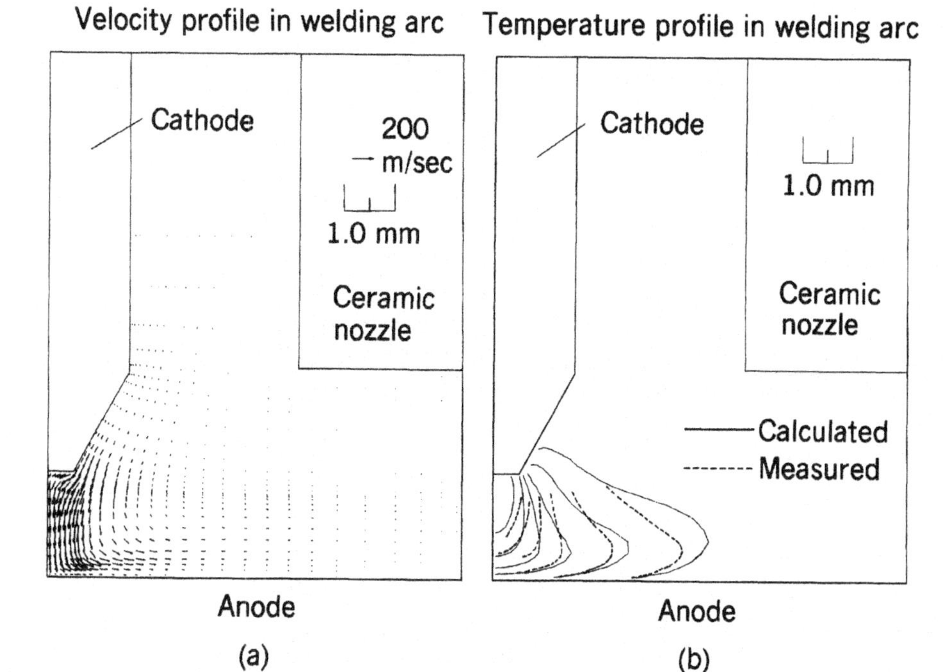

Fig. 8.5-22 Calculated results for a gas tungsten arc: (*a*) velocity field; (*b*) temperature field. (From M. C. Tsai and S. Kou, *International Journal of Heat and Mass Transfer*, **33**, 1990, p. 2089.)

Fiber Surface ($r = r_f$):

$$v_r = 0 \qquad [8.5\text{-}78]$$

$$v_z = V_f \qquad [8.5\text{-}79]$$

$$k\frac{\partial T}{\partial r} = \frac{r_f}{2}\left(\rho_f C_f V_f \frac{\partial T_f}{\partial z}\right) + q_R \qquad [8.5\text{-}80]$$

where the last condition is from Eqs. [8.5-76] and [8.5-77].

Channel Wall ($r = r_w$):

$$v_r = 0 \qquad [8.5\text{-}81]$$

$$v_z = 0 \qquad [8.5\text{-}82]$$

$$T = T_w \qquad [8.5\text{-}83]$$

where T_w is the temperature of the channel wall, which can be water-cooled.

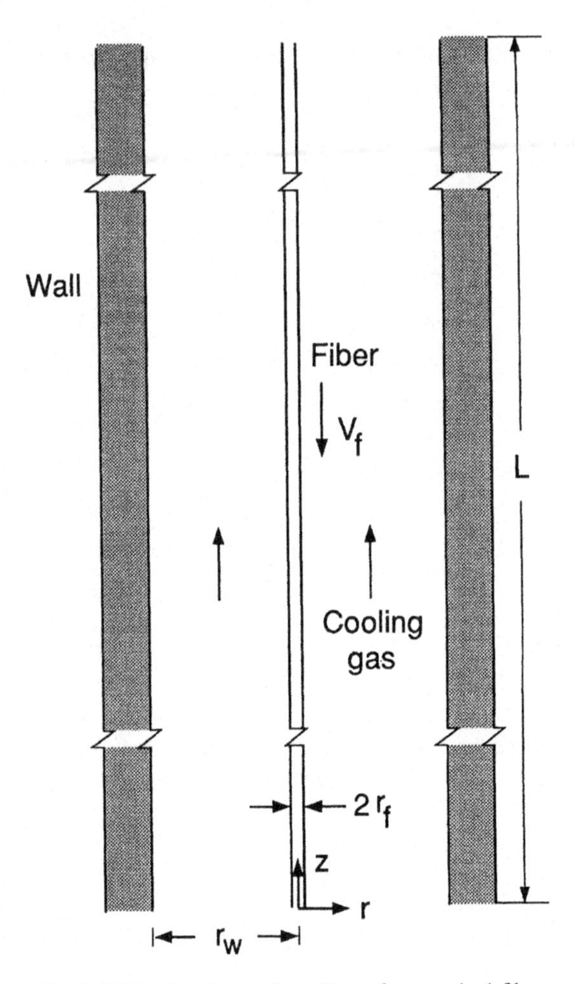

Fig. 8.5-23 Accelerated cooling of an optical fiber.

Channel Inlet ($z = 0$):

$$v_r = 0 \tag{8.5-84}$$

$$v_z = V_0 \tag{8.5-85}$$

$$T = T_0 \tag{8.5-86}$$

Other inlet boundary conditions have also been used, depending on the way the cooling gas is introduced.

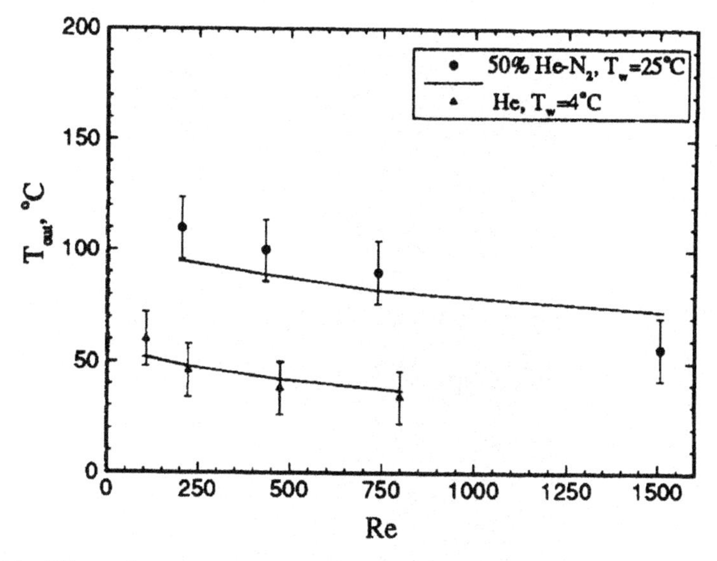

Fig. 8.5-24 Fiber exit temperature versus Reynolds number of the cooling gas. (From T. Vaskopulos, C. Polymeropoulos, and A. Zebid, "Heat Transfer from a Moving Surface Applied to Force Cooling of Optical Fiber," in *Heat and Mass Transfer in Materials Processing and Manufacturing*, D. A. Zumbrunner, ed., ASME publication HTD-vol. 261, New York, 1993, p. 21. Reprinted with permission from ASME.)

Channel Outlet ($z = L$):

$$\frac{\partial v_r}{\partial z} = 0 \qquad [8.5\text{-}87]$$

$$\frac{\partial v_z}{\partial z} = 0 \qquad [8.5\text{-}88]$$

$$\frac{\partial^2 T}{\partial z^2} = 0 \qquad [8.5\text{-}89]$$

At the outlet the gas velocity is assumed to be fully developed, and the axial temperature gradient stops changing.

Figure 8.5-24 shows the calculated and measured fiber exit temperatures for a 150-μm-radius fiber cooled by introducing the gas from near the bottom but through the channel wall.[74] The Reynolds numbers of the gas is defined as

[74] T. Vaskopulos, C. Polymeropoulos, and A. Zebid, in *Heat and Mass Transfer in Materials Processing and Manufacturing*, D. A. Zumbrunner et al., eds., ASME Publication HTD-Vol. 261, New York, 1993, p. 21.

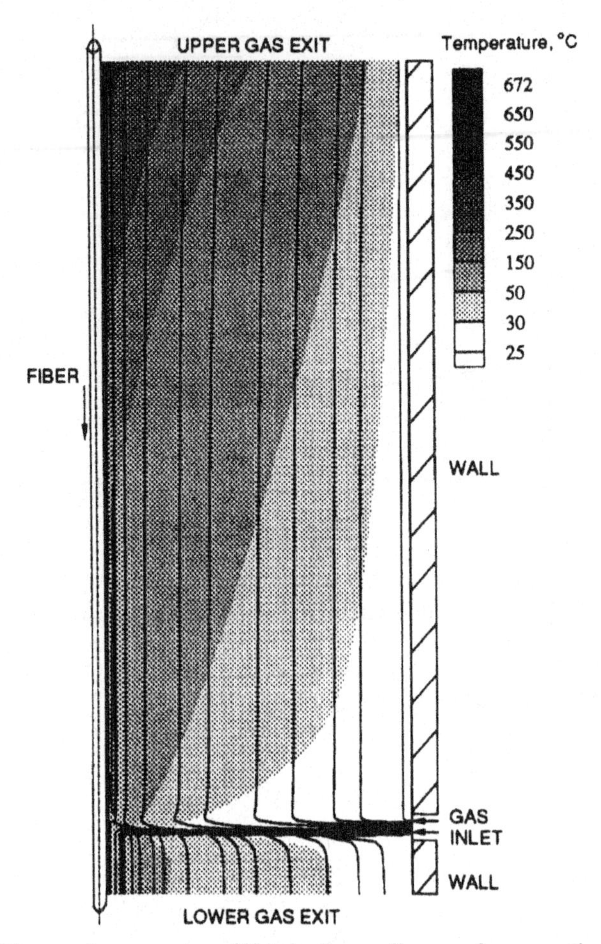

Fig. 8.5-25 Flow and temperature fields in the cooling gas for an optical fiber. (From T. Vaskopulos, C. Polymeropoulos, and A. Zebid, in "Heat Transfer from a Moving Surface Applied to Force Cooling of Optical Fiber," *Heat and Mass Transfer in Materials Processing and Manufacturing*, D. A. Zumbrunner, ed., ASME publication HTD-vol. 261, New York, 1993, p. 21. Reprinted with permission from ASME.)

$2\rho_w V_m(r_w - r_f)/\mu_w$, where ρ_w and μ_w are the gas density and viscosity at the wall temperature (25°C), respectively, and V_m is the average gas velocity. As shown, the fiber exit temperature is affected significantly by the composition and temperature of the cooling gas, but less so by the gas flow rate. Helium is known to have a higher thermal conductivity than nitrogen. The effect of other process variables will be left as exercise problems. An example of the calculated streamlines and isotherms is shown in Fig. 8.5-25 for a 150-μm-radius fiber pulled at 2 m/s and cooled by helium at Re = 210.

PROBLEMS[75]

8.1-1 It has been considered to die cast from a powder/melt mixture rather than a pure melt as usual. How much is the solidification time reduced if as much as 50% powder is used, and why?

8.1-2 Magnesium is sometimes considered as an alternative to aluminum in die casting in view of similarities in their properties. It, however, is more attractive in terms of the solidification time. Using the data given below, find the savings in the solidification time, assuming that a casting of the same dimensions is made with Mg rather than Al.

Al: $T_m = 933\,°K, \quad \rho_s = 2.70\,\mathrm{g/cm^3}, \quad H_f = 1.08\ \mathrm{kJ/cm^3}$

Mg: $T_m = 922\,°K, \quad \rho_s = 1.74\,\mathrm{g/cm^3}, \quad H_f = 0.64\ \mathrm{kJ/cm^3}$

8.4-1 Show that from Eq. [8.2-10] the maximum dimensionless temperature of the workpiece is

$$T^*_{\max} = (\sqrt{2\pi} a^{*1.5})^{-1}$$

for a uniform square laser beam of $2a \times 2a$, where $T^*_{\max}$ and a^* are defined in Section 8.4.2.

8.4-2 Consider the laser transformation hardening of a rod of radius a by a uniform ring-shaped laser beam of width $2b$, as illustrated in Fig. P8.4-2a. The beam is stationary while the rod travels at a constant

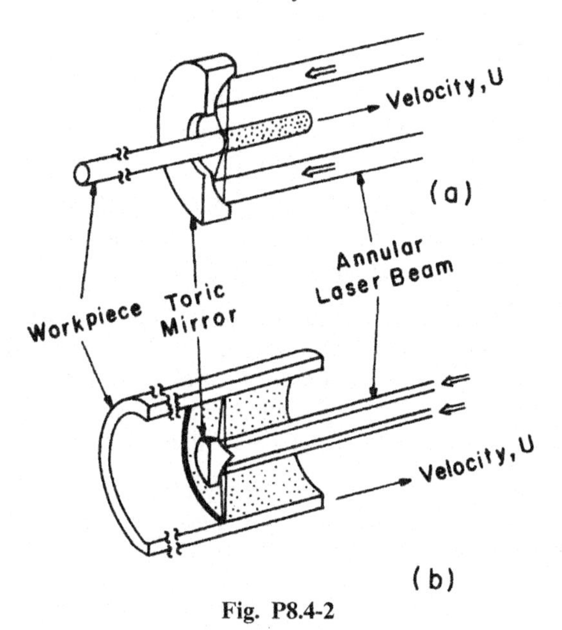

Fig. P8.4-2

[75] The first two digits in problem numbers refer to section numbers in text.

velocity U in the z direction. Give the equation of energy and boundary conditions.

8.4-3 Consider the laser transformation hardening of a tube of outer radius a' and inner radius a by a uniform ring-shaped laser beam of width $2b$, as shown in Fig. P8.4-2b. The beam is stationary while the tube travels at a constant velocity U in the z direction. Give the equation of energy and boundary conditions.

8.4-4 Consider the GTA weld shown in Fig. 8.4-12. The effective radius of the arc $a = 3$ mm. How does the transverse cross section of the weld vary if a is varied from 2 to 8 mm?

8.4-5 Consider the GTA weld shown in Fig. 8.4-12. The initial workpiece temperature T_0 is the room temperature. How does the transverse cross section of the weld vary if T_0 is varied from 0 to 150°C?

8.4-6 Consider the GTA weld shown in Fig. 8.4-12. The effective thermal conductivity $k_{eff} = 1.5\,k$, where k is the thermal conductivity of the solid. How does the peak temperature of the pool vary if k_{eff} is varied from $1k$ to $10k$?

8.4-7 Consider the GTA welding of the thin sheets shown in Fig. 8.4-13. Give the energy equation and boundary conditions for the two-dimensional heat conduction in the workpiece. The thickness and width of the sheet are b and w, respectively, and the effective radius of the arc is a. Neglect the initial and final transients of heat flow.

8.5-1 Consider the Czochralski pulling of the oxide crystal shown in Fig. P8.5-1 (see p. 523).[76] The thermal expansion coefficient β_T is $1.6 \times 10^{-8}\,°C^{-1}$ and the crystal is not rotated. How would the growth front and isotherms be affected if β_T were changed to zero?

8.5-2 Consider the Czochralski pulling of the oxide crystal shown in Fig. P8.5-1.[76] How would the growth front and streamlines be affected if the crystal were rotated at significantly high rates?

8.5-3 Consider the Czochralski pulling of the silicon crystal shown in Fig. P8.5-3[77] (see p. 523). The surface-tension temperature coefficient $\partial\gamma/\partial T = -4 \times 10^{-6}$ dyn cm^{-1} °C^{-1} and buoyancy convection is neglected. How would the growth front and streamlines be affected if $\partial\gamma/\partial T$ were increased to -4×10^{-3} dyn cm^{-1} °C^{-1}.

8.5-4 Consider the case of floating-zone crystal growth shown in Fig. 8.5-7a. How would the growth front and streamlines be affected if: (a) $\partial\gamma/\partial T = 0$ and (b) $\partial\gamma/\partial T = \beta_T = 0$?

[76] Q. Xiao and J. J. Derby, *J. Crystal Growth*, **128**, 188, 1993.

[77] R. A. Brown, T. A. Kinney, P. A. Sackinger, and D. E. Bornside, *J. Crystal Growth*, **97**, 99, 1989.

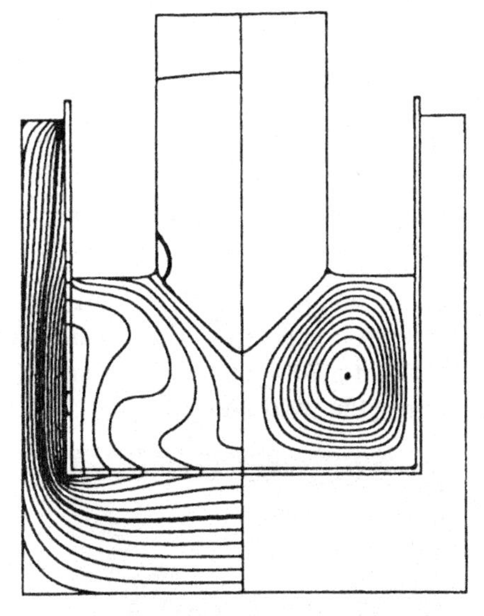

Fig. P8.5-1

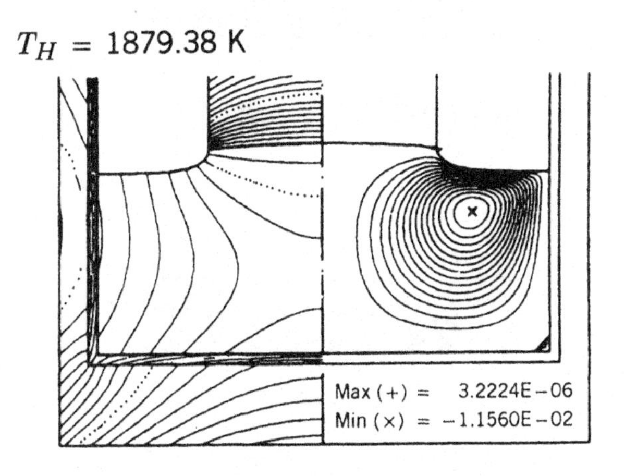

$\gamma_T = 4 \times 10^{-6}$ dyn/cm−K, Ma° = 1.16
Re° = 13.15, $\Delta H_0 = -1.75$ mm

Fig. P8.5-3

8.5-5 Consider the case of floating-zone crystal growth shown in Fig. 8.5-7a. How would the free surface and streamlines be affected if gravity $g = 0$?

8.5-6 Consider the case of floating-zone crystal growth shown in Fig. 8.5-8c. How would the growth front and streamlines be affected if the feed rod alone were rotated?

8.5-7 Consider the case of floating-zone crystal growth shown in Fig. 8.5-8*c*. How would the growth front and streamlines be affected if the crystal alone were rotated?

8.5-8 Consider the case of floating-zone crystal growth shown in Fig. 8.5-8*c*. How would the growth front and streamlines be affected if the crystal grew downward (crystal at top and feed at bottom) and if the crystal alone were rotated?

8.5-9 Consider the two gas tungsten arc welds in a thin stainless-steel sheet shown in Fig. P8.5-9.[78] The two welds were made under identical conditions, except that a 700-ppm SO_2 was added to the shielding gas in one weld. Identify the weld and explain.

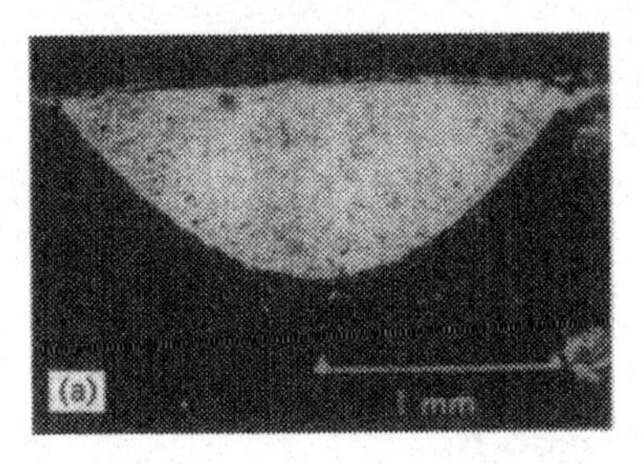

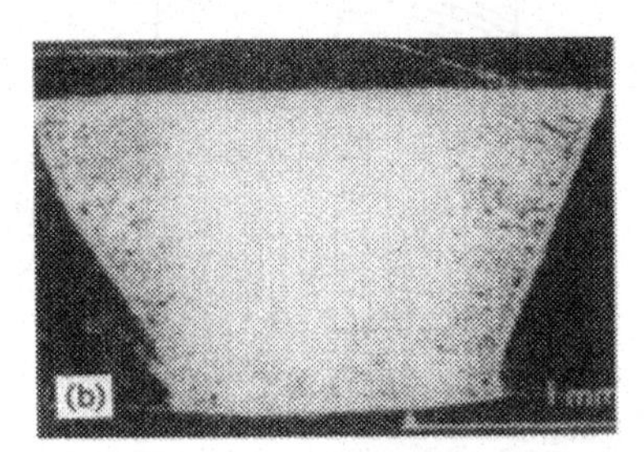

Fig. P8.5-9

8.5-10 The boundary conditions for a weld pool produced by a stationary heat source have been given in Section 8.5.3. Give similar boundary conditions for a weld pool produced by a heat source traveling at a constant speed V in the negative x direction. This is similar to the case shown in Fig. 8.4-9, except that the workpiece is now of finite thickness g and width $2l$.

8.5-11 Consider the gas tungsten welding arc shown in Fig. 8.5-22. How would the velocities and isotherms in the arc be affected if the conical surface of the electrode were extended, that is, having less truncation of the

[78] C. R. Heiple and P. Burgardt, *Welding J.*, **64**, 159s, 1985.

electrode tip? This change would allow a greater portion of the welding current to enter the inclined surface of the electrode.

8.5-12 Consider the accelerated cooling of an optical fiber. How would the fiber exit temperature be affected by the drawing speed and diameter of the fiber?

8.5-13 Consider the accelerated cooling of an optical fiber. How would the fiber exit temperature be affected by the inner diameter and length of the cooling channel?

8.5-14 Consider the accelerated cooling of an optical fiber. How should the boundary conditions in Section 8.5.5 be changed if the "fiber" has a substantial diameter so that its radial temperature gradients become significant? What about the equation of energy for the fiber?

CHAPTER 9

MASS TRANSFER IN MATERIALS PROCESSING

In our discussion on mass transfer in Chapter 3, the species overall mass-balance equation (Eq. [3.2-7] or [3.2-9]) and the species equation of continuity (Eq. [3.3-11] or [3.3-12]) were derived. In the present chapter these equations will be applied to materials processing: the former in Section 9.1; the latter, throughout the rest of the chapter.

9.1. OVERALL BALANCE

9.1.1. Crystal Growth: Vertical Bridgman

Figure 9.1-1 is an illustration of vertical Bridgman crystal growth (Section 6.1.1), the seed being neglected for simplicity. As shown in Figs. 9.1-1*a* and 9.1-1*b*, the melt tends to shrink on solidification if $\rho_S > \rho_L$ and vice versa. As described previously, in Section 6.1.1.4, at the crystal/melt interface the dopant (or solute) concentration of the crystal w_{AS}^* and that of the melt w_{AL}^* are related to each other through $w_{AS}^* = k_0 w_{AL}^*$, where k_0 is the equilibrium segregation coefficient. For simplicity, the melt composition is assumed uniform: $w_{AL} = w_{AL}^*$. This assumption of complete mixing in the melt is reasonable when convection in the melt is strong or when the melt is very short.

For simplicity we make the following four assumptions: (1) a planar crystal/melt interface, (2) negligible concentration gradients in the lateral directions, (3) negligible diffusion in the crystal, and (4) constant equilibrium segregation coefficient k_0. The criterion for a stable planar interface has been described in Section 6.1.1.6. The diffusion coefficient of the dopant (or solute) is much smaller in the solid than in the melt. For crystal growth of most electronic and

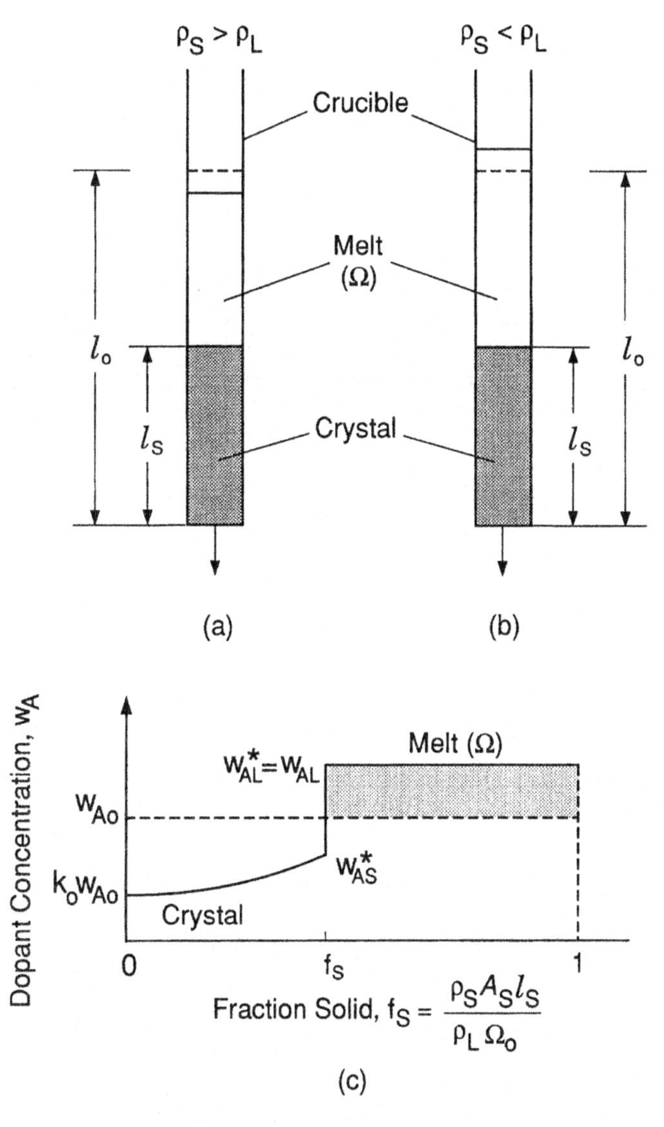

Fig. 9.1-1 Vertical Bridgman crystal growth: (*a*) $\rho_S > \rho_L$; (*b*) $\rho_S < \rho_L$; (*c*) dopant redistribution.

optical materials, the dopant concentration is low and k_0 can be considered constant.

Only the case of $k_0 < 1$ will be discussed here; that of $k_0 > 1$ follows in a similar way. Let us consider crystal growth from a melt of initial composition w_{A0}. Since for $k_0 < 1$, that is, the fact that the crystal cannot accommodate as much dopant as the melt, the dopant is rejected into the melt. The shaded area in

Fig. 9.1-1c represents the dopant increase in the melt due to such rejection. This caused w_{AL}^* to rise, which, in turn, causes w_{AS}^* to rise, since $w_{AS}^* = k_0 w_{AL}^*$.

Let the control volume Ω be the melt in the crucible and A_S the transverse cross section of the crystal and the melt. Since the total mass of the melt and the crystal equals the initial mass in the crucible

$$\rho_L \Omega + \rho_S A_S l_S = \rho_L \Omega_0 \qquad [9.1\text{-}1]$$

where l_S is the height of the crystal and Ω_0 the initial melt volume.

Let us now consider the mass balance in the melt Ω. Let the crystal/melt interface be stationary and the crucible move at the velocity of dl_S/dt, where l_S is the length of the crystal. Evidently, no material enters Ω. At the crystal/melt interface the material leaves Ω in the form of the crystal at the velocity dl_S/dt. As such, from Eq. [1.2-6]

$$\frac{d}{dt}(\rho_L \Omega) = 0 - \rho_S\left(\frac{dl_S}{dt}\right) A_S \qquad \text{(overall mass balance)} \qquad [9.1\text{-}2]$$

Since the material leaves Ω at the composition of w_{AS}^*, from Eq. [3.2-7]

$$\frac{d}{dt}(w_{AL}\rho_L \Omega) = 0 - w_{AS}^* \rho_S\left(\frac{dl_S}{dt}\right) A_S \qquad \text{(species overall mass balance)} \qquad [9.1\text{-}3]$$

because the diffusion and reaction terms are zero. Substituting Eq. [9.1-2] into Eq. [9.1-3] and multiplying by dt

$$\rho_L \Omega\, dw_{AL} - w_{AL}\rho_S A_S\, dl_S = -\, w_{AS}^* \rho_S A_S\, dl_S \qquad [9.1\text{-}4]$$

Substituting Eq. [9.1-1] and $w_{AS}^* = k_0 w_{AL}$ into Eq. [9.1-4]

$$(\rho_L \Omega_0 - \rho_S A_S l_S)\frac{dw_{AS}^*}{k_0} = \left(\frac{1}{k_0} - 1\right) w_{AS}^* \rho_S A_S\, dl_S \qquad [9.1\text{-}5]$$

Multiplying both sides by $-k_0/(\rho_S A_S)$ and rearranging

$$\frac{dw_{AS}^*}{dl_S} = \frac{(k_0 - 1)\, w_{AS}^*}{l_S - (\rho_L \Omega_0/\rho_S A_S)} \qquad [9.1\text{-}6]$$

The initial condition is

$$w_{AS}^* = k_0 w_{A0} \quad \text{at} \quad l_S = 0 \qquad [9.1\text{-}7]$$

where w_{A0} is the initial composition of the melt.

TABLE 9.1-1 Solid/Melt Density Ratios of Some Materials

	Semiconductors[a]				Oxides		Metals[f]		
	Si	Ge	GaAs	InSb	GGG[b]	$NaNO_3$[e]	Al	Fe	Cu
ρ_S (g/cm^3)	2.30	5.26	5.32	5.77	7.2[c]	2.12	2.70	7.86	8.96
ρ_L (g/cm^3)	2.53	5.51	5.71	6.47	5.7[d]	1.90	2.38	7.03	8.00
ρ_S/ρ_L	0.909	0.955	0.932	0.892	1.263	1.116	1.134	1.118	1.12
			< 1			> 1		> 1	

[a] V. M. Glazov, S. N. Chizhevskaya, and N. N. Glagoleva, *Liquid Semiconductors*, Plenum Press, New York, 1969, p. 61.
[b] Gadolinium gallium garnet (a solid state laser).
[c] D. C. Miller, A. J. Valentino, and L. K. Shick, *J. Crystal Growth*, **44**, 121, 1978.
[d] S. Geller, G. P. Espinosa, and P. B. Grandall, *J. Appl. Cryst.*, **2**, 1986, 1969.
[e] C. W. Lan and S. Kou, *J. Crystal Growth*, **108**, 351, 1991.
[f] T. Iida and R. I. L. Guthrie, *The Physical Properties of Liquid Metals*, Oxford University Press, Oxford, 1988, p. 71.

The solution, from Case B of Appendix A, is as follows:

$$\frac{w_{AS}^*}{k_0 w_{A0}} = \left(1 - \frac{\rho_S A_S l_S}{\rho_L \Omega_0}\right)^{k_0 - 1} \tag{9.1-8}$$

or

$$\boxed{\frac{w_{AS}^*}{k_0 w_{A0}} = (1 - f_S)^{k_0 - 1}} \tag{9.1-9}$$

where the fraction solid

$$f_S = \frac{\rho_S A_S l_S}{\rho_L \Omega_0} \tag{9.1-10}$$

The solute redistribution equation, Eq. [9.1-9], is valid for both $k_0 < 1$ and $k_0 > 1$. It is identical in form to the normal freezing equation or **Scheil equation**,[1] which is usually derived based on the assumption of $\rho_S = \rho_L$:

$$f_S = \frac{A_S l_S}{\Omega_0} = \frac{l_S}{l_0} \tag{9.1-11}$$

where l_0 is the initial height of the melt. This equation can lead to significant errors in f_S and hence w_{AS}^* if the ρ_S/ρ_L ratio differs from unity significantly. This ratio is listed in Table 9.1-1 for some representative materials.

[1] E. Scheil, *Z. Metallk.*, **34**, 70, 1942.

9.1.2. Crystal Growth: Horizontal Bridgman

Figure 9.1-2 is an illustration of horizontal Bridgman crystal growth (Section 6.1.1.2); the crystal is assumed uniform in cross-section for simplicity (i.e., the

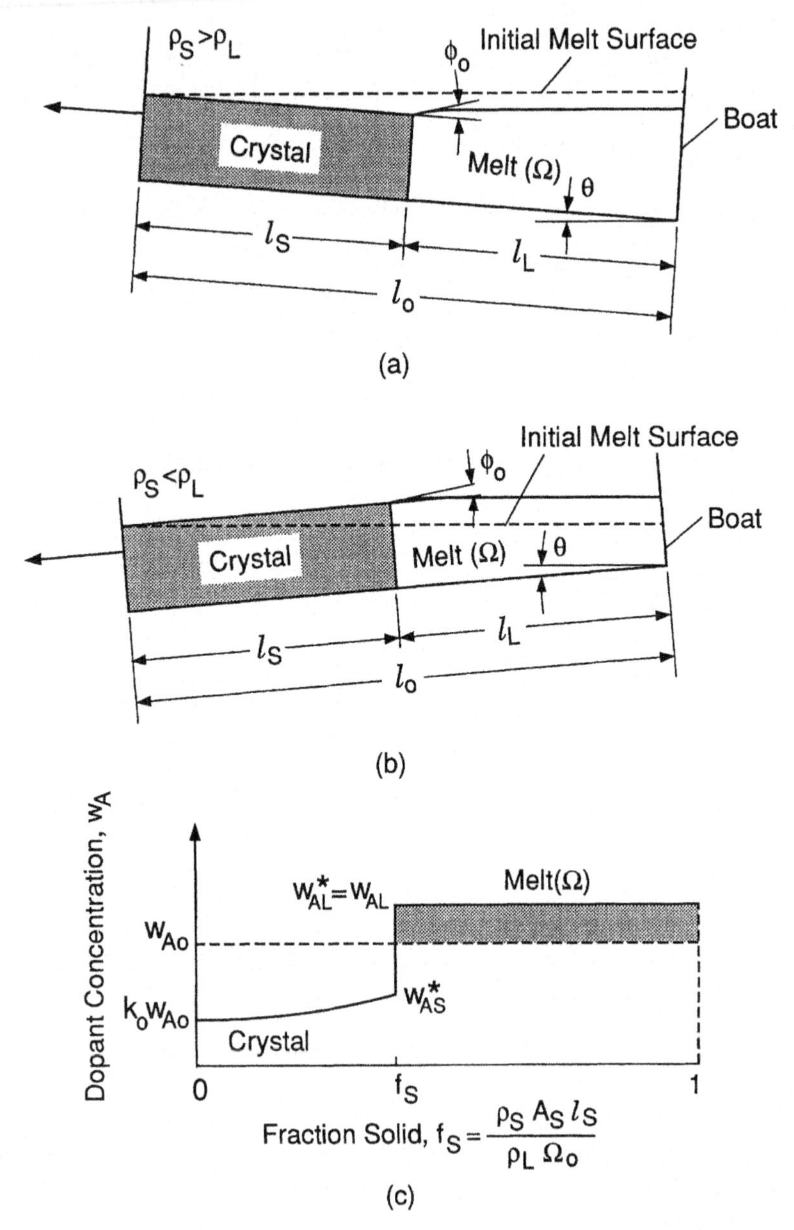

Fig. 9.1-2 Horizontal Bridgman crystal growth: (*a*) $\rho_S > \rho_L$; (*b*) $\rho_S < \rho_L$; (*c*) dopant redistribution.

growth angle $\phi = \phi_0$). In order to keep the crystal thickness uniform, the seed end of the boat needs to be raised (the tilt angle $\theta > 0$) if $\rho_S > \rho_L$ and vice versa.

Let the control volume Ω be the melt in the boat. The melt is assumed uniform in composition and the four simplifying assumptions in Section 9.1.1 are made. The derivation of the dopant (or solute) redistribution equation is identical to that described in Section 9.1.1 for vertical Bridgman. The resultant equation is identical to Eq. [9.1-9].

The resultant dopant (or solute) distribution, either vertical or horizontal Bridgman, is illustrated in Fig. 9.1-3. As indicated by the arrows on the solidus

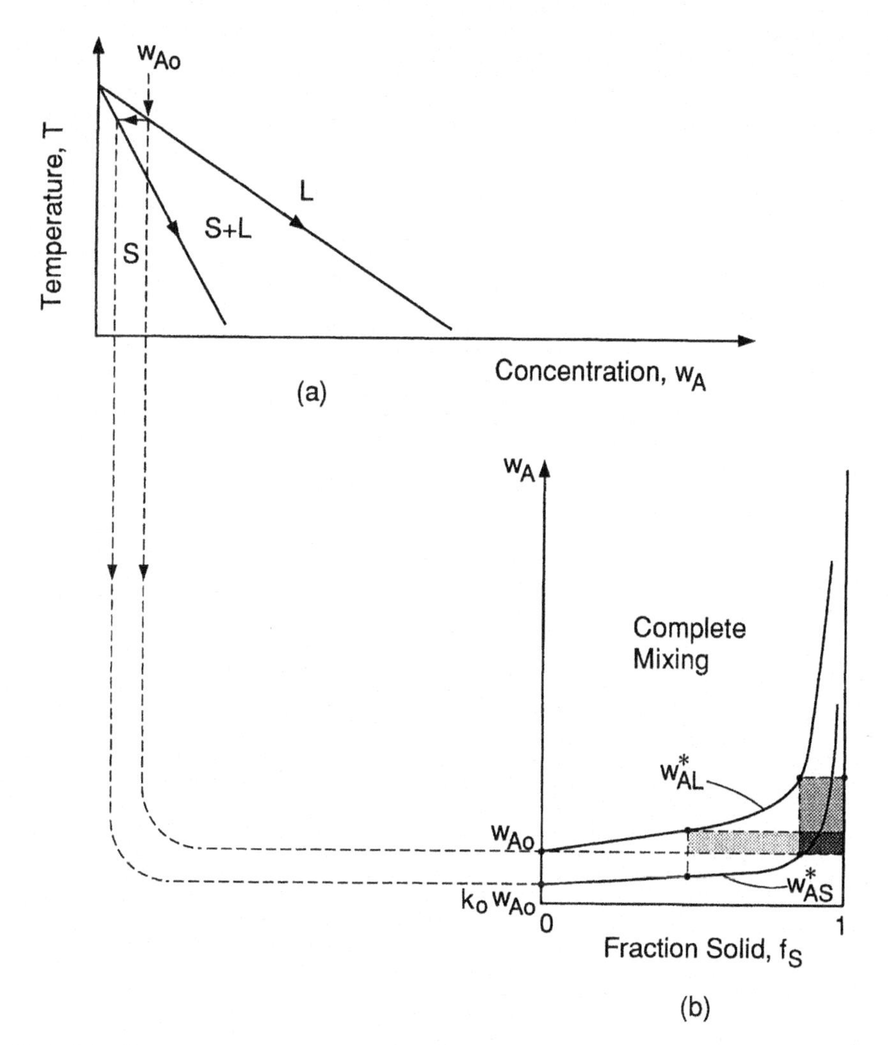

Fig. 9.1-3 Dopant segregation under complete mixing in melt: (*a*) phase diagram; (*b*) concentration profiles.

line and the liquidus line in Fig. 9.1-3*a*, w_{AS}^* and w_{AL}^* both increase during crystal growth. As already mentioned, the melt is assumed uniform in composition, that is, complete mixing in the melt. The shaded areas represent, at two different stages of crystal growth, the concentration increases in the melt due to rejection of dopant (or solute) from the crystal ($k_0 < 1$).

9.1.3. Crystal Growth: Czochralski

Figure 9.1-4 illustrates Czochralski crystal growth (Section 6.1.1.1); the crystal diameter is assumed uniform for simplicity (i.e., the growth angle $\phi = \phi_0$). As in the previous sections, we consider the crystal/melt interface as stationary. In other words, the crystal moves at the velocity dl_S/dt away from the interface while the crucible is pushed toward it at $(dl_S/dt)A_S/A_L$, where l_S is the length of the crystal and A_S and A_L are the cross-sectional areas of the crystal and the crucible, respectively.

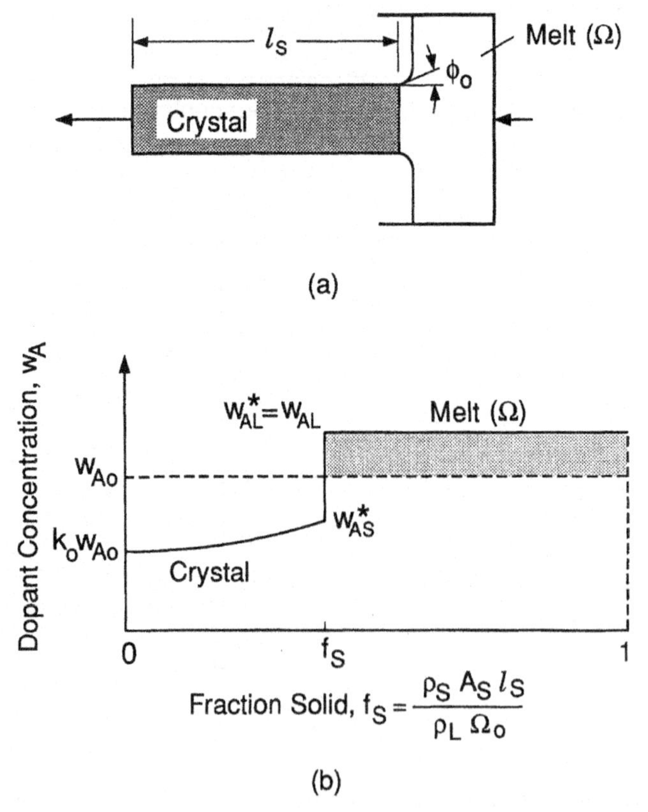

Fig. 9.1-4 Czochralski crystal growth; (*a*) control volume (Ω); (*b*) dopant redistribution.

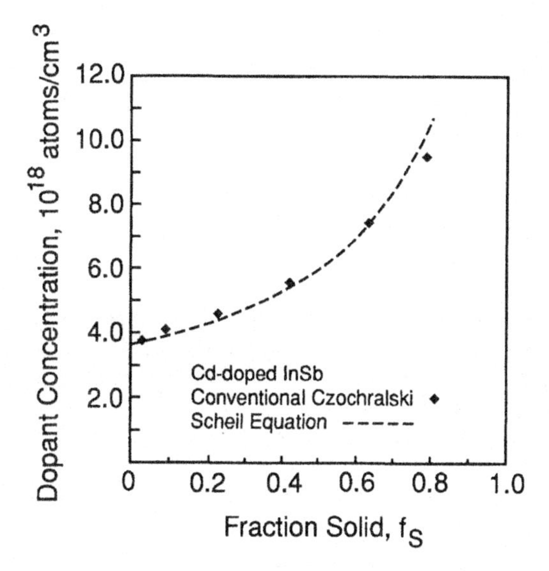

Fig. 9.1-5 Axial dopant segregation of Cd in an InSb Czochralski crystal. (From M. H. Lin and S. Kou, *Journal of Crystal Growth*, **152**, 1995, p. 256.)

Let the melt be the control volume Ω. The melt is assumed uniform in composition, and the four simplifying assumptions in Section 9.1.1 are made. The derivation of the dopant (or solute) distribution equation is identical to that described in Section 9.1.1 for vertical Bridgman. The resultant equation is identical to Eq. [9.1-9].

Figure 9.1-5 shows the dopant distribution in a Cd-doped InSb single crystal ($k_0 = 0.274$) grown by the Czochralski process.[2] The initial dopant concentration of the melt is 13.0×10^{18} atoms/cm^3. The Scheil equation appears to be a good approximation in this case.

9.1.4. Crystal Growth: Zone Melting

Figure 9.1-6 is an illustration of zone-melting crystal growth (Section 6.1.1.3), vertical and horizontal. We consider the molten zone as stationary while the crystal and the feed both move at the velocity dl_S/dt, where l_S is the length of the crystal. For simplicity, we assume that the melt is uniform in composition and that the zone length l_L is constant, except during the final transient after the molten zone reaches the end of the sample. We also assume that the crystal is uniform in cross-section, that is, the growth angle $\phi = \phi_0$. In addition, we also adopt the four simplifying assumptions in Section 9.1.1.

[2] M. H. Lin and S. Kou, *J. Crystal Growth*, **152**, 256, 1995.

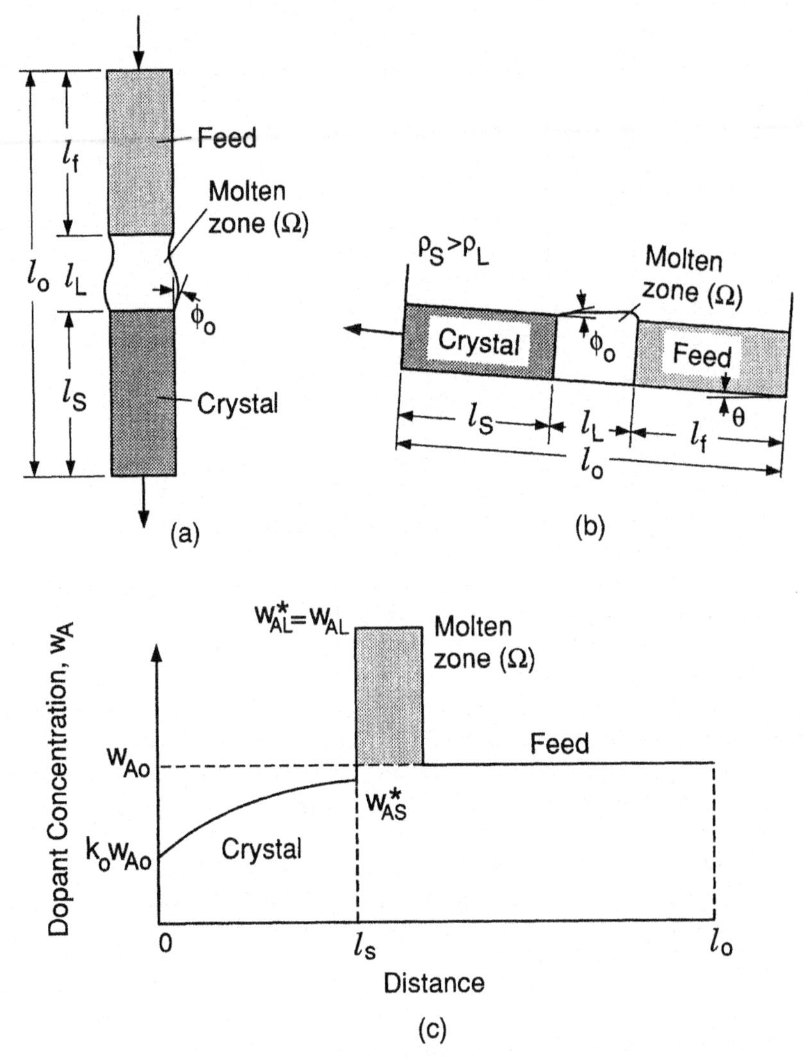

Fig. 9.1-6 Zone melting crystal growth: (*a*) floating zone; (*b*) horizontal; (*c*) dopant redistribution.

Let the control volume Ω be the molten zone and A_S the transverse cross sectional area of the crystal and the feed. Since the mass of the zone is the same before and after melting

$$\rho_L \Omega = \rho_S A_S l_L \qquad [9.1\text{-}12]$$

From Eq. [1.2-6]

$$\frac{d}{dt}(\rho_L \Omega) = \rho_S \left(\frac{dl_S}{dt}\right) A_S - \rho_S \left(\frac{dl_S}{dt}\right) A_S = 0 \qquad \text{(overall mass balance)} \qquad [9.1\text{-}13]$$

From Eq. [3.2.7]

$$\frac{d}{dt}(\rho_L w_{AL}\Omega) = \rho_S w_{A0}\left(\frac{dl_S}{dt}\right)A_S - \rho_S w_{AS}^*\left(\frac{dl_S}{dt}\right)A_S$$

(species overall mass balance) [9.1-14]

where w_{A0} is the composition of the feed. Substituting Eq. [9.1-13] into Eq. [9.1-14] and multiplying by dt

$$\rho_L\Omega\, dw_{AL} + w_{AL}\cdot 0 = \rho_S A_S(w_{A0} - w_{AS}^*)dl_S \qquad [9.1\text{-}15]$$

Substituting Eq. [9.1-12] and $w_{AS}^* = k_0 w_{AL}$ into Eq. [9.1-15]

$$\rho_S A_S l_L \frac{dw_{AS}^*}{k_0} = \rho_S A_S(w_{A0} - w_{AS}^*)\, dl_S \qquad [9.1\text{-}16]$$

Multiplying both sides by $k_0/(\rho_S A_S l_L dl_S)$

$$\frac{dw_{AS}^*}{dl_S} = \frac{-k_0(w_{AS}^* - w_{A0})}{l_L} \qquad [9.1\text{-}17]$$

The initial condition is

$$w_{AS}^* = k_0 w_{A0} \quad \text{at} \quad l_S = 0 \qquad [9.1\text{-}18]$$

The solution, from Case A of Appendix A, is as follows:

$$\boxed{\frac{w_{AS}^* - w_{A0}}{k_0 w_{A0} - w_{A0}} = \exp\left(-\frac{k_0 l_S}{l_L}\right)} \qquad [9.1\text{-}19]$$

This equation is identical to that derived by Pfann.[3] Notice that it does not require that ρ_L equal ρ_S.

9.1.5. Casting and Welding: Microsegregation

Unlike the scenario in crystal growth, where the crystal/melt interface is planar, the solid/melt interface in the casting or welding of an alloy is usually dendritic (Fig. 6.1-21). If the equilibrium segregation coefficient $k_0 \neq 1$, the solute is rejected ($k_0 < 1$) or absorbed ($k_0 > 1$) by the dendrite arms, causing w_{AL}^* and hence w_{AS}^* to vary during solidification. Since this type of solute segregation is on the microscopic scale of a dendrite arm, it is referred to as **microsegregation**. It

[3] W. G. Pfann, *Trans. AIME*, **194**, 747, 1952.

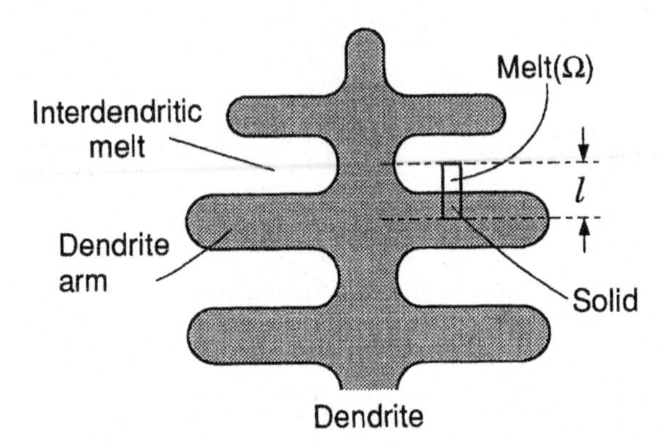

Fig. 9.1-7 Control volume (Ω) for microsegregation analysis in dendritic solidification.

differs from the macrosegregation on the scale of a crystal described in Sections 9.1.1–9.1.4.

Let us consider the macroscopic volume element shown in Fig. 9.1-7, which is equivalent to the macroscopic ones shown in Figs. 9.1-1 and 9.1-2 for Bridgman crystal growth. During solidification the dendrite arms grow thicker at the expense of the melt in between the arms.

The interdendritic melt is assumed uniform in composition. This simplifying assumption seems reasonable in view of the small dendrite arm spacing, for instance $2l \approx 5 \times 10^{-3}$ cm (i.e., 50 μm). With the help of the four simplifying assumptions in Section 9.1.1 and neglecting the curvature of dendrite arms, an equation identical to Eq. [9.1-9] can be derived to describe microsegregation in castings and welds.

9.1.6. Welding: Weld Metal Composition

The composition of a weld metal can significantly affect its properties, such as strength and resistance to cracking during welding, specifically, solidification cracking.[4] It is, therefore, of practical interest to be able to calculate the weld metal composition.

Figure 9.1-8 shows the transverse cross section of a weld $(A_1 + A_2 + A_f)$. The weight fractions of component A in the filler metal, base metal 1 and base metal 2, are w_{Af}, w_{A1}, and w_{A2}, respectively. The filler metal moves forward at the speed V (i.e., the welding speed) and feeds at the rate m_f (e.g., g/s), both of which are constant. Areas A_1 and A_2 represent the portions of the base metals 1 and 2 melted into the weld metal, respectively. Area A_f represents the filler metal added to the weld metal.

For the purpose of discussion let us assume that the weld pool is represented by the control volume Ω under the filler metal. The validity of the analysis here,

[4] S. Kou, *Welding Metallurgy*, Wiley, New York, 1987.

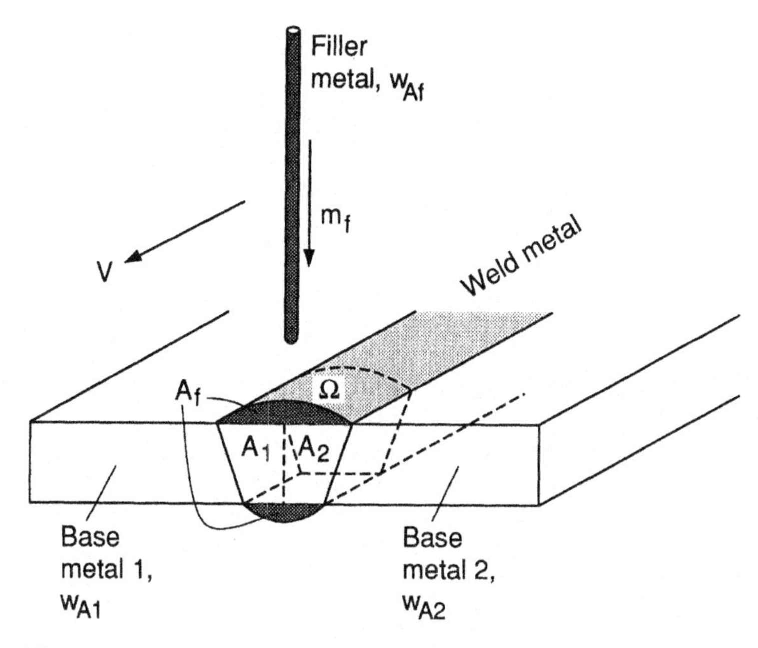

Fig. 9.1-8 Control volume (Ω) for analysis of weld metal composition.

however, is not affected by the shape of the weld pool assumed. For convenience the control volume Ω and the filler metal are considered stationary, while the workpiece moves backward at the speed V. In other words, the two base metals enter the control volume at the velocity V and the weld metal leaves at the same velocity. Because of the significant electromagnetic stirring in the weld pool, the composition of the weld metal w_{Aw} is essentially uniform.

From Eq. [1.2-6]

$$0 = \rho_1 V A_1 + \rho_2 V A_2 + m_f - \rho_w V(A_1 + A_2 + A_f) \qquad \text{(overall mass balance)} \qquad [9.1\text{-}20]$$

where ρ_1, ρ_2, and ρ_w are the densities of base metal 1, base metal 2, and the weld metal, respectively. From Eq. [3.2-7]

$$0 = w_{A1}\rho_1 V A_1 + w_{A2}\rho_2 V A_2 + w_{Af} m_f - w_{Aw}\rho_w V(A_1 + A_2 + A_f) \qquad \text{(species overall mass balance)} \qquad [9.1\text{-}21]$$

Substituting Eq. [9.1-20] into Eq. [9.1-21]

$$\boxed{w_{Aw} = \frac{w_{A1}\rho_1 V A_1 + w_{A2}\rho_2 V A_2 + w_{Af} m_f}{(\rho_1 V A_1 + \rho_2 V A_2 + m_f)}} \qquad [9.1\text{-}22]$$

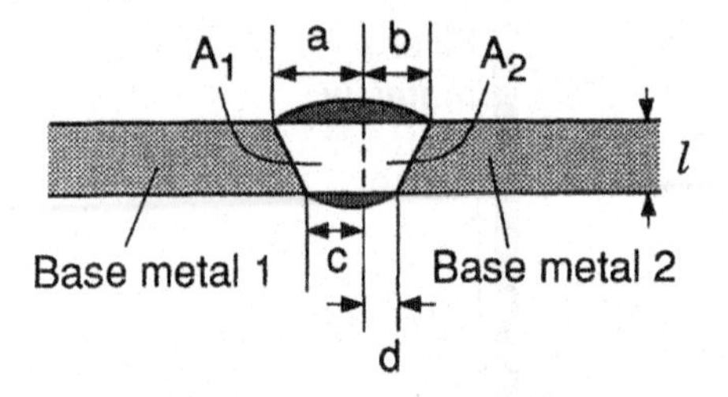

Fig. 9.1-9 Distances from the weld boundaries to the joint (dashed line) before welding.

In Eq. [9.1-22] everything needed to find the weld metal composition w_{Aw} is known except for areas A_1 and A_2. Let us consider the butt joint shown in Fig. 9.1-9. As an approximation we assume that the fusion boundaries are straight lines. This approximation works better when the thickness l is small. By marking before welding, for example, scratching straight lines on the workpiece surfaces at a short distance from and parallel to the joint, distances a, b, c, and d can be easily found. As such

$$A_1 = \left(\frac{a+c}{2}\right) l \qquad [9.1\text{-}23]$$

and

$$A_2 = \left(\frac{b+d}{2}\right) l \qquad [9.1\text{-}24]$$

which can be substituted into Eq. [9.1-22] to find the weld metal composition.

Other joint designs, such as a single-V joint, can be dealt with in a similar way.

9.2. ONE-DIMENSIONAL MASS TRANSFER

9.2.1. Surface Heat Treating: Carburizing

The surface of a carbon steel of an initial carbon level w_{Ai} is to be carburized (Section 6.3.1). The steel is heated to the desired temperature in a furnace. At time $t = 0$ the steel is exposed to a gas mixture containing CO_2 and CO, which keeps its surface at a constant carbon level w_{AS} throughout carburizing, as illustrated in Fig. 9.2-1. The carburized layer is much thinner than the steel itself and the latter can thus be considered semiinfinite.

For simplicity we shall assume that the overall density ρ and the diffusion coefficient of carbon in steel D_A are both constant. Since the steel is stationary and there are no chemical reactions in it, the species continuity equation—

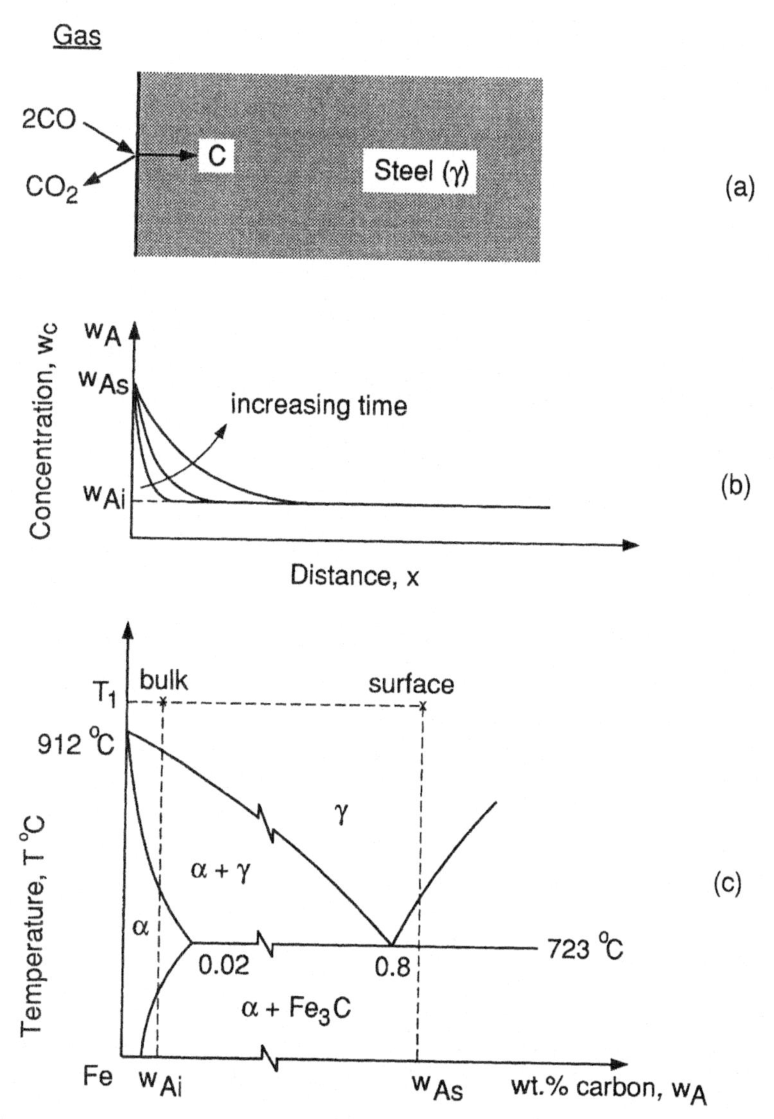

Fig. 9.2-1 Carburization of steel: (*a*) process; (*b*) concentration profile; (*c*) phase diagram.

Eq. [A] of Table 3.3-1—reduces to the following one-dimensional diffusion equation for carbon:

$$\frac{\partial \rho_A}{\partial t} = D_A \frac{\partial^2 \rho_A}{\partial x^2} \tag{9.2-1}$$

Since $\rho_A = \rho w_A$ and ρ is constant, this equation becomes

$$\frac{\partial w_A}{\partial t} = D_A \frac{\partial^2 w_A}{\partial x^2} \qquad [9.2\text{-}2]$$

The initial and boundary conditions are

$$w_A(x, 0) = w_{Ai} \qquad [9.2\text{-}3]$$

$$w_A(0, t) = w_{AS} \qquad [9.2\text{-}4]$$

$$w_A(\infty, t) = w_{Ai} \qquad [9.2\text{-}5]$$

The solution, from Case O of Appendix A, is as follows:

$$\boxed{\frac{w_A - w_{AS}}{w_{Ai} - w_{AS}} = \operatorname{erf}\left[\frac{x}{\sqrt{4D_A t}}\right]} \qquad [9.2\text{-}6]$$

9.2.2. Semiconductor Device Fabrication: Dopant Diffusion

As described in Section 6.3.2, doping by diffusion is usually conducted in two steps: (1) predeposition and (2) drive-in.

Let us first consider the predeposition of a dopant A into an initially dopant-free substrate. Assume that the diffusion coefficient of the dopant D_A and the density ρ are constant, and that the doped layer is much thinner than the substrate, that is, the substrate is semiinfinite. Since $w_{Ai} = 0$, from Eq. [9.2-6]

$$\frac{w_A - w_{AS}}{0 - w_{AS}} = \operatorname{erf}\left[\frac{x}{\sqrt{4D_A t}}\right] \qquad [9.2\text{-}7]$$

or

$$w_A = w_{AS}\left[1 - \operatorname{erf}\left(\frac{x}{\sqrt{4D_A t}}\right)\right] \qquad [9.2\text{-}8]$$

Let M be the amount of dopant predeposited per unit area:

$$M = \int_0^\infty \rho w_A \, dx \qquad [9.2\text{-}9]$$

Substituting Eq. [9.2-8] into [9.2-9]

$$M = \rho w_{AS} \int_0^\infty \left[1 - \operatorname{erf}\left(\frac{x}{\sqrt{4D_A t}}\right)\right] dx \qquad [9.2\text{-}10]$$

or

$$M = \rho w_{AS}\sqrt{4D_A t}\int_0^{\infty}\left[1 - \operatorname{erf}\left(\frac{x}{\sqrt{4D_A t}}\right)\right] d\left(\frac{x}{\sqrt{4D_A t}}\right) \quad [9.2\text{-}11]$$

To determine M, the following integration formula can be used:

$$\int_0^{\infty} [1 - \operatorname{erf}(\eta)]\, d(\eta) = \frac{1}{\sqrt{\pi}} \quad [9.2\text{-}12]$$

Substituting Eq. [9.2-12] into Eq. [9.2-11]

$$\boxed{M = \rho w_{AS}\sqrt{\frac{4D_A t}{\pi}}} \quad [9.2\text{-}13]$$

Let us now consider the drive-in of dopant A. We assume that the depth of diffusion in predeposition is much smaller than that in drive-in, and that the latter is in turn much smaller than the thickness of the substrate. From Eq. [9.2-2]

$$\frac{\partial w_A}{\partial t} = D_A \frac{\partial^2 w_A}{\partial x^2} \quad [9.2\text{-}14]$$

The initial and boundary conditions are

$$w_A(x, 0) = 0 \quad [9.2\text{-}15]$$

$$w_A(\infty, t) = 0 \quad [9.2\text{-}16]$$

and the mass conservation requirement is

$$M = \int_0^{\infty} \rho w_A\, dx \quad [9.2\text{-}17]$$

The solution, from Case Q of Appendix A, is as follows:

$$\boxed{w_A = \frac{M}{\rho\sqrt{\pi D_A t}} \exp\left(\frac{-x^2}{4D_A t}\right)} \quad [9.2\text{-}18]$$

where D_A is the diffusion coefficient of the dopant at the drive-in temperature and t is the drive-in time. Equation [9.2-18] describes the concentration profile of dopant A in the substrate. The amount of the dopant predeposited, M, can be determined from Eq. [9.2-13].

9.2.3. Semiconductor Device Fabrication: Oxidation

Let us consider the growth of a SiO_2 layer on a Si substrate by thermal oxidation (Section 6.3.2). Let c_O and c_S be the molar densities of the oxidant at the gas/SiO_2 and SiO_2/Si interfaces, respectively, as shown in Fig. 9.2.-2. Let D_A be the diffusion coefficient of the oxidant in the SiO_2 layer and l the thickness of the SiO_2 layer. When the oxide layer grows very slowly, diffusion of the oxidant through the layer is not affected by the growth rate. It can be shown (e.g., in Problem 3.3-1), that the concentration profile of the oxidant in the oxide

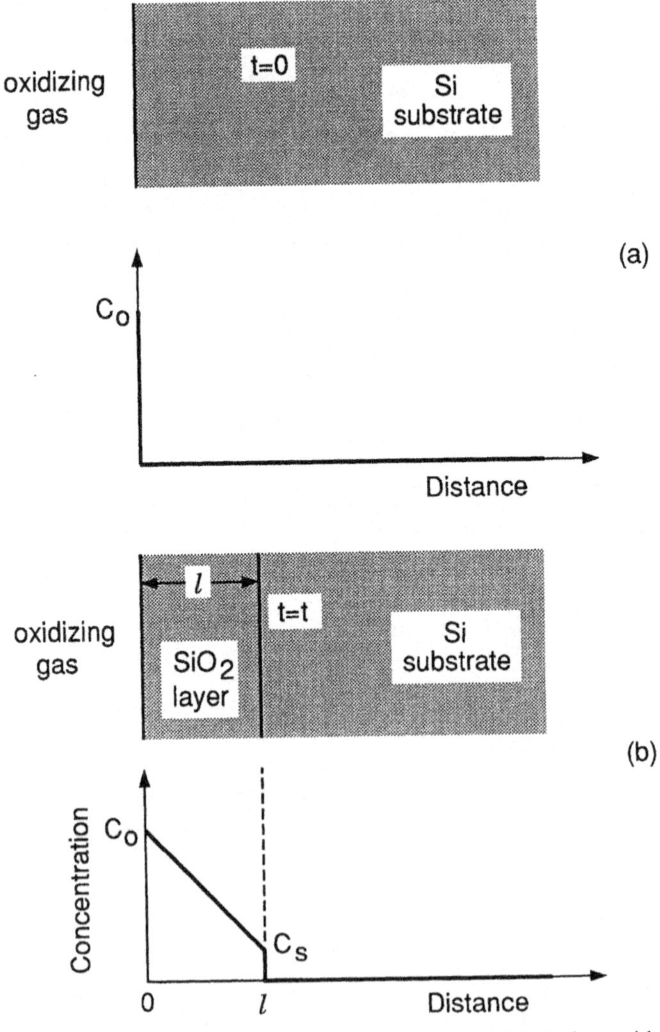

Fig. 9.2-2 Oxidation of silicon: (a) before oxidation; (b) during oxidation.

layer is linear and the diffusion flux j is as follows:

$$j = \frac{D_A(c_O - c_S)}{l} \qquad [9.2\text{-}19]$$

The mass balance of the oxidant at the SiO_2/Si interface requires that

$$\underset{\text{(diffusion rate)}}{\left[\frac{D_A(c_O - c_S)}{l}\right] A_i} = \underset{\text{(reaction rate)}}{(k_1 c_S) A_i} \qquad [9.2\text{-}20]$$

where A_i is the interface area and k_1 the reaction rate constant. The reaction at the SiO_2/Si interface is assumed to be of the first order and proportional to c_S. From Eq. [9.2-20]

$$c_S = \frac{D_A c_O}{k_1 l + D_A} \qquad [9.2\text{-}21]$$

The mass of the SiO_2 layer increases at a rate of $\rho_{SiO_2}(dl/dt)A_i$. To make this possible O_2(g) needs to be consumed, in terms of moles per unit time, at $(\rho_{SiO_2}/M_{SiO_2})(dl/dt)A_i$, where M_{SiO_2} is the molecular weight of SiO_2. If H_2O(g) is the oxidant, however, it needs to be consumed at $2(\rho_{SiO_2}/M_{SiO_2})(dl/dt)A_i$ since only 0.5 mol of SiO_2 is produced per mole of H_2O(g) consumed. Since the LHS of Eq. [9.2-20] represents the rate of consumption of the oxidant, it follows that

$$\left(\frac{m\rho_{SiO_2}}{M_{SiO_2}}\frac{dl}{dt}\right) A_i = (k_1 c_S) A_i \qquad [9.2\text{-}22]$$

where $m = 1$ for O_2(g) and 2 for H_2O(g).

Substituting Eq. [9.2-21] into Eq. [9.2-22]

$$m\frac{\rho_{SiO_2}}{M_{SiO_2}}\frac{dl}{dt} = \frac{k_1 D_A c_O}{k_1 l + D_A} \qquad [9.2\text{-}23]$$

or

$$2l\,dl + \left(\frac{2D_A}{k_1}\right) dl = (2D_A c_O)\left(\frac{M_{SiO_2}}{m\rho_{SiO_2}}\right) dt \qquad [9.2\text{-}24]$$

Integrating from $l = l_0$ (i.e., thickness of the native SiO_2 at surface) at $t = 0$

$$\boxed{l^2 + Al = B(t + \tau_0)} \qquad [9.2\text{-}25]$$

where

$$A = \frac{2D_A}{k_1} \qquad [9.2\text{-}26]$$

$$B = 2D_A c_0 \frac{M_{SiO_2}}{m\rho_{SiO_2}} \qquad [9.2\text{-}27]$$

and

$$\tau_0 = \frac{l_0^2 + 2D_A l_0/k_1}{2D_A c_0 M_{SiO_2}/(m\rho_{SiO_2})} \qquad [9.2\text{-}28]$$

Equation [9.2-25] was derived by Deal and Grove.[5]

For short oxidation times, $l \ll D_A/k_1$ and the first LHS term of Eq. [9.2-24] can be neglected. The following linear equation is obtained on integration:

$$l = \frac{B}{A}(t + \tau_0) \qquad [9.2\text{-}29]$$

and B/A is thus called the **linear rate constant**. For long oxidation times, on the other hand, $l \gg D_A/k_1$ and the second term in Eq. [9.2-24] can be neglected. The following parabolic equation is obtained on integration:

$$l^2 = B(t + \tau_0) \qquad [9.2\text{-}30]$$

and B is thus called the **parabolic rate constant**.

It is interesting to point out that the analysis in this section can also be applied to the formation of surface tarnish layers on metals. In fact, a parabolic equation similar to Eq. [9.2-30], with $\tau_0 = 0$, has also been derived. The oxide layer is thought to consist of the cations of the metal substrate and anions of oxygen. Because of the large ionic radius of the oxygen anions, the diffusing species is believed to be the metal cations. In other words, the mechanism of oxidation is that the metal cations diffuse outward to the gas/oxide interface and react with the oxygen gas.[6]

9.2.4. Semiconductor Device Fabrication: Chemical Vapor Deposition

The single-wafer reactor shown in Fig. 9.2-3*a* has been considered by Middleman and Hochberg[7] to help deposit a thin film of uniform thickness on a wafer;

[5] B. E. Deal and A. S. Grove, *J. Appl. Phys.*, **36**, 3370, 1965.

[6] G. H. Geiger and D. R. Poirier, *Transport Phenomena in Metallurgy*, Addison-Wesley, Reading, MA, 1973, p. 502.

[7] S. Middleman and A. K. Hochberg, *Process Engineering Analysis in Semiconductor Device Fabrication*, McGraw-Hill, New York, 1993, p. 560.

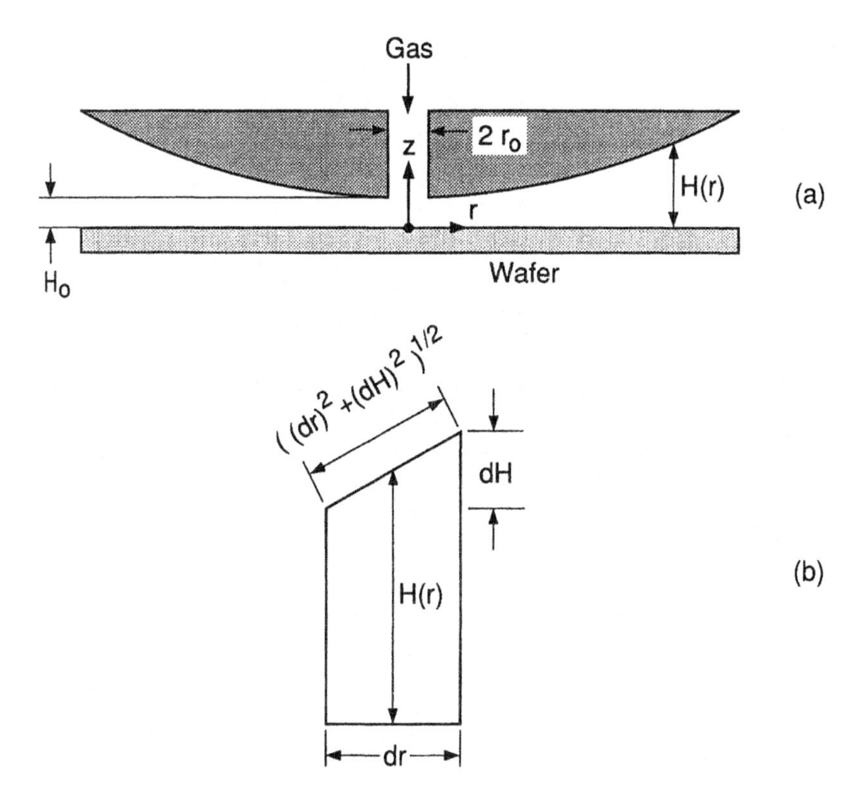

Fig. 9.2-3 Chemical vapor deposition: (*a*) single wafer reactor; (*b*) control volume.

let us assume that the reactor is isothermal and that the gas is uniform in molar density. For simplicity, let us assume that the variations in the gas composition and v_r in the z direction are negligible as compared to those in the r direction. Also, mass transfer in the r direction by diffusion is negligible as compared to that by convection.

Let us consider the following reactions, which are assumed linear in kinetics:

$$A(\text{g}) \xrightarrow{k_1} B(\text{g}) \qquad [9.2\text{-}31]$$

$$B(\text{g}) \xrightarrow{k_2} C(\text{s}) + D(\text{g}) \qquad [9.2\text{-}32]$$

The gas carrying species A at the molar concentration of c_A° is introduced into the reactor at the volume flow rate Q. It decomposes into species B at the rate of $k_1 c_A$ per unit volume of the gas. Species B then reacts at the solid surfaces to form a film of species C at the rate of $k_2 C_B$ per unit surface area. Let us consider the differential volume in Fig. 9.2-3*b*. The consumption of species B per unit volume is

$$\frac{k_2 c_B[2\pi r\,dr + 2\pi r\sqrt{(dr)^2 + (dH)^2}]}{(2\pi r\,dr)H}$$

or

$$\frac{k_2 c_B}{H}\left[1 + \sqrt{1 + \left(\frac{dH}{dr}\right)^2}\right]$$

where the first and second terms represent the reactions at the wafer and the roof, respectively.

The species continuity equations, Eq. [B] of Table 3.3-2, reduce to the following equations at the steady state:

$$v_r \frac{dc_A}{dr} = -k_1 c_A \qquad [9.2\text{-}33]$$

and

$$v_r \frac{dc_B}{dr} = k_1 c_A - \frac{k_2 c_B}{H}\left[1 + \sqrt{1 + \left(\frac{dH}{dr}\right)^2}\right] \qquad [9.2\text{-}34]$$

The first and second RHS terms of Eq. [9.2-34] represent the rates of generation and consumption of species B, respectively.

In order to have a uniform film thickness c_B must be uniform in the radial direction; in other words, $c_B = c_B^\circ$ for all r. As such, from Eq. [9.2-34]

$$c_A = \frac{k_2 c_B^\circ}{k_1 H}\left[1 + \sqrt{1 + \left(\frac{dH}{dr}\right)^2}\right] \qquad [9.2\text{-}35]$$

If dH/dr is neglected, Eq. [9.2-35] becomes identical to the equation derived by Middleman and Hochberg.[8]

Since v_r is uniform in the z direction, the volume flow rate

$$Q = 2\pi r H v_r \qquad [9.2\text{-}36]$$

Substituting Eq. [9.2-36] into Eq. [9.2-33]

$$\frac{dc_A}{dr} = -\frac{2\pi r H k_1 c_A}{Q} \qquad [9.2\text{-}37]$$

[8] S. Middleman and A. K. Hochberg, ibid.

On rearranging and integrating

$$\int_{c_A^\circ}^{c_A} \frac{dc_A}{c_A} = -\frac{\pi k_1}{Q}\int_{r_0}^{r} 2r\,H\,dr \qquad [9.2\text{-}38]$$

or

$$\ln\left(\frac{c_A}{c_A^\circ}\right) = \frac{\pi k_1}{Q}\int_{r}^{r_0} 2r\,H\,dr \qquad [9.2\text{-}39]$$

Substituting Eq. [9.2-35] into Eq. [9.2-39]

$$\boxed{\frac{k_2 c_B^\circ}{k_1 c_A^\circ H}\left[1+\sqrt{1+\left(\frac{dH}{dr}\right)^2}\right] = \exp\left[\frac{\pi k_1}{Q}\int_{r}^{r_0} 2r\,H\,dr\right]} \qquad [9.2\text{-}40]$$

This equation is nonlinear and has to be solved numerically to find H as a function of r.

9.3. ONE-DIMENSIONAL DIFFUSION WITH SOLIDIFICATION

9.3.1. Crystal Growth: Bridgman

In Section 9.1.1 we considered Bridgman crystal growth in the extreme case where the melt is uniform in composition, that is, with complete mixing in the melt. Here, we shall consider the other extreme case where mass transfer in the melt is by diffusion only, that is, where there is no convection in the melt. Melt convection can be suppressed in crystal growth under microgravity. It can also be suppressed by using a steady magnetic field if the melt is a conductor such as a semiconductor or metal.[9]

Figure 9.3-1, which is the counterpart of Fig. 9.1-3 for the case of complete mixing, illustrates the evolution of w_{AS}^* and w_{AL}^* during crystal growth. The values w_{AS}^* and w_{AL}^*, which are respectively the compositions of the crystal and the melt at the crystal/melt interface, are related to each other through the equilibrium segregation coefficient defined by $k_0 = w_{AS}^*/w_{AL}^*$.

As shown in Fig. 9.3-1*b*, the initial compositions of the melt and the crystal at the crystal/melt interface are respectively w_{A0} and $k_0 w_{A0}$, since $w_{AS}^* = k_0 w_{AL}^* = k_0 w_{A0}$. Since $k_0 < 1$, the crystal cannot accommodate as much dopant as the melt. As such, the dopant is rejected by the crystal into the melt, causing a dopant-rich (diffusion boundary) layer in the melt ahead of the crystal/melt

[9] H. P. Utech and M. C. Flemings, *J. Appl. Phys.*, **37**, 2021, 1966.

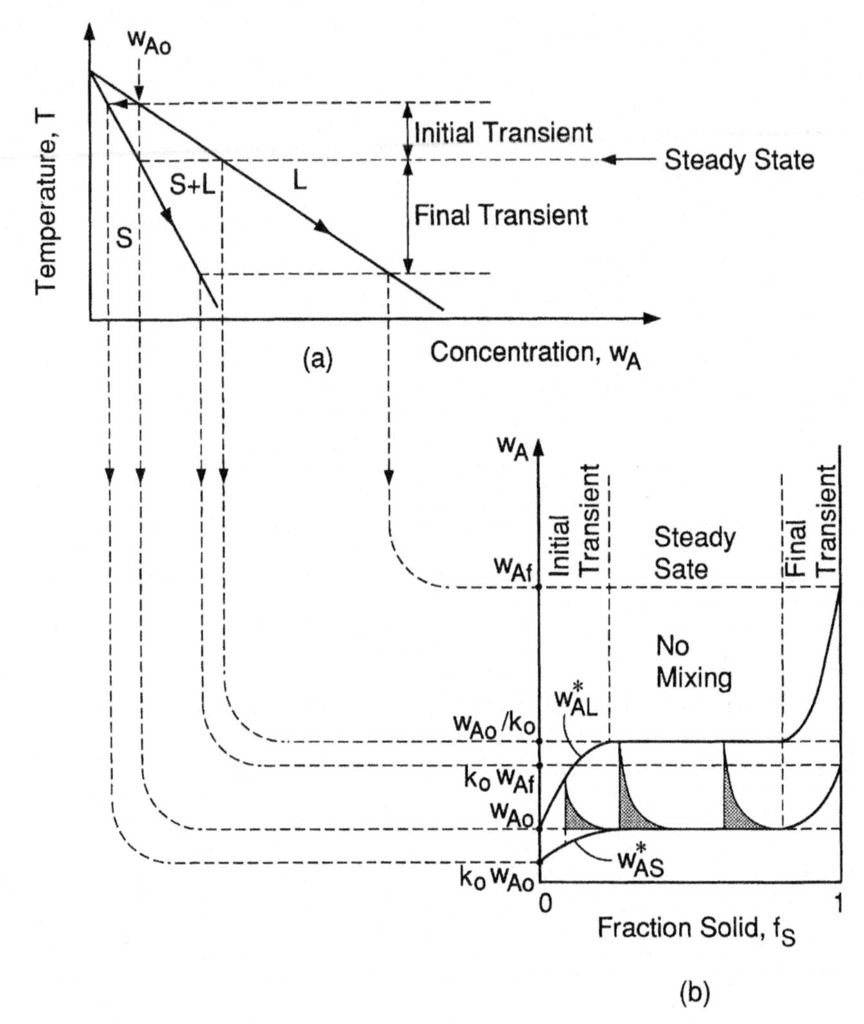

Fig. 9.3-1 Dopant segregation under no mixing in melt: (*a*) phase diagram; (*b*) concentration profiles.

interface. This causes w_{AL}^* to rise, which, in turn, causes w_{AS}^* to rise since $w_{AS}^* = k_0 w_{AL}^*$.

As w_{AL}^* reaches w_{A0}/k_0, w_{AS}^* reaches w_{A0}, which is the composition of the bulk melt from which the crystal grows. As such, w_{AL}^* and w_{AS}^* remain constant as crystal growth proceeds, that is, a steady state is reached. This continues until the dopant-rich layer impinges on the end of the melt, whereupon w_{AL}^* and hence w_{AS}^* begin to rise until crystal growth is completed.

Let us first consider the dopant distribution in the melt and the thickness of the dopant-rich layer. The steady-state composition profile of the dopant-rich layer is illustrated in Fig. 9.3-2. In addition to the four simplifying assumptions

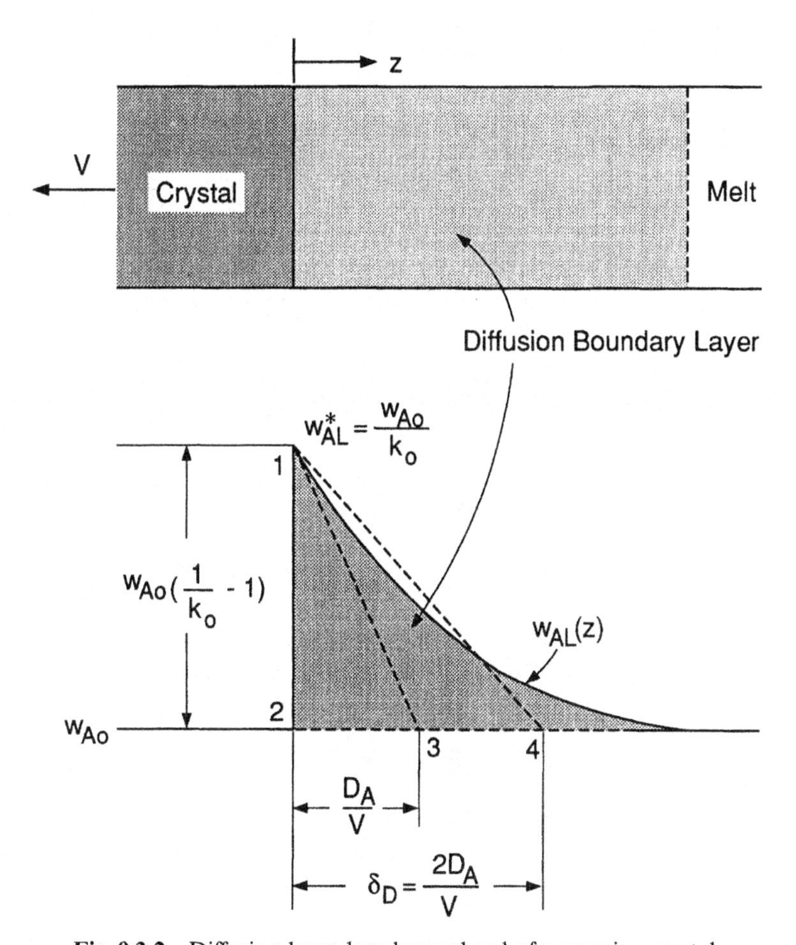

Fig. 9.3-2 Diffusion boundary layer ahead of a growing crystal.

in Section 9.1.1, it is assumed that there is no convection in the melt. For convenience, the crystal/melt interface is considered stationary while the material moves at a constant velocity V (> 0) in the negative z direction.

At the steady state and with no chemical reactions the species continuity equation, Eq. [A] of Table 3.3-1, reduces to

$$-V\frac{d\rho_{AL}}{dz} = D_A\frac{d^2\rho_{AL}}{dz^2} \tag{9.3-1}$$

Since ρ is constant and $\rho_{AL} = \rho w_{AL}$, Eq. [9.3-1] becomes

$$\frac{d^2 w_{AL}}{dz^2} = -\frac{V}{D_A}\frac{dw_{AL}}{dz} \tag{9.3-2}$$

The boundary conditions are as follows:

$$w_{AL} = w^*_{AL} = \frac{w_{A0}}{k_0} \quad \text{at} \quad z = 0 \qquad [9.3\text{-}3]$$

$$w_{AL} = w_{A0} \quad \text{for} \quad z = \infty \qquad [9.3\text{-}4]$$

The solution, from Case E of Appendix A, is as follows:

$$\frac{w_{AL} - (w_{A0}/k_0)}{w_{A0} - (w_{A0}/k_0)} = \frac{e^{-Vz/D_A} - 1}{e^{-V\infty/D_L} - 1} \qquad [9.3\text{-}5]$$

or

$$\boxed{w_{AL} = w_{A0}\left[1 + \left(\frac{1}{k_0} - 1\right)e^{-Vz/D_A}\right]} \qquad [9.3\text{-}6]$$

This equation, which describes the steady-state concentration distribution in the diffusion boundary layer, has been derived by Tiller et al.[10]

Equation [9.3-6] can be differentiated with respect to z to give

$$\frac{dw_{AL}}{dz} = w_{A0}\left(\frac{1}{k_0} - 1\right)\left(\frac{-V}{D_A}\right)e^{-Vz/D_A} \qquad [9.3\text{-}7]$$

and, at the crystal/melt interface

$$\left.\frac{dw_{AL}}{dz}\right|_{z=0} = w_{A0}\left(\frac{1}{k_0} - 1\right)\left(\frac{-V}{D_A}\right) = \frac{w_{A0}[(1/k_0) - 1]}{0 - (D_A/V)} \qquad [9.3\text{-}8]$$

Therefore, D_A/V is the length of the line segment 2–3 shown in Fig. 9.3-2; line segment 1–3 is the tangent to the concentration profile at the crystal/melt interface. As will be shown below, D_A/V appears to be half the effective thickness of the diffusion boundary layer δ_D.

The bottom shaded area in Fig. 9.3-2 represents the dopant rejected by the growing crystal. Following the approach of Kurz and Fisher,[11] the triangle $\Delta(1\text{–}2\text{–}4)$ is meant to have the same area as the shaded area, so that the line segment 2–4 can be taken as the effective thickness of the diffusion boundary

[10] W. A. Tiller, K. A. Jackson, J. W. Rutter, and B. Chalmer, *Acta Metallurgica*, **1**, 429, 1953.

[11] W. Kurz and D. J. Fisher, *Fundamentals of Solidification*, Trans Tech Publications, Switzerland, 1986, p. 225.

layer δ_D. In other words

$$\int_0^\infty (w_{AL} - w_{A0})\,dz = \Delta(1\text{–}2\text{–}4) = \frac{1}{2}\left[w_{A0}\left(\frac{1}{k_0} - 1\right)\right]\delta_D \qquad [9.3\text{-}9]$$

From Eq. [9.3-6]

$$\int_0^\infty (w_{AL} - w_{A0})\,dz = w_{A0}\left(\frac{1}{k_0} - 1\right)\int_0^\infty e^{-Vz/D_A}\,dz$$

$$= w_{A0}\left(\frac{1}{k_0} - 1\right)\left(-\frac{D_A}{V}\right)(0 - 1) \qquad [9.3\text{-}10]$$

From Eq. [9.3-9] and [9.3-10]

$$\boxed{\delta_D = \frac{2D_A}{V}} \qquad [9.3\text{-}11]$$

Therefore, the effective thickness of the diffusion boundary layer is $2D_A/V$.

Let us now consider the initial transient. The species continuity equation, Eq. [A] of Table 3.3-1, reduces to

$$\frac{\partial w_{AL}}{\partial t} - V\frac{\partial w_{AL}}{\partial z} = D_A\frac{\partial^2 w_{AL}}{\partial z^2} \qquad [9.3\text{-}12]$$

A rigorous solution for this equation has been solved by Smith et al.,[12] which is too tedious to be presented here. Instead, an approximate approach similar to that of Kurz and Fisher[13] will be considered here. In this approach, the diffusion boundary layer is considered as the control volume Ω. As shown in Fig. 9.3-3, Ω is stationary while the material moves through it at a constant velocity V in the negative z direction. Let us now apply the overall mass balance of component A given by Eq. [3.2-7]. Mass generation R_A can be dropped since no chemical reactions are involved here. Diffusion flux J_A can also be dropped. Diffusion of A across the left surface of Ω is neglected because it is assumed that there is no diffusion in the crystal. Diffusion of A across the right surface is also neglected because of the negligible local concentration gradient.

As illustrated in Fig. 9.3-3, the melt of concentration w_{A0} enters Ω through its right surface and the crystal concentration w_{AS}^* leaves Ω through its left surface, both at the velocity of V. Let the area of these surfaces, that is, the

[12] V. G. Smith, W. A. Tiller, and J. W. Rutter, *Can. J. Phys.*, **33**, 723, 1955.

[13] W. Kurz and D. J. Fisher, *Fundamentasls of Solidification*, Trans Tech Publications, Switzerland, 1986, p. 225.

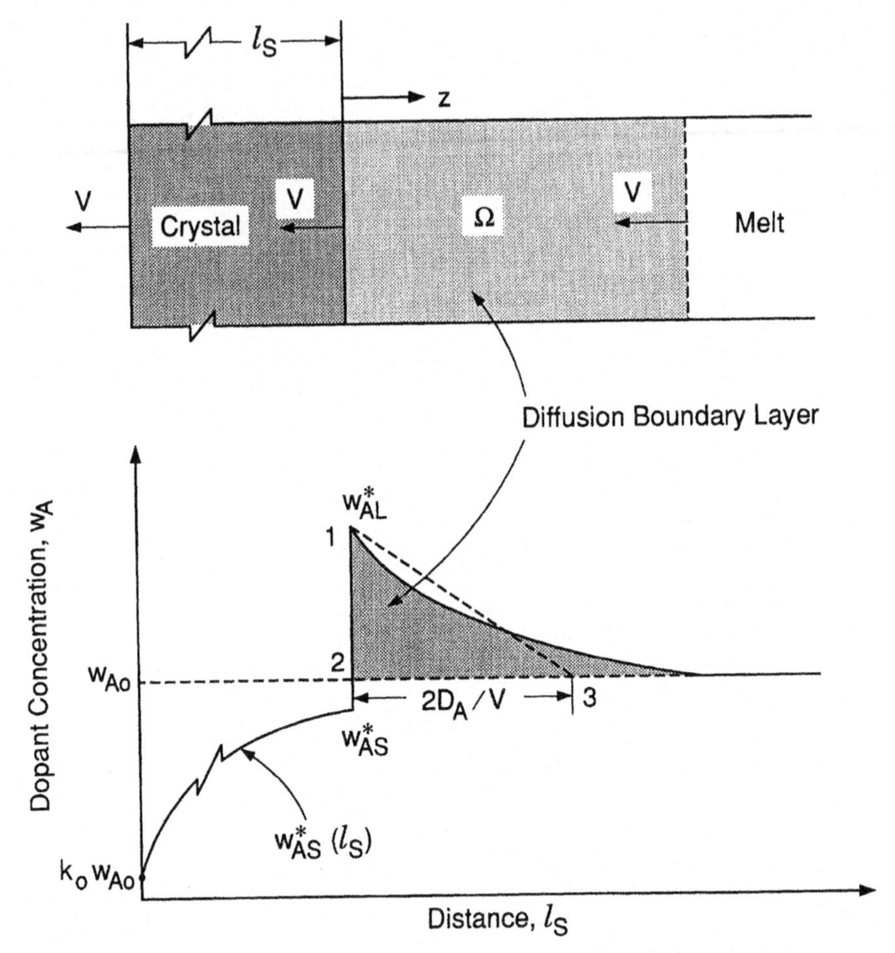

Fig. 9.3-3 Diffusion boundary layer in Bridgman crystal growth.

cross-sectional area of the crystal, be A_S. As such, from Eq. [3.2-7]

$$\frac{dM_{A\Omega}}{dt} = \frac{d}{dt}\left[\iiint_\Omega \rho w_{AL}\, d\Omega\right] = \rho w_{A0} V A_S - \rho w^*_{AS} V A_S \qquad [9.3\text{-}13]$$

Since

$$d\Omega = A_S\, dz \qquad [9.3\text{-}14]$$

Equation [9.3-13] becomes

$$\frac{d}{dt}\left[\int_0^\infty w_{AL}\, dz\right] = w_{A0} V - w^*_{AS} V \qquad [9.3\text{-}15]$$

Since w_{A0} is time-independent

$$\frac{d}{dt}\left[\int_0^\infty (w_{AL} - w_{A0})\,dz\right] = (w_{A0} - w_{AS}^*)\,V \qquad [9.3\text{-}16]$$

The growth rate V is related to the length of the growing crystal as follows:

$$V = \frac{dl_s}{dt} \qquad [9.3\text{-}17]$$

Therefore, Eq. [9.3-16] becomes

$$d\left[\int_0^\infty (w_{AL} - w_{A0})\,dz\right] = (w_{A0} - w_{AS}^*)\,dl_s \qquad [9.3\text{-}18]$$

As an approximation, we assume that the thickness of the diffusion boundary layer δ_D remains essentially constant throughout crystal growth, except, of course, for the final transient. According to Eq. [9.3-11]

$$\delta_D = \frac{2D_A}{V} \qquad [9.3\text{-}19]$$

and, as illustrated in Fig. 9.3-3

$$\int_0^\infty (w_{AL} - w_{A0})\,dz = \Delta(1\text{–}2\text{–}3) = \tfrac{1}{2}(w_{AL}^* - w_{A0})\,\delta_D \qquad [9.3\text{-}20]$$

Substituting Eq. [9.3-20] into Eq. [9.3-18]

$$\tfrac{1}{2}\,\delta_D\,dw_{AL}^* = (w_{A0} - w_{AS}^*)\,dl_s \qquad [9.3\text{-}21]$$

Since $w_{AL}^* = w_{AS}^*/k_0$

$$\frac{dw_{AS}^*}{dl_s} = -\frac{2k_0}{\delta_D}(w_{AS}^* - w_{A0}) \qquad [9.3\text{-}22]$$

The initial condition is

$$w_{AS}^* = k_0 w_{A0} \quad \text{at} \quad l_s = 0 \qquad [9.3\text{-}23]$$

The solution, from Case A of Appendix A, is as follows:

$$\frac{w_{AS}^* - w_{A0}}{k_0 w_{A0} - w_{A0}} = e^{-2k_0 l_s/\delta_D} \qquad [9.3\text{-}24]$$

or, from Eq. [9.3-19],

$$w_{AS}^* = w_{A0}[1-(1-k_0)e^{-2k_0l_S/\delta_D}] = w_{A0}[1-(1-k_0)e^{-k_0Vl_S/D_A}] \quad [9.3\text{-}25]$$

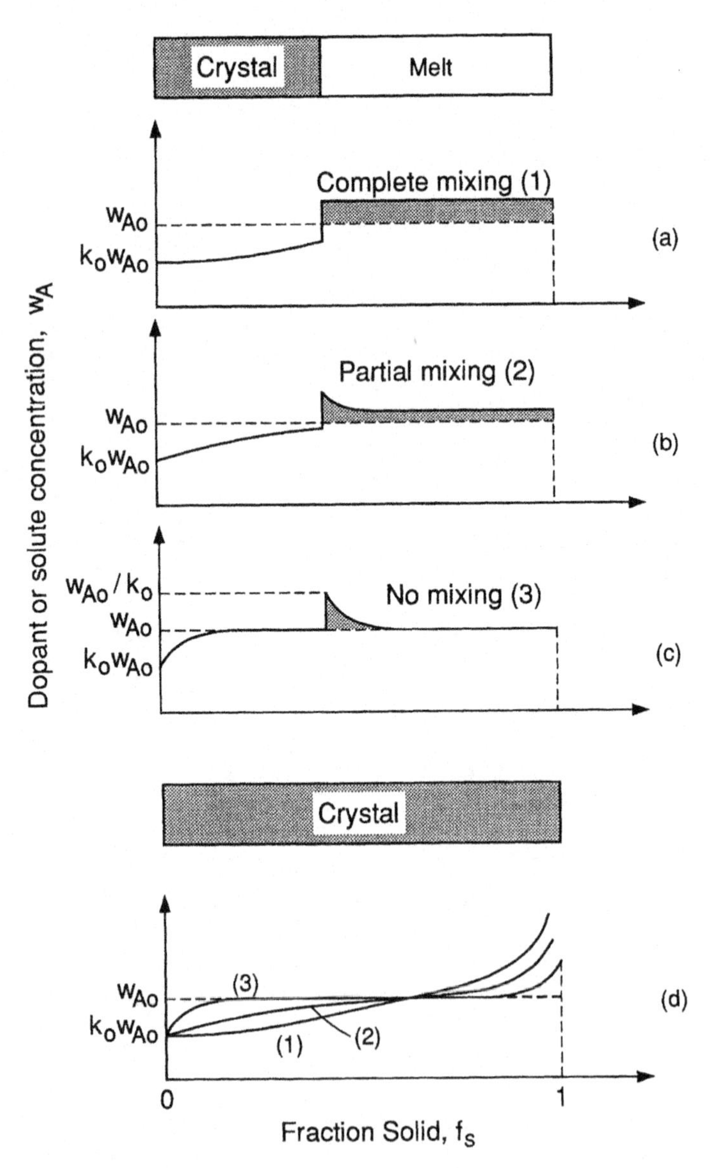

Fig. 9.3-4 Solute redistribution in crystal growth: (*a*) complete mixing in melt; (*b*) partial mixing; (*c*) no mixing; (*d*) dopant segregation in resultant crystals.

Equation [9.3-25], which is identical to that derived by Pohl,[14] describes the distribution of dopant A in the initial portion of the single crystal, where l_S is the distance from the starting point of the crystal.

Figure 9.3-4 shows complete, partial, and no mixing of the melt during Bridgman crystal growth and the dopant distributions in the resultant crystals. As shown in Fig. 9.3-4*d*, the severity of segregation increases with the extent of mixing in the melt.

9.3.2. Crystal Growth: Zone Melting with Segregation Control

Zone leveling is a well-known technique for reducing dopant segregation in zone-melting crystal growth. As shown in Fig. 9.1-6, in conventional zone-melting growth the concentration at the growth front (w_{AS}^*) starts from $k_0 w_{A0}$ and approaches w_{A0} rather slowly. With zone leveling, however, extra dopant is added to the melt to produce a uniform melt concentration of w_{A0}/k_0 at the beginning: $w_{AL}^* = w_{AL} = w_{A0}/k_0$. This is illustrated by the broken line in Fig. 9.3-5.[15] The zone length l_L should be kept constant so that the melt composition does not change during crystal growth. Since $w_{AS}^* = w_{A0}$, Eq. [9.1-17] reduces to

$$\frac{dw_{AS}^*}{dl_s} = 0 \qquad [9.3\text{-}26]$$

Therefore, as crystal growth proceeds, w_{AS}^* remains constant at w_{A0}. In other words, the initial transient region is eliminated, as illustrated by the solid line in Fig. 9.3-5. This steady-state growth condition continues until the melt zone reaches the end of the sample, whereupon w_{AL}^* and hence w_{AS}^* increase rapidly.

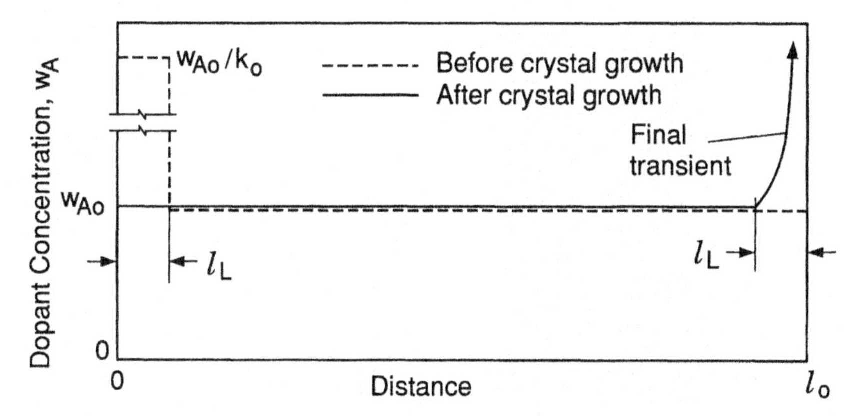

Fig. 9.3-5 Dopant segregation control in zone-melting crystal growth.

[14] R. G. Pohl, *J. Appl. Phys.*, **25**, 669, 1954.

[15] W. G. Pfann, *Zone Melting*, 3rd ed., Wiley, New York, 1966, pp. 201–204.

9.3.3. Crystal Growth: Czochralski with Segregation Control

The concept of zone leveling can be applied to Czochralski crystal growth; one example is the so-called floating-crucible method. As shown in Fig. 9.3-6*a*, the floating crucible has an orifice at the bottom. The melt inside the floating crucible – the growth melt – is predoped to w_{AO}/k_O, while the replenishing melt in the outer crucible w_{AO}. This allows the crystal to grow at the targeted

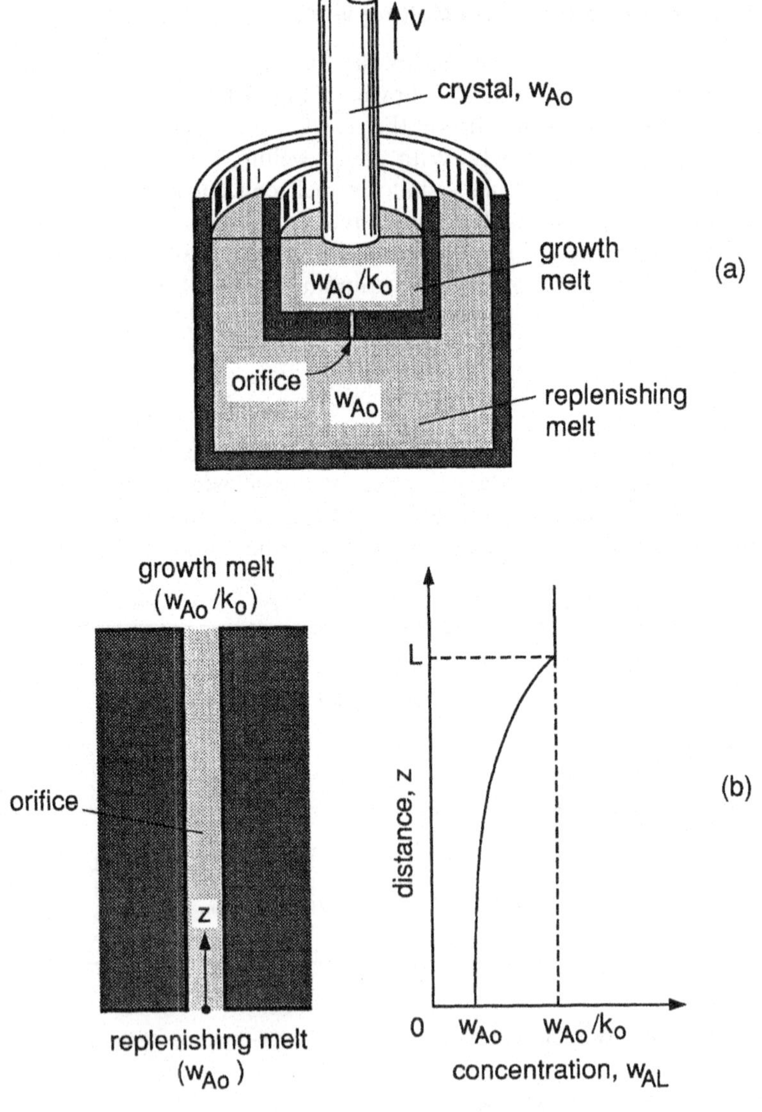

Fig. 9.3-6 Dopant segregation control in Czochralski crystal growth.

composition of w_{A0} from the beginning. As the melt inside the floating crucible is being consumed by crystal growth, it is replenished by the melt entering through the orifice. In this way the melt volume in the floating crucible remains constant during crystal growth. The growth and replenishing melts are, in fact, equivalent to the molten zone and the feed in zone-melting crystal growth, respectively.

To ensure proper segregation control, dopant backdiffusion from the growth melt into the replenishing melt must be suppressed so that their compositions can remain constant during crystal growth. In zone-melting crystal growth this backdiffusion is not a concern since the diffusion coefficient in solids (on the order of 10^{-7} cm²/s) is much smaller than that in liquids (on the order of 10^{-5} cm²/s).

For simplicity let us assume plug flow in the orifice, that is, $v_z = V_0$, where V_0 is the average fluid velocity in the orifice. Assuming constant density and the steady state, the species equation of continuity, Eq. [B] of Table 3.3-1, can be reduced to

$$\frac{d^2 w_{AL}}{dz^2} = \frac{V_0}{D_A}\frac{dw_{AL}}{dz} \qquad [9.3\text{-}27]$$

As illustrated in Fig. 9.3-6*b*, the boundary conditions are as follows:

$$w_{AL} = w_{A0} \quad \text{at} \quad z = 0 \qquad [9.3\text{-}28]$$

$$w_{AL} = \frac{w_{A0}}{k_0} \quad \text{at} \quad z = L \qquad [9.3\text{-}29]$$

These boundary conditions, in fact, require complete mixing within both melts and negligible diffusion between the two. The solution, from Case E of Appendix A, is as follows:

$$\boxed{\frac{w_{AL} - w_{A0}}{(w_{A0}/k_0) - w_{A0}} = \frac{e^{V_0 z/D_A} - 1}{e^{V_0 L/D_A} - 1}} \qquad [9.3\text{-}30]$$

and so

$$j_{Az} = -\rho_L D_A \frac{dw_{AL}}{dz} = \frac{\rho_L w_{A0} V_0 (k_0 - 1) e^{V_0 z/D_A}}{k_0 (e^{V_0 L/D_A} - 1)} \qquad [9.3\text{-}31]$$

From Eq. [3.1-13] the mass flux through the orifice

$$n_{Az} = \rho_L w_{AL} V_0 + j_{Az} \qquad [9.3\text{-}32]$$

Substituting Eqs. [9.3-30] and [9.3-31]

$$n_{Az} = \rho_L w_{A0} V_0 \frac{k_0 - e^{-V_0 L/D_A}}{k_0 (1 - e^{-V_0 L/D_A})} \qquad [9.3\text{-}33]$$

As the steady state the dopant entering the growth melt through the orifice is consumed by the growing crystal; i.e.,

$$n_{Az}A_0 = \rho_S w_{A0} A_S V_p \qquad [9.3\text{-}34]$$

where A_0 is the orifice cross-sectional area, A_S the crystal cross-sectional area, and V_p the crystal production speed. Substituting Eqs. [7.1-7] and [9.3-33] into Eq. [9.3-34]

$$\left(\frac{1}{k_0} - 1\right) e^{-V_0 L/D_A} = 0 \qquad [9.3\text{-}35]$$

Taking e^{-5} (i.e., 0.0067) as an approximation for zero (i.e., the RHS)

$$\boxed{\frac{V_0 L}{D_A} = 5 + \ln\left|\left(\frac{1}{k_0} - 1\right)\right|} \qquad [9.3\text{-}36]$$

where the melt velocity in the orifice can be calculated from Eq. [7.1-8].

Equations [9.3-35] and [9.3-36] differ from those of Blackwell[16] in that $(1/k_0 - 1)$ is not included in the latter. For a large crytal growing at a high speed, a small L (e.g., 3 mm) is sufficient to suppress dopant diffusion. For a small crystal (e.g., ≤ 1 cm diameter) growing at low speeds (e.g., 1 mm/h), L must be substantial (e.g., 2–3 cm).

In reality several facts must be considered when using the floating-crucible technique. First, the orifice must be narrow enough to suppress backdiffusion of the dopant into the replenishing melt, but wide enough to permit the necessary melt inflow. Since most semiconductor melts do not wet their crucible materials, they have difficulties going through a narrow orifice (e.g., ≤ 3 mm diameter). Second, during waiting (for thermal equilibrium in the growth system) and even during seeding and necking, backdiffusion may occur. Third, the growth melt may be forced out of the floating crucible when the seed is dipped into the melt to initiate growth, especially if the growth melt is small in volume. To avoid these problems, an extra long passageway between the two melts is preferred. This is especially true when growing alloy semiconductors (e.g., $In_xGa_{1-x}Sb$, Si_xGe_{1-x}), where the pulling rates are very slow (e.g., 1 mm/h).

Lin and Kou have designed floating crucibles with an extra long melt passageway.[17] Figure 9.3-7 shows one such floating crucible and the resultant dopant distributions in a Cd-doped InSb crystal.

[16] G. R. Blackwell, *Solid-State Electronics*, **7**, 105, 1964.
[17] M. H. Lin and S. Kou, *J. Crystal Growth*, **152**, 256, 1995.

9.3.4. Crystal Growth: Bridgman with Segregation Control

The concept of zone leveling can also be applied to vertical Bridgman crystal growth. The submerged-heater technique illustrated in Fig. 9.3-8a[18] was originally intended to reduce convection near the growth front, the submerged heater acting as a convection baffle[19]. Recently, it has been coupled with zone leveling for segregation control. The melt is separated by a submerged heater or baffle

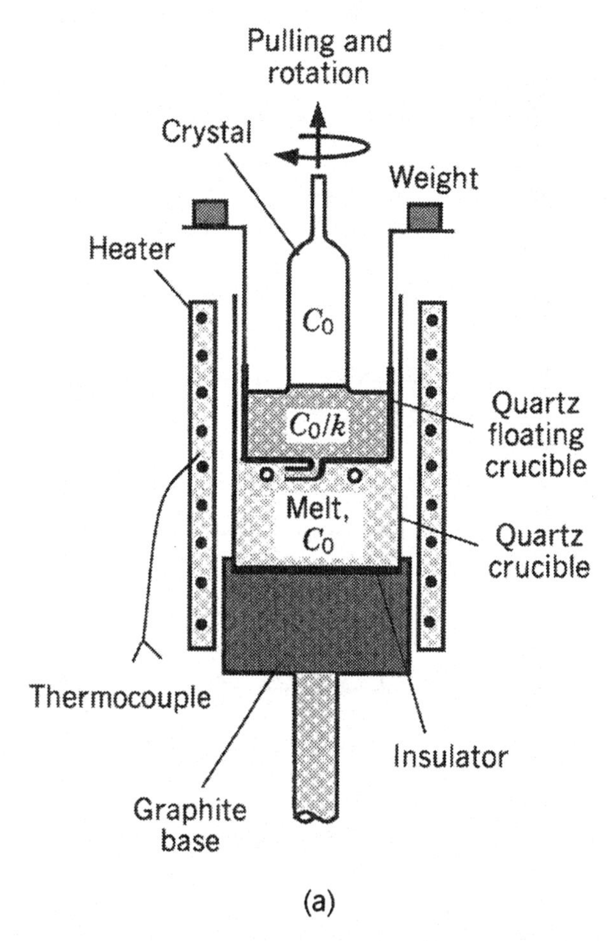

Fig. 9.3-7 Dopant segregation control in Czochralski pulling of InSb: (*a*) floating quartz crucible with a spiral inlet tube at bottom; (*b*) axial dopant distribution; (*c*) radial dopant distribution. (From M. H. Lin and S. Kou, *Journal of Crystal Growth*, **152**, 1995, p. 256.).

[18] A. G. Ostrogorsky, H. I. Sell, S. Scharl, and G. Muller, *J. Crystal Growth*, **128**, 201, 1993.
[19] A. G. Ostrogorsky, *J. Crystal Growth*, **104**, 233, 1990.

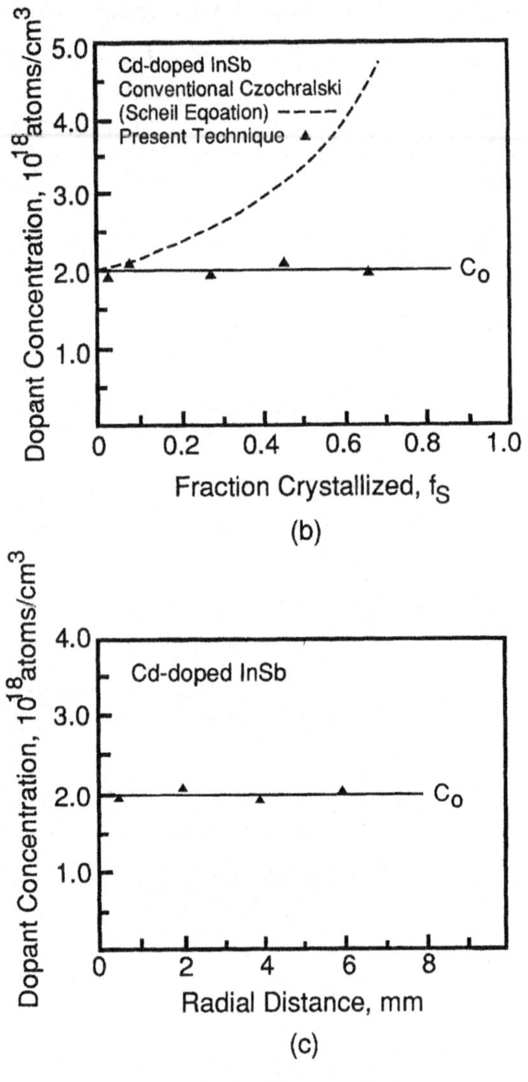

Fig. 9.3-7 *(Continued).*

into a top melt of one composition and an enclosed melt of a different composition. Reduction in dopant segregation has been observed.

To illustrate the concept of zone leveling in vertical Bridgman, let us consider the arrangement shown in Fig. 9.3-8*b*.[20] As shown, the melt is separated by a diffusion baffle into a growth melt at w_{A0}/k_0 and a replenishing melt at w_{A0}.

[20] Y. Tao and S. Kou, *J. Crystal Growth*, in press.

This arrangement has a long passageway to suppress dopant diffusion between the two melts.

The length of the passageway L required for suppression of backdiffusion can be determined using Eq. [9.3-36], which has been derived previously for floating-crucible Czochralski. To determine the velocity V_0 required in the equation, let us consider the growth melt as the control volume Ω. From Eq. [1.2-6]

$$0 = \rho_L V_0 A_0 - \rho_S V A_S \qquad [9.3\text{-}37]$$

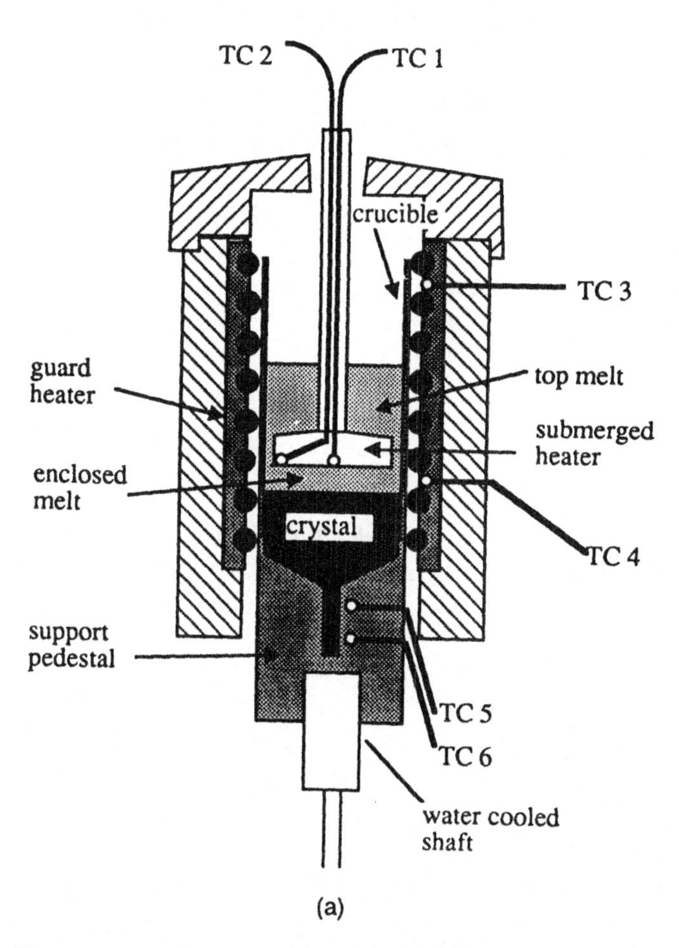

Fig. 9.3-8 Dopant segregation control in vertical Bridgman: (*a*) submerged heater or baffle. (From A. Ostrogorski, "Convection and Segregation during Growth of Ge and InSb by SHM," *Journal of Crystal Growth*, **128**, 1993, p. 201); (*b*) a similar technique (from Y. Tao and S. Kou, *Journal of Crystal Growth*, in press).

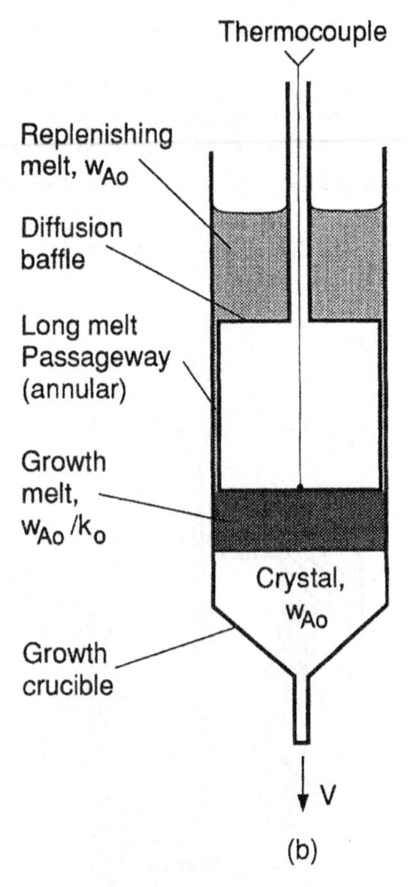

Fig. 9.3-8 (*Continued*).

and so

$$\frac{V_0}{V} = \frac{\rho_S A_S}{\rho_L A_0} \qquad [9.3\text{-}38]$$

where A_S is the crystal cross-sectional area and A_0 the annular area between the crucible and the baffle.

It is often desirable to initiate melt replenishing when the crystal grows to its shoulder, which is the top of the conical expansion. This is because the melt dopant concentration at this point can be easily predicted by Scheil's equation (Eq. [9.1-9]) and because the conical portion of a Bridgman crystal is not used, anyway. The time interval t from melting the charge in the crucible to the commencement of replenishing can range from 5 to 50 h, depending on the initial crucible position in the furnace and the crucible lowering speed V. From Eq. [9.2-6] and the fact that erf(2) = 0.99, the required length of the passageway

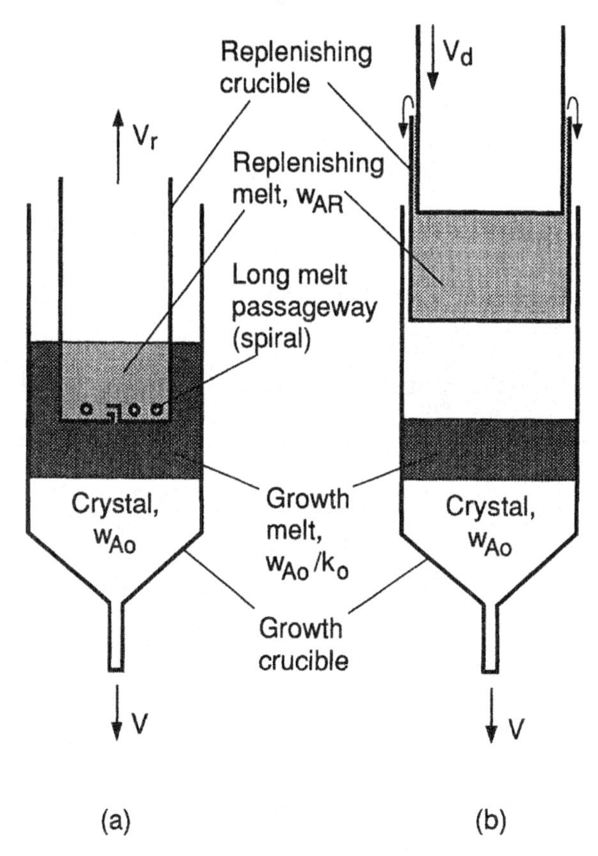

Fig. 9.3-9 Dopant segregation control in vertical Bridgman: (*a*) immersed replenishing crucible; (*b*) nonimmersed replenishing crucible. (From Y. Tao and S. Kou, *Journal of Crystal Growth*, in press.)

is $4\sqrt{Dt}$, where D is the dopant diffusion coefficient in the melt. Since D is on the order of 5×10^{-5} cm^2/s, the length is 4–12 cm.

Let us now consider two other segregation control techniques for vertical Bridgman growth.[21] In the immersed-crucible technique shown in Fig. 9.3-9*a*, backdiffusion is suppressed by a long melt passageway. In the non-immersed crucible technique shown in Fig. 9.3-9*b*, backdiffusion is eliminated altogether.

Let us determine the required dopant concentration of the replenishing melt w_{AR}. With the growth melt in the first technique as the control volume Ω, Eq. [1.2-6] becomes

$$\frac{d}{dt}(\rho_L \Omega) = \rho_L (V + V_r) A_r - \rho_S V A_S \qquad \text{(overall mass balance)} \qquad [9.3\text{-}39]$$

[21] Y. Tao and S. Kou, ibid.

where V_r and A_r are the lifting speed and inner cross-sectional area of the replenishing crucible, respectively. If the composition of the growth melt is to be uniform at w_{A0}/k_0, from Eq. [3.2-7],

$$\frac{d}{dt}\left(\frac{w_{A0}}{k_0}\rho_L\Omega\right) = w_{AR}\rho_L(V + V_r)A_r - w_{A0}\rho_S VA_S \quad \text{(species overall mass balance)} \tag{9.3-40}$$

Substituting Eq. [9.3-39] into Eq. [9.3-40]

$$\frac{w_{A0}}{k_0}\rho_L(V + V_r)A_r - \frac{w_{A0}}{k_0}\rho_S VA_S = w_{AR}\rho_L(V + V_r)A_r - w_{A0}\rho_S VA_S \tag{9.3-41}$$

On rearranging

$$\boxed{\frac{w_{AR}}{w_{A0}} = \frac{1}{k_0}\left[1 - (1 - k_0)\frac{\rho_S VA_S}{\rho_L(V + V_r)A_r}\right]} \tag{9.3-42}$$

Similarly, with the growth melt in the second technique as the control volume Ω, Eq. [1.2-6] becomes

$$\frac{d}{dt}(\rho_L\Omega) = \rho_L V_d A_d - \rho_S VA_S \qquad \text{(overall mass balance)} \tag{9.3-43}$$

where V_d and A_d respectively are the lowering speed and transverse cross-sectional area of the dummy. If the composition of the growth melt is to be uniform at w_{A0}/k_0, from Eq. [3.2-7]

$$\frac{d}{dt}\left(\frac{w_{A0}}{k_0}\rho_L\Omega\right) = w_{AR}\rho_L V_d A_d - w_{A0}\rho_S VA_S \qquad \text{(species overall mass balance)} \tag{9.3-44}$$

Substituting Eq. [9.3-43] into Eq. [9.3-44]

$$\frac{w_{A0}}{k_0}\rho_L V_d A_d - \frac{w_{A0}}{k_0}\rho_S VA_S = w_{AR}\rho_L V_d A_d - w_{A0}\rho_S VA_S \tag{9.3-45}$$

On rearranging

$$\boxed{\frac{w_{AR}}{w_{A0}} = \frac{1}{k_0}\left[1 - (1 - k_0)\frac{\rho_S VA_S}{\rho_L V_d A_d}\right]} \tag{9.3-46}$$

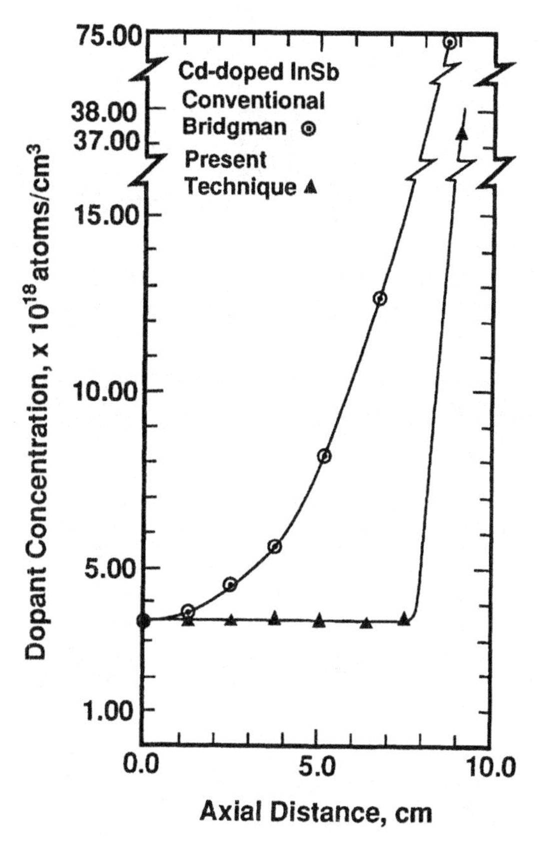

Fig. 9.3-10 Dopant distributions in two vertical Bridgman InSb single crystals grown without and with segregation control. (From Y. Tao and S. Kou, *Journal of Crystal Growth*, in press.)

Figure 9.3-10[22] shows the dopant distributions in two Cd-doped InSb single crystals grown by conventional Bridgman and by the technique shown in Fig. 9.3-9*b*. With segregation control the dopant concentration is uniform until the final stage of crystal growth, where it rises rapidly since the replenishing melt runs out. Similar results have been obtained by using the technique shown in Fig. 9.3-9*a* in a KNO_3-doped $NaNO_3$ single crystal.[22]

As for horizontal Bridgman, the use of a diffusion baffle alone is not enough. The boat must be tilted since ρ_S usually differs from ρ_L, as illustrated in Fig. 9.3-11[22] for the case of $\rho_S > \rho_L$. Boat tilting helps keep the crystal cross section constant, in order to keep the volume and hence composition of the growth melt constant. Figure 9.3-12 shows the segregation control in a horizontal Bridgman crystal of $NaNO_3$.

[22] Y. Tao and S. Kou, ibid.

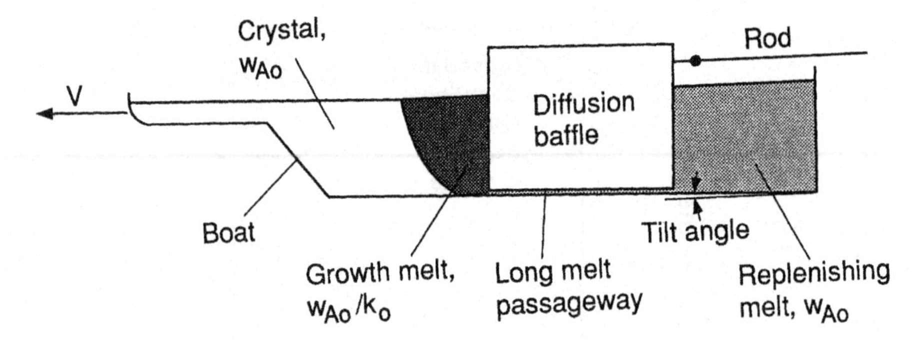

Fig. 9.3-11 Dopant segregation control in horizontal Bridgman. (From Y. Tao and S. Kou, *Journal of Crystal Growth*, in press.)

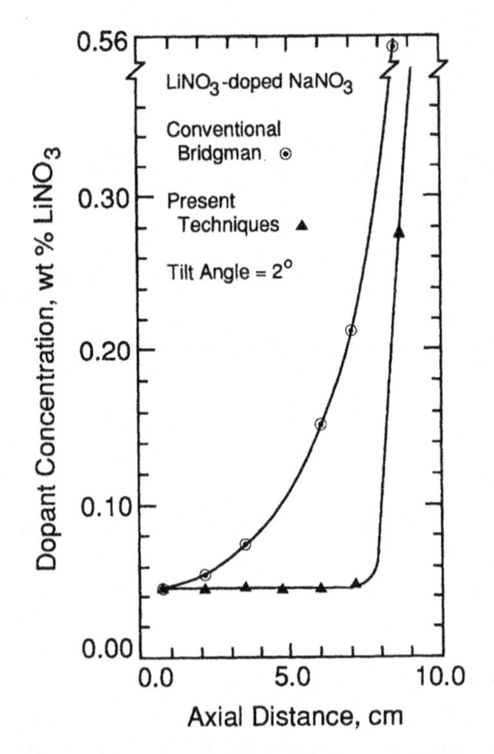

Fig. 9.3-12 Dopant segregation control in a horizontal Bridgman $NaNO_3$ crystal. (From Y. Tao and S. Kou, *Journal of Crystal Growth*, in press.)

9.4. CONVECTIVE MASS TRANSFER

9.4.1. Crystal Growth: Bridgman

Dopant segregation in Bridgman crystal growth is dominated by convection in the melt. Study of simultaneous fluid flow, heat transfer, and mass transfer in the melt can help one understand and control dopant segregation in Bridgman crystal growth. Extensive work has been conducted recently, notably by Brown and Derby and their associates. Convective mass transfer has also been studied in Czochralski crystal growth (e.g., by Kobayashi[23] and Toh[24] et al.).

The governing equations for simultaneous fluid flow, heat transfer, and mass transfer are as follows

$$\text{Continuity:} \quad \nabla \cdot \mathbf{v} = 0 \tag{9.4-1}$$

$$\text{Motion:} \quad \rho \frac{\partial \mathbf{v}}{\partial t} + \rho \mathbf{v} \cdot \nabla \mathbf{v} = -\nabla P + \mu \nabla^2 \mathbf{v} - \rho \beta_T \mathbf{g}(T - T_0) - \rho \beta_S \mathbf{g}(c_A - c_{A0}) + \sigma B_0^2 \mathbf{v} \times \mathbf{e}_B \times \mathbf{e}_B \tag{9.4-2}$$

$$\text{Energy:} \quad \frac{\partial T}{\partial t} + \mathbf{v} \cdot \nabla T = \alpha \nabla^2 T \tag{9.4-3}$$

$$\text{Species:} \quad \frac{\partial c_A}{\partial t} + \mathbf{v} \cdot \nabla c_A = D_A \nabla^2 c_A \tag{9.4-4}$$

In Eq. [9.4-2], which is Eq. [4.2-23], β_T and β_S are the thermal and solutal expansion coefficients, respectively, and B_0 is a uniform magnetic field applied in the direction $\mathbf{e}_B$. In the following discussion the magnetic field, if it is applied, is assumed to be in the z direction: $\mathbf{e}_B = \mathbf{e}_z$, where $\mathbf{e}_z$ is the unit vector in the z direction. The z direction is taken vertically upwards.

The vertical Bridgman configuration is shown schematically in Fig. 9.4-1.[25] As shown, the furnace consists of three different zones: a hot zone at the top, a cold zone at the bottom, and a gradient (or adiabatic) zone in between. The ampoule (crucible) is lowered at a constant speed V_s through the stationary furnace. Since the origin of the coordinate system is fixed to the top of the ampoule, the coordinate system and the ampoule can be considered stationary while the furnace rises at V_s. For simplicity, convection in the melt is assumed laminar and axisymmetric. In reality, misalignment of the ampoule axis from the gravitational vector or the furnace axis can cause asymmetric convection.

[23] S. Kobayashi, *J. Crystal Growth*, **75**, 301, 1986; **85**, 69, 1987.

[24] K. Toh and H. Ozoe, *J. Crystal Growth*, **130**, 645, 1993.

[25] R. A. Brown and D. H. Kim, *J. Crystal. Growth*, **109**, 50, 1991.

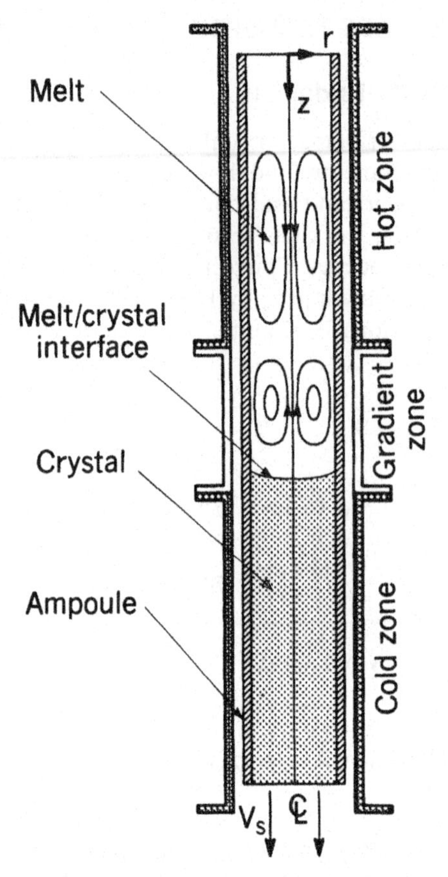

Fig. 9.4-1 A vertical Bridgman furnace. (From R. A. Brown and D. H. Kim, *Journal of Crystal Growth*, **109**, 1991, p. 50, "Modeling of Directional Solidification: From Scheil to Detailed Numerical Simulation.").

The boundary conditions are too numerous to describe here. Those for the melt are as follows:

Crystal/Melt Interface:

$$\mathbf{v} = 0 \qquad [9.4\text{-}5]$$

$$T = T_L \qquad [9.4\text{-}6]$$

$$(k\mathbf{n}\cdot\nabla T)_L + \rho H_f \mathbf{n}\cdot\mathbf{R} = (k\mathbf{n}\cdot\nabla T)_S \qquad [9.4\text{-}7]$$

$$(-D_A\mathbf{n}\cdot\nabla c_A)_L = c_A(1 - k_0)\mathbf{n}\cdot\mathbf{R} \qquad [9.4\text{-}8]$$

The first equation is the no-slip condition at the interface, Eq. [5.1-3], assuming equal densities of the melt and the crystal. In the second equation, T_L is the liquidus temperature, Eq. [5.1-12]. The third equation is the Stefan condition at the interface, Eq. [5.1-9]; **R** is the local growth rate and **n** the unit normal vector at the interface pointing into the melt. The fourth equation is the species mass balance at the interface, Eq. [5.1-11], assuming negligible diffusion in the crystal and dropping the asterisk (*) at the interface for convenience. The equilibrium segregation coefficient k_0, as defined in Section 5.1.3, is the ratio of the solute concentration of the crystal at the interface over that of the melt at the interface.

Centerline ($r = 0$):

$$\frac{\partial v_z}{\partial r} = \frac{\partial T}{\partial r} = \frac{\partial c_A}{\partial r} = 0 \qquad [9.4\text{-}9]$$

$$v_r = 0 \qquad [9.4\text{-}10]$$

These two equations follow from the axisymmetry assumption.

Ampoule Wall:

$$\mathbf{v} = 0 \qquad [9.4\text{-}11]$$

$$(T)_L = (T)_a \qquad [9.4\text{-}12]$$

$$(k\mathbf{n}\cdot\nabla T)_L = (k\mathbf{n}\cdot\nabla T)_a \qquad [9.4\text{-}13]$$

$$\frac{\partial c_A}{\partial r} = 0 \qquad [9.4\text{-}14]$$

The first equation is again the no-slip condition. The second and third equations are the continuity in temperature and the heat flux at the ampoule wall (case 6, Fig. 2.3-2). The last equation states that the solute is not assumed to penetrate through the ampoule wall (case 3, Fig. 3.3-2).

Free Surface ($z = 0$):

$$v_z = 0 \qquad [9.4\text{-}15]$$

$$\frac{\partial v_r}{\partial z} = 0 \qquad [9.4\text{-}16]$$

$$T = T_{\text{furnace}} \qquad [9.4\text{-}17]$$

$$\frac{\partial c_A}{\partial z} = 0 \qquad [9.4\text{-}18]$$

The first equation states that no melt is to cross the stationary free surface (Eq. [5.2-5]). The second equation states that there is no shear stress at the free surface (Eq. [5.2-3]), assuming that $\partial\gamma/\partial T$ and $\partial\gamma/\partial c_A$ are both zero. The third equation states that the free surface is simply assumed to be equilibrated with the furnace temperature. The last equation states that no solute evaporates from the free surface.

The governing equations and boundary conditions were converted into the dimensionless form and solved by the finite-element and Newton methods.[26] The dimensionless governing equations have been given previously (Eqs. [4.2-51]–[4.2–54]) along with the definitions of the dimensionless numbers such as the Prandtl number Pr, thermal Rayleigh number Ra_T, solutal Rayleigh number Ra_S, Hartmann number Ha, and Schmidt number Sc.

Let us now consider Ga-doped Ge where $Pr = 6.4 \times 10^{-3}$ and $Sc = 6.2$. Like most other dopants, the Ga concentration is very low: 10^{14}–10^{19} atoms/cm^3 or 10^{-9}–10^{-4} mole fraction. Solutal convection can be ignored (i.e., $Ra_S = 0$) at this low level of dopant concentration.

Figure 9.4-2 shows the temperature fields for the growth of the Ga-doped Ge at various times (in seconds); the growth rate V_g is 4 μm/s.[26] The results in Fig. 9.4-2*a* are for the case of no buoyancy convection, $Ra_T = 0$, while those in Fig. 9.4-2*b* are for the case of significant buoyancy convection, $Ra_T = 2 \times 10^5$. As shown, the results are not much different from each other, as expected from the very low Pr and the small ampoule size (1.36-cm ID).

The flow fields corresponding to the temperature fields in Fig. 9.4-2*b* are shown in Fig. 9.4-3*a*. As shown, there are two flow cells, an upper one in the bulk melt and a lower one near the crystal/melt interface. The upper cell is counter-clockwise – warm light melt rises along the ampoule wall, while cooler, denser melt falls along the centerline, typical of buoyancy convection. Since the crystal/melt interface is concave toward the melt, the lower cell is clockwise – the cool dense melt near the interface sinks inward along the interface. The strength of this cell is almost constant.

The corresponding Ga concentration fields are shown in Fig. 9.4-3*b*. The high Schmidt number causes the isoconcentration lines to be severely distorted by convection. The two distinct regions, corresponding to the two flow cells in Fig. 9.4-3a, are nearly well mixed and are separated by a thin diffusion layer. The calculated axial dopant distribution along the resultant crystal is very close to that predicted by the Scheil equation for complete mixing[26] (i.e., Eq. [9.1-9]).

In addition to Ga-doped Ge, Kim and Brown[27] have also studied the $Hg_{1-x}Cd_xTe$ system. The latter can be considered as a heavier HgTe alloyed with a lighter CdTe to the mole fraction of x. Since the density of the melt

[26] R. A. Brown and D. H. Kim, ibid.

[27] D. H. Kim and R. A. Brown, *J. Crystal Growth*, **114**, 411, 1991.

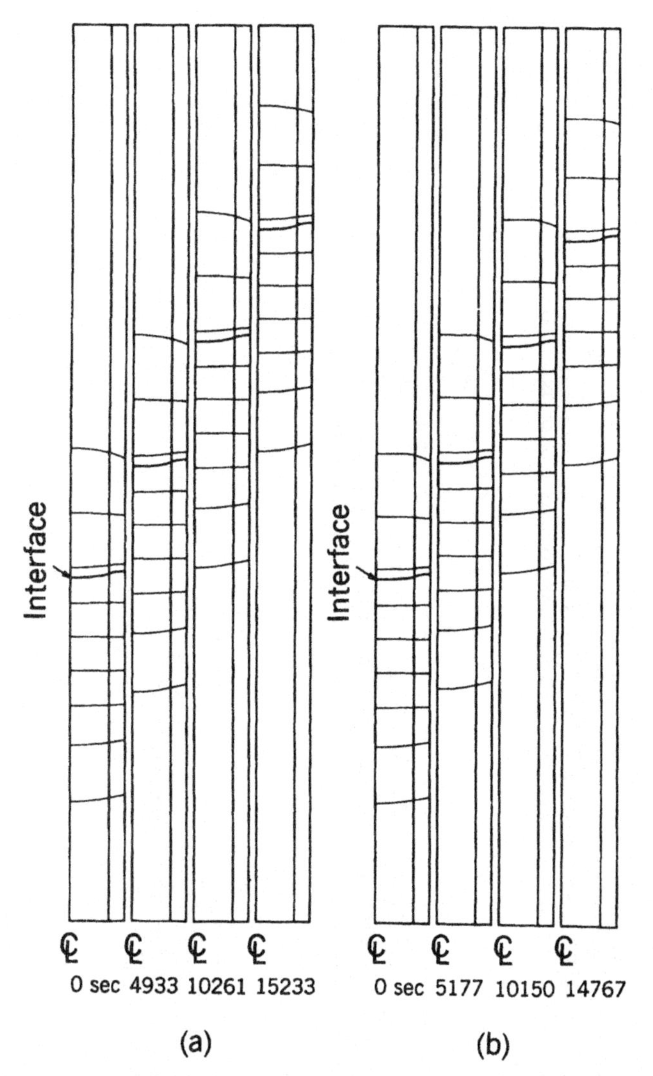

Fig. 9.4-2 Temperature fields in vertical Bridgman growth of Ga-doped Ge: (*a*) $Ra_T = 0$; (*b*) $Ra_T = 2 \times 10^5$. (From R. A. Brown and D. H. Kim, "Modeling of Directional Solidification: From Scheil to Detailed Numerical Simulation," *Journal of Crystal Growth*, **109**, 1991, p. 50.)

depends not only on temperature but also the CdTe concentration, the thermo-solutal flow fields appear significantly more complex than the buoyancy flow fields for Ga-doped Ge.

In an earlier study on Ga-doped Ge, Kim et al.[28] investigated the effect of a vertical magnetic field on convection and segregation. Figure 9.4-4*a* shows the

[28] D. H. Kim, P. M. Adornato, and R. A. Brown, *J. Crystal Growth*, **89**, 339, 1988.

flow fields at three different levels of the Hartmann number, which is proportional to the strength of the magnetic field. The thermal Rayleigh number $Ra_T = 1 \times 10^7$, the solutal Rayleigh number $Ra_S = 0$, and the ampoule translation velocity $V_g = 4\ \mu m/s$. With a relatively weak magnetic field, Ha = 100,

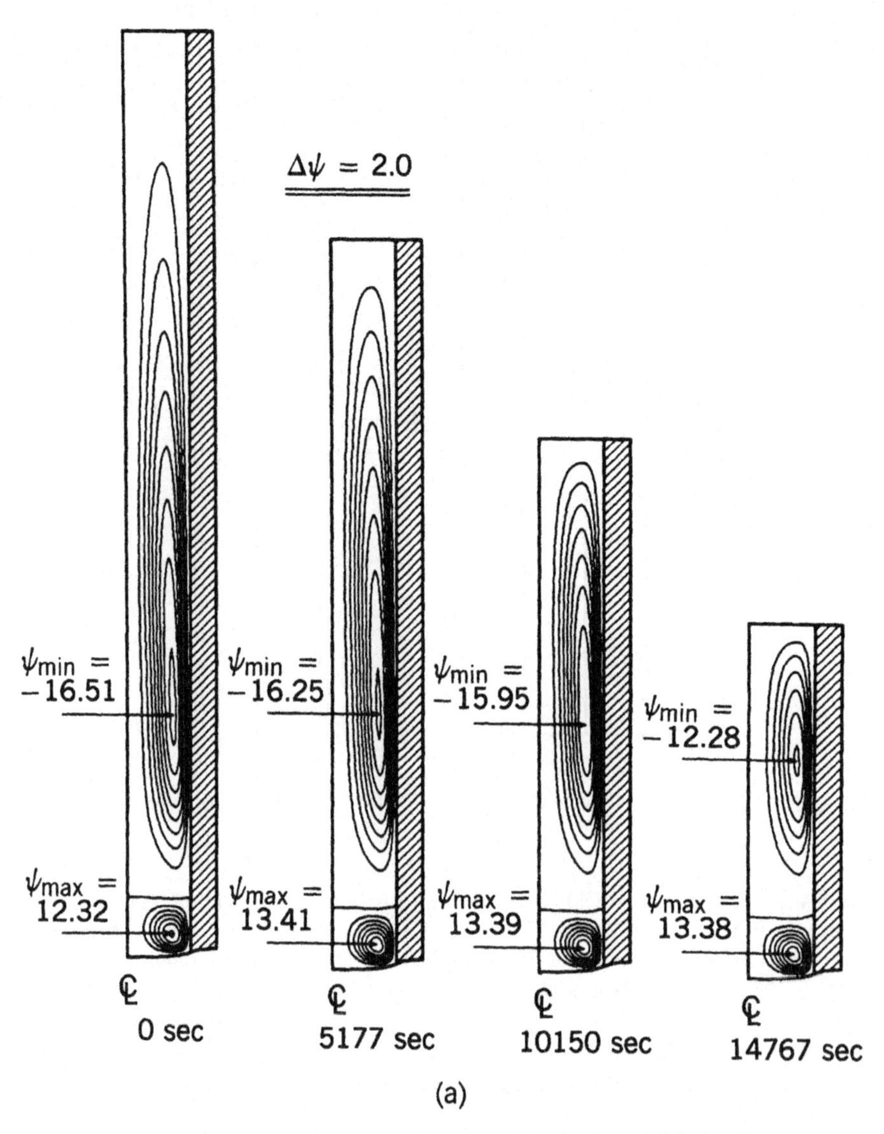

Fig. 9.4-3 Vertical Bridgman growth of Ga-doped Ge: (*a*) flow fields; (*b*) Ga concentration fields. (From R. A. Brown and D. H. Kim, "Modeling of Directional Solidification: From Scheil to Detailed Numerical Simulation," *Journal of Crystal Growth*, **109**, 1991, p. 50.)

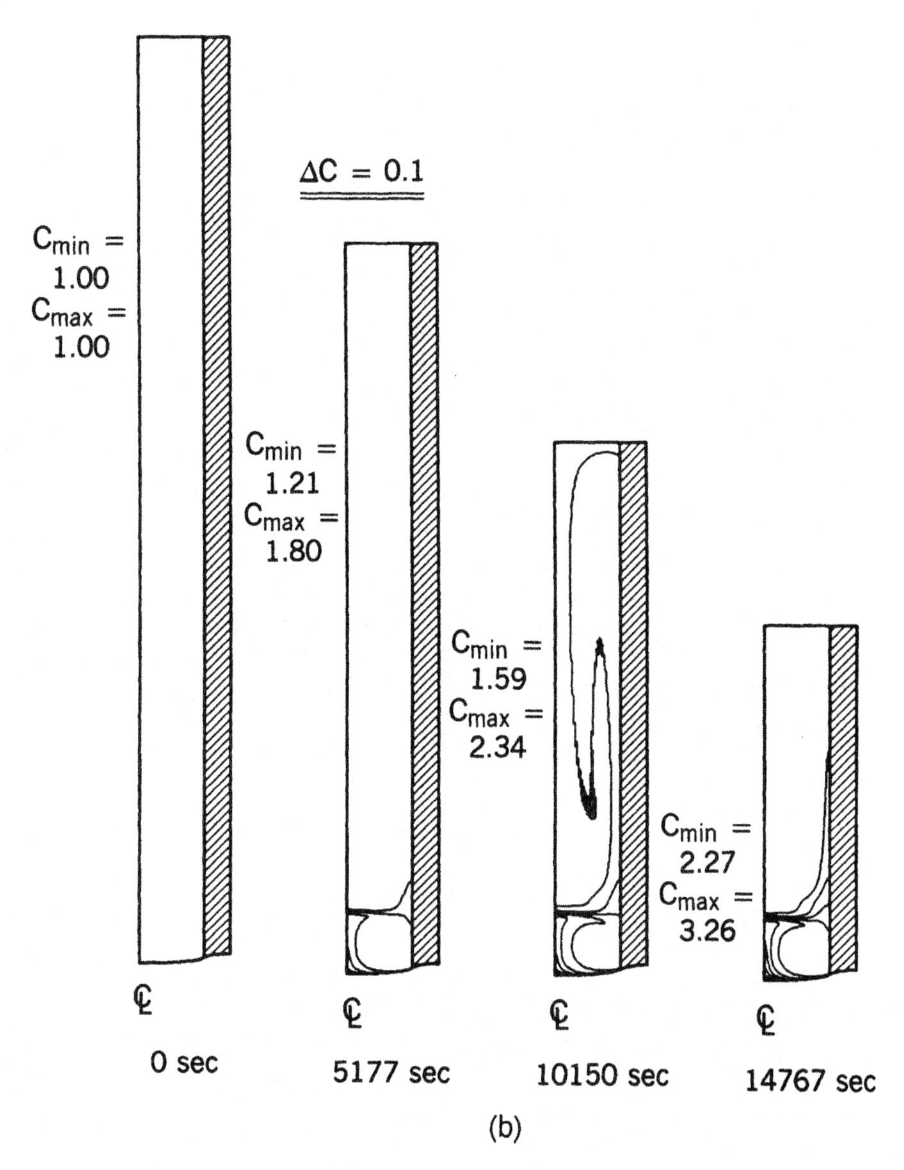

Fig. 9.4-3 (*Continued*).

the flow field is similar to those shown previously in Fig. 9.4-3*a*. As the magnetic field is increased in strength, convection is damped. First, the lower cell disappears and then the upper one, leaving a unidirectional flow field due to the ampoule translation.

The concentration fields corresponding to the flow fields in Fig. 9.4-4*a* are shown in Fig. 9.4-4*b*. As the magnetic field is increased in strength, the

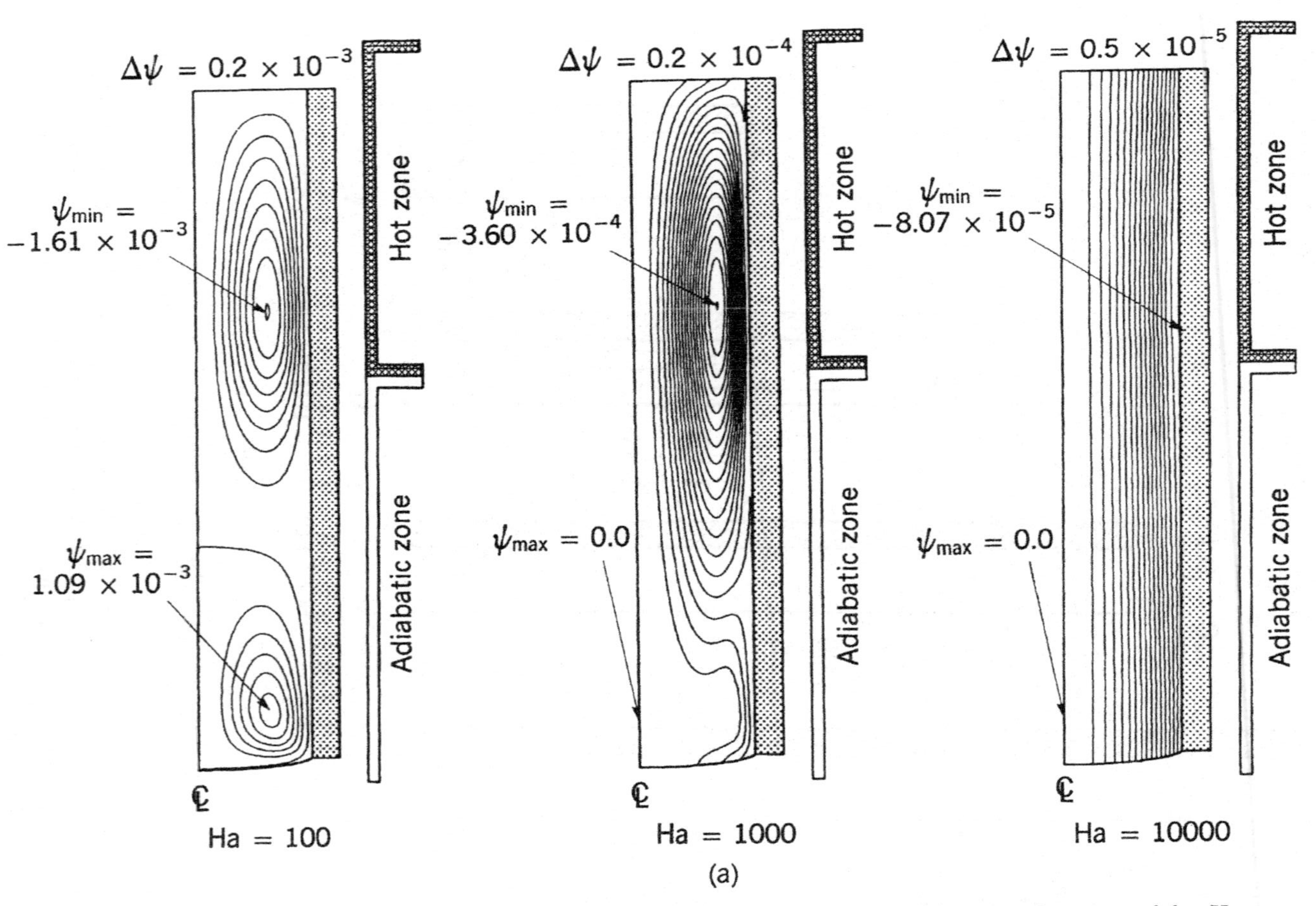

Fig. 9.4-4 Vertical Bridgman growth of Ga-doped Ge with increasing magnetic field strength measured by Ha ($Ra_T = 1 \times 10^7$): (*a*) flow fields; (*b*) Ga concentration fields. (From D. H. Kim, P. M. Adornato, and R. A. Brown, "Effect of Vertical Magnetic Field on Convection and Segregation in Vertical Bridgman Crystal Growth," *Journal of Crystal Growth*, **89**, 1988, p. 339).

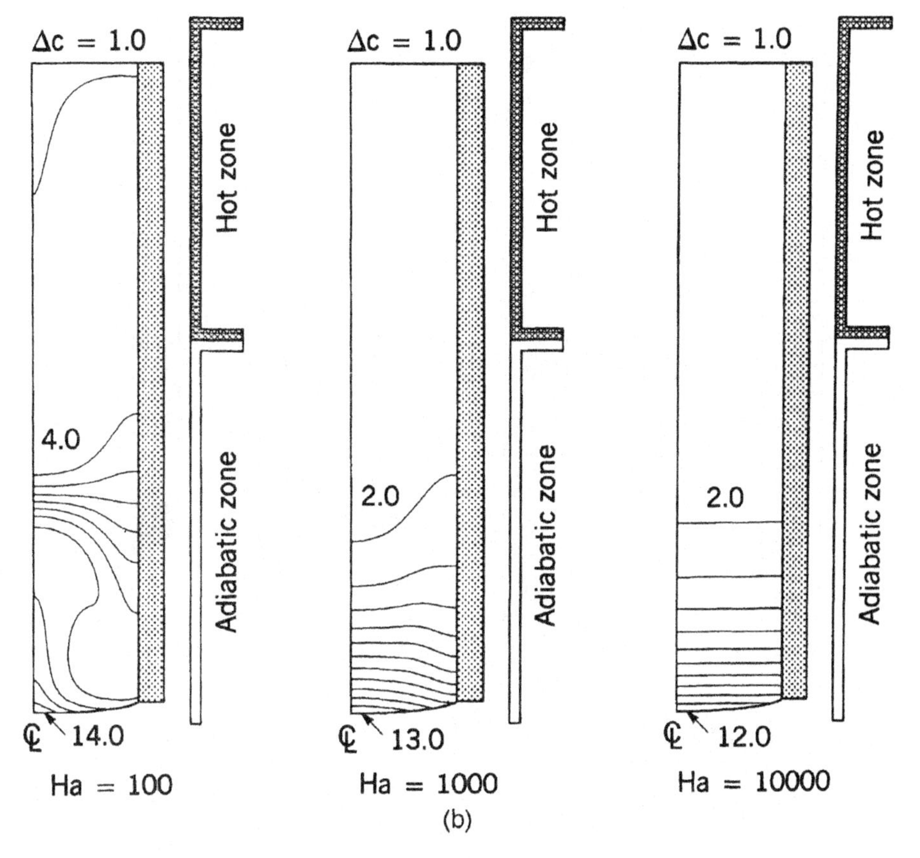

Fig. 9.4-4 (*Continued*).

concentration field becomes more like one due to diffusion alone, with flat isoconcentration lines. Axial dopant segregation is reduced. At Ha $= 10^4$ radial dopant segregation is also reduced, as can be seen from the reduced number of intersections between the isoconcentration lines and the crystal/melt interface. However, since weak mixing can cause more radial segregation than either complete or no mixing, moderate magnetic damping can, in fact, aggravate radial segregation.

Brandon and Derby[29] have considered internal radiation in the Bridgman growth of oxides. The absorption coefficient of the crystal is varied from $a = \infty$

[29] S. Brandon and J. J. Derby, *J. Crystal Growth*, **121**, 473, 1992.

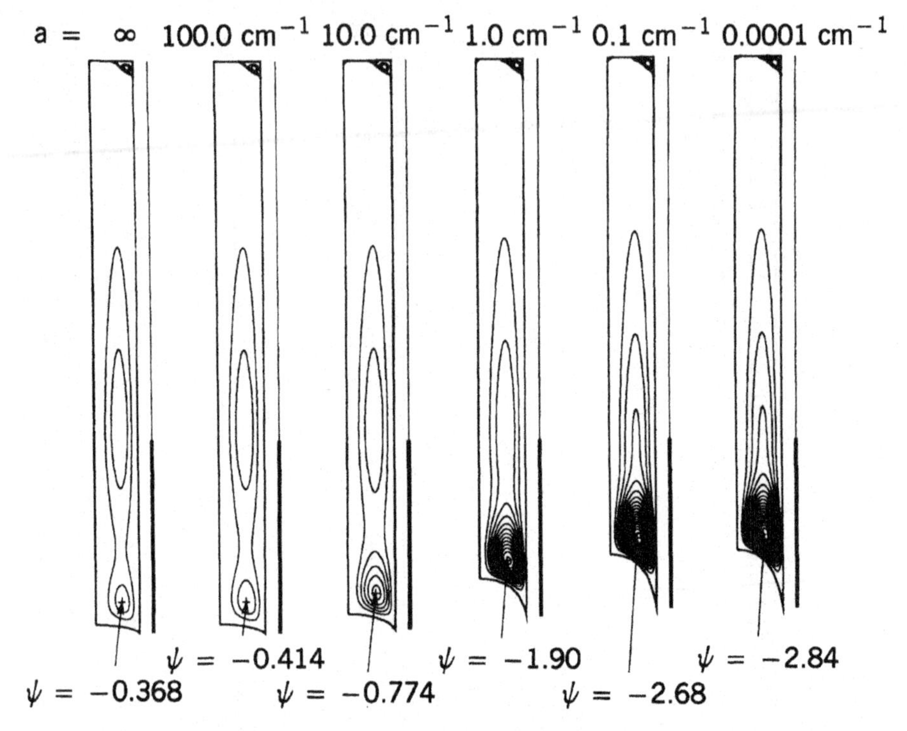

Fig. 9.4-5 Flow fields in vertical Bridgman growth of an oxide with decreasing optical absorption coefficient in the crystal. The axial positions of the hot zone and the adiabatic zone are marked by a thin line and a thick line, respectively. (From S. Brandon and J. J. Derby, "Heat Transfer in Vertical Bridgman Growth of Oxides: Effects of Conduction, Convection, and Internal Radiation," *Journal of Crystal Growth*, **121**, 1992, p. 473.)

(opaque) to 10^{-4} cm (transparent), as shown in Fig. 9.4-5. Heat loss through the crystal is increased by internal radiation. This causes the crystal/melt interface to move upward; however, the highly conductive molybdenum ampoule wall serves as a channel for heat flow, thus melting away the crystal at its edges and making the interface convex toward the melt. The degree of convexity increases with decreasing absorption coefficient (increasing transparency) of the crystal. Unlike the pattern in Fig. 9.4-3*a*, where the interface is concave, the cell near the interface is counterclockwise. The cool dense melt near the interface sinks outward along the interface. This cell grows stronger with increasing convexity of the interface.

Xiong and Kou[30] have measured velocity fields in a simulated Bridgman system of $NaNO_3$, in which the convexity of the crystal/melt interface can be

[30] B. Xiong and S. Kou, unpublished research at the University of Wisconsin, Madison, WI, 1995.

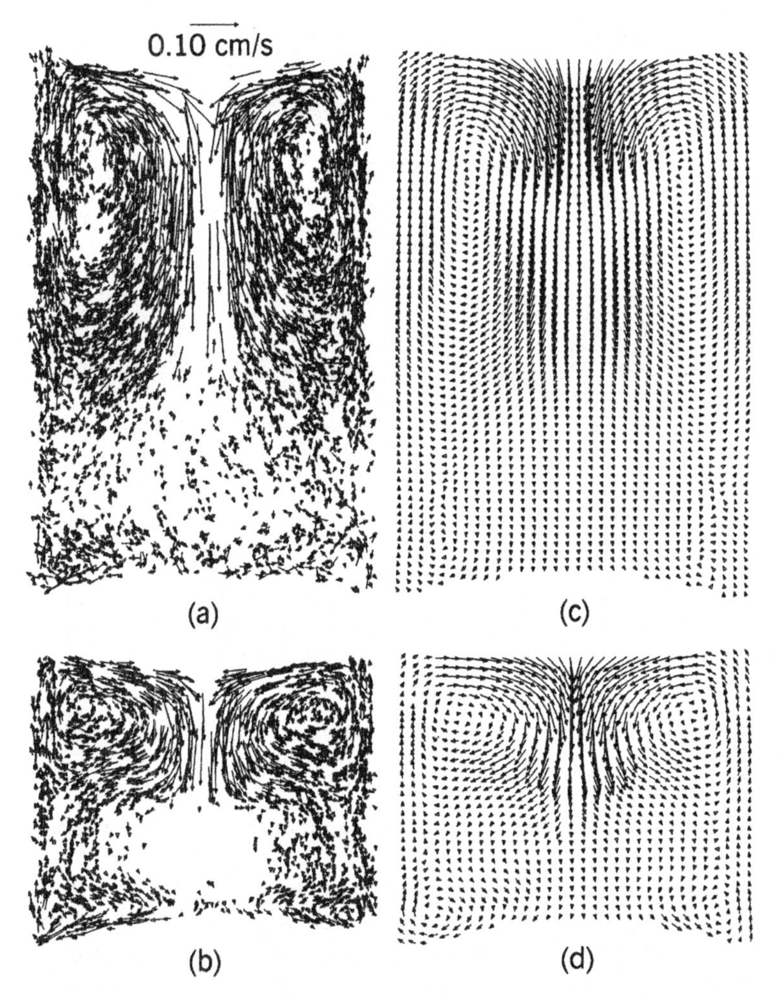

Fig. 9.4-6 Velocity fields in simulated vertical Bridgman growth of 2.5-cm-diameter $NaNO_3$: (*a*) 3.8-cm-deep melt, as measured; (*b*) 3.8-cm-deep melt, interpolated; (*c*) 2.2-cm-deep melt, as measured; (*d*) 2.2-cm-deep melt, interpolated. (From B. Xiong and S. Kou, unpublished work, University of Wisconsin, Madison, WI, 1994.)

controlled. Figure 9.4-6 shows the flow fields in two $NaNO_3$ melts, in ampoules of 25 mm ID. The deeper melt (Fig. 9.4-6*a*) represents an earlier stage of crystal growth. In either case there exists a zone of negligible convection near the crystal. The cells above this zone are induced by the heated ampoule wall; warm light melt rises along the ampoule wall and cooler denser melt falls along the centerline. The lower cells near the crystal/melt interface are induced by the radial temperature gradients caused by the convex crystal/melt interface. The cool dense melt near the interface sinks outward along the interface. The lower

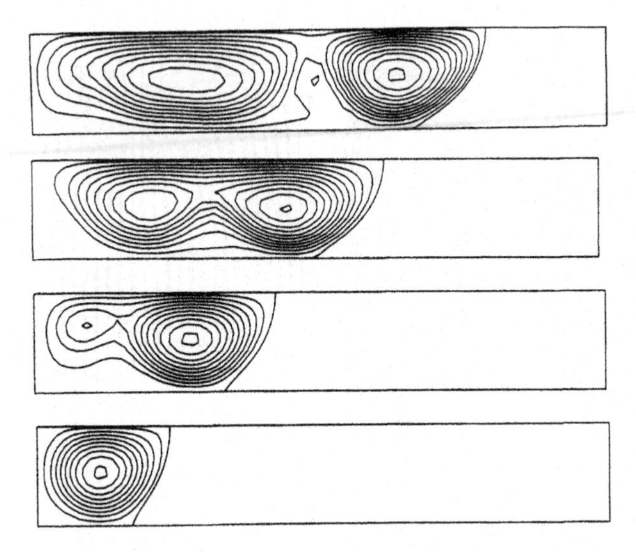

Fig. 9.4-7 Streamlines and melt/crystal interface in horizontal Bridgman crystal growth ($Ra_T = 3 \times 10^{94}$, $Pr = 0.0015$). (From M. J. Crochet, F. T. Geyling, and J. J. Van Schaftingen, "Numerical Simulation of the Horizontal Bridgman Growth of a Gallium Arsenide Single Crystal," *Journal of Crystal Growth*, **65**, 1983, p. 166.)

cells in the deeper melt, although not as clear, are similar to those in the shallower melt. This has been confirmed by additional velocity measurements focusing on the melt near the crystal.

Since the crystal is at the bottom, the region of the cool, dense melt near the crystal is very stable and difficult to penetrate, resulting in a zone of minimal convection. This is contrary to the Czochralski melt shown in Fig. 8.5-5*b*, where the downward flow from the top of the centerline penetrates all the way to the bottom. The melt near the Czochralski crystal is cool and dense, thus exhibiting a tendency to penetrate to the bottom of the melt.

Xiao et al.[31] have developed a three-dimensional computer model for vertical Bridgman growth. It was found that a slight (e.g., 1°) misalignment of the ampoule axis from the gravitational vector produces a significant three-dimensional flow and an asymmetric radial dopant distribution. Lan and Ting[32] have considered the coincal shape of the ampoule bottom and its effect on the growth front shape.

Before leaving the subject of Bridgman crystal growth, it should be mentioned that horizontal Bridgman has also been studied, notably by Crochet[33], Roux,[34] and their associates. Figure 9.4-7 is an example of the calculated flow

[31] Q. Xiao, S. Kuppurao, A. Yeckel, and J. J. Derby, *J. Crystal Growth*, in press.

[32] C. W. Lan and C. C. Ting, *J. Crystal Growth*, **149**, 175, 1995.

[33] M. J. Crochet, F. T. Geyling, and J. J. Van Schaftingen, *J. Crystal Growth*, **65**, 166, 1983.

[34] B. Roux, H. Ben Hadid, and P. Laure, *J. Crystal Growth*, **97**, 201, 1989.

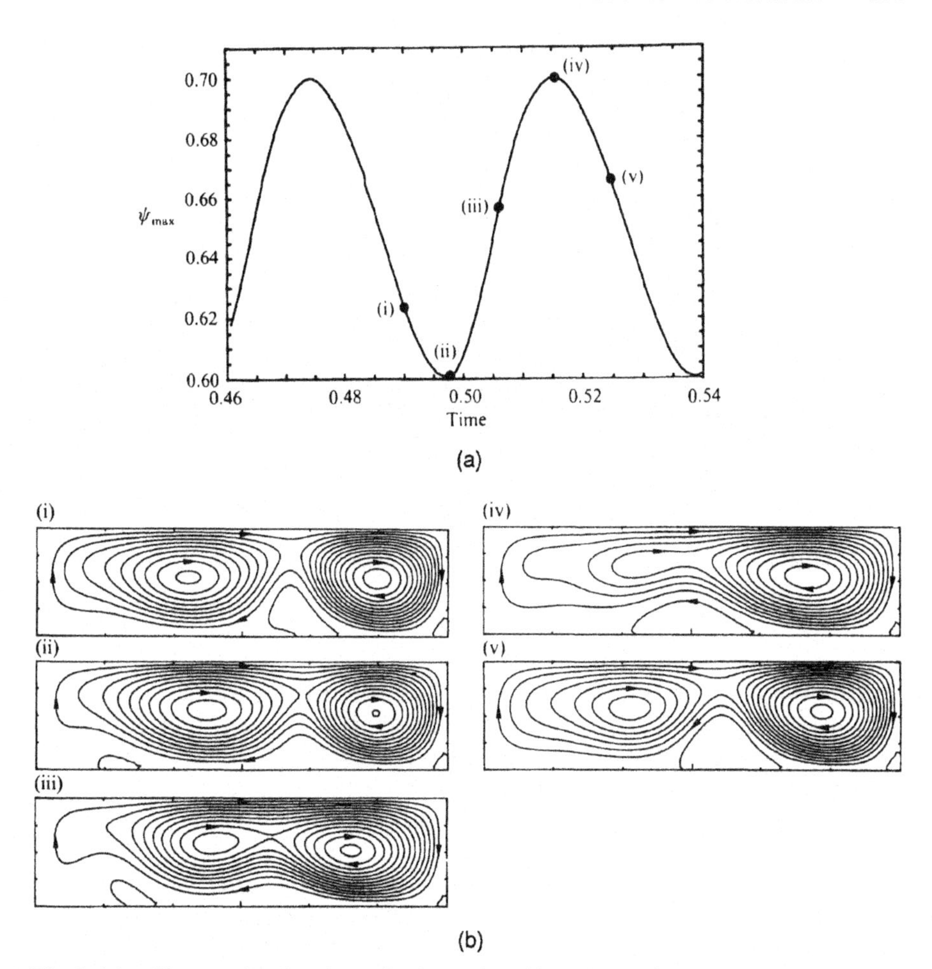

Fig. 9.4-8 Flow oscillation in a horizontal Bridgman configuration: (*a*) maximum stream function; (*b*) streamlines ($Gr = 4 \times 10^4$). (From H. Ben Hadid and B. Roux, "Buoyancy- and Thermocapillary-Driven Flows in Differentially Heated Cavities for Low-Prandtl-Number Flow," *Journal of Fluid Mechanics*, **235**, 1992, p 1. Reprinted with the permission of Cambridge University Press.)

fields. One interesting aspect of horizontal Bridgman is that flow oscillation can occur easily. As shown in Fig. 9.4-8,[35] the cells can expand and contract periodically, causing the stream function (and temperature) to oscillate. Kaddeche et al.[36] have calculated macrosegregation in horizontal Bridgman.

[35] H. Ben Hadid and B. Roux, *J. Fluid Mechanics*, **235**, 1, 1992.

[36] S. Kaddeche, H. B. Hadid, and D. Henry, *J. Crystal Growth*, **135**, 341, 1994.

9.4.2. Crystal Growth: Floating Zone

Lan and Kou[37] have studied radial dopant segregation in floating-zone crystal growth. $NaNO_3$ was selected as the model material and convection in the molten zone was assumed steady and axisymmtric, as actually observed in flow visualization.[38] As mentioned previously (Section 8.5.2), convection can, in fact, be oscillatory and asymmetric, as in Si. The crystal growth system has already been shown schematically in Fig. 8.5-6. The cylindrical coordinate (r, z) and the ambient-temperature distribution $T_a(z)$ due to the stationary ring heater (not shown) are both stationary. Since the steady-state condition is assumed, the position and shape of the molten zone remain constant. The feed and the crystal both move in the negative z direction. The governing equations for fluid flow and heat transfer are identical to Eqs. [8.5-23] through [8.5-26]. The governing equation for mass transfer, according to Eq. [3.4-8] and without chemical reactions, is as follows:

$$\text{Species:}\quad \frac{\partial c_A}{\partial t} + \left[\frac{1}{r}\frac{\partial \psi}{\partial z}\frac{\partial c_A}{\partial r} - \frac{1}{r}\frac{\partial \psi}{\partial r}\frac{\partial c_A}{\partial z}\right] = D_A\left[\frac{1}{r}\frac{\partial}{\partial r}\left(r\frac{\partial c_A}{\partial r}\right) + \frac{\partial^2 c_A}{\partial z^2}\right] \tag{9.4-19}$$

The boundary conditions for fluid flow and heat transfer are identical to Eqs. [8.5-30] through [8.5-49], except that the melting point T_m is replaced by the liquidus temperature T_L, which, according to Eq. [5.1-13], is as follows:

$$T = T_m + m_L c_A \tag{9.4-20}$$

where m_L is the slope of the liquidus line of the phase diagram.

At the steady state, the boundary conditions for mass transfer are as follows:

$$\text{Crystal/melt interface:}\quad (c_A - k_0 c_A)(-V_c)\mathbf{n}\cdot\mathbf{e}_z = (-D_A\mathbf{n}\cdot\nabla c_A)_L \tag{9.4-21}$$

where k_0 is the equilibrium segregation coefficient, V_c the crystal pulling velocity (<0), and $\mathbf{e}_z$ the unit vector in the z direction. Equation [9.4-21] is similar to Eq. [5.1-30]. Since the diffusion coefficient D_A is much smaller for solid than melt, solid-state diffusion can be neglected.

$$\text{Feed/melt interface:}\quad (c_A - c_{Af})V_f\mathbf{n}\cdot\mathbf{e}_z = (D_A\mathbf{n}\cdot\nabla c_A)_L \tag{9.4-22}$$

where c_{Af} is the dopant concentration of the feed and V_f the rod feeding rate (<0). Equation [9.4-22] is similar to Eq. [5.1-32].

[37] C. W. Lan and S. Kou, *J. Crystal Growth*, **132**, 578, 1993.

[38] Y. Tao and S. Kou, *J. Crystal Growth*, **137**, 72, 1994.

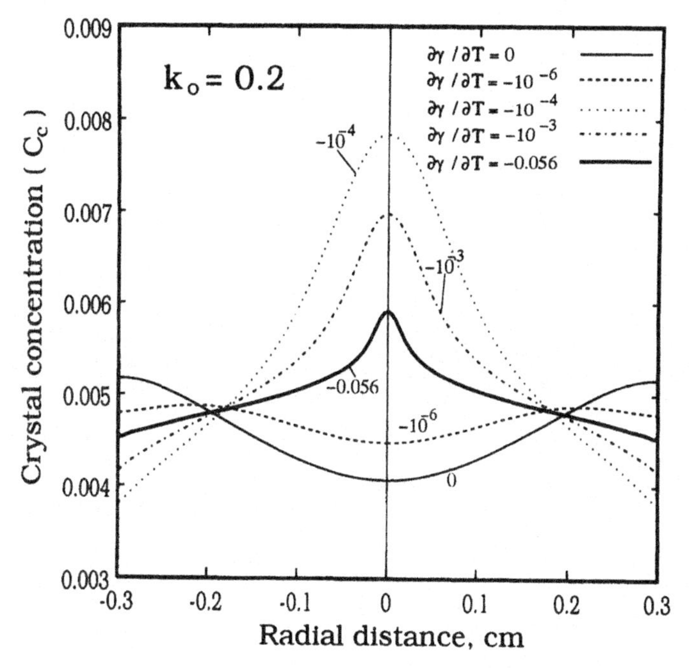

Fig. 9.4-9 Effect of $\partial\gamma/\partial T$ on radial dopant segregation in zero-gravity floating-zone crystal growth of KNO_3-doped $NaNO_3$. (From C. W. Lan and S. Kou, *Journal of Crystal Growth*, **132**, 1993, p. 578).

$$\text{Centerline:} \quad \frac{\partial c_A}{\partial r} = 0 \qquad [9.4\text{-}23]$$

due to axisymmetry.

$$\text{Free surface:} \quad \mathbf{n} \cdot \nabla c_A = 0 \qquad [9.4\text{-}24]$$

assuming no dopant evaporates from the free surface, that is, the mass flux is zero.

Let us consider the steady-state, floating-zone growth of a 0.6-cm-diameter $NaNO_3$ crystal from a feed rod doped with 0.48 wt% KNO_3 under zero gravity. The equilibrium segregation coefficient k_0 is 0.2, the diffusion coefficient D_A is $2 \times 10^{-5}\,\text{cm}^2/\text{s}$, and the surface tension temperature coefficient $\partial\gamma/\partial T$ is -0.056 dyn $\text{cm}^{-1}\,°\text{C}^{-1}$. Since the growth angle ϕ_0 is only 2.5°, the free surface of the melt is nearly cylindrical under zero gravity.

In order to study the effect on $\partial\gamma/\partial T$ on radial dopant segregation, it is varied from 0 to -0.056 dyn $\text{cm}^{-1}\,°\text{C}^{-1}$. As shown in Fig. 9.4-9,[39] with $\partial\gamma/\partial T = 0$ the

[39] C. W. Lan and S. Kou, ibid.

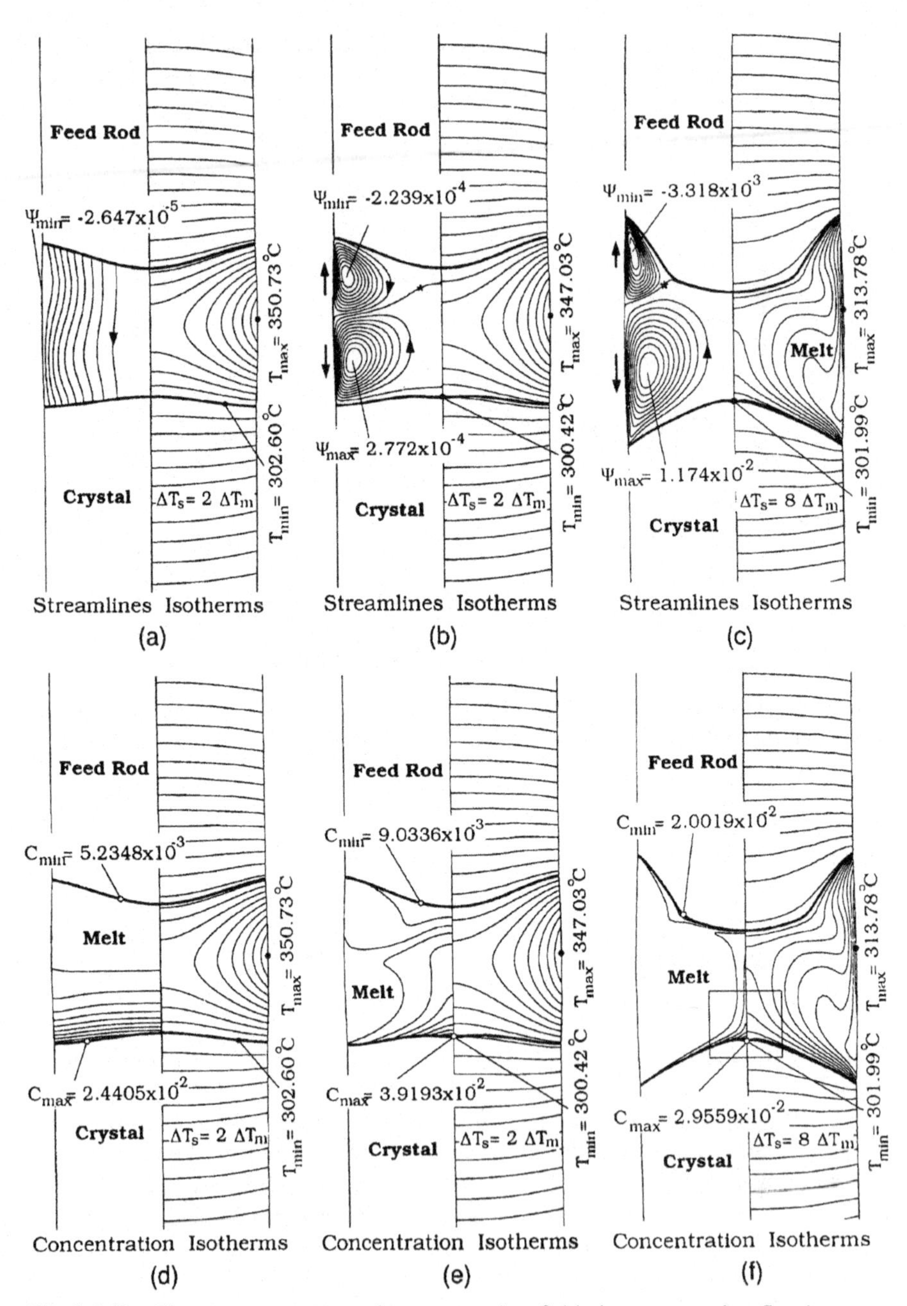

Fig. 9.4-10 Flow, temperature and concentration fields in zero-gravity, floating-zone crystal growth of KNO_3-doped $NaNO_3$: (*a*), (*d*) $\partial\gamma/\partial T = -1\times10^{-6}$ dyn cm^{-1} °C^{-1}; (*b*), (*e*) $\partial\gamma/\partial T = -1\times10^{-4}$ dyn cm^{-1} °C^{-1}; (*c*), (*f*) $\partial\gamma/\partial T = -0.056$ dyn cm^{-1} °C^{-1}. (From C. W. Lan and S. Kou, *Journal of Crystal Growth*, **132**, 1993, p. 578.)

dopant segregates toward the surface of the crystal. As $\partial\gamma/\partial T$ becomes increasingly negative, the tendency to segregate toward the centerline first increases and then decreases.

With $\partial\gamma/\partial T = 0$ there is no convection and mass transfer is due to diffusion alone. Since the crystal/melt interface is convex toward the melt and since diffusion in the melt is away from the interface, the dopant tends to segregate toward the surface.

The flow, temperature, and concentration fields are shown in Fig. 9.4-10 for $\partial\gamma/\partial T = -1\times10^{-6}$, -1×10^{-4}, and -0.056 dyn cm^{-1} °C^{-1}. The stream function here equals the ψ defined by Eqs. [8.5-27] and [8.5-28] multiplied by $-\rho_L$, which is -1.9 g/cm^3. As shown in Fig. 9.4-10*a*, with $\partial\gamma/\partial T = -1\times10^{-6}$ dyn cm^{-1} °C^{-1} the melt near the crystal/melt interface flows slightly radially inward, opposite to the slightly outward dopant diffusion induced by the convexity of the interface. The tendency for the dopant to segregate toward the surface is, therefore, slightly reduced.

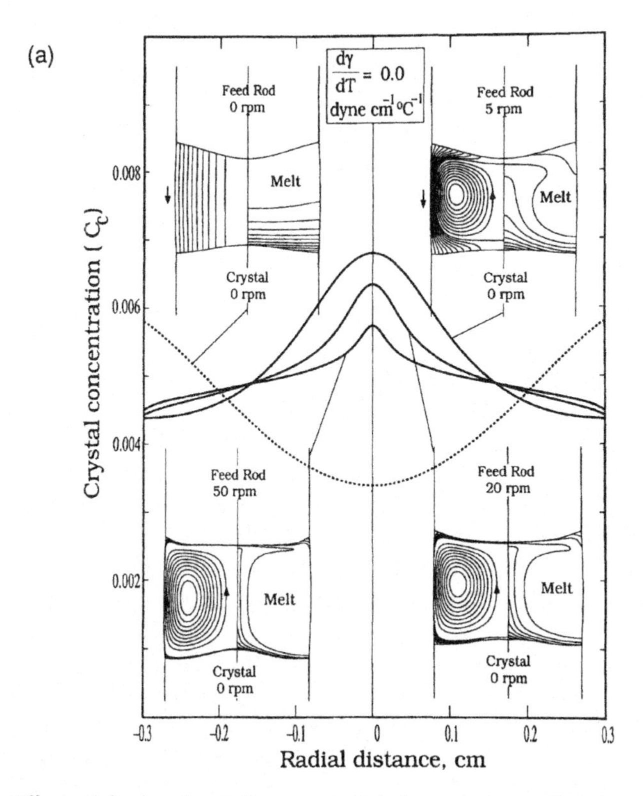

Fig. 9.4-11 Effect of feed-rod rotation on radial dopant segregation in zero-gravity, floating-zone crystal growth of KNO_3-doped $NaNO_3$: (*a*) $\partial\gamma/\partial T = 0$; (*b*) $\partial\gamma/\partial T = -0.056$ dyn cm^{-1} °C^{-1}. (From C. W. Lan and S. Kou, *Journal of Crystal Growth*, **133**, 1993, p. 309.)

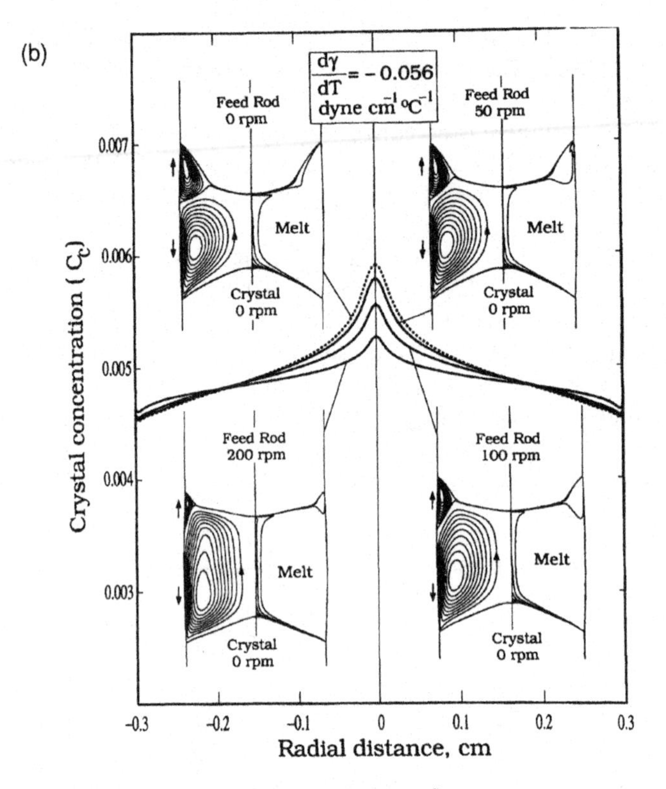

Fig. 9.4-11 *(Continued).*

As shown in Fig. 9.4-10*b*, with $\partial\gamma/\partial T = -1 \times 10^{-4}$ dyn cm^{-1} °C^{-1}, two distinct cells develop in the melt. Since $\partial\gamma/\partial T < 0$, the direction of the lower cell is such that the melt near the crystal/melt interface flows radially inward. This pushes the isoconcentration lines at the interface radially inward, as shown in Fig. 9.4-10*e*. Although the convection is weak, it is sufficient to cause the dopant to segregate toward the centerline significantly, as shown in Fig. 9.4-9.

With $\partial\gamma/\partial T$ equal to its actual value of -0.056 dyn cm^{-1} °C^{-1}, both cells in the melt grow much stronger as indicated by the ψ values of the cells in Fig. 9.4-10*c*. Since the dopant is better mixed in the melt, the extent of radial dopant segregation is reduced significantly, as shown in Fig. 9.4-9.

The effect of feed-rod rotation on radial dopant segregation is shown in Fig. 9.4-11.[40] With $\partial\gamma/\partial T = 0$ (i.e., Fig. 9.4-11*a*), the dopant diffuses toward the surface in the absence of rotation. With 5-rpm feed-rod rotation, the melt near the feed rod is push radially outward by the centrifugal force. Near the crystal

[40] C. W. Lan and S. Kou, *J. Crystal Growth*, **133**, 309, 1993.

the melt flows radially inward to feed the crystal, causing the dopant to segregate toward the centerline. As the rotation speed is increased, radial segregation is reduced because of better mixing of the dopant in the melt.

With $\partial\gamma/\partial T = -0.056$ dyn cm^{-1} °C^{-1} (Fig. 9.4-11*b*), the dopant segregates toward the centerline in the absence of rotation. As a result of strong thermocapillary convection, there is already significant mixing in the melt. Further increase in mixing, which requires substantial feed-rod rotation, results in reduction in dopant segregation.

9.4.3. Casting: Macrosegregation

One important mechanism for solute macrosegregation in casting (Section 6.1.2) is the flow of interdendritic liquid in the mushy zone during solidification. The mushy zone is a two-phase mixture of the dendrites and the interdendritic liquid.

Computation of simultaneous fluid flow, heat transfer, and mass transfer in the mushy zone as well as the liquid pool, can provide useful information for studying macrosegregation in casting. Incropera[41] and his associates have developed a dual-phase approach for such computation. Subsequently, Beckermann[42] and his associates also used a dual-phase approach for the same purpose.

According to Table 4.1-2, the general integral-balance equation for a single phase problem, that is, Eq. [B] is as follows:

$$\frac{\partial}{\partial t}\iiint_{\Omega} \rho\phi \, d\Omega = -\iint_{A} (\rho\mathbf{v}\phi)\cdot\mathbf{n}\, dA - \iint_{A} \mathbf{j}_{\phi}\cdot\mathbf{n}\, dA + \iiint_{\Omega} s_{\phi}\, d\Omega \qquad [9.4\text{-}25]$$

where ϕ is the quantity of interest, for instance, velocity, energy, or concentration, and $\mathbf{j}_\phi$ is the flux, such as momentum, heat, or species diffusion.

In order to deal with a multiphase mixture such as the mushy zone, Bennon and Incropera[41] have used an equation, which is similar to Eq. [9.4-25] in form but valid for phase k in the mixture:

$$\frac{\partial}{\partial t}\iiint_{\Omega_k} \rho_k\phi_k \, d\Omega_k = -\iint_{A_k} (\rho_k\mathbf{v}_k\phi_k)\cdot\mathbf{n}\, dA_k - \iint_{A_k} \mathbf{j}_{\phi k}\cdot\mathbf{n}\, dA_k + \iiint_{\Omega_k} s_{\phi k}\, d\Omega_k \qquad [9.4\text{-}26]$$

where ρ_k and $\mathbf{v}_k$ are density and velocity of phase k, respectively. Assume that g_k – the volume fraction of phase k in the mixture – is continuous and that

$$dA_k = g_k\, dA \qquad [9.4\text{-}27]$$

[41] W. D. Bennon and F. P. Incropera, *Int. J. Heat Mass Transfer*, **30**, 2161, 2171, 1987.

[42] M. C. Schneider and C. Beckermann, in *Transport Phenomena in Solidification*, C. Beckermann et al., eds., HTD-Vol. 284; AMD Vol. 182, ASME, New York, 1994, p. 43.

and

$$d\Omega_k = g_k \, d\Omega \qquad [9.4\text{-}28]$$

Substituting Eqs. [9.4-27] and [9.4-28] into Eq. [9.4-26]

$$\frac{\partial}{\partial t} \iiint_{\Omega_k} g_k \rho_k \phi_k \, d\Omega = -\iint_A (g_k \rho_k \mathbf{v}_k \phi_k) \cdot \mathbf{n} \, dA - \iint_A (g_k \mathbf{j}_{\phi k}) \cdot \mathbf{n} \, dA + \iiint_\Omega g_k s_{\phi k} \, d\Omega. \qquad [9.4\text{-}29]$$

From the Gauss divergence theorem (i.e., Eq. [A.4-1]) and assuming that the control volume Ω does not change with time, Eq. [9.4-29] becomes

$$\iiint_{\Omega_k} \frac{\partial}{\partial t} (g_k \rho_k \phi_k) \, d\Omega = -\iiint_A \nabla \cdot (g_k \rho_k \mathbf{v}_k \phi_k) \, d\Omega - \iiint_A \nabla \cdot (g_k \mathbf{j}_{\phi k}) \, d\Omega + \iiint_\Omega g_k s_{\phi k} \, d\Omega. \qquad [9.4\text{-}30]$$

Since this equation must hold for any arbitrary region Ω

$$\boxed{\frac{\partial}{\partial t} (g_k \rho_k \phi_k) = -\nabla \cdot (g_k \rho_k \mathbf{v}_k \phi_k) - \nabla \cdot (g_k \mathbf{j}_{\phi k}) + g_k s_{\phi k}} \qquad [9.4\text{-}31]$$

For $g_k = 1$ (i.e., a single-phase material) Eq. [9.4-31] reduces to the form of the following general differential-balance equation for single-phase materials (i.e., Eq. [B] of Table 4.1-4):

$$\frac{\partial}{\partial t} (\rho \phi) = -\nabla \cdot (\rho \mathbf{v} \phi) - \nabla \cdot \mathbf{j}_\phi + s_\phi \qquad [9.4\text{-}32]$$

Table 4.1-4 shows how the equations of continuity, motion, energy, and species continuity can be obtained from Eq. [9.4-32] for single-phase materials. In a similar way, these equations can be obtained from Eq. [9.4-31] for multiphase materials.

Prescott and Incropera[43] have considered the solidification of a Pb–19 wt% Sn alloy in an annular mold 15 cm high, as illustrated in Fig. 9.4-12. The outer surface is cooled with a coolant of temperature T_c; the heat transfer coefficient is U. All remaining surfaces are insulated. The Pb–Sn phase diagram is shown in Fig. 9.4-13. The composition of the liquid in the mushy zone, that is, the

[43] P. J. Prescott and F. P. Incropera, *Metallurigcal Trans. B*, **22B**, 529, 1991.

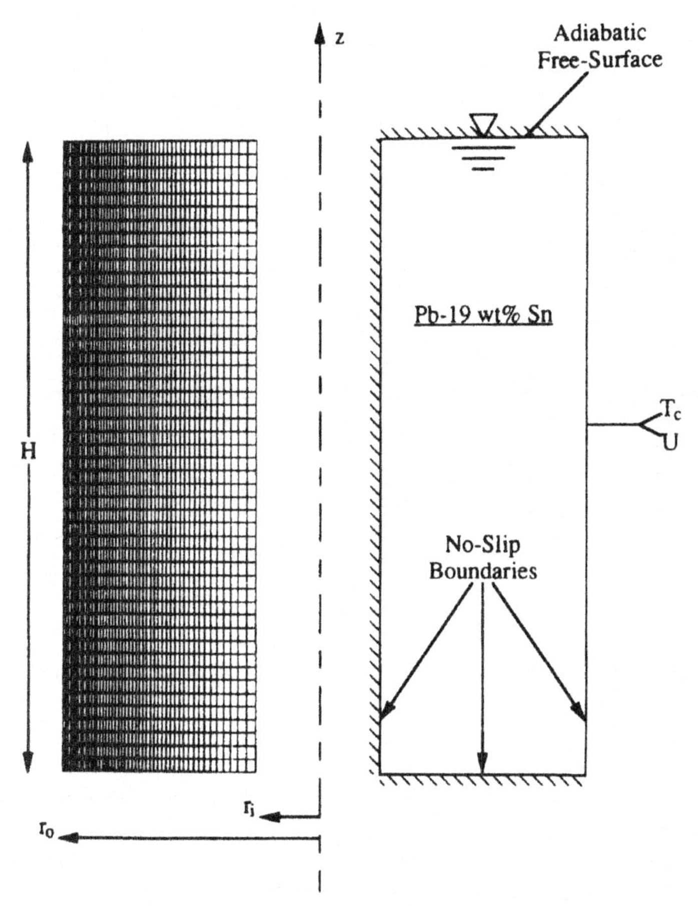

Fig. 9.4-12 An annular mold for studying macrosegregation in a Pb–Sn alloy; cooling is from the outer surface of the mold. (From P. J. Prescott and F. P. Incropera, *Metallurgical Transactions*, **22B**, 1991, p. 529.)

interdendritic liquid, is governed by the (left) liquidus line. As shown, the cooler the interdendritic liquid is, the richer it is in Sn and hence the lighter it is. This is because Sn is about one-third lighter than Pb.

The results at 110 s after cooling are shown in Fig. 9.4-14. As shown, the mushy zone (S + L) grows from the lower outer (left) corner. The small clockwise cell near the top of the mushy zone is solutally driven, that is, the cooler and lighter Sn-rich interdendritic liquid rises and, in fact, escapes from the mushy zone. The large counter clockwise cell in the liquid pool, on the other hand, is thermally driven.

As it rises, the Sn-rich interdendritic liquid can result in the remelting of the dendrites, or the formation of channels (freckles). In order to dilute the extra Sn

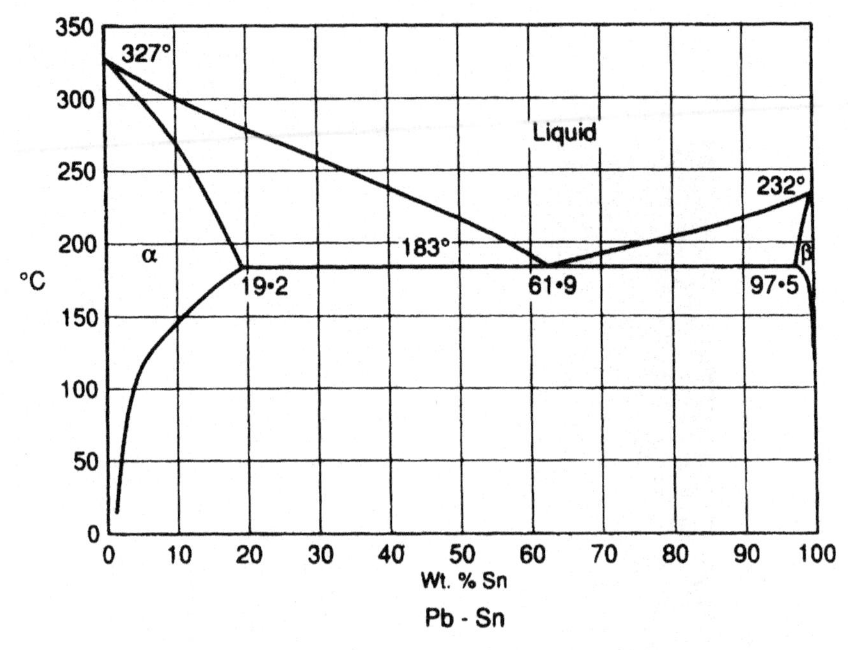

Fig. 9.4-13 A Pb–Sn phase diagram. (From P. J. Prescott and F. P. Incropera, *Metallurgical Transactions*, **22B**, 1991, p. 529.)

that arrives, the dendrites, which are poor in Sn, can in fact remelt to supply a low Sn melt.

As solidification continues, the solutally driven cell grows with the mushy zone, while the thermally driven cell shrinks with the liquid pool, as shown in Fig. 9.4-15. A channel has formed in the mushy zone.

The evolution of macrosegregation is shown in Fig. 9.4-16. As shown, Sn segregates more and more to the upper inner (right) corner. This, as already explained, is due to the fact that the lighter Sn-rich interdendritic liquid rises and escapes from the mushy zone as it grows upward and inward.

9.4.4. Semiconductor Device Fabrication: Chemical Vapor Deposition

Computation of simultaneous fluid flow, heat transfer, and mass transfer has been a widely used tool for studying chemical vapor deposition. Extensive work has been conducted by numerous investigators, notably by Jensen and his associates.[44]

Let us consider the axisymmetric vertical reactor shown in Fig. 9.4-17. The carrier gas (usually H_2 or N_2) with a low concentration of the reactant species A

[44] D. I. Fotiadis, S. Kieda, and K. F. Jensen, *J. Crystal Growth*, **102**, 411, 1990.

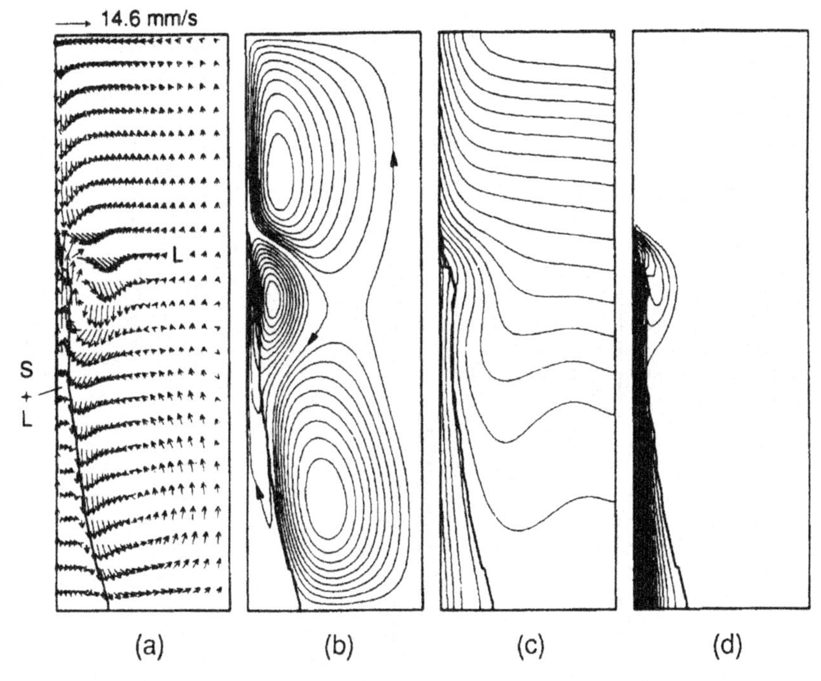

Fig. 9.4-14 Solidification of a Pb–19% Sn annular casting 110 s after cooling: (*a*) velocity vectors; (*b*) streamlines; (*c*) isotherms; (*d*) liquid isoconcentration lines. (From P. J. Prescott and F. P. Incropera, *Metallurgical Transactions*, **22B**, 1991, p. 529.)

is introduced from the top of the reactor. The wafer is heated by the graphite susceptor, which is, in turn, heated by the induction heater. The quartz tube, however, is not induction-heated. The susceptor and hence the wafer rotate at a constant angular velocity ω.

Since the flow rates are low, the flow in the reactor can be considered laminar and steady. Since the physical properties of the carrier gas can vary with temperature appreciably, it is useful to follow governing equations that are used for variable properties.

From Eqs. [4.2-7], [4.2-10] and [4.2-12], it is seen that $-\nabla p + \rho \mathbf{g} = -\nabla P - \rho\beta_T \mathbf{g}(T - T_0)$. From Table 4.1-4 and with gravity being the only body force involved, we have the following governing equations:

Continuity: $$\nabla \cdot (\rho \mathbf{v}) = 0 \qquad [9.4\text{-}33]$$

Momentum: $$\nabla \cdot (\rho \mathbf{v}\mathbf{v}) = -\nabla P - \nabla \cdot \boldsymbol{\tau} - \rho\beta_T \mathbf{g}(T - T_0) \qquad [9.4\text{-}34]$$

Energy: $$\nabla \cdot (\rho \mathbf{v} C_v T) = -\nabla \cdot \mathbf{q} + s \qquad [9.4\text{-}35]$$

Species: $$\nabla \cdot (\mathbf{v} c_A) = -\nabla \cdot \underline{\mathbf{j}}_A + \underline{r}_A \qquad [9.4\text{-}36]$$

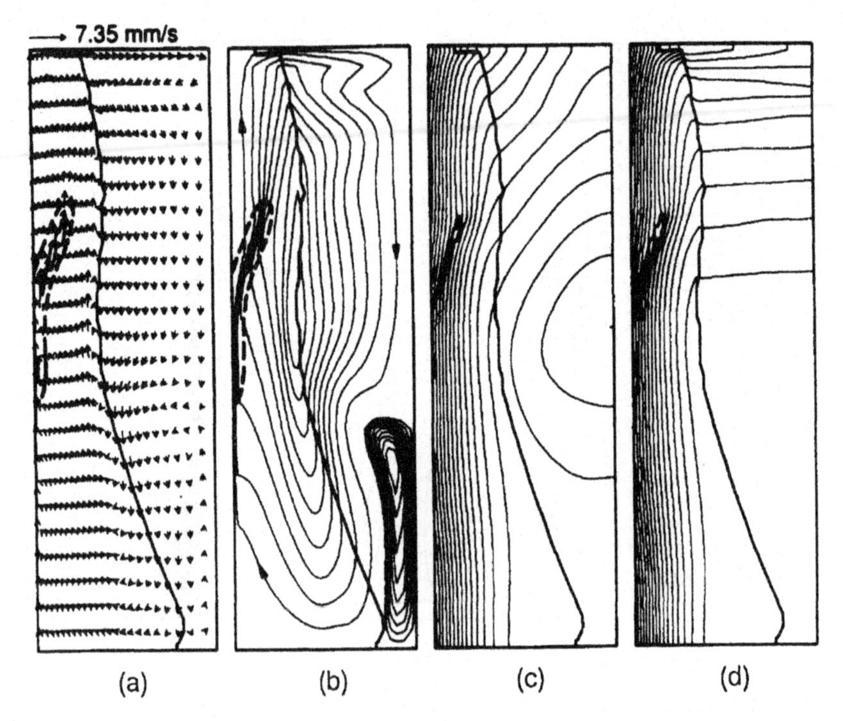

Fig. 9.4-15 Same as Fig. 9.4-14 but 165 s after cooling. (From P. J. Prescott and F. P. Incropera, *Metallurgical Transactions*, **22B**, 1991, p. 529.)

Regarding Eqn. [9.4-33], from the ideal-gas law

$$\rho = \frac{p_0 M_c}{RT} \qquad [9.4\text{-}37]$$

where p_0 is the operating pressure of the reactor, M_c the molecular weight of the carrier gas, R the gas constant, and T the temperature. Since pressure drops over the reactor are usually small, p_0 may be used instead of the local pressure.[45]

Regarding Eq. [9.4-35], in view of the very low concentration of the reactant species in the carrier gas, the heat of reactions can be neglected (i.e., $s = 0$). Viscous dissipation is also neglected.

Regarding Eq. [9.4-36], from Eq. [3.1-10]

$$\underline{\mathbf{j}}_A = -D_A\left(\nabla c_A + \frac{\alpha_T}{\rho} c_A c_C \nabla \ln T\right) \qquad [9.4\text{-}38]$$

where α_T is the thermal diffusion factor.

[45] D. I. Fotiadis, S. Kieda, and K. F. Jensen, ibid.

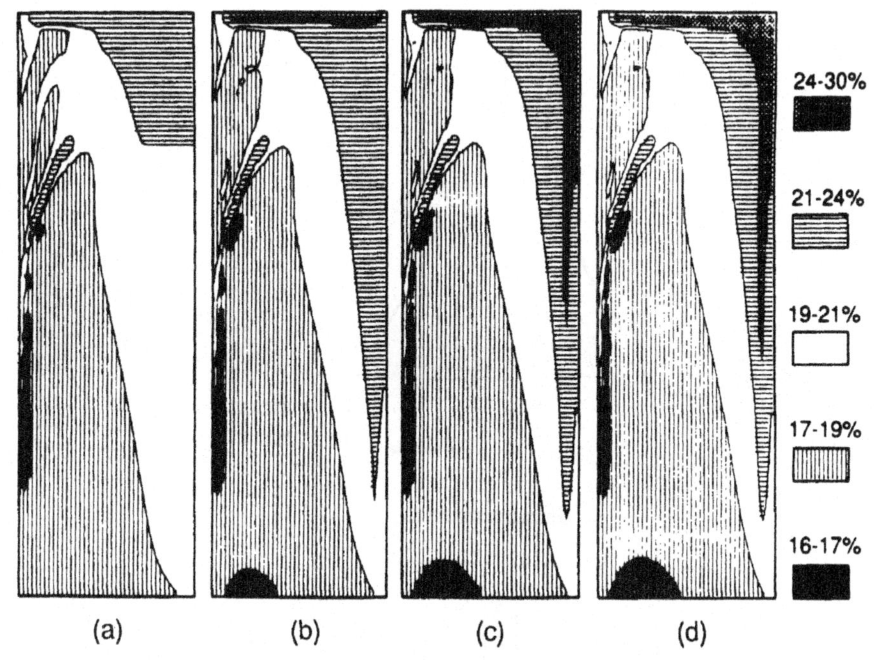

Fig. 9.4-16 Macrosegregation of Sn in a Pb–19% Sn annular casting after: (*a*) 300 s; (*b*) 450 s; (*c*) 600 s; (*d*) 1830 s. (From P. J. Prescott and F. P. Incropera, *Metallurgical Transactions*, **22B**, 1991, p. 529.)

The boundary conditions related to the gas phase are as follows:

Solid Surfaces:

$$v_r = v_z = 0 \qquad [9.4\text{-}39]$$

$$v_\theta = \omega_s r \qquad [9.4\text{-}40]$$

$$(-k\mathbf{n}\cdot\nabla T)_S = (-k\mathbf{n}\cdot\nabla T) + q_{\text{rad}} \qquad [9.4\text{-}41]$$

$$-D_A\left[\mathbf{n}\cdot\left(\nabla c_A + \frac{\alpha_T}{\rho}c_A c_C \nabla \ln T\right)\right] = k_{\text{surf}}c_{A,\text{surf}} \qquad [9.4\text{-}42]$$

In these equations ω_s is zero at the reactor wall and ω elsewhere. The constant k_{surf} is the surface reaction rate constant, assuming that the reaction kinetics are linear and that the diffusion of species A to the solid surface $(\mathbf{n}\cdot\underline{\mathbf{j}}_A)$ is the rate-determining step. The radiation heat flux

$$q_{\text{rad}} = \sigma \sum_{j=1}^{n} F_{js}[\varepsilon_{js}T_j^4 - \varepsilon_{jw}T_w^4] \qquad [9.4\text{-}43]$$

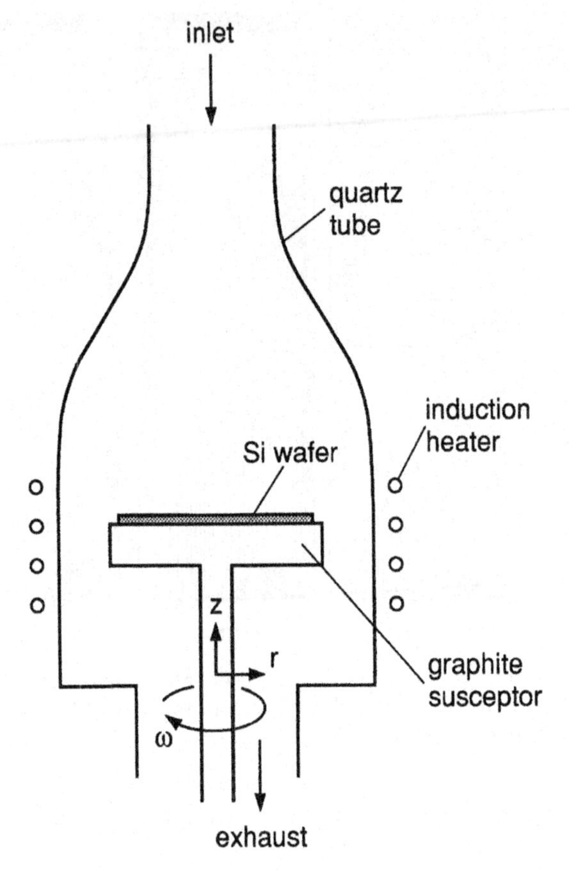

Fig. 9.4-17 An axisymmetric vertical reactor for chemical vapor deposition.

where σ is the Stefan–Boltzmann constant, F_{js} the view factor for the jth element of any solid surface to the reactor wall, and ε the effective emissivity. The solid surfaces are divided into small elements for finite-element computation. Heat flow in the solids is also calculated in order to determine $(-k\mathbf{n}\cdot\nabla T)_s$.

Centerline:

$$v_r = v_\theta = 0 \qquad [9.4\text{-}44]$$

$$\frac{\partial v_z}{\partial r} = \frac{\partial T}{\partial r} = \frac{\partial c_A}{\partial r} = 0 \qquad [9.4\text{-}45]$$

Inlet:

$$v_r = v_\theta = 0 \qquad [9.4\text{-}46]$$

$$v_z = \frac{Q}{\pi R_i^2} m \qquad [9.4\text{-}47]$$

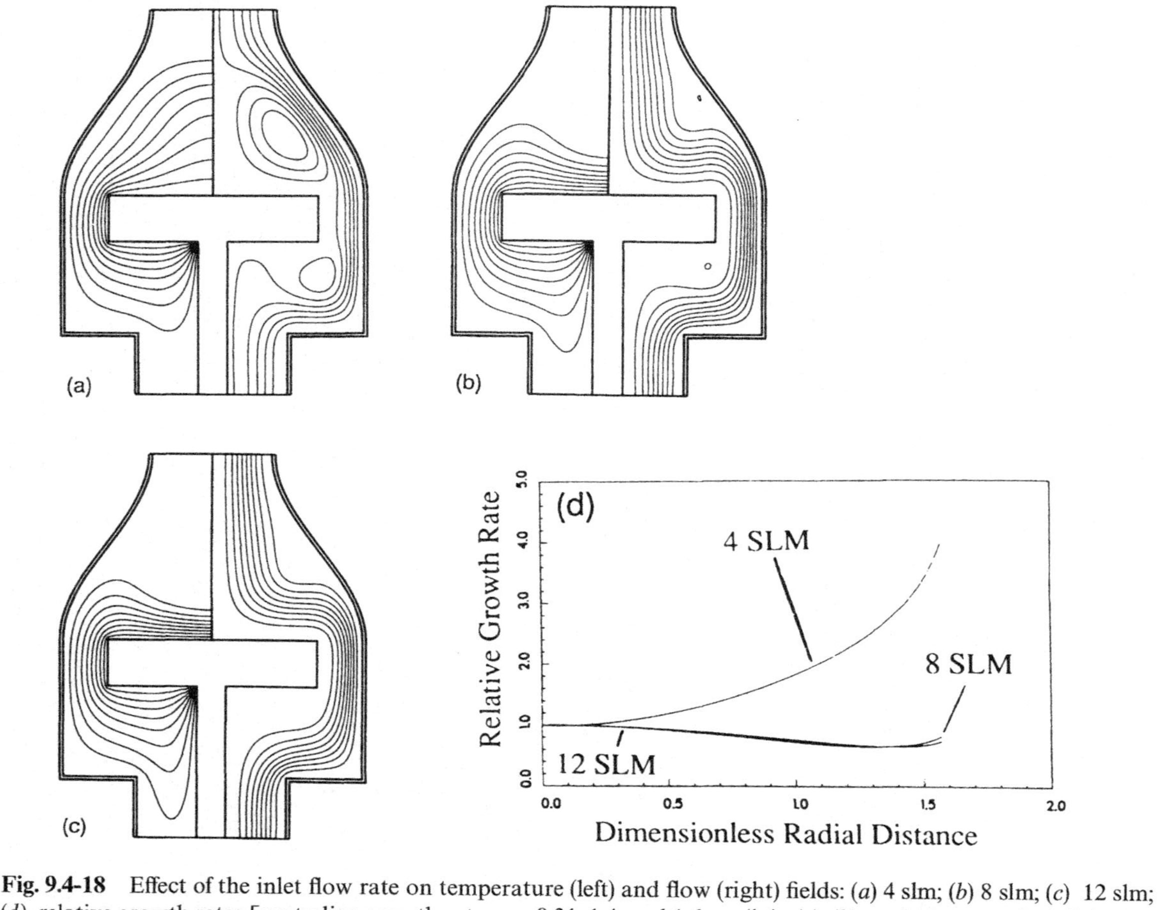

Fig. 9.4-18 Effect of the inlet flow rate on temperature (left) and flow (right) fields: (*a*) 4 slm; (*b*) 8 slm; (*c*) 12 slm; (*d*) relative growth rates [centerline growth rates are 0.31, 1.4, and 1.6 μm/h in (*a*), (*b*), and (*c*), respectively]. (From D. I. Fotiadis, S. Kieda, and K. F. Jensen, "Transpsort Phenomena in Vertical Reactors for Metal Organic Vapor Phase Epitaxy. 1. Effects of Heat Transfer Characteristics, Reactor Geometry, and Operating Conditions," *Journal of Crystal Growth*, **102**, 1990, p. 441.)

$$T = T_i \qquad [9.4\text{-}48]$$

$$c_A = c_{Ai} \qquad [9.4\text{-}49]$$

where Q is the volume flow rate of the carrier gas at the inlet; R_i, T_i and c_{Ai} are the inlet radius, temperature, and reactant concentration, respectively; and

$$m = 2\left[1 - \left(\frac{r}{R_i}\right)^2\right] \quad \text{(for parabolic velocity profile)} \qquad [9.4\text{-}50]$$

and

$$m = 1 \quad \text{(for plug velocity profile)} \qquad [9.4\text{-}51]$$

Outlet:

$$\frac{\partial v_r}{\partial z} = \frac{\partial v_\theta}{\partial z} = \frac{\partial v_z}{\partial z} = \frac{\partial T}{\partial z} = \frac{\partial c_A}{\partial z} = 0 \qquad [9.4\text{-}52]$$

assuming that the outlet tube is long enough that the velocity, temperature, and composition are fully developed.

Once the concentration field is obtained, the growth rate of the thin film on the wafer can be calculated as follows:

$$\text{Growth rate} = \bar{V}\, k_{\text{surf}} c_{A,\,\text{surf}} \qquad [9.4\text{-}53]$$

where $\bar{V}$ is the molar volume of the film material.

The results shown in Fig. 9.4-18 are for the chemical vapor deposition of GaAs films using arsine and trimethylgallium (TMG). The growth rate is considered to be limited by the diffusion of TMG to the solid surface, in the temperature range of 850–1050 K. The susceptor is 7 cm in radius and 900 K in temperature, and the carrier gas is H_2 at 0.13 atm. As shown in Fig. 9.4-18*a*, at a low flow rate such as 4 slm (standard liters per minute), the flow over the susceptor is dominated by a buoyancy convection cell recirculating in the clockwise direction. The presence of this cell reduces mass transfer to and hence the growth rate in the center of the wafer, with a thinner film in the center of the wafer. The growth rate at the centerline is 0.31 μm/h.

As the flow rate is increased to 8 and 12 slm, the buoyancy recirculation cell is eliminated, as shown in Figs. 9.4-18*b* and 9.4-14*c*. Consequently, the growth rate becomes more uniform, as shown in Fig. 9.4-18*d*. Moreover, the absolute growth rate is increased to 1.4 and 1.6 μm/h at the centerline. The separation of the streamlines from the upper reactor wall is due to the expansion (divergence) of the wall.

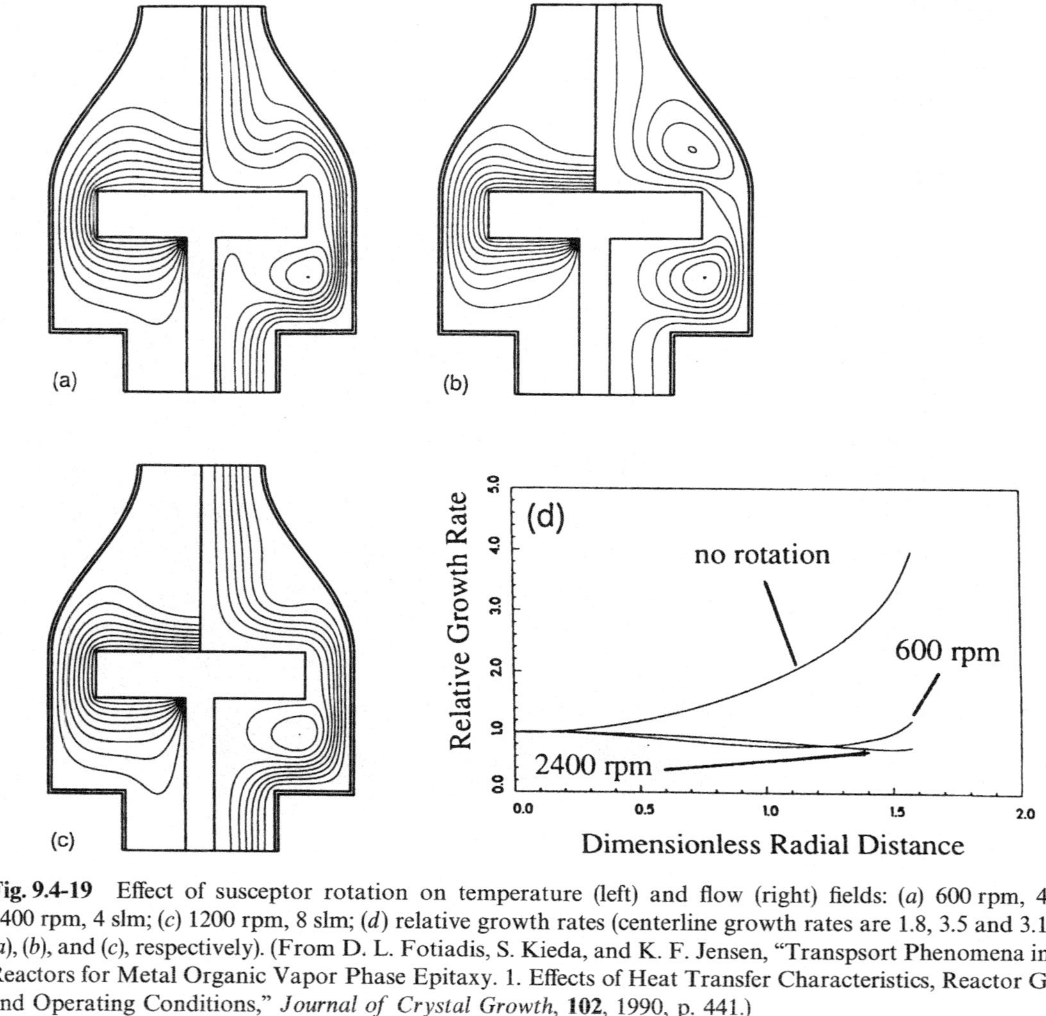

Fig. 9.4-19 Effect of susceptor rotation on temperature (left) and flow (right) fields: (*a*) 600 rpm, 4 slm; (*b*) 2400 rpm, 4 slm; (*c*) 1200 rpm, 8 slm; (*d*) relative growth rates (centerline growth rates are 1.8, 3.5 and 3.1 μm/h in (*a*), (*b*), and (*c*), respectively). (From D. L. Fotiadis, S. Kieda, and K. F. Jensen, "Transpsort Phenomena in Vertical Reactors for Metal Organic Vapor Phase Epitaxy. 1. Effects of Heat Transfer Characteristics, Reactor Geometry, and Operating Conditions," *Journal of Crystal Growth*, **102**, 1990, p. 441.)

Figure 9.4-19 shows the effect of susceptor rotation. At 600 rpm the buoyant recirculation cell is eliminated, as shown in Fig. 9.4-19*a*. As a result, the growth rate uniformity is improved, as shown in Fig. 9.4-19*d*. The gas near the rotating susceptor is subjected to a centrifugal force to flow radially outward, that is, opposite to the direction of the buoyant recirculation cell. At a very high rotation rate (e.g. 2400 rpm), the radially outward flow near the susceptor is strong enough to induce a counterclockwise recirculation cell above the susceptor, as shown in Fig. 9.4-19*b*. Although this recirculation cell does not affect the growth rate uniformity significantly, it can be reduced or even eliminated by increasing the inlet flow rate.

Figure 9.4-20 shows the comparison between calculated and observed flow patterns.[46] Forced convection due to susceptor rotation dominates in the case shown in Fig. 9.4-20*a*, whereas buoyancy convection dominates in the case in Fig. 9.4-20*b*.

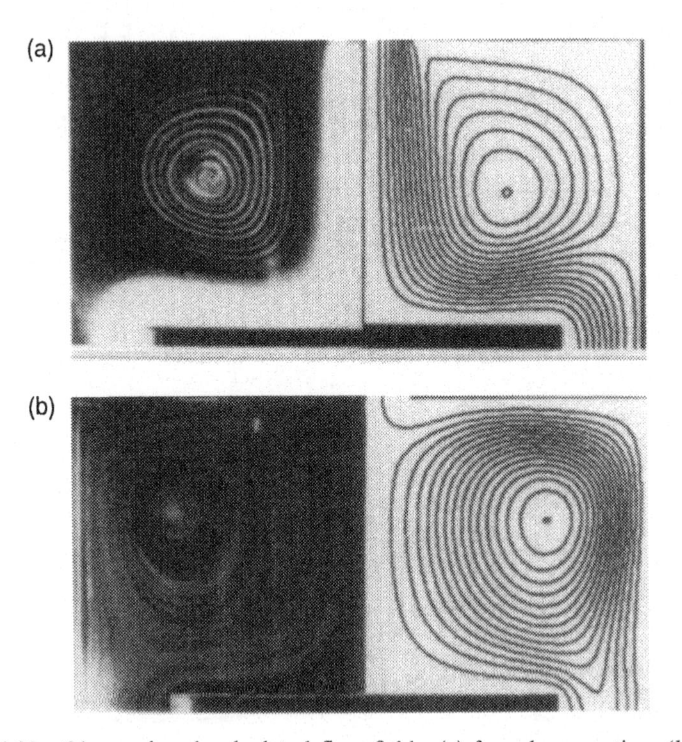

Fig. 9.4-20 Observed and calculated flow fields: (*a*) forced convection; (*b*) buoyancy convection. (From D. I. Fotiadis, A. M. Kremer, D. R. McKenna, and K. F. Jensen, "Complex Flow Phenomena in Vertical MOCVD Reactors: Effects on Deposition Uniformity and Interface Abruptness," *Journal of Crystal Growth*, **85**, 1987, p. 154.)

[46] D. I. Foitadis, A. M. Kremer, D. R. McKenna, and K. F. Jensen, *J. Crystal Growth*, **85**, 154, 1987.

PROBLEMS[47]

9.1-1 The so-called pedestal growth process is illustrated in Fig. P9.1-1. The molten zone is stationary and has a steady volume of Ω_L. The crystal pulling speed is V, and the cross-sectional areas of the crystal and the feed are A_S and A_f, respectively. The melt composition is uniform. Derive an equation similar to Eq. [9.1-19], with w_{AS}^* as a function of w_{A0}, k_0, A_s, A_f, l_S, and Ω_L.

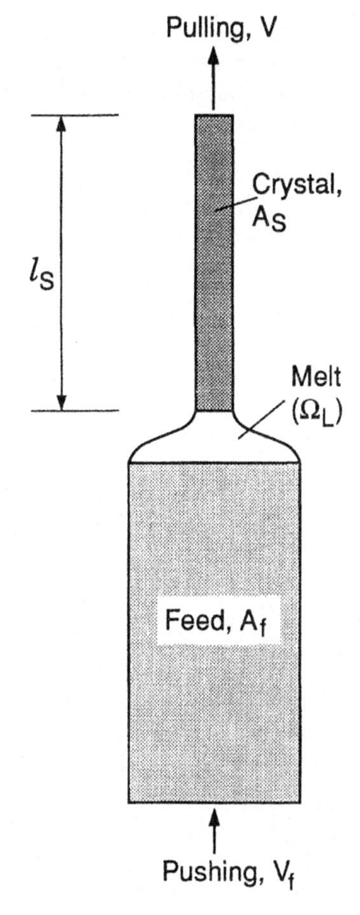

Fig. P9.1-1

9.2-1 A 1010 steel (containing 0.10 wt% carbon) is carburized at 975 °C for 5 h by exposure to a gas mixture containing about 16% CO and 5% CO_2. Determine the thickness of the carburized layer in which the

[47] The first two digits in problem numbers refer to section numbers in text.

carbon content is no less than 0.7 wt%. The average diffusion coefficient of carbon is austenite (γ)

$$D = 0.12 \times \exp\left(-\frac{32{,}000}{RT}\right) \quad \mathrm{cm^2/s}$$

where

$$R = 8.31 \quad \mathrm{J\ mol^{-1}\ K^{-1}}$$

and T is in kelvins.

9.3-1 Referring to Fig. 9.3-5, determine the concentration profile in the final transient region of the crystal. The length of the feed is l_f. The melt composition is uniform at w_{A0}/k_0 before the final transient.

9.4-1 Consider the Bridgman growth of a crystal from a slightly doped melt, in which the melt density depends on temperature only. Explain the convection patterns shown in Figs. P9.4-1*a* and P9.4-1*b*.

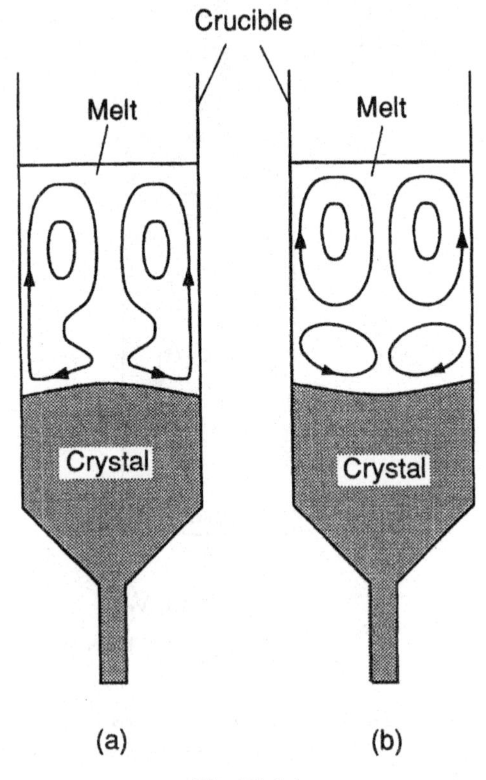

Fig. P9.4-1

9.4-2 Sketch the radial dopant distributions in the crystals shown in Fig. P9.4-1 assuming $k_0 < 1$.

9.4-3 Sketch the radial dopant distributions in the crystals shown in Fig. P9.4-1 assuming $k_0 > 1$.

9.4-4 How would the radial dopant distributions in the crystals shown in Fig. P9.4-1 be affected if convection became strong?

9.4-5 Consider the effect of $\partial\gamma/\partial T$ on the radial dopant segregation shown in Fig. 9.4-9, characterize the segregation, assuming $k_0 > 1$.

9.4-6 Consider the microgravity floating-zone crystal growth shown in Fig. 9.4-11*a* for the case of $\partial\gamma/\partial T = 0$. What would radial dopant segregation be like if the crystal alone were rotated?

9.4-7 Consider the microgravity floating-zone crystal growth shown in Fig. 9.4-11*a* for the case of $\partial\gamma/\partial T = 0$. What would radial dopant segregation be like if the feed rod and the crystal were counterrotated?

9.4-8 Consider the microgravity floating-zone crystal growth shown in Fig. 9.4-11*b* for the case of $\partial\gamma/\partial T = -0.056$ dyn cm^{-1} °C^{-1}. What would radial dopant segregation be like if the crystal alone were rotated?

9.4-9 Consider the microgravity floating-zone crystal growth shown in Fig. 9.4-11*b* for the case of $\partial\gamma/\partial T = -0.056$ dyn cm^{-1} °C^{-1}. What would radial dopant segregation be like if the feed rod and the crystal were counterrotated?

9.4-10 Consider the directional solidification of the horizontal ingot shown in Fig. P9.4-10. What would the solute segregation in the transverse direction be like if the ingot were a Sn–15% Pb alloy?

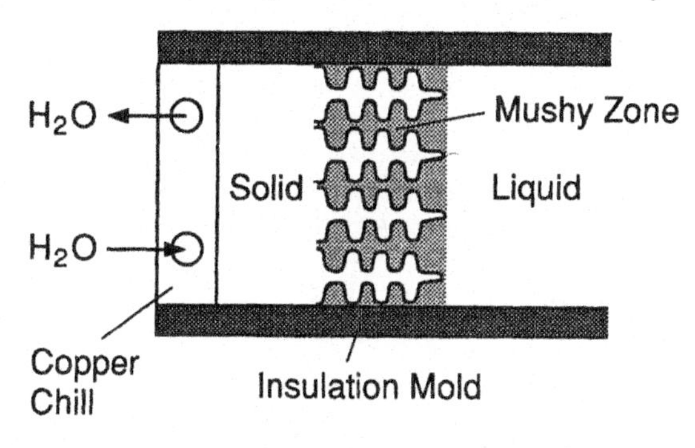

Fig. P9.4-10

9.4-11 Consider the directional solidification of the horizontal ingot shown in Fig. P9.4-10. What would the solute segregation in the transverse direction be like if the ingot were a Pb–15% Sn alloy?

9.4-12 Consider the casting of the ingot shown in Fig. 6.1-22. What would radial solute segregation be like if the ingot were a Pb–15% Sn alloy?

9.4-13 Consider the casting of the ingot shown in Fig. 6.1-22. Suppose the ingot is Pb–15% Sn. How would rotation affect radial solute segregation?

9.4-14 Consider the case of chemical vapor deposition shown in Fig. 9.4-18*a*, where the reactor wall is at 300 K. How would the flow pattern, isotherms, and growth rate be affected if the reactor wall were insulated?

9.4-15 Consider the case of chemical vapor deposition shown in Fig. 9.4-18*a*, where the reactor inlet is 4 cm in radius. How would the flow pattern, isotherms, and growth rate be affected if the inlet radius were decreased to 1 cm and increased to 6 cm?

9.4-16 Consider the case of chemical vapor deposition shown in Fig. 9.4-18*a*, where the susceptor–inlet distance is 12 cm. How would the flow pattern, isotherms, and growth rate be affected if the distance were increased to 20 cm and decreased to 4 cm?

9.4-17 Consider the case of chemical vapor deposition shown in Fig. 9.4-18*a*. How would the flow pattern, isotherms, and growth rate be affected if the whole reactor system were inverted upside down?

9.4-18 Consider the case of chemical vapor deposition shown in Fig. 9.4-18*a*, where hydrogen is the carrier gas. How would the flow pattern, isotherms, and growth rate be affected if nitrogen were used? As compared to hydrogen, nitrogen is heavier and less thermally conductive.

APPENDIX A

MATHEMATICS REVIEW: VECTORS, TENSORS, AND DIFFERENTIAL EQUATIONS

Vectors, tensors, and differential equations, which are used throughout the book, are briefly reviewed. This chapter is not intended to be a general discussion of the subjects. Instead, it is designed to provide equations of vectors and tensors, and solutions of differential equations that are used in the book. The purpose is to minimize detraction from the discussion of the main subject of transport phenomena.

In transport phenomena physical quantities can be classified into the following categories: **scalars**, such as temperature, energy, density, volume, pressure, and time; **vectors**, such as displacement, velocity, momentum, acceleration, and force; and **tensors**, such as shear stress and momentum flux. In this book, vectors and tensors are represented in boldface and scalars in lightface.

For convenience physical quantities and the multiplication signs between them are each assigned an order, as shown in Table A.1-1. According to the table, scalars, vectors, and tensors have the orders of 0, 1, and 2, respectively. From the table the resultant order of a product of two similar or different physical quantities can be determined. For example, the product $\mathbf{v} \cdot \mathbf{w}$ of two

TABLE A.1-1 Orders of Physical Quantities and Their Multiplication Signs

	Physical Quantity			Multiplication Sign			
	Scalar	Vector	Tensor	None	$\times$	$\cdot$	$:$
Order	0	1	2	0	-1	-2	-4

vectors $\mathbf{v}$ and $\mathbf{w}$ has the resultant order of $1-2+1=0$ and is, therefore, a scalar. For another example, the product $\nabla \times \mathbf{v}$ of a vector differential operator ∇ and a vector $\mathbf{v}$ has the resultant order of $1-1+1=1$ and is, therefore, a vector. For yet another example, the product $\mathbf{v} \cdot \boldsymbol{\tau}$ of a vector $\mathbf{v}$ and a tensor $\boldsymbol{\tau}$ has the resultant order of $1-2+2=1$ and is, therefore, a vector.

A.1. VECTORS

A.1.1. Definitions

A.1.1.1. Unit Vectors

Unit vectors $\mathbf{e}_x$, $\mathbf{e}_y$, and $\mathbf{e}_z$, shown in Fig. A.1-1, are vectors in the x, y, and z directions, respectively, and with unit magnitude:

$$|\mathbf{e}_x| = 1$$

$$|\mathbf{e}_y| = 1$$

$$|\mathbf{e}_z| = 1 \qquad \text{[A.1-1]}$$

A.1.1.2. Vectors

Unlike a scalar, which has only one component, a vector has three components. The velocity vector $\mathbf{v}$, for instance, has three components: v_x, v_y, and v_z in the

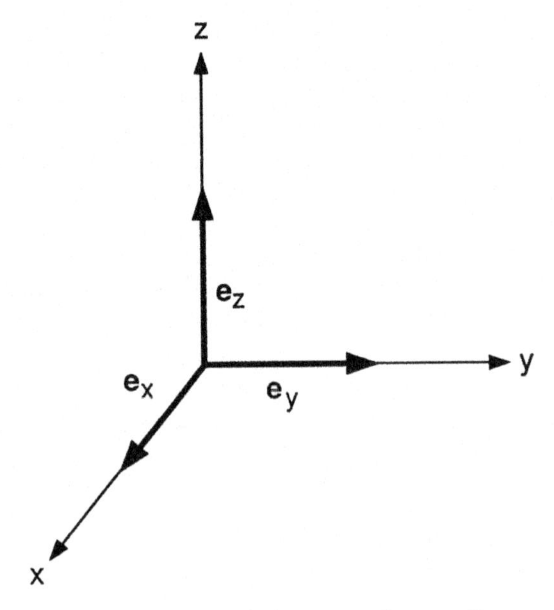

Fig. A.1-1 Unit vectors in rectangular coordinates.

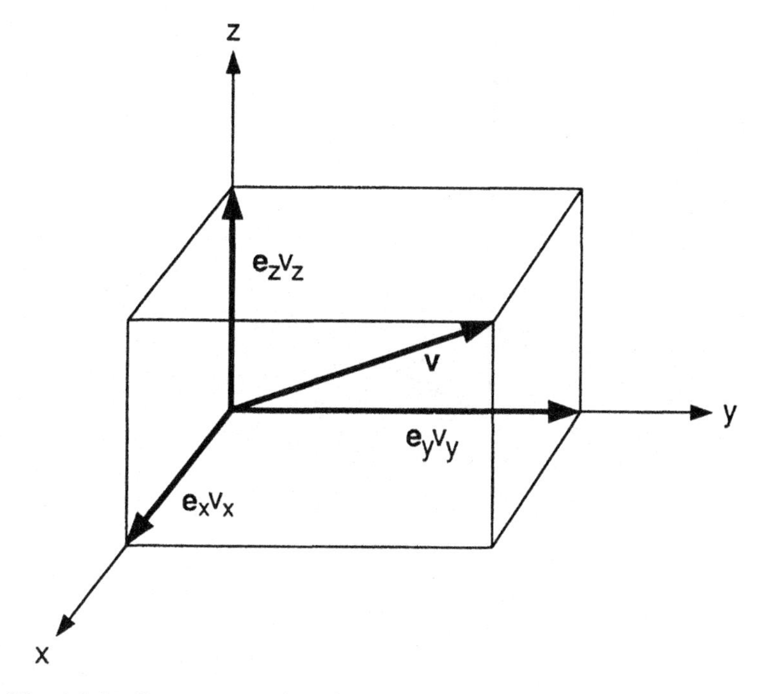

Fig. A.1-2 Components of a velocity vector **v** in rectangular coordinates.

x, y, and z directions, respectively, as illustrated in Fig. A.1-2. It can be expressed as follows:

$$\boxed{\mathbf{v} = \mathbf{e}_x v_x + \mathbf{e}_y v_y + \mathbf{e}_z v_z} \quad \text{[A.1-2]}$$

It can also be expressed in as a row matrix

$$\boxed{\mathbf{v} = (v_x \;\; v_y \;\; v_z)} \quad \text{[A.1-3]}$$

or as a column matrix

$$\boxed{\mathbf{v} = \begin{pmatrix} v_x \\ v_y \\ v_z \end{pmatrix}} \quad \text{[A.1-4]}$$

A **radial displacement vector**, shown in Fig. A.1-3, can be expressed as follows:

$$\mathbf{r} = \mathbf{e}_x x + \mathbf{e}_y y + \mathbf{e}_z z \quad \text{[A.1-5]}$$

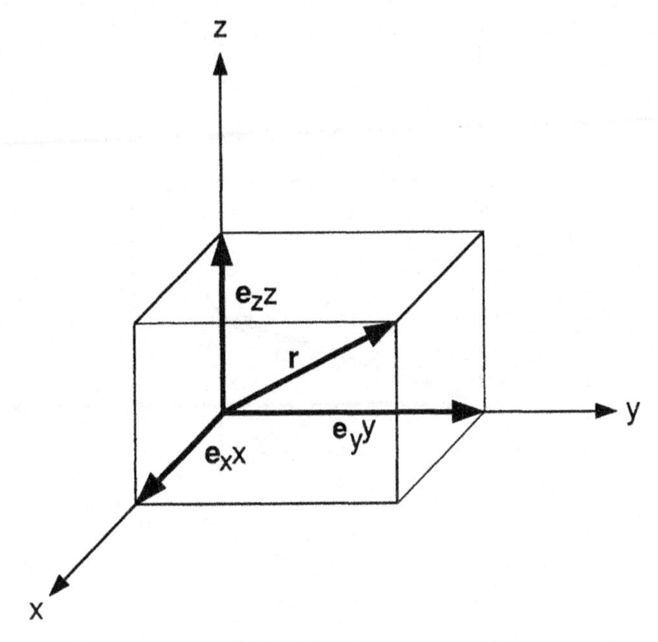

Fig. A.1-3 Components of a radial displacement vector **r** in rectangular coordinates.

From Eq. [A.1-2] the unit vector $\mathbf{e}_y$ can be expressed as follows:

$$\mathbf{e}_y = \mathbf{e}_x 0 + \mathbf{e}_y 1 + \mathbf{e}_z 0 \qquad \text{[A.1-6]}$$

or

$$\mathbf{e}_y = (0 \ \ 1 \ \ 0) \qquad \text{[A.1-7]}$$

or

$$\mathbf{e}_y = \begin{pmatrix} 0 \\ 1 \\ 0 \end{pmatrix} \qquad \text{[A.1-8]}$$

A.1.2. Products

A.1.2.1. Scalar Product

The scalar product of two vectors **v** and **w**, denoted as **v**·**w**, is a scalar quantity defined by

$$\boxed{\mathbf{v}\cdot\mathbf{w} = |\mathbf{v}|\,|\mathbf{w}|\cos(\mathbf{v}, \mathbf{w})} \qquad \text{[A.1-9]}$$

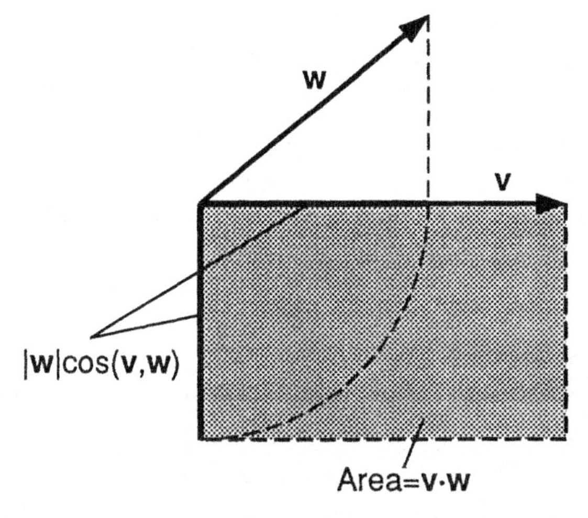

Fig. A.1-4 Scalar product $\mathbf{v} \cdot \mathbf{w}$ of two vectors, $\mathbf{v}$ and $\mathbf{w}$.

where $(\mathbf{v}, \mathbf{w})$ is the angle between vectors $\mathbf{v}$ and $\mathbf{w}$. The rectangle in Fig. A.1-4 respresents $\mathbf{v} \cdot \mathbf{w}$.

From Eq. [A.1-9] and Fig. A.1-1

$$\mathbf{e}_x \cdot \mathbf{e}_x = |\mathbf{e}_x|\,|\mathbf{e}_x| \cos 0° = 1 \qquad \text{[A.1-10]}$$

$$\mathbf{e}_x \cdot \mathbf{e}_y = |\mathbf{e}_x|\,|\mathbf{e}_y| \cos 90° = 0 \qquad \text{[A.1-11]}$$

In summary

$$\mathbf{e}_x \cdot \mathbf{e}_x = 1$$

$$\mathbf{e}_y \cdot \mathbf{e}_y = 1$$

$$\mathbf{e}_z \cdot \mathbf{e}_z = 1 \qquad \text{[A.1-12]}$$

and

$$\mathbf{e}_x \cdot \mathbf{e}_y = \mathbf{e}_y \cdot \mathbf{e}_x = 0$$

$$\mathbf{e}_y \cdot \mathbf{e}_z = \mathbf{e}_z \cdot \mathbf{e}_y = 0 \qquad \text{[A.1-13]}$$

$$\mathbf{e}_z \cdot \mathbf{e}_x = \mathbf{e}_x \cdot \mathbf{e}_z = 0$$

From Eqs. [A.1-2], [A.1-12], and [A.1-13]

$$\mathbf{v}\cdot\mathbf{w} = (\mathbf{e}_x v_x + \mathbf{e}_y v_y + \mathbf{e}_z v_z)\cdot(\mathbf{e}_x w_x + \mathbf{e}_y w_y + \mathbf{e}_z w_z)$$

$$= v_x w_x + v_y w_y + v_z w_z \qquad \text{[A.1-14]}$$

It can be easily shown that $\mathbf{v}\cdot\mathbf{w} = \mathbf{w}\cdot\mathbf{v}$. If $\mathbf{w} = \mathbf{v}$, Eq. [A.1-14] becomes

$$\mathbf{v}\cdot\mathbf{v} = v_x^2 + v_y^2 + v_z^2 = v^2 \qquad \text{[A.1-15]}$$

A.1.2.2. Vector Product

The vector product of two vectors, $\mathbf{v}$ and $\mathbf{w}$, denoted as $\mathbf{v}\times\mathbf{w}$, is a vector defined by

$$\boxed{\mathbf{v}\times\mathbf{w} = |\mathbf{v}|\,|\mathbf{w}|\sin(\mathbf{v},\mathbf{w})\,\mathbf{n}_{vw}} \qquad \text{[A.1-16]}$$

where $\mathbf{n}_{vw}$ is a unit vector normal to the plane containing $\mathbf{v}$ and $\mathbf{w}$ and pointing in the direction that a right-handed screw will move if turned from $\mathbf{v}$ to $\mathbf{w}$. The parallelogram in Fig. A.1-5 represents the magnitude of $\mathbf{v}\times\mathbf{w}$. It is clear that $\mathbf{v}\times\mathbf{w} = -\mathbf{w}\times\mathbf{v}$.

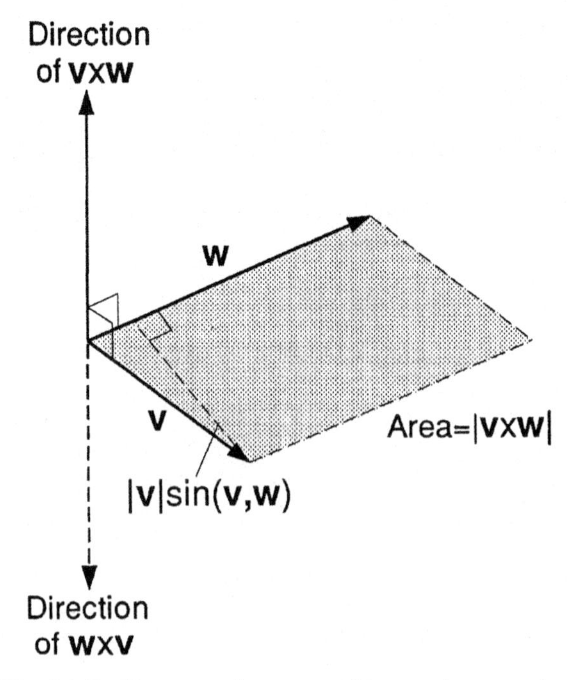

Fig. A.1-5 Vector product $\mathbf{v}\times\mathbf{w}$ of two vectors, $\mathbf{v}$ and $\mathbf{w}$.

From Eq. [A.1-16] and Fig. A.1-1

$$\mathbf{e}_x \times \mathbf{e}_x = 0$$
$$\mathbf{e}_y \times \mathbf{e}_y = 0$$
$$\mathbf{e}_z \times \mathbf{e}_z = 0 \qquad [A.1\text{-}17]$$

and

$$\mathbf{e}_x \times \mathbf{e}_y = \mathbf{e}_z$$
$$\mathbf{e}_y \times \mathbf{e}_x = -\mathbf{e}_z$$
$$\mathbf{e}_x \times \mathbf{e}_z = -\mathbf{e}_y$$
$$\mathbf{e}_z \times \mathbf{e}_x = \mathbf{e}_y$$
$$\mathbf{e}_z \times \mathbf{e}_y = -\mathbf{e}_x$$
$$\mathbf{e}_y \times \mathbf{e}_z = \mathbf{e}_x \qquad [A.1\text{-}18]$$

From Eq. [A.1-2]

$$\mathbf{v} \times \mathbf{w} = (\mathbf{e}_x v_x + \mathbf{e}_y v_y + \mathbf{e}_z v_z) \times (\mathbf{e}_x w_x + \mathbf{e}_y w_y + \mathbf{e}_z w_z) \qquad [A.1\text{-}19]$$

Substituting Eqs. [A.1-17] and [A.1-18] into Eq. [A.1-19] and rearranging

$$\mathbf{v} \times \mathbf{w} = \mathbf{e}_x(v_y w_z - v_z w_y) + \mathbf{e}_y(v_z w_x - v_x w_z) + \mathbf{e}_z(v_x w_y - v_y w_x)$$
$$= \begin{vmatrix} \mathbf{e}_x & \mathbf{e}_y & \mathbf{e}_z \\ v_x & v_y & v_z \\ w_x & w_y & w_z \end{vmatrix} \qquad [A.1\text{-}20]$$

A.2. TENSORS

A.2.1. Definitions

Unlike a vector, which has three components, a tensor (2nd order) has nine components. A tensor $\boldsymbol{\tau}$, such as a stress tensor, can be expressed as follows in rectangular coordinates:

$$\boldsymbol{\tau} = \begin{pmatrix} \tau_{xx} & \tau_{xy} & \tau_{xz} \\ \tau_{yx} & \tau_{yy} & \tau_{yz} \\ \tau_{zx} & \tau_{zy} & \tau_{zz} \end{pmatrix} \qquad [A.2\text{-}1]$$

A.2.2. Product

The **tensor product** of two vectors **v** and **w**, denoted as **vw**, is a tensor defined by

$$\mathbf{vw} = \begin{pmatrix} v_x \\ v_y \\ v_z \end{pmatrix} (w_x \;\; w_y \;\; w_z) = \begin{pmatrix} v_x w_x & v_x w_y & v_x w_z \\ v_y w_x & v_y w_y & v_y w_z \\ v_z w_x & v_z w_y & v_z w_z \end{pmatrix} \qquad \text{[A.2-2]}$$

The **vector product** of a tensor $\boldsymbol{\tau}$ and a vector **v**, denoted by $\boldsymbol{\tau}\cdot\mathbf{v}$, is a vector defined by

$$\begin{aligned}\boldsymbol{\tau}\cdot\mathbf{v} &= \begin{pmatrix} \tau_{xx} & \tau_{xy} & \tau_{xz} \\ \tau_{yx} & \tau_{yy} & \tau_{yz} \\ \tau_{zx} & \tau_{zy} & \tau_{zz} \end{pmatrix} \cdot \begin{pmatrix} v_x \\ v_y \\ v_z \end{pmatrix} \\ &= \mathbf{e}_x(v_x\tau_{xx} + v_y\tau_{xy} + v_z\tau_{xz}) + \mathbf{e}_y(v_x\tau_{yx} + v_y\tau_{yy} + v_z\tau_{yz}) \\ &\quad + \mathbf{e}_z(v_x\tau_{zx} + v_y\tau_{zy} + v_z\tau_{zz}) \end{aligned} \qquad \text{[A.2-3]}$$

For example, if $\mathbf{v} = \mathbf{e}_y$, from Eqs. [A.1-8] and [A.2-3]

$$\boldsymbol{\tau}\cdot\mathbf{e}_y = \begin{pmatrix} \tau_{xx} & \tau_{xy} & \tau_{xz} \\ \tau_{yx} & \tau_{yy} & \tau_{yz} \\ \tau_{zx} & \tau_{zy} & \tau_{zz} \end{pmatrix} \cdot \begin{pmatrix} 0 \\ 1 \\ 0 \end{pmatrix} = \mathbf{e}_x\tau_{xy} + \mathbf{e}_y\tau_{yy} + \mathbf{e}_z\tau_{zy} \qquad \text{[A.2-4]}$$

Similarly, the product between a tensor **vv** and a vector **n** is a vector as shown below:

$$\begin{aligned}\mathbf{vv}\cdot\mathbf{n} &= \begin{pmatrix} v_x v_x & v_x v_y & v_x v_z \\ v_y v_x & v_y v_y & v_y v_z \\ v_z v_x & v_z v_y & v_z v_z \end{pmatrix} \cdot \begin{pmatrix} n_x \\ n_y \\ n_z \end{pmatrix} \\ &= \mathbf{e}_x(v_x v_x n_x + v_x v_y n_y + v_x v_z n_z) + \mathbf{e}_y(v_y v_x n_x + v_y v_y n_y + v_y v_z n_z) \\ &\quad + \mathbf{e}_z(v_z v_x n_x + v_z v_y n_y + v_z v_z n_z) \\ &= (\mathbf{e}_x v_x + \mathbf{e}_y v_y + \mathbf{e}_z v_z)(v_x n_x + v_y n_y + v_z n_z) \\ &= \mathbf{v}(\mathbf{v}\cdot\mathbf{n}) \end{aligned} \qquad \text{[A.2-5]}$$

The **scalar product** of two tensors $\boldsymbol{\sigma}$ and $\boldsymbol{\tau}$, denoted as $\boldsymbol{\sigma}:\boldsymbol{\tau}$, is a scalar defined by

$$\begin{aligned}\boldsymbol{\sigma}:\boldsymbol{\tau} &= \begin{pmatrix} \sigma_{xx} & \sigma_{xy} & \sigma_{xz} \\ \sigma_{yx} & \sigma_{yy} & \sigma_{yz} \\ \sigma_{zx} & \sigma_{zy} & \sigma_{zz} \end{pmatrix} : \begin{pmatrix} \tau_{xx} & \tau_{xy} & \tau_{xz} \\ \tau_{yx} & \tau_{yy} & \tau_{yz} \\ \tau_{zx} & \tau_{zy} & \tau_{zz} \end{pmatrix} \\ &= \sigma_{xx}\tau_{xx} + \sigma_{xy}\tau_{yx} + \sigma_{xz}\tau_{zx} + \sigma_{yx}\tau_{xy} + \sigma_{yy}\tau_{yy} + \sigma_{yz}\tau_{zy} \\ &\quad + \sigma_{zx}\tau_{xz} + \sigma_{zy}\tau_{yz} + \sigma_{zz}\tau_{zz} \end{aligned} \qquad \text{[A.2-6]}$$

Similarly, the **scalar product** of two tensors **vw** and $\boldsymbol{\tau}$ is

$$\mathbf{vw}\cdot\boldsymbol{\tau} = \begin{pmatrix} v_x w_x & v_x w_y & v_x w_z \\ v_y w_x & v_y w_y & v_y w_z \\ v_z w_x & v_z w_y & v_z w_z \end{pmatrix} : \begin{pmatrix} \tau_{xx} & \tau_{xy} & \tau_{xz} \\ \tau_{yx} & \tau_{yy} & \tau_{yz} \\ \tau_{zx} & \tau_{zy} & \tau_{zz} \end{pmatrix}$$

$$= v_x w_x \tau_{xx} + v_x w_y \tau_{yx} + v_x w_z \tau_{zx} + v_y w_x \tau_{xy} + v_y w_y \tau_{yy} + v_y w_z \tau_{zy} + v_z w_x \tau_{xz} + v_z w_y \tau_{yz} + v_z w_z \tau_{zz} \qquad [A.2-7]$$

A.3. DIFFERENTIAL OPERATORS

A.3.1. Definitions

The vector differential operation ∇, called "del," (also called inverted delta "nabla") has components similar to those of a vector. However, unlike a vector, it cannot stand alone and must operate on a scalar, vector, or tensor function. In rectangular coordinates it is defined by

$$\boxed{\nabla = \mathbf{e}_x \frac{\partial}{\partial x} + \mathbf{e}_y \frac{\partial}{\partial y} + \mathbf{e}_z \frac{\partial}{\partial z}} \qquad [A.3-1]$$

It can also be expressed in matrix forms similar to those in Eqs. [A.1-3] and [A.1-4].

The **gradient** of a scalar field s, denoted as ∇s, is a vector defined by

$$\nabla s = \mathbf{e}_x \frac{\partial s}{\partial x} + \mathbf{e}_y \frac{\partial s}{\partial y} + \mathbf{e}_z \frac{\partial s}{\partial z} \qquad [A.3-2]$$

For the product rs between two scalars r and s, $\nabla(rs)$ is a vector defined by

$$\nabla(rs) = \mathbf{e}_x \frac{\partial}{\partial x}(rs) + \mathbf{e}_y \frac{\partial}{\partial y}(rs) + \mathbf{e}_z \frac{\partial}{\partial z}(rs)$$

$$= \mathbf{e}_x\left(r\frac{\partial s}{\partial x} + s\frac{\partial r}{\partial x}\right) + \mathbf{e}_y\left(r\frac{\partial s}{\partial y} + s\frac{\partial r}{\partial y}\right) + \mathbf{e}_z\left(r\frac{\partial s}{\partial z} + s\frac{\partial r}{\partial z}\right)$$

$$= r\left(\mathbf{e}_x\frac{\partial s}{\partial x} + \mathbf{e}_y\frac{\partial s}{\partial y} + \mathbf{e}_z\frac{\partial s}{\partial z}\right) + s\left(\mathbf{e}_x\frac{\partial r}{\partial x} + \mathbf{e}_y\frac{\partial r}{\partial y} + \mathbf{e}_z\frac{\partial r}{\partial z}\right) \qquad [A.3-3]$$

Therefore

$$\boxed{\nabla(rs) = r\nabla s + s\nabla r} \qquad [A.3-4]$$

A.3.2. Products

The **divergence** of a vector field $\mathbf{v}$, denoted as $\nabla \cdot \mathbf{v}$, is a scalar similar to the scalar product between two vectors. From Eq. [A.1-14]

$$\nabla \cdot \mathbf{v} = \left(\mathbf{e}_x \frac{\partial}{\partial x} + \mathbf{e}_y \frac{\partial}{\partial y} + \mathbf{e}_z \frac{\partial}{\partial z} \right) \cdot (\mathbf{e}_x v_x + \mathbf{e}_y v_y + \mathbf{e}_z v_z)$$

$$= \frac{\partial v_x}{\partial x} + \frac{\partial v_y}{\partial y} + \frac{\partial v_z}{\partial z} \qquad \text{[A.3-5]}$$

Similarly

$$\nabla \cdot (a\mathbf{v}) = \left(\mathbf{e}_x \frac{\partial}{\partial x} + \mathbf{e}_y \frac{\partial}{\partial y} + \mathbf{e}_z \frac{\partial}{\partial z} \right) \cdot (\mathbf{e}_x a v_x + \mathbf{e}_y a v_y + \mathbf{e}_z a v_z)$$

$$= \frac{\partial}{\partial x}(a v_x) + \frac{\partial}{\partial y}(a v_y) + \frac{\partial}{\partial z}(a v_z)$$

$$= a \left(\frac{\partial v_x}{\partial x} + \frac{\partial v_y}{\partial y} + \frac{\partial v_z}{\partial z} \right) + \left(v_x \frac{\partial a}{\partial x} + v_y \frac{\partial a}{\partial y} + v_z \frac{\partial a}{\partial z} \right) \qquad \text{[A.3-6]}$$

In other words

$$\boxed{\nabla \cdot (a\mathbf{v}) = a(\nabla \cdot \mathbf{v}) + \mathbf{v} \cdot \nabla a} \qquad \text{[A.3-7]}$$

From Eqs. [A.3-1] and [A.3-2]

$$\nabla \cdot \nabla s = \left(\mathbf{e}_x \frac{\partial}{\partial x} + \mathbf{e}_y \frac{\partial}{\partial y} + \mathbf{e}_z \frac{\partial}{\partial z} \right) \cdot \left(\mathbf{e}_x \frac{\partial s}{\partial x} + \mathbf{e}_y \frac{\partial s}{\partial y} + \mathbf{e}_z \frac{\partial s}{\partial z} \right)$$

$$= \frac{\partial^2 s}{\partial x^2} + \frac{\partial^2 s}{\partial y^2} + \frac{\partial^2 s}{\partial z^2} \qquad \text{[A.3-8]}$$

In other words

$$\nabla \cdot \nabla s = \nabla^2 s \qquad \text{[A.3-9]}$$

where the differential operator ∇^2, called **Laplace operator**, is defined as

$$\boxed{\nabla^2 = \frac{\partial^2}{\partial x^2} + \frac{\partial^2}{\partial y^2} + \frac{\partial^2}{\partial z^2}} \qquad \text{[A.3-10]}$$

The **curl** of a vector field v, denoted by $\nabla \times \mathbf{v}$, is a vector like the vector product of two vectors. From Eq. [A.1-20]

$$\nabla \times \mathbf{v} = \begin{vmatrix} \mathbf{e}_x & \mathbf{e}_y & \mathbf{e}_z \\ \dfrac{\partial}{\partial x} & \dfrac{\partial}{\partial y} & \dfrac{\partial}{\partial z} \\ v_x & v_y & v_z \end{vmatrix}$$

$$= \mathbf{e}_x\left(\frac{\partial v_z}{\partial y} - \frac{\partial v_y}{\partial z}\right) + \mathbf{e}_y\left(\frac{\partial v_x}{\partial z} - \frac{\partial v_z}{\partial x}\right) + \mathbf{e}_z\left(\frac{\partial v_y}{\partial x} - \frac{\partial v_x}{\partial y}\right) \qquad \text{[A.3-11]}$$

Like the tensor product of two vectors, $\nabla \mathbf{v}$ is a tensor as shown below:

$$\nabla \mathbf{v} = \begin{pmatrix} \dfrac{\partial}{\partial x} \\ \dfrac{\partial}{\partial y} \\ \dfrac{\partial}{\partial z} \end{pmatrix} (v_x \ v_y \ v_z)$$

$$= \begin{pmatrix} \dfrac{\partial v_x}{\partial x} & \dfrac{\partial v_y}{\partial x} & \dfrac{\partial v_z}{\partial x} \\ \dfrac{\partial v_x}{\partial y} & \dfrac{\partial v_y}{\partial y} & \dfrac{\partial v_z}{\partial y} \\ \dfrac{\partial v_x}{\partial z} & \dfrac{\partial v_y}{\partial z} & \dfrac{\partial v_z}{\partial z} \end{pmatrix} \qquad \text{[A.3-12]}$$

Also, like the vector product of a vector and a tensor, $\nabla \cdot \boldsymbol{\tau}$ is a vector. From Eq. [A.2-3]

$$\nabla \cdot \boldsymbol{\tau} = \left(\frac{\partial}{\partial x} \ \frac{\partial}{\partial y} \ \frac{\partial}{\partial z}\right) \cdot \begin{pmatrix} \tau_{xx} & \tau_{xy} & \tau_{xz} \\ \tau_{yx} & \tau_{yy} & \tau_{yz} \\ \tau_{zx} & \tau_{zy} & \tau_{zz} \end{pmatrix}$$

$$= \mathbf{e}_x\left(\frac{\partial \tau_{xx}}{\partial x} + \frac{\partial \tau_{yx}}{\partial y} + \frac{\partial \tau_{zx}}{\partial z}\right) + \mathbf{e}_y\left(\frac{\partial \tau_{xy}}{\partial x} + \frac{\partial \tau_{yy}}{\partial y} + \frac{\partial \tau_{zy}}{\partial z}\right)$$

$$+ \mathbf{e}_z\left(\frac{\partial \tau_{xz}}{\partial x} + \frac{\partial \tau_{yz}}{\partial y} + \frac{\partial \tau_{zz}}{\partial z}\right) \qquad \text{[A.3-13]}$$

From Eq. [A.2-2]

$$\nabla\cdot(\rho\mathbf{v}\mathbf{v}) = \left(\frac{\partial}{\partial x}\ \frac{\partial}{\partial y}\ \frac{\partial}{\partial z}\right)\cdot\begin{pmatrix} \rho v_x v_x & \rho v_x v_y & \rho v_x v_z \\ \rho v_y v_x & \rho v_y v_y & \rho v_y v_z \\ \rho v_z v_x & \rho v_z v_y & \rho v_z v_z \end{pmatrix}$$

$$= \mathbf{e}_x\left(\frac{\partial}{\partial x}(\rho v_x v_x) + \frac{\partial}{\partial y}(\rho v_y v_x) + \frac{\partial}{\partial z}(\rho v_z v_x)\right)$$

$$+ \mathbf{e}_y\left(\frac{\partial}{\partial x}(\rho v_x v_y) + \frac{\partial}{\partial y}(\rho v_y v_y) + \frac{\partial}{\partial z}(\rho v_z v_y)\right)$$

$$+ \mathbf{e}_z\left(\frac{\partial}{\partial x}(\rho v_x v_z) + \frac{\partial}{\partial y}(\rho v_y v_z) + \frac{\partial}{\partial z}(\rho v_z v_z)\right) \qquad [A,3\text{-}14]$$

It can be shown that

$$\boxed{\nabla\cdot(\rho\mathbf{v}\mathbf{v}) = \mathbf{v}(\nabla\cdot\rho\mathbf{v}) + \rho\mathbf{v}\cdot\nabla\mathbf{v}} \qquad [A.3\text{-}15]$$

This equation resembles Eq. [A.3-4] for a scalar product and, in fact, can be derived in a similar manner.

The following two equations are also useful:[1]

$$(\mathbf{v}\cdot\nabla)\,\mathbf{v} = \tfrac{1}{2}\nabla v^2 - \mathbf{v}\times(\nabla\times\mathbf{v}) \qquad [A.3\text{-}16]$$

and

$$(\boldsymbol{\tau}:\nabla\mathbf{v}) = (\nabla\cdot[\boldsymbol{\tau}\cdot\mathbf{v}]) - (\mathbf{v}\cdot[\nabla\cdot\boldsymbol{\tau}]) \qquad [A.3\text{-}17]$$

A.4. DIVERGENCE THEOREM

A.4.1. Vectors

Let Ω be a closed region in space surrounded by a surface A and $\mathbf{n}$ the outward-directed unit vector normal to the surface.

For a vector $\mathbf{v}$

$$\boxed{\iiint_\Omega \nabla\cdot\mathbf{v}\,d\Omega = \iint_A \mathbf{v}\cdot\mathbf{n}\,dA} \qquad [A.4\text{-}1]$$

This equation, called the **Gauss divergence theorem**, is useful for converting from a surface integral to a volume integral or vice versa. Three related theorems for scalars and tensors are also available as shown below.

[1] R. B. Bird, W. E. Stewart, and E. N. Lightfood, *Transport Phenomena*, Wiley, New York, 1960, pp. 727 and 731.

A.4.2. Scalars

For a scalar s

$$\iiint_{\Omega} \nabla s \, d\Omega = \iint_{A} s\mathbf{n} \, dA \qquad \text{[A.4-2]}$$

A.4.3. Tensors

For a tensor $\boldsymbol{\tau}$ or $\mathbf{vv}$

$$\iiint_{\Omega} [\nabla \cdot \boldsymbol{\tau}] \, d\Omega = \iint_{A} [\boldsymbol{\tau} \cdot \mathbf{n}] \, dA \qquad \text{[A.4-3]}$$

$$\iiint_{\Omega} [\nabla \cdot \mathbf{vv}] \, d\Omega = \iint_{A} [\mathbf{vv} \cdot \mathbf{n}] \, dA \qquad \text{[A.4-4]}$$

A.5. CURVILINEAR COORDINATES

Besides the rectangular coordinate system that we have used so far, curvilinear coordinate systems, especially cylindrical and spherical, are also often used. For many problems in transport phenomena, such as those involving a cylinder or sphere, these curvilinear coordinates are more natural than rectangular coordiantes.

As illustrated in Fig. A.5-1, a point P in space can be represented by $P(x, y, z)$ in rectangular coordinates, $P(r, \theta, z)$ in cylindrical coordinates, or $P(r, \theta, \phi)$ in spherical coordinates. Like unit vectors $\mathbf{e}_x$, $\mathbf{e}_y$, and $\mathbf{e}_z$ in rectangular coordinates, unit vectors $\mathbf{e}_r$, $\mathbf{e}_\theta$, and $\mathbf{e}_z$ in cylindrical coordinates are perpendicular to one another, so are unit vectors $\mathbf{e}_r$, $\mathbf{e}_\theta$, and $\mathbf{e}_\phi$ in spherical coordinates.

A.5.1. Cylindrical Coordinates

As shown in Fig. A.5-1*b*, for cylindrical coordinates the variables r, θ, and z are related to x, y, and z as follows

$$x = r\cos\theta \qquad \text{[A.5-1]}$$

$$y = r\sin\theta \qquad \text{[A.5-2]}$$

$$z = z \qquad \text{[A.4-3]}$$

A vector $\mathbf{v}$ and a tensor $\boldsymbol{\tau}$ can be expressed as follows:

$$\mathbf{v} = \mathbf{e}_r v_r + \mathbf{e}_\theta v_\theta + \mathbf{e}_z v_z \qquad [A.5\text{-}4]$$

and

$$\boldsymbol{\tau} = \begin{pmatrix} \tau_{rr} & \tau_{r\theta} & \tau_{rz} \\ \tau_{\theta r} & \tau_{\theta\theta} & \tau_{\theta z} \\ \tau_{zr} & \tau_{z\theta} & \tau_{zz} \end{pmatrix} \qquad [A.5\text{-}5]$$

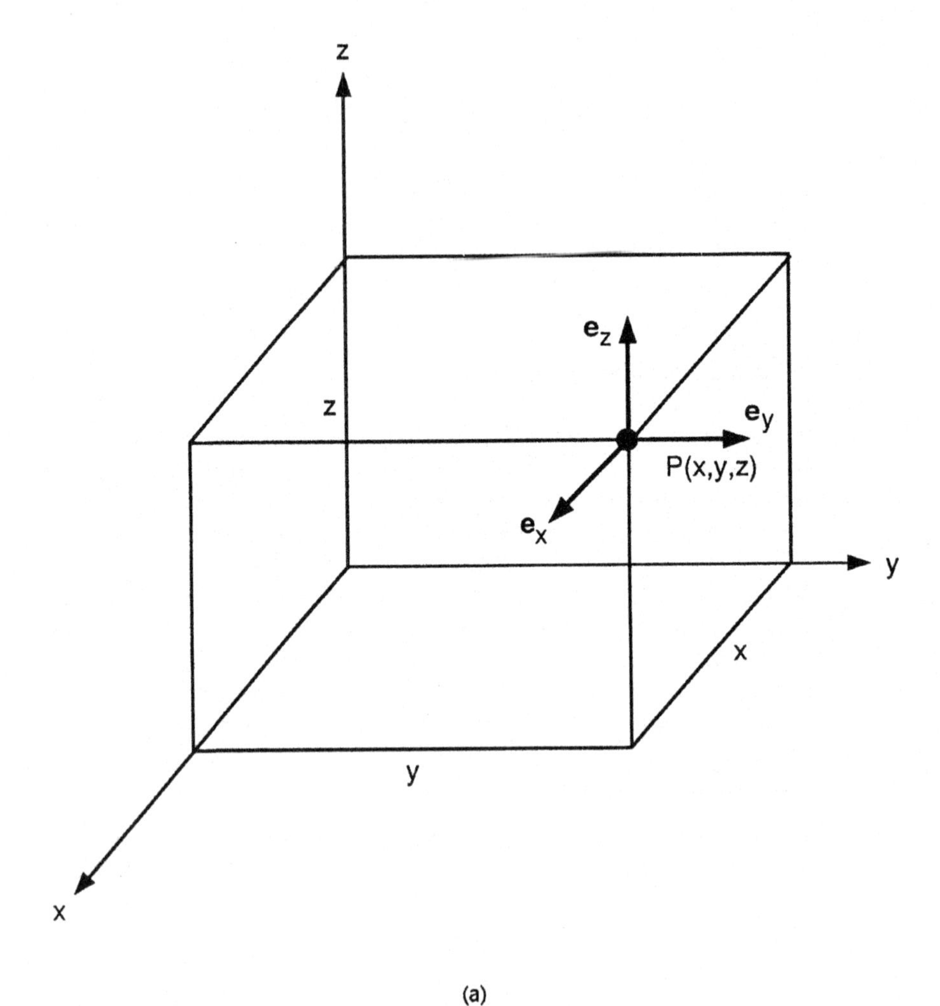

Fig. A.5-1 Three types of commonly used coordinates: (*a*) rectangular; (*b*) cylindrical; (*c*) spherical.

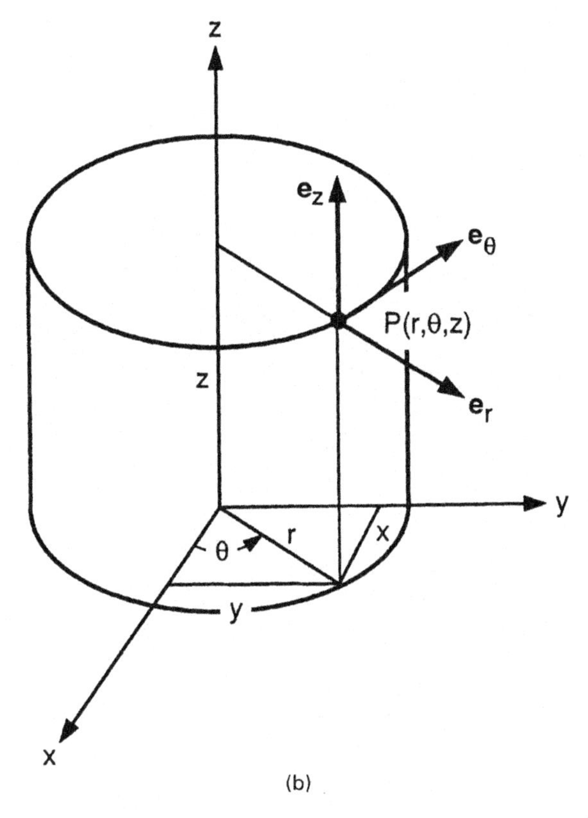

Fig. A.5-1 (*Continued*).

A.5.2. Spherical Coordinates

As shown in Fig. A.5-1*c*, for spherical coordinates

$$x = r \sin\theta \cos\phi \quad \text{[A.5-6]}$$

$$y = r \sin\theta \sin\phi \quad \text{[A.5-7]}$$

$$z = r \cos\theta \quad \text{[A.5-8]}$$

A vector $\mathbf{v}$ and a tensor $\boldsymbol{\tau}$ can be expressed as follows:

$$\mathbf{v} = \mathbf{e}_r v_r + \mathbf{e}_\theta v_\theta + \mathbf{e}_\phi v_\phi \quad \text{[A.5-9]}$$

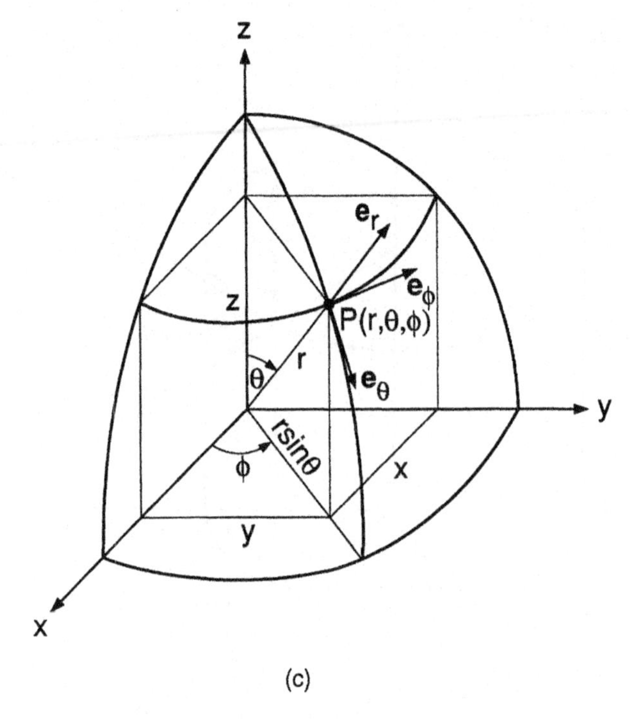

(c)

Fig. A.5-1 (*Continued*).

and

$$\boldsymbol{\tau} = \begin{pmatrix} \tau_{rr} & \tau_{r\theta} & \tau_{r\phi} \\ \tau_{\theta r} & \tau_{\theta\theta} & \tau_{\theta\phi} \\ \tau_{\phi r} & \tau_{\phi\theta} & \tau_{\phi\phi} \end{pmatrix} \qquad \text{[A.5-10]}$$

A.5.3. Differential Operators

Vectors, tensors, and their products in curvilinear coordinates are similar in form to those in curvilinear coordinates. For example, if $\mathbf{v} = \mathbf{e}_r$ in cylindrical coordinates

$$\boldsymbol{\tau} \cdot \mathbf{e}_r = \begin{pmatrix} \tau_{rr} & \tau_{r\theta} & \tau_{rz} \\ \tau_{\theta r} & \tau_{\theta\theta} & \tau_{\theta z} \\ \tau_{zr} & \tau_{z\theta} & \tau_{zz} \end{pmatrix} \cdot \begin{pmatrix} 1 \\ 0 \\ 0 \end{pmatrix}$$

$$= \mathbf{e}_r \tau_{rr} + \mathbf{e}_\theta \tau_{\theta r} + \mathbf{e}_z \tau_{zr} \qquad \text{[A.5-11]}$$

Similarly, if $\mathbf{v} = \mathbf{e}_r$ in spherical coordinates

$$\boldsymbol{\tau} \cdot \mathbf{e}_r = \begin{pmatrix} \tau_{rr} & \tau_{r\theta} & \tau_{r\phi} \\ \tau_{\theta r} & \tau_{\theta\theta} & \tau_{\theta\phi} \\ \tau_{\phi r} & \tau_{\phi\theta} & \tau_{\phi\phi} \end{pmatrix} \cdot \begin{pmatrix} 1 \\ 0 \\ 0 \end{pmatrix}$$

$$= \mathbf{e}_r \tau_{rr} + \mathbf{e}_\theta \tau_{\theta r} + \mathbf{e}_\phi \tau_{\phi r} \qquad \text{[A.5-12]}$$

Equations [A.5-11] and [A.5-12] are similar in form to Eq. [A.2-4].

The situation of the vector differential operator ∇, however, is quite different. In fact, ∇ is in the form of Eq. [A.3-1] only in rectangular coordinates. In curvilinear coordinates, ∇ assumes different forms depending on the orders of the physical quantity and the multiplication sign involved.

For example, in cylindrical coordinates

$$\nabla s = \mathbf{e}_r \frac{\partial s}{\partial r} + \mathbf{e}_\theta \frac{1}{r} \frac{\partial s}{\partial \theta} + \mathbf{e}_z \frac{\partial s}{\partial z} \qquad \text{[A.5-13]}$$

whereas in spherical coordinates

$$\nabla s = \mathbf{e}_r \frac{\partial s}{\partial r} + \mathbf{e}_\theta \frac{1}{r} \frac{\partial s}{\partial \theta} + \mathbf{e}_\phi \frac{1}{r \sin\theta} \frac{\partial s}{\partial \phi} \qquad \text{[A.5-14]}$$

Equations [A.5-13] and [A.5-14] differ in form from Eq. [A.3-2].

The equations for ∇s, $\nabla \cdot \mathbf{v}$, $\nabla \times \mathbf{v}$, and $\nabla^2 s$ in rectangular, cylindrical, and spherical coordinates are given in Tables A.5-1, A.5-2, and A.5-3, respectively. The equations in the last two tables can be derived from those in the first by coordinate transformation with the help of the chain rule.

TABLE A.5-1 ∇ Operations in Rectangular Coordinates

$$\nabla s = \mathbf{e}_x \frac{\partial s}{\partial x} + \mathbf{e}_y \frac{\partial s}{\partial y} + \mathbf{e}_z \frac{\partial s}{\partial z} \qquad \text{[A]}$$

$$\nabla \cdot \mathbf{v} = \frac{\partial v_x}{\partial x} + \frac{\partial v_y}{\partial y} + \frac{\partial v_z}{\partial z} \qquad \text{[B]}$$

$$\nabla \times \mathbf{v} = \mathbf{e}_x \left(\frac{\partial v_z}{\partial y} - \frac{\partial v_y}{\partial z} \right) + \mathbf{e}_y \left(\frac{\partial v_x}{\partial z} - \frac{\partial v_z}{\partial x} \right) + \mathbf{e}_z \left(\frac{\partial v_y}{\partial x} - \frac{\partial v_x}{\partial y} \right) \qquad \text{[C]}$$

$$\nabla^2 s = \frac{\partial^2 s}{\partial x^2} + \frac{\partial^2 s}{\partial y^2} + \frac{\partial^2 s}{\partial z^2} \qquad \text{[D]}$$

TABLE A.5-2 ∇ Operations in Cylindrical Coordinates

$$\nabla s = \mathbf{e}_r \frac{\partial s}{\partial r} + \mathbf{e}_\theta \frac{1}{r}\frac{\partial s}{\partial \theta} + \mathbf{e}_z \frac{\partial s}{\partial z} \quad [A]$$

$$\nabla \cdot \mathbf{v} = \frac{1}{r}\frac{\partial (r v_r)}{\partial r} + \frac{1}{r}\frac{\partial v_\theta}{\partial \theta} + \frac{\partial v_z}{\partial z} \quad [B]$$

$$\nabla \times \mathbf{v} = \mathbf{e}_r \left[\frac{1}{r}\frac{\partial v_z}{\partial \theta} - \frac{\partial v_\theta}{\partial z}\right] + \mathbf{e}_\theta \left[\frac{\partial v_r}{\partial z} - \frac{\partial v_z}{\partial r}\right] + \mathbf{e}_z \left[\frac{1}{r}\frac{\partial}{\partial r}(r v_\theta) - \frac{1}{r}\frac{\partial v_r}{\partial \theta}\right] \quad [C]$$

$$\nabla^2 s = \frac{1}{r}\frac{\partial}{\partial r}\left(r \frac{\partial s}{\partial r}\right) + \frac{1}{r^2}\frac{\partial^2 s}{\partial \theta^2} + \frac{\partial^2 s}{\partial z^2} \quad [D]$$

TABLE A.5-3 ∇ Operations in Spherical Coordinates

$$\nabla s = \mathbf{e}_r \frac{\partial s}{\partial r} + \mathbf{e}_\theta \frac{1}{r}\frac{\partial s}{\partial \theta} + \mathbf{e}_\phi \frac{1}{r\sin\theta}\frac{\partial s}{\partial \phi} \quad [A]$$

$$\nabla \cdot \mathbf{v} = \frac{1}{r^2}\frac{\partial}{\partial r}(r^2 v_r) + \frac{1}{r\sin\theta}\frac{\partial}{\partial \theta}(v_\theta \sin\theta) + \frac{1}{r\sin\theta}\frac{\partial v_\phi}{\partial \phi} \quad [B]$$

$$\nabla \times \mathbf{v} = \frac{\mathbf{e}_r}{r\sin\theta}\left[\frac{\partial}{\partial \theta}(v_\phi \sin\theta) - \frac{\partial v_\theta}{\partial \phi}\right] + \mathbf{e}_\theta \left[\frac{1}{r\sin\theta}\frac{\partial v_r}{\partial \phi} - \frac{1}{r}\frac{\partial (r v_\phi)}{\partial r}\right] + \mathbf{e}_\phi \left[\frac{1}{r}\frac{\partial (r v_\theta)}{\partial r} - \frac{1}{r}\frac{\partial v_r}{\partial \theta}\right] \quad [C]$$

$$\nabla^2 s = \frac{1}{r^2}\frac{\partial}{\partial r}\left(r^2 \frac{\partial s}{\partial r}\right) + \frac{1}{r^2\sin\theta}\frac{\partial}{\partial \theta}\left(\sin\theta \frac{\partial s}{\partial \theta}\right) + \frac{1}{r^2\sin^2\theta}\frac{\partial^2 s}{\partial \phi^2} \quad [D]$$

A.6. ORDINARY DIFFERENTIAL EQUATIONS

Solutions are provided for 14 different cases as follows.

A.6.1. Rectangular Coordinates

Case A

$$\frac{d\phi}{du} = a(\phi - b) \quad [A.6\text{-}1]$$

$$\phi = \phi_0 \quad \text{at} \quad u = 0 \qquad \text{Boundary (or Initial) Condition} \quad [A.6\text{-}2]$$

□

The variable ϕ, such as, temperature, is a function of u, such as position x, y, or z (or time t), whereas a and b are constants. Since $d\phi = d(\phi - b)$, from Eq. [A.6-1]

$$\frac{d(\phi - b)}{(\phi - b)} = a\,du \qquad \text{[A.6-3]}$$

On integration

$$\ln(\phi - b) = au + I \qquad \text{[A.6-4]}$$

Substituting Eq. [A.6-2] into Eq. [A.6-4]

$$\ln(\phi_0 - b) = I \qquad \text{[A.6-5]}$$

Substituting Eq. [A.6-5] into Eq. [A.6-4]

$$\ln(\phi - b) = au + \ln(\phi_0 - b) \qquad \text{[A.6-6]}$$

Therefore

$$\ln\left(\frac{\phi - b}{\phi_0 - b}\right) = au \qquad \text{[A.6-7]}$$

or

$$\boxed{\frac{\phi - b}{\phi_0 - b} = e^{au}} \qquad \text{[A.6-8]}$$

Equation [A.6-2] becomes an initial condition if u is time t.

Case B

$$\frac{d\phi}{du} = \frac{a(\phi - b)}{c(u - d)} \qquad \text{[A.6-9]}$$

$$\phi = \phi_0 \quad \text{at} \quad u = u_0 \qquad \text{Boundary (or Initial) Condition} \qquad \text{[A.6-10]}$$

□

From Eq. [A.6-9] and the fact that $d\phi = d(\phi - b)$ and $du = d(u - d)$

$$\frac{cd(\phi - b)}{(\phi - b)} = \frac{ad(u - d)}{(u - d)} \qquad \text{[A.6-11]}$$

On integration

$$\ln(\phi - b)^c = \ln(u - d)^a + I \qquad \text{[A.6-12]}$$

Substituting Eq. [A.6-10] into Eq. [A.6-12]

$$\ln(\phi_0 - b)^c = \ln(u_0 - d)^a + I \qquad \text{[A.6-13]}$$

So

$$I = \ln(\phi_0 - b)^c - \ln(u_0 - d)^a \qquad \text{[A.6-14]}$$

Substituting Eq. [A.6-14] into Eq. [A.6-12]

$$\ln(\phi_0 - b)^c = \ln(u - d)^a + \ln(\phi_0 - b)^c - \ln(u_0 - d)^a \qquad \text{[A.6-15]}$$

or

$$\ln\left(\frac{\phi - b}{\phi_0 - b}\right)^c = \ln\left(\frac{u - d}{u_0 - d}\right)^a \qquad \text{[A.6-16]}$$

So

$$\boxed{\left(\frac{\phi - b}{\phi_0 - b}\right)^c = \left(\frac{u - d}{u_0 - d}\right)^a} \qquad \text{[A.6-17]}$$

Case C

$$\frac{d^2\phi}{du^2} = a \qquad \text{[A.6-18]}$$

$$\phi = \phi_1 \quad \text{at} \quad u = u_1 \qquad \text{Boundary Condition 1} \qquad \text{[A.6-19]}$$

$$\phi = \phi_2 \quad \text{at} \quad u = u_2 \qquad \text{Boundary Condition 2} \qquad \text{[A.6-20]}$$

□

Integrating Eq. [A.6-18]

$$\frac{d\phi}{du} = au + I_1 \qquad \text{[A.6-21]}$$

Integrating Eq. [A.6-21]

$$\phi = \tfrac{1}{2} au^2 + I_1 u + I_2 \quad \text{[A.6-22]}$$

Substituting Eq. [A.6-19] into Eq. [A.6-22]

$$\phi_1 = \tfrac{1}{2} au_1^2 + I_1 u_1 + I_2 \quad \text{[A.6-23]}$$

Similarly, substituting Eq. [A.6-20] into Eq. [A.6-22]

$$\phi_2 = \tfrac{1}{2} au_2^2 + I_1 u_2 + I_2 \quad \text{[A.6-24]}$$

From Eqs. [A.6-23] and [A.6-24]

$$I_1 = \frac{\phi_2 - \phi_1}{u_2 - u_1} - \frac{1}{2} a(u_1 + u_2) \quad \text{[A.6-25]}$$

and

$$I_2 = \frac{\phi_1 u_2 - \phi_2 u_1}{u_2 - u_1} + \frac{1}{2} a u_1 u_2 \quad \text{[A.6-26]}$$

Substituting Eqs. [A.6-25] and [A.6-26] into Eq. [A.6-22]

$$\boxed{\phi = \frac{1}{2} a u^2 + \left[\frac{\phi_2 - \phi_1}{u_2 - u_1} - \frac{1}{2} a(u_1 + u_2)\right] u + \left(\frac{\phi_1 u_2 - \phi_2 u_1}{u_2 - u_1} + \frac{1}{2} a u_1 u_2\right)} \quad \text{[A.6-27]}$$

Case D

$$\frac{d^2\phi}{du^2} = a \quad \text{[A.6-28]}$$

$$\frac{d\phi}{du} = b \quad \text{at} \quad u = u_1 \quad \text{Boundary Condition 1} \quad \text{[A.6-29]}$$

$$\phi = \phi_2 \quad \text{at} \quad u = u_2 \quad \text{Boundary Condition 2} \quad \text{[A.6-30]}$$

□

Integrating Eq. [A.6-28]

$$\frac{d\phi}{du} = au + I_1 \qquad [A.6-31]$$

Substituting Eq. [A.6-29] into Eq. [A.6-31]

$$b = au_1 + I_1 \qquad [A.6-32]$$

Substituting Eq. [A.6-32] into Eq. [A.6-31]

$$\frac{d\phi}{du} = au + b - au_1 \qquad [A.6-33]$$

Integrating Eq. [A.6-33]

$$\phi = \tfrac{1}{2}au^2 + (b - au_1)u + I_2 \qquad [A.6-34]$$

Substituting Eq. [A.6-30] into Eq. [A.6-34]

$$\phi_2 = \tfrac{1}{2}au_2^2 + (b - au_1)u_2 + I_2 \qquad [A.6-35]$$

So

$$I_2 = \phi_2 - \tfrac{1}{2}au_2^2 - (b - au_1)u_2 \qquad [A.6-36]$$

Substituting Eq. [A.6-36] into Eq. [A.6-34]

$$\boxed{\phi = \phi_2 + \tfrac{1}{2}a(u^2 - u_2^2) + (b - au_1)(u - u_2)} \qquad [A.6-37]$$

Case E

$$\frac{d^2\phi}{du^2} = a\frac{d\phi}{du} + b \qquad [A.6-38]$$

$$\phi = \phi_0 \quad \text{at} \quad u = 0 \quad \text{Boundary Condition 1} \qquad [A.6-39]$$

$$\phi = \phi_1 \quad \text{at} \quad u = u_1 \quad \text{Boundary Condition 2} \qquad [A.6-40]$$

□

Define

$$\phi' = \frac{d\phi}{du} \qquad \text{[A.6-41]}$$

From Eqs. [A.6-41] and [A.6-38]

$$\frac{d\phi'}{du} = a\phi' + b \qquad \text{[A.6-42]}$$

or

$$\frac{d[\phi' + (b/a)]}{\phi' + (b/a)} = a\,du \qquad \text{[A.6-43]}$$

Integrating Eq. [A.6-43]

$$\ln\left(\phi' + \frac{b}{a}\right) = au + I_1' = \ln e^{au + I_1'} \qquad \text{[A.6-44]}$$

So

$$\phi' + \frac{b}{a} = \frac{d\phi}{du} + \frac{b}{a} = I_1\, e^{au} \qquad \text{[A.6-45]}$$

where

$$I_1 = e^{I_1'} \qquad \text{[A.6-46]}$$

Integrating Eq. [A.6-45]

$$\phi = \frac{I_1}{a} e^{au} - \frac{b}{a} u + I_2 \qquad \text{[A.6-47]}$$

Substituting Eq. [A.6-39] into Eq. [A.6-47]

$$\phi_0 = \frac{I_1}{a} + I_2 \qquad \text{[A.6-48]}$$

Similarly, substituting Eq. [A.6-40] into Eq. [A.6-47]

$$\phi_1 = \frac{I_1}{a} e^{au_1} - \frac{b}{a} u_1 + I_2 \qquad \text{[A.6-49]}$$

From Eqs. [A.6-48] and [A.6-49]

$$I_1 = \frac{a(\phi_1 - \phi_0) + bu_1}{e^{au_1} - 1} \qquad \text{[A.6-50]}$$

and

$$I_2 = \phi_0 - \frac{(\phi_1 - \phi_0) + (b/a)u_1}{e^{au_1} - 1} \qquad \text{[A.6-51]}$$

Substituting Eqs. [A.6-50] and [A.6-51] into Eq. [A.6-47]

$$\phi = \left(\frac{\phi_1 - \phi_0 + (b/a)u_1}{e^{au_1} - 1}\right) e^{au} - \frac{b}{a}u + \phi_0 - \frac{\phi_1 - \phi_0 + (b/a)u_1}{e^{au_1} - 1} \qquad \text{[A.6-52]}$$

So

$$\phi = \phi_0 + \left(\phi_1 - \phi_0 + \frac{b}{a}u_1\right)\left(\frac{e^{au} - 1}{e^{au_1} - 1}\right) - \frac{b}{a}u \qquad \text{[A.6-53]}$$

or

$$\boxed{\frac{\phi - \phi_0}{\phi_1 - \phi_0} = \frac{e^{au} - 1}{e^{au_1} - 1} + \frac{b}{a}\left[u_1\left(\frac{e^{au} - 1}{e^{au_1} - 1}\right) - u\right]} \qquad \text{[A.6-54]}$$

Case F

$$\frac{d^2\phi}{du^2} = au \qquad \text{[A.6-55]}$$

$$\phi = 0 \quad \text{at} \quad u = -u_1 \qquad \text{Boundary Condition 1} \qquad \text{[A.6-56]}$$

$$\phi = 0 \quad \text{at} \quad u = u_1 \qquad \text{Boundary Condition 2} \qquad \text{[A.6-57]}$$

□

Integrating Eq. [A.6-55]

$$\frac{d\phi}{du} = \frac{1}{2}au^2 + I_1 \qquad \text{[A.6-58]}$$

Integrating Eq. [A.6-58]

$$\phi = \tfrac{1}{6} a u^3 + I_1 u + I_2 \qquad \text{[A.6-59]}$$

Substituting Eq. [A.6-56] into Eq. [A.6-59]

$$0 = \frac{-1}{6} a u_1^3 - I_1 u_1 + I_2 \qquad \text{[A.6-60]}$$

Similarly, substituting Eq. [A.6-57] into Eq. [A.6-59]

$$0 = \tfrac{1}{6} a u_1^3 + I_1 u_1 + I_2 \qquad \text{[A.6-61]}$$

From Eqs. [A.6-60] and [A.6-61]

$$I_2 = 0 \qquad \text{[A.6-62]}$$

and

$$I_1 = \frac{-1}{6} a u_1^2 \qquad \text{[A.6-63]}$$

Substituting Eqs. [A.6-62] and [A.6-63] into Eq. [A.6-59]

$$\phi = \tfrac{1}{6} a u^3 - \tfrac{1}{6} a u_1^2 u \qquad \text{[A.6-64]}$$

So

$$\phi = \frac{a}{6}(u^3 - u_1^2 u) \qquad \text{[A.6-65]}$$

Case G

$$\frac{d^2\phi}{du^2} = a(\phi - b) \qquad (a > 0) \qquad \text{[A.6-66]}$$

$$\frac{d\phi}{du} = c \quad \text{at} \quad u = u_1 \qquad \text{Boundary Condition 1} \qquad \text{[A.6-67]}$$

$$\phi = \phi_2 \quad \text{at} \quad u = u_2 \qquad \text{Boundary Condition 2} \qquad \text{[A.6-68]}$$

□

Let

$$\psi = \phi - b \qquad [A.6\text{-}69]$$

Since b is a constant, Eqs. [A.6-66] through [A.6-68] become

$$\frac{d^2\psi}{du^2} = a\psi \qquad [A.6\text{-}70]$$

$$\frac{d\psi}{du} = c \quad \text{at} \quad u = u_1 \qquad [A.6\text{-}71]$$

$$\psi = \phi_2 - b = d \quad \text{at} \quad u = u_2 \qquad [A.6\text{-}72]$$

The solution of Eq. [A.6-70] is of the form $\psi = e^{mu}$, which, when substituted into Eq. [A.6-70], results in

$$m^2 e^{mu} = a e^{mu} \qquad [A.6\text{-}73]$$

that is,

$$m = \pm\sqrt{a} \qquad [A.6\text{-}74]$$

Since Eq. [A.6-70] is linear

$$\psi = Ae^{\sqrt{a}u} + Be^{-\sqrt{a}u} \qquad [A.6\text{-}75]$$

Substituting into Eqs. [A.6-71] and [6.7-72]

$$Ae^{\sqrt{a}u_1} + Be^{-\sqrt{a}u_1} = \frac{c}{\sqrt{a}} = f \qquad [A.6\text{-}76]$$

$$Ae^{\sqrt{a}u_2} + Be^{-\sqrt{a}u_2} = d \qquad [A.6\text{-}77]$$

Solving the two equations simultaneously

$$A = \frac{(c/\sqrt{a})\,e^{-\sqrt{a}(u_1 + 2u_2)} + de^{-\sqrt{a}(2u_1 + u_2)}}{e^{-2\sqrt{a}u_1} + e^{-2\sqrt{a}u_2}} \qquad [A.6\text{-}78]$$

$$B = \frac{-fe^{-\sqrt{a}u_1} + de^{-\sqrt{a}u_2}}{e^{-2\sqrt{a}u_1} + e^{-2\sqrt{a}u_2}} \qquad [A.6\text{-}79]$$

Substituting Eqs. [A.6-78], [A.6-79], and [A.6-72] into Eq. [A.6-75]

$$\phi = b + \frac{1}{e^{-2\sqrt{a}u_1} + e^{-2\sqrt{a}u_2}} \left\{ \frac{c}{\sqrt{a}} \left[e^{\sqrt{a}(u - u_1 - 2u_2)} - e^{-\sqrt{a}(u + u_1)} \right] \right.$$

$$\left. + (\phi_2 - b)\left[e^{\sqrt{a}(u - 2u_1 - u_2)} + e^{-\sqrt{a}(u + u_2)} \right] \right\} \qquad \text{[A.6-80]}$$

Case H

$$\frac{d^2\phi}{du^2} = a\frac{d\phi}{du} + b\phi + c \qquad \text{[A.6-81]}$$

$$\phi = \phi_1 \quad \text{at} \quad u = u_1 \quad \text{Boundary Condition 1} \qquad \text{[A.6-82]}$$

$$\phi = \phi_2 \quad \text{at} \quad u = u_2 \quad \text{Boundary Condition 2} \qquad \text{[A.6-83]}$$

□

Let

$$\psi = \phi + \frac{c}{b} \qquad \text{[A.6-84]}$$

Since c/b is a constant, Eqs. [A.6-81] through [A.6-83] become

$$\frac{d^2\psi}{du^2} = a\frac{d\psi}{du} + b\psi \qquad \text{[A.6-85]}$$

$$\psi = \psi_1 = \phi_1 + \frac{c}{b} \quad \text{at} \quad u = u_1 \qquad \text{[A.6-86]}$$

$$\psi = \psi_2 = \phi_2 + \frac{c}{b} \quad \text{at} \quad u = u_2 \qquad \text{[A.6-87]}$$

The solution of Eq. [A.6-85] is of the form $\psi = e^{mu}$, which, when substituted into Eq. [A.6-85], results in

$$m^2 e^{mu} - ame^{mu} - be^{mu} = 0 \qquad \text{[A.6-88]}$$

or

$$m^2 - am - b = 0 \qquad \text{[A.6-89]}$$

The roots are

$$m_1 = \frac{a + \sqrt{a^2 + 4b}}{2} \qquad [A.6\text{-}90]$$

$$m_2 = \frac{a - \sqrt{a^2 + 4b}}{2} \qquad [A.6\text{-}91]$$

Since Eq. [A.6-85] is linear

$$\psi = Ae^{m_1 u} + Be^{m_2 u} \qquad [A.6\text{-}92]$$

Substituting Eq. [A.6-92] into Eqs. [A.6-86] and [6.7-87]

$$Ae^{m_1 u_1} + Be^{m_2 u_1} = \psi_1 \qquad [A.6\text{-}93]$$

$$Ae^{m_1 u_2} + Be^{m_2 u_2} = \psi_2 \qquad [A.6\text{-}94]$$

Solving Eqs. [A.6-93] and [A.6-94] simultaneously

$$A = \frac{\psi_1 e^{m_2 u_2} - \psi_2 e^{m_2 u_1}}{e^{m_2 u_2 + m_1 u_1} - e^{m_2 u_1 + m_1 u_2}} \qquad [A.6\text{-}95]$$

and

$$B = \frac{\psi_1 e^{-m_1 u_1} - \psi_2 e^{-m_1 u_2}}{e^{(m_2 - m_1)u_1} - e^{(m_2 - m_1)u_2}} \qquad [A.6\text{-}96]$$

Substituting Eqs. [A.6-95] and [A.6-96] into Eq. [A.6-92]

$$\boxed{\psi = \left(\frac{\psi_1 e^{m_2 u_2} - \psi_2 e^{m_2 u_1}}{e^{m_2 u_2 + m_1 u_1} - e^{m_2 u_1 + m_1 u_2}} \right) e^{m_1 u} + \left(\frac{\psi_1 e^{-m_1 u_1} - \psi_2 e^{-m_1 u_2}}{e^{(m_2 - m_1)u_1} - e^{(m_2 - m_1)u_2}} \right) e^{m_2 u}} \qquad [A.6\text{-}97]$$

where

$$\psi_1 = \phi_1 + \frac{c}{b} \qquad [A.6\text{-}98]$$

$$\psi_2 = \phi_2 + \frac{c}{b} \qquad [A.6\text{-}99]$$

$$m_1 = \tfrac{1}{2}(a + \sqrt{a^2 + 4b}) \qquad [A.6\text{-}100]$$

$$m_2 = \tfrac{1}{2}(a - \sqrt{a^2 + 4b}) \qquad [A.6\text{-}101]$$

A.6.2. Curvilinear Coordinates

Case I

$$\frac{1}{r}\frac{d}{dr}\left(r\frac{d\phi}{dr}\right) = a + br^2 \qquad [A.6-102]$$

$$\frac{d\phi}{dr} = 0 \quad \text{at} \quad r = r_1 \qquad \text{Boundary Condition 1} \qquad [A.6-103]$$

$$\phi = \phi_2 \quad \text{at} \quad r = r_2 \qquad \text{Boundary Condition 2} \qquad [A.6-104]$$

□

The variable ϕ is a function of radius r, whereas a is a constant. From Eq. [A.6-102]

$$\frac{d}{dr}\left(r\frac{d\phi}{dr}\right) = ar + br^3 \qquad [A.6-105]$$

Integrating once

$$r\frac{d\phi}{dr} = \frac{1}{2}ar^2 + \frac{b}{4}r^4 + I_1 \qquad [A.6-106]$$

or

$$\frac{d\phi}{dr} = \frac{1}{2}ar + \frac{b}{4}r^3 + \frac{I_1}{r} \qquad [A.6-107]$$

Integrating one more time

$$\phi = \frac{1}{4}ar^2 + \frac{b}{16}r^4 + I_1 \ln r + I_2 \qquad [A.6-108]$$

Substituting Eq. [A.6-103] into Eq. [A.6-106]

$$0 = \frac{1}{2}ar_1^2 + \frac{b}{4}r_1^4 + I_1 \qquad [A.6-109]$$

So

$$I_1 = \frac{-1}{2}ar_1^2 - \frac{b}{4}r_1^4 \qquad [A.6-110]$$

Substituting Eqs. [A.6-104] and [A.6-110] into Eq. [A.6-108]

$$\phi_2 = \frac{1}{4}ar_2^2 + \frac{b}{16}r_2^4 - \left(\frac{1}{2}ar_1^2 + \frac{b}{4}r_1^4\right)\ln r_2 + I_2 \qquad \text{[A.6-111]}$$

So

$$I_2 = \phi_2 - \frac{1}{4}ar_2^2 - \frac{b}{16}r_2^4 + \left(\frac{1}{2}ar_1^2 + \frac{b}{4}r_1^4\right)\ln r_2 \qquad \text{[A.6-112]}$$

Substituting Eqs. [A.6-110] and [A.6-112] into Eq. [A.6-108]

$$\phi = \frac{1}{4}ar^2 + \frac{b}{16}r^4 - \left(\frac{1}{2}ar_1^2 + \frac{b}{4}r_1^4\right)\ln r + \phi_2 - \frac{1}{4}ar_2^2 - \frac{b}{16}r_2^4 + \left(\frac{1}{2}ar_1^2 + \frac{b}{4}r_1^4\right)\ln r_2 \qquad \text{[A.6-113]}$$

or

$$\boxed{\phi = \phi_2 + \frac{a}{4}(r^2 - r_2^2) + \frac{b}{16}(r^4 - r_2^4) - \left(\frac{a}{2}r_1^2 + \frac{b}{4}r_1^4\right)\ln\left(\frac{r}{r_2}\right)} \qquad \text{[A.6-114]}$$

Case J

$$\frac{1}{r}\frac{d}{dr}\left(r\frac{d\phi}{dr}\right) = a \qquad \text{[A.6-115]}$$

$$\phi = \phi_1 \quad \text{at} \quad r = r_1 \qquad \text{Boundary Condition 1} \qquad \text{[A.6-116]}$$

$$\phi = \phi_2 \quad \text{at} \quad r = r_2 \qquad \text{Boundary Condition 2} \qquad \text{[A.6-117]}$$

□

From Eq. [1.6-108] of Case I with $b = 0$

$$\phi = \tfrac{1}{4}ar^2 + I_1 \ln r + I_2 \qquad \text{[A.6-118]}$$

Substituting Eq. [A.6-116] into Eq. [A.6-118]

$$\phi_1 = \tfrac{1}{4}ar_1^2 + I_1 \ln r_1 + I_2 \qquad \text{[A.6-119]}$$

Similarly, substituting Eq. [A.6-117] into Eq. [A.6-118]

$$\phi_2 = \tfrac{1}{4} a r_2^2 + I_1 \ln r_2 + I_2 \qquad \text{[A.6-120]}$$

From Eqs. [A.6-119] and [A.6-120]

$$I_1 = \frac{(\phi_2 - \phi_1) - (a/4)(r_2^2 - r_1^2)}{\ln(r_2/r_1)} \qquad \text{[A.6-121]}$$

and

$$I_2 = \phi_1 - \frac{1}{4} a r_1^2 - \left[\frac{(\phi_2 - \phi_1) - (a/4)(r_2^2 - r_1^2)}{\ln(r_2/r_1)}\right] \ln r_1 \qquad \text{[A.6-122]}$$

Substituting Eqs. [A.6-121] and [A.6-122] into Eq. [A.6-118]

$$\boxed{\phi_2 = \phi_1 + \frac{a}{4}(r^2 - r_1^2) + \left[(\phi_2 - \phi_1) - \frac{a}{4}(r_2^2 - r_1^2)\right] \frac{\ln(r/r_1)}{\ln(r_2/r_1)}} \qquad \text{[A.6-123]}$$

Case K

$$\frac{1}{r}\frac{d}{dr}\left(r \frac{d\phi}{dr}\right) = a \qquad \text{[A.6-124]}$$

$$\frac{d\phi}{dr} = 0 \qquad \text{at} \quad r = 0 \qquad \text{Boundary Condition 1} \qquad \text{[A.6-125]}$$

$$\frac{d\phi}{dr} = b(\phi - c) \quad \text{at} \quad r = r_1 \qquad \text{Boundary Condition 2} \qquad \text{[A.6-126]}$$

□

Let

$$\psi = \phi - c \qquad \text{[A.6-127]}$$

Since c is a constant

$$\frac{d\psi}{dr} = \frac{d\phi}{dr} \qquad \text{[A.6-128]}$$

So, Eqs. [A.6-124] through [A.6-126] become

$$\frac{d}{dr}\left(r\frac{d\psi}{dr}\right) = ar \tag{A.6-129}$$

$$\frac{d\psi}{dr} = 0 \quad \text{at} \quad r = 0 \tag{A.6-130}$$

$$\psi = \frac{1}{b}\frac{d\psi}{dr} \quad \text{at} \quad r = r_1 \tag{A.6-131}$$

Integrating Eq. [A.6-129]

$$r\frac{d\psi}{dr} = \frac{1}{2}ar^2 + I_1 \tag{A.6-132}$$

or

$$\frac{d\psi}{dr} = \frac{1}{2}ar + \frac{I_1}{r} \tag{A.6-133}$$

Substituting into Eq. [A.6-130]

$$I_1 = 0 \tag{A.6-134}$$

and Eq. [A.6-132] becomes

$$\frac{d\psi}{dr} = \frac{1}{2}ar \tag{A.6-135}$$

Integrating this equation

$$\psi = \tfrac{1}{4}ar^2 + I_2 \tag{A.6-136}$$

Substituting Eqs. [A.6-135] and [A.6-136] into Eq. [A.6-131]

$$\frac{1}{4}ar_1^2 + I_2 = \frac{1}{b}\left(\frac{1}{2}ar_1\right) \tag{A.6-137}$$

or

$$I_2 = \frac{ar_1}{2b} - \frac{ar_1^2}{4} \tag{A.6-138}$$

Substituting into Eq. [A.6-136]

$$\psi = \frac{a}{4}(r^2 - r_1^2) + \frac{ar_1}{2b} \qquad \text{[A.6-139]}$$

Substituting into Eq. [A.6-127]

$$\boxed{\phi = \frac{a}{4}(r^2 - r_1^2) + \frac{ar_1}{2b} + c} \qquad \text{[A.6-140]}$$

Case L

$$\frac{d}{dr}\left[\frac{1}{r}\frac{d}{dr}(r\phi)\right] = 0 \qquad \text{[A.6-141]}$$

$$\phi = \phi_1 \quad \text{at} \quad r = r_1 \qquad \text{Boundary Condition 1} \qquad \text{[A.6-142]}$$

$$\phi = \phi_2 \quad \text{at} \quad r = r_2 \qquad \text{Boundary Condition 2} \qquad \text{[A.6-143]}$$

□

On integration

$$\frac{1}{r}\frac{d}{dr}(r\phi) = I_1 \qquad \text{[A.6-144]}$$

or

$$\frac{d}{dr}(r\phi) = I_1 r \qquad \text{[A.6-145]}$$

On integration again

$$r\phi = \tfrac{1}{2} I_1 r^2 + I_2 \qquad \text{[A.6-146]}$$

or

$$\phi = \frac{1}{2} I_1 r + \frac{I_2}{r} \qquad \text{[A.6-147]}$$

Substituting Eq. [A.6-142] into Eq. [A.6-147]

$$\phi_1 = \frac{1}{2} I_1 r_1 + \frac{I_2}{r_1} \qquad \text{[A.6-148]}$$

Similarly, substituting Eq. [A.6-143] into Eq. [A.6-147]

$$\phi_2 = \frac{1}{2} I_1 r_2 + \frac{I_2}{r_2} \qquad \text{[A.6-149]}$$

From Eqs. [A.6-148] and [A.6-149]

$$I_1 = \frac{2(r_2 \phi_2 - r_1 \phi_1)}{(r_2^2 - r_1^2)} \qquad \text{[A.6-150]}$$

$$I_2 = \frac{r_1 r_2 (r_2 \phi_1 - r_1 \phi_2)}{(r_2^2 - r_1^2)} \qquad \text{[A.6-151]}$$

Substituting Eqs. [A.6-150] and [A.6-151] into Eq. [A.6-147]

$$\boxed{\phi = \frac{r^2 (r_2 \phi_2 - r_1 \phi_1) + r_1 r_2 (r_2 \phi_1 - r_1 \phi_2)}{r(r_2^2 - r_1^2)}} \qquad \text{[A.6-152]}$$

Case M

$$\frac{d}{dr}\left(r^2 \frac{d\phi}{dr} \right) = 0 \qquad \text{[A.6-153]}$$

$\phi = \phi_1$ at $r = r_1$ Boundary Condition 1 [A.6-154]

$\phi = \phi_2$ at $r = r_2$ Boundary Condition 2 [A.6-155]

□

Integrating Eq. [A.6-153]

$$r^2 \frac{d\phi}{dr} = I_1 \qquad \text{[A.6-156]}$$

and

$$\frac{d\phi}{dr} = \frac{I_1}{r^2} \qquad \text{[A.6-157]}$$

Integrating Eq. [A.6-157]

$$\phi = -\frac{I_1}{r} + I_2 \qquad \text{[A.6-158]}$$

Substituting Eq. [A.6-154] into Eq. [A.6-158]

$$\phi_1 = -\frac{I_1}{r_1} + I_2 \qquad \text{[A.6-159]}$$

Similarly, substituting Eq. [A.6-155] into Eq. [A.6-158]

$$\phi_2 = -\frac{I_1}{r_2} + I_2 \qquad \text{[A.6-160]}$$

From Eqs. [A.6-159] and [A.6-160]

$$I_1 = \frac{\phi_2 - \phi_1}{(1/r_1) - (1/r_2)} \qquad \text{[A.6-161]}$$

and

$$I_2 = \frac{\phi_2 - \phi_1}{r_1[(1/r_1) - (1/r_2)]} + \phi_1 \qquad \text{[A.6-162]}$$

Substituting Eqs. [A.6-161] and [A.6-162] into Eq. [A.6-158]

$$\phi = -\frac{\phi_2 - \phi_1}{[(1/r_1) - (1/r_2)]\, r} + \frac{\phi_2 - \phi_1}{r_1\, [(1/r_1) - (1/r_2)]} + \phi_1 \qquad \text{[A.6-163]}$$

or

$$\boxed{\frac{\phi - \phi_1}{\phi_2 - \phi_1} = \frac{(1/r_1) - (1/r)}{(1/r_1) - (1/r_2)}} \qquad \text{[A.6-164]}$$

Case N

$$\frac{1}{r^2}\frac{d}{dr}\left(r^2 \frac{d\phi}{dr}\right) = a\phi \qquad \text{[A.6-165]}$$

$$\frac{d\phi}{dr} = 0 \quad \text{at} \quad r = 0 \qquad \text{Boundary Condition 1} \qquad \text{[A.6-166]}$$

$$\phi = \phi_1 \quad \text{at} \quad r = r_1 \qquad \text{Boundary Condition 2} \qquad \text{[A.6-167]}$$

□

Let

$$\psi = r\phi \qquad [A.6\text{-}168]$$

and so

$$\frac{d\psi}{dr} = r\frac{d\phi}{dr} + \phi = r\frac{d\phi}{dr} + \frac{\psi}{r} \qquad [A.6\text{-}169]$$

or

$$r^2\frac{d\phi}{dr} = r\frac{d\psi}{dr} - \psi \qquad [A.6\text{-}170]$$

Substituting into Eq. [A.6-165]

$$\frac{d}{dr}\left(r\frac{d\psi}{dr} - \psi\right) = ar\psi \qquad [A.6\text{-}171]$$

or

$$r\frac{d^2\psi}{dr^2} + \frac{d\psi}{dr} - \frac{d\psi}{dr} = ar\psi \qquad [A.6\text{-}172]$$

which is

$$\frac{d^2\psi}{dr^2} = a\psi \qquad [A.6\text{-}173]$$

Substituting Eqs. [A.6-168] and [A.6-170] into Eqs. [A.6-166] and [A.6-167]

$$\psi = 0 \quad \text{at} \quad r = 0 \qquad [A.6\text{-}174]$$

$$\psi = r_1\phi_1 \quad \text{at} \quad r = r_1 \qquad [A.6\text{-}175]$$

The solution of Eq. [A.6-173] is of the form $\psi = e^{mr}$, which, when substituted into Eq. [A.6-173], results in

$$m^2 e^{mr} = ae^{mr} \qquad [A.6\text{-}176]$$

that is

$$m = \pm\sqrt{a} \qquad [A.6\text{-}177]$$

Since Eq. [A.6-173] is linear

$$\psi = Ae^{\sqrt{a}r} + Be^{-\sqrt{a}r} \qquad [A.6\text{-}178]$$

Substituting into Eqs. [A.6-174] and [A.6-175]

$$0 = A + B \qquad [A.6\text{-}179]$$

$$r_1 \phi_1 = Ae^{\sqrt{a}r_1} + Be^{-\sqrt{a}r_1} \qquad [A.6\text{-}180]$$

Solving the two equations simultaneously

$$A = \frac{r_1 \phi_1}{e^{\sqrt{a}r_1} - e^{-\sqrt{a}r_1}} \qquad [A.6\text{-}181]$$

$$B = \frac{-r_1 \phi_1}{e^{\sqrt{a}r_1} - e^{-\sqrt{a}r_1}} \qquad [A.6\text{-}182]$$

Substituting into Eq. [A.6-178]

$$\psi = r_1 \phi_1 \left(\frac{e^{\sqrt{a}r} - e^{-\sqrt{a}r}}{e^{\sqrt{a}r_1} - e^{-\sqrt{a}r_1}} \right) \qquad [A.6\text{-}183]$$

Substituting Eq. [A.6-168]

$$\boxed{\frac{\phi}{\phi_1} = \frac{r_1}{r} \left(\frac{e^{\sqrt{a}r} - e^{-\sqrt{a}r}}{e^{\sqrt{a}r_1} - e^{-\sqrt{a}r_1}} \right)} \qquad [A.6\text{-}184]$$

A.7. PARTIAL DIFFERENTIAL EQUATIONS

Solutions are provided for nine different cases as follows.

A.7.1. Rectangular Coordinates

Case O

$$\frac{\partial \phi}{\partial t} = a \frac{\partial^2 \phi}{\partial u^2} \qquad [A.7\text{-}1]$$

$$\phi(u, 0) = \phi_i, \quad \text{Initial Condition} \qquad [A.7\text{-}2]$$

$$\phi(0, t) = \phi_S, \quad \text{Boundary Conditions 1} \tag{A.7-3}$$

$$\phi(\infty, t) = \phi_i, \quad \text{Boundary Condition 2} \tag{A.7-4}$$

□

The variable ϕ (e.g., temperature or concentration) is a function of both position u and time t [i.e., $\phi = \phi(u, t)$], whereas a is a constant. Let us define

$$\eta = \frac{u}{\sqrt{4at}} \tag{A.7-5}$$

$$\frac{\partial \phi}{\partial t} = \frac{d\phi}{d\eta}\frac{\partial \eta}{\partial t} = \frac{-u}{2\sqrt{4at^3}}\frac{d\phi}{d\eta} = \frac{-1}{2}\frac{\eta}{t}\frac{d\phi}{d\eta} \tag{A.7-6}$$

$$\frac{\partial \phi}{\partial u} = \frac{d\phi}{d\eta}\frac{\partial \eta}{\partial u} = \frac{1}{\sqrt{4at}}\frac{d\phi}{d\eta} \tag{A.7-7}$$

$$\frac{\partial^2 \phi}{\partial u^2} = \frac{\partial}{\partial u}\left(\frac{\partial \phi}{\partial u}\right) = \frac{1}{\sqrt{4at}}\frac{d}{d\eta}\left(\frac{\partial \phi}{\partial u}\right) = \frac{1}{\sqrt{4at}}\frac{d}{d\eta}\left[\frac{1}{\sqrt{4at}}\frac{d\phi}{d\eta}\right] = \frac{1}{4at}\frac{d^2\phi}{d\eta^2} \tag{A.7-8}$$

Substituting Eqs. [A.7-6] and [A.7-8] into Eq. [A.7-1]

$$\frac{-\eta}{2t}\frac{d\phi}{d\eta} = \frac{1}{4t}\frac{d^2\phi}{d\eta^2} \tag{A.7-9}$$

or

$$\frac{d^2\phi}{d\eta^2} = -2\eta\frac{d\phi}{d\eta} \tag{A.7-10}$$

Notice that the partial differential equation, Eq. [A.7-1], has been converted into an ordinary differential equation, Eq. [A.7-10]. The corresponding initial and boundary conditions are

$$\text{BC1:} \quad \phi = \phi_S \quad \text{at} \quad \eta = 0 \tag{A.7-11}$$

$$\text{IC, BC2:} \quad \phi = \phi_i \quad \text{at} \quad \eta = \infty \tag{A.7-12}$$

Let us define

$$\phi' = \frac{d\phi}{d\eta} \tag{A.7-13}$$

Substituting Eq. [A.7-13] into Eq. [A.7-10]

$$\frac{d\phi'}{d\eta} = -2\eta\phi' \qquad \text{[A.7-14]}$$

or

$$\frac{d\phi'}{\phi'} = -d(\eta^2) \qquad \text{[A.7-15]}$$

On integration

$$\ln \phi' = -\eta^2 + I_1' \qquad \text{[A.7-16]}$$

Therefore

$$\phi' = e^{-\eta^2 + I_1'} = e^{I_1'} e^{-\eta^2} = I_1 e^{-\eta^2} \qquad \text{[A.7-17]}$$

Substituting Eq. [A.7-13] into Eq. [A.7-17]

$$\frac{d\phi}{d\eta} = I_1 e^{-\eta^2} \qquad \text{[A.7-18]}$$

On integration

$$\phi = I_1 \int_0^{\eta} e^{-\eta^2} d\eta + I_2 \qquad \text{[A.7-19]}$$

Arbitrarily selecting the lower limit of this integral as zero does not affect the validity of the equation. Rather, it affects only the value of the integration constant I_2.

Substituting Eq. [A.7-11] into Eq. [A.7-19]

$$I_2 = \phi_S \qquad \text{[A.7-20]}$$

Substituting Eqs. [A.7-12] and [A.7-20] into Eqn. [A.7-19]

$$I_1 = \frac{\phi_i - \phi_S}{\int_0^{\infty} e^{-\eta^2} d\eta} = \frac{2}{\sqrt{\pi}}(\phi_i - \phi_S) \qquad \text{[A.7-21]}$$

Substituting Eq. [A.7-20] and [A.7-21] into Eq. [A.7-19]

$$\frac{\phi - \phi_S}{\phi_i - \phi_S} = \text{erf}(\eta) \qquad \text{[A.7-22]}$$

where the **error function** erf(η) is defined as

$$\operatorname{erf}(\eta) = \frac{\int_0^\eta e^{-\eta^2}\,d\eta}{\int_0^\infty e^{-\eta^2}\,d\eta} = \frac{2}{\sqrt{\pi}}\int_0^\eta e^{-\eta^2}\,d\eta \qquad \text{[A.7-23]}$$

Substituting Eq. [A.7-5] into Eq. [A.7-22]

$$\boxed{\frac{\phi - \phi_S}{\phi_i - \phi_S} = \operatorname{erf}\left[\frac{u}{\sqrt{4at}}\right]} \qquad \text{[A.7-24]}$$

From Eqs. [A.7-7], [A.7-18], and [A.7-21]

$$\frac{\partial \phi}{\partial u} = \frac{1}{\sqrt{\pi a t}}(\phi_i - \phi_s)\, e^{-u^2/(4at)} \qquad \text{[A.7-25]}$$

Case P

$$\frac{\partial \phi}{\partial t} = a\frac{\partial^2 \phi}{\partial u^2} \qquad \text{[A.7-26]}$$

$\phi(u, 0) = 0$ Initial Condition [A.7-27]

$\phi(\pm\infty, t) = 0$ Boundary Condition [A.7-28]

$M = \int_{-\infty}^{\infty} \phi\, du$ Conservation Requirement [A.7-29]

□

It can be shown that Eq. [A.7-26] can be satisfied by[2]

$$\phi = \frac{I}{\sqrt{t}}\exp\left(\frac{-u^2}{4at}\right) \qquad \text{[A.7-30]}$$

where I is an arbitrary constant. It is seen that Eq. [A.7-30] suggests that ϕ is symmetric with respect to $u = 0$. Furthermore, $\phi(\pm\infty, t) = 0$ and $\phi(u, 0) = 0$ for any values of u other than zero. Substituting Eq. [A.7-30] into Eq. [A.7-29]

$$M = I\int_{-\infty}^{\infty} \frac{1}{\sqrt{t}}\exp\left(\frac{-u^2}{4at}\right) du \qquad \text{[A.7-31]}$$

[2] J. Crank, The *Mathematics of Diffusion*, 2nd ed., Oxford University Press, London, 1975, p. 11.

or

$$M = I\sqrt{4a}\int_{-\infty}^{\infty} \exp\left(\frac{-u^2}{4at}\right) d\left(\frac{u}{\sqrt{4at}}\right) \qquad \text{[A.7-32]}$$

The integration in Eq. [A.7-32] can be carried out with the help of the following integration formula:

$$\int_{-\infty}^{\infty} e^{-\eta^2}\, d\eta = \sqrt{\pi} \qquad \text{[A.7-33]}$$

Substituting Eq. [A.7-33] into Eq. [A.7-32]

$$M = 2I\sqrt{\pi a} \qquad \text{[A.7-34]}$$

or

$$I = \frac{M}{2\sqrt{\pi a}} \qquad \text{[A.7-35]}$$

Substituting Eq. [A.7-35] into Eq. [A.7-30]

$$\boxed{\phi = \frac{M}{2\sqrt{\pi a t}} \exp\left(\frac{-u^2}{4at}\right)} \qquad \text{[A.7-36]}$$

Case Q

$$\frac{\partial \phi}{\partial t} = a\frac{\partial^2 \phi}{\partial u^2} \qquad \text{[A.7-37]}$$

$\phi(u, 0) = 0$ Initial Condition [A.7-38]

$\phi(\infty, t) = 0$ Boundary Condition [A.7-39]

$M = \int_0^{\infty} \phi\, du$ Conservation Requirement [A.7-40]

□

As already mentioned, Eq. [A.7-37] can be satisfied by

$$\phi = \frac{I}{\sqrt{t}} \exp\left(\frac{-u^2}{4at}\right) \qquad \text{[A.7-41]}$$

Substituting Eq. [A.7-41] into Eq. [A.7-40]

$$M = I \int_0^\infty \frac{1}{\sqrt{t}} \exp\left(\frac{-u^2}{4at}\right) du \qquad \text{[A.7-42]}$$

or

$$M = I\sqrt{4a} \int_0^\infty \exp\left(\frac{-u^2}{4at}\right) d\left(\frac{u}{\sqrt{4at}}\right) \qquad \text{[A.7-43]}$$

The integration in Eq. [A.7-43] can be carried out with the help of the following integration formula:

$$\int_0^\infty e^{-\eta^2}\, d\eta = \frac{\sqrt{\pi}}{2} \qquad \text{[A.7-44]}$$

Substituting Eq. [A.7-44] into Eq. [A.7-43]

$$M = I\sqrt{\pi a} \qquad \text{[A.7-45]}$$

or

$$I = \frac{M}{\sqrt{\pi a}} \qquad \text{[A.7-46]}$$

Substituting Eq. [A.7-46] into Eq. [A.7-41]

$$\boxed{\phi = \frac{M}{\sqrt{\pi a t}} \exp\left[\frac{-u^2}{4at}\right]} \qquad \text{[A.7-47]}$$

Case R

$$\frac{\partial \phi}{\partial t} = a \frac{\partial^2 \phi}{\partial u^2} \qquad \text{[A.7-48]}$$

$$\phi(u, 0) = \phi_i \qquad \text{Initial Condition} \qquad \text{[A.7-49]}$$

$$-b \frac{\partial \phi(0, t)}{\partial t} = \psi_0 \qquad \text{Boundary Condition 1} \qquad \text{[A.7-50]}$$

$$\phi(u, t) = \phi_i \quad \text{as} \quad u \to \infty \qquad \text{Boundary Condition 2} \qquad \text{[A.7-51]}$$

□

Let us define

$$\psi = -b\frac{\partial\phi}{\partial u} \qquad [A.7\text{-}52]$$

and substitute it into Eq. [A.7-48]. Then

$$\frac{\partial\phi}{\partial t} = a\frac{\partial}{\partial u}\left(\frac{-\psi}{b}\right) = \frac{-a}{b}\frac{\partial\psi}{\partial u} \qquad [A.7\text{-}53]$$

This equation can be differentiated with respect to u to give

$$\frac{\partial^2\phi}{\partial t\,\partial u} = \frac{-a}{b}\frac{\partial^2\psi}{\partial u^2} \qquad [A.7\text{-}54]$$

Substituting Eq. [A.7-52] into Eq. [A.7-54]

$$\frac{-1}{b}\frac{\partial\psi}{\partial t} = \frac{-a}{b}\frac{\partial^2\psi}{\partial u^2} \qquad [A.7\text{-}55]$$

and so

$$\frac{\partial\psi}{\partial t} = a\frac{\partial^2\psi}{\partial u^2} \qquad [A.7\text{-}56]$$

The initial boundary conditions are

$$\text{IC:} \quad \psi(u, 0) = 0 \qquad [A.7\text{-}57]$$

$$\text{BC1:} \quad \psi(0, t) = \psi_0 \qquad [A.7\text{-}58]$$

$$\text{BC2:} \quad \psi(\infty, t) = 0 \qquad [A.7\text{-}59]$$

Note that from Eq. [A.7-49] $\partial\phi(u,0)/\partial u = 0$ and from Eq. [A.7-51] $\partial\phi(u,t)/\partial u = 0$ as $u \to \infty$, thus leading to Eqs. [A.7-57] and [A.7-59], respectively. Since Eqs. [A.7-56] through [A.7-59] are similar to Eqs. [A.7-1] through [A.7-4] of Case O, in light of Eq. [A.7-24] we can write

$$\frac{\psi - \psi_0}{0 - \psi_0} = \operatorname{erf}\left[\frac{u}{\sqrt{4at}}\right] \qquad [A.7\text{-}60]$$

Substituting Eq. [A.7-52] into Eq. [A.7-60]

$$-b\frac{\partial\phi}{\partial u} = \psi = \psi_0\left\{1 - \mathrm{erf}\left[\frac{u}{\sqrt{4at}}\right]\right\} \quad \text{[A.7-61]}$$

which can be integrated, from $\phi = \phi$ at $u = u$ to $\phi = \phi_i$ at $u = \infty$ (Eq. [A.7-51]), to give

$$-(\phi_i - \phi) = \frac{\psi_0}{b}\int_u^\infty \left\{1 - \mathrm{erf}\left[\frac{u}{\sqrt{4at}}\right]\right\} du \quad \text{[A.7-62]}$$

or

$$\phi - \phi_i = \frac{\psi_0}{b}\sqrt{4at}\int_{u/\sqrt{4at}}^\infty \left\{1 - \mathrm{erf}\left[\frac{u}{\sqrt{4at}}\right]\right\} d\left[\frac{u}{\sqrt{4at}}\right] \quad \text{[A.7-63]}$$

So

$$\boxed{\phi - \phi_i = \frac{\psi_0}{b}\sqrt{4at}\, i\,\mathrm{erf}\,c\left[\frac{u}{\sqrt{4at}}\right]} \quad \text{[A.7-64]}$$

where

$$i\,\mathrm{erf}\,c(\xi) = \int_\xi^\infty [1 - \mathrm{erf}(\xi)]\, d\xi = \frac{1}{\sqrt{\pi}}e^{-\xi^2} - \xi[1 - \mathrm{erf}(\xi)] \quad \text{[A.7-65]}$$

Case S

$$\frac{\partial\phi}{\partial t} = a\frac{\partial^2\phi}{\partial u^2} \quad \text{[A.7-66]}$$

$\phi(u, 0) = \phi_i$ Initial Condition [A.7-67]

$b\dfrac{\partial\phi(0, t)}{\partial u} = c[\phi(0, t) - \phi_f]$ Boundary Condition 1 [A.7-68]

$\phi(u, t) = \phi_i$ as $u \to \infty$ Boundary Condition 2 [A.7-69]

□

Let us define

$$\theta = \frac{\phi - \phi_f}{\phi_i - \phi_f} \qquad [A.7-70]$$

Equation [A.7-66] becomes

$$\frac{\partial \theta}{\partial t} = a \frac{\partial^2 \theta}{\partial u^2} \qquad [A.7-71]$$

The initial and boundary conditions are

$$\text{IC:} \quad \theta(u, 0) = 1 \qquad [A.7-72]$$

$$\text{BC1:} \quad \frac{\partial \theta(0, t)}{\partial u} - \frac{c}{b}\theta(0, t) = 0 \qquad [A.7.73]$$

$$\text{BC2:} \quad \theta(u, t) = 1 \quad \text{as} \quad u \to \infty \qquad [A.7-74]$$

Let us further define

$$\psi = \theta - \frac{b}{c}\frac{\partial \theta}{\partial u} \qquad [A.7-75]$$

and Eqs. [A.7-71] through [A.7-74] become

$$\frac{\partial \psi}{\partial t} = a \frac{\partial^2 \psi}{\partial u^2} \qquad [A.7-76]$$

$$\text{IC:} \quad \psi(u, 0) = 1 \qquad [A.7-77]$$

$$\text{BC1:} \quad \psi(0, t) = 0 \qquad [A.7-78]$$

$$\text{BC2:} \quad \psi(\infty, t) = 1 \qquad [A.7-79]$$

Note that from Eq. [A.7-72] $\partial\theta(u, 0)/\partial u = 0$ and from Eq. [A.7-74] $\partial\theta(u, t)/\partial u = 0$ as $u \to \infty$, thus leading to Eqs. [A.7-77] and [A.7-79], respectively. Since Eqs. [A.7-76] through [A.7-79] are similar to Eqs. [A.7-1] through [A.7-4] of Case O, in light of Eq. [A.7-24] we can write

$$\frac{\psi - 0}{1 - 0} = \text{erf}\left[\frac{u}{\sqrt{4at}}\right] \qquad [A.7-80]$$

thus

$$\psi = \operatorname{erf}\left[\frac{u}{\sqrt{4at}}\right] \qquad \text{[A.7-81]}$$

where the error function, as in Eq. [A.7-23] of Case K, is defined as

$$\operatorname{erf}(\eta) = \frac{2}{\sqrt{\pi}} \int_{p=0}^{p=\eta} \exp(-p^2)\, dp \qquad \text{[A.7-82]}$$

Substituting Eq. [A.7-81] into Eq. [A.7-75] and rearranging

$$\frac{\partial \theta}{\partial u} - \frac{c}{b}\theta = -\frac{c}{b}\operatorname{erf}\left[\frac{u}{\sqrt{4at}}\right] \qquad \text{[A.7-83]}$$

Multiplying this equation by an integration factor $\exp(-cu/b)$

$$\frac{\partial}{\partial u}\left[\theta \exp\left(\frac{-cu}{b}\right)\right] = -\frac{c}{b}\operatorname{erf}\left[\frac{u}{\sqrt{4at}}\right]\exp\left(\frac{-cu}{b}\right) \qquad \text{[A.7-84]}$$

On integration from ∞ to u

$$\theta = \left[\exp\left(\frac{cu}{b}\right)\right]\left(\frac{-c}{b}\right)\int_{u=\infty}^{u=u} \operatorname{erf}\left[\frac{u}{\sqrt{4at}}\right]\exp\left(\frac{-cu}{b}\right) du \qquad \text{[A.7-85]}$$

The following integration formula can be used:

$$\int_{\infty}^{u} \operatorname{erf}\left(\frac{\xi}{m}\right)\exp\left(-\frac{\xi}{n}\right) d\xi = -n\exp\left(-\frac{u}{n}\right)\left\{\operatorname{erf}\left(\frac{u}{m}\right) + \exp\left(\frac{u}{n} + \frac{m^2}{4n^2}\right)\right.$$
$$\left.\times\left[1 - \operatorname{erf}\left(\frac{u}{m} + \frac{m}{2n}\right)\right]\right\} \qquad \text{[A.7-86]}$$

Substituting this equation into Eq. [A.7-85]

$$\boxed{\theta \equiv \frac{\phi - \phi_f}{\phi_i - \phi_f} = \operatorname{erf}\left(\frac{u}{\sqrt{4at}}\right) + \exp\left(\frac{cu}{b} + \frac{c^2 at}{b^2}\right)\left[1 - \operatorname{erf}\left(\frac{u}{\sqrt{4at}} + \frac{c}{b}\sqrt{at}\right)\right]}$$

[A.7-87]

Case T

$$\frac{\partial \phi}{\partial t} = a \frac{\partial^2 \phi}{\partial u^2} \quad \text{[A.7-88]}$$

$$\phi(u, 0) = \phi_i \qquad \text{Initial Condition} \quad \text{[A.7-89]}$$

$$\frac{\partial \phi(0, t)}{\partial u} = 0 \qquad \text{Boundary Condition 1} \quad \text{[A.7-90]}$$

$$-b \frac{\partial \phi(L, t)}{\partial u} = c[\phi(L, t) - \phi_f] \qquad \text{Boundary Condition 2} \quad \text{[A.7-91]}$$

□

Let us define

$$\theta = \phi - \phi_f \quad \text{[A.7-92]}$$

$$\frac{\partial \theta}{\partial t} = a \frac{\partial^2 \theta}{\partial u^2} \quad \text{[A.7-93]}$$

with the initial and boundary conditions as follows

$$\text{IC:} \quad \theta(u, 0) = \phi_i - \phi_f = \theta_i \quad \text{[A.7-94]}$$

$$\text{BC1:} \quad \frac{\partial \theta(0, t)}{\partial u} = 0 \quad \text{[A.7-95]}$$

$$\text{BC2:} \quad \frac{\partial \theta(L, t)}{\partial u} + \frac{c}{b} \theta(L, t) = 0 \quad \text{[A.7-96]}$$

Since Eqs. [A.7-93] through [A.7-96] are linear and homogeneous, we can apply the method of **separation of variables**. In other words, we assume that

$$\theta(u, t) = F(u) G(t) \quad \text{[A.7-97]}$$

and convert Eq. [A.7-93] into

$$\frac{1}{F} \frac{d^2 F}{du^2} = \frac{1}{aG} \frac{dG}{dt} = -\lambda^2 \quad \text{[A.7-98]}$$

In this equation $(d^2F/du^2)/F$ is either a function of u or a constant, and $(dG/dt)/(aG)$ is either a function of t or a constant. Therefore, they both must be equal to a constant, and this constant is arbitrarily assigned as $-\lambda^2$.

From Eq. [A.7-98]

$$\frac{d^2F}{du^2} + \lambda^2 F = 0 \tag{A.7-99}$$

and the solution to this linear homogeneous second-order ordinary differential equation is well known:

$$F = c_1 \cos \lambda u + c_2 \sin \lambda u \tag{A.7-100}$$

Also from Eq. [A.7-98]

$$\frac{dG}{dt} + a\lambda^2 G = 0 \tag{A.7-101}$$

and the solution is

$$G = c_3 e^{-a\lambda^2 t} \tag{A.7-102}$$

Now let us determine the constants c_1, c_2, and c_3. From Eqs. [A.7-95], [A.7-97], and [A.7-100]

$$G(t)\frac{dF(0)}{du} = G(t)\,\lambda c_2 \cos(0) = 0 \tag{A.7-103}$$

thus suggesting that $c_2 = 0$ and Eq. [A.7-100] reduces to

$$F = c_1 \cos \lambda u \tag{A.7-104}$$

From Eqs. [A.7-96], [A.7-97], and [A.7-104]

$$G(t)\frac{dF(L)}{du} + \frac{c}{b}G(t)\,F(L) = G(t)\,(-\lambda c_1 \sin \lambda L) + \frac{c}{b}G(t)\,(c_1 \cos \lambda L) = 0 \tag{A.7-105}$$

thus requiring that

$$(\lambda L)\tan(\lambda L) = \mathrm{Bi} \tag{A.7-106}$$

where

$$\mathrm{Bi} = \frac{cL}{b} \quad \text{[A.7-107]}$$

Since $\tan \lambda L$ is a periodic function, an infinite number of λ values can satisfy Eq. [A.7-106]. Let λ_n be one of such λ values, where $n = 1, 2, 3, \ldots, \infty$. This λ_n is called the **eigenvalue**. Now, from Eqs. [A.7-97], [A.7-102], and [A.7-104], the general solution is as follows:

$$\theta = \sum_{n=1}^{\infty} A_n (\cos \lambda_n u) e^{-a\lambda_n^2 t} \quad \text{[A.7-108]}$$

where $A_n = c_{1,n} c_{3,n}$.

Substituting Eq. [A.7-94] into Eq. [A.7-108]

$$\theta_i = \sum_{n=1}^{\infty} A_n \cos \lambda_n u \quad \text{[A.7-109]}$$

To determine A_n, let us multiply Eq. [A.7-109] by

$$\int_0^L (\cos \lambda_m u)\, du$$

to become

$$\theta_i \int_0^L (\cos \lambda_m u)\, du = \int_0^L \sum_{n=1}^{\infty} A_n (\cos \lambda_n u)(\cos \lambda_m u)\, du \quad \text{[A.7-110]}$$

when $m \neq n$ all integrands on the RHS of Eq. [A.7-110] are zero, and when $m = n$ the integrand has the following nonzero values:

$$A_n \left[\frac{L}{2} + \frac{1}{2\lambda_n} (\sin \lambda_n L)(\cos \lambda_n L) \right]$$

The LHS, with $m = n$, equals $(\theta_i/\lambda_n)\sin(\lambda_n L)$. As such

$$A_n = \frac{2\theta_i \sin \lambda_n L}{\lambda_n L + (\sin \lambda_n L)(\cos \lambda_n L)} \quad \text{[A.7-111]}$$

Therefore, the final solution is

$$\frac{\theta}{\theta_i} = \frac{\phi - \phi_f}{\phi_i - \phi_f} = 2 \sum_{n=1}^{\infty} \frac{\sin \lambda_n L}{\lambda_n L + (\sin \lambda_n L)(\cos \lambda_n L)} \exp(-\lambda_n^2 a t) \cos(\lambda_n u) \quad \text{[A.7-112]}$$

where λ_n is the nth root of Eq. [A.7-106]. This equation can be rewritten as follows:

$$\frac{\phi - \phi_f}{\phi_i - \phi_f} = 2 \sum_{n=1}^{\infty} \frac{\sin(\lambda_n L)}{(\lambda_n L) + \sin(\lambda_n L)\cos(\lambda_n L)} \times \exp\left[-(\lambda_n L)^2 \left(\frac{at}{L^2}\right)\right] \cos\left[(\lambda_n L)\left(\frac{u}{L}\right)\right] \quad \text{[A.7-113]}$$

Case U

$$\frac{\partial \phi}{\partial t} = a\left[\frac{\partial^2 \phi}{\partial x^2} + \frac{\partial^2 \phi}{\partial y^2} + \frac{\partial^2 \phi}{\partial z^2}\right] \quad \text{[A.7-114]}$$

$\phi(x, y, z, 0) = 0$ expect at $(x', y', 0, 0)$ Initial Condition [A.7-115]

$\phi(\pm\infty, \pm\infty, \infty, t) = 0$ Boundary Condition [A.7-116]

$\int_0^{\infty} \int_{-\infty}^{\infty} \int_{-\infty}^{\infty} \phi \, dx\, dy\, dz = b$ Conservation Requirement [A.7-117]

□

Let us prove that the following equation is a solution:

$$\phi = \frac{2b}{(4\pi a t)^{3/2}} e^{-[(x-x')^2 + (y-y')^2 + z^2]/4at} \quad \text{[A.7-118]}$$

From Eq. [A.7-118]

$$\frac{\partial \phi}{\partial t} = \frac{2b}{(4\pi a)^{3/2}}\left(\frac{-3}{2t^{5/2}}\right) e^{-[(x-x')^2 + (y-y')^2 + z^2]/4at} + \frac{2b}{(4\pi a t)^{3/2}}\left(\frac{[(x-x')^2 + (y-y')^2 + z^2]}{4at^2}\right) e^{-[(x-x')^2 + (y-y')^2 + z^2]/4at} \quad \text{[A.7-119]}$$

$$\frac{\partial \phi}{\partial x} = \frac{2b}{(4\pi a t)^{3/2}}\left(\frac{-2(x-x')}{4at}\right) e^{-[(x-x')^2 + (y-y')^2 + z^2]/4at} \quad \text{[A.7-120]}$$

$$\frac{\partial^2 \phi}{\partial x^2} = \frac{2b}{(4\pi a t)^{3/2}}\left(\frac{-2}{4at}\right) e^{-[(x-x')^2 + (y-y')^2 + z^2]/4at} + \frac{2b}{(4\pi a t)^{3/2}}\left(\frac{-2(x-x')}{4at}\right) e^{-[(x-x')^2 + (y-y')^2 + z^2]/4at} \quad \text{[A.7-121]}$$

Similarly

$$\frac{\partial^2\phi}{\partial y^2} = \frac{2b}{(4\pi at)^{3/2}}\left(\frac{-2}{4at}\right)e^{-[(x-x')^2+(y-y')^2+z^2]/4at} + \frac{2b}{(4\pi at)^{3/2}}\left(\frac{-2(y-y')}{4at}\right)^2 e^{-[(x-x')^2+(y-y')^2+z^2]/4at} \quad \text{[A.7-122]}$$

$$\frac{\partial^2\phi}{\partial z^2} = \frac{2b}{(4\pi at)^{3/2}}\left(\frac{-2}{4at}\right)e^{-[(x-x')^2+(y-y')^2+z^2]/4at} + \frac{2b}{(4\pi at)^{3/2}}\left(\frac{-2z}{4at}\right)^2 e^{-[(x-x')^2+(y-y')^2+z^2]/4at} \quad \text{[A.7-123]}$$

Substituting Eqs. [A.7-119], [A.7-121], [A.7-122], and [A.7-123] into Eq. [A.7-114], we see that it is satisfied. Furthermore, Eq. [A.7-118] satisfies the conditions in Eqs. [A.7-115] and [A.7-116]. Finally, from Eq. [A.7-118]

$$\int_{z=0}^{z=\infty}\int_{y=-\infty}^{y=\infty}\int_{x=-\infty}^{x=\infty} \phi\, dx\, dy\, dz = \frac{2b}{\pi^{3/2}}\int_{-\infty}^{\infty} e^{-(x-x')^2/4at}\, d\left[\frac{(x-x')}{\sqrt{4at}}\right]\int_{-\infty}^{\infty} e^{-(y-y')^2/4at}\, d\left[\frac{(y-y')}{\sqrt{4at}}\right] \times \int_0^{\infty} e^{-z^2/4at}\, d\left[\frac{z}{\sqrt{4at}}\right] \quad \text{[A.7-124]}$$

With the help of the following formula

$$\int_{-\infty}^{\infty} e^{-\xi^2}\, d\xi = 2\int_0^{\infty} e^{-\xi^2}\, d\xi = \sqrt{\pi} \quad \text{[A.7-125]}$$

Eq. [A.7-124] becomes Eq. [A.7-117]. Therefore, Eq. [A.7-118] is a solution.

A.7.2. Curvilinear Coordinates

Case V

$$\frac{\partial C}{\partial t} = \frac{D}{r}\frac{\partial}{\partial r}\left(r\frac{\partial C}{\partial r}\right) \quad \text{[A.7-126]}$$

$$C(r, 0) = C_i \quad \text{Initial Condition} \quad \text{[A.7-127]}$$

$$C(a, t) = C_s \quad \text{Boundary Condition 1} \tag{A.7-128}$$

$$\frac{\partial C(0, t)}{\partial r} = 0 \quad \text{Boundary Condition 2} \tag{A.7-129}$$

□

By following the method of separating the variables shown in Case P, it can be shown that

$$C = ue^{-D\alpha^2 t} \tag{A.7-130}$$

is a solution of Eq. [A.7-126], provided u is a function of r only, satisfying

$$\frac{d^2u}{dr^2} + \frac{1}{r}\frac{du}{dr} + \alpha^2 u = 0 \tag{A.7-131}$$

which is the **Bessel equation of zero order.**

The solution of Eq. [A.7-131] satisfying the conditions in Eqs. [A.7-127] through [A.7-129] is such that[3]

$$\boxed{\frac{C - C_i}{C_s - C_i} = 1 - \frac{2}{a}\sum_{n=1}^{\infty} e^{-D\alpha_n^2 t}\frac{J_0(r\alpha_n)}{\alpha_n J_1(a\alpha_n)}} \tag{A.7-132}$$

where the α_n value are the positive roots of the following **Bessel function of the first kind of zero order**:

$$J_0(\alpha_n) = 0 \tag{A.7-133}$$

and the **Bessel function of the first kind of the first order**

$$J_1(a\alpha_n) = \frac{1}{a}\left[2\int_0^a r\{J_0(\alpha r)\}^2\, dr\right]^{1/2} \tag{A.7.134}$$

Case W

$$\frac{\partial C}{\partial t} = D\left(\frac{\partial^2 C}{\partial r^2} + \frac{2}{r}\frac{\partial C}{\partial r}\right) \tag{A.7-135}$$

$$C(r, 0) = C_i \quad \text{Initial Condition} \tag{A.7-136}$$

[3] J. Crank, *Mathematics of Diffusion*, 2nd ed., Oxford University Press, London, 1975.

$$C(a, t) = C_s \quad \text{Boundary Condition 1} \tag{A.7-137}$$

$$\frac{\partial C(0, t)}{\partial r} = 0 \quad \text{Boundary Condition 2} \tag{A.7-138}$$

□

Let us define

$$u = Cr \tag{A.7-139}$$

Substituting Eq. [A.7-139] into Eq. [A.7-135]

$$\frac{\partial u}{\partial t} = D\frac{\partial^2 u}{\partial r^2} \tag{A.7-140}$$

The solution of Eq. [A.7-140] satisfying the conditions in Eqs. [A.7-136] through [A.7-138] is such that[4]

$$\boxed{\frac{C - C_i}{C_s - C_i} = 1 + \frac{2a}{\pi}\sum_{n=1}^{\infty}\frac{(-1)^n}{n}\sin\left(\frac{n\pi r}{a}\right)e^{-Dn^2\pi^2 t/a^2}} \tag{A.7-141}$$

PROBLEMS[5]

A.1-1 Expand $\mathbf{e}_x \cdot \mathbf{v}$ in rectangular coordinates.

A.1-2 Expand $\mathbf{e}_x \times \mathbf{v}$ in rectangular coordinates.

A.1-3 Expand $\mathbf{u} \cdot (\mathbf{v} \times \mathbf{w})$ in rectangular coordinates.

A.2-1 Expand $\mathbf{e}_x \cdot \boldsymbol{\tau}$ in rectangular coordinates.

A.3-1 Show that $\mathbf{v} \cdot (\nabla \mathbf{v})$ is a vector and expand it in rectangular coordinates.

A.3-2 Show that $\boldsymbol{\tau} : (\nabla \mathbf{v})$ is a scalar and expand it in rectangular coordinates.

A.3-3 Show that $\nabla^2 \mathbf{v}$ is a vector and expand it in rectangular coordinates.

A.3-4 Expand $\mathbf{v} \cdot (\nabla \times \mathbf{v})$ is rectangular coordinates.

[4] J. Crank, ibid.

[5] The first two digits in problem numbers refer to section numbers in text.

A.4-1 Convert the following surface integrals into volume intergals.

$$\iint_A (\nabla \mathbf{v}) \cdot \mathbf{n}\, dA$$

and

$$\iint_A (\rho \mathbf{v}) \cdot \mathbf{n}\, dA$$

A.4-2 Convert the following surface integral into a volume integral

$$\iint_A (\nabla \times \mathbf{v}) \cdot \mathbf{n}\, dA$$

A.6-1 Find the solution for

$$\frac{d^2T}{dz^2} = 0$$

with

$$T = T_1 \quad \text{at} \quad z = z_1$$

and

$$T = 0 \quad \text{at} \quad z = z_2$$

A.6-2 Solve the following problem

$$\frac{d^2T}{dz^2} = \frac{V}{\alpha}\frac{dT}{dz}$$

$$T = T_L \quad \text{at} \quad z = 0$$

$$T = T_m \quad \text{at} \quad z = 1$$

where V/α is a constant.

A.6-3 Solve the following problem

$$\frac{1}{r}\frac{d}{dr}\left[r\frac{dv_z}{dr}\right] = \frac{P'_L - P'_0}{\mu L} \quad (= \text{constant})$$

$$\frac{dv_z}{dr} = 0 \quad \text{at} \quad r = 0 \qquad \text{Boundary Condition 1}$$

$$v_z = 0 \quad \text{at} \quad r = R \qquad \text{Boundary Condition 2.}$$

APPENDIX B

SOFTWARE USEFUL FOR TRANSPORT PHENOMENA IN MATERIALS PROCESSING[1]

1. Adina-F
   Adina R&D Inc.
   71 Elton Avenue
   Watertown, MA 02172
   (617) 926-5194
2. Airplane
   Aesop Inc.
   128 Broadmead
   Princeton, NJ 08540
   (609) 924-6602
3. APPL/3D
   INCA/3D
   Amtec Engineering Inc.
   Box 3633
   Bellevue, WA 98009
4. ASTEC
   Harwell-Flow3D
   Harwell Laboratory
   U.K. Atomic Energy Authority
   Oxfordshire OX11 ORA
   United Kingdom
5. Autodyn
   Century Dynamics
   903 Paramount Road
   Oakland, CA 94610
   (415) 763-9074
6. C-Flow
   Advanced CAE Technology Inc.
   Warren Rd. Business Park
   31 Dutch Mill Road
   Ithaca, NY 14850
   (607) 257-4280
7. Cape
   Cape Simulations, Inc.
   888 Worcester Street
   Wellesley, MA 02181
   (617) 237-3460
8. Commix-1 AR/P
   Argonne National Laboratory
   National Energy Software Center
   9700 South Cass Avenue
   Argonne, IL 60439
   (708) 972-7250

[1] Partially from A. Wolfe, *Mechanical Eng.*, p. 48, Jan. 1991.

9. COMPACT
   Innovative Research, Inc.
   2800 University Avenue SE
   Minneapolis, MN 55414
10. Eagle
    Air Force Armament Laboratory
    Eglin AFB, FL 32543-5434
    (904) 882-3124
11. Easyflow
    Phoenics
    Cham of North America Inc.
    1525-A Sparkman Drive
    Huntsville, AL 35816
    (205) 830-2620
12. FDL3D-E
    Wright Research and
    Development Center
    WRDC/FIMM
    Wright-Patterson AFB,
    OH 45433-6553
    (513) 255-2455
13. FEAP
    NISA/3D-Fluid
    Engineering Mechanics
    Research Corp.
    Box 696
    Troy, MI 48099
    (313) 689-0077
14. Fidap
    Fluid Dynamics International
    500 Davis Street, Suite 600
    Evanston, IL 60201
    (708) 491-0200
15. Fire
    AVL List GmbH
    Kleiststrasse/48
    A-8020 Graz
    Austria
    43 316 987 441
16. Flotran
    PC-Flotran
    Compulfo
    1575 State Farm Boulevard
    Charlottesville, VA 22901-8611
    (804) 977-3569
17. Flow3D
    Flow Science Inc.
    1325 Trinity Drive
    Los Alamos, NM 87544
    (505) 662-2636
18. Fluent
    Nekton
    Creare.x Inc.
    Box 71
    Etna Road
    Hanover, NH 03755
    (603) 643-3800
19. INS3D
    NASA Ames Research Center
    Moffett Field, CA
20. MAGMASoft (mold filling)
    MAGMA GmbH
    Aachen, Germany
21. Moldflow
    Moldflow Pty Ltd.
    Colchester Rd.
    Kilsyth, Victoria 3137
    Australia
    03 720 2088
22. N3S
    Laboratoire National
    d'Hydraulique
    6 Quai Watier
    BP 49 78401 Chatou Cedex 01
    France
    (33) 1 30 71 72 55
23. PAM-Fluid
    Engineering System
    International GmbH
    20 rue Sarrinen, Silic 270
    94578 Rungis Cedex
    France
    (1) 46 87 25 72
24. Panair
    Cosmic
    The University of Georgia
    382 East Broad Street
    Athens, GA 30602
    (404) 542-3265

25. Passage
Techanalysis
7120 Waldemar Drive
Indianapolis, IN 46268
(317) 291-1985
26. Poly3D
Rheotek Inc.
Box 833
Cap Rouge, Quebec G1Y 3E2
Canada
(418) 656-3565
27. PolyCAD
McMaster University
Hamilton, Ontario L8S 4L7
Canada
(416) 525-9140, ext. 4954
28. PolyFlow
PolyFlow S.A.
Fonds Jean Paques, 8
B-1348 Louvain-La-Neuve
Belgium
32-(0) 10-45-28 61
29. PROCAST (mold filling)
UES, Inc.
1595 Chickasaw Road
Arnold, MD 21012
(410) 757-9961
30. Quadpan
Lockheed-California Co.
Burbank, CA 91520-7552
(818) 847-4145
31. SCRYU
Software Cradle Co. Ltd.
6-1-15 Nishinakajima
Yodogawa-ku, Osaka 532
Japan
816-300-5641
32. SIMULOR (mold filling)
Aluminium Pechiney
France
33. Star-CD
Computational Dynamics Ltd.
317 Latimer Road
London W10 6RA England
01 969 9639
34. TAP-A
Watchem
Westinghouse Electric Corp.
Box 355
Pittsburgh, PA 15230
(800) 284-WCCS
35. TascFlow
Advanced Scientific Computing Ltd.
554 Parkside Drive
Unit 4
Waterloo, Ontario N2L 5Z4
(519) 886-8435
36. 3D Radioss Fluid
Mecalog
Zone de Courtaboeu
6 Avenue des Andes Bat. 6
91952 Les Ulis Cedex B
France
(1) 45 88 25 50
37. Trio VF, EF
CEN Saclay
91191 Gif-sur Yvette Cedex
France
(33) 67 08 24 64
38. Visiun
Northern Research and Engineering Corp.
39 Olympia Avenue
Woburn, MA 01901
(617) 935-9050
39. VSAERO
Analytical Methods, Inc.
2133 152d Avenue NE
Redmond, WA 98052
(206) 643-9090

APPENDIX C

PUBLICATIONS RELATED TO TRANSPORT PHENOMENA IN MATERIALS PROCESSING

C.1. TMS SERIES

1. H. D. Brody and D. Apelian, eds., *Modeling of Casting and Welding Processes*, The Metallurgical Society of AIME, Warrendale, PA, 1981.
2. J. A. Dantzig and J. T. Berry, eds., *Modeling of Casting and Welding Processes II*, The Metallurgical Society of AIME, Warrendale, PA, 1984.
3. S. Kou, and R. Mehrabian, eds., *Modeling and Control of Casting and Welding Processes*, The Metallurgical Society, Warrendale, PA, 1986.
4. A. F. Giamei and G. J. Abbaschian, eds., *Modeling and Control of Casting and Welding Processes IV*, The Minerals, Metals and Materials Society, Warrendale, PA, 1988.
5. M. Rappaz, M. R. Ozgu, and K. W. Mahin, eds., *Modeling of Casting, Welding and Advanced Solidification Processes V*, The Minerals, Metals and Materials Society, Warrendale, PA, 1991.
6. T. S. Pionka, V. Voller, and L. Katgerman, eds., *Modeling of Casting, Welding and Advanced Solidification Processes VI*, The Minerals, Metals and Materials Society, Warrendale, PA, 1993.

C.2. ASME SERIES

1. M. M. Chen, J. Mazumder, and C. L. Tucker III, eds., *Transport Phenomena in Materials Processing*, PED-vol. 10, HTD-vol. 29, ASME, New York, 1983.

2. R. K. Shah, ed., *Heat Transfer in Manufacturing and Materials Processing*, HTD-vol. 113, ASME, New York, 1989.
3. M. Charmichi, M. K. Chyu, Y. Joshi, and S. M. Walsh, eds., *Transport Phenomena in Materials Processing*, HTD-vol. 132, ASME, New York, 1990.
4. P. J. Bishop et al., eds., *Transport Phenomena in Materials Processing*, HTD-vol. 146, ASME, New York, 1990.
5. M. Charmachi et al., eds., *Transport Phenomena in Materials Processing and Manufacturing*, HTD-vol. 196, ASME, New York, 1992.
6. C. Beckermann, L. A. Bertram, S. J. Pien, and R. E. Smelser, eds., *Micro/Macro Scale Phenomena in Solidification*, HTD-vol. 218, AMD-vol. 139, ASME, New York, 1992.
7. J. C. Khanpara and P. Bishop, eds., *Heat Transfer in Materials Processing*, HTD-vol. 224, ASME, New York, 1992.
8. C. L. Chan, F. P. Incropera, and V. Prasad, eds., *Transport Phenomena in Nonconventional Manufacturing and Materials Processing*, HTD-vol. 259, ASME, New York, 1993.
9. D. A. Zumbrunnen, ed., *Heat and Mass Transport in Materials Processing and Manufacturing*, HTD-vol. 261, ASME, New York, 1993.
10. C. Beckerman, H. P. Wang, and L. A. Bertram, eds., *Transport Phenomena in Solidification*, HTD-vol. 284, AMD-vol. 182, ASME, New York, 1994.

C.3. JOURNALS RELATED TO MATERIALS PROCESSING

1. *Acta Metallurgia et Materialia*, Elsevier Science, Inc., Tarrytown, NY.
2. *Advanced Materials and Processes*, ASM International, Materials Park, OH.
3. *High Temperature Materials and Processes*, Fruend Publishing House, London, UK.
4. *International Journal of Powder Metallurgy*, APMI International, Princeton, NJ.
5. *International Materials Reviews*, Institute of Materials, London, UK and ASM International, Materials Park, OH.
6. *Iron and Steel Engineer*, Association of Iron and Steel Engineers, Pittsburgh, PA.
7. *Iron and Steelmaker*, The Iron and Steel Society, Warrendale, PA.
8. *Journal of Crystal Growth*, North-Holland, Amsterdam, The Netherlands.
9. *Journal of Materials Engineering and Performance*, ASM International, Materials Park, OH.
10. *Journal of Materials Research*, Materials Research Society, Pittsburgh, PA.
11. *Journal of Materials Science*, Chapman & Hall, London, UK.

12. *Journal of Materials Synthesis and Processing*, Plenum Press, New York.
13. *Journal of Minerals, Metals and Materials*, The Minerals, Metals and Materials Society, Warrendale, PA.
14. *Journal of Thermal Spray Technology*, ASM International, Materials Park, OH.
15. *Materials and Design*, Elsevier Science, Oxford, UK.
16. *Materials Science and Technology*, Institute of Materials, London, UK.
17. *Materials Science and Engineering*, Elsevier Science, Oxford, UK.
18. *Materials and Manufacturing Processes*, Marcel Dekker, New York.
19. *Metal Heat Treating*, Penton Publishing, Cleveland, OH.
20. *Metallurgical and Materials Transactions*, ASM International, Materials Park, OH and The Minerals, Metals and Materials Society, Warrendale, PA.
21. *Powder Metallurgy*, The Institute of Materials, London, UK.
22. *Welding Journal*, American Welding Society, Miami, FL.
23. *Welding in the World*, International Institute of Welding, Pergamon, Oxford, UK.

INDEX

CPSIA information can be obtained
at www.ICGtesting.com
Printed in the USA
BVOW06*0937090217
475701BV00012B/83/P

9 780471 076674